SYMBOLS, PREFIXES, AND ABBREVIATIONS

See Appendixes A–1 (Units) and A–2 (constants and conversions) for more detailed information.

A	ampere
Å	angstrom $= 10^{-10}$ m
A	vector potential, Wb m^{-1}
A, a	area, m^2
AR	axial ratio
AU	astronomical unit
a	atto $= 10^{-18}$ (prefix)
â	unit vector
B, B	magnetic flux density, T = Wb m^{-2}
B	susceptance, $\mho$
B	susceptance/unit length, $\mho$ m^{-1}
BWFN	beam width, first nulls
C	coulomb
C	capacitance, F
C	capacitance/unit length, F m^{-1}
C, c	a constant, c = velocity of light
cc	cubic centimeter
°C	degree Celsius
D, D	electric flux density, C m^{-2}
d	distance, m
deg	degree, angle
dB	decibel $= 10 \log (P_2/P_1)$
dBi	decibels over isotropic
dl	element of length (scalar), m
$d\mathbf{l}$	element of length (vector), m
ds	element of surface (scalar), m^2
$d\mathbf{s}$	element of surface (vector), m^2
dv	element of volume (scalar), m^3
E, E	electric field intensity, V m^{-1}
E	exa $= 10^{18}$ (prefix)
emf	electromotive force, V
e	electric charge, C
F	farad
F, F	force, N
f	femto $= 10^{-15}$ (prefix)
f	force per volume, N m^{-3}
f	frequency, Hz
G	giga $= 10^9$ (prefix)
G	conductance, $\mho$
G	conductance/unit length, $\mho$ m^{-1}
g	gram
H	henry
H, H	magnetic field, A m^{-1}
HPBW	half-power beam width
Hz	hertz = 1 cycle per second
I, I, i	current, A
J	joule
J, J	current density, A m^{-2}
Jy	jansky, 10^{-26} W m^{-2} Hz^{-1}
K	kelvin
K, K	sheet-current density, A m^{-1}
K, k	a constant
k	kilo $= 10^3$ (prefix)
kg	kilogram
L	inductance, H
L	inductance/unit length, H m^{-1}
l	liter
l, L	length (scalar), m
l	length (vector), m
LCP	left circularly polarized
LEP	left elliptically polarized
ln	natural logarithm (base e)
log	common logarithm (base 10)
M	mega $= 10^6$ (prefix)
M, M	magnetization, A m^{-1}
M	polarization state $= M(\epsilon, \tau)$
m, m	magnetic (dipole) moment, A m^2
m	meter
m	milli $= 10^{-3}$ (prefix)
min	minute
N	newton
N, n	number (integer)
Np	neper
n	nano $= 10^{-9}$ (prefix)
$\hat{\mathbf{n}}$	unit vector normal to a surface
P, P	polarization of dielectric, C m^{-2}
p	electric dipole moment, C m^{-3}
P	peta $= 10^{15}$ (prefix)
P	polarization state $= P(\gamma, \delta)$
P	power, W
P	power per solid angle, W rad^{-2}
P_n	normalized power pattern, dimensionless
$\mathscr{P}$	permeance, H
$\mathscr{P}$	radiation pressure, N m^{-2}
p	pico $= 10^{-12}$ (prefix)
Q, q	charge, C
R	resistance, Ω
R	resistance/unit length, Ω m^{-1}
$\mathscr{R}$	reluctance, H^{-1}
RCP	right circular polarization
REP	right elliptical polarization
r	revolution
r	radius, m; also coordinate direction
$\hat{\mathbf{r}}$	unit vector in r direction
rad	radian
rad^2	square radian = steradian = sr
S, S	Poynting vector, W m^{-2}

S	flux density, $\text{W m}^{-2}\,\text{Hz}^{-1}$
S	resistivity, $\Omega\,\text{m}$
S, s	distance, m; also surface area, m^2
s	second (of time)
sr	steradian = square radian = rad^2
T	tesla = Wb m^{-2}
T	tera = 10^{12} (prefix)
T	torque, N m
t	time, s
U	magnetostatic potential, A
V	volt
V	voltage (also emf), V
$\mathscr{V}$	emf (electromotive force), V
v	velocity, m s^{-1}
W	watt
W	energy, J
Wb	weber = 10^4 gauss
w	energy density, J m^{-3}
X	reactance, Ω
X	reactance/unit length, $\Omega\,\text{m}^{-1}$
$\hat{\mathbf{x}}$	unit vector in x direction
x	coordinate direction
Y	admittance, $\mho$
Y	admittance/unit length, $\mho\,\text{m}^{-1}$
$\hat{\mathbf{y}}$	unit vector in y direction
y	coordinate direction
Z	impedance, Ω
Z	impedance/unit length, $\Omega\,\text{m}^{-1}$
Z_c	intrinsic impedance, conductor, Ω per square
Z_d	intrinsic impedance, dielectric, Ω per square
Z_L	load impedance, Ω
Z_{yz}	transverse impedance, rectangular waveguide, Ω
$Z_{r\phi}$	transverse impedance, cylindrical waveguide, Ω
Z_0	intrinsic impedance, space, Ω per square
Z_0	characteristic impedance, transmission line, Ω
$\hat{\mathbf{z}}$	unit vector in z direction
z	coordinate direction, also red shift
α	(alpha) angle, deg or rad
α	attenuation constant, nep m^{-1}
β	(beta) angle, deg or rad; also phase constant $= 2\pi/\lambda$
γ	(gamma) angle, deg or rad
δ	(delta) angle, deg or rad
ϵ	(epsilon) permittivity, F m^{-1}
ϵ_0	permittivity of vacuum, F m^{-1}
η	(eta) index of refraction
θ	(theta) angle, deg or rad
$\hat{\boldsymbol{\theta}}$	(theta) unit vector in θ direction
κ	(kappa) constant
Λ	(capital lambda) flux linkage, Wb turn
λ	(lambda) wavelength, m
μ	(mu) permeability, H m^{-1}
μ_0	permeability of vacuum, H m^{-1}
μ	(mu) mobility, $\text{m}^2\,\text{s}^{-1}\,\text{V}^{-1}$
μ	(mu) micro = 10^{-6} (prefix)
ν	(nu)
ξ	(xi)
π	(pi) = 3.1416
ρ	(rho) electric charge density, C m^{-3}; also mass density, kg m^{-3}
ρ	reflection coefficient, dimensionless
ρ_s	surface charge density, C m^{-2}
ρ_L	linear charge density, C m^{-1}
σ	(sigma) conductivity, $\mho\,\text{m}^{-1}$
τ	(tau) tilt angle, polarization ellipse, deg or rad
τ	transmission coefficient, deg or rad
ϕ	(phi) angle, deg or rad
$\hat{\boldsymbol{\phi}}$	(phi) unit vector in ϕ direction
χ	(chi) susceptibility, dimensionless
ψ	(psi) angle, deg or rad
ψ_m	magnetic flux, Wb
Ω	(capital omega) ohm
Ω	(capital omega) solid angle, sr or deg^2
$\mho$	(upsidedown capital omega) mho ($\mho = 1/\Omega = \mathbf{S}$, siemens)
ω	(omega) angular frequency ($= 2\pi f$), rad s^{-1}

ELECTROMAGNETICS

McGraw-Hill Series in Electrical Engineering

Consulting Editor
Stephen W. Director, Carnegie-Mellon University

Networks and Systems
Communications and Information Theory
Control Theory
Electronics and Electronic Circuits
Power and Energy
Electromagnetics
Computer Engineering
Introductory and Survey
Radio, Television, Radar, and Antennas

Previous Consulting Editors

Ronald M. Bracewell, Colin Cherry, James F. Gibbons, Willis W. Harman, Hubert Heffner, Edward W. Herold, John G. Linvill, Simon Ramo, Ronald A. Rohrer, Anthony E. Siegman, Charles Susskind, Frederick E. Terman, John G. Truxal, Ernst Weber, and John R. Whinnery

Electromagnetics

Consulting Editor
Stephen W. Director

Combination earth station, radio-telescope, and solar collector designed and built by the author. The 2.6-m dish antenna, of low-cost design, receives Clarke-orbit TV satellites, serves as a radio-telescope antenna, and, with a water tank at the focus, as a solar power collector.

The antenna is on an al-az (altitude-azimuth) mount operated by 12-V automotive-type electric jacks. The 4- or 12-GHz low-noise amplifier (LNA) at the focus is connected by a short length of 50-Ω coaxial cable to a down-converter behind the dish. The 70-MHz IF output of the down-converter is connected by a longer 50-Ω coaxial cable to the main satellite receiver and TV set. As a radio-telescope, the entire LNA bandwidth of nearly 1 GHz can be used to increase sensitivity for the detection of celestial sources. As a solar power collector, the aperture can gather in over 5 kW of power from the sun. (See Sec. 10-22 and also problems 10-6-1, 14-5-13, 14-6-8, 14-6-9, and 14-6-10.)

ELECTROMAGNETICS

Third Edition

John D. Kraus

Director, Radio Observatory
Taine G. McDougal Professor Emeritus of
Electrical Engineering and Astronomy
The Ohio State University

McGRAW-HILL BOOK COMPANY

Auckland Bogotá Guatemala Hamburg Lisbon
London Madrid Mexico New Delhi Panama Paris
San Juan São Paulo Singapore Sydney Tokyo

Other books by John D. Kraus:

ANTENNAS (McGraw-Hill)
RADIO ASTRONOMY (Cygnus-Quasar)
BIG EAR (Cygnus-Quasar)
OUR COSMIC UNIVERSE (Cygnus-Quasar)

ELECTROMAGNETICS

INTERNATIONAL EDITION

This book was set in Times Roman.
The editors were T. Michael Slaughter and Madelaine Eichberg;
The production supervisor was Leroy A. Young.
New drawings were done by J & R Services, Inc.

Library of Congress Cataloging in Publication Data

Kraus, John Daniel, date
    Electromagnetics.

    (McGraw-Hill series in electrical engineering.
Electromagnetics)
    Bibliography: p.
    Includes index.
    1. Electromagnetic theory. I. Title. II. Series.
QC661.K72 1984   530.1'41   83-11351
ISBN 0-07-035423-5
ISBN 0-07-035424-3 (solutions manual)

When ordering this title use ISBN 0-07-Y66380-7

Printed and Bound by KIN KEONG PRINTING CO. PTE. LTD. – Republic of Singapore.

CONTENTS

Preface   xv

Chapter 1 Introduction   1

1-1 Electromagnetics: Its History and Importance   1
1-2 Dimensions and Units   5
1-3 Fundamental and Secondary Units   6
1-4 How to Read the Symbols and Notation   7
1-5 Equation Numbering   8
1-6 Dimensional Analysis   8
     Problems   9

Chapter 2 The Static Electric Field   11

2-1 Introduction   11
2-2 The Force between Point Charges and Coulomb's Law   11
2-3 Idealness and Staticness   14
2-4 Electric Field Intensity   14
2-5 Positiveness, Right-Handedness, and Outwardness   16
2-6 The Electric Field of Several Point Charges and the
     Principle of Superposition of Fields   17
2-7 The Electric Scalar Potential   20
2-8 The Electric Scalar Potential as a Line Integral of the Electric Field 22
2-9 Relation of Electric Field Lines and Equipotential Contours:
     Orthogonality   25
2-10 Field of Two Equal Point Charges of Opposite Sign and of
     Same Sign   26
2-11 Charge Density and Continuous Distributions of Charge   28
2-12 Electric Potential of Charge Distributions and the Principle of
     Superposition of Potential   29
2-13 The Electric Field as the Gradient of the Electric Potential   32
2-14 Gradient in Rectangular Coordinates   32
2-15 The Electric Dipole and Electric-Dipole Moment   35

2-16 Electric Flux 38
2-17 Electric Flux over a Closed Surface; Gauss' Law 40
2-18 Single Shell of Charge 42
2-19 Conductors and Induced Charges 44
2-20 Conducting Shell 44
Problems 47

Chapter 3 The Static Electric Field in Dielectrics 56

3-1 Introduction 56
3-2 Homogeneity, Linearity, and Isotropy 56
3-3 Dielectrics and Permittivity 57
3-4 The Electric Field in a Dielectric 59
3-5 Polarization 59
3-6 Boundary Relations 63
3-7 Table of Boundary Relations 68
3-8 Capacitors and Capacitance 69
3-9 Dielectric Strength 72
3-10 Energy in a Capacitor and Energy Density 74
3-11 Field Distributions 77
3-12 Field of a Finite Line of Charge 78
3-13 Field of an Infinite Line of Charge 80
3-14 Infinite Cylinder of Charge 80
3-15 Infinite Coaxial Transmission Line 81
3-16 Two Infinite Lines of Charge 82
3-17 Infinite Two-Wire Transmission Line 84
3-18 Infinite Single-Wire Transmission Line 85
3-19 Field Maps and Field Cells 86
3-20 Heart Dipole Field 93
3-21 Divergence of the Flux Density **D** 94
3-22 Maxwell's Divergence Equation from Gauss' Law 97
3-23 Divergence Theorem 99
3-24 Divergence of **D** and **P** in a Capacitor 99
3-25 The Laplacian Operator; Poisson's and Laplace's Equations 102
Problems 103

Chapter 4 The Steady Electric Current 112

4-1 Introduction 112
4-2 The Electric Current; Current Density 112
4-3 Resistance and Ohm's Law 114
4-4 Power Relations and Joule's Law 115
4-5 The Electric Circuit 116
4-6 Resistivity and Conductivity 116
4-7 Ohm's Law at a Point 119
4-8 Dielectrics, Conductors, and Semiconductors 120
4-9 Table of Conductivities 124
4-10 Kirchhoff's Voltage Law and the Difference between
Potential and EMF 124
4-11 Tubes of Current 128

4-12 Kirchhoff's Current Law 129
4-13 Divergence of **J** and Continuity Relations for Current 130
4-14 Current and Field at a Conductor-Insulator Boundary 131
4-15 Current and Field at a Conductor-Conductor Boundary 134
4-16 Current Mapping and the Resistance of Simple Geometries;
    Conductor Cells 135
4-17 Laplace's Equations for Conducting Media 138
    Problems 140

Chapter 5 The Static Magnetic Field of Steady Electric Currents

147

5-1 Introduction 147
5-2 Effect of a Magnet on a Current-Carrying Wire 148
5-3 The Magnetic Field of a Current-Carrying Element;
    The Biot-Savart Law 151
5-4 The Magnetic Field of an Infinite Linear Conductor 153
5-5 The Force between Two Parallel Linear Conductors;
    Definition of the Ampere 154
5-6 The Magnetic Field of a Current-Carrying Loop 155
5-7 Magnetic Flux ψ_m and Magnetic Flux Density **B** 156
5-8 Magnetic Flux over a Closed Surface 157
5-9 Magnetic Field Relations in Vector Notation; Cross Product 158
5-10 Torque of a Loop; Magnetic Moment 161
5-11 The Solenoid 162
5-12 Inductors and Inductance 164
5-13 Inductance of Simple Geometries 165
5-14 Ampère's Law and **H** 169
5-15 Ampère's Law Applied to a Conducting Medium and
    Maxwell's Equation 171
5-16 Magnetostatic Potential U and MMF F 172
5-17 Field Cells and Permeability 176
5-18 Curl 179
5-19 Maxwell's First Curl Equation 185
5-20 Summary of Operations Involving ∇ 185
5-21 A Comparison of Divergence and Curl 186
5-22 The Vector Potential 188
5-23 Charged Particles in Electric and Magnetic Fields 192
    Problems 198

Chapter 6 The Static Magnetic Field of Ferromagnetic Materials

210

6-1 Introduction 210
6-2 Magnetic Dipoles, Loops, and Solenoids 210
6-3 Magnetic Materials 214
6-4 Relative Permeability 215
6-5 Magnetic Dipoles and Magnetization 216

6-6   Uniformly Magnetized Rod and Equivalent Air-Filled Solenoid   218
6-7   The Magnetic Vectors **B, H,** and **M**   220
6-8   Energy in an Inductor and Energy Density   231
6-9   Boundary Relations   234
6-10 Table of Boundary Relations for Magnetic Fields   237
6-11 Ferromagnetism   238
6-12 Magnetization Curves   240
6-13 Hysteresis   245
6-14 Energy in a Magnet   248
6-15 Permanent Magnets   250
6-16 Table of Permanent Magnetic Materials   251
6-17 Demagnetization   252
6-18 The Magnetic Circuit; Reluctance and Permeance   254
6-19 Magnetic Field Mapping; Magnetic Field Cells   258
6-20 Comparison of Field Maps in Electric, Magnetic, and Current Cases   263
6-21 Gapless Circuit   267
6-22 Magnetic Circuit with Air Gap   269
6-23 Magnetic Gap Force   270
6-24 Permanent Magnet with Gap   272
6-25 A Comparison of Static Electric and Magnetic Fields   273
6-26 Comparison of Electric and Magnetic Relations Involving Polarization and Magnetization   275
6-27 Do Magnetic Monopoles Exist?   276
      Problems   276

Chapter 7   Bounded Fields and Laplace's Equation   282

7-1   Introduction   282
7-2   Laplace's Equation in Rectangular Coordinates; Separation of Variables   283
7-3   Example 1: The Parallel-Plate Capacitor   284
7-4   Uniqueness   286
7-5   Point-by-Point, or Iterative, Method   287
7-6   Example 2: The Infinite Square Trough   289
7-7   Example 3: Infinite Square Trough, Digital-Computer Solution   292
7-8   Analog-Computer Solution with Laplace's and Kirchoff's Laws   294
7-9   Example 4: Conducting Sheet between Two Conducting Planes   297
7-10 Solution of Laplace's Equation in Cylindrical and Spherical Coordinates   303
7-11 Example 5: Coaxial Line   304
7-12 Poisson's Equation   307
7-13 Example 6: Parallel-Plate Capacitor with Space Charge   308
7-14 The Theory of Images   311
7-15 Example 7: Charged Cylindrical Conductor over an Infinite Metal Ground Plane   312
7-16 Example 8: Current-Carrying Conductor over Infinite Metal Ground Plane   314
      Problems   315

Chapter 8   Time-Changing Electric and Magnetic Fields   322

8-1   Introduction   322
8-2   Faraday's Law   322
8-3   Moving Conductor in a Magnetic Field   326
8-4   General Case of Induction   327
8-5   Examples of Induction   328
8-6   Stokes' Theorem   332
8-7   Maxwell's Equation from Faraday's Law: Differential Form   334
8-8   Inductance, Self and Mutual   335
8-9   Alternating-Current Behavior of Ferromagnetic Materials   338
8-10   Eddy Currents   340
8-11   Displacement Current   340
8-12   Maxwell's Equation from Ampère's Law: Complete Expression   342
8-13   Phasors   344
8-14   Dielectric Hysteresis   346
8-15   Boundary Relations   350
8-16   General Field Relations   351
8-17   Comparison of Electric and Magnetic Field Relations   353
  Problems   353

Chapter 9   The Relation between Field and Circuit Theory; Maxwell's Equations   365

9-1   Introduction   365
9-2   Applications of Circuit and Field Theory   366
9-3   The Series Circuit; Comparison of Field and Circuit Theory   368
9-4   Maxwell's Equations as Generalizations of Circuit Equations   371
9-5   Maxwell's Equations in Free Space   374
9-6   Maxwell's Equations for Harmonically Varying Fields   374
9-7   Tables of Maxwell's Equations   374
  Problems   376

Chapter 10   Waves and Transmission Lines   378

10-1   Introduction   378
10-2   The Wave Equation for Waves in Space and on Transmission Lines   380
10-3   Coaxial, Two-Wire, and Field-Cell Transmission Lines   391
10-4   The Infinite Uniform Transmission Line: Characteristic Impedance   392
10-5   Impedance of Transmission Lines and of Media   395
10-6   The Terminated Uniform Transmission Line   402
10-7   Reflection Coefficient, Slotted Line, and Smith Chart   408
10-8   Directional Coupler   419
10-9 $\lambda/4$ Transformers and Bandwidth   420
10-10.1   Pulses and Transients   424
10-10.2   Wave Reflections on a $\lambda/4$ Transformer   428
10-10.3 S, or Scattering Parameters   431
10-11   Relative Phase Velocity and Index of Refraction   434
10-12   Group Velocity   436

10-13 Traveling Waves and Standing Waves 439
10-14 Conductors and Dielectrics 445
10-15 Conducting Media and Lossy Lines 447
10-16 Plane Waves at Interfaces 454
10-17 Power and Energy Relations 464
10-18 Power Flow on a Transmission Line 470
10-19 Circuit Application of the Poynting Vector 472
10-20 The Axon: An Active, Lossless, Noiseless Transmission Line 475
10-21 Shielding of Transmission Lines 476
10-22 Radio Link Transmission Line 480
10-23 General Development of the Wave Equation 482
10-24 Table of Circuit and Field Relations 483
10-25 Table of Impedance, Velocity, Attenuation, Phase, and other
      Constants for Transmission Lines and Waves 484
      Problems 485

Chapter 11   Wave Polarization 495

11-1 Introduction 495
11-2 Linear, Elliptical, and Circular Polarization 495
11-3 Poynting Vector for Elliptically or Circularly Polarized Waves 498
11-4 The Polarization Ellipse and the Poincaré Sphere 500
11-5 Partial Polarization and the Stokes Parameters 503
     Problems 509

Chapter 12   Reflection, Refraction, and Diffraction 511

12-1 Introduction 511
12-2 Oblique Incidence 511
12-3 Elliptically Polarized Plane Wave, Oblique Incidence 519
12-4 Huygens' Principle and Physical Optics 524
12-5 Geometrical-Optics Concepts 529
     Problems 531

Chapter 13   Waveguides and Resonators 534

13-1 Introduction 534
13-2 Circuits, Lines, and Guides: A Comparison 534
13-3 TE-Mode Wave in the Infinite-Parallel-Plane Transmission Line
     or Guide 535
13-4 The Hollow Rectangular Waveguide 543
13-5 The Hollow Cylindrical Waveguide 561
13-6 Hollow Waveguides of Other Cross Section 569
13-7 Attenuation at Frequencies Less than Cutoff 571
13-8 Attenuation at Frequencies Greater than Cutoff 572
13-9 Waveguide Devices 578
13-10 Waveguide Iris Theory 579
13-11 Intrinsic, Characteristic, and Wave Impedances 581
13-12 Waves Traveling Parallel to a Plane Boundary 582
13-13 Open Waveguides 586

13-14 Dielectric Sheet Waveguide   589
13-15 Dielectric Fiber and Rod Waveguides   593
13-16 Retinal Optic Fibers   596
13-17 Cavity Resonators   597
13-18 Modes   604
       Problems   605

Chapter 14 Antennas and Radiation   612

14-1   Introduction   612
14-2   Retarded Potentials   618
14-3   The Short Dipole Antenna   620
14-4   Radiation Resistance of a Short Dipole   629
14-5   Effective Aperature, Directivity, and Gain   633
14-6   Array Theory   640
       14-6.1 Two Isotropic Point Sources   640
       14-6.2 Pattern Multiplication   641
       14-6.3 Binomial Array   642
       14-6.4 Array with n Sources of Equal Amplitude and Spacing   643
       14-6.5 Array with n Sources of Equal Amplitude and Spacing:
               Broadside Case   645
       14-6.6 Array with n Sources of Equal Amplitude and Spacing:
               End-Fire Case   646
       14-6.7 Graphical Representation of Phasor Addition of Fields   648
       14-6.8 Simple Two-Element Interferometer   650
14-7   Continuous Aperature Distribution   653
14-8   Fourier Transform Relations between the Far-Field Pattern and
       the Aperture Distribution   655
14-9   Linear Antennas   657
14-10 Fields of $\lambda/2$ Dipole Antenna   659
14-11 Traveling-Wave Antennas   661
14-12 The Small Loop Antenna   665
14-13 The Helical-Beam Antenna   666
14-14 Beam Width and Directivity of Arrays   670
14-15 Scanning Arrays   670
14-16 Frequency-Independent Antennas   671
14-17 Reciprocity   673
14-18 Self and Mutual Impedance and Arrays of Dipoles   675
14-19 Reflector and Lens Antennas   683
       14-19.1 Infinite-Flat-Sheet Reflector   684
       14-19.2 Finite-Flat-Sheet Reflector   684
       14-19.3 Thin Reflectors and Directors   686
       14-19.4 Corner Reflectors   688
       14-19.5 Parabolic Reflectors and Lens Antennas   689
       14-19.6 Some Comments on Corner Reflectors vs.
                Parabolic Reflectors   692
14-20 Slot and Complementary Antennas   693
14-21 Horn Antennas   694
14-22 Aperture Concept   695

14-23 Friis Formula and Radar Equation 697
14-24 Radio Telescopes, Antenna Temperature, System Temperature,
        Remote Sensing, and Resolution 702
14-25 Table of Antenna and Antenna System Relations 712
        Problems 713

Appendix A Units, Constants, and Other Useful Relations 725

A-1 Units 725
A-2 Constants and Conversions 732
A-3 Trigonometric Relations 733
A-4 Hyperbolic Relations 734
A-5 Logarithmic Relations 735
A-6 Approximation Formulas for Small Quantities 735
A-7 Series 736
A-8 Solution of Quadratic Equation 736
A-9 Vector Identities 736
A-10 Recurrence Relations for Bessel Functions 738
A-11 Coordinate Diagrams 738
A-12 Table of Dielectric Materials 741
A-13 Smith Chart 742
A-14 Noise-Temperature–Noise-Figure Chart 743
A-15 Fundamental Force Laws 744
A-16 Unit Vector, Scalar (Dot) Products between Coordinate Systems 744
A-17 Component Transformations 746

Appendix B Bibliography 747

Appendix C Answers to Starred Problems 749

        Index 755

PREFACE

Many changes and new features have been incorporated in this edition to better fit needs expressed by users of the previous edition. The aim of the book, as before, is to present the basic elements of electromagnetic theory for an introductory course. Topics are developed in easy steps from the simplest special cases to more general ones with applications throughout to numerous practical situations.

The introduction (Chapter 1) sets the stage for the book and provides a perspective for the terms, names, and units which appear thousands of times throughout the book by relating them to a dynamic history. Chapters 2 to 6 on static electric and magnetic fields and steady currents form the foundation of field theory. Chapters 7 to 9 discuss boundary-value problems, time-changing fields, and the relation of field and circuit theory, culminating in Maxwell's equations. Problem solutions by a variety of methods including analytical, graphical, and computer (both digital and analog) are discussed.

Chapters 10 to 12 cover transmission lines, plane waves, polarization, reflection, refraction, and diffraction. Chapters 13 and 14 explain waveguides, resonators, antennas, and radiation with examples of many of the latest innovations including fiber optics, remote sensing, and geostationary satellite television.

Transmission lines have such important applications that many instructors place special emphasis on them and desire to teach them as soon as possible. Accordingly, transmission lines are introduced earlier and are discussed in more detail than in the previous edition. The field-cell transmission line concept is used to relate transmission lines and plane waves, the equations for both being identical in form.

In response to a growing interest by engineers in biological areas, brief discussions are included on the heart as an electric dipole, the axon (nerve fiber) as a noiseless-lossless transmission line, and the sensors of the retina of the eye as polyrod antennas.

There are many new illustrations in this edition totaling nearly 1000 diagrams, charts, and photographs. The problem sets have been completely redone and problems are now grouped according to both chapter and section, facilitating assignments. Many of the problems are adapted for solution with computers. Sample solutions in both BASIC and FORTRAN are given. Many problems pertain to modern real-world engineering situations, as, for example, problems involving studies of the engineering feasibility and/or the design of practical devices. Some of these have multiple or indefinite solutions (few real-world problems have a single exact answer) and illustrate the compromises and trade-offs required in practice. Including multiple parts, this edition contains nearly 1500 problems covering all levels of difficulty.

The problems constitute an important extension of the text, many topics being covered that are not included in the main text. Many "to prove" or "to show" problems contain valuable supplemental information helpful in design work or in the solution of other problems. Reading the problems, even without solving them, is highly recommended especially the ones in the Practical Applications groups (last group of each set). Many topics of current interest are discussed, some of which could serve as subjects for term papers.

Appendix A contains tables of important constants, equations, formulas, and other useful information. There is also a very complete tabulation of units and their equivalents. Appendix B is a bibliography and Appendix C gives answers to many of the problems. There is a complete listing of symbols inside the front cover and some frequently used vector relations inside the back cover.

The chapters on particles and plasmas and moving systems and relativity which I wrote for the second edition have been omitted since most instructors did not have the time to include them.

Particular attention has been given to arranging topics so that the book can be used not only for a two-semester (or three-quarter) course but also for shorter courses. Thus, a one-semester course might cover about half the book (most of the first five chapters, parts of the next two, and a few introductory sections of the last four). In a brief course, most of Chapter 6 and all of Chapter 7 could be omitted.

As prerequisites the student is assumed to have a knowledge of introductory physics and mathematics through differential and integral calculus. Also a course in vector analysis is desirable either beforehand or concurrently.

Although great care has been exercised, some errors in the text or figures are inevitable. Anyone finding them will do me a great service by writing me so that they can be corrected in subsequent printings.

It would not have been possible to prepare this edition without the dedicated assistance of Dr. Erich Pacht, of the Ohio State University Radio Observatory, who has been involved in all aspects of the process. Dr. Keith R. Carver, who assisted me on the second edition, was on extended leave from his institution and not able to help on this edition.

I have had the benefit of innumerable useful comments and suggestions from many persons at the Ohio State University and elsewhere and, in particular, I gratefully acknowledge the help of Professors Louis Bailin, Walter Burnside, Stuart Collins, John Cooper, John Cowan, Jr., Charles Fell, Walter Flood, Robert Garbacz, Daniel Hodge, Carl Ingling, Edward Kennaugh, Hsien Ching Ko, Robert Kouyoumjian, Curt Levis, William Peake, Leon Peters, Jack Richmond, Clarence Schultz, Thomas Seliga, Chen-To Tai, and Herman Weed.

Finally, I express my sincere appreciation to my wife, Alice, for her assistance, patience, and encouragement during all the years of the book's preparation.

John D. Kraus

INTRODUCTION

1-1 ELECTROMAGNETICS: ITS HISTORY AND IMPORTANCE

Electric and magnetic forces, gravity, and the so-called "weak" and "strong" forces are the five known forces of physics. Gravity is dominant on a planetary-to-stellar scale, while the weak and strong forces are important inside the nucleus of atoms. Electric and magnetic forces are important in between. In fact, most of the forces of our everyday experience that are not gravitational are electric or magnetic.

Electromagnetics embraces both electricity and magnetism and is basic to everything electric and magnetic. Although a resistor, capacitor, or inductor may be regarded as a simple two-terminal circuit element without regard to electromagnetic field theory, an understanding of what goes on inside these circuit elements requires a knowledge of *electromagnetic fields*, a field being any region in which electric and magnetic forces act. As discussed further in Chap. 9, it is important to note that the power supplied from a generator to a load flows not through the wires connecting them but rather by the fields surrounding them. And when it comes to waves in waveguides, waves radiating from antennas, or waves traveling in space, electromagnetic field theory offers the only answers. Although circuit theory may be adequate in many situations, electromagnetic field theory is required for a full understanding.

Some further insights into electromagnetics and its scope and importance become evident through the following brief history tracing its development.

Six hundred years before Christ, a Greek mathematician, astronomer, and philosopher, Thales of Miletus, noted that when amber is rubbed with silk it produces sparks and has a seemingly magical power to attract particles of fluff and straw. The Greek word for amber is *elektron* and from this we get our words *electricity, electron*, and *electronics*. Thales also noted the attractive power between pieces of a natural magnetic rock called loadstone found at a place called *Magnesia*, from which is derived the words *magnet* and *magnetism*. Thales was a pioneer in both electricity and magnetism but his interest, like that of others of this time, was philosophical rather than practical, and it was 22 centuries before these phenomena were investigated in a serious experimental way.

It remained for William Gilbert of England in about 1600 A.D. to perform the first systematic experiments of electric and magnetic phenomena, describing his experiments in his celebrated book, *De Magnete*. Gilbert invented the electroscope for measuring electrostatic effects. He was also the first to recognize that the earth itself is a huge magnet, thus providing new insights into the principles of the compass and dip needle.

In experiments with electricity made about 1750 that led to his invention of the lightning rod, Benjamin Franklin, the American scientist-statesman, established the law of conservation of charge and determined that there are both *positive* and *negative* charges. Later, Charles Augustin de Coulomb of France measured electric and magnetic forces with a delicate torsion balance he invented. During this period Karl Friedrich Gauss, a German mathematician and astronomer, formulated his famous divergence theorem relating a volume and its surface.

By 1800 Alessandro Volta of Italy had invented the voltaic cell and, connecting several in series, the electric battery. With batteries, electric currents could be produced, and in 1819 the Danish professor of physics Hans Christian Oersted found that a current-carrying wire caused a nearby compass needle to deflect, thus discovering that *electricity could produce magnetism*. Before Oersted, electricity and magnetism were considered as entirely independent phenomena.

The following year, André Marie Ampère, a French physicist, extended Oersted's observations. He invented the solenoidal coil for producing magnetic fields and theorized correctly that the atoms in a magnet are magnetized by tiny electric currents circulating in them. About this time Georg Simon Ohm of Germany published his now famous law relating current, voltage, and resistance. However, it initially met with ridicule and a decade passed before scientists began to recognize its truth and importance.

Then in 1831, Michael Faraday of London demonstrated that a changing magnetic field could produce an electric current. Whereas Oersted found that electricity could produce magnetism, Faraday discovered that *magnetism could produce electricity*. At about the same time, Joseph Henry of Albany, New York, observed the effect independently. Henry also invented the electric telegraph and relay.

Faraday's extensive experimental investigations enabled James Clerk Maxwell, a professor at Cambridge University, England, to establish in a profound and elegant manner the interdependence of electricity and magnetism. In his classic treatise of 1873, he published the first unified theory of electricity and magnetism and founded the science of electromagnetics. He postulated that light was electromagnetic in nature and that electromagnetic radiation of other wavelengths should be possible.

Although Maxwell's equations are of great importance and, with boundary, continuity, and other auxiliary relations, form the basic tenets of modern electromagnetics, many scientists of Maxwell's time were skeptical of his theories. It was not until 15 years later, in 1888, that his theories were vindicated by Heinrich Hertz, a physics professor at Karlsruhe, Germany. Hertz generated and detected radio waves of about 5 meters wavelength. With a spark transmitter and receiver, Hertz demonstrated that, except for a difference in wavelength, the polarization, reflection, and refraction of radio waves were identical with light.

Hertz was the father of radio, but his invention remained a laboratory curiosity until Guglielmo Marconi of Italy adapted Hertz's spark system to send messages

Pioneers of electromagnetics

Name	Dates	Role played	Units (if any)
Thales of Miletus	636–546 B.C.	Pioneered in electricity and magnetism	
William Gilbert	1540–1603 A.D.	Recognized that the earth is a huge magnet	gilbert (Gb)
Benjamin Franklin	1706–1790	Established conservation of charge	
Charles A. de Coulomb	1736–1806	Measured electric and magnetic forces	coulomb (C)
Karl F. Gauss	1777–1855	Enunciated divergence theorem	gauss (G)
Alessandro Volta	1745–1827	Invented voltaic cell	volt (V)
Hans C. Oersted	1777–1851	Discovered that electricity could produce magnetism	oersted (Oe)
André M. Ampère	1775–1836	Invented solenoid	ampere (A)
Joseph Henry	1797–1878	Made experiments leading to electric telegraph	henry (H)
Georg S. Ohm	1787–1854	Formulated Ohm's law	ohm (Ω)
Michael Faraday	1791–1867	Demonstrated that magnetism could produce electricity	farad (F)
James P. Joule	1818–1889	Established that heating is proportional to the square of the current	joule (J)
James C. Maxwell	1831–1879	Founded electromagnetic theory	maxwell (Mx)
Heinrich Hertz	1857–1894	Father of radio	hertz (Hz)
Guglielmo Marconi	1874–1937	Made radio practical	
Thomas A. Edison	1847–1931	Invented incandescent light bulb and built first electric power systems	
Nikola Tesla	1856–1943	Demonstrated value of alternating current	tesla (T)
Albert Einstein	1879–1955	Made Maxwell's equations universal through his theory of relativity	

Other pioneers honored with SI units

Name	Dates	Role played	Units
Isaac Newton	1642–1727 A.D.	Formulated laws of motion and of universal gravitation. Newton's laws are to mechanics what Maxwell's equations are to electromagnetics.	newton (N)
James Watt	1736–1819	Pioneered applications of steam power.	watt (W)
Wilhelm E. Weber	1804–1891	Did pioneering work on terrestrial magnetism.	weber (Wb)

across space. Marconi added tuning, a large antenna, and ground systems and at longer wavelengths was able to signal over long distances. In 1901 he created a sensation by sending radio signals across the Atlantic Ocean. Marconi pioneered in developing radio communication for ships. Prior to radio, or "wireless," as it was then called, complete isolation enshrouded a ship at sea. Disaster could strike without anyone on the shore or other ships being aware that anything had happened. Marconi changed all that, and radio began to develop great commercial importance.

Thomas Alva Edison, the prolific American inventor, put electricity and magnetism to practical applications in telegraphy, telephony, lighting, and power generation and transmission. Whereas Edison was partial to direct current, Nikola Tesla developed alternating-current transmission of power and invented the induction motor. He designed the great power generating system at Niagara Falls. When it started operating in 1895, it generated as much power as all of the other generating stations in the United States combined. As a young man, Tesla had emigrated to the United States from what is now Yugoslavia.

Early thinkers believed a given instant of time had the same meaning for all observers, moving or stationary. However, about 1905, a critical analysis of these concepts by Albert Einstein, an examiner in the Swiss patent office, resulted in a new formulation of ideas about space and time. Einstein's relativity, or space-time, concept tells us that there is no such thing as a pure electric or a pure magnetic field which retains its identity for all observers. Thus, what appears to be a static electric field to a stationary observer appears, at least partially, as a magnetic field to a moving observer.

Until Einstein, gravity and electromagnetics were regarded as completely unrelated, but his prediction that electromagnetic waves passing near a massive object like a star would be bent, or refracted, by the object's gravitational field has been amply confirmed.

Einstein and others have sought to relate the five forces of physics in a *Grand Unified Theory* in which Maxwell's equations would be a special case. Such a unification has not yet been achieved, but its realization is a major goal of modern physics.

There are now few subjects understood as thoroughly as electromagnetics and few that have had greater practical application. Electric motors and generators, electric lighting and heating, telephones, radio, television, data links, medical electronics, radar, and remote sensing have completely changed our way of life. Hundreds of stationary communication satellites now ring the earth as though mounted on towers 36,000 kilometers high. And even now our probes are exploring the solar system to Saturn and beyond, responding to our commands and sending back pictures and data even though it takes more than an hour for the radio waves to travel the distance one way. With telescopes, both ground-based and in orbit, the universe is being explored to its limits at all electromagnetic wavelengths—from the shortest gamma rays to the longest radio waves. See Fig. 1-1. Our telescopes receive electromagnetic waves from objects so distant that the waves have been traveling for many billions of years.

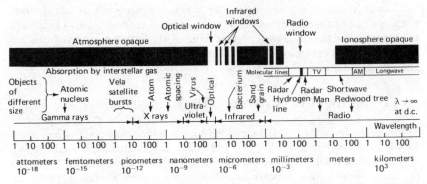

Figure 1-1 The electromagnetic spectrum with wavelength on a logarithmic scale from the shortest gamma rays to the longest radio waves. Wavelengths are expressed in metric units. The atmospheric opacity is shown at the top with the optical and radio windows in evidence.

Our civilization has been revolutionized by electromagnetics. We are, in fact, an electromagnetic society. But we would be very smug to think that we now know it all. H. G. Wells once wrote, "The past is but a beginning of a beginning, and all that is and has been is but the twilight of the dawn."

1-2 DIMENSIONS AND UNITS

Lord Kelvin is reported to have said:

When you can measure what you are speaking about and express it in numbers you know something about it; but when you cannot measure it, when you cannot express it in numbers your knowledge is of a meagre and unsatisfactory kind; it may be the beginning of knowledge but you have scarcely progressed in your thoughts to the stage of science whatever the matter may be.

To this it might be added that before we can measure something, we must define its dimensions and provide some standard, or reference unit, in terms of which the quantity can be expressed numerically.

A *dimension* defines some physical characteristic. For example, length, mass, time, velocity, and force are dimensions. The dimensions of length, mass, time, electric current, temperature, and luminous intensity are considered as the *fundamental dimensions* since other dimensions can be defined in terms of these six. This choice is arbitrary but convenient. Let the letters L, M, T, I, $\mathcal{T}$, and $\mathcal{I}$ represent the dimensions of length, mass, time, electric current, temperature, and luminous intensity. Other dimensions are then secondary dimensions. For example, area is a secondary dimension which can be expressed in terms of the fundamental dimension of length squared (L^2). As other examples, the fundamental dimensions of velocity are L/T and of force are ML/T^2.

A *unit* is a standard or reference by which a dimension can be expressed numerically. Thus, the meter is a unit in terms of which the dimension of length can be expressed, and the kilogram is a unit in terms of which the dimension of mass can be expressed. For example, the length (dimension) of a steel rod might be 2 meters and its mass (dimension) 5 kilograms.

1-3 FUNDAMENTAL AND SECONDARY UNITS

The units for the fundamental dimensions are called the *fundamental or base units.* In this book the International System of Units, abbreviated SI, is used.† In this system the *meter, kilogram, second, ampere, kelvin,* and *candela* are the base units for the six fundamental dimensions of length, mass, time, electric current, temperature, and luminous intensity. The definitions for these fundamental units are:

Meter (m). Length equal to 1,650,763.73 wavelengths in vacuum corresponding to the $2p_{10}-5d_5$ transition of krypton 86.

Kilogram (kg). Equal to mass of international prototype kilogram, a platinum-iridium mass preserved at Sèvres, France. This standard kilogram is the only artifact among the SI base units.

Second (s). Equal to duration of 9,192,631,770 periods of radiation corresponding to the transition between two hyperfine levels of the ground state of cesium 133. The second was formerly defined as 1/86,400 part of a mean solar day. The earth's rotation rate is gradually slowing down, but the atomic (cesium 133) transition is much more constant and is now the standard. The two standards differ by about 1 second per year.

Ampere (A). Electric current which if flowing in two infinitely long parallel wires in vacuum separated by 1 meter produces a force of 200 nanonewtons per meter of length (200 nN m^{-1} = 2 × 10^{-7} N m^{-1}).

Kelvin (K). Temperature equal to 1/273.16 of the triple point of water (or triple point of water equals 273.16 kelvins).‡

Candela (cd). Luminous intensity equal to that of 1/600,000 square meter of a perfect radiator at the temperature of freezing platinum.

The units for other dimensions are called *secondary* or *derived* units and are based on these fundamental units (see Table 2, Sec. A-1, in Appendix A).

The material in this book deals almost exclusively§ with the four fundamental dimensions *length, mass, time,* and *electric current* (dimensional symbols *L, M, T,* and *I*). The four fundamental units for these dimensions are the basis of what was

---

† The International System of Units is the modernized version of the metric system. The abbreviation SI is from the French name *Système Internationale d'Unités.* For the complete official description of the system see *U.S. Natl. Bur. Stand. Spec. Pub.* 330, 1971.

‡ Note that the symbol for degree is not used with kelvins. Thus, the boiling temperature of water is 373 kelvins (373 K), *not* 373°K. However, the degree sign is retained with degrees Celsius.

§ A couple of equations involve temperature. None involves luminous intensity.

formerly called the meter-kilogram-second-ampere (mksa) system, now a subsystem of the SI.

The complete SI involves not only units but also other recommendations, one of which is that multiples and submultiples of the SI units be stated in steps of 10^3 or 10^{-3}. Thus, the kilometer (1 km $= 10^3$ m) and the millimeter (1 mm $= 10^{-3}$ m) are preferred units of length, but the centimeter ($= 10^{-2}$ m) is not. For example, the proper SI designation for the width of motion-picture film is 35 mm, not 3.5 cm. For a list of the preferred units see Appendix A, Sec. A-1, Table 1. This table also gives the pronounciation, abbreviation, and derivation of these units.

In this book *rationalized* SI units are used. The rationalized system has the advantage that the factor 4π does not appear in Maxwell's equations although it does appear in certain other relations. A complete table of units in this system is given as Table 2, Sec. A-1, in Appendix A. The table lists dimensions, or quantities, alphabetically under each of the following headings: Fundamental, Mechanical, Electrical, and Magnetic. For each quantity the mathematical symbol (as used in equations), description, SI unit and abbreviation, equivalent units, and fundamental dimensions are listed.

It is suggested that as each new quantity and unit is discussed the student refer to the table and, in particular, become familiar with the fundamental dimensions for the quantity.

1-4 HOW TO READ THE SYMBOLS AND NOTATION

In this book *quantities*, or *dimensions*, which are scalars, like charge Q, mass M, or resistance R, are always in italics. Quantities which may be vectors *or* scalars are boldface as vectors and italics as scalars, e.g., electric field **E** (vector) or E (scalar). Unit vectors are always boldface with a hat (circumflex) over the letter, e.g., $\hat{\mathbf{x}}$ or $\hat{\mathbf{r}}$.†

Units are in roman type, i.e., *not* italic; for example, H for henry, s for second, or A for ampere.‡ The abbreviation for a unit is capitalized if the unit is derived from a proper name; otherwise it is lowercase (small letter). Thus, we have C for coulomb but m for meter. Note that when the unit is written out, it is always lowercase even though derived from a proper name. *Prefixes* for units are also roman, like n in nC for nanocoulomb or M in MW for megawatt. See Table 1, Sec. A-1, in Appendix A for a complete list of prefixes.

Example 1 $\mathbf{D} = \hat{\mathbf{x}}200 \text{ pC m}^{-2}$

means that the electric flux density **D** is a vector in the positive x direction with a magnitude of 200 picocoulombs per square meter ($= 2 \times 10^{-10}$ coulomb per square meter).

---

† In longhand notation a vector may be indicated by a bar over the letter and hat (^) over the unit vector.

‡ In longhand notation no distinction is usually made between quantities (italics) and units (roman). However, it can be done by placing a bar under the letter to indicate italics or writing the letter with a distinct slant.

Example 2 $$V = 10 \text{ V}$$

means that the voltage V equals 10 volts. Distinguish carefully between V (italics) for voltage, V (roman) for volts, **v** (lowercase, boldface) for velocity, and v (lowercase, italics) for volume.

Example 3 $$S = 4 \text{ W m}^{-2} \text{ Hz}^{-1}$$

means that the flux density S (a scalar) equals 4 watts per square meter per hertz. This can also be written $S = 4$ W/m^2/Hz or 4 W/(m^2 Hz), but the form W m^{-2} Hz^{-1} is more direct and less ambiguous.

Note that for conciseness, prefixes are used where appropriate instead of exponents. Thus, the velocity of light would be given as $c = 300$ Mm s^{-1} (300 megameters per second) and *not* 3×10^8 m s^{-1}. However, in solving a problem the exponential form (3×10^8 m s^{-1}) would be used.

The modernized metric (SI) units and the conventions used herein combine to give a concise, exact, and unambiguous notation, and if one is attentive to the details, it will be seen to possess both elegance and beauty.

1-5 EQUATION NUMBERING

Important equations and those referred to in the text are numbered consecutively beginning with each section. When reference is made to an equation in a different section, its number is preceded by the chapter and section number. Thus, (14-15-3) refers to Chap. 14, Sec. 15, Eq. (3). A reference to this same equation within Sec. 15 of Chap. 14 would read simply (3). Note that chapter and section numbers are printed at the top of each page.

1-6 DIMENSIONAL ANALYSIS

It is a necessary condition for correctness that every equation be balanced dimensionally. For example, consider the hypothetical formula

$$\frac{M}{L} = DA$$

where M = mass
L = length
D = density (mass per unit volume)
A = area

The dimensional symbols for the left side are M/L, the same as those used. The dimensional symbols for the right side are

$$\frac{M}{L^3} L^2 = \frac{M}{L}$$

Therefore, both sides of this equation have the dimensions of mass per length, and the equation is balanced dimensionally. This is not a guarantee that the equation is correct; i.e., it is not a *sufficient* condition for correctness. It is, however, a *necessary* condition for correctness, and it is frequently helpful to analyze equations in this way to determine whether or not they are dimensionally balanced.

Such *dimensional analysis* is also useful for determining what the dimensions of a quantity are. For example, to find the dimensions of force, we make use of Newton's second law that

$$\text{Force} = \text{mass} \times \text{acceleration}$$

Since acceleration has the dimensions of length per time squared, the dimensions of force are

$$\frac{\text{Mass} \times \text{length}}{\text{Time}^2}$$

or in dimensional symbols

$$\text{Force} = \frac{ML}{T^2}$$

PROBLEMS†

Group 1-1: Sec. 1-1. Electromagnetic history

1-1-1. Units Each of the following units is named after a person. What were their contributions? (*a*) Coulomb; (*b*) ampere; (*c*) volt; (*d*) ohm; (*e*) farad; (*f*) hertz; (*g*) newton; (*h*) joule; (*i*) watt.

1-1-2. Units List at least four famous scientists or inventors for which no units are (as yet) named.

1-1-3. Electromagnetic spectrum (*a*) How many octaves does the electromagnetic spectrum of Fig. 1-1 cover? (*b*) For how many octaves is the earth's atmosphere transparent?

Group 1-2: Secs. 1-2 through 1-6. Dimensions, units, and dimensional analysis

★**1-2-1. Dimensions and units** Give (*a*) the dimensional description, (*b*) the dimensional formulas in terms of the symbols M, L, T, and I, and (*c*) the SI units for each of the following:

$$\frac{dl}{dt} \quad \text{where } l = \text{length and } t = \text{time}$$

$$\int F \, dl \quad \text{where } F = \text{force}$$

$$\frac{dl}{dx}$$

★**1-2-2. Dimensions and units** Give the information requested in Prob. 1-2-1 for each of the following:

(*a*) $\iiint \rho \, dv$; (*b*) V; (*c*) E; (*d*) $\int \mathbf{E} \cdot d\mathbf{l}$; (*e*) $\dfrac{1}{4\pi\epsilon_0}$; (*f*) $\dfrac{Q^2}{4\pi\epsilon_0 r^2}$; (*g*) J; (*h*) BIl.

† Answers to starred problems are given in Appendix C.

★**1-2-3. Energy** Energy $\mathscr{E}$ (dimensional symbols ML^2/T^2) is a fundamental physical quantity. If $\mathscr{E}$ is used along with M, L, T, and I or Q (charge $= IT$) for the fundamental dimensional symbols of quantities, we note, for example, that

$$\text{Force} = \frac{\mathscr{E}}{L}$$

$$\text{Emf} = \frac{\mathscr{E}}{Q}$$

$$\text{Magnetic flux} = \frac{\mathscr{E}}{I}$$

Find the most concise $\mathscr{E}$, M, L, T, I, and Q notation for (a) work, (b) moment, (c) momentum, (d) power, (e) electric field E, (f) magnetic flux density B, and (g) energy density.

★**1-2-4. Dimensional balance** (a) Is dimensional balance of an equation a necessary condition for correctness? (b) A sufficient condition for correctness?

THE STATIC ELECTRIC FIELD

2-1 INTRODUCTION

In this chapter the basic relations for static electric fields in free space are discussed. They include Coulomb's law and the force per charge, Gauss' law and the electric flux, and the superposition of fields. The equivalence of the electric field to the gradient of the electric potential and of the electric potential to the line integral of the electric field are explained. The orthogonality of field lines and equipotential contours is pointed out. Shells of charge are also considered. Also mentioned are the simplifying concepts of staticness and idealness and the conventions of positiveness, right-handedness, and outwardness.

2-2 THE FORCE BETWEEN POINT CHARGES
AND COULOMB'S LAW

In his early electrical experiments, Thales noted that pieces of amber rubbed with silk attracted pieces of lint or straw. Amber, or matter in general in its normal, or neutral, state is regarded as containing equal amounts of positive and negative electricity (or electric charge). Rubbing transfers some charge from one object to the other so they are no longer neutral, one becoming positive and the other negative.

The charge transferred consists of electrons mechanically stripped from the outer atomic shells of the amber. Since electrons are considered to have a *negative* charge, the electron-deficient amber is said to be *positively* charged. Amber tends to retain such a charge because charges are not free to migrate through it. This behavior is typical of *insulators* (or *dielectrics*). Metals, on the other hand, are able to conduct or permit easy migration of charges and are classed as *conductors*.

(a)

(b) **Figure 2-1** Two point charges of same sign repel (a), but of opposite sign attract (b).

The basic experiment of electrostatics, reported by Coulomb in 1785,[†] involved small charged bodies.[‡] The results of this experiment are given by *Coulomb's law*, which states that the force F between two point charges Q_1 and Q_2 is proportional to the product of the charges and inversely proportional to the square of the distance r between them, i.e.,

$$F = k \frac{Q_1 Q_2}{r^2} \quad \text{(N)} \tag{1}$$

where k is a constant of proportionality. Because of the inverse-square effect of distance, this law is said to be an inverse-square law. The force is in the direction of the line connecting the charges. As suggested in Fig. 2-1, the force is outward (repulsive force) if the two charges are of the same sign, but inward (attractive force) if the two charges are of opposite sign.

In the International System the constant of proportionality is given by

$$k = \frac{1}{4\pi\epsilon}$$

where ϵ is the permittivity of the medium in which the charges are situated. By dimensional analysis of (1) we find that ϵ has the dimensions of capacitance per length, or, in dimensional symbols, $T^4 I^2 / M L^3$. (Capacitance is discussed in Sec. 3-8.) The SI unit for permittivity is the farad per meter (F m^{-1}). The permittivity of vacuum is

$$\epsilon_0 = 8.85 \times 10^{-12} \text{ F m}^{-1} = 8.85 \text{ pF m}^{-1}$$

$$\approx \frac{1}{36\pi} \times 10^{-9} \text{ F m}^{-1} = \frac{1}{36\pi} \text{ nF m}^{-1}$$

The permittivity of air is substantially the same as for vacuum. In this chapter it is assumed that the medium is air or vacuum. Thus, to be explicit, the permittivity is given as ϵ_0, although it is to be noted that the equations of this chapter can also apply for $\epsilon \neq \epsilon_0$ in an *unbounded* medium which is isotropic, homogeneous, and linear. The general situation of media for which $\epsilon \neq \epsilon_0$ is discussed in detail in Chap. 3.

† Charles A. de Coulomb, *History Royal Acad. Sci. (France)*, 1785, pp. 569 and 579.

‡ Ideally, it is convenient to regard them as *point charges*. This assumption leads to no appreciable error provided the size of the volume occupied by the charged particles is small compared with the other distances involved.

Force is a vector; i.e., it has both magnitude and direction. Rewriting (1) as a vector equation and also substituting the value of k, we have

$$\boxed{\mathbf{F} = \hat{\mathbf{r}}\, \frac{Q_1 Q_2}{4\pi\epsilon_0 r_{12}^2}} \tag{2}$$

where $\mathbf{F}$ = force, N†

$\hat{\mathbf{r}}$ = unit vector (see Fig. 2-1) pointing in direction of line joining charges; thus $\mathbf{F} = \hat{\mathbf{r}}F$

Q_1 = charge 1, C

Q_2 = charge 2, C

ϵ_0 = permittivity of vacuum, F m^{-1}

r_{12} = distance between point charges, m

This is the complete vector expression for Coulomb's law stated in rationalized SI units. To demonstrate the application of this law let us consider the following problems.

Example 1 A negative point charge of 1 μC is situated in air at the origin of a rectangular coordinate system. A second negative point charge of 100 μC is situated on the positive x axis at the distance of 500 mm from the origin. What is the force on the second charge?

SOLUTION By Coulomb's law the force

$$\mathbf{F} = \hat{\mathbf{x}}\, \frac{(-10^{-6})(-10^{-4})}{(4\pi \times 0.5^2)(10^{-9}/36\pi)}$$

$$= +\,\hat{\mathbf{x}}3.6 \text{ N}$$

That is, there is a force of 3.6 N (0.8 lb) in the positive x direction on the second charge.

Example 2 Two point charges of 1 C each and of the same sign are placed 1 mm apart in air. What is the magnitude of the repulsive force?

SOLUTION From Coulomb's law the force is

$$F = \frac{1 \times 1}{4\pi\epsilon_0 \times 10^{-6}} = 9 \times 10^{15} \text{ N}$$

$$\approx \text{ weight of } 9.2 \times 10^{11} \text{ metric tons}$$

$$\approx \text{ weight of } 10^{12} \text{ U.S. tons}$$

† One *newton* equals the force required to accelerate 1 kg 1 m s^{-2}.

This force is sufficient to lift millions of Empire State Buildings, or roughly all the buildings in the United States simultaneously. [Note that the force N at the earth's surface due to gravity equals the mass (kg) times 9.8 m s^{-2}.]

The Coulomb force between charged particles extends beyond such macroscopic situations even to the forces that bind the electrons of an atom to its nucleus or the forces that bind atoms into molecules or into larger aggregations of liquid or solid form. It is literally the force that holds solids, including you and me, together.

2-3 IDEALNESS AND STATICNESS

The concept of a point charge employed in Sec. 2-2 is an idealization. In many situations discussed in science and technology *idealness* is invoked in this manner in order to simplify the problem.

In Chaps. 2 and 3 it is important to note that it is the *static* electric field which is discussed. The implication here is that all charges and objects are at rest with respect to one another. This is also something of an idealization, but one which is useful in developing the basic concepts of field theory. In later chapters the situation under conditions of motion will be treated.

2-4 ELECTRIC FIELD INTENSITY

Consider a positive electric point charge Q_1 placed rigidly at the origin of a polar coordinate system. If another positive point charge Q_2 is brought into the vicinity of Q_1, it is acted upon by a force. This force is directed radially outward and becomes greater as Q_2 approaches Q_1. It may be said that Q_1 has a field around it where forces act. Thus, a *field* is a region where forces act.

The nature of the field around the point charge is suggested by the vector diagram of Fig. 2-2, the length of the vector being proportional to the force at the point.

Dividing (2-2-2) by Q_2 puts the equation in the dimensional form of force per charge, i.e.,

$$\frac{F}{Q_2} = \frac{\text{force}}{\text{charge}}$$

which has the dimensional symbols

$$\frac{ML}{T^2 Q} = \frac{ML}{T^3 I}$$

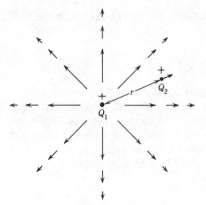

Figure 2-2 Point charge Q_1 with vectors indicating magnitude and direction of associated electric field.

If Q_2 is a *positive* test charge, the resulting *force per unit charge* is defined as the *electric field intensity* **E**. Thus,

$$\mathbf{E} = \frac{\mathbf{F}}{Q_2} = \hat{\mathbf{r}}\,\frac{Q_1}{4\pi\epsilon_0\,r^2} \qquad (1)$$

where Q_2 = positive test charge
r = distance of Q_2 from Q_1

The SI unit of electric field intensity is the newton per coulomb (N C^{-1}). As will appear after the discussion of electric potential (Sec. 2-7), an equivalent unit for the electric field intensity is the volt per meter (V m^{-1}).

According to (1), the point charge Q_1 is surrounded by an *electric field* of intensity **E** which is proportional to Q_1 and is inversely proportional to r^2. The electric field intensity **E** is a vector having the same direction as the force **F** but differing in dimensions and in numerical magnitude.

It is not implied by (1) that the positive test charge has a value of 1 C. It may have any convenient value since the ratio of the force (newtons) to the test charge (coulombs) is independent of the size of the charge provided the test charge does not disturb the field being measured. As noted in Example 2 in Sec. 2-2, 1 C represents a much larger charge than is ordinarily encountered in static problems. Thus, if we attempted to use a test charge of 1 C, we would tend to disturb the charges whose field we seek to measure. Therefore, it is necessary to use small test charges; in fact, the test charge should be small enough to ensure that it does not appreciably disturb the charge configuration whose field is to be measured.

If the test charge is made small enough, it may be regarded as of infinitesimal size, so that the ultimate value of the electric field intensity at a point becomes the force $\Delta\mathbf{F}$ on a positive test charge ΔQ divided by the charge (ΔQ) with the limit taken as the charge approaches zero; i.e.,

$$\mathbf{E} = \lim_{\Delta Q \to 0} \frac{\Delta\mathbf{F}}{\Delta Q} \tag{2}$$

Actually the smallest available test charge is an electron. Since this is a finite charge, it follows that $\mathbf{E}$ cannot be measured with unlimited accuracy. Although this is of importance in atomic problems, it need not concern us in the large-scale, or macroscopic, problems treated in this book. In practice, $\mathbf{E}$ would be measured with a small but finite test charge, and if this charge is small enough, $\mathbf{E}$ would differ inappreciably from that measured with an infinitesimal, or vanishingly small, test charge as implied in (2).

Example A negative point charge 10 nC is situated in air at the origin of a rectangular coordinate system. What is the electric field intensity at a point on the positive x axis 3 m from the origin?

SOLUTION By (1) the field intensity is given by

$$\mathbf{E} = -\hat{\mathbf{x}} \frac{10^{-8}}{(4\pi \times 3^2)(10^{-9}/36\pi)}$$

$$= -\hat{\mathbf{x}}10 \qquad \text{N C}^{-1} \text{ (or V m}^{-1})$$

That is, the electric field intensity is 10 N C^{-1} (or 10 V m^{-1}) and is in the negative x direction.

2-5 POSITIVENESS, RIGHT-HANDEDNESS, AND OUTWARDNESS

It will be convenient at the outset to adopt several conventions.

1. It is convenient to define the electric field, as in Sec. 2-4, in terms of *positive* charge. Thus, the electric field at any point is in the direction of the force on a *positive* test charge placed at that point.
2. It is convenient to standardize on right-handed coordinate systems. For example, the rectangular coordinate system in Fig. 2-3b is *right-handed* since in turning the *positive x* axis into the *positive y* axis and proceeding in the direction of a *right-handed* screw, we move in the *positive z* direction. This set of coordinate axes may accordingly be termed a *positive set*. The rectangular coordinate system in Fig. 2-3a is left-handed since in turning the positive x axis into the positive y axis, we must proceed like a left-handed screw to move in the positive z direction.
3. It is a consequence of the positiveness convention that the electric field around a positive charge is *outward* (see Fig. 2-2). A further consequence is that we

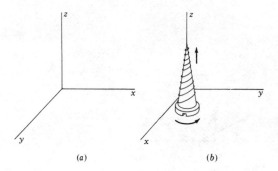

Figure 2-3 (*a*) Left-handed and (*b*) right-handed coordinate systems. Right-handed screw turned as shown moves in $+z$ direction.

take the *positive* direction of the normal at any point on a closed surface as the *outward* normal.

The concept of positiveness is inherent in our definitions and, as discussed above, is also associated with the definitions of right-handedness and outwardness.

2-6 THE ELECTRIC FIELD OF SEVERAL POINT CHARGES AND THE PRINCIPLE OF SUPERPOSITION OF FIELDS

Since the electric field of a point charge is a linear function of the value of the charge, it follows that the fields of more than one point charge are linearly superposable by vector addition. As a generalization, this fact may be stated as the *principle of superposition* applied to electric fields as follows:

The total or resultant field at a point is the vector sum of the individual component fields at the point.

Thus, referring to Fig. 2-4, the field intensity of the charge Q_1 at the point P is $\mathbf{E}_1$ and of the charge Q_2 is $\mathbf{E}_2$. The total field at P due to both charges is the vector sum of $\mathbf{E}_1$ and $\mathbf{E}_2$, or $\mathbf{E}$, as indicated in the figure.

Example 1 A charge Q_1 of 1 nC is situated at the origin ($x = 0, y = 0$), and a charge Q_2 of -2 nC is situated on the y axis 1 m from the origin ($x = 0, y = 1$), as shown in Fig. 2-5. Find the total electric field intensity at the point P on the x axis 2 m from the origin ($x = 2, y = 0$).

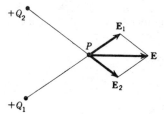

Figure 2-4 Vector addition of fields due to two equal point charges of the same sign to give resultant, or total field **E**.

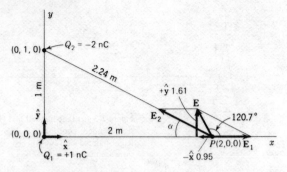

Figure 2-5 Vector addition of fields due to two unequal point charges of opposite sign to give resultant, or total field **E**.

SOLUTION The vector value of the electric field $\mathbf{E}_1$ at P at the point $(2, 0)$ due to the charge Q_1 at the point $(0, 0)$ is, from (2-4-1),

$$\mathbf{E}_1 = \hat{\mathbf{x}} \frac{10^{-9}}{(4\pi \times 2^2)(10^{-9}/36\pi)}$$

$$= \hat{\mathbf{x}} 2.25 \ \text{N} \ \text{C}^{-1}$$

The magnitude of the field $\mathbf{E}_2$ at P due to the charge Q_2 at $(0, 1)$ is

$$\mathbf{E}_2 = \frac{-2 \times 10^{-9}}{(4\pi \times 2.24^2)(10^{-9}/36\pi)}$$

$$= -3.59 \ \text{N} \ \text{C}^{-1}$$

The vector value of $\mathbf{E}_2$ is given by

$$\mathbf{E}_2 = -\hat{\mathbf{x}} 3.59 \cos \alpha + \hat{\mathbf{y}} 3.59 \sin \alpha$$

$$= -\hat{\mathbf{x}} 3.59 \frac{2}{2.24} + \hat{\mathbf{y}} 3.59 \frac{1}{2.24}$$

$$= -\hat{\mathbf{x}} 3.2 + \hat{\mathbf{y}} 1.6 \ \text{N} \ \text{C}^{-1}$$

where $\hat{\mathbf{x}}$ = unit vector in x direction
$\hat{\mathbf{y}}$ = unit vector in y direction

The total vector field **E** at P can be obtained by graphical vector addition of $\mathbf{E}_1$ and $\mathbf{E}_2$ or analytically as follows:

$$\mathbf{E} = \hat{\mathbf{x}}(2.25 - 3.2) + \hat{\mathbf{y}} 1.6$$

or in both rectangular and polar forms

$$\mathbf{E} = -\hat{\mathbf{x}} 0.95 + \hat{\mathbf{y}} 1.61 = 1.86 \underline{/120.7^\circ} \ \text{N} \ \text{C}^{-1}$$

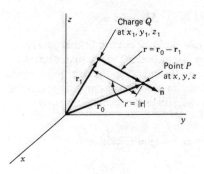

Figure 2-6 Geometry for full vector notation of field $\mathbf{E}$ at point $P(x, y, z)$ due to charge Q at (x_1, y_1, z_1).

In full vector notation, we have, in general, that $\mathbf{E}$ at a point P (x, y, z) due to a charge Q at a point (x_1, y_1, z_1) (see Fig. 2-6) is given by

$$\mathbf{E} = \frac{\mathbf{r}}{|\mathbf{r}|} \frac{Q}{4\pi\epsilon_0 r^2} = \hat{\mathbf{r}} \frac{Q}{4\pi\epsilon_0 r^2} \qquad (\text{N C}^{-1}) \tag{1}$$

where $\hat{\mathbf{r}} = \dfrac{\mathbf{r}}{|\mathbf{r}|} = \dfrac{\mathbf{r}}{r}$ = unit vector in direction of vector $\mathbf{r}$ where $\mathbf{r}(=|\mathbf{r}|\hat{\mathbf{r}})$ has the length r and direction of P from the charge Q and where, in rectangular coordinates,

$$\mathbf{r} = \hat{\mathbf{x}}(x - x_1) + \hat{\mathbf{y}}(y - y_1) + \hat{\mathbf{z}}(z - z_1)$$

$$r = \sqrt{(x - x_1)^2 + (y - y_1)^2 + (z - z_1)^2}$$

and

$$\mathbf{E} = \frac{Q[\hat{\mathbf{x}}(x - x_1) + \hat{\mathbf{y}}(y - y_1) + \hat{\mathbf{z}}(z - z_1)]}{4\pi\epsilon_0[(x - x_1)^2 + (y - y_1)^2 + (z - z_1)^2]^{3/2}} \tag{2}$$

Example 2 Repeat Example 1 of Fig. 2-5 using full vector notation.

SOLUTION We have

$$Q_1 \ (1 \text{ nC}) \text{ at } x_1 = 0, y_1 = 0, z_1 = 0$$

$$Q_2 \ (-2 \text{ nC}) \text{ at } x_2 = 0, y_2 = 1, z_2 = 0$$

$$\text{Point } P \text{ at } x = 2, y \ = 0, z \ = 0$$

Therefore, from (2),

$$\mathbf{E}_1 = \frac{9[\hat{\mathbf{x}}2 + \hat{\mathbf{y}}0 + \hat{\mathbf{z}}0]}{[4]^{3/2}} = \hat{\mathbf{x}}2.25 + \hat{\mathbf{y}}0 + \hat{\mathbf{z}}0 \text{ N C}^{-1}$$

$$\mathbf{E}_2 = \frac{-18[\hat{\mathbf{x}}2 - \hat{\mathbf{y}}1 + \hat{\mathbf{z}}0]}{[5]^{3/2}} = -\hat{\mathbf{x}}3.2 + \hat{\mathbf{y}}1.61 + \hat{\mathbf{z}}0 \text{ N C}^{-1}$$

Adding $\mathbf{E}_1$ and $\mathbf{E}_2$ yields the total $\mathbf{E}$ at P as follows:

$$\mathbf{E} = \mathbf{E}_1 + \mathbf{E}_2 = -\hat{\mathbf{x}}0.95 + \hat{\mathbf{y}}1.61 + \hat{\mathbf{z}}0 \text{ N C}^{-1}$$

2-7 THE ELECTRIC SCALAR POTENTIAL

Consider two points, x_1 and x_2, situated in a uniform electric field **E** parallel to the x direction. Let a *positive* test charge at x_2 be moved in the negative x direction to x_1, as in Fig. 2-7. The field exerts a force on the charge so that it requires work to move the charge against the force. The amount of work per unit charge is equal to the force per unit charge (or field intensity **E**) times the distance through which the charge is moved. Thus,

$$E(x_2 - x_1) = \text{work per unit charge} \qquad \text{(joules/coulomb, J C}^{-1}) \qquad (1)$$

The dimensions are

$$\frac{\text{Force}}{\text{Charge}} \times \text{length} = \frac{\text{work}}{\text{charge}}$$

or

$$\frac{ML}{T^2}\frac{L}{Q} = \frac{ML^2}{T^3 I}$$

In SI units the relation is

$$\frac{\text{Newtons}}{\text{Coulomb}} \times \text{meters} = \frac{\text{joules}}{\text{coulomb}}$$

The dimensions of work per charge are those of potential. In our example (Fig. 2-7), the work or energy per unit charge required to transport the test charge from x_2 to x_1 is called the difference in *electric potential*† of the points x_2 and x_1. The point x_1 has the higher potential since it requires work to reach it from point x_2. Thus, moving from x_2 to x_1 (opposite to **E**), we experience a *rise* in potential. The unit of electric potential V is the volt (V) and is equal to 1 J C^{-1}. Hence, electric potential is expressible either in joules per coulomb or in volts.

$$\frac{\text{Newtons}}{\text{Coulomb}} \times \text{meters} = \frac{\text{joules}}{\text{coulomb}} = \text{volts}$$

Dividing by meters, we obtain

$$\frac{\text{Newtons}}{\text{Coulomb}} = \frac{\text{volts}}{\text{meter}} = \text{electric field intensity}$$

Thus, the electric field intensity **E** is expressible in either newtons per coulomb or in volts per meter.

---

† *Potential*, in general, is a measure of energy per some kind of unit quantity. For example, the difference in gravitational potential at sea level and 100 m above sea level is equal to the work required to raise a 1-kg mass from sea level to a height of 100 m against the earth's gravitational field. Potential is a scalar quantity; i.e., it has magnitude but no direction.

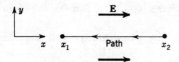

Figure 2-7 Linear path in uniform electric field.

Example Let the uniform electric field in Fig. 2-7 have an intensity **E** of 10 V m^{-1}. If the distance $x_2 - x_1$ is 100 mm, what is the potential difference of the two points?

SOLUTION From (1) the electric potential is given by

$$V = 10 \times 0.1 = 1 \text{ V}$$

That is, the potential of x_1 is 1 V higher than the potential of x_2.

Consider next the case of a nonuniform field such as exists in the vicinity of the positive point charge Q (Fig. 2-8a). The electric field **E** is radial and is inversely proportional to the square of the distance r from the charge Q. The energy per coulomb required to move a *positive* test charge from r_2 to r_1 along a radial path equals the potential difference, or rise, V_{21} between the points. This is given by

$$V_{21} = \int_{r_2}^{r_1} dV = - \int_{r_2}^{r_1} E \, dr \qquad \text{(V)}$$

The negative sign takes into account the fact that the motion from r_2 to r_1 is opposite to the direction of the field. Substituting the value of E from (2-4-1) yields

$$V_{21} = V_1 - V_2 = - \int_{r_2}^{r_1} \frac{Q}{4\pi\epsilon_0 r^2} \, dr$$

$$= - \frac{Q}{4\pi\epsilon_0} \int_{r_2}^{r_1} \frac{dr}{r^2} = \frac{Q}{4\pi\epsilon_0} \left(\frac{1}{r_1} - \frac{1}{r_2} \right) \qquad (2)$$

where $V_1 =$ potential at point r_1
        $V_2 =$ potential at point r_2

The potential difference in (2) is positive since work must be expended to move the positive test charge from r_2 to r_1 against the field. However, if the motion is

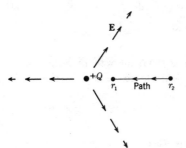

Figure 2-8a Linear path in nonuniform electric field.

from r_1 to r_2, the field does work on the charge and there is a fall in potential (negative potential difference).

If the point r_2 (Fig. 2-8a) is removed to infinity, we can arbitrarily define it to be at zero potential. Thus, (2) becomes

$$V_1 = \frac{Q}{4\pi\epsilon_0 r_1} \quad \text{(V)} \tag{3}$$

This potential is called the *absolute potential* of the point r_1 due to the charge Q. It is inversely proportional to the distance from Q to the point r_1 and is, by definition, the work per coulomb required to bring a positive test charge from infinity to the point r_1.

2-8 THE ELECTRIC SCALAR POTENTIAL AS A LINE INTEGRAL OF THE ELECTRIC FIELD

In Sec. 2-7 the test charge is moved via the shortest path between two points. Actually, the path followed is immaterial since the potential difference is determined solely by the difference in potential of the two end points of the path. Thus, referring to Fig. 2-8a, the potential at the point r_1 with respect to the potential at r_2 is said to be *single-valued*; i.e., it can have only one value regardless of the path taken from r_2 to r_1. When the path of the test charge is not parallel to **E** but at an angle θ, as in Fig. 2-8b, the potential difference V_{21} between the points x_2 and x_1 is equal to the path length $x_2 - x_1$ multiplied by the component of **E** parallel to it. Thus, $V_{21} = (x_2 - x_1)E \cos\theta$.

If the test charge is moved perpendicular to the direction of the field ($\theta = 90°$), no work is performed and hence this path is said to be an *equipotential* line. It is an important property of fields that equipotential and field lines are orthogonal.

Let us consider next the case where the path of the test charge is curved. Then the potential difference between the end points of the path is given by the product of the infinitesimal element of path length dl and the component of **E** parallel to it, integrated over the length of the path from a to b. For the path in the uniform field **E** in Fig. 2-9, the infinitesimal potential rise dV between the ends of the path element dl is given by

$$dV = -E \cos\theta \, dl = -\mathbf{E} \cdot d\mathbf{l} \tag{1}$$

where $dl = |d\mathbf{l}|$ = magnitude of path element $d\mathbf{l}$
θ = angle between $d\mathbf{l}$ and field **E**

A potential rise (positive potential difference dV) requires that the component of the motion parallel to **E** be opposed to the field. Hence the negative sign in (1).

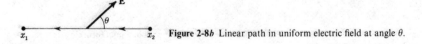

Figure 2-8b Linear path in uniform electric field at angle θ.

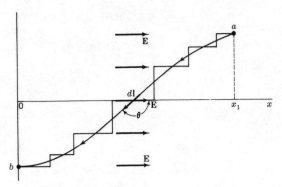

Figure 2-9 Curved path in a uniform electric field.

By integrating (1) between the points a and b we obtain the potential rise V_{ab} between the points a and b. Thus,

$$V_{ab} = \int_a^b dV = V_b - V_a = - \int_a^b E \cos \theta \, dl = - \int_a^b \mathbf{E} \cdot d\mathbf{l} \qquad (2)$$

Referring to the zigzag path in Fig. 2-9, we note that contributions to the work occur only when the path has a component parallel to $\mathbf{E}$. Hence, the work in going from a to b is the sum of the work increments along the zigzag-path elements parallel to $\mathbf{E}(\theta = 180°)$ taken in the limit as the increments approach zero with no work contribution from the zigzag-path elements perpendicular to $\mathbf{E}$ ($\theta = 90°$).

The integral involving dl in (2) is called a *line integral*. Hence, the potential rise between a and b equals the line integral of $\mathbf{E}$ along the curved path from a to b. As written in (2), the result can be expressed either in scalar form (with cos θ) or in vector form as a scalar or dot product.

Example 1 In Fig. 2-9 let $\mathbf{E}$ be everywhere in the $+x$ direction and equal to 10 V m^{-1} (a uniform field). Let $x_1 = 1$ m. Find V_{ab}.

SOLUTION From (2)

$$V_{ab} = - \int_a^b E \cos \theta \, dl = - \int_{x_1}^0 E \, dx = E \, x_1 = +10 \text{ V}$$

Consider now the situation where the path of the test charge is curved and the electric field is nonuniform. Let the nonuniform field be produced by a point charge $+Q$ as in Fig. 2-10. The field intensity due to a point charge is given by (2-4-1). Substituting this in (2) and also putting $dr = \cos \theta \, dl$, where dr is an infinitesimal element of radial distance, gives

$$V_{ab} = - \frac{Q}{4\pi\epsilon_0} \int_{r=a}^{r=b} \frac{dr}{r^2} = \frac{Q}{4\pi\epsilon_0} \left(\frac{1}{b} - \frac{1}{a} \right) \qquad (\text{V}) \qquad (3)$$

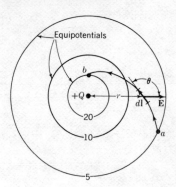

Figure 2-10 Curved path in a nonuniform electric field.

Putting $b = r_1$ and $a = r_2$ makes this result identical with (2-7-2), where the path is along a radial line.

Example 2 Let the positive charge Q, Fig. 2-10, be equal to 223 pC. Also let $a = 400$ mm and $b = 100$ mm. The medium is air. Find the absolute potential V_a at a, the absolute potential V_b at b, and the potential rise V_{ab}

SOLUTION

$$V_a = \frac{Q}{4\pi\epsilon_0} \frac{1}{a} \approx 5 \text{ V}$$

$$V_b = \frac{Q}{4\pi\epsilon_0} \frac{1}{b} \approx 20 \text{ V}$$

$$V_{ab} = V_b - V_a \approx 15 \text{ V}$$

The work to move a test charge along an equipotential contour or surface is zero ($\theta = 90°$). The maximum amount of work per unit distance is performed by moving normal to an equipotential surface. This coincides with the direction of the field.

The work to transport a test charge around any *closed path* in a static field is zero since the path starts and ends at the same point. Thus, the upper and lower limits of the integrals in (2) become the same, and the result is zero. A property of the *static electric field* is, then, that the *line integral of this field around a closed path is zero*, that is,

$$\oint \mathbf{E} \cdot d\mathbf{l} = 0 \qquad (4)$$

A field for which (4) holds is called a *conservative or lamellar field*. It follows that the potential difference between any two points of a conservative field is independent of the path.

2-9 RELATION OF ELECTRIC FIELD LINES AND EQUIPOTENTIAL CONTOURS; ORTHOGONALITY

A field line indicates the direction of the force on a positive test charge introduced into the field. If the test charge is released, it moves in the direction of the field line.

In a uniform field, the field lines are parallel, as in Fig. 2-11. A single field line gives no information about the intensity of the field. It indicates only the direction. However, by measuring the work per coulomb required to move a positive test charge along a field line the potential differences along the line can be determined. The larger the potential difference between two points a unit distance apart, the stronger the field. In a uniform field the potential difference per unit length is constant, so that the equipotential lines (which are orthogonal to the field lines) are equally spaced. In the example of Fig. 2-11, the electric field intensity is 0.2 V mm^{-1}, so that the equipotential contours at 1-V intervals are parallel lines spaced 5 mm apart. One of the lines is arbitrarily taken as having a zero potential so that the potentials shown are relative to this line.

Consider now the case of a nonuniform field such as exists in the vicinity of the positive point charge Q in Fig. 2-12. If a positive test charge is released in this field, it moves radially away from Q, so that the field lines are radial. The field intensity varies inversely as the square of the distance. In Fig. 2-12 this is suggested by the fact that the field lines become more widely separated as the distance from Q increases. The absolute potential is inversely proportional to the distance from Q. If $Q = 10$ pC, the equipotential contours for 20, 10, 5, and 3 V are then as shown by the concentric circles in Fig. 2-12.

It is to be noted that a potential *rise* is always in the opposite direction to **E**.

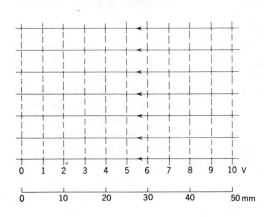

Figure 2-11 Field lines (solid) and equipotential lines (dashed) of a uniform electric field.

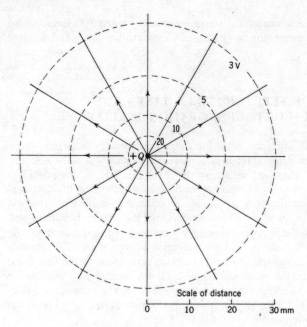

Figure 2-12 Field lines (solid) and equipotential lines (dashed) of a nonuniform electric field.

2-10 FIELD OF TWO EQUAL POINT CHARGES OF OPPOSITE SIGNS AND OF SAME SIGN

The electric field at a point P due to two point charges $+Q$ and $-Q$ is equal to the vector sum of the fields at P due to each of the charges alone. This is illustrated in Fig. 2-13. The potential V at P is equal to the algebraic sum of the potentials at P due to each charge alone.

A map of the field lines (solid) and equipotential contours (dashed) is shown in Fig. 2-13 for point charges $+Q$ and $-Q$ separated by 127 mm. The equipotential contours are given in volts for $Q = 140$ pC. The charge configuration in Fig. 2-13 constitutes an electric dipole with a charge separation of 127 mm.

In contrast to the above let us consider two equal point charges of the same sign (positive) as in Fig. 2-14. A map of the field lines (solid) and equipotential contours (dashed) is shown for a charge separation of 127 mm. The equipotential contours are given in volts for $Q = 140$ pC. The only difference between the charge configuration of Fig. 2-14 and that in Fig. 2-13 is that the lower charge is positive.

Near each charge the effect of the other charge is small, and the equipotentials are circles like those around an isolated point charge. For intermediate distances the

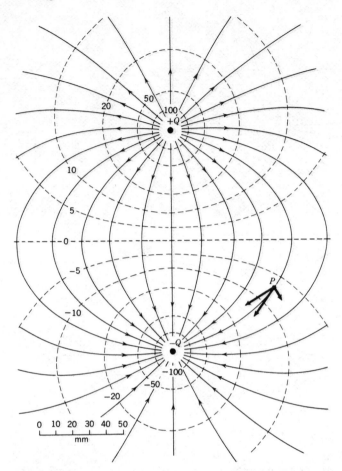

Figure 2-13 Electric field and potential variation around an electric dipole consisting of positive and negative charges of 140 pC separated by 127 mm. The solid lines are field lines, and the dashed lines are equipotential contours, with their potential level indicated in volts.

equipotentials have the shapes shown in Fig. 2-14. Of particular interest is the figure-eight-shaped equipotential ($V = 39.5$ V) that crosses itself at the point P, where $\mathbf{E} = 0$. This point is called a *singular point*. At such a point, field and equipotential lines are not perpendicular.

Note that in three dimensions the (dashed) equipotential lines are surfaces generated by rotating Figs. 2-13 and 2-14 about the (vertical) axis through the charges.

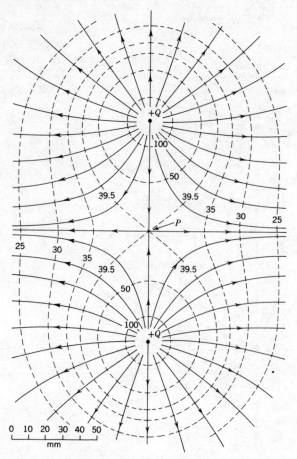

Figure 2-14 Electric field and potential variation around two equal positive charges of 140 pC separated by 127 mm. The solid lines are field lines, and the dashed lines are equipotential contours, with their potential level indicated in volts.

2-11 CHARGE DENSITY AND CONTINUOUS DISTRIBUTIONS OF CHARGE

The (average) electric *charge density* ρ is equal to the total charge Q in a volume v divided by the volume. Thus,

$$\rho = \frac{Q}{v} \tag{1}$$

Electric charge density has the dimensions of charge per unit volume, or in dimensional symbols $Q/L^3 = IT/L^3$. The SI unit of charge density is the *coulomb per cubic meter* $(C\ m^{-3})$.

By assuming that electric charge can be continuously distributed throughout a region we can also define the value of the charge density ρ at a point P as the charge ΔQ in a small volume element Δv divided by the volume, with the limit of this ratio taken as the volume shrinks to zero around the point P. In symbols,

$$\rho = \lim_{\Delta v \to 0} \frac{\Delta Q}{\Delta v} \tag{2}$$

This gives the value of ρ at a point and hence defines ρ as a point function.

It will be convenient to use this definition of ρ, but it is to be noted that it is based on the assumption that the electric charge is continuously distributed. Actually electric charge is not continuously distributed but is associated with discrete particles (electrons or atoms) separated by finite atomic distances. Nevertheless, the assumption of a continuous charge distribution (an idealization) leads to no appreciable error provided the region contains many atoms or electrons and the distances involved are large compared with atomic dimensions. The assumption of continuous charge distribution can be applied to the large-scale, or macroscopic, problems treated in this book but would not be applicable to problems on atomic structure, where the noncontinuous nature of the charge distribution must be taken into account.

The charge density ρ, discussed above, is sometimes called a *volume charge density* to distinguish it from surface charge density and linear charge density. The *surface charge density* ρ_s gives the charge per unit area (coulombs per square meter, $C\ m^{-2}$) at a point on a continuous surface distribution of charge. The *linear charge density* ρ_L gives the charge per unit length (coulombs per meter, $C\ m^{-1}$) at a point on a continuous line distribution of charge. Both ρ_s and ρ_L are point functions which can be defined as in (2), with a surface or line element substituted for the volume element.

2-12 ELECTRIC POTENTIAL OF CHARGE DISTRIBUTIONS AND THE PRINCIPLE OF SUPERPOSITION OF POTENTIAL

Since the electric scalar potential due to a single point charge is a linear function of the value of its charge, it follows that the potentials of more than one point charge are linearly superposable by scalar (algebraic) addition. As a generalization, this fact may be stated as the *principle of superposition* applied to electric potential† as follows:

The total electric potential at a point is the algebraic sum of the individual component potentials at the point.

† Although electric *scalar* potential is implied, the word *scalar* will usually be omitted for brevity.

Thus, if only the three point charges Q_1, Q_2, and Q_3 are present in Fig. 2-15, the total electric potential (work per unit charge) at the point P is given by

$$V_p = \frac{1}{4\pi\epsilon_0}\left(\frac{Q_1}{r_1} + \frac{Q_2}{r_2} + \frac{Q_3}{r_3}\right) \tag{1}$$

where r_1 = distance from Q_1 to P
r_2 = distance from Q_2 to P
r_3 = distance from Q_3 to P

This can also be expressed with a summation sign. Thus,

$$V_p = \frac{1}{4\pi\epsilon_0} \sum_{n=1}^{3} \frac{Q_n}{r_n} \tag{2}$$

If the charge is not concentrated at a point but is distributed along a line as in Fig. 2-15, the potential at P due to this linear charge distribution is

$$V_L = \frac{1}{4\pi\epsilon_0} \int \frac{\rho_L}{r}\, dl \quad \text{(V)} \tag{3}$$

where ρ_L = linear charge density, C m^{-1}
dl = element of length of line, m

The integration is carried out over the entire line of charge.

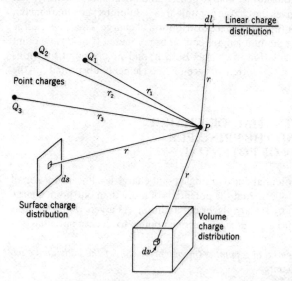

Figure 2-15 Electric potential at P is the algebraic sum of the potentials due to the point, line, surface, and volume distributions of charge.

When the charge is distributed over a surface, as in Fig. 2-15, the potential at P caused by this surface charge distribution is

$$V_s = \frac{1}{4\pi\epsilon_0} \iint \frac{\rho_s}{r} ds \quad \text{(V)} \tag{4}$$

where ρ_s = surface charge density, C m^{-2}
ds = element of surface, m^2

The integration is carried out over the entire surface of charge.
For a volume charge distribution, as in Fig. 2-15,

$$V_v = \frac{1}{4\pi\epsilon_0} \iiint \frac{\rho}{r} dv \quad \text{(V)} \tag{5}$$

where ρ = (volume) charge density, C m^{-3}
dv = element of volume, m^3

The integration is taken throughout the volume containing charge.
If the point charges, the line charge distribution, the surface charge distribution, and the volume charge distribution of Fig. 2-15 are all present simultaneously, the total electric potential at the point P due to all these distributions is by the superposition principle the algebraic sum of the individual component potentials. Thus, in general,

$$V = V_p + V_L + V_s + V_v \tag{6}$$

or

$$\boxed{V = \frac{1}{4\pi\epsilon_0} \left(\sum_1^N \frac{Q_n}{r_n} + \int \frac{\rho_L}{r} dl + \iint \frac{\rho_s}{r} ds + \iiint \frac{\rho}{r} dv \right)} \tag{7}$$

Example As shown in Fig. 2-16 a square that is 1 m on a side in air has a point charge $Q_1 = +1$ pC at the upper right corner, a point charge $Q_2 = -10$ pC at the lower right corner, and a line distribution of charge of uniform density $p_L = +10$ pC m^{-1} along the left edge. Find the potential at the point P at the center of the square.

SOLUTION The potential at P due to the point charges is

$$V_p = \frac{1}{4\pi\epsilon_0} \left(\frac{10^{-12}}{0.707} - \frac{10^{-11}}{0.707} \right) = -0.115 \text{ V}$$

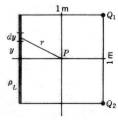

Figure 2-16 Line and point charges for example illustrating superposition of electric potential.

The potential at P caused by the line distribution of charge is

$$V_L = \frac{1}{4\pi\epsilon_0} \int_{y=-0.5}^{y=0.5} \frac{10^{-11}}{\sqrt{0.5^2 + y^2}} \, dy = +0.158 \text{ V}$$

The total potential at P is then

$$V = V_p + V_L = +43 \text{ mV}$$

The principle of superposition stated for the special cases of potential in this section and for fields in Sec. 2-6 can be applied, in general, to any quantity which is linearly related to its cause. The electric fields or potentials at a point are linear functions of the charge producing them and hence are superposable (by vector addition for fields and scalar addition for potential).

2-13 THE ELECTRIC FIELD AS THE GRADIENT OF THE ELECTRIC POTENTIAL

The potential rise between two points along an electric field line is a measure of the *gradient* of the potential in the same way that the elevation rise between two points on a slope is a measure of the gradient of the slope. More specifically the gradient of the potential at a point is defined as the potential rise ΔV across an element of length Δl *along a field line* divided by Δl, with the limit of this ratio taken as Δl approaches zero. Thus,

$$\text{Gradient of } V = \lim_{\Delta l \to 0} \frac{\Delta V}{\Delta l} = \frac{dV}{dl} \qquad (1)$$

More completely, the gradient of V is a vector whose direction is along a field line with magnitude as given in (1). Since a potential *rise* occurs when moving *against* the electric field (see Sec. 2-8), the *direction of the gradient is opposite to that of the field*. Thus,

$$\text{Gradient of } V = -\mathbf{E}$$

In more symbolic notation we can write

$$\boxed{\text{Gradient of } V = \text{grad } V = \nabla V = -\mathbf{E}} \qquad (2)$$

where grad V stands for the gradient of V. As will be discussed in the next section, the gradient of V can also be expressed with the operator del, or nabla, ∇, as ∇V.

2-14 GRADIENT IN RECTANGULAR COORDINATES

In this section a relation for gradient will be developed in rectangular coordinates. To do this, consider the electric potential distribution of Fig. 2-17. The work per coulomb to bring a positive test charge to the point P (at origin of coordinates) is

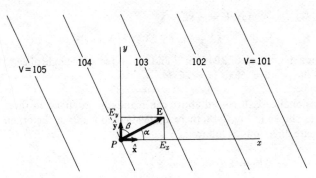

Figure 2-17 Potential distribution with electric field **E** at a point P.

104 V. This is the absolute potential V at P. The potential elsewhere is a function of both x and y, and its variation is indicated by the equipotential contours. The field is uniform. Thus, the contours are straight, parallel, and equally spaced. There is no variation with respect to z (normal to page). At P the electric field is as indicated by the vector **E**, perpendicular to the equipotential line.

Consider now the change in potential along an infinitesimal element of path length in the x direction ($y = $ constant). Then

$$-\frac{\partial V}{\partial x} = E \cos \alpha = E_x$$

where $\alpha = $ angle between **E** and x axis
$E_x = $ component of **E** in x direction

Likewise, for an infinitesimal element of path length in the y direction,

$$-\frac{\partial V}{\partial y} = E \cos \beta = E_y$$

where $\beta = $ angle between **E** and y axis
$E_y = $ component of **E** in y direction

By the principle of superposition the total field **E** at the point P is the vector sum of the component fields at the point. Hence,

$$\mathbf{E} = \hat{\mathbf{x}}E_x + \hat{\mathbf{y}}E_y = -\left(\hat{\mathbf{x}}\frac{\partial V}{\partial x} + \hat{\mathbf{y}}\frac{\partial V}{\partial y}\right) = -\text{grad } V = -\nabla V$$

Thus, the gradient in this rectangular two-dimensional case is equal to the x and y derivatives of the potential added vectorially.

Example Suppose that in Fig. 2-17 the potential decreases by 2 V m^{-1} in the x direction and by 1 V m^{-1} in the y direction. Find the electric field **E**.

SOLUTION
$$\text{grad } V = -\hat{\mathbf{x}}2 - \hat{\mathbf{y}}1$$

and
$$\mathbf{E} = -\text{grad } V = \hat{\mathbf{x}}2 + \hat{\mathbf{y}}1 = 2.24\underline{/26.6°} \text{ V m}^{-1}$$

Therefore, **E** has a magnitude of 2.24 V m^{-1} and is directed at an angle of 26.6° with respect to the positive x axis ($\alpha = 26.6°$).

The two-dimensional case discussed above can readily be extended to three dimensions. Thus, as shown in Fig. 2-18, there are field components at the origin in the three coordinate directions as follows:

$$\hat{\mathbf{x}}E_x = \hat{\mathbf{x}}E \cos \alpha = -\hat{\mathbf{x}}\frac{\partial V}{\partial x}$$

$$\hat{\mathbf{y}}E_y = \hat{\mathbf{y}}E \cos \beta = -\hat{\mathbf{y}}\frac{\partial V}{\partial y}$$

$$\hat{\mathbf{z}}E_z = \hat{\mathbf{z}}E \cos \gamma = -\hat{\mathbf{z}}\frac{\partial V}{\partial z}$$

By the principle of superposition the total field **E** at the origin is the vector sum of the component fields, or

$$\mathbf{E} = -\left(\hat{\mathbf{x}}\frac{\partial V}{\partial x} + \hat{\mathbf{y}}\frac{\partial V}{\partial y} + \hat{\mathbf{z}}\frac{\partial V}{\partial z}\right) \tag{1}$$

where the relation in the parentheses is the complete expression in rectangular coordinates for the gradient of V. It is often convenient to consider that this expression is the product of V and the operator del (∇). Thus, in rectangular coordinates

$$\nabla = \hat{\mathbf{x}}\frac{\partial}{\partial x} + \hat{\mathbf{y}}\frac{\partial}{\partial y} + \hat{\mathbf{z}}\frac{\partial}{\partial z} \tag{2}$$

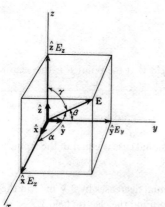

Figure 2-18 Components of electric field in rectangular coordinates.

The quantity ∇ is a vector operator. It is meaningless until applied. Taking the product of ∇ and V yields the gradient of V. That is,

$$\nabla V = \hat{\mathbf{x}}\frac{\partial V}{\partial x} + \hat{\mathbf{y}}\frac{\partial V}{\partial y} + \hat{\mathbf{z}}\frac{\partial V}{\partial z} = \text{grad } V = -\mathbf{E} \tag{3}$$

Expressions for the gradient in other coordinate systems are given inside the back cover.

2-15 THE ELECTRIC DIPOLE AND ELECTRIC-DIPOLE MOMENT

The combination of two equal point charges Q of opposite sign separated by a small distance l is called an *electric dipole*, and the product Ql is called the *electric-dipole moment*. If the two charges were superposed, the resultant field would be zero. However, when the two charges are separated by even a small distance, there is a finite resultant field (see Fig. 2-13).

By regarding the separation between the charges as a vector $\mathbf{l}$, pointing from the negative to the positive charge as in Fig. 2-19, the dipole moment can be expressed as a vector $Q\mathbf{l}$ with the magnitude Ql and the direction of $\mathbf{l}$.

Referring to Fig. 2-19, the potential of the positive charge at a point P is

$$V_1 = \frac{Q}{4\pi\epsilon_0 r_1}$$

The potential of the negative charge at P is

$$V_2 = \frac{-Q}{4\pi\epsilon_0 r_2}$$

The total potential V at P is then

$$V = V_1 + V_2 = \frac{Q}{4\pi\epsilon_0}\left(\frac{1}{r_1} - \frac{1}{r_2}\right)$$

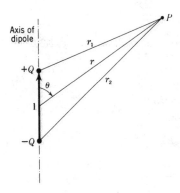

Figure 2-19 Electric dipole.

If the point P is at a large distance compared with the separation l, so that the radial lines r_1, r, and r_2 are essentially parallel, we have very nearly that

$$r_1 = r - \frac{l}{2} \cos \theta \quad \text{and} \quad r_2 = r + \frac{l}{2} \cos \theta$$

where r = distance from center of dipole to point P
  θ = angle between axis of dipole and r

Hence, the potential V at a distance r from an electric dipole is

$$\boxed{V = \frac{Ql \cos \theta}{4\pi\epsilon_0 r^2} \quad \text{(V)}}$$ (1)

where it is assumed that r is much greater than l ($r \gg l$) so that terms in l^2 can be neglected compared with those in r^2.

It is instructive to consider (1) as the product of four factors involving the dipole moment, the angle, the distance, and a constant, characteristic of the system of units employed. Thus,

$$V = \underset{\substack{\text{Dipole} \\ \text{moment}}}{Ql} \quad \underset{\substack{\text{Angle} \\ \text{factor}}}{\cos \theta} \quad \underset{\substack{\text{Distance} \\ \text{factor}}}{\frac{1}{r^2}} \quad \underset{\text{Constant}}{\frac{1}{4\pi\epsilon_0}}$$ (2)

Expressions for the potential and also the electric field of dipoles (and quadrupoles and higher-order configurations) always contain these four kinds of factors.

To find the electric field of the dipole of Fig. 2-19 it is convenient to make use of the gradient discussed in the preceding section. Thus, let us take the gradient of the potential given by (1), obtaining†

$$\mathbf{E} = -\hat{\mathbf{r}}\frac{\partial V}{\partial r} - \hat{\boldsymbol{\theta}}\frac{1}{r}\frac{\partial V}{\partial \theta} = \hat{\mathbf{r}}\frac{Ql \cos \theta}{2\pi\epsilon_0 r^3} + \hat{\boldsymbol{\theta}}\frac{Ql \sin \theta}{4\pi\epsilon_0 r^3}$$ (3)

where $\hat{\mathbf{r}}$ = unit vector in r direction (see Fig. 2-20)
  $\hat{\boldsymbol{\theta}}$ = unit vector in θ direction
  l = separation of dipole charges Q

According to (3), the electric field has two components, as shown in Fig. 2-20, one in the r direction (E_r) and one in the θ direction (E_θ). Thus

$$\mathbf{E} = \hat{\mathbf{r}}E_r + \hat{\boldsymbol{\theta}}E_\theta$$ (4)

or

$$E_r = \frac{Ql \cos \theta}{2\pi\epsilon_0 r^3} \quad \text{and} \quad E_\theta = \frac{Ql \sin \theta}{4\pi\epsilon_0 r^3} \quad \text{(V m}^{-1}\text{)}$$ (5)

In these equations the restriction applies that $r \gg l$.

---

† See inside back cover for gradient in spherical coordinates.

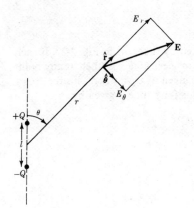

Figure 2-20 Component fields and total field **E** at a distance r from an electric dipole.

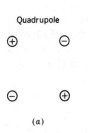

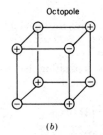

Figure 2-21 Charge configuration: for a quadrupole (a) and for an octopole (b).

By a similar procedure to that followed above it is possible to obtain the potential and field relations for more complex charge configurations, e.g., the quadrupole or octopole shown in Fig. 2-21 (see Prob. 2-3-17). The potential and field variations with distance for these configurations are listed in Table 2-1, along with the distance factors for a dipole and a single charge. We note that the higher the order of the charge configuration, the more rapidly the potential or field falls off with distance. We also note that for any given configuration the field falls off more rapidly than the potential.

Table 2-1 Distance factor of electric potential and electric field for different charge configurations

Configuration	Electric potential V	Electric field **E**
Single charge	r^{-1}	r^{-2}
Dipole	r^{-2}	r^{-3}
Quadrupole	r^{-3}	r^{-4}
Octopole	r^{-4}	r^{-5}

2-16 ELECTRIC FLUX

Suppose field lines, as in Fig. 2-13, are replaced by tubes, as in Fig. 2-22. If each tube represents a constant amount of charge or *electric flux* ψ, then at any point there is a *flux density* D (proportional to E) as given by ψ/A, where A is the cross-sectional area of the tube. Thus, the electric flux for a tube is given by

$$\psi = DA \quad \text{(C)}$$

where D = average flux density, C m^{-2}
A = cross-sectional area of tube, m^2

More generally we have

$$\psi = \iint \mathbf{D} \cdot d\mathbf{s} = \iint D_n \, ds \quad \text{(C)} \tag{1}$$

This relation states that the electric flux through any surface equals the integral of the flux density over the surface. In Fig. 2-22 ψ is constant along any tube. If all space is filled with flux tubes connecting $+Q$ and $-Q$, the total flux through all the tubes passing through the infinite plane separating the charges is equal to Q. Likewise integrating $\mathbf{D}$ over the infinite plane yields Q, or

$$\psi = \iint_{\substack{\text{infinite} \\ \text{plane}}} \mathbf{D} \cdot d\mathbf{s} = Q \quad \text{(C)} \tag{2}$$

Consider next the electric field around a single isolated positive point charge Q as in Fig. 2-23.† This is given by

$$\mathbf{E} = \hat{\mathbf{r}} \frac{Q}{4\pi\epsilon_0 r^2} \quad \text{or} \quad \epsilon_0 \mathbf{E} = \hat{\mathbf{r}} \frac{Q}{4\pi r^2}$$

The second equation has the dimensions of charge per unit area, or electric flux density. It follows that

$$\mathbf{D} = \hat{\mathbf{r}} \frac{Q}{4\pi r^2} \tag{3}$$

or

$$\boxed{\mathbf{D} = \epsilon_0 \mathbf{E}} \tag{4}$$

where $\mathbf{D}$ = electric flux density, C m^{-2}
ϵ_0 = permittivity of vacuum, F m^{-1}
$\mathbf{E}$ = electric field intensity, V m^{-1}

---

† We imagine that the field lines end on an equal negative charge distributed over the inside of a sphere at infinity.

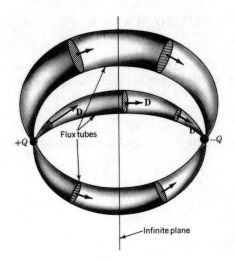

Flux tubes

+Q   -Q

Infinite plane

Figure 2-22 Electric field between two point charges of opposite sign with flux tubes joining them. The tubes follow the electric field lines. Each tube has a constant amount of flux.

According to (4), the flux density and field intensity are vectors with the same direction. This is true for all isotropic media, i.e., media whose properties do not depend on direction.

Since $4\pi r^2$ equals the area of a sphere of radius r, it follows that the magnitude of **D** at the radius r is identical with the surface charge density which would occur if the charge Q were distributed uniformly over a sphere of radius r instead of concentrated at the center (see Fig. 2-23).

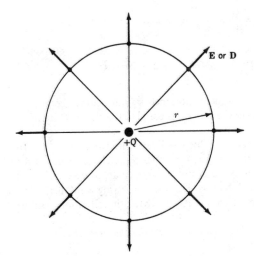

E or D

r

+Q

Figure 2-23 Electric field of an isolated positive point charge Q. The flux density **D** at a radius r is the same as the surface charge density on the sphere if Q were uniformly distributed over the sphere.

2-17 ELECTRIC FLUX OVER A CLOSED SURFACE; GAUSS' LAW

Referring to Fig. 2-24, let a positive point charge Q be situated at the center of an imaginary sphere of radius r. The infinitesimal amount of electric flux through the surface element ds is given by

$$d\psi = \mathbf{D} \cdot d\mathbf{s}$$

Integrating this over the sphere of radius r then gives the total flux through the sphere

$$\psi = \iint \mathbf{D} \cdot d\mathbf{s} \tag{1}$$

We note that $\mathbf{D}$ is everywhere normal to the sphere, so that

$$\mathbf{D} \cdot d\mathbf{s} = D \, ds$$

where $D = |\mathbf{D}|$ = scalar magnitude of the vector $\mathbf{D}$
$ds = |d\mathbf{s}|$ = scalar magnitude of the vector $d\mathbf{s}$

From (1) and (2-16-3)

$$\psi = \iint \frac{Q}{4\pi r^2} \, ds \tag{2}$$

From Fig. 2-24 $ds = (r \, d\theta)(r \, d\phi \sin \theta) = r^2 \sin \theta \, d\theta \, d\phi$, and the solid angle $d\Omega$ subtended by the spherical element of surface area ds is $ds/r^2 = d\Omega = \sin \theta \, d\theta \, d\phi$; therefore

$$\psi = \frac{Q}{4\pi} \iint d\Omega = \frac{Q}{4\pi} \int_0^{2\pi} \int_0^{\pi} \sin \theta \, d\theta \, d\phi$$

$$= \frac{Q}{4\pi} [-\cos \theta]_0^{\pi} \int_0^{2\pi} d\phi = \frac{Q}{4\pi} \times 2 \times 2\pi = Q \tag{3}$$

Thus, the total electric flux over the sphere (obtained by integrating the normal component of the flux density $\mathbf{D}$ over the sphere) is equal to the charge Q enclosed by the sphere.

We could have obtained the result for this special case more simply by multiplying $D = Q/4\pi r^2$ by the area of the sphere ($4\pi r^2$). But if the charge is not at the center of the sphere, or if there is a distribution of charge enclosed by a surface of arbitrary shape, the result is not as obvious. However, applying a mathematical relation formulated by Karl Freidrich Gauss in 1813, the situation can be generalized as follows:

The electric flux through any closed surface equals the charge enclosed.

This is Gauss' law for electric fields.

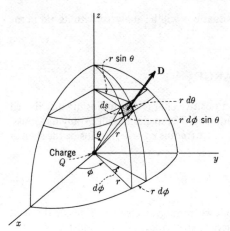

Figure 2-24 Point charge Q at origin of spherical coordinate system.

Since the surface integral of the normal component of the electric flux density **D** gives the flux through the surface, we have

$$\iint \mathbf{D} \cdot d\mathbf{s} = \iint D \cos \theta \, ds = Q \qquad \text{(C)} \qquad (4)$$

where Q is the total or net charge enclosed. This charge may also be expressed as the volume integral of the charge density ρ, so that Gauss' law is given by

$$\iint \mathbf{D} \cdot d\mathbf{s} = \iiint \rho \, dv = Q \qquad (5)$$

where the surface integration is carried out over a closed surface and the volume integration throughout the region enclosed. An alternative notation of Gauss' law is

$$\boxed{\oint_s \mathbf{D} \cdot d\mathbf{s} = \oint_v \rho \, dv = Q} \qquad (6)$$

where $\oint_s$ = double, or surface, integral over closed surface
$\quad \oint_v$ = triple, or volume, integral throughout region enclosed

Gauss' law is the basic theorem of electrostatics. It is a necessary consequence of the inverse-square law (Coulomb's law). Thus, if **D** for a point charge did not vary as $1/r^2$, the total flux over a surface enclosing it would not equal the charge.

In later chapters we will discuss a set of four fundamental electromagnetic relations known as *Maxwell's equations*. These equations can be stated in either integral or differential (point) form. Equation (6) is one of these relations in its integral form. More explicitly, it is *Maxwell's equation from Gauss' law*. In Sec. 3-22 the same equation is discussed in its point, or differential form.

To illustrate Gauss' law, several situations will be analyzed with its aid in the following sections.

2-18 SINGLE SHELL OF CHARGE

Referring to Fig. 2-25a, suppose that a positive charge Q is uniformly distributed over an imaginary spherical shell of radius r_1. The medium is air. Applying Gauss' law by integrating $\mathbf{D}$ over a spherical surface (radius $r_1 - dr$) just inside the shell of charge, we have

$$\epsilon_0 \oint_s \mathbf{E} \cdot d\mathbf{s} = 0 \tag{1}$$

since the charge enclosed is zero. It follows from symmetry that $\mathbf{E}$ inside the shell is zero. Applying Gauss' law to a spherical shell (radius $r_1 + dr$) just outside the shell of charge, we have, neglecting infinitesimals,

$$\epsilon_0 \oint_s \mathbf{E} \cdot d\mathbf{s} = \epsilon_0 E \, 4\pi r_1^2 = Q \tag{2}$$

or
$$E = \frac{Q}{4\pi\epsilon_0 r_1^2} \tag{3}$$

This value of field intensity is identical with that at a radius r_1 from a point charge Q. We can therefore conclude that the field outside the shell of charge is the same as if the charge Q were concentrated at the center. Summarizing, the field everywhere due to a spherical shell of charge is

$$\mathbf{E} = 0 \text{ inside} \qquad r \leq r_1 \tag{4}$$

$$\mathbf{E} = \hat{\mathbf{r}} \frac{Q}{4\pi\epsilon_0 r^2} \text{ outside} \qquad r \geq r_1 \tag{5}$$

The variation of $\mathbf{E}$ as a function of r is illustrated by Fig. 2-25b.†

The absolute potential at a radius r outside the shell is given by

$$V = -\int_\infty^r \mathbf{E} \cdot d\mathbf{r} \tag{6}$$

Introducing the value of $\mathbf{E}$ from (5),

$$V = -\frac{Q}{4\pi\epsilon_0} \int_\infty^r \frac{dr}{r^2} = \frac{Q}{4\pi\epsilon_0 r} \qquad (V) \tag{7}$$

---

† Note that a point charge at the origin gives an infinite $\mathbf{E}$ as $r \to 0$ but a surface charge of finite area at a radius r_1 gives a finite $\mathbf{E}$ as $r \to r_1$. This is because the volume charge density ρ of a point charge is infinite, whereas the surface charge density ρ_s of the shell of charge is finite. In the present case $\rho_s = Q/4\pi r_1^2$.

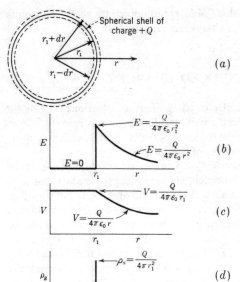

Figure 2-25 Uniformly charged spherical shell with graphs showing variation of electric field intensity E, electric potential V, and surface charge density ρ_s as a function of radial distance r.

At the shell, where $r = r_1$, we have

$$V = \frac{Q}{4\pi\epsilon_0 r_1} \tag{8}$$

Since **E** inside the shell is zero, it requires no work to move a test charge inside and therefore the potential is constant, being equal to the value at the shell. Summarizing, the electric potential everywhere due to a spherical shell of charge of radius r_1 is

$$V = \frac{Q}{4\pi\epsilon_0 r_1} \text{ inside } \qquad r \leq r_1 \tag{9}$$

$$V = \frac{Q}{4\pi\epsilon_0 r} \text{ outside } \qquad r \geq r_1 \tag{10}$$

The variation of V as a function of r is illustrated by Fig. 2-25c. The variation of the surface charge density ρ_s is shown by Fig. 2-25d. The surface density is zero everywhere except at $r = r_1$, where it has the value $Q/4\pi r_1^2$, as indicated by the vertical line, or spike.

It is to be noted that the potential is continuous, both (9) and (10) being equal *at* the shell ($r = r_1$). However, the electric field is discontinuous, jumping abruptly from zero just inside the shell to a value $Q/4\pi\epsilon_0 r_1^2$ just outside the shell. This results from the assumption that the shell of charge has zero thickness.

2-19 CONDUCTORS AND INDUCED CHARGES

A conductor can conduct, or convey, electric charge. In static situations a conductor may be defined as a medium in which the electric field is always zero. It follows that all parts of a conductor must be at the same potential. Metals such as copper, brass, aluminum, and silver are examples of conductors.

When a metallic conductor is brought into an electric field, different parts of the conductor would assume different potentials were it not for the fact that electrons flow in the conductor until a surface charge distribution is built up that reduces the total field in the conductor to zero.† This surface charge distribution is said to consist of *induced charges*. The field in which the conductor is placed may be called the *applied field* $\mathbf{E}_a$, while the field produced by the surface charge distribution may be called the *induced field* $\mathbf{E}_i$. The sum of the applied and induced fields yields a total field in the conductor equal to zero. Although the total field inside the conductor is zero after the static situation has been reached, the total field is not zero while the induced charges are in motion, i.e., while currents are flowing.

To summarize, under static conditions the electric field in a conductor is zero, and its potential is a constant. Charge may reside on the surface of the conductor, and, in general, the surface charge density need not be constant.

2-20 CONDUCTING SHELL

An initially uncharged conducting shell of inner radius a and outer radius b (wall thickness $b - a$) is shown in cross section in Fig. 2-26. Let a point charge $+Q$ be placed at the center of the shell. This might be done by introducing the charge through a hole in the shell which is plugged after the charge is inside.‡ The point charge has a radial electric field. Let this be called the applied field $\mathbf{E}_a$. For the total field $\mathbf{E}$ in the conducting wall to be zero requires an induced field $\mathbf{E}_i$ inside the wall such that

$$\mathbf{E}_a + \mathbf{E}_i = \mathbf{E} = 0 \tag{1}$$

or

$$\mathbf{E}_i = -\mathbf{E}_a \tag{2}$$

The induced field $\mathbf{E}_i$ is produced by a distribution of induced negative charges on the inner shell wall and induced positive charges on the outer shell wall, as suggested in

---

† The electrons in the outermost shell of the atoms of a conductor are so loosely held that they migrate readily from atom to atom under the influence of an electric field.

‡ This is an idealized version of an experiment first performed by Faraday, using an ice pail.

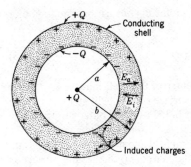

Figure 2-26 Conducting shell of wall thickness $b - a$ with point charge Q at center.

Fig. 2-26. Let us apply Gauss' law to this situation to determine quantitatively the magnitude of these induced charges.

Suppose that an imaginary sphere designated S_1 with a radius $a - dr$ is situated just inside the inner wall of the shell as in Fig. 2-27. By Gauss' law the surface integral of the normal component of **D** over this sphere must equal $+Q$. That is,

$$\oint_{S_1} \mathbf{D} \cdot d\mathbf{s} = +Q \tag{3}$$

Applying Gauss' law to the sphere S_2 of radius $a + dr$ just inside the conductor, we have, since the total field **E** in the conductor is zero,

$$\oint_{S_2} \mathbf{D} \cdot d\mathbf{s} = \epsilon_0 \oint_{S_2} \mathbf{E} \cdot d\mathbf{s} = 0 \tag{4}$$

Thus, the total charge inside the sphere S_2 must be zero. It follows that a charge $-Q$ is situated on the inner surface of the shell wall. Since the shell was originally uncharged, this negative charge Q, produced by a migration of electrons to the inner surface, must leave a deficiency of electrons, or positive charge Q, on the outer surface of the shell. It is assumed that the surface charges reside in an infinitesimally thin layer.

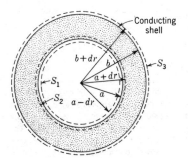

Figure 2-27 Conducting shell of wall thickness $b - a$ with surfaces of integration.

Applying Gauss' law to the sphere S_3 of radius $b + dr$ just outside the outer surface of the shell, we then have

$$\oint_{S_3} \mathbf{D} \cdot d\mathbf{s} = +Q \qquad (5)$$

To summarize, the charge $+Q$ at the center of the shell induces an exactly equal but negative charge $-Q$ on the inner surface of the shell, and this in turn results in an equal positive charge $+Q$ distributed over the outer surface of the shell. The flux tubes originating on $+Q$ at the center end on the equal negative charge on the inside of the shell. There is no total field and no flux in the shell wall. Outside the shell the flux tubes continue from the charge $+Q$ on the outer surface as though no shell were present. The variation of the component fields $\mathbf{E}_a$ (applied) and $\mathbf{E}_i$ (induced) as a function of r is illustrated by Fig. 2-28b. The variation of the total field $\mathbf{E}$ is shown in Fig. 2-28c.

If a conducting wire is connected from the inner surface of the shell to the charge $+Q$ at the center, electrons will flow and reduce the charge at the center and on the inner surface to zero. However, the charge $+Q$ remains on the outer surface of the shell. This results in an applied field only external to the shell ($r \geq b$) and of the same value as before. There is no induced field whatsoever. Thus, the total

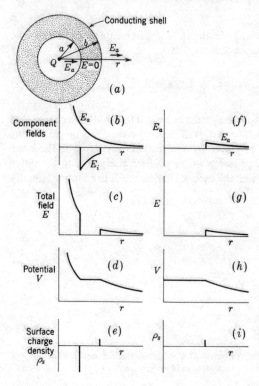

Figure 2-28 Conducting shell of wall thickness $b - a$ with graphs showing variation of applied field E_a, induced field E_i, total field E, potential V, surface charge density ρ_s, with charge Q at center (b, c, d, and e), and with charge only on outside of shell (f, g, h, and i).

field is identical with the applied field and is zero for $r \leq b$ as shown by Fig. 2-28f and g. This final result might have been achieved more simply in the first place by applying the charge $+Q$ to the outside of the originally uncharged conducting sphere.

The variation of V and ρ_s as a function of r when the charge $+Q$ is at the center of the sphere is indicated in Fig. 2-28d and e, while the variation when the charge is only on the outside of the shell is shown by Fig. 2-28h and i.

PROBLEMS†

Note: A carefully made sketch is a good first step in the solution of any problem.

Group 2-1: Secs. 2-1 through 2-6. Coulomb's law, electric field intensity, and superposition of fields

★**2-1-1. Two point charges and Coulomb's law** A positive point charge of 5 pC is located at (0, 20) mm ($x = 0, y = 20$ mm) and another such charge at (20, 0) mm. Find the magnitude and direction of the force **F** on a positive charge of 20 nC at (20, 20) mm.

2-1-2. Electric field of 2 point charges With point charges of 10 pC at (0, 50) mm ($x = 0, y = 50$ mm) and (50, 0) mm, find the electric field intensity **E** at (50, 50) mm.

★**2-1-3. Levitation of charged ball** Two small-diameter 5-g dielectric balls can slide freely on a vertical nonconducting thread. Each ball carries a negative charge of 2 μC. Find the separation if the lower ball is restrained from moving.

★**2-1-4. Electric field of 3 point charges in a line** A charge of $+1$ C is situated at $(-1, 0)$, a charge of -2 C at (0, 0), and a charge of $+1$ C at (1, 0) m. Find the electric field intensity **E** at $(0, -1)$ m.

2-1-5. Electric field of 3 point charges at corners of a square A charge of $1Q$ is situated at $(\sqrt{2}, \sqrt{2})$ m, a charge of $2Q$ at $(-\sqrt{2}, \sqrt{2})$ m, and a charge of $3Q$ at $(-\sqrt{2}, -\sqrt{2})$ m. If $Q = 4\pi \times 8.85$ pC, find the electric field intensity **E** at (0, 0) m.

★**2-1-6. Electric field on axis of 4 charges in a plane** With point charges of 5 nC at $(-2, 0, 0)$, $(0, -2, 0)$, $(2, 0, 0)$, and $(0, 2, 0)$ m, find the electric field **E** at (0, 0, 3) m ($x = 0, y = 0, z = 3$ m).

2-1-7. Zero field from 3 point charges at corners of a triangle Two point charges $+Q$ are situated at $(10, 0)$ and $(-10, 0)$ m. A third charge $-Q$ is situated at $(0, -15)$ m. Find the coordinates of the point where the electric field **E** $= 0$.

★**2-1-8. Electric field at center of cube face with 8 charges at cube corners** A cube 1 m square has charges of 6 nC at its 8 corners. Find the electric field at the center of any face.

2-1-9. Field of point charge Find **E** at the point (1, 2, 3) m due to a point charge of 2 nC at (3, 2, 1) m.

2-1-10. Field of point charge Find **E** at the point (2, 4, 3) m due to a point charge of 5 nC at (1, 2, 4) m.

2-1-11. Coulomb's law and Newton's law It is interesting to note that Coulomb's law has the same form as the Law of Universal Gravitation formulated by Isaac Newton in 1687. Thus, according to Newton, the force between two point masses is given by

$$\mathbf{F} = -\hat{\mathbf{r}} G \frac{M_1 M_2}{r^2}$$

where **F** = force, N
    $\hat{\mathbf{r}}$ = unit vector, dimensionless
    r = distance between masses, m
    M_1 = mass 1, kg
    M_2 = mass 2, kg
    G = universal gravitational constant = 6.7×10^{-11} N m^2 kg^{-2}

† Answers to starred problems are given in Appendix C.

Note that whereas electric charges of the same sign repel (Coulomb's law), masses which are all of the same kind or sign attract (Newton's law). Hence the negative sign in Newton's law (above).

Use Newton's law to verify that a mass at the earth's surface and attracted to it by 9×10^{15} N is equal to 9.2×10^{11} metric tons (tonnes), as in the worked example of Sec. 2-2. Let M_1 equal the mass of the earth considered as a point mass at the center of the earth and set r equal to the radius of the earth. See Appendix A-2.

2-1-12. Electrostatic attraction and repulsion Explain why a piece of paper released near an electrostatic charge (as on the electrode of an electrostatic generator) is attracted to it but, after touching it, is repelled. Are Coulomb forces involved?

★**2-1-13. Earth-moon force** Find the electric charges required on the earth and moon to balance their gravitational attraction if the charge on the earth is 10 times that on the moon.

2-1-14. V and E for quarter ring A charge $+Q$ is uniformly distributed over a ring section of radius r_1 extending between angles of 90 and 180° (i.e., a quarter ring situated in the 2d quadrant). Find V and E at the origin.

2-1-15. Potential of 3 point charges Point charges of 1 nC are situated at the points $(-4, 6)$, $(-3, -7)$, and $(6, -5)$ m. Find V at the point $(5, 3)$ m.

Group 2-2: Secs. 2-7 through 2-14. Electric scalar potential, line integral, charge distributions, field maps, and gradient

★**2-2-1. Electric scalar potential of 2 point charges** A point charge of 20 pC is situated at $(1, 1)$ and another point charge of -20 pC at $(-1, 1)$ m. Find the electric scalar potential V at $(1, 0)$ m.

2-2-2. Potential and field of 2 positive point charges Point charges of 5 nC are situated at $(1, 1)$ and $(-1, -1)$ m. (a) Find the electric scalar potential V at the origin $(0, 0)$. (b) Find the electric field intensity **E** at the same point.

★**2-2-3. Potential and field of 2 equal point charges of opposite sign** A point charge of 250 pC is situated at $(1, 1)$ and a point charge of -250 pC at $(-1, -1)$ m. (a) Find the electric scalar potential V at the origin $(0, 0)$. (b) Find the electric field intensity **E** at the same point.

2-2-4. Potential of 2 point charges on the x axis Point charges of 500 pC are situated at $(-500, 0)$ and $(500, 0)$ mm. Find the electric scalar V at $(0, 1000)$ mm.

2-2-5. Potential of 3 point charges on the x axis Point charges of 333 pC are situated at $(-500, 0)$, $(0, 0)$, and $(500, 0)$ mm. Find the electric scalar potential V at $(0, 1000)$ mm.

2-2-6. Potential of 5 point charges on the x axis Point charges of 200 pC are situated at $(-500, 0)$, $(-250, 0)$, $(0, 0)$, $(250, 0)$, and $(500, 0)$ mm. Find the electric scalar potential V at $(0, 1000)$ mm.

★**2-2-7. Potential of line charge on x axis** A uniform line charge of 1 nC is situated along the x axis between the points $(-500, 0)$ and $(500, 0)$ mm. Find the electric scalar potential V at $(0, 1000)$ mm.

2-2-8. Equivalence of point and line charges Probs. 2-2-4, 2-2-5, 2-2-6, and 2-2-7 all involve the same total charge (1 nC) distributed either uniformly or at discrete points along the x axis between $(-500, 0)$ and $(500, 0)$ mm. How many uniformly spaced equal point charges totaling 1 nC would be required between the same points on the x axis to give a potential V at $(0, 1000)$ mm that is within 1 percent of the value for the continuous uniform line charge of Prob. 2-2-7?

★**2-2-9. Two line charges along x axis** A uniform line charge of $-4\pi \times 8.85$ pC m^{-1} is situated between the points $(-5, 0)$ and $(-2, 0)$ m and another such line of positive charge between $(2, 0)$ and $(5, 0)$ m. (a) Find the electric scalar potential V at $(1, 0)$ m. (b) Find the electric field intensity **E** at the same point.

2-2-10. Line integral Calculate the line integral of **A** from $(2, 1)$ to $(4, 2)$ where

$$\mathbf{A} = \hat{\mathbf{r}} \frac{1}{r}$$

★**2-2-11. Normal to a plane** Find the unit normal to the plane defined by the equation $2x + 3y + 6z = 12$.

2-2-12. Potential of a cosine line distribution compared with point charge and uniform distribution (*a*) A line charge extends between the points (1, 0) and (3, 0) m with a distribution given by

$$\rho_L = 200 \cos \frac{\pi}{2} x \quad (\text{pC m}^{-1})$$

Find the potential V at the origin (0, 0). (*b*) Find the potential V at the origin if all of the line charge is concentrated at the point (2, 0). (*c*) Find the potential V at the origin if the line charge is uniformly distributed between (1, 0) and (3, 0) (same total charge). (*d*) If (*b*) and (*c*) are regarded as approximations to (*a*), what are the percentage differences [(*b*) to (*a*) and (*c*) to (*a*)]?

2-2-13. Potential and field of ring of charge A circular ring of 1-m radius centered on the origin and situated in the *xy* plane has a charge of 1 nC uniformly distributed over it. (*a*) Find the electric scalar potential V on the ring axis (*z* axis) at a distance of 1 m from the ring [point (0, 0, 1)]. (*b*) Find the electric field intensity **E** at the same point.

2-2-14. Potential and field of square loop of charge A square loop 2 m on a side situated in the *xy* plane has a charge of 1 nC uniformly distributed over it. (*a*) Find the electric scalar potential V at a distance of 1 m perpendicular from the center of the loop [point (0, 0, 1)]. (*b*) Find the electric field intensity **E** at the same point.

★**2-2-15. Potential and field of disk** A flat disk of 1-m radius centered on the origin and situated in the *xy* plane has a charge of 1 nC uniformly distributed over its surface. (*a*) Find the electric scalar potential V on the disk axis (*z* axis) at a distance of 1 m from the disk [point (0, 0, 1)]. (*b*) Find the electric field intensity **E** at the same point.

2-2-16. Potential and field of square A square sheet 2 m on a side situated in the *xy* plane and centered on the origin has a charge of 1 nC uniformly distributed over its surface. (*a*) Find the electric scalar potential V at a distance of 1 m perpendicular from the center of the square [point (0, 0, 1)]. (*b*) Find the electric field intensity **E** at the same point.

★**2-2-17. Potential and field of sphere** A sphere of 1-m radius has a charge of 1 nC uniformly distributed throughout its volume. (*a*) Find the electric scalar potential V at a point 2 m from the center of the sphere. (*b*) Find the electric field intensity **E** at the same point.

2-2-18. Potential and field of cube A cube 2 m on a side has a charge of 1 nC uniformly distributed throughout its volume. (*a*) Find the electric scalar potential V at a point 2 m from the center of the cube and perpendicular to a face of the cube. (*b*) Find the electric field intensity **E** at the same point.

2-2-19. Comparison of potential and field of sphere, cube, and point charge Compare the values of V and **E** of Probs. 2-2-17 and 2-2-18 with those at 2 m from a point charge of 1 nC.

2-2-20. Gradient A potential distribution is given by $V = 3x + 6$ V. (*a*) What is the expression for the field **E**? (*b*) Find **E** (magnitude and direction) at (0, 10) and (10, 10) m.

2-2-21. Gradient A potential distribution is given by $V = 8x + 6y^2 - 3z^{1/2}$ V. (*a*) What is the expression for the field **E**? (*b*) Find **E** (magnitude and direction) at (3, 4, 9) m.

2-2-22. Gradient A potential distribution is given by $V = 2/(x^2 - y^3 + z^4)$ V. (*a*) What is the expression for the field **E**? (*b*) Find **E** (magnitude and direction) at (2, 1, 2) m.

★**2-2-23. Gradient** If $V = -x^2y^2z^3$ V, find **E**.

2-2-24. Gradient If $V = r\theta\phi$ V, find **E**.

2-2-25. Gradient If $V = 2r\phi z^2$ V, find **E**.

★**2-2-26. Potential and field of 3 charges** A point charge $-Q$ is situated at (0, 1) and a point charge $+Q$ at (0, -1) m. A third point charge $+2Q$ is situated at (-1, 0) m. If $Q = 4\pi \times 8.85$ pC, find V and **E** at the origin.

2-2-27. Field map of 3 point charges Three equal point charges are arranged as in Fig. P2-2-27. Charges 1 and 2 are positive with charge 3 negative. All charges are $\pm\frac{1}{9}$ nC so that the potential at any point is

$$V = \sum_{n=1}^{n=3} \frac{1}{r} \quad (\text{V})$$

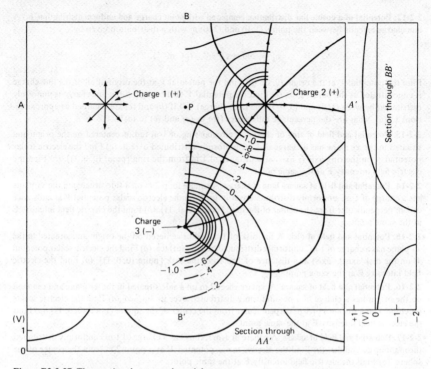

Figure P2-2-27 Three point charges and partial map.

where r is the distance from each charge to the point. The distance between charges 1 and 2 is 4 m. Charge 3 is midway between charges 1 and 2 and 3 m below. Thus,

$$V \text{ (at } P) = \frac{1}{2} + \frac{1}{2} - \frac{1}{3} = \frac{2}{3} \quad \text{(V)}$$

(a) Complete potential contours at 0.2 V intervals from $+1.6$ V to -1.6 V and label each. Note that the left and right halves of the map are mirror images. (b) Complete E lines from charges 1 and 2 in 8 directions, as indicated in the figure. (c) Draw graphs of V along the lines AA' and BB'. (d) Find the point where $E = 0$. (See Prob. 2-1-7.) A partially completed map is given in Fig. P2-2-27 with the graph for V along the line AA' also shown. Note that the field E is inversely proportional to the spacing between the equipotential contours. This problem is ideally suited for solution with a computer.

2-2-28. Field map of 8 point charges Eight equal positive or negative point charges are located at points (x, y) of (100, 100), (100, 250), (100, 400), (250, 100), (250, 400), (400, 100), (400, 250), and (400, 400) mm. Using a computer program and one of the charge configurations of Fig. P2-2-28a (or other combinations), obtain a computer-plotted contour map of the area $0 \leq x \leq 500$ and $0 \leq y \leq 500$ mm and label all contours with the proper potential (voltage). The actual map can be drawn to some convenient size smaller than 500 mm by 500 mm by applying a scale factor. All charges are $\pm\frac{1}{9}$ nC, so that the potential at any point is

$$V = \frac{1}{4\pi\epsilon_0} \sum \frac{Q}{r} = \sum \frac{1}{r} \quad \text{(V)}$$

Make the contour interval 1 or 2 V. In Fig. P2-2-28a contours are shown only at 5- or 10-V intervals except in interesting regions, where the interval is 1 V. The heavy lines in Fig. P2-2-28a are the zero-potential contours. It is suggested that the program be written to calculate the potential at the points of a 50 × 50 grid (2500 points). The program then calculates the location of the equipotential contour lines. To avoid infinite potentials the map may be terminated within 5 grid units of each charge or whenever the total potential exceeds a specified value.

Complete the map by drawing a suitable number of electric field lines with arrows. Note that these lines must always cross the equipotentials at right angles. Note that where equipotentials are close together, the field is strong, and where they are far apart, the field is weak.

Finally, prepare a voltage graph of the potential along several map bisectors (vertical, horizontal, or diagonal) as done for the sample map of Fig. P2-2-28b in Fig. P2-2-28c.

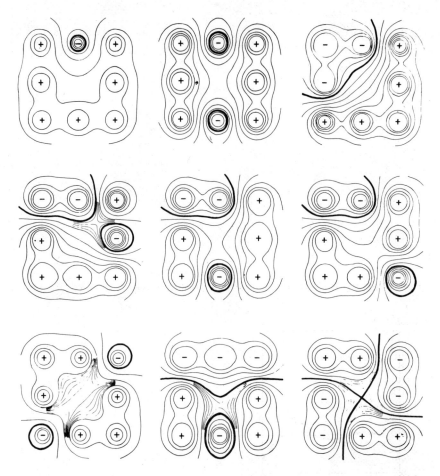

Figure P2-2-28 (a) Computer-generated potential contour maps for different configurations of eight charges of the same magnitude with signs as indicated. The heavy line is the zero-potential contour.

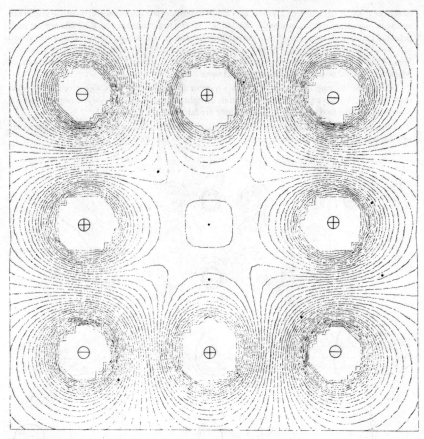

Figure P2-2-28 (*b*) Detail map for eight charges to 1-V contour interval.

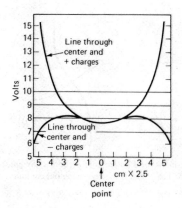

Figure P2-2-28 (*c*) Voltage graph for eight-charge field map (Figure P2-2-28*b*).

2-2-29. Line charge A line charge of length l is situated coincident with the x axis with one end at the origin and the other at $x = l$. If the line charge density varies as x^2, find V at a point x_1 where $x_1 > l$.

2-2-30. Sheet of charge with hole An infinite sheet of uniform charge density ρ_s is situated coincident with the xy plane at $z = 0$. The sheet has a hole of radius r centered on the origin. Find (a) V and (b) E for points along the z axis.

2-2-31. Parabolic line charge A parabolic line of charge with vertex at the origin and focus at $x = 1$ m extends 4 m in both $+y$ and $-y$ directions. If $\rho_L = 2$ nC m^{-1}, find (a) V and (b) E at the focus.

2-2-32. Potential plots The potential variation from a point charge at the origin is shown in Fig. P2-2-32 along lines parallel to the x axis but displaced from it. A sample program in BASIC used to generate the figure is given. In a similar manner generate figures for (a) 2, (b) 3, and (c) 4 point charges situated with uniform spacing along the x axis. Note that the Y in the program corresponds to the potential V. The displacement from the x axis is given by the terms involving 100 ($= 10^2$), 400 ($= 20^2$), etc.

```
]10 HGR
]20 HCOLOR=7
]30 FOR X = 20 TO 256 STEP 1
]40 Y = 1000 / ((138 − X)^2 + 100)^.5
]50 HPLOT X + 20 , Y + 40
]60 Y = 1000 / ((138 − X)^2 + 400)^.5
]70 HPLOT X + 15 , Y + 30
]80 Y = 1000 / ((138 − X)^2 + 900)^.5
]90 HPLOT X + 10 , Y + 20
]100 Y = 1000 / ((138 − X)^2 + 1600)^.5
]110 HPLOT X + 5 , Y + 10
]111 Y = 1000 / ((138 − X)^2 + 2500)^.5
]112 HPLOT X , Y
]120 NEXT X
```

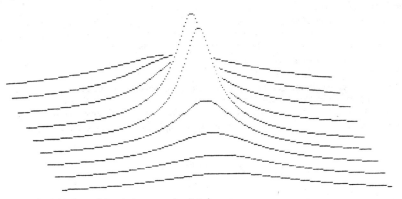

Figure P2-2-32 Potential variation around point charge.

Group 2-3: Secs. 2-15 through 2-20. Electric dipole, electric flux, Gauss' law, shell of charge, and conductors

2-3-1. Force on dipole An electric dipole of moment 6 pC m is situated 100 mm from an isolated point charge of 2 nC. Find the maximum net force on the dipole.

★2-3-2. Torque on dipole In a uniform field a dipole experiences no net (translational) force but it does experience a torque. Develop the expression for this torque on a dipole of moment ql in a uniform field **E**. Let θ be the angle between the dipole axis and the field.

2-3-3. Electric flux over sphere If $E = \hat{x}x + \hat{y}y + \hat{z}z$ V m^{-1}, find the total electric flux over a sphere of 2-m radius.

2-3-4. Scalar product Find $A \cdot B$, where $A = \hat{x}3 + \hat{y}2 + \hat{z}$ and $B = \hat{r}$ at $\theta = 45°$, $\phi = 30°$. Solve two ways: (1) using rectangular coordinates and (2) using spherical coordinates.

2-3-5. Scalar product If $A = \hat{x}3 + \hat{y}4 - \hat{z}5$ and $B = \hat{x}3 + \hat{y}4 + \hat{z}5$, find the angle between A and B.

★2-3-6. Scalar product If $A = \hat{x}2 + \hat{y}0 + \hat{z}2$ and $B = \hat{r}2$ at $\theta = 45°$, $\phi = 0°$, find the angle between A and B.

2-3-7. Sphere and Gauss' law A spherical distribution of charge has a density $\rho = kr$ C m^{-3} where $k = $ constant. Find D inside the sphere.

★2-3-8. Parallel unit vector Find the unit vector which is parallel to the resultant of the vectors $A = \hat{x} + \hat{y}2 + \hat{z}3$ and $B = \hat{x}2 - \hat{y}8 - \hat{z}2$.

2-3-9. Potential and field of disk A thin flat disk of charge of 1-m radius has a surface charge density $\rho_s = \frac{1}{9}\pi$ nC m^{-2}. Find E and V at the following distances perpendicular to the center of the disk: (a) 1 μm; (b) 1 m; (c) 2 m; and (d) 20 m.

2-3-10. Potential and field of shell of charge An infinitesimally thin shell of charge of 1-m radius has a surface charge density $\rho_s = \frac{1}{36}\pi$ nC m^{-2}. Find E and V at the following distances from its center: (a) 1 μm; (b) 1 m $\pm$ 1 μm; (c) 2 m; and (d) 20 m.

★2-3-11. Water drops A spherical drop of water with a charge of 10 pC has a surface potential of 100 V. (a) Find the radius of the drop. (b) If two such drops coalesce into a larger spherical drop, find the surface potential of the new larger drop.

2-3-12. Comparison of disk, shell, and point charge Find E and V at the following distances from a $\frac{1}{9}$-nC point charge: (a) 1 μm; (b) 1 m; (c) 2 m; and (d) 20 m. Compare these results with the E and V values obtained for a disk and shell of charge in the preceding problems. How good an approximation are the point charge values to the disk and shell values?

2-3-13. Gravitational field of earth shell (a) If the mass of the earth were uniformly distributed in a thin spherical shell of the same radius as the earth, what would the gravitational field be outside the shell? (b) Inside the shell?

2-3-14. Concentric shells of charge Two concentric shells have radii r_1 and r_2. Charge $+Q_1$ is uniformly distributed over the shell of radius r_1, and charge $+Q_2$ is uniformly distributed over the shell of radius r_2. Apply Gauss' law to find (a) the field E everywhere and (b) the potential V everywhere.

2-3-15. Gauss' law and inverse-square relation Show that Gauss' law depends on an inverse-square relation between field and distance; that is, $E \propto 1/r^2$ for a point charge.

2-3-16. Flux from opposite charges Two equal point charges of opposite sign are enclosed by a surface. (a) Is it possible for electric flux from these charges to cross this surface? (b) What is the total flux through the surface due to these charges and any other charges which may exist outside?

2-3-17. Quadrupole (a) Show that at a large distance from a quadrupole (as in Fig. 2-21a) the potential is given by

$$V = \frac{Ql^2 \sin\theta \cos\theta}{2\pi\epsilon_0 r^3}$$

where $Q = $ individual charge
$l = $ spacing between charges
$r = $ radial distance
$\theta = $ angle from x or y axis to radial line (quadrupole oriented as in Fig. 2-21a)

(b) Find the electric field E.

2-3-18. Flux from line charge A uniform line charge of $\rho_L = 6$ nC m^{-1} is situated coincident with the x axis. Find the electric flux per unit length of line passing through a plane strip extending in the x direction with edges at $y = 1$, $z = 0$, and $y = 1$, $z = 5$ m.

2-3-19. Flux through cylinder A cylinder of 1-m radius with axis coincident with the z axis extends between z values of ± 3 m. Find the electric flux passing through the surface of the cylinder from a point charge of 3 μC situated at $(0, 0, 1)$.

★**2-3-20. Flux through sphere** A sphere of 2-m radius has a point charge of 8 nC at its center. Find the electric flux passing through that part of the sphere between $\pm 60°$ latitude and $\pm 20°$ longitude.

2-3-21. Charged balloon A spherical surface such as that of a rubber balloon has a uniform electric charge density. (a) What is the electric field $\mathbf{E}$ inside? (b) If the balloon is flattened, does $\mathbf{E}$ inside change? (c) If the balloon remains spherical but the surface charge is nonuniform, does $\mathbf{E}$ inside change?

Group 2-4: Practical applications

2-4-1. Aircraft bank indication using atmospheric potential A person standing in an open field on a clear day has a potential difference of about 300 V between head and feet due to the electric field E of earth's atmosphere. This $\mathbf{E} \simeq 180$ V m^{-1} is directed downward so that the person's head is 300 V more positive than the feet. Can this potential gradient be used as a bank and pitch indicator on an aircraft? Explain. (See M. L. Hill, Introducing the Electrostatic Autopilot, *Astronautics and Aeronautics*, November 1972, pp. 22–31.) When a thundercloud comes along, the potential difference between a person's head and feet may rise to 20,000 V. (See A. D. Moore, "Electrostatics," Doubleday & Company, Inc., Garden City, N.Y., 1968; J. A. Chalmers, "Atmospheric Electricity," Pergamon Press, New York, 1967.)

2-4-2. Xerographic machine Describe how electrostatic fields are employed in xerographic machines. There are books on the subject.

THREE

THE STATIC ELECTRIC FIELD IN DIELECTRICS

3-1 INTRODUCTION

This chapter extends the theory introduced in Chap. 2 to include material media. The conditions of homogeneity, linearity, and isotropy are pointed out. The influence of a dielectric medium on field distributions is discussed, and the concepts of electrostatic energy and energy density are introduced. The fields of several charge distributions are found. Introduction of the divergence concept leads to Maxwell's divergence equation in **D**, which is then applied to the parallel-plate capacitor.

3-2 HOMOGENEITY, LINEARITY, AND ISOTROPY

A medium is *homogeneous* if its physical characteristics (mass density, molecular structure, etc.) do not vary from point to point. If the medium is not homogeneous, it may be described as inhomogeneous, nonhomogeneous, or heterogeneous.

A medium is *linear* with respect to an electrostatic field if the flux density **D** is proportional to the electric field intensity **E**. This is the case in free space, where $\mathbf{D} = \epsilon_0 \mathbf{E}$. Here the factor ϵ_0, or permittivity, is a constant. In material media the permittivity ϵ may not always be constant. If it is not, the material is said to be nonlinear.

An *isotropic* material is one whose properties are independent of direction. Generally, materials whose molecular structure is randomly oriented will be isotropic. However, crystalline media or certain plasmas may have directional characteristics. Such materials are said to be nonisotropic or anisotropic.

In this book concepts are usually developed first for the case where the medium is *homogeneous*, *linear*, and *isotropic*.† Later the ideas may be extended to cases where one or more of these restrictions no longer hold.

3-3 DIELECTRICS AND PERMITTIVITY

In a *conductor* the outer electrons of an atom are easily detached and migrate readily from atom to atom under the influence of an electric field. In a *dielectric*, on the other hand, the electrons are so well bound or held near their equilibrium positions that they cannot be detached by the application of ordinary electric fields. Hence, an electric field produces no migration of charge in a dielectric, and, in general, this property makes dielectrics act as good insulators. Paraffin, glass, and mica are examples of dielectrics.

An important characteristic of a dielectric is its *permittivity* ϵ.‡ Since the permittivity of a dielectric is always greater than the permittivity of vacuum, it is often convenient to use the relative permittivity ϵ_r of the dielectric, i.e., the ratio of its permittivity to that of vacuum. Thus

$$\epsilon_r = \frac{\epsilon}{\epsilon_0}$$

(1)

where ϵ_r = relative permittivity of dielectric
ϵ = permittivity of dielectric
ϵ_0 = permittivity of vacuum = 8.85 pF m^{-1}

Whereas ϵ or ϵ_0 is expressed in farads per meter (F m^{-1}), the relative permittivity ϵ_r is a dimensionless ratio.

The relative permittivity is the value ordinarily given in tables. The relative permittivity of a few media is shown in Table 3-1, where the values are for static (or low-frequency) fields and, except for vacuum or air, are approximate. Note that ϵ_r for air is so close to unity that for most problems we can consider air equivalent to vacuum.

---

† Dielectrics of this type are sometimes designated *Class A dielectrics*.

‡ Also called the *dielectric constant*. However, the permittivity is not always a constant, as one might be led to infer by this term (see Chap. 8).

Table 3-1 Permittivities of dielectric media for static fields†

Medium	Relative permittivity ϵ_r
Vacuum	1‡
Air (atmospheric pressure)	1.0006
Styrofoam	1.03
Paraffin	2.1
Plywood	2.1
Polystyrene	2.7
Amber	3
Rubber	3
Birch	3
Plexiglas	3.4
Dry sandy soil	3.4
Nylon (solid)	3.8
Sulfur	4
Quartz	5
Bakelite	5
Formica	6
Lead glass	6
Mica	6
Marble	8
Flint glass	10
Ammonia (liquid)	22
Glycerin	50
Water (distilled)	81
Rutile (TiO_2)	89–173§
Barium titanate ($BaTiO_3$)	1,200¶
Barium strontium titanate ($2\,BaTiO_3 : 1\,SrTiO_3$)	10,000¶
Barium titanate zirconate ($4\,BaTiO_3 : 1\,BaZrO_3$)	13,000††
Barium titanate stannate ($9\,BaTiO_3 : 1\,BaSnO_3$)	20,000††

† For a comprehensive tabulation of permittivities, see A. R. Von Hippel, "Dielectric Materials and Applications," pp. 301–370, The M.I.T. Press, Cambridge, Mass., 1954.

‡ By definition.

§ Crystals, in general, are nonisotropic; i.e., their properties vary with direction. Rutile is an example of such a nonisotropic crystalline substance. Its relative permittivity depends on the direction of the applied electric field with relation to the crystal axes, being 89 when the field is perpendicular to a certain crystal axis and 173 when the field is parallel to this axis. For an aggregation of randomly oriented rutile crystals $\epsilon_r = 114$. All crystals, except those of the cubic system, are nonisotropic to electric fields; i.e., their properties vary with direction. Thus, the permittivity of many other crystalline substances may vary with direction. However, in many cases the difference is slight. For example, a quartz crystal has a relative permittivity of 4.7 in one direction and 5.1 at right angles. The average value is 4.9. The nearest integer is 5, and this is the value given in the table.

¶ The permittivity of these titanates is highly temperature-sensitive. The above values are for 25°C. See, for example, E. Wainer, High Titania Dielectrics, *Trans. Electrochem. Soc.*, **89** (1946).

†† K. W. Plessner and R. West, High-permittivity Ceramics for Capacitors, in J. B. Birks and J. H. Schulman (eds.), "Progress in Dielectrics," vol. 2, John Wiley & Sons, Inc., New York, 1960.

3-4 THE ELECTRIC FIELD IN A DIELECTRIC

In free space, the electric field is defined as *force per unit charge*. This implies that the electric field in free space is a *measurable quantity*. However, to measure the electric field inside a dielectric or other material medium may be very difficult or impractical. But, if we confine our attention to the external effects of the dielectric, such internal measurements become unnecessary provided a theory can be formulated for the behavior of the dielectric which produces agreement with external measurements. Thus, a distinction should be made between an electric field as a *measurable quantity* (as in free space) and an electric field as a *theoretical quantity* (as in a dielectric).† In this chapter a theory for the electrostatic field in a dielectric is developed and then related to the external field by means of boundary conditions.

3-5 POLARIZATION

Although there is no migration of charge when a dielectric is placed in an electric field, there does occur a slight displacement of the negative and positive charges of the dielectric's atoms or molecules so that they behave like very small *dipoles*. The dielectric is said to be *polarized* or in a state of *polarization* when the dipoles are present. For most materials,‡ the removal of the field results in the return of the atoms or molecules to their normal, or unpolarized, state and the disappearance of the dipoles.

As a simple example, consider that a polarized atom of a dielectric material is represented by an electric dipole, i.e., a positive point charge representing the nucleus and a negative point charge representing the electronic charge, the two charges being separated by a small distance. The electrons orbit the nucleus and act like a negatively charged cloud surrounding the nucleus. When the atom is unpolarized, the cloud surrounds the nucleus symmetrically, as in Fig. 3-1a, and the dipole moment is zero (the equivalent positive and negative point charges have zero displacement). Under the influence of an electric field the electron cloud becomes slightly displaced or asymmetrical, as in Fig. 3-1b, and the atom is polarized. According to our simple picture, the atom may then be represented by the equivalent point-charge dipole of Fig. 3-1c (dipole moment $p = ql$).

Consider the dielectric slab of permittivity ϵ in Fig. 3-2a situated in vacuum. Let a uniform field **E** be applied normal to the slab. This polarizes the dielectric, i.e., induces atomic dipoles throughout the slab. In the interior the positive and negative charges of adjacent dipoles annul each other's effects. The net result of the polarization is to produce a layer of negative charge on one surface of the slab and a layer of positive charge on the other, as suggested in Fig. 3-2a.

---

† The Teaching of Electricity and Magnetism at the College Level (Report of the Coulomb's Law Committee of the American Association of Physics Teachers), *Am. J. Phys.*, **18**(1):5 (January 1950).

‡ When polarization in a dielectric persists in the absence of an applied electric field, the substance is permanently polarized and is called an *electret*. A strained piezoelectric crystal is an example of an electret.

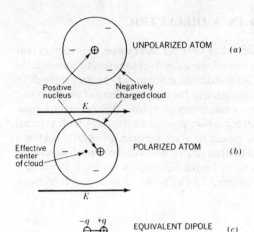

UNPOLARIZED ATOM (a)

Positive nucleus

Negatively charged cloud

E

Effective center of cloud

POLARIZED ATOM (b)

E

$-q$ $+q$

EQUIVALENT DIPOLE (c)
(Moment $= ql$)

Figure 3-1 Unpolarized atom as in (a) becomes polarized as in (b) when electric field is applied. Equivalent dipole is shown at (c).

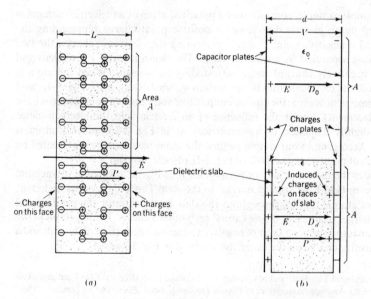

Figure 3-2 (a) Dielectric slab in uniform electric field. (b) Parallel-plate capacitor with dielectric slab in lower part.

The effect of the atomic dipoles can be described by the polarization P, or *dipole moment per unit volume.* Thus,

$$P = \frac{n}{v} ql = \frac{Ql}{v} \tag{1}$$

where n = number of dipoles in volume v
$\quad Q = nq$ = charge of all dipoles
$\quad Ql$ = net dipole moment in volume v

For example, consider the rectangular volume of surface area A and thickness $L (v = AL)$ in Fig. 3-2a. For this volume

$$P = \frac{QL}{AL} = \frac{Q}{A} = \rho_{sp} \qquad (\text{C m}^{-2}) \tag{2}$$

where ρ_{sp} is the surface charge density of polarization charge appearing on the slab faces. Thus, P has the dimensions of charge per area, the same as D.

The value of P in (2) is an average for the volume v. To define the meaning of P at a point, it is convenient to assume that a dielectric in an electric field has a continuous distribution of infinitesimal dipoles, i.e., a continuous polarization, whereas the dipoles actually are discrete polarized atoms. The assumption of a continuous distribution leads to no appreciable error provided that we consider only volumes containing many atoms or dipoles, i.e., macroscopic regions. Assuming now a continuously polarized dielectric, the value of P at a point can be defined as the net dipole moment Ql of a small volume Δv divided by the volume Δv, with the limit taken as Δv shrinks to zero around the point. Thus,

$$P = \lim_{\Delta v \to 0} \frac{Ql}{\Delta v} \tag{3}$$

Consider now a uniform electric field in a parallel-plate capacitor with plates separated by a distance d, as in the cross-sectional view of Fig. 3-2b. There is a voltage V between the plates so that the electric field $E = V/d$ everywhere. The medium in the upper part of the capacitor is vacuum (or air) with permittivity ϵ_0. The lower part is filled with a dielectric of permittivity ϵ. The dielectric completely fills the space between the plates, but in Fig. 3-2b there is a gap in order to show the charges on the plates.

In the upper part of the capacitor (Fig. 3-2b) we have

$$D_0 = \epsilon_0 E \tag{4}$$

where D_0 = electric flux density in vacuum (or air-filled) part of capacitor, C m^{-2}
$\quad \epsilon_0$ = permittivity of vacuum = 8.85 pF m^{-1}
$\quad E = V/d$ = electric field intensity, V m^{-1}

In the lower, dielectric-filled part of the capacitor the electric field polarizes the dielectric, causing a surface charge density ρ_{sp} to appear on both faces of the dielectric slab. These bound charges induce free charges of opposite sign on the capacitor plates (compare upper and lower parts of Fig. 3-2b). As a result the free-charge surface density on the plates is increased by ρ_{sp}. Therefore, in the dielectric we have

$$D_d = \epsilon_0 E + \rho_{sp} \tag{5}$$

but from (2) $\rho_{sp} = P$, and so

$$D_d = \epsilon_0 E + P \tag{6}$$

where D_d = electric flux density in dielectric, C m^{-2}
ϵ_0 = permittivity of vacuum = 8.85 pF m^{-1}
E = electric field intensity, V m^{-1}
P = polarization (of dielectric), C m^{-2}

Equation (6) implies the presence of dielectric (because of the P term), and so the subscript to D is redundant. Therefore, (6) can be written

$$\boxed{D = \epsilon_0 E + P} \tag{7}$$

Although developed for the special case of a parallel-plate capacitor, (7) is a (vector) relation which applies in general.

In the dielectric we can also write

$$\boxed{D = \epsilon E} \quad \text{or} \quad D = D_d = \epsilon E \tag{8}$$

where ϵ is the permittivity of the dielectric material in farads per meter. Equating (6) and (8) gives

$$\epsilon E = \epsilon_0 E + P \tag{9}$$

or $$\boxed{\epsilon = \epsilon_0 + \frac{P}{E}} \quad \text{or} \quad \frac{P}{E} = \epsilon - \epsilon_0 \tag{10}†$$

Example Find the relative permittivity ϵ_r and polarization P for the dielectric slab of Fig. 3-2b. Assume each plus or minus sign represents 1 nC m^{-2} and $A = 1$ m^2.

SOLUTION $$\epsilon_r = \frac{\epsilon}{\epsilon_0} = \frac{\epsilon E}{\epsilon_0 E} = \frac{D_d}{D_0} = \frac{5}{3} \quad \text{(dimensionless)}$$

---

† The ratio P/E is also sometimes written $P/E = \chi\epsilon_0$, where χ is the *electric susceptibility* (dimensionless). Comparing this with (10) gives $\chi = \epsilon_r - 1$. Thus, the susceptibility $\chi = 0$ for vacuum, for which $\epsilon_r = 1$.

By inspection of Fig. 3-2*b* we can write

$$P = \rho_{sp} = 2 \, \text{nC m}^{-2}$$

This result can also be obtained as follows. From (6)

$$P = D_d - \epsilon_0 E$$

Therefore,

$$P = 5 - 3 = 2 \, \text{nC m}^{-2}$$

the same as before.

In isotropic media **P** and **E** are in the same direction, so that their quotient is a scalar and hence ϵ is a scalar. In nonisotropic media, such as crystals, **P** and **E** are, *in general*, not in the same direction, so that ϵ is no longer a scalar. Thus, **D** $= \epsilon_0 \mathbf{E} + \mathbf{P}$ is a general relation, while **D** $= \epsilon \mathbf{E}$ is a more concise expression, which, however, has a simple significance only for isotropic media (or certain special cases in nonisotropic media).

In practice it is simpler, whenever possible, to describe a dielectric by its permittivity ϵ $[= \epsilon_0 + (P/E)]$, thus including implicity the effect of any polarization. However, if we want to consider what is going on in the dielectric, we must deal with the polarization P explicitly.

3-6 BOUNDARY RELATIONS

In a single medium the electric field is continuous. That is, the field, if not constant, changes only by an infinitesimal amount in an infinitesimal distance. However, at the boundary between two different media the electric field may change abruptly both in magnitude and direction. It is of great importance in many problems to know the relations of the fields at such boundaries. These boundary relations are discussed in this section.

It is convenient to analyze the boundary problem in two parts, considering first the relation between the fields *tangent* to the boundary and second the fields *normal* to the boundary.

Taking up first the relation of the fields tangent to the boundary, let two dielectric media of permittivities ϵ_1 and ϵ_2 be separated by a plane boundary as in Fig. 3-3. It is assumed that both media are perfect insulators, i.e., the conductivities† σ_1 and σ_2 of the two media are zero. Consider a rectangular path, half in each medium, of length Δx parallel to the boundary and of length Δy normal to the boundary. Let the average electric field intensity tangent to the boundary in medium 1 be E_{t1} and the average field intensity tangent to the boundary in medium 2 be E_{t2}. The work per unit charge required to transport a positive test charge around this closed path is the line integral of **E** around the path ($\oint \mathbf{E} \cdot d\mathbf{l}$). By

† For discussion of conductivity see Sec. 4-6.

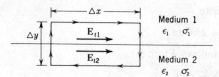

Figure 3-3 The tangential electric field is continuous across a boundary.

making the path length Δy approach zero the work along the segments of the path normal to the boundary is zero even though a finite electric field may exist normal to the boundary. The line integral of **E** around the rectangle in the direction of the arrows is then

$$E_{t1} \Delta x - E_{t2} \Delta x = 0 \tag{1}$$

or

$$\boxed{E_{t1} = E_{t2}} \tag{2}$$

According to (2) *the tangential components of the electric field are the same on both sides of a boundary between two dielectrics.* In other words, the tangential electric field is *continuous* across such a boundary.

If medium 2 is a conductor ($\sigma_2 \neq 0$), the field E_{t2} in medium 2 must be zero under static conditions and hence (2) reduces to

$$E_{t1} = 0 \tag{3}$$

According to (3), the *tangential electric field at a dielectric-conductor boundary is zero.*†

Turning our attention next to the fields normal to the boundary, consider two dielectric media of permittivities ϵ_1 and ϵ_2 separated by the xy plane as shown in Fig. 3-4. It is assumed that both media are perfect insulators ($\sigma_1 = \sigma_2 = 0$). Suppose that an imaginary box is constructed, half in each medium, of area $\Delta x \, \Delta y$ and height Δz. Let D_{n1} be the average flux density normal to the top of the box in medium 1 and D_{n2} the average flux density normal to the bottom of the box in medium 2. D_{n1} is an outward normal (positive), while D_{n2} is an inward normal (negative). By Gauss' law the electric flux or surface integral of the normal component of **D** over a closed surface equals the charge enclosed. By making the height of the box Δz approach zero the contribution of the sides to the surface integral is zero. The total flux over the box is then due entirely to flux over the top and bottom surfaces. If the

---

† This assumes that no currents are flowing. If currents are present, **E** in the conductor is not zero, unless the conductivity is infinite, and (2) applies rather than (3). In Chap. 8 the relations of (2) and (3) are extended to include *time-changing fields*, and it is shown that the relation $E_{t1} = E_{t2}$ of (2) holds with static or changing fields for the boundary between *any* two media of permittivities, permeabilities, and conductivities ϵ_1, μ_1, and σ_1 and ϵ_2, μ_2, and σ_2. Furthermore, for changing fields the relation $E_{t1} = 0$ of (3) is restricted to the case where the conductivity of medium 2 is infinite ($\sigma_2 = \infty$). This follows from the fact that a time-changing electric field in a conductor is zero only if the conductivity is infinite.

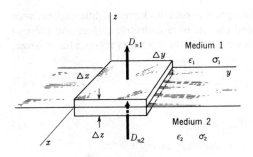

Figure 3-4 The normal component of the flux density is continuous across a charge-free boundary.

average surface charge density on the boundary is ρ_s, we have on applying Gauss' law

$$D_{n1}\,\Delta x\,\Delta y - D_{n2}\,\Delta x\,\Delta y = \rho_s\,\Delta x\,\Delta y$$

or
$$\boxed{D_{n1} - D_{n2} = \rho_s}\qquad(4)$$

According to (4) *the normal component of the flux density changes at a charged boundary between two dielectrics by an amount equal to the surface charge density.* This is usually zero at a dielectric-dielectric boundary unless charge has been placed there by mechanical means, as by rubbing.

If the boundary is free from charge, $\rho_s = 0$ and (4) reduces to

$$D_{n1} = D_{n2}\qquad(5)$$

According to (5), *the normal component of the flux density is continuous across the charge-free boundary between two dielectrics.*

If medium 2 is a conductor, $D_{n2} = 0$ and (4) reduces to

$$D_{n1} = \rho_s\qquad(6)$$

According to (6), *the normal component of the flux density at a dielectric-conductor boundary is equal to the surface charge density on the conductor.*[†]

It is important to note that ρ_s in these relations refers to actual electric charge separated by finite distances from equal quantities of opposite charge and *not* to surface charge ρ_{sp} due to polarization. The polarization surface charge is produced by atomic dipoles having equal and opposite charges separated by what is assumed to be an infinitesimal distance. It is not permissible to separate the positive and negative charges of such a dipole by a surface of integration, and hence the volume must always contain an integral (whole) number of dipoles and, therefore, zero net charge. Only when the positive and negative charges are separated by a macroscopic distance (as on the opposite surfaces of a conducting sheet) can we separate

---

† In Chap. 8 it is pointed out that the relation $D_{n1} - D_{n2} = \rho_s$ of (4) and $D_{n1} = D_{n2}$ of (5) hold with static *or* time-changing fields for *any* two media of permittivities, permeabilities, and conductivities ϵ_1, μ_1, and σ_1 and ϵ_2, μ_2, and σ_2.

them by a surface of integration. This emphasizes a fundamental difference between the polarization, or so-called *bound* charge, on a dielectric surface and the *true* charge on a conductor surface. In a similar way the boundary relation for polarization is

$$P_{n1} - P_{n2} = -\rho_{sp} \tag{7}$$

If medium 2 is free space,

$$P_{n1} = -\rho_{sp} \tag{8}$$

Equations (6) and (8) are written more generally as

$$\mathbf{D} \cdot \hat{\mathbf{n}} = \rho_s \quad \text{and} \quad -\mathbf{P} \cdot \hat{\mathbf{n}} = \rho_{sp} \tag{9}$$

The minus sign in the polarization relation results from the fact that with positive polarization charge at a dielectric surface the polarization is directed *inward* while the surface normal is directed *outward*.

Example 1 *Boundary between two dielectrics.* Let two isotropic dielectric media 1 and 2 be separated by a charge-free plane boundary as in Fig. 3-5. Let the permittivities be ϵ_1 and ϵ_2, and let the conductivities $\sigma_1 = \sigma_2 = 0$. Referring to Fig. 3-5, the problem is to find the relation between the angles α_1 and α_2 of a static field line or flux tube which traverses the boundary. For example, given α_1, find α_2.

SOLUTION Let

$$D_1 = \text{magnitude of } \mathbf{D} \text{ in medium 1}$$

$$D_2 = \text{magnitude of } \mathbf{D} \text{ in medium 2}$$

$$E_1 = \text{magnitude of } \mathbf{E} \text{ in medium 1}$$

$$E_2 = \text{magnitude of } \mathbf{E} \text{ in medium 2}$$

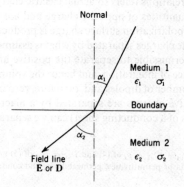

Figure 3-5 Boundary between two dielectric media showing change in direction of field line.

In an isotropic medium, **D** and **E** have the same direction. According to the boundary relations,

$$D_{n1} = D_{n2} \quad \text{and} \quad E_{t1} = E_{t2} \tag{10}$$

Referring to Fig. 3-5,

$$D_{n1} = D_1 \cos \alpha_1 \quad \text{and} \quad D_{n2} = D_2 \cos \alpha_2 \tag{11}$$

while $E_{t1} = E_1 \sin \alpha_1 \quad \text{and} \quad E_{t2} = E_2 \sin \alpha_2$ (12)

Substituting (11) and (12) into (10) and dividing the resulting equations yields

$$\frac{D_1 \cos \alpha_1}{E_1 \sin \alpha_1} = \frac{D_2 \cos \alpha_2}{E_2 \sin \alpha_2} \tag{13}$$

But $D_1 = \epsilon_1 E_1$ and $D_2 = \epsilon_2 E_2$; so that (13) becomes

$$\frac{\tan \alpha_1}{\tan \alpha_2} = \frac{\epsilon_1}{\epsilon_2} = \frac{\epsilon_{r1}\epsilon_0}{\epsilon_{r2}\epsilon_0} = \frac{\epsilon_{r1}}{\epsilon_{r2}} \tag{14}$$

where ϵ_{r1} = relative permittivity of medium 1
ϵ_{r2} = relative permittivity of medium 2
ϵ_0 = permittivity of vacuum

Suppose, for example, that medium 1 is air ($\epsilon_{r1} = 1$), while medium 2 is a slab of sulfur ($\epsilon_{r2} = 4$). Then when $\alpha_1 = 30°$, the angle α_2 in medium 2 is 66.6°.

Example 2 *Boundary between a conductor and a dielectric.* Suppose that medium 2 in Fig. 3-5 is a conductor. Find α_1.

SOLUTION   Since medium 2 is a conductor, $D_2 = E_2 = 0$ under static conditions. According to the boundary relations,

$$D_{n1} = \rho_s \quad \text{or} \quad E_{n1} = \frac{\rho_s}{\epsilon_1}$$

and $E_{t1} = 0$

Therefore

$$\alpha_1 = \tan^{-1}\frac{E_{t1}}{E_{n1}} = \tan^{-1} 0 = 0$$

It follows that a static electric field line or flux tube at a dielectric-conductor boundary is always perpendicular to the conductor surface (when no currents are present). This fact is of fundamental importance in field mapping (see Sec. 3-19).

If a thin conducting sheet is introduced normal to an electric field, surface charges are induced on the sheet so that the original field external to the sheet is

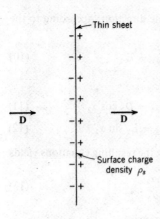

Figure 3-6 Thin conducting sheet placed normal to field has an induced surface charge density ρ_s, equal to the flux density D of the field at the sheet. The surface charge densities on the two sides of the sheet are equal in magnitude but opposite in sign.

undisturbed. The value of the induced surface charge density ρ_s is equal to the flux density D at the sheet. Hence, one can interpret the flux density D at a point as equal to the charge density ρ_s which would appear on a thin conducting sheet introduced normal to $\mathbf{D}$ at the point. Referring, for example, to the thin conducting sheet normal to the field in Fig. 3-6, the relation of $\mathbf{D}$ and ρ_s is as follows:

$$\text{On left side:} \quad \mathbf{D} = -\hat{\mathbf{n}}\rho_s$$

$$\text{On right side:} \quad \mathbf{D} = +\hat{\mathbf{n}}\rho_s$$

where $\hat{\mathbf{n}}$ is the unit vector normal to the surface. Thus $\mathbf{D}$ is normally inward on the left side and normally outward on the right. The magnitude of the flux density on each side is equal to the charge density ρ_s.

3-7 TABLE OF BOUNDARY RELATIONS

Table 3-2 summarizes the boundary relations for static fields developed in the preceding section.

Table 3-2 Boundary relations for static electric fields†

Field component	Boundary relation		Condition
Tangential	$E_{t1} = E_{t2}$	(1)	Any two media
Tangential	$E_{t1} = 0$	(2)	Medium 1 is a dielectric; medium 2 is a conductor
Normal	$D_{n1} - D_{n2} = \rho_s$	(3)	Any two media with charge at boundary
Normal	$D_{n1} = D_{n2}$	(4)	Any two media with no charge at boundary
Normal	$D_{n1} = \rho_s$	(5)	Medium 1 is a dielectric; medium 2 is a conductor with surface charge

† Relations (1), (3), and (4) apply in the presence of currents and also for time-varying fields (Chap. 8). The other relations, (2) and (5), also apply for time-changing situations provided $\sigma_2 = \infty$.

3-8 CAPACITORS AND CAPACITANCE

A *capacitor†* is an electrical device consisting of two conductors separated by an insulating or dielectric medium.

By definition the *capacitance* of a capacitor is the ratio of the charge on one of its conductors to the potential difference between them. Thus, the capacitance C of a capacitor is

$$\frac{Q}{V} = C \tag{1}$$

where Q = charge on one conductor

V = potential difference of conductors

Dimensionally (1) is

$$\frac{\text{Charge}}{\text{Potential}} = \frac{\text{charge}}{\text{energy/charge}} = \frac{\text{charge}^2}{\text{energy}} = \text{capacitance}$$

or in dimensional symbols

$$\frac{Q^2}{ML^2/T^2} = \frac{I^2 T^4}{ML^2}$$

The SI unit of capacitance is the *farad* (F). Thus, 1 C V^{-1} equals 1 F, or

$$\frac{\text{Coulombs}}{\text{Volt}} = \text{farads}$$

In other words, a capacitor that can store 1 C charge with a potential difference of 1 V has a capacitance of 1 F. Since a capacitor of 1 F capacitance is much larger than is ordinarily used in practice, microfarads (μF) and picofarads (pF) are commonly used.

Example A *parallel-plate capacitor* consists of two flat metal sheets of area A separated by distance d, as shown in cross section in Fig. 3-7. If the medium has a permittivity ϵ, find the capacitance.

SOLUTION If a voltage V is applied between the plates, we have

$$V = Ed \qquad \text{(V)}$$

where E is the electric field between the plates. The charge on one plate is given by

$$Q = DA \qquad \text{(C)}$$

† Formerly called a *condenser*.

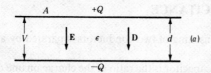

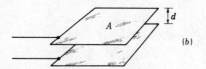

Figure 3-7 (a) Parallel-plate capacitor with voltage difference V and plate charge Q. The area of each plate is A, and the plate separation is d. (b) Parallel-plate capacitor.

where D is the flux density between the plates. But $D = \epsilon E$; so $Q = \epsilon EA$, and the capacitance is

$$C = \frac{Q}{V} = \frac{\epsilon EA}{Ed} = \frac{\epsilon A}{d} \quad \text{(F)}$$

Thus the capacitance is proportional to the permittivity ϵ and the plate area A and is inversely proportional to the plate-separation distance d. It was assumed that the field was uniform between the plates and zero outside (no fringing). Actually any fringing effects will be small if the plates are large compared to d.

Consider now the parallel-plate capacitor shown in Fig. 3-7b. From the above example we have

$$C = \frac{\epsilon A}{d} \quad (2)$$

where ϵ = permittivity of medium between capacitor plates, F m^{-1}
A = area of plates, m^2
d = separation of plates, m

Introducing the relative permittivity ϵ_r and the value of ϵ_0, we obtain for the capacitance of the parallel-plate capacitor

$$C = 8.85 \frac{A\epsilon_r}{d} \quad \text{(pF)} \quad (3)$$

where ϵ_r is the relative permittivity of the medium between the plates.

Example 1 Referring to Fig. 3-8, a parallel-plate capacitor consists of two square metal plates 500 mm on a side separated by 10 mm. A slab of sulfur ($\epsilon_r = 4$) 6 mm thick is placed on the lower plate, leaving an air gap 4 mm thick between it and the upper plate. Find the capacitance of the capacitor. Neglect fringing of the field at the edges of the capacitor.

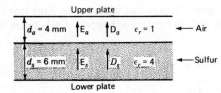

Figure 3-8 Cross section through capacitor with sulfur slab on lower plate and an air space above.

SOLUTION Imagine that a thin metal foil is placed on the upper surface of the sulfur slab. The foil is not connected to either plate. Since the foil is normal to **E** (and assuming that it is of negligible thickness) the field in the capacitor is undisturbed. The capacitor may now be regarded as two capacitors in series, an air capacitor of 4 mm plate spacing and capacitance C_a, and a sulfur-filled capacitor of 6 mm plate spacing and capacitance C_s. The capacitance of the air capacitor is, from (3),

$$C_a = \frac{8.85 A \epsilon_r}{d} = \frac{8.85 \times 0.5^2 \times 1}{0.004} = 553 \text{ pF}$$

The capacitance of the sulfur-filled capacitor is

$$C_s = \frac{8.85 \times 0.5^2 \times 4}{0.006} = 1,475 \text{ pF}$$

The total capacitance of two capacitors in parallel is the sum of the individual capacitances. However, the total capacitance of two capacitors in series, as here, is the reciprocal of the sum of the reciprocals of the individual capacitances. Thus, the total capacitance C is given by

$$\frac{1}{C} = \frac{1}{C_a} + \frac{1}{C_s}$$

or

$$C = \frac{C_a C_s}{C_a + C_s} = \frac{553 \times 1,475}{553 + 1,475} = 402 \text{ pF}$$

Example 2 If 100 V is applied across the capacitor plates, find the: electric field (E_a) in air, electric field (E_s) in sulfur, flux density (D_a) in air, flux density (D_s) in sulfur, potential V_a across air gap, and potential V_s across sulfur slab.

SOLUTION Let the field quantities and dimensions be as given in Fig. 3-8. Since **D** is continuous at the air-sulfur boundary,

$$D_a = D_s \quad \text{or} \quad \epsilon_0 E_a = \epsilon E_s$$

Thus

$$E_a = \frac{\epsilon}{\epsilon_0} E_s = \epsilon_r E_s = 4 E_s$$

Thus, the electric field in the air gap is 4 times that in the sulfur. Note that the potential V across the capacitor is given by the line integral of $\mathbf{E}$ between the plates, or

$$V = E_a d_a + E_s d_s = \epsilon_r E_s d_a + E_s d_s$$

$$= (\epsilon_r d_a + d_s) E_s$$

Hence $\qquad E_s = \dfrac{V}{\epsilon_r d_a + d_s} \qquad$ and $\qquad E_a = \dfrac{\epsilon_r V}{\epsilon_r d_a + d_s}$

Introducing numerical values,

$$E_s = \frac{100}{4 \times 0.004 + 0.006} = 4545 \text{ V m}^{-1}$$

and $\qquad\qquad E_a = 4 \times E_s = 18{,}182 \text{ V m}^{-1}$

We also have

$$V_a = E_a d_a = 72.7 \text{ V} \qquad \text{and} \qquad V_s = E_s d_s = 27.3 \text{ V}$$

and $\qquad\qquad V = V_a + V_s = 72.7 + 27.3 = 100 \text{ V (as given)}$

The flux density

$$D_a = D_s = \epsilon_0 E_a = \epsilon_r \epsilon_0 E_s = 161 \text{ nC m}^{-2}$$

3-9 DIELECTRIC STRENGTH

The field intensity $\mathbf{E}$ in a dielectric cannot be increased indefinitely. If a certain value is exceeded, sparking occurs and the dielectric is said to break down. The maximum field intensity that a dielectric can sustain without breakdown is called its *dielectric strength*.

In the design of capacitors it is important to know the maximum potential difference that can be applied before breakdown occurs. For a given plate spacing this breakdown is proportional to the dielectric strength of the medium between the plates. The radius of curvature of the conducting surface is another factor. The electric field E adjacent to a conductor is proportional to the electric charge density ρ_s on the conductor surface (Sec. 3-6), and this charge density tends to be higher on surfaces with small radii of curvature and less on surfaces with large radii of curvature (Sec. 3-19).

As E is gradually increased, sparking occurs in air almost immediately when a critical value of field is exceeded if the field is uniform (E everywhere parallel), but a discharge may occur first if the field is nonuniform (diverging) with sparkover following as E is increased further.

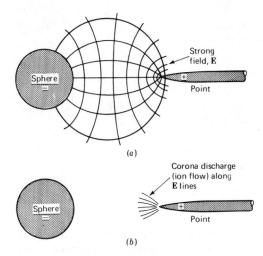

Figure 3-9 Spherical and point conductors with field map (a) showing strong field E near point resulting in corona discharge in air (b).

Near sharp points E may become very high (see Fig. 3-9a), and although air is usually regarded as a nonconducting medium, it does contain ions† produced by cosmic rays from above and radioactivity from the earth below. Thus, a charged point will attract ions of opposite polarity with a force $F(=qE$, where q = charge of ion), and if E is large enough, the ions will be accelerated to such velocities as to produce more ions by collision. This flow of ions constitutes an electric current or *corona discharge* with a visible glow and an audible hissing sound (see Fig. 3-9b). In this process the charge on the sharp point tends to be neutralized by the ions attracted to it, reducing the potential difference between the point and the surrounding space. However, if the potential difference is maintained or increased, the air may break down completely and a spark or electric discharge will occur.

Between a thundercloud and the ground there may be a potential difference of more than 100 million V, and when a discharge, or lightning, occurs, currents of 100,000 A or more may flow momentarily between the cloud and the ground. Although the lightning current is of short duration, the stroke energy may be 1 billion (10^9) J!

The lightning rod with its sharp point and conductor to ground, invented by Benjamin Franklin, provides a low-resistance path for the lightning current so that, if the rod is hit, the damage will be minimal as compared to that which a typical composite poorly conducting, mostly dielectric structure (such as a wood or masonry building) would experience (the I^2R heating may produce steam or other gases with explosive effects). In principle, a lightning rod should help to reduce the electric field above it by corona discharge and thereby reduce the chances of a hit,

---

† Ions are atoms or molecules with an electric charge. Thus, when an electron is detached from a molecule, the electron-deficient molecule becomes a positive ion, while a molecule which acquires an extra electron becomes a negative ion.

Table 3-3 Dielectric strengths

Material	Dielectric strength MV m^{-1}
Air (atmospheric pressure)	3
Oil (mineral)	15
Paper (impregnated)	15
Polystyrene	20
Rubber (hard)	21
Bakelite	25
Glass (plate)	30
Paraffin	30
Quartz (fused)	30
Mica	200

but the important factor is that, if a hit does occur, the rod with its cable to the ground and the adequate ground connection can handle the lightning stroke without damage.

Electrostatic discharge phenomena have many important applications. Electrojet printing, spray painting around corners, and precipitation of particulate matter from smokestack exhausts are based on electrostatic principles, as are also all Xerox and similar copying machines (see Probs. 2-4-1, 3-5-5, 3-5-7, 3-5-13, and 3-5-14).

Many capacitors have air as the dielectric. These types have the advantage that if breakdown occurs, the capacitor is not permanently damaged. For applications requiring large capacitance or small physical size or both, other dielectrics are employed. The dielectric strengths of a number of common dielectric materials are listed in Table 3-3. The dielectric strengths are for a uniform field, and the materials are arranged in order of increasing strength.

3-10 ENERGY IN A CAPACITOR AND ENERGY DENSITY

It requires work to charge a capacitor. Hence energy is stored by a charged capacitor.

To determine the magnitude of this energy, consider a capacitor of capacitance C charged to a potential difference V between the two conductors. Then from (3-8-1)

$$q = CV \tag{1}$$

where q is the charge on each conductor. Potential is work per charge. In terms of infinitesimals it is the infinitesimal work dW per infinitesimal charge dq. That is,

$$V = \frac{dW}{dq} \tag{2}$$

Introducing the value of V from (2) in (1) we have

$$dW = \frac{q}{C}\,dq \qquad (3)$$

If the charging process starts from a zero charge and continues until a final charge Q is delivered, the total work W is the integral of (3), or

$$W = \frac{1}{C} \int_0^Q q\,dq = \frac{1}{2}\frac{Q^2}{C} \qquad (4)$$

This is the energy stored in the capacitor. By (3-8-1) this relation can be variously expressed as

$$\boxed{W = \frac{1}{2}\frac{Q^2}{C} = \frac{1}{2}CV^2 = \frac{1}{2}QV} \qquad (5)$$

where W = energy, J
$\quad C$ = capacitance, F
$\quad V$ = potential difference, V
$\quad Q$ = charge on one conductor, C

Consider the parallel-plate capacitor of capacitance C shown in Fig. 3-10. When it is charged to a potential difference V between the plates, the energy stored is

$$W = \tfrac{1}{2}CV^2 = \tfrac{1}{2}QV \qquad (6)$$

The question may now be asked: In what part of the capacitor is the energy stored? The answer is: The energy is stored in the electric field *between* the plates. To demonstrate this, let us proceed as follows. Consider the small cubical volume $\Delta v\,(=\Delta l^3)$ between the plates as indicated in Fig. 3-10. This volume is shown to a larger scale in Fig. 3-11. The length of each side is Δl, and the top and bottom faces (of area Δl^2) are parallel to the capacitor plates (normal to the field **E**). If thin sheets of metal foil are placed coincident with the top and bottom faces of the volume, the field will be undisturbed provided the sheets are sufficiently thin. The volume Δv now constitutes a small capacitor of capacitance

$$\Delta C = \frac{\epsilon\,\Delta l^2}{\Delta l} = \epsilon\,\Delta l \qquad (7)$$

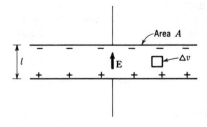

Figure 3-10 Energy is stored in the electric field between the capacitor plates.

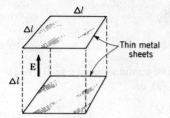

Figure 3-11 Small cubical volume or cell of capacitance $\epsilon \, \Delta l$.

The potential difference ΔV of the thin sheets is given by

$$\Delta V = E \, \Delta l \qquad (8)$$

Now the energy ΔW stored in the volume Δv is, from (5),

$$\Delta W = \tfrac{1}{2}\Delta C \, \Delta V^2 \qquad (9)$$

Substituting (7) for ΔC and (8) for ΔV in (9), we have

$$\Delta W = \tfrac{1}{2}\epsilon E^2 \, \Delta v \qquad (10)$$

Dividing (10) by Δv and taking the limit of the ratio $\Delta W/\Delta v$ as Δv approaches zero, we obtain the energy per volume, or *energy density w*, at the point around which Δv shrinks to zero. Thus†

$$w = \lim_{\Delta v \to 0} \frac{\Delta W}{\Delta v} = \tfrac{1}{2}\epsilon E^2 \qquad (\text{J m}^{-3}) \qquad (12)$$

No material medium need be present for energy to be stored by a field. Thus, energy is present even in vacuum. This energy is equivalent to that required to charge the capacitor cell like the energy stored in a lifted weight. However, if a dielectric material is present, the amount of energy is increased in proportion to the permittivity ϵ. From (3-5-7) or (3-5-9), we can express the energy density as

$$w = \tfrac{1}{2}(\epsilon_0 E^2 + PE) = \tfrac{1}{2}\epsilon E^2 \qquad (\text{J m}^{-3}) \qquad (13)$$

where $\tfrac{1}{2}\epsilon_0 E^2$ = energy density in vacuum
$\tfrac{1}{2}PE$ = additional energy density in dielectric medium

The additional energy in the dielectric is due to the polarization of its molecules in the electric field like the energy stored in a stretched spring.

With no dielectric medium, there is no polarization and the energy density is simply $\tfrac{1}{2}\epsilon_0 E^2$. With a dielectric it is increased by $\tfrac{1}{2}PE$ to a total energy density of $\tfrac{1}{2}\epsilon E^2$.

† For the more general case of a nonisotropic medium in which **D** and **E** may not be in the same direction,

$$w = \tfrac{1}{2}\mathbf{D} \cdot \mathbf{E} \qquad (11)$$

Thus, an electric field contains energy. It will be shown later that a magnetic field also contains energy and that a moving electromagnetic field, or wave, transports energy.

Now the total energy W stored by the capacitor of Fig. 3-10 is the integral of the energy density w over the entire region in which the electric field $\mathbf{E}$ has a value,

$$W = \int_v w \, dv = \frac{1}{2} \int_v \epsilon E^2 \, dv \qquad (14)$$

Assuming that the field is uniform between the plates and that there is no fringing of the field at the edges of the capacitor, we have on evaluating (14)

$$W = \tfrac{1}{2}\epsilon E^2 Al = \tfrac{1}{2}DA\, El = \tfrac{1}{2}QV \qquad \text{(J)} \qquad (15)$$

where A = area of one capacitor plate, m^2
l = spacing between capacitor plates, m

This result, obtained by integrating the energy density throughout the volume between the plates of the capacitor, is identical with the relation given by (5).

3-11 FIELD DISTRIBUTIONS

In many problems it is desirable to know the distribution of the electric field and the associated potential. For example, if the field intensity exceeds the breakdown value for the dielectric medium, sparking, or corona, can occur. From a knowledge of the field distribution, the charge surface density on conductors bounding the field and the capacitance between them can also be determined.

In the following sections the field and potential distributions for a number of simple geometric forms are discussed. The field and potential distributions around point charges, charged spheres, line charges, and charged cylinders are considered first. The field and potential distributions of these configurations can be expressed by relatively simple equations. The extension of these relations by the method of images to situations involving large conducting sheets or ground planes is then considered. Finally, the field and potential distributions for some conductor configurations which are not easily treated mathematically are found by a simple graphical method known as *field mapping*. Many field problems are conveniently solved by the use of computers; such methods are discussed in Chap. 7.

Field and potential distributions may be presented in various ways. For example, a graph of the variation of the magnitude of the electric field $\mathbf{E}$ and of the electric potential V along a reference line may give the desired information. This is illustrated by the curves for E and V in Fig. 3-12a for the field and potential along a radial line extending from the center of a charged conducting sphere of radius r_1. Or the field and potential distributions may be indicated by a contour map, or graph, as in Fig. 3-12b. In this map the radial lines indicate the direction of the electric field, while the circular contours are equipotential lines. In this diagram the potential difference between contours is a constant.

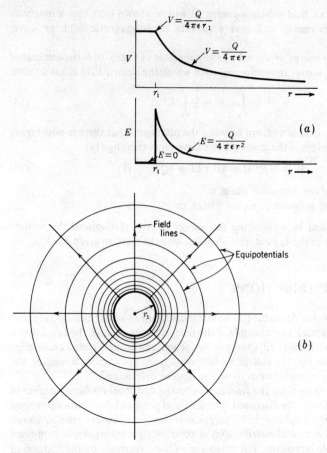

Figure 3-12 Variation of electric field E and potential V of an isolated charged conducting sphere of radius r_1.

3-12 FIELD OF A FINITE LINE OF CHARGE

Consider now the field produced by a thin line of electric charge. Let a positive charge Q be distributed uniformly as an infinitesimally thin line of length $2a$ with center at the origin, as in Fig. 3-13. The linear charge density ρ_L (charge per unit length) is then

$$\rho_L = \frac{Q}{2a} \tag{1}$$

where ρ_L is in coulombs per meter when Q is in coulombs and a is in meters.

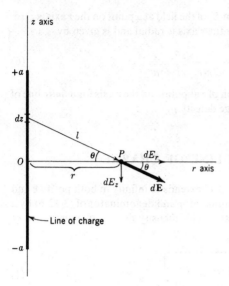

Figure 3-13 Thin line of charge of length $2a$.

At the point P on the r axis, the infinitesimal electric field $d\mathbf{E}$ due to an infinitesimal length of wire dz is the same as from a point charge of magnitude $\rho_L\,dz$. Thus,

$$d\mathbf{E} = \hat{\mathbf{l}}\,\frac{\rho_L\,dz}{4\pi\epsilon l^2} \tag{2}$$

where $l = \sqrt{r^2 + z^2}$
$\hat{\mathbf{l}}$ = unit vector in direction of l

Since the z axis in Fig. 3-13 is an axis of symmetry, the field has only z and r components. These are

$$dE_r = dE\cos\theta = dE\,\frac{r}{l} \tag{3}$$

and

$$dE_z = dE\sin\theta = dE\,\frac{r}{l} \tag{4}$$

The resultant or total r component E_r of the field at a point on the r axis is obtained by integrating (3) over the entire line of charge. That is,

$$E_r = \frac{\rho_L r}{4\pi\epsilon}\int_{-a}^{+a}\frac{dz}{l^3} = \frac{\rho_L r}{4\pi\epsilon}\int_{-a}^{+a}\frac{dz}{\sqrt{(r^2 + z^2)^3}}$$

and

$$E_r = \frac{\rho_L a}{2\pi\epsilon r\sqrt{r^2 + a^2}} \tag{5}$$

By symmetry the resultant z component E_z of the field at a point on the r axis is zero. Hence the total field $\mathbf{E}$ at points along the r axis is radial and is given by

$$|\mathbf{E}| = E_r = \frac{\rho_L a}{2\pi\epsilon r\sqrt{r^2 + a^2}} \qquad (6)$$

This relation gives the field as a function of r at points on the r axis for a *finite* line of charge of length $2a$ and uniform charge density ρ_L.

3-13 FIELD OF AN INFINITE LINE OF CHARGE

Consider that the line of charge in Fig. 3-13 extends to infinity in both positive and negative z directions. By dividing the numerator and denominator of (3-12-6) by a and letting a become infinite the electric field intensity due to an *infinite line of positive charge* is found to be

$$\boxed{|\mathbf{E}| = E_r = \frac{\rho_L}{2\pi\epsilon r}} \qquad (1)$$

The potential difference V_{21} between two points at radial distances r_2 and r_1 from the infinite line of charge is then the work per unit charge required to transport a positive test charge from r_2 to r_1. Assume that $r_2 > r_1$. This potential difference is given by the line integral of E_r from r_2 to r_1, the potential at r_1 being higher than at r_2 if the line of charge is positive. Thus

$$V_{21} = -\int_{r_2}^{r_1} E_r \, dr = \frac{\rho_L}{2\pi\epsilon} \int_{r_1}^{r_2} \frac{dr}{r}$$

or

$$V_{21} = \frac{\rho_L}{2\pi\epsilon} \ln r \Big]_{r_1}^{r_2} = \frac{\rho_L}{2\pi\epsilon} \ln \frac{r_2}{r_1} \qquad (2)\dagger$$

3-14 INFINITE CYLINDER OF CHARGE

If the charge is distributed uniformly along a cylinder of radius r_1 instead of concentrated along an infinitesimally thin line, the field external to the cylinder is given by (3-13-1) for $r \geq r_1$. Inside the cylinder, $\mathbf{E} = 0$.

The potential difference between the cylinder and points outside the cylinder is given by (3-13-2), where $r_2 > r_1$ and ρ_L is the charge per unit length of the cylinder. Inside the cylinder the potential is the same as the potential at the surface ($r = r_1$).

---

† If we put $r = r_1$, then $V = 0$ at $r = r_2$. That is, the surface for which $r = r_2$ becomes a reference surface with respect to which the potential values are measured.

3-15 INFINITE COAXIAL TRANSMISSION LINE

The concentric cylindrical conductors arranged as in Fig. 3-14a form a useful current-carrying arrangement called a *coaxial transmission line*. Much can be learned about its properties from a consideration of its behavior under static conditions. Let a fixed potential difference be applied between the inner and outer conductors of an infinitely long coaxial line so that the charge Q per unit length l of one line is ρ_L. The field is confined to the space between the two conductors. The field lines are radial, and the equipotential lines are concentric circles, as indicated in Fig. 3-14b. The magnitude of the field at a radius r is given by (3-13-1), where $a \leq r \leq b$, and where ρ_L is the charge per unit length on the inner conductor. The potential difference V between the conductors is, from (3-13-2),

$$V = \frac{\rho_L}{2\pi\epsilon} \ln \frac{b}{a} \qquad (1)$$

Since capacitance is given by the ratio of charge to potential, $C = Q/V$. Dividing by length l, we have $C/l = (Q/l)/V$. The ratio Q/l equals the linear charge density ρ_L (C m^{-1}). Hence, the capacitance per unit length C/l of the coaxial line is

$$\frac{C}{l} = \frac{\rho_L}{V} = \frac{2\pi\epsilon}{\ln(b/a)} \qquad \text{(F m}^{-1}) \qquad (2)$$

where ϵ is the permittivity of the medium between conductors.

Since $\epsilon = \epsilon_0 \epsilon_r$, where $\epsilon_0 = 8.85$ pF m^{-1}, (2) can be expressed more conveniently as

$$\frac{C}{l} = \frac{55.6\epsilon_r}{\ln(b/a)} = \frac{24.2\epsilon_r}{\log(b/a)} \qquad \text{(pF m}^{-1}) \qquad (3)$$

where ϵ_r = relative permittivity of medium between conductors
$\quad b$ = inside radius of outer conductor
$\quad a$ = radius of inner conductor (in same units as b)

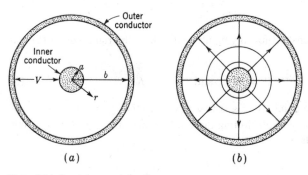

(a)   (b)

Figure 3-14 Coaxial transmission line.

3-16 TWO INFINITE LINES OF CHARGE

Let two infinite parallel lines of charge be separated by a distance $2s$ as in Fig. 3-15. Assume that the linear charge density of the two lines is equal but of opposite sign. The resultant electric field **E** at a point P, distant r_1 from the negative line and r_2 from the positive line, is then the vector sum of the field of each line taken alone.

Let the origin of the coordinates in Fig. 3-15 be the reference for potential. Imagine that only the positively charged line is present. Then from (3-13-2) the potential difference between P and the origin is

$$V_+ = \frac{\rho_L}{2\pi\epsilon} \ln \frac{s}{r_2} \tag{1}$$

Similarly for the negatively charged line

$$V_- = -\frac{\rho_L}{2\pi\epsilon} = \frac{s}{r_1} \tag{2}$$

With both lines present the total potential difference V between P and the origin is the algebraic sum of (1) and (2), or

$$V = V_+ + V_- = \frac{\rho_L}{2\pi\epsilon} \ln \frac{r_1}{r_2} \tag{3}$$

If V in (3) is a constant, (3) is the equation of an equipotential line. The form of the equipotential line will be more apparent if (3) is transformed in the following manner. From (3)

$$\ln \frac{r_1}{r_2} = \frac{2\pi\epsilon V}{\rho_L} \tag{4}$$

and

$$\frac{r_1}{r_2} = e^{2\pi\epsilon V/\rho_L} \tag{5}$$

Since $2\pi\epsilon V/\rho_L$ is a constant for any equipotential line, the right side of (5) is a constant K. Thus

$$e^{2\pi\epsilon V/\rho_L} = K \quad \text{and} \quad r_1 = Kr_2 \tag{6}$$

The coordinates of the point P in Fig. 3-15 are (x, y), so that $r_1 = \sqrt{(s + x)^2 + y^2}$ and $r_2 = \sqrt{(s - x)^2 + y^2}$. Substituting these values of r_1 and r_2 in (6), squaring, and rearranging yields

$$x^2 - 2xs \frac{K^2 + 1}{K^2 - 1} + s^2 + y^2 = 0 \tag{7}$$

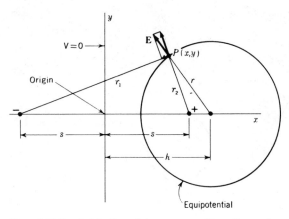

Figure 3-15 Two infinite lines of charge separated by a distance $2s$.

Adding $s^2(K^2 + 1)^2/(K^2 - 1)^2$ to both sides of (7) to complete the square on the left side, we have

$$\left(x - s\frac{K^2 + 1}{K^2 - 1}\right)^2 + y^2 = \left(\frac{2Ks}{K^2 - 1}\right)^2 \tag{8}$$

This is the equation of a circle having the form

$$(x - h)^2 + y^2 = r^2 \tag{9}$$

where x, y = coordinates of point on circle
$\quad\quad h$ = x coordinate of center of circle
$\quad\quad r$ = radius of circle

Comparing (8) and (9), it follows that the equipotential curve passing through the point (x, y) is a circle of radius

$$r = \frac{2Ks}{K^2 - 1} \tag{10}$$

with its center on the x axis at a distance from the origin

$$h = s\frac{K^2 + 1}{K^2 - 1} \tag{11}$$

An equipotential line of radius r with center at $(h, 0)$ is shown in Fig. 3-15. As K increases, corresponding to larger equipotentials, r approaches zero and h approaches s, so that the equipotentials are smaller circles with their centers more nearly at the line of charge. This is illustrated by the additional equipotential circles in Fig. 3-16. The potential is zero along the y axis. That is, $V = 0$ at $x = 0$. Thus, the plane $x = 0$ is the reference plane for potential.

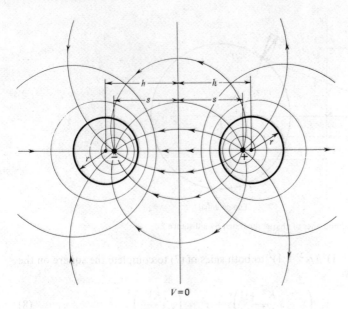

Figure 3-16 Field and equipotential lines around two infinite parallel lines of charge or around an infinite two-conductor transmission line.

Field lines are also shown in Fig. 3-16. These are everywhere orthogonal to the potential circles and also are circles with their centers on the y axis.

3-17 INFINITE TWO-WIRE TRANSMISSION LINE

The discussion of two infinite lines of charge in the previous section is easily extended to the case of an infinite line consisting of two parallel current-carrying wires. This arrangement constitutes a *two-wire transmission line*, a form in very common use. Much can be learned about its properties from a consideration of its behavior under static conditions. Let a fixed potential difference be applied between the conductors so that the charge per unit length of each conductor is ρ_L.

The surface of the wires is an equipotential surface, and therefore an equipotential circle in Fig. 3-16 will coincide with the wire surface. Thus, the heavy circles of radius r and center-to-center spacing $2h$ can represent the two wires. The field and potential distributions external to the wire surfaces are the same as if the field were produced by two infinitesimally thin lines of charge with a spacing of $2s$. The field inside the wires is, of course, zero, and the potential is the same as on the surface. The charge is not uniformly distributed on the wire surface but has higher density on the adjacent sides of the conductors.

The potential difference V_c between one of the conductors and a point midway between them is, from (3-16-3) and (3-16-6),

$$V_c = \frac{\rho_L}{2\pi\epsilon} \ln K \tag{1}$$

The value of K can be expressed in terms of the radius r and half the center-to-center spacing h by eliminating s from (3-16-10) and (3-16-11) and solving for K, obtaining

$$K = \frac{h}{r} + \sqrt{\frac{h^2}{r^2} - 1} \tag{2}$$

The potential difference V_{2c} between the two conductors is then

$$V_{2c} = 2V_c = \frac{\rho_L}{\pi\epsilon} \ln\left(\frac{h}{r} + \sqrt{\frac{h^2}{r^2} - 1}\right) \tag{3}$$

To find the *capacitance per unit length*, C/l, of the two-conductor line we take the ratio of the charge per unit length on one conductor to the difference of potential between the conductors, i.e.,

$$\frac{C}{l} = \frac{\rho_L}{V_{2c}} = \frac{12.1\epsilon_r}{\log[(h/r) + \sqrt{(h/r)^2 - 1}]} \qquad (\text{pF m}^{-1}) \tag{4}$$

where　ϵ_r = relative permittivity of the medium surrounding the conductors, dimensionless

　　　　h = half center-to-center spacing

　　　　r = conductor radius (same units as h)

3-18 INFINITE SINGLE-WIRE TRANSMISSION LINE

A single-wire transmission line with ground return is another form sometimes used. Let the conductor radius be r and the height of the center of the conductor above ground be h. Assume that the conductor has a positive charge ρ_L per unit length and that the ground is at zero potential.

The field and potential distribution of this type of line is readily found by the *theory of images*.† Thus, if the ground is removed and an identical conductor with charge $-\rho_L$ per unit length placed as far below ground level as the other conductor is above, the situation is the same as for a two-conductor line (Fig. 3-16). The conductor which replaces the ground is called the *image* of the upper conductor. The field and potential distribution for the single conductor line is then as illustrated by Fig. 3-17.

---

† The theory of images is discussed more fully in Sec. 7-14.

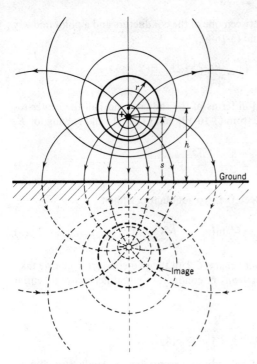

Figure 3-17 Conductor over a metallic ground plane.

The difference in potential between the single conductor and the ground is as given by (3-17-1) or by one-half of (3-17-3). The capacitance per unit length C/l is twice the value given by (3-17-4), or

$$\frac{C}{l} = \frac{24.2\epsilon_r}{\log[(h/r) + \sqrt{(h/r)^2 - 1}]} \quad \text{(pF m}^{-1}) \tag{1}$$

A strip transmission line is a variant of the single-wire transmission line (see Sec. 10-5).

3-19 FIELD MAPS AND FIELD CELLS

Not all conductor configurations can be treated mathematically as readily as those above. Other methods are discussed in Chap. 7. However, for two-dimensional problems† a very effective *graphical-field-mapping‡* method is applicable. Some

---

† By a two-dimensional problem is meant one in which the conductor configuration can be shown by a single cross section, all cross sections parallel to it being the same.

‡ A. D. Moore, "Fundamentals of Electrical Design," McGraw-Hill Book Company, New York, 1927.

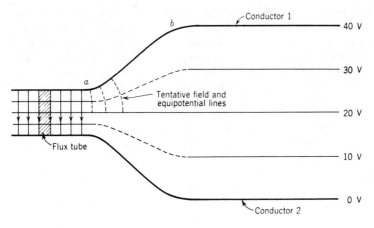

Figure 3-18 Cross section of two sheet conductors with partially completed field map.

experience in graphical field mapping is valuable because it is based on and emphasizes such important fundamental properties of static electric fields as the following:

1. Field and potential lines intersect at right angles.†
2. The surface of a conductor is an equipotential surface.
3. The field meets a conducting surface normally.
4. In a uniform field, the potential varies linearly with distance.
5. A flux tube is parallel to the field,‡ and the electric flux is constant over any cross section of a flux tube.
6. A tube of flux originates on a positive charge and ends on an equal negative charge.

   Graphical field mapping will be introduced with the aid of an example. Consider two charged sheet conductors 1 and 2 as shown in cross section in Fig. 3-18. The sheets extend infinitely far to the left and right and also normal to the page. This is a two-dimensional problem, all cross sections parallel to the page being the same. Therefore, the field and potential distribution everywhere between the sheets will be known if it can be found for a two-dimensional cross section such as shown in Fig. 3-18. Let the potential difference between the conductors be 40 V, with the upper conductor positive and the lower conductor at zero potential. To the left of a and to the right of b the field is uniform, so that equipotential surfaces 10 V apart are equally spaced as indicated, the conductor surfaces being equipotentials at 0 and 40 V. Between a and b the conductor spacing changes, and the equipotentials may be drawn tentatively as shown by the dashed lines.

---

† Except at singular points, as in a corner.
‡ The sidewalls of a flux tube are field lines.

The next step in the mapping procedure is to draw field lines from conductor 1 to conductor 2 in the uniform field region to the left of *a* with the spacing equal to that between the equipotentials. In this way the region is divided into squares. Each square is the end surface of a rectangular volume, or cell, of depth *d* into the page. A stack, or series, of squares bounded by the same field lines represents a rectangular flux tube extending between the positive charge on one conductor and the negative charge on the other. The field map is next extended to the right by drawing *field lines as nearly normal to the equipotentials as possible*, with the field lines spaced so that the areas formed are as nearly square as possible. After one or two revisions of the tentative equipotentials between *a* and *b* and also of the field lines, it should be possible to remap the region to the right of *a* so that field and equipotential lines are everywhere orthogonal and the areas between the lines are all *squares* or *curvilinear squares*. The completed field map is shown in Fig. 3-19.

By a *curvilinear square* is meant *an area that tends to yield true squares as it is subdivided into smaller and smaller areas by successive halving of the equipotential interval and the flux per tube*. A partially subdivided curvilinear square is illustrated in Fig. 3-20.

A field map, such as shown in Fig. 3-19, divides the field into many squares each of which represents a side of a *field cell*. These field cells have a depth *d* (into the page) as suggested by the three-dimensional view of the typical field cell in Fig. 3-19. The cell has a length *l* (parallel to the field) and a width *b*. The sidewalls of a field

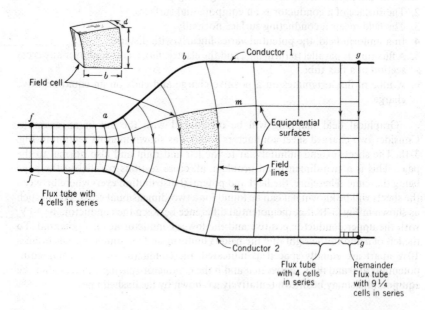

Figure 3-19 Cross section of two sheet conductors with completed field map. Inset shows three-dimensional view of a field cell.

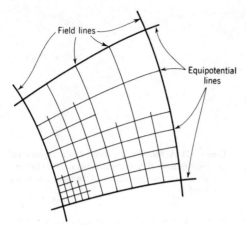

Figure 3-20 Partially subdivided curvilinear square.

cell are the walls of a flux tube (parallel to the field), while the top and bottom coincide with equipotential surfaces. As curvilinear cells are subdivided into smaller cells, their end areas tend to become true squares. The subdivided cells are always of depth d (into the page), the same as the larger cells. Thus, a *field cell*, or simply a *cell*, may be defined as a *curvilinear square volume*.

If thin sheets of metal foil are applied to the equipotential surfaces of a field cell, we have a *field-cell capacitor*. The capacitance C of a parallel-plate capacitor is

$$C = \frac{\epsilon A}{l} \qquad (1)$$

where ϵ = permittivity of medium
  A = area of plates
  l = spacing of plates

Applying this relation to a field-cell capacitor with a square end ($b = l$), we have for the capacitance C_0 of the field cell

$$C_0 = \frac{\epsilon b d}{l} = \epsilon d \qquad (2)$$

Dividing by d, we obtain the capacitance per unit depth of a field cell as

$$\boxed{\frac{C_0}{d} = \epsilon} \qquad (3)$$

where ϵ is the permittivity of the medium in farads per meter.

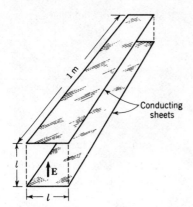

Figure 3-21 A field-cell capacitor has a capacitance per unit depth equal to the permittivity ϵ of the medium. For air ($\epsilon = \epsilon_0$) the capacitance of the capacitor shown is 8.85 pF.

Thus, the significance of the value of ϵ is that it is the capacitance per unit depth of a field-cell capacitor. For example, a field-cell capacitor of unit depth in a medium of air (or vacuum) has a capacitance of 8.85 pF. Such a capacitor is illustrated in Fig. 3-21.†

Any field cell can be subdivided into smaller square-ended cells with as many cells in parallel as in series. Hence the capacitance‡ per unit depth of *any* field cell, large or small, exactly square or curvilinear is equal to ϵ.

In a field map, such as in Fig. 3-19, most of the area is divided into "regular" cells with four in series for each flux tube. These cells all have the same potential difference across them (10 V). Hence these cells may be defined as *cells of the same kind*. The remaining area of the map (at the right) consists of a fractional, or remainder, flux tube. This tube is also divided into cells. These cells are of two kinds, both different from those in the rest of the map. One kind of cell in the remainder flux tube has about 4.3 V across it and the other kind about 1 V across it. There are nine 4.3-V cells and four 1-V cells.

Any field cell has the same capacitance per unit depth. Many additional properties are common to *field cells of the same kind*. These cells of the same kind have the same potential differences across them. In uniform fields the areas of the ends of those cells are the same, but in nonuniform fields the areas will not be the same.

Since the capacitance per unit depth of any cell of the same kind is the same, it follows that the electric flux through any cell of the same kind is the same ($Q/d = VC/d$). Thus, the 10-V cells in Fig. 3-19 have a flux of 10ϵC/unit depth, while the 4.3- and 1-V cells have 4.3ϵC and ϵC/unit depth, respectively.

---

† The capacitance of an isolated capacitor such as shown in Fig. 3-21 is somewhat greater than 8.85 pF because of fringing of the field. However, a field cell represents only a portion of a more extensive field, and its sides are parallel to the field (no fringing).

‡ It is understood that this capacitance is that which would be obtained if the *field cell* were made into a *field-cell capacitor* by placing thin sheets of metal foil coincident with its equipotential surfaces (if no conductor is already present).

Now the average flux density D at the equipotential surface of a field cell is given by

$$D = \frac{Q}{bd} = \rho_s \qquad (\text{C m}^{-2}) \qquad (4)$$

where Q = total charge on foil at equipotential surface of field cell = total flux ψ through cell, C

b = width of cell, m

d = depth of cell, m

ρ_s = average surface charge density on foil at equipotential surface, C m^{-2}

Hence, the average flux density is inversely proportional to the field-cell or flux-tube width. Also the average surface charge density ρ_s at a conducting surface is inversely proportional to the width of the field cell or flux tube at the surface. For example, the spacing of conductors 1 and 2 to the right of b in Fig. 3-19 is 4 times that to the left of a; so in the uniform field region to the left of a the surface charge density ρ_s is 4 times the value of ρ_s in the uniform field region to the right of b. The surface charge density is even smaller than to the right of b in the region of concave conductor curvature near b and somewhat larger than to the left of a in the region of convex conductor curvature near a.

Since $E = D/\epsilon$, the field intensity is also inversely proportional to the cell width, or length ($E = V/l$). Furthermore, the energy $W\ (=\frac{1}{2}QV)$ stored in any cell of the same kind is the same. It follows that the average energy density w is inversely proportional to the area of the end of the cell ($= bl$ for a square-ended cell). For example, the energy density in the uniform field region to the left of a in Fig. 3-19 is 16 times the energy density in the uniform field region to the right of b.

To summarize, *the properties of an accurate electric field map†* are as follows:

1. The capacitance of *any* field cell is the same.
2. The capacitance C_0 per unit depth of *any* field cell is equal to the permittivity ϵ of the medium.
3. The potential difference across any field cell of the same kind is the same.
4. The flux ψ through any field cell of the same kind is the same.
5. The flux ψ over any cross section of a flux tube is the same.
6. The average flux density D in any cell of the same kind is inversely proportional to the width of the cell or flux tube.
7. The average charge density ρ_s at the conducting boundary of any cell of the same kind is inversely proportional to the width of the cell at the boundary.
8. The average field intensity E in any cell of the same kind is inversely proportional to the cell width.
9. The energy stored in any cell of the same kind is the same.
10. The average energy density w in any cell of the same kind is inversely proportional to the area of the end of the cell. (This is the area that appears in the field map.)

---

† In a single medium of uniform permittivity.

In addition to providing quantitative information about a field, a field map gives a simple easily interpreted visual picture of a field. Thus, on a map with all cells of the same kind, the field is strong where the squares are small and weak where the squares are large.

In order to test the accuracy of a field map, and hence the accuracy with which the above properties hold for a particular map, the curvilinear squares of the map can be further subdivided by halving the equipotential interval and halving the flux per tube, as in Fig. 3-20. If the smaller regions so produced tend to become more nearly true squares, the field is accurately mapped. However, if the regions tend to become rectangles, the map is inaccurate and another attempt should be made. Also field and equipotential lines should intersect orthogonally. It is especially important that this rule be observed at all stages of making a field map. Often it is better to erase and begin again than to try to revise an inaccurate map. *In graphical field mapping an eraser is as important as a pencil.*

With analog and digital computers and methods discussed in Chap. 7, a process equivalent to graphical field mapping is carried out. With the digital computer, iteration is continued until no further change occurs in the map. Experience in graphical field mapping is a valuable prelude to computer mapping.

Example Referring to Fig. 3-19, let the conductor separation at ff be 10 mm and at gg be 40 mm, and let the conductors have a depth (into the page) of 200 mm. If the conductors end at ff and gg, and if fringing of the field is neglected, find the capacitance C of the resulting capacitor. The medium in the capacitor is air.

SOLUTION The method of solution will be to evaluate the series-parallel combination of capacitors formed by the individual cells.

Each cell has a capacitance

$$C_0 = \epsilon_0 d = 8.85 \times 0.2 = 1.77 \text{ pF}$$

The capacitance between the ends of each flux tube with 4 cells in series is then

$$\frac{1.77}{4} = 0.443 \text{ pF}$$

The capacitance between the ends of the remainder flux tube with 9.25 cells in series is

$$\frac{1.77}{9.25} = 0.191 \text{ pF}$$

There are fifteen 4-cell tubes and one remainder (9.25-cell) tube. Hence the total capacitance C between ff and gg is the sum of the capacitances of all the flux tubes, or

$$C = (15 \times 0.443) + 0.191 = 6.83 \text{ pF}$$

The above calculation is somewhat simplified if each cell is arbitrarily assigned a capacitance of unity. On this basis the total capacitance in arbitrary units is given by

$$\frac{15}{4} + \frac{1}{9.25} = 3.86 \text{ units}$$

and the total actual capacitance C is the product of this result and the actual capacitance of a cell, or

$$C = 3.86 \times (8.85 \times 0.2) = 6.83 \text{ pF}$$

Yet another method of calculation is to use the relation that the total capacitance C is given by

$$C = \frac{N}{n} C_0 \tag{5}$$

where N = number of cells (or flux tubes) in parallel
n = number of cells in series
C_0 = capacitance of one cell

and where all cells are of the same kind. Thus in the above example, counting in terms of the 10-V cells, we have

$$C = \frac{15.43}{4} \times (8.85 \times 0.2) = 6.83 \text{ pF}$$

Note that if we had wanted the capacitance of a capacitor with conductors coinciding with the equipotentials m and n (Fig. 3-19) and of 200 mm depth, the cells in series would be reduced to two and the capacitance doubled. In this way the capacitance of any conductor configuration conforming to the equipotential surfaces of a field map can be easily calculated.

3-20 HEART DIPOLE FIELD

The heart of all mammals contracts or beats in response to an electric potential difference across it which reaches a maximum value just prior to the start of the blood-pumping contraction. The potentials measured on the skin of the animal at this instant of maximum voltage have a distribution like that from a dipole aligned with the heart. The similarity of the field measured on a human chest (Fig. 3-22a) to the field of a dipole in an isotropic medium (Fig. 3-22b) is apparent. (See also Fig. 2-13.) The electric fields of animals are of great diagnostic value. For example, a different (or abnormal) heart position would be evident from a field map. An advantage of this mapping technique is that it is harmless, in contrast to x-ray or certain other techniques for acquiring similar information.

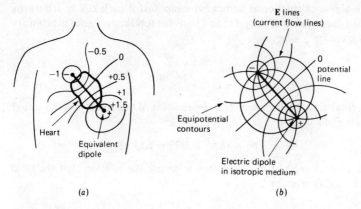

(a) (b)

Figure 3-22 (a) Equipotential contours of electric field from the heart measured on a human chest just prior to heart contraction with approximate position of the equivalent heart dipole. The field shown is a somewhat simplified version of an actual measured field. (b) Equipotential and field lines of electric dipole in an isotropic medium. Note that the heart dipole is within an imperfect (conducting) dielectric medium and the potentials are measured on a surface displaced from it. The dipole in (b), however, is in a uniform medium and the map is in a plane parallel to and coincident with the dipole. Thus, one should not expect the two maps to be identical. The equipotential contours in (a) are in millivolts. [(a) is after B. Taccardi, Circ. Res., 1963.]

3-21 DIVERGENCE OF THE FLUX DENSITY D

In Sec. 2-17 Gauss' law is applied to surfaces enclosing finite volumes, and it is shown that the normal component of the flux density **D** integrated over a closed surface equals the electric charge enclosed. By an extension of this relation to surfaces enclosing infinitesimal volumes we are led to a useful relation called *divergence*.

Let Δv be a small but finite volume. If we assume a uniform charge density throughout the volume, the charge ΔQ enclosed is the product of the volume charge density ρ and the volume Δv. By Gauss' law the charge enclosed is also equal to the integral of the normal component D_n of the flux density over the surface of the volume Δv. Thus,

$$\oint_s D_n \, ds = \Delta Q = \rho \, \Delta v \qquad (1)$$

and

$$\frac{\oint_s D_n \, ds}{\Delta v} = \rho \qquad (2)$$

If the charge density is not uniform throughout Δv, we may take the limit of (2) as Δv shrinks to zero, obtaining the charge density ρ at the point around which Δv

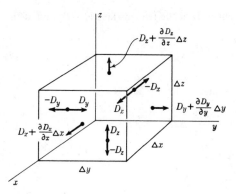

Figure 3-23 Construction used to develop differential expression for divergence of **D**.

collapses. The limit of (2) as Δv approaches zero is called the divergence of **D**, written div **D** or $\nabla \cdot$ **D**. Hence

$$\lim_{\Delta v \to 0} \frac{\oint_s D_n \, ds}{\Delta v} = \text{div } \mathbf{D} = \rho \qquad (\text{C m}^{-3}) \tag{3}$$

Whereas the integral of the normal component of **D** over a finite volume yields the *free charge* enclosed, the divergence of **D** gives the *free charge density at a point*. If the charge is zero at a point, it follows that the charge density is zero and also that the divergence of **D** is zero at that point. It is important to note that the divergence of **D** is a scalar point function.

Let us now discuss divergence in a more formal way, developing it as a differential expression. A small volume $\Delta x \, \Delta y \, \Delta z = \Delta v$ is placed in an electric field with flux density **D**, having components D_x, D_y, and D_z in the three coordinate directions as shown in Fig. 3-23. The total flux density **D** is related to its components by

$$\mathbf{D} = \hat{x}D_x + \hat{y}D_y + \hat{z}D_z \tag{4}$$

The normal outward component of **D** at the back face is $-D_x$ since the field is directed inward. If the field changes between the back and front faces, the normal component of **D** at the front face can, by Taylor's theorem, be represented by an infinite series,

$$D_x + \frac{\partial D_x}{\partial x} \frac{\Delta x}{1} + \frac{\partial^2 D_x}{\partial x^2} \frac{\Delta x^2}{2!} + \frac{\partial^3 D_x}{\partial x^3} \frac{\Delta x^3}{3!} + \cdots \tag{5}$$

When Δx is very small, the square and higher-order terms may be neglected, so that at the front face we have for the normal component of **D**

$$D_x + \frac{\partial D_x}{\partial x} \Delta x \tag{6}$$

In like manner the normal component of **D** at the left side face is $-D_y$ and at the right side face is

$$D_y + \frac{\partial D_y}{\partial y} \Delta y \tag{7}$$

Similarly at the bottom face it is $-D_z$ and at the top face is

$$D_z + \frac{\partial D_z}{\partial z} \Delta z \tag{8}$$

The outward flux of **D** over the back face is

$$-D_x \, \Delta y \, \Delta z \tag{9}$$

and over the front face is

$$\left(D_x + \frac{\partial D_x}{\partial x} \Delta x\right)\Delta y \, \Delta z \tag{10}$$

Adding up the outward flux of **D** over the entire volume, we obtain for the total flux

$$\Delta \psi = \left(-D_x + D_x + \frac{\partial D_x}{\partial x} \Delta x\right)\Delta y \, \Delta z$$

$$+ \left(-D_y + D_y + \frac{\partial D_y}{\partial y} \Delta y\right)\Delta x \, \Delta z + \left(-D_z + D_z + \frac{\partial D_z}{\partial z} \Delta z\right)\Delta x \, \Delta y \tag{11}$$

which simplifies to

$$\Delta \psi = \left(\frac{\partial D_x}{\partial x} + \frac{\partial D_y}{\partial y} + \frac{\partial D_z}{\partial z}\right)\Delta x \, \Delta y \, \Delta z \tag{12}$$

From Gauss' law we know that the total electric flux over the surface of the volume (or integral of the normal component of **D** over the surface of the volume) is equal to the charge enclosed. The charge enclosed is also equal to the integral of the charge density ρ over the volume. Therefore

$$\Delta \psi = \oint_s D_n \, ds = \left(\frac{\partial D_x}{\partial x} + \frac{\partial D_y}{\partial y} + \frac{\partial D_z}{\partial z}\right)\Delta v = \oint_v \rho \, dv \tag{13}$$

Dividing by Δv and taking the limit as Δv approaches zero, we obtain the divergence of **D**. Thus,

$$\lim_{\Delta v \to 0} \frac{\oint_s D_n \, ds}{\Delta v} = \frac{\partial D_x}{\partial x} + \frac{\partial D_y}{\partial y} + \frac{\partial D_z}{\partial z} = \rho \tag{14}$$

and

$$\operatorname{div} \mathbf{D} = \frac{\partial D_x}{\partial x} + \frac{\partial D_y}{\partial y} + \frac{\partial D_z}{\partial z} = \rho \tag{15}$$

The center member of (15) is a differential relation for the divergence of **D** expressed in rectangular coordinates. The divergence of **D** can also be written as the scalar, or dot product of the operator ∇ and **D**. That is,

$$\text{div } \mathbf{D} = \nabla \cdot \mathbf{D} \tag{16}$$

This is readily seen by expanding (16) into the expressions for ∇ and **D**. Then

$$\nabla \cdot \mathbf{D} = \underbrace{\left(\hat{\mathbf{x}} \frac{\partial}{\partial x} + \hat{\mathbf{y}} \frac{\partial}{\partial y} + \hat{\mathbf{z}} \frac{\partial}{\partial z} \right)}_{\nabla} \cdot \underbrace{(\hat{\mathbf{x}} D_x + \hat{\mathbf{y}} D_y + \hat{\mathbf{z}} D_z)}_{\mathbf{D}} \tag{17}$$

Performing the multiplication indicated in (17) gives nine dot-product terms as follows:

$$\nabla \cdot \mathbf{D} = \hat{\mathbf{x}} \cdot \hat{\mathbf{x}} \frac{\partial D_x}{\partial x} + \hat{\mathbf{y}} \cdot \hat{\mathbf{x}} \frac{\partial D_x}{\partial y} + \hat{\mathbf{z}} \cdot \hat{\mathbf{x}} \frac{\partial D_x}{\partial z} + \hat{\mathbf{x}} \cdot \hat{\mathbf{y}} \frac{\partial D_y}{\partial x} + \hat{\mathbf{y}} \cdot \hat{\mathbf{y}} \frac{\partial D_y}{\partial y}$$

$$+ \hat{\mathbf{z}} \cdot \hat{\mathbf{y}} \frac{\partial D_y}{\partial z} + \hat{\mathbf{x}} \cdot \hat{\mathbf{z}} \frac{\partial D_z}{\partial x} + \hat{\mathbf{y}} \cdot \hat{\mathbf{z}} \frac{\partial D_z}{\partial y} + \hat{\mathbf{z}} \cdot \hat{\mathbf{z}} \frac{\partial D_z}{\partial z} \tag{18}$$

The dot product of a unit vector on itself is unity, but the dot product of a vector with another vector at right angles is zero. Accordingly, six of the nine dot products in (18) vanish, but the three involving $\hat{\mathbf{x}} \cdot \hat{\mathbf{x}}$, $\hat{\mathbf{y}} \cdot \hat{\mathbf{y}}$, and $\hat{\mathbf{z}} \cdot \hat{\mathbf{z}}$ do not, and the product indicated by (17) becomes

$$\nabla \cdot \mathbf{D} = \frac{\partial D_x}{\partial x} + \frac{\partial D_y}{\partial y} + \frac{\partial D_z}{\partial z} \tag{19}$$

The dot product of the operator ∇ with a vector function is the divergence of the vector. The quantity $\nabla \cdot$ may be considered as a *divergence operator*. Thus the divergence operator applied to a vector function yields a scalar function. For example, $\nabla \cdot \mathbf{D}$ (divergence of **D**) is given in rectangular coordinates by the right side of (19) and is a scalar, being equal to the charge density ρ.

If **D** is known everywhere, taking the divergence of **D** enables us to find the sources (positive charge regions) and sinks (negative charge regions) responsible for the electric flux, and, hence, for **D**.

3-22 MAXWELL'S DIVERGENCE EQUATION FROM GAUSS' LAW

The relation (3-21-15) that

$$\boxed{\nabla \cdot \mathbf{D} = \rho} \tag{1}$$

was developed by an application of Gauss' law to an infinitesimal volume. It is the fundamental differential relation for static electric fields. This relation is one of a set of four known as *Maxwell's equations.* We considered it in its integral form in Sec. 2-17. The other three relations are developed in later chapters.

In a region free from charge, $\rho = 0$, and

$$\mathbf{V} \cdot \mathbf{D} = 0 \qquad (2)$$

Example As a simple nonelectrical example of divergence consider that a long hollow cylinder is filled with air under pressure. If the cover over one end of the cylinder is removed quickly, the air rushes out. It is apparent that the velocity of the air will be greatest near the open end of the cylinder, as suggested by the arrows representing the velocity vector **v** in Fig. 3-24a. Suppose that the flow of air is free from turbulence, so that **v** has only an x component. Let us also assume that the velocity **v** in the cylinder is independent of y but is directly proportional to x, as indicated by

$$|\mathbf{v}| = v_x = Kx \qquad (3)$$

where K is a constant of proportionality. The question is: What is the divergence of **v** in the cylinder?

SOLUTION Apply the divergence operator to (3), giving

$$\mathbf{V} \cdot \mathbf{v} = \frac{\partial v_x}{\partial x} = K \qquad (4)$$

Hence, the divergence of **v** is equal to the constant K.

A velocity field can be represented graphically by lines showing the direction of **v** with the density of the lines indicating the magnitude of **v**. The velocity field in the cylinder represented in this way is illustrated in Fig. 3-24b. We note that **v** lines originate, i.e., have their source, throughout the cylinder,

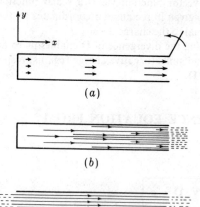

(a)

(b)

(c)

Figure 3-24 The velocity **v** of air rushing from a tube has divergence (a) and (b). When air flows with uniform velocity through a tube open at both ends, as at (c), the divergence of v is zero.

the number increasing with x. This indicates that $\mathbf{v}$ increases as a function of x. This situation is concisely expressed by div $\mathbf{v} = K$. That is, the divergence of $\mathbf{v}$ has a constant value K throughout the cylinder, and this tells us that, assuming (3) to be correct, the source of the velocity field provided by the expanding air is uniformly distributed throughout the cylinder.

If, on the other hand, both ends of the cylinder are open and air passes through with the same velocity everywhere, v_x equals a constant and the divergence of $\mathbf{v}$ is zero in the cylinder. In this case, the source of the velocity field must be somewhere external to the cylinder, and the velocity field diagram would be as shown in Fig. 3-24c.

If more lines enter a small volume† than leave it or more leave it than enter, the field has divergence. If the same number enter as leave the volume, the field has a zero divergence.

3-23 DIVERGENCE THEOREM

From Gauss' law (2-17-6) we have

$$\oint_s \mathbf{D} \cdot d\mathbf{s} = \oint_v \rho \, dv \tag{1}$$

where $\mathbf{D}$ is integrated over the surface s and ρ is integrated throughout the volume v enclosed by s.

From (3-22-1) let us introduce $\nabla \cdot \mathbf{D}$ for ρ in (1), obtaining

$$\oint_s \mathbf{D} \cdot d\mathbf{s} = \oint_v \nabla \cdot \mathbf{D} \, dv \tag{2}$$

The relation stated in (2) is the *divergence theorem* as applied to the flux density $\mathbf{D}$, or *Gauss' theorem* (as distinguished from Gauss' law). This relation holds not only for $\mathbf{D}$, as in (2), but also for any vector function. In words, the divergence theorem states that *the integral of the normal component of a vector function over a closed surface s equals the integral of the divergence of that vector throughout the volume v enclosed by the surface s.*

3-24 DIVERGENCE OF D AND P IN A CAPACITOR

As further illustrations of the significance of divergence let us consider the charged parallel-plate capacitor of Fig. 3-25. A slab of paraffin fills the space between the plates except for the small air gaps. True charge of surface density ρ_s is present on

† In the limit an infinitesimal volume.

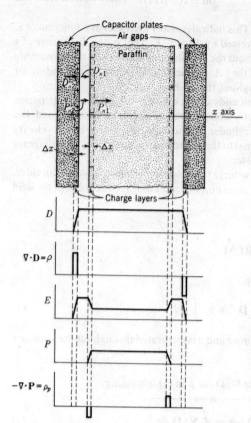

Figure 3-25 Cross section through parallel-plate capacitor with paraffin slab showing the variation of the flux density **D**, charge density ρ, electric field E, polarization P, and polarization charge density ρ_p along axis between the plates. The thickness Δx of the charge layers is greatly exaggerated.

the surface of the plates. Polarization charge of surface density ρ_{sp} is present on the surface of the paraffin.

According to (3) of Table 3-2, the relation of **D** at a boundary is given by

$$D_{n1} - D_{n2} = \rho_s \tag{1}$$

where D_{n1} = flux density in air gap

D_{n2} = flux density in conducting plate = 0

ρ_s = true surface charge density

Suppose that the true charge ρ_s is distributed *uniformly* throughout a thin layer of thickness Δx, as suggested in Fig. 3-25. Then the total change ΔD_n in flux density from one side of the layer to the other is given by

$$D_{n1} - D_{n2} = \Delta D_n = \Delta D_x \tag{2}$$

But when Δx is small,

$$\Delta D_x = \frac{dD_x}{dx} \Delta x \tag{3}$$

Therefore, (1) becomes

$$\frac{dD_x}{dx} = \frac{\rho_s}{\Delta x} = \rho \tag{4}$$

where ρ is the volume charge density in the charge layer. Since $\mathbf{D}$ has only an x component, $dD_x/dx = \operatorname{div} \mathbf{D}$. Thus,

$$\frac{dD_x}{dx} = \mathbf{V} \cdot \mathbf{D} = \rho \tag{5}$$

Hence the change of $\mathbf{D}$ with distance (in the charge layer) equals the divergence of $\mathbf{D}$ and also the volume charge density. It follows that if the charge layer is infinitesimally thin ($\Delta x \to 0$), then $\mathbf{V} \cdot \mathbf{D}$ and ρ approach infinity. However, it is more reasonable to consider that the charge layer is of small but finite thickness, so that although $\mathbf{V} \cdot \mathbf{D}$ and ρ may be large, they are not infinite. The variation of $\mathbf{D}$ and $\mathbf{V} \cdot \mathbf{D}$ along the x axis of the capacitor is shown graphically in Fig. 3-25.

At the paraffin surface $\mathbf{D}$ is constant, but both $\mathbf{E}$ and $\mathbf{P}$ change. From (3-5-7)

$$\mathbf{P} = \mathbf{D} - \epsilon_0 \mathbf{E} \tag{6}$$

Now the change in polarization $\mathbf{P}$ is equal to the surface charge density ρ_{sp} due to polarization. Thus, from (3-6-7),

$$P_{n1} - P_{n2} = -\rho_{sp} \tag{7}$$

where P_{n1} = polarization in paraffin
P_{n2} = polarization in air gap ≈ 0
ρ_{sp} = polarization surface-charge density

Assume that the polarization surface charge is uniformly distributed throughout a thin layer of thickness Δx at the paraffin surface, as suggested in Fig. 3-25. Then the total change ΔP_n in polarization from one side of the layer to the other is given by

$$P_{n1} - P_{n2} = \Delta P_n = \Delta P_x \tag{8}$$

But when Δx is small,

$$\Delta P_x = \frac{dP_x}{dx} \Delta x \tag{9}$$

Therefore, (7) becomes

$$\frac{dP_x}{dx} = -\frac{\rho_{sp}}{\Delta x} = -\rho_p \tag{10}$$

where ρ_p is the volume density of polarization charge in the layer at the paraffin surface in coulombs per cubic meter. Polarization charge differs from true charge ρ in that it cannot be isolated, whereas true charge can. In this sense it is a fictitious charge. Since $\mathbf{P}$ has only an x component, which is a function only of x, $dP_x/dx =$ div $\mathbf{P}$. Thus,

$$\frac{dP_x}{dx} = \boxed{\nabla \cdot \mathbf{P} = -\rho_p} \tag{11}$$

Hence the change of $\mathbf{P}$ with distance (in charge layer) equals the divergence of $\mathbf{P}$ and also the volume density ρ_p of polarization charge. The assumption of a polarization charge layer that is of small but finite thickness results in a value of $\nabla \cdot \mathbf{P}$ that may be large but not infinite.

The divergence of $\mathbf{D}$ yields the sources of the $\mathbf{D}$ field (true charge), while the divergence of $\mathbf{P}$ yields the sources of the polarization field.

The variation of $\mathbf{E}$, $\mathbf{P}$, and $-\nabla \cdot \mathbf{P}$ along the x axis of the capacitor is illustrated graphically in Fig. 3-25.

It may be shown that the potential V_p in an unbounded volume due to a polarization distribution is given by

$$V_p = -\frac{1}{4\pi\epsilon_0} \int \frac{\nabla \cdot \mathbf{P}}{r} \, dv \tag{12}$$

Thus, when both true charge and polarization are present and the distribution of both are fixed, the total potential V_T in the unbounded volume is

$$V_T = \frac{1}{4\pi\epsilon_0} \int_v \frac{\rho}{r} \, dv - \frac{1}{4\pi\epsilon_0} \int_v \frac{\nabla \cdot \mathbf{P}}{r} \, dv$$

or

$$V_T = \frac{1}{4\pi\epsilon_0} \int_v \frac{\rho - \nabla \cdot \mathbf{P}}{r} \, dv \tag{13}$$

where ρ = true volume charge density, C m^{-3}

$\mathbf{P}$ = polarization, C m^{-2}

ϵ_0 = permittivity of vacuum = 8.85 pF m^{-1}

r = distance from volume element containing charge or polarization to point at which V_T is to be calculated, m

The volume integration is taken over all regions containing charge or polarization. The field intensity $\mathbf{E}$ is then

$$\mathbf{E} = -\nabla V_T \tag{14}$$

3-25 THE LAPLACIAN OPERATOR; POISSON'S AND LAPLACE'S EQUATIONS

As an extension of the divergence operator we are led to the Laplacian (la-plah'-si-an) operator. From (3-22-1)

$$\nabla \cdot \mathbf{D} = \rho \tag{1}$$

Since $\mathbf{D} = \epsilon\mathbf{E}$ and $\mathbf{E} = -\nabla V$, we have

$$\mathbf{D} = -\epsilon\,\nabla V \tag{2}$$

Introducing (2) in (1) gives

$$\nabla\cdot\nabla V = -\frac{\rho}{\epsilon} \tag{3}$$

This is *Poisson's* (pwas'-awns) *equation.* The double operator (divergence of the gradient) is also written as ∇^2 (del squared) and is called the *Laplacian operator.* Thus Poisson's equation can be written

$$\boxed{\nabla^2 V = -\frac{\rho}{\epsilon}} \tag{4}$$

If $\rho = 0$, (4) reduces to

$$\boxed{\nabla^2 V = 0} \tag{5}$$

which is known as *Laplace's equation.*

In rectangular coordinates

$$\nabla^2 V = \frac{\partial^2 V}{\partial x^2} + \frac{\partial^2 V}{\partial y^2} + \frac{\partial^2 V}{\partial z^2} \tag{6}$$

The static potential distribution for any conductor configuration can be determined if a solution to Laplace's equation can be found which also satisfies the boundary conditions.

As will be discussed in Chap. 7, such a solution is *unique*; i.e., this is the way the field must be; no other configuration is possible. Actually, the solution need not be a mathematical one. Thus, an accurate field map obtained by graphical methods, as discussed in Sec. 3-19, is also unique, constituting a solution which satisfies Laplace's equation and the boundary conditions.

PROBLEMS†

Group 3-1: Secs. 3-1 through 3-7. Dielectrics, permittivity, polarization, and boundary relations

*3-1-1. Polarization in dielectric slab A plane slab of dielectric ($\epsilon_r = 6$) is situated normal to a uniform field $D = 2$ C m^{-2}. If the slab occupies a volume of 0.1 m^3 and is uniformly polarized, find (a) P in the slab and (b) total dipole moment of the slab.

3-1-2. Sulfur slab A flat slab of sulfur ($\epsilon_r = 4$) is placed normal to a uniform field. If $\rho_{sp} = 0.5$ C m^{-2} on the slab surfaces, find (a) P in the slab, (b) D in the slab, (c) D outside the slab (in air), (d) E in the slab, and (e) E outside the slab (in air).

† Answers to starred problems are given in Appendix C.

3-1-3. Two cavities in dielectric Two cavities are cut in a dielectric medium ($\epsilon_r = 4$) of large extent. Cavity 1 is a thin disk-shaped cavity with flat faces perpendicular to the direction of **D** in the dielectric. Cavity 2 is a long needle-shaped cavity with its axis parallel to **D**. The cavities are filled with air. If $D = 6$ pC m^{-2}, what is the magnitude of **E** at (a) center of cavity 1 and (b) the center of cavity 2?

★**3-1-4. D at angle to dielectric surface** In a dielectric, **D** makes an angle of 45° with respect to the dielectric surface. Find E_n (normal) and E_t (tangential) in the adjacent medium (air) if $\epsilon_r = 3$ for the dielectric.

3-1-5. Dielectric sandwich Three plane dielectric slabs of equal thickness with $\epsilon_r = 2$, 3, and 4 are sandwiched together. If **E** in air is at an angle of 30° with respect to a normal to the plane surface of the $\epsilon_r = 2$ slab, find the angle of **E** with respect to the normal inside the slabs. Draw a figure showing the path of the **E** line through the sandwich.

3-1-6. Artificial dielectric An artificial dielectric consists of a uniform lattice structure of conducting spheres extending in the x, y, and z directions with sphere diameters one-third their center-to-center spacing. Find ϵ_r. *Hint*: Consider that each sphere has a dipole moment equivalent to the dipole moment of a polarized atom of dielectric.

Group 3-2: Secs. 3-8 through 3-10. Capacitors, capacitance, dielectric strength, energy, and energy density

3-2-1. Capacitor with air gap A 50 mm thick slab of dielectric with $\epsilon_r = 100$ does not fit tightly in a parallel-plate capacitor, there being an air gap of 1 mm. If the plate area is 0.4 m^2 and 300 V is applied across the capacitor plates, find (a) capacitance, (b) E in air space, and (c) E in dielectric. (d) How much is the capacitance reduced by the air gap? Assume no fringing.

★**3-2-2. Three-section capacitor** A sandwich (parallel-plate) capacitor has a 3-stack dielectric filling with relative permittivities $\epsilon = \epsilon_0$, $2\epsilon_0$, and $3\epsilon_0$. If each section has a thickness of 3 mm and 12 V is applied across the capacitor, find E in the section with $\epsilon = 3\epsilon_0$.

3-2-3. Capacitance of earth What is the capacitance of the earth?

3-2-4. Earth-ionosphere capacitance The earth is surrounded by an ionized shell, or ionosphere. Assuming the ionosphere is equivalent to a conducting shell at a height of 600 km, what is the capacitance of the earth-ionosphere combination?

3-2-5. Capacitor with variable permittivity Find the capacitance of a parallel-plate capacitor of area A and plate separation x, with dielectric between the plates of permittivity $\epsilon_0(a + bx^2)$, where a and b are constants.

3-2-6. Wedge-shaped capacitor A capacitor with square plates of area A has a separation d along one edge and $2d$ along the opposite edge. Find the capacitance. Neglect fringing. $\epsilon = \epsilon_0$.

3-2-7. Capacitor plate attraction Find the force of attraction of a parallel-plate capacitor of area A and charge Q.

★**3-2-8. Capacitor partially filled with dielectric slab** A parallel-plate capacitor has a dielectric slab which is half the thickness of the plate separation. Is the capacitance dependent on the location of the slab (i.e., all air gap on one side of slab, or half on each side)?

3-2-9. Capacitor voltage before and after dielectric introduced A parallel-plate capacitor of area A and plate separation d has a voltage V_0 applied by a battery. The battery is then disconnected and a dielectric slab of permittivity ϵ and thickness d_1 inserted. (a) Find the new voltage V_1 between the plates. (b) Find the capacitance C_0 before and its value C_1 after the slab is introduced ($d_1 \leq d$).

★**3-2-10. Numerical quantities for Prob. 3-2-9** If $d_1 = \frac{1}{2}d$ and $\epsilon = 4\epsilon_0$, find (a) ratio of V_1 to V_0 and (b) ratio of C_1 to C_0 for the preceding problem.

3-2-11. Energy in capacitor Find the energy stored in a capacitor of 1,000 μF at 1,000 V.

3-2-12. Two-part capacitor The left half of a horizontal parallel-plate capacitor is filled with a dielectric of permittivity ϵ while the right half is air-filled. The plate separation is 20 mm and there is 80-V potential difference between the plates. Find E, D, and P in both halves. $\epsilon = 4\epsilon_0$.

3-2-13. Cylindrical capacitor A capacitor consists of 2 concentric thin-walled cylinders of radius a and b and length l. Find the capacitance. $\epsilon = \epsilon_0$.

★3-2-14. High-voltage bushing A high-voltage conductor is brought through a grounded metal panel by means of the double concentric capacitor bushing shown in Fig. P3-2-14. The medium is air (dielectric strength = 3 MV m^{-1}). Neglect fringing and thickness of sleeves. (a) Find the outer sleeve length L which equalizes the voltage across each space. (b) Find L which results in the same value of maximum field in each space. (c) What is the maximum working voltage for each condition? (d) If the inner sleeve (200 mm long) were removed, what would be the maximum working voltage? (e) If the number of concentric sleeves is increased in number so that the spacing between sleeves becomes smaller, what is the ultimate working voltage?

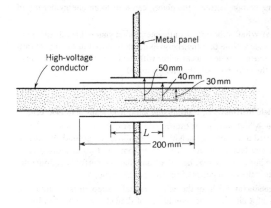

Figure P3-2-14 Concentric capacitor bushing.

3-2-15. Capacitor with nonuniform permittivity A parallel-plate capacitor of area A is filled with a dielectric of permittivity $\epsilon = \epsilon_0[1 + (\epsilon_r y/d)]$, where $y = 0$ at one plate and $y = d$ at the other plate. Neglect fringing. (a) Find E, D, P, and V as a function of distance y between the plates. (b) Make a graph showing the variation of ϵ, E, D, P, and V as a function of y. (c) Find the capacitance. (d) Find the polarization-volume-charge density ρ_p as a function of y and the polarization-surface-charge density ρ_{sp} at $y = 0$ and $y = d$. (e) Make a graph of ρ_p vs. y.

3-2-16. Problem 3-2-15 modification Repeat Prob. 3-2-15 for the case where the plate at $y = 0$ remains in contact with the slab but the other plate is moved to a distance $y = 2d$ with the space between $y = d$ and $y = 2d$ air-filled.

3-2-17. Electrostatic press Design an electrostatic press for applying pressures of 1 N mm^{-2} to 1-m^2 plywood sheets 10 mm thick (see Prob. 3-2-7).

★3-2-18. Energy in capacitor A typical 12-V automobile battery can store 1 kWh of energy. How large (in volume) must a capacitor be to store an equal amount of energy? Consult a radio parts catalog for capacitance, voltage rating, and dimensions of capacitors such as electrolytic types. Under what conditions will a battery be preferable and under what conditions will a capacitor (or capacitor bank) be preferable?

★3-2-19. Breakdown for sphere What is the maximum voltage to which a 200-mm-diameter sphere can be charged if it is situated (a) in air or (b) in mineral oil? (c) Compare these voltages with those for a 400-mm sphere.

3-2-20. n unequal capacitors Show that the total capacitance of n unequal capacitors is (a) more than the capacitance of the largest when all are in parallel and (b) less than the capacitance of the smallest when all are in series.

3-2-21. Capacitance defined in terms of energy Show that the capacitance of a capacitor can be expressed as

$$C = \frac{\int \mathbf{E} \cdot \mathbf{D} \, dv}{V^2} \quad \text{(F)}$$

where the numerator equals the energy stored and V equals the voltage between the capacitor plates.

3-2-22. Capacitor energy difference A parallel-plate capacitor is charged by a battery to an energy of 1 J. The battery is disconnected and the plate separation is then doubled. Is the stored energy changed? If there is a difference where did the energy go or come from?

3-2-23. Capacitor energy A parallel-plate capacitor of plate area A, plate separation h, and dielectric permittivity ϵ has a plate surface charge density ρ_s. Starting with A, h, and ρ_s, show how $\frac{1}{2}QV$, where Q is the charge on one plate and V the voltage between the plates, can lead to an energy density of $\frac{1}{2}\epsilon E^2$ in the dielectric.

★**3-2-24. Energy stored around sphere** (a) What is the electrostatic energy of a charged metal sphere of radius R and charge Q? The sphere is in air. Note that the energy is actually stored in the region outside the sphere as though the sphere is one electrode of a capacitor with the other electrode a sphere at infinity. (b) Within what radius from the sphere is half the energy stored?

Group 3-3: Secs. 3-11 through 3-20. Field distributions, transmission lines, field maps, and field cells

3-3-1. High-voltage dc transmission line A high-voltage dc transmission line consists of a single 25-mm diameter cylindrical conductor supported at an average height of 15 m above the ground. What is E at ground level (a) directly under the conductor, (b) at a distance of 25 m, and (c) at a distance of 50 m perpendicular to the line. Assume line at an average height, ground flat and perfectly conducting, and line operating at 500 kV. (d) What is the capacitance of the line per kilometer?

★**3-3-2. Coaxial line with square outer conductor** (a) Map the field lines and equipotential contours of a coaxial transmission line consisting of a circular inner conductor of diameter d symmetrically located inside an outer conductor of square cross section with an inner side dimension of $3d$. Note that thanks to symmetry only one octant (45° sector) need be mapped. (b) If the inner conductor is at $+100$ V and the outer conductor at zero, what is the potential midway between inner conductor and a corner? (c) What is the potential halfway between the inner conductor and the center of one side? (d) What is the capacitance of the line per meter length?

3-3-3. Coaxial line with triangular outer conductor A coaxial transmission line consists of an inner conductor of diameter d situated symmetrically inside a triangular outer conductor of equilateral cross section with inner side lengths all equal to $3d$. (a) Map the field lines and equipotential contours. (b) What is the ratio of surface charge density at the center point of a side of the outer conductor to that at a point on the side midway from the center to a corner? (c) If 100 V is applied to the line, what is the potential at a point midway from the inner conductor to a corner? The outer conductor is at zero potential. (d) What is the potential halfway between the inner conductor and the center of one side? (e) What is the capacitance of the line per unit length?

★**3-3-4. Asymmetric coaxial line** If 100 V is applied to the asymmetric coaxial line of Fig. P3-3-4, find V at the point P halfway between inner and outer conductors. The potential of the outer conductor is zero.

3-3-5. Strip line with half-cylinders attached A flat-strip transmission line with conducting half-cylinders attached is shown in cross section in Fig. P3-3-5. (a) If the upper strip is at $+100$ V and the lower strip at zero, find E at the points A and B. The radius $r = 10$ mm. (b) What is the capacitance of the line per meter length? (c) What is the capacitance per meter length of the strip line without the half-cylinders? Assume no fringing of the field at the edges of the strips in both cases.

★**3-3-6. Nonplanar strip line** For a strip line with the cross section and field map shown in Fig. P3-3-6 find (a) capacitance per unit length (into page), (b) E at A, and (c) E at B. The potential across the plates is 10 V with lower plate at zero potential. $\epsilon = \epsilon_0$ everywhere. Assume no fringing.

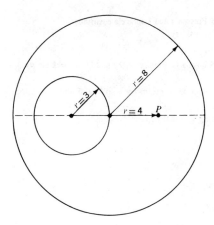

Figure P3-3-4 Cross section of asymmetrical coaxial transmission line.

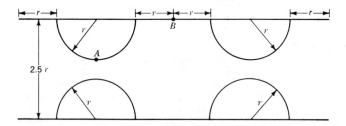

Figure P3-3-5 Parallel plane capacitor with metal half-cylinders attached.

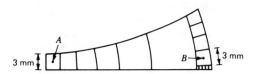

Figure P3-3-6 Nonplanar strip line.

Group 3-4: Secs. 3-21 through 3-25. Divergence and Poisson's and Laplace's equations

★3-4-1. Divergence If $\mathbf{D} = \hat{\mathbf{x}}3 + \hat{\mathbf{y}}6y + \hat{\mathbf{z}}z^2$ C m^{-2}, find $\nabla \cdot \mathbf{D}$.

3-4-2. Divergence If $\mathbf{D} = \hat{\mathbf{x}}y + \hat{\mathbf{y}}z + \hat{\mathbf{z}}x$ C m^{-2}, find $\nabla \cdot \mathbf{D}$.

3-4-3. Divergence Find the divergence of (a) $\mathbf{F} = \hat{\mathbf{x}} \cos az + \hat{\mathbf{y}} \sin ax + \hat{\mathbf{z}}y$ and (b) $\mathbf{G} = \hat{\mathbf{x}} \sin ay + \hat{\mathbf{y}}x - \hat{\mathbf{z}} \cos az$.

★3-4-4. Divergence for sphere Find $\mathbf{D}$ and $\nabla \cdot \mathbf{D}$ as a function of radius for a sphere of uniform charge density ρ and radius r_1.

3-4-5. Divergence Find the divergence of the following vector functions:
  (a) $\mathbf{A} = \hat{\mathbf{x}}3x + \hat{\mathbf{y}}4y^3 + \hat{\mathbf{z}}5z^2$
  (b) $\mathbf{B} = \hat{\mathbf{x}}2xy^2z - \hat{\mathbf{y}}3xz + \hat{\mathbf{z}}4x^2y^3$
  (c) $\mathbf{C} = \hat{\mathbf{x}}2 + \hat{\mathbf{y}}7x^2y^2z^2 - \hat{\mathbf{z}}z^{-1}$
  (d) $\mathbf{D} = \hat{\mathbf{x}}y + \hat{\mathbf{y}}z + \hat{\mathbf{z}}xy$
  (e) $\mathbf{E} = \hat{\mathbf{r}} \sin \phi + \hat{\boldsymbol{\phi}} \cos \phi + \hat{\mathbf{z}}z^2$
  (f) $\mathbf{F} = \hat{\mathbf{r}}2r^{1/2} - \hat{\boldsymbol{\phi}} \sin \phi \cos \phi + \hat{\mathbf{z}} \tan \phi$
  (g) $\mathbf{G} = \hat{\mathbf{r}}r^{-1} + \hat{\boldsymbol{\theta}} \sin \theta \sin \phi - \hat{\boldsymbol{\phi}} \sin \theta \sin \phi$
  (h) $\mathbf{H} = \hat{\mathbf{r}}z + \hat{\boldsymbol{\theta}} \cos \theta \sin \phi + \hat{\boldsymbol{\phi}}r^{1/2}$

3-4-6. Gradient and laplacian Find the gradient and laplacian of the following scalar functions: (a) $U = 1/(x^2 + y^2)$, (b) $U = x^3y^2z$, (c) $V = 1/r$ (cylindrical coordinates), (d) $V = 1/r$ (spherical coordinates), (e) $V = 1/e^r$ (cylindrical coordinates), (f) $V = 1/e^r$ (spherical coordinates), (g) $W = x^2 - yz$, and (h) $W = xy + xz$.

3-4-7. Cube of charge. Divergence Find the total charge inside a cubical volume 1 m on a side situated in the positive octant with three edges coincident with the x, y, and z axes and one corner at the origin if (a) $\mathbf{D} = \hat{\mathbf{x}}(x + 3)$, (b) $\mathbf{D} = \hat{\mathbf{y}}(y^2 + 4)$, (c) $\mathbf{D} = \hat{\mathbf{z}}2x^2$, and (d) $\mathbf{D} = \hat{\mathbf{x}}xyz$. Solve by first taking the divergence of $\mathbf{D}$ and then integrating this result *over the volume* of the cube. Compare these results with those obtained by integrating $\mathbf{D}$ *over the surface* of the cube.

Group 3-5: Practical applications

★3-5-1. Aluminum plate capacitor for radio transmitter A capacitor in the final amplifier of a radio transmitter has a stack of 8 aluminum plates 2 mm thick and 200 mm square as one electrode. The other electrode consists of 7 plates of the same dimensions interleaved with the first stack so that there is a uniform spacing between all plates of 8 mm. (a) Find the capacitance. (b) Find the dc breakdown voltage. Neglect fringing. The edges of the plates are rounded.

3-5-2. Twin-line A 2-wire transmission line called *twin-line* used for TV reception consists of 2 parallel 1-mm-diameter conductors spaced 10 mm on centers symmetrically imbedded in a flexible dielectric rod 20 mm in diameter. If $\epsilon_r = 1.8$ for the rod, find the capacitance per meter of the twin-line.

3-5-3. Flexible coaxial line with air core A flexible coaxial transmission line consists of an outer conductor of wire braid enclosing a dielectric tube (Fig. P3-5-3). The core of the tube is air-filled with the inner conductor positioned at the center by a thin radial fin of dielectric which spirals along inside the tube. The dielectric tube has an outside diameter of 10 mm and an inside diameter of 6 mm. (a) If the inner-conductor diameter is 1 mm and $\epsilon_r = 2$ for the dielectric, find the capacitance of the line per meter length. (b) What is the capacitance per meter length if the dielectric completely fills the line?

3-5-4. Field under thundercloud. Field mapping If a thundercloud is above the ridge and ravine of Fig. P3-5-4, what is E (a) on the ridge and (b) in the ravine? Take E under the cloud as 10 kV m^{-1}.

3-5-5. Field under high-voltage line Many 60-Hz high-voltage transmission lines operate at 765 kV. This is an ac rms value so the peak voltage is 1.08 MV. (a) What is the peak E at ground level under a conductor of such a line if the conductor is 12 m above the ground? (b) What is the peak potential difference between the head and feet of a 1.8-m person standing under the conductor? *Note*: The field is sufficient to ignite standard fluorescent lamps to brilliance even though the lamps have no wires connected to them.

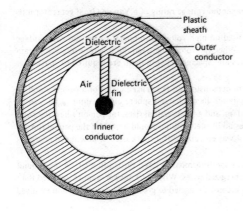

Figure P3-5-3 Cross section of air-core coaxial transmission line.

Figure P3-5-4 Ridge and ravine with thundercloud over head for field map.

★3-5-6. Millikan oil-drop experiment In Robert Millikan's famous oil-drop experiment, in which he measured the charge of an electron, a tiny charged droplet of oil of charge q and mass m is suspended in equilibrium with gravity (of acceleration g) by an electric field E between 2 horizontal capacitor plates. If the droplet has N electron charges, find the charge q of an electron in terms of m, g, N, and E.

3-5-7. Electrostatic paint spraying A flat metal sheet can be readily painted on its left-hand side with a spray can as suggested in Fig. P3-5-7. How may electrostatics be utilized to spray the right-hand or back side of the sheet without turning or moving the sheet and with the spray can still at A?

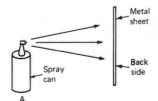

Figure P3-5-7 Arrangement for spraying behind metal sheet using electrostatic fields.

3-5-8. Attraction and repulsion When a small piece of paper is brought near an electrically charged sphere, it is attracted to the sphere, but after touching it, the paper is repelled. Why?

3-5-9. Van de Graaf generator What is the maximum voltage rating of a Van de Graaf generator if its top spherical electrode is 2 m in diameter?

★**3-5-10. Sphere gaps** Ball or sphere gaps are used on high-voltage equipment to prevent over-voltage produced, for example, by a lightning stroke. When the voltage between the spheres exceeds a certain value, sparking occurs. What diameter spheres and spacing between them are required to spark over at or above 100 kV?

3-5-11. Van de Graaf demonstration A frequent demonstration at science fairs is for a person standing on an insulated pedestal to place a hand firmly on the discharged sphere of a medium sized Van de Graaf generator. The generator is then turned on, and as its potential rises, the person's hair rises and stands straight out, but the person experiences no ill effects. What would happen if the person (on the insulated pedestal) touched the sphere *after* it was energized? If a spark occurs, how much energy is transferred?

3-5-12. Birds on high-voltage lines On high-voltage transmission lines, birds often sit on the ground wire (highest wire of the line) but not on the energized wires. Why? The answer is more involved than saying they may receive a shock because they are not connected to ground. The phenomenon involved is electrostatic.

★**3-5-13. Electrostatic energy in a thundercloud** Electrostatically, a typical thundercloud may be represented by a capacitor model with horizontal plates 10 km^2 in area separated by a vertical distance of 5 km. The upper plate has a positive charge of 200 C and the lower plate an equal negative charge. (*a*) Find the electrostatic energy stored in the cloud. (*b*) What is the potential difference V between the top and bottom of the cloud? (*c*) What is the average electric field E in the cloud? (*d*) How close is this value to the dielectric strength of dry air?

3-5-14. Shielding from lightning Explain why the danger of injury from lightning is greater standing in the open, under a power line or an isolated tree than it is in a depression, in a metal enclosed vehicle, or in a rod-protected building. Note that all 6 situations require different reasons. Note that if the depression is an arroyo creek bed, it's important to get out of it before the flash flood that follows the storm.

3-5-15. Heart dipole field The heart dipole in Fig. 3-22 is approximately parallel to the surface of the chest. How would the potential contours on the chest appear if the dipole were perpendicular to the surface of the chest?

3-5-16. Field in bone Quartz is *piezoelectric*; i.e., a pressure applied produces a potential difference while a potential difference applied produces a mechanical stress and deformation. Bone is also piezoelectric to some extent. In simplistic terms, exercise stresses a bone, producing an electric potential

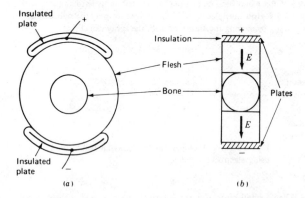

Figure P3-5-16 Capacitor plates across arm for application of electric field to bone.

difference, which in turn promotes calcium deposit to strengthen the bone. Persons with a broken arm in a cast can't exercise it, but if a potential difference is applied across the arm, it can produce an electric field at the bone, promoting calcium deposit and more rapid mending.

If 50 V is applied by insulated "capacitor plates" across a 75-mm-diameter arm as in Fig. P3-5-16a, what is the electric field E in the bone? The bone diameter is 25 mm. Take ϵ_r (insulation) = 1.5, ϵ_r (bone) = 2, and ϵ_r (flesh) = 4. Note that since the plates are insulated, there is no current flow between them. The insulation thickness is 500 μm. *Hint:* Consider that the field at the bone is essentially uniform so that the problem can be treated approximately as a parallel-plate capacitor with different dielectrics, as in Fig. P3-5-16b.

FOUR

THE STEADY ELECTRIC CURRENT

4-1 INTRODUCTION

Electric charge in motion constitutes an *electric current* and any current-carrying medium may be called a *conductor*. In metallic conductors the charge is carried by electrons. In plasmas or gaseous conductors the charge is carried by (negative) electrons and positive ions (electron deficient atoms or molecules). A plasma contains equal numbers of positive and negative charges (total charge equals zero). The earth's ionosphere is a highly rarefied gas which may also be called a plasma. In liquid conductors (electrolytes) the charge is carried by ions, both positive and negative. In semiconductors the charge is carried by electrons and holes, the holes behaving like positive charges.

In this chapter the important relations governing the behavior of steady electric currents in conductors are discussed.†

The fields associated with steady currents are constant with time and hence are static fields. In Chaps. 2 and 3 the discussion was almost entirely concerned with static fields having all associated charges stationary, i.e., with no currents present. In this chapter the fields are also static, but steady direct currents may be present.

4-2 THE ELECTRIC CURRENT; CURRENT DENSITY

Referring to Fig. 4-1, a test charge q introduced into an electric field $\mathbf{E}$ experiences a force $\mathbf{F}$ which, from (2-4-1), is given by

$$\mathbf{F} = q\mathbf{E} \quad \text{(N)} \tag{1}$$

† Specifically, a *steady direct current* is meant. This should not be confused with a steady-state current, which may imply a time-changing current that repeats itself periodically.

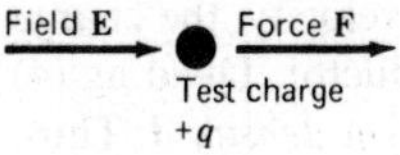

Field E   ●   Force F
Test charge
+q

Figure 4-1 Test charge q in electric field E experiences a force F.

If the charge is free to move, it will receive an acceleration which, from Newton's second law (**F** = m**a**), is

$$a = \frac{F}{m} \quad (\text{m s}^{-2}) \tag{2}$$

where m is the mass of the charged particle in kilograms. In the absence of restraints, the particle's velocity v ($= at$) will increase indefinitely with the time t provided the electric field **E** is constant. However, in gaseous, liquid, or solid conductors the particle collides repeatedly with other particles, losing some of its energy and causing random changes in its direction of motion. If **E** is constant and the medium is homogeneous, the net effect of the collisions is to restrain the charged particle to a constant average velocity called the *drift velocity* $\mathbf{v}_d$. This drift velocity has the same direction as the electric field and is related to it by a constant called the *mobility* μ. Thus,

$$\mathbf{v}_d = \mu\mathbf{E} \quad (\text{m s}^{-1}) \tag{3}$$

where μ = mobility, m^2 V^{-1} s^{-1}
     E = electric field intensity, V m^{-1}

If a medium of uniform cross section A, as in Fig. 4-2, contains many free-to-move charged particles of volume density ρ, then these moving charges will form a *current* **I** in coulombs per second passing a given reference point as given by

$$\mathbf{I} = \mathbf{v}_d \rho A \tag{4}$$

where I = current, C s^{-1}
    $\mathbf{v}_d$ = drift velocity, m s^{-1}
    ρ = charge density, C m^{-3}
    A = area of conducting medium, m^2

The SI unit for current (coulombs per second) is the *ampere*.

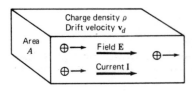

Figure 4-2 Medium of cross-sectional area A containing free-to-move charged particles has a current **I** in the presence of an electric field **E**. The current consists of charged particles of volume density ρ drifting with a velocity $\mathbf{v}_d$. Charges may be both positive and negative. Positive charges move in direction of field, negative charges in opposite direction but both add to the total current. For simplicity only positive charges are shown here.

According to (4), the current is proportional to the drift velocity, the charge density, and the area of the current-carrying medium or conductor. Dividing (4) by the area A, we obtain the current per unit area or the *current density* **J**. Thus,

$$\mathbf{J} = \frac{\mathbf{I}}{A} = \mathbf{v}_d \rho \qquad (5)$$

Current density has the dimensions of current per area and in SI units is expressed in amperes per square meter ($A\ m^{-2}$).

If the current is not uniformly distributed, (5) gives the average current density. However, it is often of interest to consider the current density at a point. This is defined as the current $\Delta\mathbf{I}$ through a small area Δs divided by Δs, with the limit of this ratio taken as Δs approaches zero. Hence, the current density at a point is given by

$$\mathbf{J} = \lim_{\Delta s \to 0} \frac{\Delta\mathbf{I}}{\Delta s} \qquad (A\ m^{-2}) \qquad (6)$$

It is assumed that the surface Δs is normal to the current direction. The current density **J** is a vector point function having a magnitude equal to the current density at the point and the direction of the current at the point.

4-3 RESISTANCE AND OHM'S LAW

In 1826 Georg Simon Ohm experimentally determined the relations between the voltage V over a length of conductor and the current I through the conductor in terms of a parameter characteristic of the conductor (see Fig. 4-3). This parameter, called the resistance R, is defined as the ratio of the voltage V to the current I. Thus,

$$R = \frac{V}{I} \qquad \text{or} \qquad \boxed{V = IR} \qquad (1)$$

The latter form is the usual statement of *Ohm's law*, which states that *the potential difference or voltage V between the ends of a conductor is equal to the product of its resistance R and the current I.*

It is assumed that R is independent of I; that is, R is a constant. Conversely, such a resistance is said to obey Ohm's law. There exist circuit elements, however, such as rectifiers, whose resistance is not a constant. Such elements are said to be *nonlinear*, and a V-versus-I diagram is required to display their behavior. Nevertheless, the resistance of the nonlinear element is still defined by $R = V/I$, but R is not independent of I and the resistance does not obey Ohm's law.

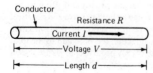

Figure 4-3 A voltage V across a conductor of length d produces a current I. The ratio of V to I is a measure of the resistance R of the conductor.

Resistance has the dimensions of potential divided by current or, in dimensional symbols, ML^2/I^2T^3. The SI unit of resistance is the ohm (Ω). Thus

$$\frac{\text{Volts}}{\text{Ampere}} = \text{ohms}$$

That is, the resistance of a conductor is 1 Ω if a current of 1 A flows when a potential difference of 1 V is applied between the ends of the conductor.

4-4 POWER RELATIONS AND JOULE'S LAW

Referring again to Fig. 4-3, the potential difference or voltage V across the length d of the conductor is equal to the work per unit charge (current $\times$ time) required to move a charge through the distance d. Multiplying by the current I yields the work per unit time, or power P. Thus,

$$\frac{\text{Work}}{\text{Current} \times \text{time}} \times \text{current} = \frac{\text{work}}{\text{time}} = \text{power}$$

or
$$\boxed{VI = P} \tag{1}$$

This is the power dissipated in the length d of the conductor. The SI unit of power is the watt (W). Hence,

$$\text{Volts} \times \text{amperes} = \text{watts} \quad (\text{W})$$

or in dimensional symbols

$$\frac{ML^2}{IT^3} I = \frac{ML^2}{T^3}$$

Introducing the value of V from Ohm's law (4-3-1) into (1) yields

$$\boxed{P = I^2R} \tag{2}$$

Thus, the work or energy dissipated per unit time in the conductor is given by the product of its resistance R and the square of the current I. This energy appears as heat in the conductor.

The energy W dissipated in the conductor in a time T is then

$$W = PT = I^2RT \tag{3}$$

where W = energy, J
P = power, W
I = current, A
R = resistance, Ω
T = time, s

This relation is known as *Joule's law*. It is assumed that P is constant over the time T. If it is not constant, I^2R is integrated over the time interval T.

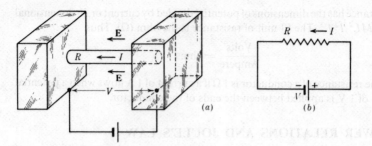

Figure 4-4 (*a*) Cylindrical conductor of length *d* between end blocks, and (*b*) schematic diagram.

4-5 THE ELECTRIC CIRCUIT

The discussion in the preceding sections concerns an infinitely long conductor along which a field **E** is applied. Consider now a cylindrical conductor of finite length *d* as in Fig. 4-4*a* between two large conducting blocks of negligible resistance maintained at a constant potential difference *V* by a battery. This produces a uniform field **E** along the conductor. This field is given by $E = V/d$. As long as this field is maintained in the conductor, current flows that has a value

$$I = \frac{V}{R} = \frac{Ed}{R}$$

Assuming that the resistance of the wires connected to the battery is negligible compared with *R*, the potential difference *V* is equal to the voltage appearing across the terminals of the battery. The arrangement of Fig. 4-4*a* may then be represented by the schematic diagram of Fig. 4-4*b*.

This is a diagram of a closed *electric circuit* of the most elementary form consisting of a *resistor* of resistance *R* and a battery of voltage *V*.

4-6 RESISTIVITY AND CONDUCTIVITY

The resistance of a conductor depends not only on the type of material of which the conductor is made but also on its shape and size. To facilitate comparisons between different types of substances, it is convenient to define a quantity which is characteristic only of the substance. The *resistivity S* is such a quantity. The resistivity of a material is numerically equal to the resistance of a homogeneous unit cube of the material with a uniform current distribution. This condition may be produced by clamping the cube between two heavy blocks of negligible resistance, as in Fig. 4-5. With a current *I* through the cube, the resistivity *S* of the material is given by $S = V/I$, where *V* is the potential between the blocks.

In SI units, this measurement is in ohms for a cube of material 1 m on a side. If two cubes are placed in series between the blocks, the resistance measured is 2*S*,

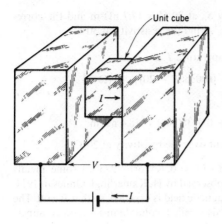

Figure 4-5 Unit cube between end blocks.

while if two cubes are placed in parallel, the resistance is $\frac{1}{2}S$. It follows that the resistance R of a rectangular block of length l and cross section a, as in Fig. 4-6, is

$$R = \frac{Sl}{a} \qquad \text{or} \qquad \boxed{S = \frac{Ra}{l}} \tag{1}$$

where S is the resistivity of the block material. From (1), resistivity has the dimensions

$$\frac{\text{Resistance} \times \text{area}}{\text{Length}} = \text{resistance} \times \text{length}$$

Thus, resistivity has the dimensions of resistance times length and in SI units is expressed in ohm-meters (Ω m).

The resistivity is a function of the temperature. In metallic conductors it varies nearly linearly with absolute temperature. Over a considerable temperature range from a reference or base temperature T_0 the resistivity S is given approximately by

$$S = S_0[1 + \alpha(T - T_0)]$$

where T = temperature of material, °C

T_0 = reference temperature (usually 20°C)

S_0 = resistivity at temperature T_0, Ω m

α = temperature coefficient of resistivity, numerical units °C^{-1}

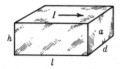

Figure 4-6 Block of conducting material.

Example For copper the resistivity S_0 at 20°C is 17.7 nΩ m and the corresponding coefficient $\alpha = 0.0038°C^{-1}$. Find the resistivity at 30°C.

SOLUTION The resistivity S at a temperature T is

$$S = 1.77 \times 10^{-8}[1 + 0.0038(T - 20)] \qquad (\Omega \text{ m})$$

At a temperature of 30°C,

$$S = 1.77 \times 10^{-8}[1 + 0.0038(10)] = 18.4 \text{ nΩ m}$$

This is an increase of nearly 4 percent over the resistivity at 20°C.

At temperatures near absolute zero ($T = 0$ K† or $-273°C$) some metals become superconducting, an effect first observed by H. Kamerlingh Onnes in 1911. The resistivity drops to zero, and the magnetic field is expelled, so that $B = 0$. The transition is very abrupt. Niobium (also called columbium) becomes superconducting at 9.2 K, aluminum at 1.2 K, but copper and gold are not superconducting, at least not at temperatures down to 0.05 K. Some intermetallic compounds like $Nb_3(Al-Ge)$ become superconducting at the relatively high temperature of about 21 K. If materials are developed which become superconducting at 25 K or more, a technological revolution will occur since this will permit use of superconductors cooled by relatively inexpensive liquid hydrogen (boiling temperature 20.4 K).‡

The reciprocal of resistance R is *conductance* G. That is, $G = 1/R$. Since resistance is expressed in ohms, conductance is expressed in reciprocal ohms. A reciprocal ohm is called a *mho* (ohm spelled backward); conductance is given in mhos, and the SI symbol is an upside-down capital omega (℧). An alternative unit for conductance is the *siemens* (S), i.e., 1 siemens = 1 mho.

The reciprocal of resistivity S is *conductivity* σ. That is,

$$\sigma = \frac{1}{S} = \frac{l}{Ra} \qquad (1a)$$

Although the resistivity is convenient in certain applications, it is often more convenient to deal with its reciprocal, the conductivity, e.g., where parallel circuits are involved. Since resistivity is expressed in ohm-meters, the conductivity is expressed in mhos per meter ($℧ \text{m}^{-1}$).

The resistance R of a rectangular block, as in Fig. 4-6, of material of conductivity σ is

$$R = \frac{l}{\sigma a} \qquad (\text{ohms}, \Omega) \qquad (2)$$

---

† The SI unit for absolute temperature is the kelvin (K) (see Sec. 1-3).

‡ B. T. Matthias, The Search for High-temperature Superconductors, *Phys. Today*, **24**: 23–28 (August 1971).

or the conductance G of the block is

$$G = \frac{1}{R} = \frac{\sigma a}{l} \quad (\text{mhos}, \mho) \tag{3}$$

where σ = conductivity of block material, $\mho\,\text{m}^{-1}$
  a = cross-sectional area of block, m^2
  l = length of block, m

4-7 OHM'S LAW AT A POINT

Consider a block of conducting material as indicated in Fig. 4-7. Let a small imaginary rectangular cell of length l and cross section a be constructed around a point P in the interior of the block with a normal to **J** as indicated. Then on applying Ohm's law to this cell we have $V = IR$, where V is the potential difference between the ends of the cell. But $V = El$ and $I = Ja$; so $El = JaR$ and

$$J = \frac{l}{aR} E \tag{1}$$

By making the cell enclosing P as small as we wish, this relation can be made to apply at the point P, and we can write

$$\boxed{J = \sigma E} \tag{2}$$

Equation (2) is *Ohm's law at a point* and relates the current density **J** at a point to the total field **E** at the point and the conductivity σ of the material. In the above discussion it is assumed that the conducting material is homogeneous (same material throughout), isotropic (resistance between opposite faces of a cube independent of the pair chosen), and linear (resistance independent of current).

From (2) the conductivity is the ratio of the current density J to the applied field E or

$$\sigma = \frac{J}{E} \quad (\mho\,\text{m}^{-1}) \tag{3}$$

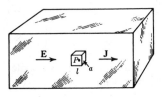

Figure 4-7 Block of conducting material with small imaginary cell enclosing the point P.

but from (4-2-5) the current density is the product of the drift velocity v_d and the charge density ρ, or

$$J = \rho v_d \tag{4}$$

Introducing this value of J in (3), the conductivity is given by

$$\sigma = \rho \frac{v_d}{E} = \rho \mu \quad (\mho \, m^{-1}) \tag{5}$$

where ρ = charge density, C m^{-3}
μ = mobility, m^2 V^{-1} s^{-1}

4-8 DIELECTRICS, CONDUCTORS, AND SEMICONDUCTORS

In a *dielectric*, or *insulator*, charges are bound and not free to migrate so that the conductivity is ideally zero. Although a field applied to an insulator may produce no migration of charge, it can produce a polarization of the insulator, or dielectric (Sec. 3-5), i.e., a displacement of the electrons with respect to their equilibrium positions.

In *plasmas, gases, and liquids*, charges of both signs are usually present and free to migrate. Assuming that all negative particles are of the same kind and all positive particles (or ions) are of the same kind, the conductivity will have two terms as follows:

$$\sigma = \rho_- \mu_- + \rho_+ \mu_+ \quad (\mho \, m^{-1}) \tag{1}$$

where ρ_- = density of negatively charged particles, C m^{-3}
μ_- = mobility of negatively charged particles, m^2 V^{-1} s^{-1}
ρ_+ = density of positively charged particles (or ions), C m^{-3}
μ_+ = mobility of positively charged particles (or ions), m^2 V^{-1} s^{-1}

The first term represents the contribution to the conductivity from negatively charged particles moving opposite to the field **E** while the second term represents the contribution from positively charged particles moving with **E**.

In ordinary (metallic) conductors, such as copper and aluminum, the outer, or valence, electrons are readily detached from their atoms by an applied electric field and are free to migrate. The atoms, however, remain fixed in the conductor's lattice so only the electrons have mobility. Therefore, the conductivity for conductors has only one term as given by

$$\sigma = \rho_e \mu_e \quad (\mho \, m^{-1}) \tag{2}$$

where ρ_e = density of electrons, C m^{-3}
μ_e = mobility of the electrons, m^2 V^{-1} s^{-1}

In *semiconductors* the normal conduction by valence electrons is supplemented by another charge carrier called a *hole*, which represents a vacant space in the lattice structure of the semiconductor. Vacant spaces left by valence electrons can migrate from atom to atom in a semiconductor so that the hole tends to behave like a positively charged ion but its mobility is more like that of an electron. Thus, the conductivity of a semiconductor is given by

$$\sigma = \rho_e \mu_e + \rho_h \mu_h \qquad (\mho \, m^{-1}) \qquad (3)$$

where ρ_e = charge density of electrons, C m^{-3}
μ_e = mobility of electrons, m^2 s^{-1} V^{-1}
ρ_h = charge density of holes, C m^{-3}
μ_h = mobility of holes, m^2 s^{-1} V^{-1}

Germanium and silicon are typical semiconductors. In their intrinsic or pure form, the electrons and vacancies (or electron-hole pairs) have only a short lifetime, disappearing as electrons and holes recombine. But new pairs are continually being formed, and so some are always present. Increasing the temperature accelerates pair formation, increasing the density of electrons and holes and, hence, the conductivity. This behavior with temperature is opposite to that for ordinary conductors, which decrease in conductivity (increase in resistance) with temperature.

If small amounts of certain impurities are added to the semiconductor, either during crystal growth or by diffusion, the carrier density and conductivity can be greatly increased. Impurities such as phosphorus provide more electrons and are called *donors*, forming *n-type* semiconductors in which electrons constitute most of the carriers, holes being in the minority. Impurities such as boron introduce more holes and are called *acceptors*, forming *p-type* semiconductors with holes predominating. The procedure of introducing impurities is called *doping*. Donor concentrations of less than one part per million can increase the conductivity by a factor of nearly a million. Thus, although the normal behaviour of pure, or intrinsic, semiconductors continues in the presence of impurities, it is a minor effect in comparison with the extra electrons in *n*-type and extra holes in *p*-type semiconductors.

Semiconductor mobilities
(at 300 K in m^2 V^{-1} s^{-1})

	Electrons	Holes
Pure germanium	0.37	0.18
Pure silicon	0.13	0.03

By way of comparison, the mobilities of conductors are much less, being typically 0.0014 for aluminum, 0.0040 for copper, and 0.0050 for silver.

The boundary between n-type and p-type regions of a single semiconductor crystal forms a junction region utilized in diodes and transistors.

Six different conditions of conductivity are illustrated in Fig. 4-8. These are for (1) a dielectric, or insulator; (2) a plasma, gas or liquid; (3) an ordinary (metallic) conductor; (4) a pure semiconductor; (5) an n-type semiconductor; and (6) a p-type semiconductor.

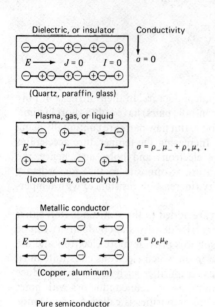

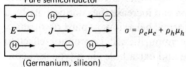

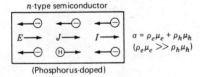

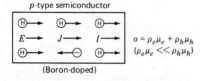

Figure 4-8 Six different conditions of conductivity. Note that negatively charged particles (electrons) travel opposite to **E** while positively charged particles or holes (H) travel with **E** but that both *add* to the total current.

Table 4-1 Table of conductivities

Substance	Type	Conductivity, $\mho\,m^{-1}$
Quartz, fused	Insulator	$\sim 10^{-17}$
Ceresin wax	Insulator	$\sim 10^{-17}$
Sulfur	Insulator	$\sim 10^{-15}$
Mica	Insulator	$\sim 10^{-15}$
Paraffin	Insulator	$\sim 10^{-15}$
Rubber, hard	Insulator	$\sim 10^{-15}$
Porcelain	Insulator	$\sim 10^{-14}$
Glass	Insulator	$\sim 10^{-12}$
Bakelite	Insulator	$\sim 10^{-9}$
Distilled water	Insulator	$\sim 10^{-4}$
Dry, sandy soil	Poor insulator	$\sim 10^{-3}$
Marshy soil	Poor insulator	$\sim 10^{-2}$
Fresh water	Poor insulator	$\sim 10^{-2}$
Animal fat†	Poor insulator	4×10^{-2}
Animal body (ave)†	Poor conductor	0.2
Animal muscle†	Poor conductor	0.4
Animal blood†	Poor conductor	0.7
Germanium (pure)	Semiconductor	~ 2
Seawater	Conductor	~ 4
Tellurium	Conductor	$\sim 5 \times 10^2$
Carbon	Conductor	$\sim 3 \times 10^4$
Graphite	Conductor	$\sim 10^5$
Cast iron	Conductor	$\sim 10^6$
Mercury	Conductor	10^6
Nichrome	Conductor	10^6
Constantan	Conductor	2×10^6
Silicon steel	Conductor	2×10^6
German silver	Conductor	3×10^6
Lead	Conductor	5×10^6
Tin	Conductor	9×10^6
Phosphor bronze	Conductor	10^7
Brass	Conductor	1.1×10^7
Zinc	Conductor	1.7×10^7
Tungsten	Conductor	1.8×10^7
Duralumin	Conductor	3×10^7
Aluminum, hard-drawn	Conductor	3.5×10^7
Gold	Conductor	4.1×10^7
Copper	Conductor	5.7×10^7
Silver	Conductor	6.1×10^7
$Nb_3(Al\text{-}Ge)$	Superconductor	$\sim \infty$

† Typical of human beings.

4-9 TABLE OF CONDUCTIVITIES

The conductivities σ (for direct current) of some common materials are listed in Table 4-1 for a temperature of 20°C and for a superconductor at temperatures below 21 K. See page 123.

4-10 KIRCHHOFF'S VOLTAGE LAW AND THE DIFFERENCE BETWEEN POTENTIAL AND EMF

Consider the simple electric circuit shown by the schematic diagram in Fig. 4-9. The circuit consists of a resistor R_0 and the battery. The current is I at all points in the circuit. At any point in the conducting material of the circuit we have from Ohm's law at a point that $\mathbf{J}/\sigma = \mathbf{E}$, where $\mathbf{E}$ is the total field at the point.

In general the total field $\mathbf{E}$ may be due not only to static charges but also to other causes such as the chemical action in a battery. To indicate this explicitly, we write

$$\mathbf{E} = \mathbf{E}_c + \mathbf{E}_e \tag{1}$$

where $\mathbf{E}_c$ = static electric field due to charges; the subscript c indicates explicitly that the field is due to *charges*

$\mathbf{E}_e$ = electric field generated by other causes as by a battery; the subscript e indicates explicitly that it is an *emf-producing* field

Whereas $\mathbf{E}_c$ is derivable as the gradient of a scalar potential due to charges ($\mathbf{E}_c = -\nabla V$), this is not the case for $\mathbf{E}_e$. Substituting (1) in (4-7-2), writing $\mathbf{J} = \mathbf{I}/a$, where a is the cross-sectional area of the conductor, and noting the value of σ from (4-6-2), we have

$$\frac{\mathbf{J}}{\sigma} = \mathbf{I}\frac{R}{l} = \mathbf{E}_c + \mathbf{E}_e \tag{2}$$

where R/l is the resistance per unit length in ohms per meter. This applies at any point in the circuit. Integrating around the complete circuit gives

$$\oint \mathbf{E}_c \cdot d\mathbf{l} + \oint \mathbf{E}_e \cdot d\mathbf{l} = I \oint \frac{R}{l} dl \tag{3}$$

The first term is zero; i.e., the line integral of a field due to charges is zero around a closed circuit (see Sec. 2-8, last paragraph). However, the second term involving the line integral of $\mathbf{E}_e$ around the circuit is not zero but is equal to a voltage called

Figure 4-9 Series circuit of battery and external resistance.

the total *electromotive force*, or *emf*, $\mathscr{V}_T$ of the circuit. The field $\mathbf{E}_e$ is produced, in the present example, by chemical action in the battery. If it were absent, no current would flow since an electric field $\mathbf{E}_c$ due to charges is not able to maintain a steady current. The right-hand side of (3) equals the total IR drop around the circuit. Hence (3) becomes

$$\mathscr{V}_T = IR_T \tag{4}$$

where R_T is the total resistance of the circuit (equals R_0 if the internal resistance of the battery is zero).

In general, for a closed circuit containing many resistors and sources of emf,

$$\boxed{\sum \mathscr{V} = I \sum R} \tag{5}$$

This is *Kirchhoff's voltage law*. In words it states that *the algebraic sum of the emfs around a closed circuit equals the algebraic sum of the ohmic, or IR, drops around the circuit.*† Kirchhoff's voltage law applies not only to an isolated electric circuit as in Fig. 4-9 but to any single mesh (closed path) of a network.

To distinguish emf from the scalar potential V, the symbol $\mathscr{V}$ (script V) is used for emf. Both V and $\mathscr{V}$ are expressed in volts, so that either may be referred to as a voltage if one does not wish to make a distinction between potential and emf.

It is to be noted that the scalar potential V is equal to the negative of the line integral of the static field $\mathbf{E}_c$, while the emf $\mathscr{V}$ equals the line integral of $\mathbf{E}_e$. Thus, between any two points a and b,‡

$$V_{ab} = V_b - V_a = -\int_a^b \mathbf{E}_c \cdot d\mathbf{l} \tag{6}$$

and

$$\mathscr{V}_{ab} = \int_a^b \mathbf{E}_e \cdot d\mathbf{l} \tag{7}$$

In (6) V_{ab} is independent of the path of integration between a and b, but $\mathscr{V}_{ab}$, in (7), is not.

For closed paths, $\oint \mathbf{E}_c \cdot d\mathbf{l} = 0$ and $\oint \mathbf{E}_e \cdot d\mathbf{l} = \mathscr{V}_T$, where $\mathscr{V}_T$ is the total emf around the circuit.

Example Let the circuit of Fig. 4-9 be redrawn as in Fig. 4-10*a*. The battery has an internal resistance R_1, and it will be convenient in this example to assume that the field $\mathbf{E}_e$ in the battery is uniform between the terminals

---

† In time-varying situations, where the circuit dimensions are small compared with the wavelength, Kirchhoff's law is modified: The algebraic sum of the *instantaneous* emfs around a closed circuit equals the algebraic sum of the *instantaneous* ohmic drops around the circuit.

‡ An open-circuited battery (no current flowing) has a terminal potential difference V equal to its emf $\mathscr{V}$. The potential V is as given by (6). As explained in the examples that follow, $\mathbf{E}_c$ and $\mathbf{E}_e$ have opposite directions in the battery. Therefore, in order that $V_{ab} = \mathscr{V}_{ab}$ for an open-circuited battery, (7) has no negative sign.

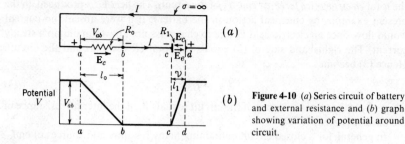

(a)

(b)

Figure 4-10 (a) Series circuit of battery and external resistance and (b) graph showing variation of potential around circuit.

(c and d). The point b (or c) is taken arbitrarily to be at zero potential. The resistor has a uniform resistance R_0, and the wires connecting the resistor and the battery are assumed to have infinite conductivity ($\sigma = \infty$). Hence, in the wire, $\mathbf{E}_c = 0$. The field $\mathbf{E}_e$ has a value only in the battery, being zero elsewhere. Find the variation of the potential V around the circuit.

SOLUTION By Kirchhoff's voltage law the sum of the emfs around the circuit equals the sum of the IR drops. Thus

$$\mathscr{V} = IR_0 + IR_1$$

or

$$I = \frac{\mathscr{V}}{R_0 + R_1}$$

In the resistor $\mathbf{E}_e = 0$, but $\mathbf{E}_c$ has a value (as discussed in connection with Fig. 4-4). Applying Ohm's law (4-7-2) to the resistor (between a and b), we have

$$\mathbf{E}_c = \frac{\mathbf{J}}{\sigma_0} = \mathbf{I}\frac{R_0}{l_0}$$

where σ_0 = conductivity of resistor material (assumed uniform)
l_0 = distance from a to b

Integrating from a to b yields

$$\int_a^b \mathbf{E}_c \cdot d\mathbf{l} = I\frac{R_0}{l_0}\int_a^b dl$$

or

$$V_{ab} = -IR_0$$

where V_{ab} is the potential difference between a and b. Since point a is connected to d and b to c with perfectly conducting wires, $V_{cd} = -V_{ab}$, where V_{cd} is the potential difference appearing across the terminals of the battery. Therefore, we have

$$\mathscr{V} = V_{cd} + IR_1$$

or

$$V_{cd} = \mathscr{V} - IR_1$$

Thus, the potential difference appearing between the terminals of the battery is equal to the emf $\mathscr{V}$ of the battery minus the IR_1 drop due to the internal resistance of the battery. The variation of the potential V around the circuit is as indicated in Fig. 4-10b.

It is instructive to compare the electric circuits of the above examples with the analogous hydraulic systems. Thus, a hydraulic system analogous to the circuit of the above example (Fig. 4-10) is shown in Fig. 4-11. Between b and c there is an open horizontal trough at what may be called a lower level, corresponding to the ground potential in Fig. 4-10. Between c and d there is a pump which raises the water or other liquid against the gravitational field in the same manner as the battery in Fig. 4-10 raises positive charge against the static electric field $\mathbf{E}_c$. Thus, the water in the upper trough has a higher potential energy than the water in the lower trough in the same way as the charge in the wire between d and a in the electric circuit of Fig. 4-10 is at a higher potential than the charge in the wire from b to c. From d to a the water moves in a horizontal frictionless trough at an upper level corresponding to the perfectly conducting wire between these points in Fig. 4-10. From a to b the water falls through a pipe to the lower level and in so doing gives up the energy it acquired in being pumped to the upper level. The pipe offers resistance to the flow of water, and the energy given up by the water appears as heat. This energy is analogous to that appearing as heat in the resistor of Fig. 4-10 owing to charge falling in potential from a to b. In this analogy the pump does work, raising the water against the gravitational field in the same manner as the chemical action in the battery does work per unit charge (against the electrostatic field $\mathbf{E}_c$ and internal resistance R_1) equal to the emf of the battery.

In a single-cell battery with two electrodes the field $\mathbf{E}_e$ is largely confined to a thin layer at the surface between the electrode and the electrolyte and is zero in the electrolyte between the two electrodes. Thus, the potential variation assumed in the

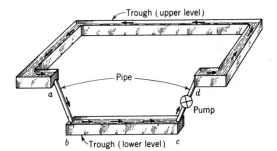

Figure 4-11 Hydraulic analog for electric circuit of Fig. 4-10.

preceding example is not representative of an actual two-electrode cell although it could be approached if each battery consisted of a large number of cells of small emf connected in series between c and d in Fig. 4-10.

4-11 TUBES OF CURRENT

In Chap. 2 we discussed tubes of flux. Let us now consider an analogous concept, namely, that of *tubes of current*. A tapered section of a long conductor is shown in Fig. 4-12. Let all the space in the conductor be filled with current tubes. Each tube is everywhere parallel to the current direction and hence, from the relation $\mathbf{J} = \sigma\mathbf{E}$, is also parallel to the electric field. Since no current passes through the wall of a current tube, the total current I_0 through any cross section of a tube is a constant. Thus

$$I_0 = \iint_a \mathbf{J} \cdot d\mathbf{s} = \text{constant} \qquad (1)$$

where $\mathbf{J}$ = current density, A m^{-2}

a = cross section of tube (over which $\mathbf{J}$ is integrated), m^2

If $\mathbf{J}$ is constant over the cross section and normal to it, then $I_0 = Ja$ or, referring to the current tube of rectangular cross section in Fig. 4-12,

$$I_0 = Jbd$$

where b and d are the tube cross-sectional dimensions (see Fig. 4-12).

If all of the conductor is divided up into current tubes, each with the same current I_0, the total current I through the conductor is $I = I_0 n$, where n is the number of current tubes.

Surfaces normal to the direction of the current (or field) are equipotential surfaces. The potential difference V between two equipotential surfaces separated by a distance l is by Ohm's law equal to the current I_0 in a current tube times the

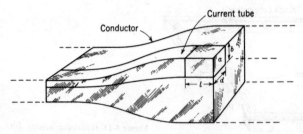

Figure 4-12 Tapered section of a long conductor showing current tube.

resistance R of the section of tube of length l. Thus, $V = I_0 R$. If the current density is uniform (field uniform), the resistance R is, from (4-6-2), given by

$$R = \frac{l}{\sigma a} \quad (\Omega) \tag{2}$$

where l = length of tube section, m
     σ = conductivity of conducting medium, $\mho\ m^{-1}$
     a = cross-sectional area of tube, m^2

4-12 KIRCHHOFF'S CURRENT LAW

Whereas flux tubes in a static electric field begin and end on electric charge and hence are *discontinuous*, the tubes of a steady current form closed circuits on themselves and hence are *continuous*. To describe this continuous nature of steady currents, it is said that the current is *solenoidal*. That is, it has no sources or sinks (ending places) as do the flux tubes, which start and end on electric charges in a static electric field. As a consequence as much current must flow into a volume as leaves it. Thus, in general, the integral of the normal component of the current density **J** over a closed surface s must equal zero, or

$$\oint_s \mathbf{J} \cdot d\mathbf{s} = 0 \tag{1}$$

This relation is for steady currents and applies to any volume. For example, the volume may be entirely inside a conducting medium, or it may be only partially filled with conductors. The conductors may form a network inside the volume, or they may meet at a point. As an illustration of this latter case, the surface S in Fig. 4-13 encloses a volume that contains five conductors meeting at a junction point P. Taking the current flowing away from the junction as positive and the current flowing toward the junction as negative, we have from (1) that

$$I_1 - I_2 - I_3 + I_4 - I_5 = 0$$

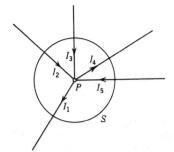

Figure 4-13 Junction point of several conductors.

In other words, *the algebraic sum of the currents at a junction is zero.* This is *Kirchhoff's current law,* which can be expressed in general by the relation

$$\sum I = 0 \tag{2}$$

4-13 DIVERGENCE OF J AND CONTINUITY RELATIONS FOR CURRENT

Consider the small volume element Δv shown in Fig. 4-14 located inside a conducting medium. The current density **J** is a vector having the direction of the current flow. In general, it has three rectangular components that vary with position. The development here is formally the same as for **D** in Sec. 3-21 and yields

$$\nabla \cdot \mathbf{J} = 0 \tag{1}$$

This is a point relation. It states that steady currents have no sources or sinks. Any vector function whose divergence is everywhere zero is said to be solenoidal.

Equation (4-12-1) is an *integral* relation involving **J** for a finite region. Equation (1) is a *differential* relation involving **J** at a point. Both equations are expressions of the continuous nature of **J**. Both are statements of Kirchhoff's current law.

Let us digress briefly to consider the situation if the current is not steady as assumed above. Then (4-12-1) does not necessarily hold, and the difference between the total current flowing out of and into a volume must equal the rate of change of electric charge inside the volume. Specifically, a *net* flow of current *out of* the volume (positive current flow) must equal the *negative* rate of change of charge with time (rate of decrease of charge). Now the total charge in the volume Δv of Fig. 4-14 is $\rho \, \Delta v$, where ρ is the average charge density. Therefore

$$\oint_s \mathbf{J} \cdot d\mathbf{s} = -\frac{\partial \rho}{\partial t} \Delta v \tag{2}$$

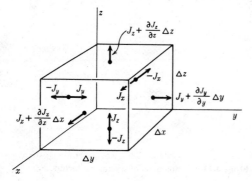

Figure 4-14 Construction used to develop differential expression for divergence of **J**.

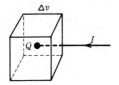

Figure 4-15 Construction for the continuity relation between current and charge.

Dividing by Δv and taking the limit as Δv approaches zero, we obtain

$$\boxed{\nabla \cdot \mathbf{J} = -\frac{\partial \rho}{\partial t}} \qquad (3)$$

This is the general *continuity relation*† between current density $\mathbf{J}$ and the charge density ρ at a point. For steady currents $\partial \rho / \partial t = 0$ and (3) reduces to (1).

Consider now the situation shown in Fig. 4-15, where a wire carrying a current I terminates inside a small volume Δv. Applying (2) to this situation, the integral of $\mathbf{J}$ over the volume yields the net current entering or leaving the volume. Assuming that I is entering the volume, we have

$$\oint \mathbf{J} \cdot d\mathbf{s} = -I \qquad (4)$$

Since $\rho \, \Delta v$ equals the total charge Q inside the volume,

$$\frac{\partial \rho}{\partial t} \Delta v = \frac{dQ}{dt} \qquad (5)$$

Substituting (4) and (5) in (2) yields

$$I = \frac{dQ}{dt} \qquad (6)$$

This is the continuity relation between the current and charge in a wire.

4-14 CURRENT AND FIELD AT A CONDUCTOR-INSULATOR BOUNDARY

The relation between the current density $\mathbf{J}$ and the electric field intensity $\mathbf{E}$ in a conductor is $\mathbf{J} = \sigma \mathbf{E}$. Consider now the situation at a conductor-insulator boundary as in Fig. 4-16. Assuming that the conductivity of the insulator is zero, $\mathbf{J} = 0$ in the

---

† Also called the equation of *conservation of charge*.

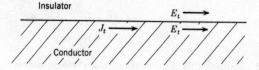

Figure 4-16 Insulator-conductor boundary.

insulator. At the boundary, current in the conductor must flow tangentially to the boundary surface. Thus, on the conductor side of the boundary we have

$$E_t = \frac{J_t}{\sigma} \tag{1}$$

where E_t = component of electric field tangential to boundary = $|\mathbf{E}|$
$\quad J_t$ = component of current density tangential to boundary = $|\mathbf{J}|$
$\quad \sigma$ = conductivity of conducting medium

By the continuity of the tangential electric field at a boundary, the tangential field on the insulator side of the boundary is also E_t.

When current flows, a conductor of finite conductivity is not an equipotential body as it is in the static case with no currents present. For example, the potential varies along a current-carrying wire with uniform current density as suggested in Fig. 4-17. The arrows indicate the field and current directions, while the transverse lines are equipotentials. Since $\mathbf{E}$ is uniform, the potential difference V of two points separated by a distance l along the wire is El. This potential difference is also equal to the IR drop, that is, $V = IR$, where I is the current in the wire and R is the resistance of a length l of the wire. The field is the same both inside and outside the wire and is entirely tangential (and parallel to the axis of the wire).

If superimposed on this situation there is a static electric charge distribution at the boundary surface due to the proximity of other conductors at a different potential, a component of the electric field E_n normal to the conductor-insulator boundary is also present on the insulator side of the boundary. The total field in the insulator is then the vector sum of the normal component E_n and the tangential component E_t. In the conductor, $E_n = 0$, and the field is entirely tangential to the boundary. For instance, consider the longitudinal cross section shown in Fig. 4-18 through a part of a long coaxial cable. Current flows to the right in the inner conductor and returns through the outer conductor. The field in the conductor is

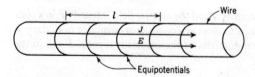

Figure 4-17 Section of long wire.

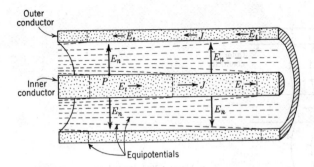

Figure 4-18 Longitudinal cross section of coaxial transmission line. Equipotentials are shown by the dashed lines. The arrows indicate the direction of the normal and tangential field components, E_n and E_t, and the current density J.

entirely tangential (and parallel to the axis of the cable) and is indicated by E_t. Since the conductivity of the conductor is large, this field is relatively weak, as suggested by the short arrows for E_t. In the insulating space between the inner and outer conductors there may exist a relatively strong field due to the voltage applied at the end of the cable. This field is a static electric field. It is entirely normal to the surfaces and is indicated by E_n. It is relatively strong, as suggested by the long arrows for E_n. At a point P at the surface of the inner conductor (Fig. 4-18) the total field **E** is then the sum of the two components E_n and E_t added vectorially as in Fig. 4-19. If the conductivity of the metal in the cable is high, E_t may be so small that **E** is substantially normal to the surface and equal to E_n. However, the size of E_t has been exaggerated in Fig. 4-19 in order to show the slant of the total field more clearly. The shape of the total field lines across the entire insulating space between the inner and outer conductors is suggested in Fig. 4-20, with the slant of the field at the conductors greatly exaggerated. Equipotential surfaces are indicated by the dashed lines.

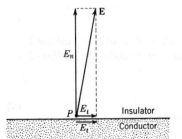

Figure 4-19 Total field **E** at insulator-conductor boundary (point P in Fig. 4-18) resolved into normal and tangential components. The tangential component E_t is due to the current in the conductor ($=J_t/\sigma$) while the normal component E_n is due to the surface charge induced by the voltage between the inner and outer conductors.

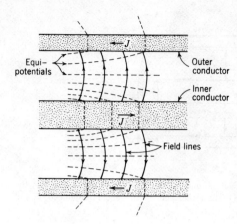

Figure 4-20 Longitudinal cross section of coaxial transmission line showing equipotentials (dashed) and total field lines (solid).

4-15 CURRENT AND FIELD AT A CONDUCTOR-CONDUCTOR BOUNDARY

Consider the conductor-conductor boundary shown in Fig. 4-21 between two media of constants σ_1 and ϵ_1 and σ_2 and ϵ_2. In general, the direction of the current changes in flowing from one medium to the other.

For steady currents we have the boundary relation

$$J_{n1} = J_{n2} \tag{1}$$

where J_{n1} = component of current density normal to boundary in medium 1
J_{n2} = component of current density normal to boundary in medium 2

From relation (1) of Table 3-2 we also have

$$E_{t1} = E_{t2} \tag{2}$$

where E_{t1} = component of field tangent to boundary in medium 1
E_{t2} = component of field tangent to boundary in medium 2

It follows that

$$\frac{J_{t1}}{\sigma_1} = \frac{J_{t2}}{\sigma_2} \tag{3}$$

where J_{t1} = component of current density tangent to boundary in medium 1
J_{t2} = component of current density tangent to boundary in medium 2

Dividing (3) by (1) gives

$$\frac{J_{t1}}{\sigma_1 J_{n1}} = \frac{J_{t2}}{\sigma_2 J_{n2}} \tag{4}$$

or

$$\frac{\tan \alpha_1}{\tan \alpha_2} = \frac{\sigma_1}{\sigma_2} \tag{5}$$

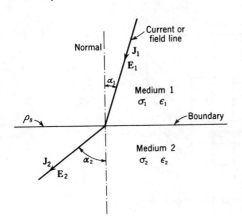

Figure 4-21 Boundary between two different conducting media showing change in direction of current or field line.

where α_1 and α_2 are as shown in Fig. 4-21. This relation is similar to Snell's law of refraction (see Chap. 12).

4-16 CURRENT MAPPING AND THE RESISTANCE OF SIMPLE GEOMETRIES; CONDUCTOR CELLS

If the current density is uniform throughout a conductor, its resistance is easily calculated from its dimensions and conductivity. For example, consider the homogeneous rectangular bar of conductivity σ shown in Fig. 4-22. It has $l' = 1$ m, $b' = 400$ mm, and $d = 600$ mm. If the end faces of the bar are clamped against high-conductivity blocks, as in Fig. 4-5, the field and current density throughout the bar will be uniform. The resistance R of the bar is

$$R = \frac{l'}{\sigma d b'} = \frac{1}{0.24\sigma} \quad (\Omega) \qquad (1)$$

where σ is the conductivity of the bar in mhos per meter.

The resistance of the bar can also be calculated by dividing the side of the bar into square areas each representing the end surface of a *conductor cell*. The sides of the cells are equipotentials. The top and bottom surfaces of the cell are parallel to the current direction. The resistance R_0 of such a cell is given by

$$R_0 = \frac{l}{\sigma d b} = \frac{1}{\sigma d} = \frac{S}{d} \qquad (1a)$$

where S is the resistivity of the bar material. Hence the product of R_0 and the depth d equals the resistivity S of the bar material, or

$$R_0 d = S \qquad (2)$$

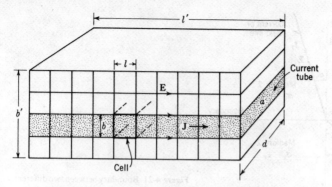

Figure 4-22 Conductor divided into current tubes. Vertical lines are equipotentials.

Taking the reciprocal of (2) yields

$$\boxed{\frac{G_0}{d} = \sigma} \qquad (3)$$

That is, the conductance per unit depth of a conductor cell is equal to the conductivity of the medium. Referring to Fig. 4-22,

$$\frac{\text{Number of cells (or current tubes) in parallel}}{\text{Number of cells in series}} = \frac{4}{10}$$

From (3) the conductance per unit depth of a conductor cell is σ, so that the actual conductance G_0 of a cell of bar material is given by

$$G_0 = d\sigma = 0.6\sigma \quad (\text{℧}) \qquad (4)$$

The total conductance of the bar is then

$$G = \tfrac{4}{10}G_0 = \tfrac{4}{10}0.6\sigma = 0.24\sigma \quad (\text{℧})$$

The total resistance is the reciprocal, or

$$R = \frac{1}{0.24\sigma} = 4.17S \quad (\Omega) \qquad (5)$$

The method of calculating the resistance or conductance of the bar by means of evaluating the series-parallel combination of conductor cells is more general than the method used in arriving at (5) since it can be applied not only to uniform current distributions (as here) but also to the more general situation where the current distribution is nonuniform. In a nonuniform distribution the sides of many or all of the conductor cells will be curvilinear squares. Their area and arrangement may be determined by graphical current-mapping techniques that are like the field-mapping procedures discussed in Sec. 3-19.

Graphical current-mapping techniques can be applied to any two-dimensional problem, i.e., to a conductor whose shape can be described by a single cross section with all other cross sections parallel to this one being identical to it. Current mapping is actually electric field mapping *in a conducting medium* since the current and the field have the same direction in isotropic media ($\mathbf{J} = \sigma\mathbf{E}$).

The following fundamental properties are useful in current mapping:

1. Current lines and equipotentials intersect at right angles.
2. Current flows tangentially to an insulating boundary.
3. The total current through any cross section of a continuous current tube is a constant.
4. In a uniform current distribution the potential varies linearly with distance.
5. Current tubes are continuous.

With these properties in mind one divides a conductor cross section into current tubes and then by equipotentials into *conductor cells* with sides that are squares or curvilinear squares, using the same trial-and-error method described in Sec. 3-19 in connection with field mapping in an insulating medium.

In calculating the conductance of a conductor with a nonuniform current distribution a current map is first made, as discussed above. The conductance G is then given by

$$G = \frac{N}{n}\, G_0 \tag{6}$$

where N = number of cells (or current tubes) in parallel
n = number of cells in series = number of cells per tube
G_0 = conductance of each cell as given by (4)

The accuracy of the conductance (or its reciprocal, the resistance) depends primarily on the accuracy with which the curvilinear squares are mapped.

Example A homogeneous rectangular bar of conductivity σ has the dimensions shown in Fig. 4-23a. This bar is identical with the one of Fig. 4-22 except that two cuts have been made across the full width of the bar, as indicated. Find the resistance of the bar when its ends are clamped between high-conductivity blocks as in Fig. 4-5.

SOLUTION A longitudinal cross section of the bar is drawn to scale and a current map made with the result shown in Fig. 4-23b.† A portion of one quadrant has been further subdivided to test the accuracy of the curvilinear squares. From (1a) the resistance R_0 of one conductor cell is $R_0 = 1/0.6\sigma$ (Ω). There are 13

---

† Although the entire cross section of the bar has been mapped, the symmetry is such that a map of only one quadrant is required.

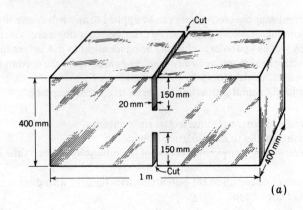

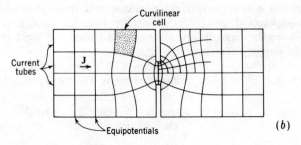

Figure 4-23 (a) Conducting bar with notch and (b) current map. Resistance of bar equals ratio of cells in series to cells in parallel multiplied by the resistance of each cell.

cells in series in a tube, and there are 4 tubes in parallel. Hence, from (6) the resistance R of the bar is

$$R = \frac{13R_0}{4} = \frac{13}{2.4\sigma} = 5.42S \quad (\Omega)$$

Thus, comparing this result with (5) for the uniform bar, the slots in the bar produce an increase of 30 percent in its resistance.

4-17 LAPLACE'S EQUATION FOR CONDUCTING MEDIA

For steady currents $\mathbf{V} \cdot \mathbf{J} = 0$, but $\mathbf{J} = \sigma \mathbf{E}$; so $\sigma \mathbf{V} \cdot \mathbf{E} = 0$. Recalling that $\mathbf{E} = -\mathbf{V}V$, we get $\sigma \mathbf{V} \cdot (\mathbf{V}V) = 0$, or

$$\boxed{\mathbf{V}^2 V = 0} \tag{1}$$

This is Laplace's equation. It was derived previously in Sec. 3-25 for static electric fields, and since it also applies here, it follows that problems involving distributions of steady currents in conducting media can be handled in the same way as problems involving static field distributions in insulating media. If we have a conductor with an unknown current distribution, and if a solution to Laplace's equation can be found that also satisfies the boundary conditions, we can obtain the potential and current distribution in the conductor. If this is not possible, we can nevertheless find the approximate potential and current distribution in two-dimensional problems by graphical current mapping as discussed in Sec. 4-16. From a knowledge of the current distribution we can determine the resistance, the maximum current density, and other items of practical importance for a given conductor configuration.

In conducting media, current tubes and the conductivity σ are analogous to the flux tubes and permittivity ϵ in insulating media. Thus in conducting media we have

$$J = \sigma E \qquad (\text{A m}^{-2}) \tag{2}$$

while in insulating media we have

$$D = \epsilon E \qquad (\text{C m}^{-2}) \tag{3}$$

It is also to be noted that in a conducting medium the *conductance per unit depth* of a conductor cell equals the conductivity σ of the medium, or

$$\frac{G_0}{d} = \sigma \qquad (\text{℧ m}^{-1}) \tag{4}$$

where d is the depth of the cell (see Fig. 4-22), while in an insulating medium the *capacitance per unit depth* of a dielectric field cell equals the permittivity ϵ of the medium, or

$$\frac{C_0}{d} = \epsilon \qquad (\text{F m}^{-1}) \tag{5}$$

For a static electric field in a dielectric medium of permittivity ϵ there are no currents, but there is a flux density $D = \epsilon E$. For a static electric field in a conducting medium of conductivity σ there is current of density $J = \sigma E$. Since both fields obey Laplace's equation, a solution in the conductor situation is also a solution for the analogous dielectric situation, and vice versa. For example, if the medium between conductors 1 and 2 in Fig. 3-19 is a conductor of conductivity σ, the conductance per unit depth between ff and gg is given by

$$\frac{G}{d} = \frac{15.43}{4}\sigma = 3.86\sigma \qquad (\text{℧ m}^{-1})$$

A further discussion of Laplace's equation is given in Chap. 7.

PROBLEMS†

4-1-1. Resistance defined in terms of power Show that the resistance of a resistor can be expressed as

$$R = \frac{\int \frac{J^2}{\sigma} \, dv}{I^2} \quad (\Omega)$$

where the numerator equals the power dissipated in the resistor (integral over volume of resistor) and I equals the current through the resistor. J = current density and σ = conductivity.

4-1-2. Electron drift velocity on dc line A transmission line 2 km long consists of 2 parallel copper wires of 2-mm diameter. When 120 V is applied at one end of the line and a 20-Ω resistance is connected at the other end of the line, find (a) current I through the resistor, (b) current density J in the line, and (c) electron drift velocity v_d in the line. Note that the drift velocity v_d is *not* the velocity with which a change (or information) propagates along the line. Thus, a change in applied voltage results in a change in resistor current within a time $t = d/v$ where d = length of line and v may be equal or close to the velocity of light (not the drift velocity). The situation is analogous to a city water system. It may take many hours for water from the pumping station (analogous to battery or generator) to reach a distant hydrant (analogous to load resistor), but if the pumps break down and there are no water towers (analogous to capacitors across the line), the pressure (analogous to voltage) at the hydrant may go to zero in seconds. A water pressure wave travels rapidly, like a voltage wave on a transmission line.

4-1-3. Electron excursion distance on ac line (a) If a standard 60-Hz sinusoidal voltage of 120 V rms is applied to the line of the preceding problem, what is the rms excursion distance (to and fro) of the electrons in the copper wires of the line? Would you have presumed such a microscopic excursion distance? (b) How can the excursion distance be increased?

4-1-4. Charge density for conductor Assuming one free electron per atom of a conductor, such as aluminum, copper, silver, or gold, show that the electric charge density of free electrons is given by

$$\rho = Nq = \frac{A_0 \rho_m q}{W} \quad (\text{C m}^{-3})$$

where N = number of electrons, m^{-3}
  q = electron charge, C
  A_0 = Avogadro's number, $= 6.0 \times 10^{23}$ atoms/mole
  ρ_m = mass density, kg m^{-3}
  W = atomic weight, kg/mole

★**4-1-5. Charge density and mobility for copper** (a) Referring to the previous problem, find the electric charge density of free electrons in copper if the mass density $\rho_m = 9.0 \times 10^3$ kg m^{-3} and the atomic weight $W = 63.5$ g/mole. (b) Using the result of (a), find the mobility μ for copper.

4-1-6. Copper bar A copper bar 30 by 80 mm in cross section by 2 m in length has 50 mV applied between its ends. Find the following quantities and give units in each case: (a) resistance R of the bar; (b) conductance G of the bar; (c) current I; (d) current density J; (e) electric field E; (f) power loss P in the bar; (g) power loss per unit volume; (h) energy loss W per hour; (i) drift velocity v of electrons. Take $T = 20°C$ and $\rho = 20$ GC m^{-3}.

4-1-7. Current density If $\mathbf{J} = \hat{\mathbf{x}} 3yz$ A m^{-2}, find the current I through a square 2 m on a side with one corner at the origin and other corners at (0, 2, 0), (0, 0, 2), and (0, 2, 2).

---

† Answers to starred problems are given in Appendix C.

★4-1-8. Leakage resistance to ground A 2-mm-diameter copper wire is enclosed in an insulating sheath of 4-mm outside diameter. If the wire is buried in a highly conducting ground, what is the leakage resistance of the sheath to ground per kilometer? The sheath conductivity is 10^{-8} $\mho$ m^{-1}.

4-1-9. Graphite washer A flat graphite washer of thickness t has an inner radius r_1 and outer radius r_2. Determine the resistance (*a*) between the inner and outer edges, (*b*) between the flat surfaces, and (*c*) around the washer (same as resistance between the edges of an infinitesimally thick saw cut through one radius of the washer). Take $\sigma = 10^5$ $\mho$ m^{-1}, $r_1 = 10$ mm, $r_2 = 20$ mm, and $t = 3$ mm.

4-1-10. Graphite cone A truncated graphite cone 30 mm long has end diameters of 6 and 18 mm. Find the resistance between the ends.

★4-1-11. Wire and sheet A straight wire is connected at right angles to the center of a large, thin conducting sheet. If the wire carries 200 A, find the sheet current density (A m^{-1}) at a distance of 100 mm from the point of connection.

Group 4-2: Secs. 4-8 through 4-12. Semiconductors, conductivities, Kirchhoff's voltage law, emf, and Kirchhoff's current law

4-2-1. Conductivity of wire A 3-mm diameter wire 1 km long has a potential drop between its ends of 18.2 V. If the current is 2 A, is the wire made of aluminum or copper?

4-2-2. Semiconductor germanium For intrinsic semiconductor germanium at 30°C, the electron-hole pair density is 2.5×10^{19} m^{-3}. Find (*a*) electron and hole charge density; (*b*) germanium conductivity if $\mu_e = 0.4$ and $\mu_h = 0.2$ m^2 V^{-1} s^{-1}; (*c*) drift velocity of electrons and holes if $E = 10$ V m^{-1}.

4-2-3. Copper-germanium comparison Give the copper to semiconductor germanium ratio for (*a*) mobility μ; (*b*) charge density ρ; (*c*) conductivity σ.

4-2-4. Cube of silicon A 1-mm cube of semiconductor silicon has 3 V applied across it. At 30°C find (*a*) J, (*b*) I, and (*c*) electron and hole drift velocities.

★4-2-5. Series circuit. Kirchhoff's voltage law (*a*) A resistance R_0 and three batteries are connected in series as shown in Fig. P4-2-5. For the first battery the emf $\mathscr{V}_1 = 1.5$ V and the electrolyte or internal resistance $R_1 = 1$ Ω; for the second battery the emf $\mathscr{V}_2 = 2$ V and the internal resistance $R_2 = 0$; and for the third battery the emf $\mathscr{V}_3 = 3$ V and the internal resistance $R_3 = 1$ Ω. The first two batteries have single cells, while the third has three cells in series, each cell of 1 V emf and $\frac{1}{3}$ Ω internal resistance. Assume that half the total emf of a cell occurs at each electrode, and assume that all connections between cells have negligible resistance. Draw a graph such as in Fig. 4-10, showing the variation of potential with position between points *a* and *c* when $R_0 = 4.5$ Ω and when $R_0 = 0$. Take $V = 0$ at the point *b*. (*b*) Referring to the circuit of Fig. P4-2-5, let the emfs be as indicated, and let $R_1 = 1.5$ Ω, $R_2 = 2$ Ω, and $R_3 = 3$ Ω. Draw a graph of the variation of potential with position when $R_0 = 6.5$ Ω and also when $R_0 = 0$.

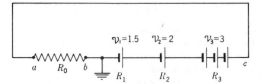

Figure P4-2-5 Series circuit.

4-2-6. Junction and Kirchhoff's current law Four wires meet at a common junction point. The current in wires 1 and 2 is 5 A each and flowing away from the junction, while the current in wire 3 is 6 A flowing toward the junction. What are the current magnitude and direction in the fourth wire?

Group 4-3: Secs. 4-13 through 4-17. Continuity of current, divergence of J, boundary conditions, current mapping, conductor cells, and Laplace's equation for conducting media. *Note*: Where exact solutions are not evident, try mapping methods or best approximations possible.

4-3-1. Spherical current density If the current density has a spherical distribution given by

$$\mathbf{J} = \hat{\mathbf{r}} r_0^2 r^{-1} \quad (\text{A m}^{-2})$$

find (a) $\nabla \cdot \mathbf{J}$ at $r = 2r_0$ and (b) I through the surface $r = 2r_0$.

4-3-2. Shunt resistance of transmission line (a) A two-conductor transmission line consists of copper wires 4 mm in diameter spaced 10 mm on center and encased in a symmetrical insulating sheath 24 mm in diameter. If the sheath insulation $\sigma = 10^{-8} \, \mho \, \text{m}^{-1}$, find the shunt (or leakage) resistance between conductors per kilometer with the line in dry air. (b) Find the shunt resistance if the line is immersed in salt water.

4-3-3. L-shaped copper bar A 40-mm-thick L-shaped copper bar has the dimensions shown in Fig. P4-3-3. At 20°C find the resistance R and conductance G (a) between faces 1 and 2, (b) between sides 3 and 4, and (c) between front and back surfaces (parallel to page). (d) If 3 mV is applied between faces 1 and 2, find the current density J and electric field E at the points P_1 and P_2. The point P_1 is 10 percent of the distance along the diagonal from the inside corner, and P_2 is 10 percent of the distance from the outside corner. (e) What is the ratio of the current density J at P_1 to that at P_2?

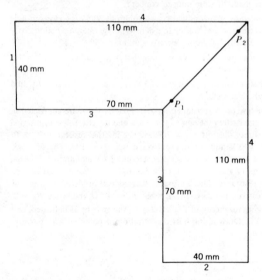

Figure P4-3-3 L-shaped bar.

4-3-4. Half bar If the bar of the preceding problem is cut in half along the diagonal line through P_1 and P_2, are the resistance values halved and conductance values doubled for each half with respect to those of parts (a), (b), and (c)? If they are not, what are the values?

★4-3-5. Straight bar If the L-shaped block of Fig. P4-3-3 is cut along the diagonal at the bend and one-half turned over, the two halves can be joined to form a straight bar 40 by 40 mm in cross section by 180 mm long. (a) What is the resistance between the ends of this bar? (b) What is the ratio of the resistance between the faces 1 and 2 of the L-shaped bar [part (a) of Prob. 4-3-3] and the resistance of the straight bar in [part (a) of this problem]? (c) Why is this ratio not unity?

4-3-6. Graphite slab A graphite slab 75 mm square and 15 mm thick has flat perfectly conducting electrodes 15 mm square in contact with the center of two opposite edges. Find the resistance between the electrodes.

4-3-7. Graphite sphere with polar caps A solid graphite sphere 200 mm in diameter has perfectly conducting caps in perfect contact on opposite poles. Each cap extends from its pole over an angle of 30° in latitude. Find the resistance measured between the caps.

4-3-8. Graphite shell with polar caps This problem is identical with the preceding problem except that the graphite sphere is a hollow shell with a thickness of 20 mm. Find the resistance measured between the caps.

★**4-3-9. Pipeline resistance** Two parallel steel pipelines have a spacing of 4 m between centers. The pipes are half buried in the ground as indicated in Fig. P4-3-9. The diameter of the pipes is 500 mm. The conductivity of the ground (sandy soil) is 100 $\mu \mho \, m^{-1}$. Find the resistance between the two pipes per kilometer. *Hint*: Note the analogy between this situation and the static electric field between two parallel cylindrical conductors.

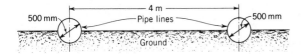

Figure P4-3-9 Half-buried pipelines.

★**4-3-10. Bar with hole** A nichrome bar 40 by 40 mm in cross section by 100 mm long has a hole through it which is symmetrically situated at the center of a long side. The hole is 30 mm in diameter. At 20°C (*a*) what is the resistance between the ends of the bar? (*b*) What is the resistance of a solid bar (without hole)? (*c*) What is the length of a solid bar having the same resistance as in (*a*)?

★**4-3-11. Bar and strip** (*a*) A bar and strip are connected as shown in Fig. P4-3-11*a*. The bar has finite conductivity, while the strip conductivity is assumed to be infinite. If the end of the bar is clamped against a large, infinitely conducting block as in Fig. P4-3-11*b* instead of connected to the strip as in Fig. P4-3-11*a*, determine by what length *l* the bar would need to be lengthened for its resistance to be the same as when connected to the strip. (*b*) Why is the resistance of the bar larger when it is connected to the strip than when contact is made with the block?

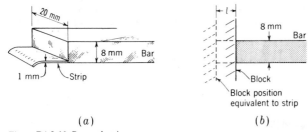

(*a*)   (*b*)

Figure P4-3-11 Bar and strip.

4-3-12. Surface charge In general, a surface charge is present on the boundary between two conductors (conductivities σ_1 and σ_2 and permittivities ϵ_1 and ϵ_2, respectively) across which current is flowing. Show that the surface charge density ρ_s is given by

$$\rho_s = J_n \left(\frac{\epsilon_1}{\sigma_1} - \frac{\epsilon_2}{\sigma_2} \right)$$

where J_n is the normal component of current density.

4-3-13. Conductor stack. Continuity of J A resistor consists of a stack of two conducting slabs 150 mm square. The upper slab is 25 mm thick and has a relative permittivity of 2 and a conductivity of $10^3 \, \mho \, m^{-1}$ while the lower slab is 50 mm thick and has a relative permittivity of 3 and a conductivity of $10^4 \, \mho \, m^{-1}$. Electrodes 150 mm square are in contact with the top of the upper slab and bottom of the lower slab. If 5 V is applied between the electrodes, find (a) E in both slabs, (b) J in both slabs, (c) the free surface charge density ρ_s at the interface between the two slabs, (d) the resistance between electrodes, and (e) the current.

4-3-14. Divergence theorem Deduce the relation $\nabla \cdot J = 0$ by applying the divergence theorem to

$$\oint_s J \cdot ds = 0$$

4-3-15. Conductor-conductor boundary Show that at a conductor-conductor boundary

$$\frac{\sigma_1}{\sigma_2} = \frac{E_{n2}}{E_{n1}} = \frac{J_{t1}}{J_{t2}}$$

★4-3-16. E at wire surface (a) A wire 2 mm in diameter has a resistance of 1 Ω per 100 m. A current of 10 A is flowing in the wire. What is the electric field E in the wire? (b) If a uniform static surface charge of 8 pC m^{-2} is applied to the wire, what is the electric field $\mathbf{E}$ (magnitude and direction) just outside the surface of the wire? The medium outside the wire is air.

4-3-17. E between coaxial conductors Demonstrate that the source of the emf energizing the coaxial line of Fig. 4-18 is at the left end by showing that if the source were at the right end, the field lines would be bowed in the opposite direction.

4-3-18. Current direction at boundary (a) The current direction at the flat boundary surface between two media makes an angle of 45° with respect to the surface in medium 1. What is the angle between the current direction and the surface in medium 2? The constants for the media are $\sigma = 1 \, M\mho \, m^{-1}$ and $\epsilon = \epsilon_0$ in medium 1 and $\sigma = 1 \, k\mho \, m^{-1}$ and $\epsilon = 3\epsilon_0$ in medium 2. (b) If the current density $\mathbf{J}$ in medium 1 is 20 A m^{-2}, what is the surface charge density at the boundary?

4-3-19. Cylindrical resistor with 2 layers A resistor consists of a cylindrical rod of length l and diameter d_1 and conductivity σ_1 inside a conducting tube of the same length, and of inner diameter d_1, outer diameter d_2, and conductivity σ_2. Find (a) the resistance between the ends of the rod and tube combination, (b) E in rod, (c) E in tube, (d) J in rod, (e) J in tube, and (f) the ratio of J in rod to J in tube. (g) Is this ratio dependent on the dimensions d_1 or d_2? If so, how?

★4-3-20. Resistance of shoe-shaped block. Field map A conducting block 20 mm thick (into page) is shown with field map in Fig. P4-3-20. If $\sigma = 10 \, \mho \, m^{-1}$, find (a) resistance between sides A and B and (b) resistance between sides C and D.

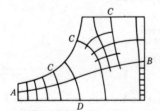

Figure P4-3-20 Shoe-shaped block.

★4-3-21. Resistance of block. Field map A conducting block 20 mm thick (into page) is shown with field map in Fig. P4-3-21. If $\sigma = 100 \, \mho \, m^{-1}$, find (a) resistance between the ends and (b) resistance between the top and bottom surfaces. Note that the top surface consists of one vertical and two horizontal surfaces.

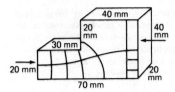

Figure P4-3-21 Conducting block with step.

4-3-22. Conductor parallel to sheet. Field map A horizontal cylindrical conductor is mounted parallel to a large flat conducting sheet as shown in the cross section with field map in Fig. P4-3-22. The cylinder and sheet have a potential difference of 300 V and the medium has a conductivity $\sigma = 10^{-4}\ \mho\ m^{-1}$. Using the field map, determine the current density J at (a) point A, (b) point B, (c) point C, (d) point D, and (e) point E.

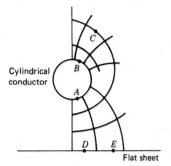

Figure P4-3-22 Field of cylindrical conductor parallel to conducting sheet.

4-3-23. Conductor parallel to sheet. Analytical method Determine the current density J at the five points of Fig. P4-3-22 as done in the preceding Prob. 4-3-22 but instead of using the field map proceed by analytical methods and compare these results with those of Prob. 4-3-22.

*4-3-24. Square resistor with cylindrical core The square resistor shown in cross section with field map in Fig. P4-3-24 has a cylindrical core which is either a perfect insulator ($\sigma = 0$) or a perfect conductor ($\sigma = \infty$). The resistor material has a conductivity σ. If the resistor has a length l (into page), find the resistance between electrodes placed in full contact (a) with faces A and B and (b) with faces C and D. (c) In each case specify also the conductivity of the core.

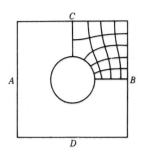

Figure P4-3-24 Square resistor with cylindrical core.

★4-3-25. Overlapping bars Two aluminum bars 20 mm by 20 mm in cross section by 100 mm long are joined by being overlapped 20 mm. At 20°C (*a*) what is the resistance between the ends of this combination (field map required)? (*b*) What is the resistance of a continuous bar 20 mm by 20 mm in cross section by 180 mm long [same length as overlapped bars in (*a*)]? (*c*) What is the resistance of a continuous bar 20 mm by 20 mm in cross section by 200 mm long [same length as two bars of (*a*) placed end to end]? (*d*) Why is resistance in (*a*) greater than in (*b*) but less than in (*c*)?

Group 4-4. Practical applications

4-4-1. Ground resistance calculation Two copper-plated steel grounding rods separated by 50 m are driven 2 m into the ground in a flat open area. If the rods are 40 mm in diameter, what is the resistance between the rods? The average ground conductivity is $5 \times 10^{-3} \, \text{\textmho} \, \text{m}^{-1}$.

★4-4-2. Ground resistance measurement To measure the resistance of the ground path between two metal grounding rods, a 1.5-V battery was connected in series with a 10-Ω resistor and a milliammeter between the rods. With the battery connected one way, the meter read 38 mA, but with battery polarity reversed the meter read 23 mA. (*a*) Find the resistance of the ground path. Assume that the difference in current readings is due to an emf in series with the ground path as might be caused by electrolytic action between the rods and ground. (*b*) How much is the ground emf?

4-4-3. Thundercloud current and charge If a thundercloud discharges by a lightning stroke once a minute, what is the total charge that must be moving in the cloud in order to maintain this discharge rate if it corresponds to the capacitor cloud model of Prob. 3-5-13? Take the velocity of charge transport $v = 20 \, \text{m s}^{-1}$.

4-4-4. Lightning stroke current A typical lightning stroke involves an energy of 1 TJ, a potential of 100 MV, and an average duration of 500 ms. Find the average current.

4-4-5. Chest resistance What is the resistance of the average adult human body measured across the chest from side to side with 100 mm square electrodes placed just under the armpits? See Table 4-1 for conductivities.

4-4-6. Heart defibrillator The heart is a mechanical pump which is bioelectrically controlled. During a heart attack the action of the heart muscle deteriorates from a regular periodic contraction to a convulsive quiver called *fibrillation*. To restore normal heart action to a human subject a capacitor storing typically 400 J may be discharged through electrodes, called *paddles*, placed across the chest as in Prob. 4-4-5. See also Fig. 3-22 showing the heart dipole field. If the current pulse lasts 3 ms, find (*a*) average current and (*b*) applied voltage. Refer to Prob. 4-4-5 for the chest resistance. See P. I. Bennett and V. C. Jones, Portable Defibrillator-Monitor for Cardiac Resuscitation, *Hewlett-Packard Journal*, February 1982, pp. 22–28.

4-4-7. Electric eel A freshwater eel may develop a potential of 500 V between electrodes on its body spaced 750 mm apart. If the eel's internal battery resistance is 15 Ω, find (*a*) resulting current and (*b*) power developed.

THE STATIC MAGNETIC FIELD OF STEADY ELECTRIC CURRENTS

5-1 INTRODUCTION

A static electric charge has an electric field, as discussed in Chaps. 2 and 3. A moving charge constitutes an electric current and possesses a *magnetic field*. Thus, a wire carrying a current I is surrounded by a region in which forces act on a magnet or compass needle, as suggested in Fig. 5-1, a discovery made by Hans Christian Oersted in 1819. Exploring this field with a compass needle, we find that the needle always turns so as to be perpendicular to the wire and to a radial line extending out from the wire. This alignment is parallel to the magnetic field. Proceeding in the direction of the needle, we find that the magnetic field forms *closed* circular loops around the wire.

The direction of the magnetic field is taken to be the direction indicated as "north" (N) by the compass needle, as in Fig. 5-1b. The relation of the magnetic field direction to the current direction can be easily remembered by means of the *right-hand rule*. With the thumb pointing in the direction of the current, as in Fig.

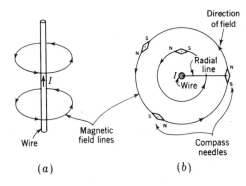

Wire

Magnetic
field lines

Compass
needles

(a)   (b)

Figure 5-1 (*a*) Magnetic field around wire carrying a current, (*b*) Cross section perpendicular to the wire. The current is flowing out of the page.

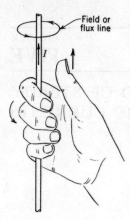

Figure 5-2 Right-hand rule relating direction of field or flux lines (fingers) to direction of current I (thumb).

5-2, the fingers of the right hand encircling the wire point in the direction of the magnetic field, or lines of magnetic flux.

5-2 EFFECT OF A MAGNET ON A CURRENT-CARRYING WIRE

If a wire is placed in the uniform field of a permanent magnet, as shown in Fig. 5-3a, there will be a force on the wire as soon as the switch is closed and current flows.† This force is basic to the operation of electric motors and is called the *motor force*.

The inward current of the wire produces a clockwise magnetic field as suggested in Fig. 5-3b, reinforcing the magnet's field above the wire and weakening it below, as suggested in Fig. 5-3c. If we now imagine, as did Michael Faraday, that the field lines are like stretched rubber bands, we may conclude that the force on the wire is *downward*.

The force F on the wire with current I is given by

$$F = LIB \qquad \text{(N)} \tag{1}$$

where L = length of wire in magnetic field (width of pole pieces), m
I = current in wire, A
B = a factor involving the magnetic field

From (1)

$$B = \frac{F}{IL} = \frac{\text{force}}{\text{current moment}} \qquad \text{(N A}^{-1}\text{ m}^{-1}\text{ or T)} \tag{2}$$

† Because of fringing effects the field will not be uniform at the edges. However, if the pole cross section is large compared to the gap spacing, the effects will be small.

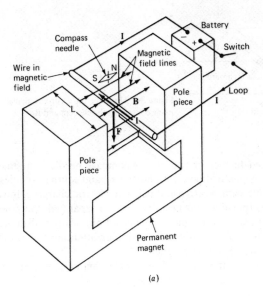

(a)

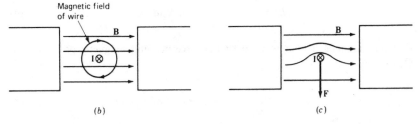

(b)   (c)

Figure 5-3 (a) When the switch is closed and a current flows in the loop, there is a force (downward) on the wire in the magnetic field (between the poles). This is the field-current interaction basic to the operation of electric motors. Note that the compass needle indicates the direction of the magnetic field. (b) The wire's current produces a field in the clockwise direction, as indicated, reinforcing the magnetic field above and opposing (weakening) it below, resulting in a downward force as suggested in (c).

Thus, B can be described as a *force per current moment*. However, as will be discussed later, it is customary to call it the *magnetic flux density*. The unit is the weber per square meter (Wb m^{-2}), or tesla (T) (1 Wb m^{-2} = 1 T).

In Fig. 5-3, **I**, **B**, and **F** are mutually perpendicular. If **I** is not perpendicular to **B**, we find that **F** is a function of ϕ (see Fig. 5-4). In general, for any infinitesimal current-carrying element dl

$$dF = BI\, dl \sin \phi \tag{3}$$

This equation and $F = LIB$ are the basic *motor equations*.

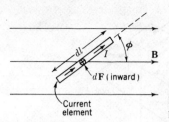

Figure 5-4 The force on a current element is normal to the plane containing the element and **B**.

As discussed further in Sec. 6-2, a compass needle is a small bar magnet which forms a magnetic dipole with north ($+$) and south ($-$) poles, analogous to an electric dipole with positive and negative electric charges. However, although electric charges (positive and negative) can be separated, the poles of a magnet cannot. Thus, an isolated magnetic pole is not physically realizable, but it may be approximated in some situations by confining attention to regions close to one end of a very long needlelike magnet.

Assuming an isolated magnetic pole of strength Q_m, we can write by analogy to the electric field as a force F per charge Q

$$E = \frac{F}{Q} \qquad (\text{N C}^{-1} \text{ or V m}^{-1}) \tag{4}$$

that the magnetic field B is given by the force per unit magnetic pole strength Q_m as

$$B = \frac{F}{Q_m} \qquad [\text{N (A m)}^{-1} \text{ or T}] \tag{5}$$

where the force on the pole is in the direction of the field B.

Comparing (5) with (2), we note that the magnitude of the pole strength Q_m is equal to a current moment ($IL = $ current $\times$ distance). Since current is electric charge per unit time t, we have (for the magnitudes) that

$$B = \frac{F}{Q_m} = \frac{F}{IL} = \frac{F}{\dfrac{Q}{t} L} = \frac{F}{Qv} \tag{6}$$

where $v = L/t = $ velocity of the charge, m s^{-1}.

Thus, if a charge Q moving with a velocity v experiences a force F, there must be a magnetic field $B = F/Qv$, the force being perpendicular to both the field and the direction of motion of the charge.

If Q were stationary (static), there would be no force. A moving charge (or current) is required for an interaction.

5-3 THE MAGNETIC FIELD OF A CURRENT-CARRYING ELEMENT; THE BIOT–SAVART LAW

Let the aligning *torque* on an arbitrarily small perfectly mounted magnetic needle be used to measure the field B produced by an incremental current-carrying element of length Δl, as in Fig. 5-5. From these measurements we find (for $r \gg \Delta l$) that the incremental B is a function of I, Δl, r, and θ, as given by

$$\Delta B = k \frac{I \, \Delta l \sin \theta}{r^2} \tag{1}$$

where k is a constant of proportionality given by

$$k = \frac{\mu}{4\pi} \tag{2}$$

where μ is the *permeability* of the medium. By dimensional analysis of (1) we find that μ has the dimensions of flux per current divided by length. It will be shown in

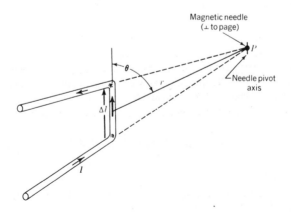

Figure 5-5 Arrangement for measuring **B** produced by short current-carrying element Δl as a function of radius r, angle θ, current **I**, and length Δl. **B** is determined by measuring the aligning torque on an arbitrarily small magnetic needle (normal to page) at P. The wires supplying current to the element Δl are always arranged so that they are radial with respect to P, and thus do not affect **B** at P.

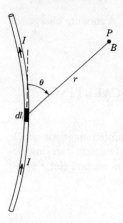

Figure 5-6 Construction for calculating flux density **B** at a point P due to a current **I** in a long conductor.

Sec. 5-12 that inductance has the dimensions of flux per current. Therefore permeability has the dimensions of inductance divided by length. The SI unit for permeability is the henry per meter.† The permeability of vacuum is

$$\mu_0 = 4\pi \times 10^{-7} \text{ H m}^{-1} = 400\pi \text{ nH m}^{-1}$$

Introducing (2) in (1) and writing infinitesimals instead of incrementals, we obtain the more fundamental relation

$$dB = \frac{\mu}{4\pi} \frac{I \, dl \sin \theta}{r^2} \tag{3}$$

The direction of dB is normal to the page (inward at P).

In case we wish to know B at a point P, as in Fig. 5-6, due to a current I in a long, straight, or curved conductor contained in the plane of the page, we assume that the conductor is made up of elements or segments of infinitesimal length dl connected in series. The total flux density B at the point P is then the sum of the contributions from all these elements and is expressed by the integral of (3). Thus

$$B = \frac{\mu I}{4\pi} \int \frac{\sin \theta}{r^2} \, dl \tag{4}$$

† Recall that permittivity ϵ has the dimensions of *capacitance* per length and is expressed in farads per meter. Note that

$$\frac{1/4\pi\epsilon_0}{\mu_0/4\pi} = \frac{1}{\mu_0\epsilon_0} = c^2$$

where c is the velocity of light or radio waves equal to 300 Mm s^{-1}.

where B = flux density at P, N A^{-1} m^{-1} or T
   μ = permeability of medium, H m^{-1}
   I = current in conductor, A
   dl = length of current element, m
   r = distance from element to P, m
   θ = angle measured clockwise from positive direction of current along dl to direction of radius vector r extending from dl to P

The integration is carried out over the length of the conductor. Equations (3) and (4) are statements of the *Biot-Savart law.*

5-4 THE MAGNETIC FIELD OF AN INFINITE LINEAR CONDUCTOR

The magnetic field or flux density B at a radius R from a thin linear conductor of infinite length with a current I can be obtained from (5-3-4).

   The geometry is shown in Fig. 5-7. With the current I as indicated, **B** at the right of the wire is into the page. This is according to the right-hand rule. Since $dl \sin \theta = r \, d\theta$ and $R = r \sin \theta$, we have from the Biot-Savart law (5-3-4) that

$$B = \frac{\mu I}{4\pi} \int_0^\pi \frac{1}{r} \, d\theta = \frac{\mu I}{4\pi R} \int_0^\pi \sin \theta \, d\theta \tag{1}$$

where the integration is between the angles $\theta = 0$ and $\theta = \pi$, that is, over the entire length of an infinite wire. Integrating gives

$$B = \frac{\mu I}{4\pi R} \left[-\cos \theta \right]_0^\pi = \frac{\mu I}{4\pi R} \, 2$$

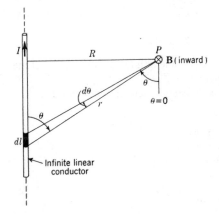

Figure 5-7 Construction for finding flux density **B** near a long straight wire. The symbol $\otimes$ (tail of an arrow) indicates a direction (in this case of **B**) into the page. The symbol $\odot$ (head of an arrow) indicates a direction out of the page.

or
$$B = \frac{\mu I}{2\pi R} \tag{2}$$

where B = flux density, N A^{-1} m^{-1} or T
μ = permeability of medium, H m^{-1}
I = current in conductor, A
R = radial distance, m

5-5 THE FORCE BETWEEN TWO PARALLEL LINEAR CONDUCTORS; DEFINITION OF THE AMPERE

Consider a length l of two long parallel linear conductors in air spaced a distance R as in Fig. 5-8a. Assume that conductor 1 carries a current I and conductor 2 a current I' in the opposite direction. The resulting magnetic field is stronger between the wires than outside, as suggested in Fig. 5-8b. Hence, using Michael Faraday's reasoning, the wires are repelled. If the currents were in the same direction, the forces would be reversed and the conductors would be attracted.

The magnitude F of the force on a length l of conductor 2 is

$$F = I'B \int_0^l dl = I'Bl \tag{1}$$

where I' = current in conductor 2
B = flux density at conductor 2 produced by current I in conductor 1

Introducing the value of B from (5-4-1) gives

$$F = \frac{\mu_0 II'}{2\pi R} l \tag{2}$$

where F = force on length l of conductor 2, N
I = current in conductor 1, A
I' = current in conductor 2, A
R = separation of conductors, m
μ_0 = permeability of vacuum or air = 400π nH m^{-1}

Since (2) is symmetrical in I and I', the force on a length l of conductor 1 is of the same magnitude as the force F on conductor 2.

Dividing (2) by l yields the *force per unit length* on either conductor as

$$\frac{F}{l} = \frac{\mu_0 II'}{2\pi R} \tag{3}$$

If $I' = I$, and introducing the value for μ_0,

$$F = 2 \times 10^{-7} \frac{I^2 l}{R} \quad \text{(N)} \tag{4}$$

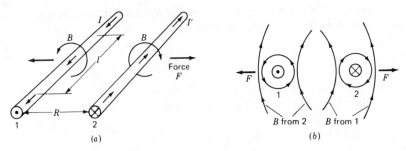

Figure 5-8 Two parallel wires carrying opposite currents as in (a) are repelled since the magnetic fields are reinforced between the wires but are opposed (weakened) outside as suggested in (b).

If $l = R = 1$ m and $F = 2 \times 10^{-7}$ N, then $I = 1$ A. This measurement is used to define the ampere in the SI system (see Sec. 1-3).

5-6 THE MAGNETIC FIELD OF A CURRENT-CARRYING LOOP

Let the loop be in the xy plane with its center at the origin, as in Fig. 5-9, so that the z axis coincides with the loop axis. The loop has a radius R and current I. At the point P on the loop axis the contribution dB produced by an element of length dl of the loop is, from (5-3-3),

$$dB = \frac{\mu I \, dl \sin \theta}{4\pi r^2} \tag{1}$$

where θ is the angle between dl and the radius vector of length r. It is assumed that the loop is in a medium of uniform permeability μ. The direction of $d\mathbf{B}$ is normal to

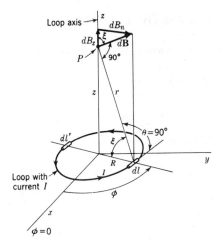

Figure 5-9 Construction for finding flux density B on axis of current loop.

the radius vector of length r, that is, at an angle ξ with respect to the loop or z axis. The component dB_z in the direction of the z axis is given by

$$dB_z = dB \cos \xi = dB \frac{R}{r} \qquad (2)$$

From Fig. 5-9 we note that $\theta = 90°$, $dl = R\, d\phi$, and $r = \sqrt{R^2 + z^2}$. Introducing these values into (1) and substituting this value for dB in (2), we have

$$dB_z = \frac{\mu I R^2}{4\pi (R^2 + z^2)^{3/2}}\, d\phi \qquad (3)$$

The total flux density B_z in the z direction is then the integral of (3) around the entire loop. The element dl also produces a component of dB_n normal to the axis of the loop. Integrating this component for all elements around the loop yields zero because of symmetry. Hence, B_z equals the total flux density B at the point P as given by

$$B = B_z = \frac{\mu I R^2}{4\pi (R^2 + z^2)^{3/2}} \int_0^{2\pi} d\phi = \frac{\mu I R^2}{2(R^2 + z^2)^{3/2}} \qquad (4)$$

At the center of the loop, $z = 0$, and

$$B = \frac{\mu I}{2R} \qquad (5)$$

where B = flux density at center of loop, N A^{-1} m^{-1} or T
   μ = permeability of medium, H m^{-1}
   I = current in loop, A
   R = radius of loop, m

5-7 MAGNETIC FLUX ψ_m AND MAGNETIC FLUX DENSITY B

The magnetic field quantity **B** introduced above as a force per current moment can also be regarded as a *magnetic flux density*, and this term is in more common use. Thus, we describe **B** as the magnetic flux per unit area or

$$B = \frac{\psi_m}{A} \qquad (1)$$

In (1) it is assumed that the magnetic field lines are perpendicular to A. More generally we have (see Fig. 5-10)

$$\psi_m = BA \cos \alpha \qquad (2)$$

where ψ_m = magnetic flux through area A
   B = magnitude of magnetic flux density **B**
   α = angle between normal to area A and direction of **B**

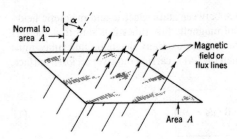

Normal to area A

Magnetic field or flux lines

Area A

Figure 5-10 Flux lines and area A.

Dimensionally

$$\psi_m = \frac{\text{force}}{\text{current moment}} \times \text{area}$$

or in dimensional symbols

$$\frac{ML}{T^2}\frac{L^2}{IL} = \frac{ML^2}{IT^2}$$

The SI unit of magnetic flux ψ_m is the weber (Wb). The unit of magnetic flux density **B** is the weber per square meter (Wb m^{-2}) or tesla (T) (dimensional symbols M/IT^2).

If **B** is not uniform over an area, the simple product (2) must be replaced by a surface integral, so that, in general, we have

$$\psi_m = \iint B \cos \alpha \, ds \tag{3}$$

where ds = element of surface area

B = magnitude of **B**

α = angle between normal to ds and direction of **B**

Equation (3) can also be written as a scalar, or dot, product. Thus,

$$\boxed{\psi_m = \iint \mathbf{B} \cdot d\mathbf{s}} \tag{4}$$

where ψ_m = magnetic flux, Wb

B = magnetic flux density, Wb m^{-2} or T or N A^{-1} m^{-1}

$d\mathbf{s}$ = vector with direction normal to surface element ds and magnitude equal to area of ds, m^2

5-8 MAGNETIC FLUX OVER A CLOSED SURFACE

The flux tubes of a static electric field originate and end on electric charges. On the other hand, tubes of magnetic flux are continuous; i.e., they have no sources or sinks. This is a fundamental difference between static electric and magnetic fields. To describe this continuous nature of magnetic flux tubes it is said that the flux density **B** is *solenoidal*. Since it is continuous, as many magnetic flux tubes must enter a volume as leave it. Hence, when (5-7-4) is carried out over a *closed* surface, the result must be zero, or

$$\oint_s \mathbf{B} \cdot d\mathbf{s} = 0$$

(1)†

This relation may be regarded as Gauss' law applied to magnetic fields [compare with (2-17-6) for electric fields].

It follows that the divergence of **B** equals zero. That is,

$$\nabla \cdot \mathbf{B} = 0$$

(2)

Both (1) and (2) are expressions of the continuous nature of **B**. Both are also members of the group known as Maxwell's equations, being expressions of *Maxwell's equation from Gauss' law for magnetic fields*. In Secs. 2-17 and 3-22 we discussed Maxwell's equation from Gauss' law for electric fields.

5-9 MAGNETIC FIELD RELATIONS IN VECTOR NOTATION; CROSS PRODUCT

A linear current-carrying conductor of length L placed in a uniform magnetic field experiences a force F that is given, from (5-2-3), by

$$F = IB \sin \phi \int_0^L dl = IBL \sin \phi$$

(1)

where F = force, N
$\quad I$ = current in conductor, A
$\quad B$ = flux density of field, T
$\quad L$ = length of conductor, m
$\quad \phi$ = angle between **I** and **B**

Equation (1) is a scalar equation and relates only the magnitudes of the quantities involved. The force **F** is perpendicular to both **I** and **B**. For example, let the conductor be normal to a uniform magnetic field of flux density **B** as in Fig. 5-11a. If the current in the conductor flows out of the page, it produces a magnetic field as indicated, which reinforces the field below the wire and weakens it above. Hence, there will be an *upward* force on the wire, as suggested in Fig. 5-11b.

---

† The symbol $\oint_s$ indicates an integral over a closed surface.

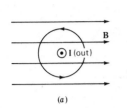

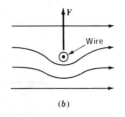

Figure 5-11 (*a*) Current-carrying wire in uniform magnetic field. (*b*) Field of current in wire decreases the field above the wire and increases it below resulting in upward force on wire.

Relating the directions to the coordinate axes as in Fig. 5-12*a*, **F** is in the positive *z* direction when **I** is in the positive *x* direction and **B** in the positive *y* direction. If the direction of **I** is not perpendicular to the direction of **B** but is as shown in Fig. 5-12*b*, the force **F** is still in the positive *z* direction with a magnitude given by (1), where ϕ equals the angle measured from the positive direction of **I** to the positive direction of **B** (counterclockwise in Fig. 5-12*b*). With ϕ measured in this way, the force **F** is in the positive *z* direction if sin ϕ is positive.

A more concise method of expressing the relation is by means of the *vector* or *cross product*.

The cross product of two vectors is defined as a third vector whose magnitude is equal to the product of the vector magnitudes and the sine of the angle between them. The direction of the third vector is perpendicular to the plane of the two vectors and in such a sense that the three vectors form a right-handed set.

Using the vector product, we can state the relation as

$$\mathbf{F} = (\mathbf{I} \times \mathbf{B})L \qquad (2)$$

For an elemental length of conductor this becomes†

$$\boxed{d\mathbf{F} = (\mathbf{I} \times \mathbf{B})\,dl} \qquad (3)$$

where $d\mathbf{F}$ = vector indicating magnitude and direction of force on element of conductor, N

$\quad$ **I** = vector indicating magnitude and direction of current in conductor, A

$\quad dl$ = length of conductor, m

$\quad$ **B** = vector indicating magnitude and direction of the flux density, T

Equations (2) and (3) relate both the magnitudes and the directions of the quantities involved, whereas (1) related only the magnitudes.

A point charge Q moving with velocity v is equivalent to a current moment IL. Thus,

$$IL = Qv$$

---

† For a volume distribution of current we have

$$d\mathbf{F} = (\mathbf{J} \times \mathbf{B})\,dv \qquad (4)$$

where $d\mathbf{F}$ is the force on the volume element dv at which the current density is **J**.

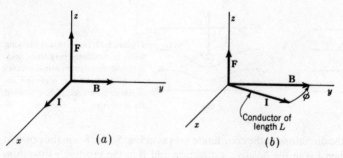

Figure 5-12 Relation between current direction **I**, field direction **B**, and force **F**.

and (2) can also be written

$$\mathbf{F} = Q(\mathbf{v} \times \mathbf{B}) \qquad (\text{N}) \qquad (5)$$

If, in addition to the magnetic field there is also an electric field **E**, we can write that the total force on the particle is given by

$$\mathbf{F} = Q\mathbf{E} + Q(\mathbf{v} \times \mathbf{B}) \qquad (\text{N}) \qquad (6)$$

where **F** is the *Lorentz force* after H. A. Lorentz. This relation, together with Newton's second law (force = mass × acceleration), is basic to calculations of charged-particle motion.

Equation (5-3-4) gives the magnitude of **B** at a point as produced by a current *I* in a straight or curved conductor contained in a single plane. A more general relation applying to a conducting wire of any shape, as in Fig. 5-13, can be stated with the aid of the vector product:

$$\mathbf{B} = \frac{\mu}{4\pi} \int \frac{\mathbf{I} \times \hat{\mathbf{r}}}{r^2} \, dl \qquad (7)$$

where **B** = flux density at *P*, T
 μ = permeability of medium, H m^{-1}
 I = current in conductor (vector pointing in positive direction of current at element *dl*), A
 $\hat{\mathbf{r}}$ = unit vector pointing from element *dl* to point *P*, dimensionless
 r = distance from *dl* to *P*, m
 dl = infinitesimal element of length of conductor, m

The integration in (7) is carried out over the length of conductor under consideration.

If the current is distributed throughout a volume, the flux density **B** is given by

$$\mathbf{B} = \frac{\mu}{4\pi} \iiint \frac{\mathbf{J} \times \hat{\mathbf{r}}}{r^2} \, dv \qquad (8)$$

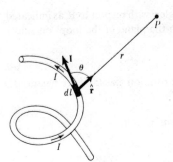

Figure 5-13 Relation for finding **B** at a point P due to a current I in a conductor of any shape.

where **J** is the current density in a volume element dv at a distance r. Equations (3), (7), and (8) are basic magnetic field relations.

5-10 TORQUE ON A LOOP; MAGNETIC MOMENT

When a current loop is placed parallel to a magnetic field, forces act on the loop that tend to rotate it. The tangential force times the radial distance at which it acts is called the *torque*, or mechanical moment, on the loop. Torque (or mechanical moment) has the dimensions of force times distance and is expressed in newton-meters (N m).

Consider the rectangular loop shown in Fig. 5-14 with sides of length l and d situated in a magnetic field of uniform flux density **B**. The loop has a steady current I. According to (5-9-3), the force on any element of the loop is

$$d\mathbf{F} = (\mathbf{I} \times \mathbf{B})\, dl \qquad \text{(N)} \qquad\qquad (1)$$

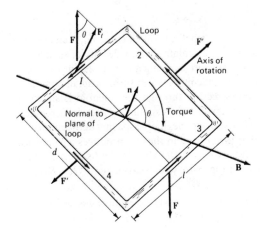

Figure 5-14 Rectangular loop in uniform field **B** experiences a torque tending to align its normal **n̂** with **B**.

If the normal $\hat{\mathbf{n}}$ to the plane of the loop is at an angle θ with respect to $\mathbf{B}$, as indicated in Fig. 5-14, the turning force (perpendicular to the plane of the loop) on side 1 (or side 3) of the loop is

$$F_t = IBl \sin \theta \qquad (2)$$

Since the forces F' on sides 2 and 4 are along the rotation axis (and are balanced), the total torque on the loop is

$$T = 2F_t \frac{d}{2} = IBld \sin \theta \qquad (3)$$

But ld equals the area A of the loop; so

$$T = IAB \sin \theta \qquad (\text{N m}) \qquad (4)$$

According to (4), the torque is proportional to the current in the loop, to its area, and to the flux density of the field in which the loop is situated.

The product IA in (4) has the dimensions of current times area and is the *magnetic moment* of the loop. Its dimensional symbols are IL^2, and it is expressed in amperes times square meters (A m^2). Let us designate magnetic moment by the letter m. Then† $m = IA$, and

$$T = mB \sin \theta \qquad (5)$$

where θ is the angle between the normal to the plane of the loop and the direction of $\mathbf{B}$ (see Fig. 5-14). If the loop has N turns, the magnetic moment $m = NIA$.

The magnetic moment may be regarded as a vector $\mathbf{m} = \hat{\mathbf{n}}m$, where the unit vector $\hat{\mathbf{n}}$ is perpendicular to the plane of the loop in the direction given by the right-hand rule (fingers in the direction of the current, thumb in the direction of $\hat{\mathbf{n}}$). Then, using the cross product, (5) can be expressed

$$\mathbf{T} = \hat{\mathbf{n}}m \times \mathbf{B} = \mathbf{m} \times \mathbf{B} \qquad (6)$$

The torque on a loop is compared with the torque on a magnetic dipole in Sec. 6-2.

5-11 THE SOLENOID

Invented in 1820 by André Marie Ampère, a helical coil, or solenoid, is an effective way of producing a magnetic field. Let us calculate the flux density of such a coil.

Let the coil consist of N turns of thin wire carrying a current I. The coil has a length l and radius R (Fig. 5-15a). The spacing between turns is small compared with the radius R of the coil. A cross section through the solenoid is shown in Fig. 5-15b. If the spacing between turns is sufficiently small, or if the wire is replaced by a thin conducting strip of width l/N and with negligible spacing between turns as in Fig. 5-15c, one may consider that the current in the coil produces a cylindrical current sheet with a linear current density $K = NI/l\,(\text{A m}^{-1})$. This is equivalent to a single turn of a conducting sheet, as in Fig. 5-16, with total current $NI(=Kl)$.

---

† Although the loop in Fig. 5-14 has a rectangular area, (4) and (5) apply regardless of the shape of the loop area.

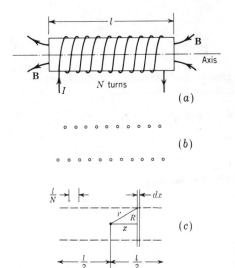

Figure 5-15 (a) Solenoid coil, (b) cross section, and (c) coil with strips.

To find the flux density B at the center of the solenoid, let a section of the coil of length dx, as in Fig. 5-15c, be regarded as a single-turn loop with a current equal to

$$K \, dx = \frac{NI}{l} \, dx \tag{1}$$

From (5-6-4) the flux density dB at the center of the solenoid due to this loop of length dx at a distance x from the center is

$$dB = \frac{\mu NIR^2}{2l(R^2 + x^2)^{3/2}} \, dx \tag{2}$$

The total flux density B at the center of this coil is then equal to this expression integrated over the length of the coil. That is,

$$B = \frac{\mu NIR^2}{2l} \int_{-l/2}^{+l/2} \frac{dx}{(R^2 + x^2)^{3/2}} \tag{3}$$

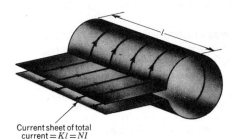

Current sheet of total
current $= Kl = NI$

Figure 5-16 Equivalent current sheet of solenoid.

Performing the integration gives

$$B = \frac{\mu NI}{\sqrt{4R^2 + l^2}} \tag{4}$$

If the length of the solenoid is much greater than its radius ($l \gg R$), (4) reduces to

$$B = \frac{\mu NI}{l} = \mu K \quad \text{and} \quad \frac{NI}{l} = K \tag{5}$$

where B = flux density, T
  μ = permeability of medium, H m^{-1}
  N = number of turns on solenoid, dimensionless
  I = current through solenoid, A
  l = length of solenoid, m
  K = sheet current density, A m^{-1}

Equations (4) and (5) give the flux density at the center of the solenoid. By changing the limits of integration in (3) to 0 and l we obtain the flux density at one end of the coil (on the axis) as

$$B = \frac{\mu NI}{2\sqrt{R^2 + l^2}} \tag{6}$$

For $l \gg R$ this reduces to

$$B = \frac{\mu NI}{2l} = \frac{1}{2}\mu K \tag{7}$$

which is one-half the value at the center of the coil as given by (5).

It may be shown (see Prob. 5-1-12) that the torque on a solenoid in a uniform magnetic field B is

$$T = NIAB \sin \theta \tag{8}$$

where θ is the angle between the solenoid axis and **B**. The torque on a solenoid is compared with the torque on a magnetic dipole in Sec. 6-2.

5-12 INDUCTORS AND INDUCTANCE

An inductor† is a device for storing energy in a magnetic field. It may be regarded as the magnetic counterpart of a capacitor, which stores energy in an electric field. As examples, loops, coils, and solenoids are inductors.

---

† An *inductor* is sometimes called an inductance. However, it is usual practice to refer to a coil or solenoid as an inductor. This makes for uniform usage when we speak, for example, of an *inductor* of 1 H *inductance*, a *capacitor* of 1 μF *capacitance*, or a *resistor* of 1 Ω *resistance*.

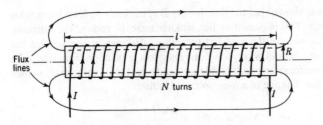

Figure 5-17 Solenoid and magnetic flux lines.

The magnetic lines produced by a current in a solenoidal coil form closed loops, as suggested in Fig. 5-17. Each line that passes through the entire solenoid as in the figure links the current N times. If all the lines link all the turns, the total *magnetic flux linkage* Λ (capital lambda) of the coil is equal to the total magnetic flux ψ_m through the coil times the number of turns, or

$$\text{Flux linkage} = \Lambda = N\psi_m \quad \text{(Wb turns)}$$

By definition the *inductance L* is the ratio of the total magnetic flux linkage to the current I through the inductor, or

$$L = \frac{N\psi_m}{I} = \frac{\Lambda}{I} \qquad (1)$$

This definition is satisfactory for a medium with a constant permeability, such as air. As discussed in Chap. 6, however, the permeability of ferrous media is not constant, and in this case the inductance is defined as the ratio of the infinitesimal change in flux linkage to the infinitesimal change in current producing it, or

$$L = \frac{d\Lambda}{dI} \qquad (2)$$

Inductance has the dimensions of magnetic flux (linkage) divided by current. The unit of inductance is the *henry* (H). Thus,

$$\text{Henrys} = \frac{\text{webers}}{\text{ampere}}$$

The dimensional symbols for inductance are ML^2/I^2T^2.

5-13 INDUCTANCE OF SIMPLE GEOMETRIES

The inductance of many inductors can be readily calculated from their geometry. As examples, expressions for the inductance of a long solenoid, a toroid, a coaxial line, and a two-wire line will be derived in this section.

In Sec. 5-11 it was shown that the flux density B at the end of a long solenoid is less than at the center. This is caused by flux leakage near the ends of the solenoid. However, this leakage is mostly confined to a short distance at the ends of the solenoid, so that if the solenoid is very long, one may, to a good approximation, take B constant over the entire interior of the solenoid and equal to its value at the center (5-11-5). The total flux linkage of a long solenoid is then

$$\Lambda = N\psi_m = NBA = \frac{\mu N^2 I A}{l} \tag{1}$$

Thus, the inductance of a long solenoid (see Fig. 5-17) is

$$L = \frac{\Lambda}{I} = \frac{\mu N^2 A}{l} \tag{2}$$

where L = inductance of solenoid, H
   Λ = flux linkage, Wb turns
   I = current through solenoid, A
   μ = permeability of medium, H m^{-1}
   N = number of turns on solenoid, dimensionless
   A = cross-sectional area of solenoid, m^2
   l = length of solenoid, m

Example Calculate the inductance of a solenoid of 2,000 turns wound uniformly over a length of 500 mm on a cylindrical paper tube 40 mm in diameter. The medium is air ($\mu = \mu_0$).

SOLUTION From (2) the inductance of the solenoid is

$$L = \frac{4\pi \times 10^{-7} \times 4 \times 10^6 \times \pi \times 4 \times 10^{-4}}{0.5} = 12.6 \text{ mH}$$

If a long solenoid is bent into a circle and closed on itself, a toroidal coil, or toroid, is obtained. When the toroid has a uniform winding of many turns, the magnetic lines of flux are almost entirely confined to the interior of the winding, B being substantially zero outside. If the ratio R/r (see Fig. 5-18) is large, one may calculate B as though the toroid were straightened out into a solenoid. Thus, the flux linkage is

$$\Lambda = N\psi_m = NBA = N\frac{\mu N I}{2\pi R} \pi r^2 = \frac{\mu N^2 I \pi r^2}{2\pi R} = \frac{\mu N^2 r^2 I}{2R} \tag{3}$$

The inductance of the toroid is then

$$L = \frac{\Lambda}{I} = \frac{\mu N^2 r^2}{2R} \tag{4}$$

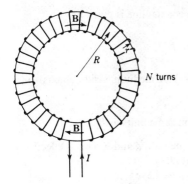

N turns

Figure 5-18 Toroid.

where L = inductance of toroid, H
   μ = permeability (uniform and constant) of medium inside coil, H m^{-1}
   N = number of turns of toroid, dimensionless
   r = radius of coil (see Fig. 5-18), m
   R = radius of toroid, m

Consider next a coaxial transmission line constructed of conducting cylinders of radius a and b, as in Fig. 5-19. The current on the inner conductor is I. The return current on the outer conductor is of the same magnitude. The flux density B at any radius r is the same as at this radius from a long straight conductor with the same current, or

$$B \text{ (at } r) = \frac{\mu I}{2\pi r} \tag{5}$$

The total flux linkage for a length d of line is then d times the integral of (5) from the inner to the outer conductor, or

$$\Lambda = d \int_a^b B \, dr = \frac{d\mu I}{2\pi} \int_a^b \frac{dr}{r} = \frac{d\mu I}{2\pi} \ln \frac{b}{a} \tag{6}$$

Hence, the inductance of a length d of the coaxial line is

$$L = \frac{\Lambda}{I} = \frac{\mu d}{2\pi} \ln \frac{b}{a} \quad \text{(H)} \tag{7}$$

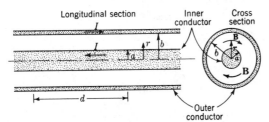

Longitudinal section   Inner   Cross
                       conductor   section

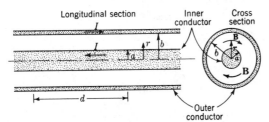

Outer
conductor

Figure 5-19 Coaxial transmission line.

or the inductance per unit length (L/d) for the coaxial line is given by

$$\frac{L}{d} = \frac{\mu}{2\pi} \ln \frac{b}{a} \quad \text{(H m}^{-1}\text{)} \tag{8}$$

where μ = permeability (uniform and constant) of medium inside coaxial line, H m^{-1}

b = inside radius of outer conductor

a = outside radius of inner conductor (in same units as b)

It is assumed that the currents are confined to the radii a and b. This is effectively the case when the walls of the conductors are thin.

Evaluating (8) for an air-filled line ($\mu = \mu_0$), we have

$$\frac{L}{d} = 0.2 \ln \frac{b}{a} = 0.46 \log \frac{b}{a} \quad (\mu\text{H m}^{-1}) \tag{9}$$

Let us consider finally a two-wire transmission line as illustrated in Fig. 5-20. The conductor radius is a, and the spacing between centers is D. At any radius r from one of the conductors the flux density B due to that conductor is given by (5). The total flux linkage due to both conductors for a length d of line is then d times twice the integral of (5) from a to D, or

$$\Lambda = 2d \int_a^D B \, dr = \frac{\mu I d}{\pi} \int_a^D \frac{dr}{r} = \frac{\mu I d}{\pi} \ln \frac{D}{a} \tag{10}$$

Hence, the inductance of a length d of the two-conductor line is

$$L = \frac{\Lambda}{I} = \frac{\mu d}{\pi} \ln \frac{D}{a} \tag{11}$$

or the inductance per unit length of line (L/d) is given by

$$\frac{L}{d} = \frac{\mu}{\pi} \ln \frac{D}{a} \quad \text{(H m}^{-1}\text{)} \tag{12}$$

where μ = permeability (uniform and constant) of medium, H m^{-1}

D = spacing between centers of conductors

a = radius of conductors (in same units as D)

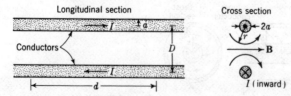

Figure 5-20 Two-wire transmission line.

It is assumed that the current is confined to a radius a. This is effectively the case when the walls of the conductors are thin.

Evaluating (12) for a medium of air ($\mu = \mu_0$), we have

$$\frac{L}{d} = 0.4 \ln \frac{D}{a} = 0.92 \log \frac{D}{a} \qquad (\mu\text{H m}^{-1}) \tag{13}$$

5-14 AMPÈRE'S LAW AND H

According to (5-4-2), the flux density B at a distance R from a long straight conductor (Fig. 5-21) is given by

$$B = \frac{\mu I}{2\pi R} \tag{1}$$

where μ = permeability of medium, H m^{-1}
 I = current in wire, A

If **B** is now integrated around a path of radius R enclosing the wire once, we have

$$\oint \mathbf{B} \cdot d\mathbf{l} = \frac{\mu I}{2\pi R} \oint dl = \frac{\mu I}{2\pi R} 2\pi R = \mu I \tag{2}$$

or
$$\oint \mathbf{B} \cdot d\mathbf{l} = \mu I \tag{3}$$

Relation (3) holds not only in the example considered but also in all cases where the integration is over a singly closed path. Equation (3) may be made independent of the medium by introducing the vector **H** defined as follows:

$$\boxed{\mathbf{H} = \frac{\mathbf{B}}{\mu}} \tag{4}$$

According to (4), **H** and **B** are vectors having the same direction. This is true for all isotropic media.

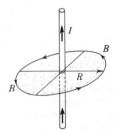

Figure 5-21 Relation of flux density **B** to current **I**.

The quantity **H** is called the magnetic field **H**, the vector **H**, or simply **H**. It has the dimensions of

$$\frac{\text{Flux density}}{\text{Permeability}} = \frac{\text{current}}{\text{length}}$$

The dimensional symbols for **H** are I/L. In SI units **H** is expressed in

$$\frac{\text{Webers/meter}^2}{\text{Webers/ampere-meter}} = \frac{\text{amperes}}{\text{meter}}$$

Introducing (4) into (3) yields

$$\oint \mathbf{H} \cdot d\mathbf{l} = I \tag{5}$$

where **H** = magnetic field, A m^{-1}

$d\mathbf{l}$ = infinitesimal element of path length, m

I = current enclosed, A

This relation is known as *Ampère's law*. In words it states that *the line integral of* **H** *around a single closed path is equal to the current enclosed.*

In the case of a single wire the integration always yields the current I in the wire regardless of the path of integration provided only that the wire is completely enclosed by the path. As illustrations, integration around the two paths at (*a*) and (*b*) in Fig. 5-22 yields I, while integration around the paths at (*c*) and (*d*) yields zero since these paths do not enclose the wire.

Example 1 The magnitude of **H** at a radius of 1 m from a long linear conductor is 1 A m^{-1}. Find the current in the wire.

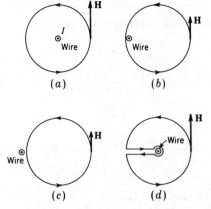

Figure 5-22 Line integral of **H** around closed paths equals current in wire when paths enclose the wire (*a*) and (*b*), but is zero when the paths do not enclose the wire (*c*) and (*d*).

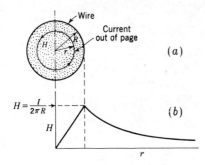

Figure **5-23** H inside and outside current-carrying wire (Example 2).

SOLUTION According to (5), the current in the wire is given by

$$I = \oint \mathbf{H} \cdot d\mathbf{l} = H \times 2\pi R = 2\pi \text{ A}$$

Example 2 A solid cylindrical conductor of radius R has a uniform current density. Derive expressions for H both inside and outside the conductor. Plot the variation of H as a function of radial distance from the center of the wire.

SOLUTION See Fig. 5-23a. Outside the wire ($r \geq R$)

$$H = \frac{I}{2\pi r} \tag{6}$$

Inside the wire the value of H at a radius r is determined solely by the current inside the radius r. Thus, inside the wire ($r \leq R$)

$$H = \frac{I'}{2\pi r} \tag{7}$$

where $I' = I(r/R)^2$ is the current inside radius r. Therefore, inside the wire

$$H = \frac{I}{2\pi R^2} r \tag{8}$$

At the surface of the wire $r = R$, and (8) equals (6). A graph of the variation of H with r is presented in Fig. 5-23b.

5-15 AMPÈRE'S LAW APPLIED TO A CONDUCTING MEDIUM AND MAXWELL'S EQUATION

Ampère's law as discussed in the preceding section may be applied to the more general situation of a path inside a conducting medium. Thus, suppose that the origin of the coordinates in Fig. 5-24 is situated inside a conducting medium of

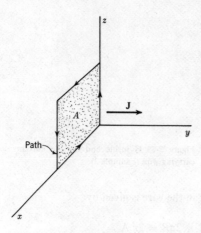

Figure 5-24 Rectangular path in medium with current density $\mathbf{J}$.

large extent. Let the current density in the medium be $\mathbf{J}$ (amperes per square meter) in the positive y direction as shown. According to Ampère's law, the line integral of $\mathbf{H}$ around the rectangular path enclosing the area A (Fig. 5-24) is equal to the current enclosed. In this case, the current I enclosed by the path is given by the integral of the normal component of $\mathbf{J}$ over the surface A, or

$$\oint \mathbf{H} \cdot d\mathbf{l} = \iint_A \mathbf{J} \cdot d\mathbf{s} = I \tag{1}$$

This expression is a generalization of Ampère's law and constitutes one of Maxwell's equations in integral form.

5-16 MAGNETOSTATIC POTENTIAL U AND MMF F

In the absence of emfs, we have from (2-8-4) that the line integral of the static electric field $\mathbf{E}$ around a closed path is zero. That is,

$$\oint \mathbf{E} \cdot d\mathbf{l} = 0 \tag{1}$$

Fields of this type are called *lamellar* and can be derived from a related scalar potential function. Thus, $\mathbf{E}$, due to static charges (written $\mathbf{E}_c$ to be explicit), is derivable as the negative gradient of a scalar electric potential V, or

$$\mathbf{E}_c = -\nabla V \tag{2}$$

Between any two points along a path in the field we have

$$\int_1^2 \mathbf{E}_c \cdot d\mathbf{l} = V_1 - V_2 \tag{3}$$

Although the static magnetic field is not lamellar (since magnetic flux lines form closed loops), it can be treated like a lamellar field if paths of integration are entirely outside current regions and do not enclose any current. Thus, *when no current is enclosed*,

$$\oint \mathbf{H} \cdot d\mathbf{l} = 0 \tag{4}$$

Under this condition, $\mathbf{H}$ can then be derived from a scalar magnetic potential function (or magnetostatic potential) U in the same way that $\mathbf{E}$ is derivable from V. That is,†

$$\mathbf{H} = -\nabla U \tag{5}$$

Between any two points along a path in the field we have

$$\int_1^2 \mathbf{H} \cdot d\mathbf{l} = U_1 - U_2 \tag{6}$$

The scalar potential U has the dimensions of

$$\frac{\text{Current}}{\text{Distance}} \times \text{distance} = \text{current}$$

Hence, U is expressed in amperes.

Returning now to a further consideration of electric fields, we have learned in Sec. 4-10 that if emfs exist in a path of integration,

$$\oint \mathbf{E} \cdot d\mathbf{l} = \mathscr{V} \tag{7}$$

where $\mathbf{E}$ = total field, V m^{-1}
$\mathscr{V}$ = total emf around path, V

In a magnetic field we may write an analogous relation, based on Ampère's law, that *when current is enclosed* by a path of integration,

$$\oint \mathbf{H} \cdot d\mathbf{l} = I = F \qquad \text{(A)} \tag{8}$$

where the quantity F, called the *magnetomotance* or *magnetomotive force* (mmf), is equal to the current enclosed. If the path of integration in (8) encloses a number of turns of wire each with a current I in the same direction, (8) may be written

$$\oint \mathbf{H} \cdot d\mathbf{l} = NI = F \qquad \text{(A turns)} \tag{9}$$

where N = number of turns of wire enclosed, dimensionless
I = current in each turn, A

---

† Since $\mathbf{V} \cdot \mathbf{D} = 0$ in charge-free regions, we obtain Laplace's equation $\nabla^2 V = 0$. In a magnetic field we always have $\mathbf{V} \cdot \mathbf{B} = 0$; so if no current is enclosed, we may write Laplace's equation in the magnetostatic potential as $\mathbf{V}^2 U = 0$.

Table 5-1 Comparison of electric and magnetic field relations

Relation	Electrostatic fields	Magnetostatic fields
Closed path	$\oint \mathbf{E}_c \cdot d\mathbf{l} = 0$ (no emfs)	$\oint \mathbf{H} \cdot d\mathbf{l} = 0$; no current enclosed (Fig. 5-25)
Gradient of scalar potential	$\mathbf{E}_c = -\nabla V$ (V m^{-1}) (no emfs)	$\mathbf{H} = -\nabla U$ (A m^{-1}); in current-free region
Integral between two points	$\int_1^2 \mathbf{E}_c \cdot d\mathbf{l} = V_1 - V_2$ (V) (no emfs)	$\int_1^2 \mathbf{H} \cdot d\mathbf{l} = U_1 - U_2$ (A); path avoids all currents
Closed path	$\oint \mathbf{E} \cdot d\mathbf{l} = \mathscr{V}$ (V) (emfs enclosed)	$\oint \mathbf{H} \cdot d\mathbf{l} = I = F$ (A); path encloses current (Fig. 5-26a) or $\oint \mathbf{H} \cdot d\mathbf{l} = NI = F$ (A turns); path encloses current N times (Fig. 5-26b)

The product NI is expressed in *ampere-turns*, and the mmf in this case has the same units.

The above relations for electric and magnetic fields are summarized in Table 5-1.

When the integration is restricted to current-free regions and to paths that are not closed, the potential U and mmf F are the same. The requirement that the path not link the current can be met by introducing a hypothetical barrier surface in the magnetic field through which the path is not allowed to pass. For example, imagine that a long conductor normal to the page as in Fig. 5-27 carries a current I. Let a barrier surface be constructed that extends from the wire an infinite distance to the left, as suggested in the figure. Now the integral of **H** from points 1 to 2 yields the current I provided 2 and 1 are separated by an infinitesimal distance. Thus

$$\int_1^2 \mathbf{H} \cdot d\mathbf{l} = U_1 - U_2 = I \quad \text{(A)} \tag{10}$$

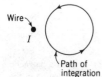

Wire
I
Path of integration

Figure 5-25 Path of integration enclosing no current (see Table 5-1).

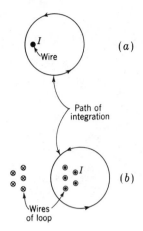

Figure 5-26 (*a*) Path of integration enclosing current **I**. (*b*) Cross section through five-turn loop showing path linking the five turns (see Table 5-1).

The requirement of (4) is still satisfied since the line integral of **H** around the closed path 1231 that avoids crossing the barrier is zero. That is,

$$\oint_{1231} \mathbf{H} \cdot d\mathbf{l} = 0 \tag{11}$$

Both U and F are scalar functions. The potential U is independent of the path of integration; that is, U is a single-valued function of position. This follows from the fact that the path of integration never completely encloses the current and is restricted to current-free regions. If a current-carrying wire is encircled more than once by the path of integration (multiple linking), the result is called the mmf F. It is multiple-valued since its magnitude depends on the number of times the path encircles the wire. Hence F is *not*, in general, independent of the path of integration.

In Fig. 5-27 the barrier surface represents a magnetic equipotential plane. If point 1 is taken arbitrarily as zero potential, then the potential of point 2 on the other side of the barrier is I. Hence, we may construct two surfaces as in Fig. 5-28,

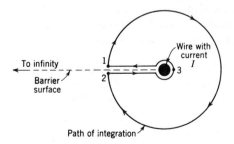

Figure 5-27 Conductor and barrier surface.

176 ELECTROMAGNETICS [CHAP. 5

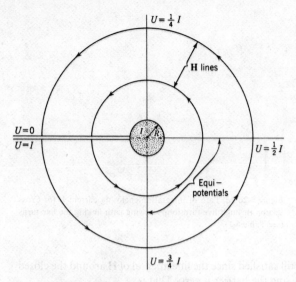

Figure 5-28 Current-carrying wire showing magnetic equipotentials (radial) and field lines (circles).

one with $U = 0$ and the other $U = I$. Other equipotential surfaces are also drawn in Fig. 5-28 for $U = I/4$, $U = I/2$, and $U = 3I/4$. The equipotential surfaces are everywhere normal to **H** and extend from the surface of the wire to infinity. They *do not extend into the interior of the wire.*

5-17 FIELD CELLS AND PERMEABILITY

Consider a transmission line of two flat parallel conducting strips as in Fig. 5-29a. The strips have a width w and a separation l. Each strip carries a current I. The transmission line is shown in cross section in Fig. 5-29b. The field between the strips is uniform, except near the open sides. If equipotentials are drawn in the uniform field region with a spacing equal to the separation l of the line, we may regard the line as being composed of a number of *field-cell transmission lines* (or *transmission-line cells*) arranged in parallel. Each transmission-line cell has a square cross section, as in Fig. 5-29c.

The current in each strip of a line cell is $I' = (l/w)I$, where I is the current in the entire line. Also, across one cell $Hl = I'$. Now the total flux linkage per length d of line is given by $\Lambda = Bld$. The inductance of this length of line is then

$$L_0 = \frac{\Lambda}{I'} = \frac{Bld}{Hl} = \mu d \qquad (1)$$

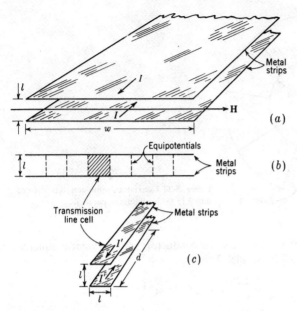

Figure 5-29 Parallel-strip transmission line (*a*) in perspective and (*b*) in cross section. (*c*) Magnetic field cell (or transmission-line cell) with strips of width equal to spacing.

or the inductance per unit length is

$$\boxed{\frac{L_0}{d} = \mu} \tag{2}$$

For air $\mu = \mu_0 = 400\pi$ nH m^{-1}, so that a field-cell transmission line with air as the medium has an inductance per unit length of 400π nH m^{-1}, or 1.26 μH m^{-1}.

Thus, the permeability μ of a medium may be interpreted as the inductance per unit length of a transmission-line cell filled with this medium.

Example Using the field-cell concept, calculate the inductance and also capacitance per unit length of the coaxial transmission line shown in cross section in Fig. 5-30. The line is air-filled.

SOLUTION The inductance per unit length of the coaxial line is given by

$$\frac{L}{d} = \frac{1}{n}\frac{L_0}{d} = \frac{\mu_0}{n} \quad \text{(H m}^{-1}) \tag{3}$$

where L_0/d = inductance per unit length of transmission-line cell
$\quad n$ = number of line cells in parallel
$\quad \mu_0$ = permeability of air = 1.26 μH m^{-1}

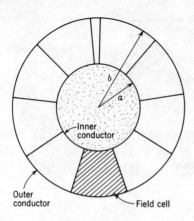

Field cell

Figure 5-30 Coaxial transmission line divided into 9.15 field-cell lines in parallel.

Dividing the space between the coaxial conductors into curvilinear squares, we obtain 9.15 line cells in parallel. Thus

$$\frac{L}{d} = \frac{1.26}{9.15} = 0.138 \ \mu\text{H m}^{-1}$$

As a check, we note that the radius of the outer conductor is twice the radius of the inner so that from (5-13-9) we get

$$\frac{L}{d} = 0.46 \log 2 = 0.138 \ \mu\text{H m}^{-1}$$

which is the same result as obtained above.

The capacitance per unit length of the coaxial line of Fig. 5-30 is given by

$$\frac{C}{d} = n \frac{C_0}{d} = n\epsilon_0 \qquad (\text{F m}^{-1}) \tag{4}$$

where C_0/d = capacitance per unit length of line cell (same as capacitance per unit length of field-cell capacitor; see Sec. 3-19)

n = number of line cells in parallel

ϵ_0 = permittivity of air = 8.85 pF m^{-1}

Thus

$$\frac{C}{d} = 9.15 \times 8.85 = 81 \text{ pF m}^{-1}$$

Using the relation of (3-15-3),

$$\frac{C}{d} = \frac{24.2}{\log 2} = 81 \text{ pF m}^{-1}$$

which is the same as obtained by the cell method.

5-18 CURL

Equation (5-14-5) relates the line integral of **H** around a *finite closed path*, or *loop*, to *I*, where *I* is the *current* enclosed by the loop.

Although relations involving finite paths are useful in circuit theory, it is frequently desirable in field theory to be able to relate quantities *at a point* in space. Curl is such a point relation and can be regarded as an extension of Ampère's law so that it applies at a point.

Consider an incremental plane area ΔS in a conducting medium with a current ΔI flowing through the area and normal to it, as in Fig. 5-31a. Integrating **H** around the periphery of the area ΔS, we have from Ampère's law that

$$\oint \mathbf{H} \cdot d\mathbf{l} = I \tag{1}$$

Now dividing by the area ΔS enclosed by the path, we have

$$\frac{\oint \mathbf{H} \cdot d\mathbf{l}}{\Delta S} = \frac{\Delta I}{\Delta S} \tag{2}$$

and taking the limit of this equation as the area ΔS approaches zero, the left-hand member is called the *curl* of **H** while the right-hand member equals the current density **J** at the point around which the area ΔS shrinks to zero. Thus,

$$\lim_{\Delta S \to 0} \frac{\oint \mathbf{H} \cdot d\mathbf{l}}{\Delta S} = \text{curl } \mathbf{H} \quad \text{and} \quad \lim_{\Delta S \to 0} \frac{\Delta \mathbf{I}}{\Delta S} = \mathbf{J} \tag{3}$$

or $$\text{curl } \mathbf{H} = \mathbf{J} \tag{4}$$

Note that curl **H** is a vector normal to ΔS and in the direction of **J**. If ΔS is not normal to **J** (as assumed above), we get only the component of the curl of **H** (and **J**) normal to ΔS.

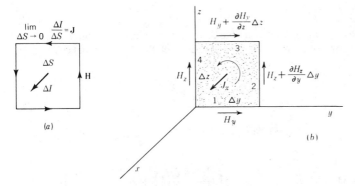

Figure 5-31 (*a*) Current density as limiting ratio of ΔI to Δs. (*b*) Construction for finding *x* component of curl of **H**.

Referring to Fig. 5-31b, let us calculate in rectangular coordinates the value of a component of $\mathbf{H}$ in the x direction by taking the integral of $\mathbf{H}$ around the periphery of the area $\Delta y \, \Delta z$ in four incremental steps. Let $\mathbf{H}$ along side 1 equal H_y and along side 4 equal H_z. If the field is not uniform, the values of $\mathbf{H}$ along sides 2 and 3 are given (to a first approximation) by

$$H_z + \frac{\partial H_z}{\partial y} \Delta y \quad \text{and} \quad H_y + \frac{\partial H_y}{\partial z} \Delta z \tag{5}$$

Thus, according to Ampère's law, we have that

$$\oint \mathbf{H} \cdot d\mathbf{l} = H_y \, \Delta y + H_z \, \Delta z + \frac{\partial H_z}{\partial y} \Delta y \, \Delta z - H_y \, \Delta y$$

$$- \frac{\partial H_y}{\partial z} \Delta z \, \Delta y - H_z \, \Delta z = J_x \, \Delta y \, \Delta z = \Delta I \tag{6}$$

Dividing by the area $\Delta y \, \Delta z$, we obtain the component of curl $\mathbf{H}$ in the x direction as

$$\text{curl}_x \, H = \lim_{\Delta y \, \Delta z \to 0} \frac{\oint \mathbf{H} \cdot d\mathbf{l}}{\Delta y \, \Delta z} = \frac{\partial H_z}{\partial y} - \frac{\partial H_y}{\partial z} = J_x \tag{7}$$

If the current has components flowing in the y and z directions, curl $\mathbf{H}$ also has components in these directions. Deriving them in an identical manner as above and adding them vectorially, we obtain the complete expression for curl $\mathbf{H}$ as

$$\text{curl } \mathbf{H} = \hat{\mathbf{x}} \left[\frac{\partial H_z}{\partial y} - \frac{\partial H_y}{\partial z} \right] + \hat{\mathbf{y}} \left[\frac{\partial H_x}{\partial z} - \frac{\partial H_z}{\partial x} \right] + \hat{\mathbf{z}} \left[\frac{\partial H_y}{\partial x} - \frac{\partial H_x}{\partial y} \right]$$

$$= \hat{\mathbf{x}} J_x + \hat{\mathbf{y}} J_y + \hat{\mathbf{z}} J_z = \mathbf{J} \tag{8}$$

or

$$\text{curl } \mathbf{H} = \mathbf{J} \tag{9}$$

Thus, curl $\mathbf{H}$ has a value wherever a current is present. From Chap. 3 we recall from

$$\text{div } \mathbf{D} = \rho \tag{9a}$$

that div $\mathbf{D}$ has a value wherever a charge is present. Curl $\mathbf{H}$ is conveniently expressed in vector notation as the cross product of the operator del ($\mathbf{\nabla}$) and $\mathbf{H}$, that is,

$$\mathbf{\nabla} = \hat{\mathbf{x}} \frac{\partial}{\partial x} + \hat{\mathbf{y}} \frac{\partial}{\partial y} + \hat{\mathbf{z}} \frac{\partial}{\partial z}$$
$$\times$$
$$\mathbf{H} = \hat{\mathbf{x}} H_x + \hat{\mathbf{y}} H_y + \hat{\mathbf{z}} H_z \tag{10}$$

which yields

$$\mathbf{\nabla} \times \mathbf{H} = \hat{\mathbf{x}} \left(\frac{\partial H_z}{\partial y} - \frac{\partial H_y}{\partial z} \right) + \hat{\mathbf{y}} \left(\frac{\partial H_x}{\partial z} - \frac{\partial H_z}{\partial x} \right) + \hat{\mathbf{z}} \left(\frac{\partial H_y}{\partial x} - \frac{\partial H_x}{\partial y} \right)$$

$$= \hat{\mathbf{x}} J_x + \hat{\mathbf{y}} J_y + \hat{\mathbf{z}} J_z = \mathbf{J} \tag{11}$$

or

$$\boxed{\nabla \times \mathbf{H} = \mathbf{J}}$$ (12)

$\nabla \times \mathbf{H}$ can also be conveniently expressed in determinant form as

$$\nabla \times \mathbf{H} = \begin{vmatrix} \hat{\mathbf{x}} & \hat{\mathbf{y}} & \hat{\mathbf{z}} \\ \dfrac{\partial}{\partial x} & \dfrac{\partial}{\partial y} & \dfrac{\partial}{\partial z} \\ H_x & H_y & H_z \end{vmatrix}$$ (13)

Example 1 A rectangular trough carries water in the x direction. A section of the trough is shown in Fig. 5-32a, the vertical direction coinciding with the z axis. The width of the trough is b. Find the curl of the velocity $\mathbf{v}$ of the water

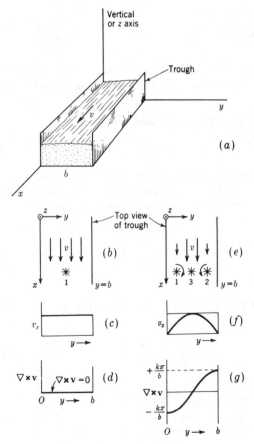

Figure 5-32 Water trough for Example 1.

for two assumed conditions. (*a*) The velocity is everywhere uniform and equal to a constant, i.e.,

$$\mathbf{v} = \hat{\mathbf{x}} K \tag{14}$$

where $\hat{\mathbf{x}}$ = unit vector in positive x direction, dimensionless
$\quad K$ = a constant, m s^{-1}

A top view of the trough is shown in Fig. 5-32*b*, with the positive x direction downward. The fact that the velocity $\mathbf{v}$ is constant is suggested by the arrows of uniform length and also by the graph of v_x as a function of y in Fig. 5-32*c*.
    (*b*) The velocity varies from zero at the edges of the trough to a maximum at the center, the quantitative variation being given by

$$\mathbf{v} = \hat{\mathbf{x}} K \sin \frac{\pi y}{b} \tag{15}$$

where K = a constant, m s^{-1}
$\quad b$ = width of trough, m

The sinusoidal variation of $\mathbf{v}$ is suggested by the arrows in the top view of the trough in Fig. 5-32*e* and also by the graph of v_x as a function of y in Fig. 5-32*f*.

SOLUTION (*a*) Equation (14) can be reexpressed

$$\mathbf{v} = \hat{\mathbf{x}} v_x \tag{16}$$

where v_x is the component of velocity in x direction. Thus $v_x = K$. The curl of $\mathbf{v}$ has two terms involving v_x, namely, $\partial v_x/\partial z$ and $\partial v_x/\partial y$. Since v_x is a constant, both terms are zero and hence $\nabla \times \mathbf{v} = 0$ everywhere in the trough (see Fig. 5-32*d*).
    (*b*) From (15)

$$v_x = K \sin \frac{\pi y}{b} \tag{17}$$

Since v_x is not a function of z, the derivative $\partial v_x/\partial z = 0$. However, v_x is a function of y so that

$$\frac{\partial v_x}{\partial y} = \frac{K\pi}{b} \cos \frac{\pi y}{b} \tag{18}$$

and we have for the curl of $\mathbf{v}$

$$\nabla \times \mathbf{v} = -\hat{\mathbf{z}} \frac{K\pi}{b} \cos \frac{\pi y}{b} \tag{19}$$

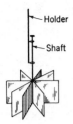

Figure 5-33 Paddle wheel for measuring curl.

where $\hat{\mathbf{z}}$ is the unit vector in the positive z direction; i.e., at the left of the center of the trough the curl of $\mathbf{v}$ is in the negative z direction (downward in Fig. 5-32a), while to the right of the center it is in the positive z direction. The variation of the curl of $\mathbf{v}$ across the trough is presented graphically in Fig. 5-32g.

A physical interpretation of the curl of $\mathbf{v}$ in the above example can be obtained with the aid of the curl-meter, or paddle-wheel, device† of Fig. 5-33. If this device is inserted with its shaft vertical into the trough with the assumed sinusoidal variation for the velocity of the water [part (b) of Example 1], it spins clockwise when it is at the left of the center of the trough (position 1 in Fig. 5-32e) and counterclockwise when it is at the right of the center of the trough (position 2 in Fig. 5-32e), corresponding to negative and positive values of curl. At the center of the trough (position 3 in Fig. 5-32e) the curl meter does not rotate since the forces on the paddles are balanced. This corresponds to the curl of $\mathbf{v}$ being zero. The rate of rotation of the paddle-wheel shaft is proportional to the curl of $\mathbf{v}$ at the point where it is inserted. Thus, it rotates fastest near the edges of the trough. At any point the rate of rotation is also a maximum with the shaft vertical (rather than inclined to the vertical), indicating that $\nabla \times \mathbf{v}$ is in the z direction. It is assumed that the paddle wheel is small enough to avoid affecting the flow appreciably and to indicate closely the conditions at a point.

If the curl meter with shaft vertical is inserted in water with uniform velocity, as assumed in part (a) of Example 1, it will not rotate (curl $\mathbf{v}$ equals zero).

Example 2 Consider a current-carrying conductor of radius R as shown in cross section in Fig. 5-34a. The current is uniformly distributed so that the current density $\mathbf{J}$ is a constant. Taking the axis of the wire in the z direction,

$$\mathbf{J} = \hat{\mathbf{z}}J = \hat{\mathbf{z}}\frac{I}{\pi R^2} \qquad (\text{A m}^{-2}) \qquad (20)$$

---

† H. H. Skilling, "Fundamentals of Electric Waves," 2d ed., p. 24, John Wiley & Sons, Inc., New York, 1948.

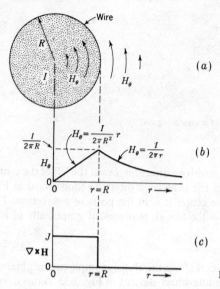

Figure 5-34 Conducting wire (Example 2).

where I is the total current in the conductor in amperes. Find the curl of $\mathbf{H}$ both inside and outside the wire.

SOLUTION The variation of H as a function of radius was worked out for this case in Example 2 of Sec. 5-14. The variation found for H is shown in the graph of Fig. 5-34b. Since $\mathbf{H}$ is entirely in the θ direction, we have

$$\mathbf{H} = \hat{\boldsymbol{\theta}} H_\theta \qquad (21)$$

where

$$H_\theta = \begin{cases} \dfrac{I}{2\pi r} & \text{outside conductor} \\[2mm] \dfrac{I}{2\pi R^2}\, r & \text{inside conductor} \end{cases}$$

Using the expression for curl in cylindrical coordinates (see inside back cover), we have

$$\nabla \times \mathbf{H} = 0 \qquad \text{outside conductor} \qquad (22)$$

$$\nabla \times \mathbf{H} = \hat{\mathbf{z}} \frac{I}{\pi R^2} = \mathbf{J} \qquad \text{inside conductor} \qquad (23)$$

Hence, the curl of $\mathbf{H}$ has a value only where there is current, being a constant in the conductor and zero outside (see Fig. 5-34c).

5-19 MAXWELL'S FIRST CURL EQUATION

The relation derived from Ampère's law in Sec. 5-18 that

$$\nabla \times \mathbf{H} = \mathbf{J} \qquad (1)\dagger$$

is one of Maxwell's equations. Equation (1) is a differential expression and relates the field $\mathbf{H}$ to the current density $\mathbf{J}$ *at a point*. The corresponding expression in integral form, as given by (5-15-1), relates $\mathbf{H}$ around a *finite closed path* to the total current passing through the area enclosed. Both are Maxwell's equations from Ampère's law.

Thus far, we have encountered three of Maxwell's four equations applying at a point. They are $\nabla \cdot \mathbf{D} = \rho$, $\nabla \cdot \mathbf{B} = 0$, and (1). The fourth relation, (8-7-2), is also an equation involving curl, so that (1) may be referred to as Maxwell's first curl equation and (8-7-2) as the second.

5-20 SUMMARY OF OPERATIONS INVOLVING ∇

We have discussed four operations involving the operator ∇ (del), namely, the gradient, divergence, Laplacian, and curl. Although the Laplacian can be resolved into the divergence of the gradient ($\nabla^2 f = \nabla \cdot \nabla f$), this operation is of such importance as to warrant listing it separately. These operators with their differential equivalents in rectangular coordinates are listed below with f representing a scalar function and $\mathbf{F}$ a vector function.

Gradient: $\qquad \operatorname{grad} f = \nabla f = \hat{\mathbf{x}} \dfrac{\partial f}{\partial x} + \hat{\mathbf{y}} \dfrac{\partial f}{\partial y} + \hat{\mathbf{z}} \dfrac{\partial f}{\partial z}$ $\qquad (1)$

Gradient operates on a scalar function to yield a vector function.

Divergence: $\qquad \operatorname{div} \mathbf{F} = \nabla \cdot \mathbf{F} = \dfrac{\partial F_x}{\partial x} + \dfrac{\partial F_y}{\partial y} + \dfrac{\partial F_z}{\partial z}$ $\qquad (2)$

Divergence operates on a vector function to yield a scalar function.

Laplacian: $\qquad \operatorname{div}(\operatorname{grad} f) = \nabla \cdot (\nabla f) = \nabla^2 f = \dfrac{\partial^2 f}{\partial x^2} + \dfrac{\partial^2 f}{\partial y^2} + \dfrac{\partial^2 f}{\partial z^2}$ $\qquad (3)$

The Laplacian operates on a scalar function to yield another scalar function.‡

---

† Equation (1) is a special form of the more general relation given in (8-12-4). The more general equation has an additional term involving the displacement current density. However, a displacement current is present only for time-changing fields so that for steady fields, as considered here, (8-12-4) reduces to (1).

‡ In rectangular coordinates it is also possible to interpret the Laplacian of a vector function as the vector sum of the Laplacians of the three scalar components of the vector. Thus

$$\nabla^2 \mathbf{F} = \hat{\mathbf{x}}\nabla^2 F_x + \hat{\mathbf{y}}\nabla^2 F_y + \hat{\mathbf{z}}\nabla^2 F_z$$

However, in no other coordinate system is this simple interpretation possible.

Curl: $\text{curl } \mathbf{F} = \nabla \times \mathbf{F} = \hat{\mathbf{x}}\left(\frac{\partial F_z}{\partial y} - \frac{\partial F_y}{\partial z}\right) + \hat{\mathbf{y}}\left(\frac{\partial F_x}{\partial z} - \frac{\partial F_z}{\partial x}\right) + \hat{\mathbf{z}}\left(\frac{\partial F_y}{\partial x} - \frac{\partial F_x}{\partial y}\right)$ (4)

Curl operates on a vector function to yield another vector function.

5-21 A COMPARISON OF DIVERGENCE AND CURL

Whereas divergence operates on a vector function to yield a scalar function, curl operates on a vector function to yield a vector function. There is another important difference. Referring to the differential relation for the divergence in (5-20-2), we note that the differentiation with respect to x is on the x component of the field, the differentiation with respect to y is on the y component, etc. Therefore, to have divergence the field must vary in magnitude along a line having the same direction as the field.†

Referring to the relation for curl in (5-20-4), we note, on the other hand, that the differentiation with respect to x is on the y and z components of the field, the differentiation with respect to y is on the x and z components, etc. Therefore, to have curl the field must vary in magnitude along a line normal to the direction of the field.‡

This comparison is illustrated in Fig. 5-35. The field at (*a*) is everywhere in the y direction. It has no variation in the x or z directions but varies in magnitude as a function of y. Therefore this field has divergence but no curl. The field at (*b*) is also everywhere in the y direction. It has no variation in the x and y directions but does vary in magnitude as a function of z. Therefore, this field has curl but no divergence.

Let us now discuss the significance of operations involving ∇ twice. First consider the divergence of the curl of a vector function; i.e.,

$$\nabla \cdot (\nabla \times \mathbf{F}) \tag{1}$$

where $\mathbf{F}$ is any vector function given in rectangular coordinates by

$$\mathbf{F} = \hat{\mathbf{x}}F_x + \hat{\mathbf{y}}F_y + \hat{\mathbf{z}}F_z$$

If we first take the curl of $\mathbf{F}$, we obtain another vector. Next taking the divergence of this vector, the result is identically zero. Thus

$$\nabla \cdot (\nabla \times \mathbf{F}) \equiv 0 \tag{2}$$

---

† This is a necessary but not a sufficient condition that a vector field have divergence. For example, the **D** field due to a point charge is radial and varies as $1/r^2$ but has no divergence except at the charge. If, however, the field is *everywhere* in the y direction, as in Fig. 5-35*a*, and varies only as a function of y, then this field does have divergence.

‡ This is a necessary but not a sufficient condition that a vector field have curl. For example, the **H** field outside of a long wire varies in magnitude as $1/r$ and has a direction normal to the radius vector; yet **H** has no curl in this region. If, however, the field is *everywhere* in the y direction, as in Fig. 5-35*b*, and varies only as a function of z, then this field does have curl.

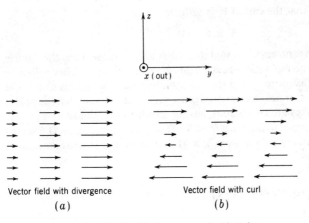

Vector field with divergence   Vector field with curl

(a)   (b)

Figure 5-35 Examples of fields with divergence and with curl.

In words (2) states that *the divergence of the curl of a vector function is zero.* As a corollary we may say that if the divergence of a vector function is zero, then it must be the curl of some other vector function.

For example, the divergence of **B** is always zero everywhere; i.e.,

$$\mathbf{V} \cdot \mathbf{B} = 0 \tag{3}$$

Therefore, **B** can be expressed as the curl of some other vector function. Let us designate this other vector function by **A**. Then

$$\boxed{\mathbf{B} = \mathbf{V} \times \mathbf{A}} \tag{4}$$

The function **A** in (4) is called the *vector potential* and is discussed in more detail in the next section.

Let us consider another operation involving **V** twice, namely, the curl of the gradient of a scalar function *f*. That is,

$$\mathbf{V} \times (\mathbf{V}f) \tag{5}$$

Taking first the gradient of *f* and then the curl of the resulting vector function, the result is found to be identically zero. Thus

$$\mathbf{V} \times (\mathbf{V}f) \equiv 0 \tag{6}$$

In words (6) states that *the curl of the gradient of a scalar function is zero.* As a corollary, any vector function which is the gradient of some scalar function has no curl.

For example, we recall from (2-13-2) that the static electric field due to charges $\mathbf{E}_c$ is derivable as the gradient of a scalar potential V. Thus

$$\mathbf{E}_c = -\mathbf{V}V \tag{7}$$

It follows, therefore, that the curl of $\mathbf{E}_c$ is zero, or

$$\nabla \times \mathbf{E}_c = 0 \tag{8}$$

If a vector field has no curl, it is said that the field is *lamellar*. Thus the electric field $\mathbf{E}_c$ is lamellar. The flux tubes of such fields are discontinuous. They originate on positive charges (as sources) and terminate on negative charges (as sinks). On the other hand, if a vector field has no divergence such as $\mathbf{B}$, it is said that the field is *solenoidal*. Its flux tubes are continuous, having no sources or sinks.

Finally, it is important to note that from $\nabla \cdot \mathbf{D} = \rho$ *the divergence of* $\mathbf{D}$ *finds the sources* (ρ) *of the electric field* and from $\nabla \times \mathbf{H} = \mathbf{J}$ *the curl of* $\mathbf{H}$ *finds the sources* $(\mathbf{J})$ *of the magnetic field.*

5-22 THE VECTOR POTENTIAL

According to (5-9-8), the magnetic flux density $\mathbf{B}$ at a point P produced by a current distribution, as in Fig. 5-36, is given by

$$\mathbf{B} = \frac{\mu}{4\pi} \iiint \frac{\mathbf{J} \times \hat{\mathbf{r}}}{r^2} \, dv \tag{1}$$

where $\mathbf{B} =$ flux density, T
$\mu =$ permeability of medium (uniform), H m^{-1}
$\mathbf{J} =$ current density at volume element, A m^{-2}
$\hat{\mathbf{r}} =$ unit vector in direction of radius vector r, dimensionless
$r =$ radius vector from volume element to point P, m
$dv =$ volume element, m^3

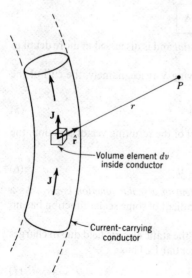
Volume element dv
inside conductor

Current-carrying
conductor

Figure 5-36 Construction for finding flux density $\mathbf{B}$ at P. $\mathbf{B}$ is into the page at P.

Carrying out the integration over the entire volume occupied by the current-carrying conductor gives the total flux density **B** at P due to the current.

In Sec. 5-21 we noted that since the divergence of **B** is always zero, it should be possible to express **B** as the curl of some other vector. Thus from (5-21-4) we can write

$$\mathbf{B} = \nabla \times \mathbf{A} \tag{2}$$

where **A** is called a *vector potential* since it is a potential function that is also a vector.† If we also make

$$\nabla \cdot \mathbf{A} = 0 \tag{3}$$

A is completely defined. Taking the curl of (2) yields

$$\nabla \times \nabla \times \mathbf{A} = \nabla \times \mathbf{B} = \mu \mathbf{J} \tag{4}$$

By the vector identity for the curl of the curl of a vector (see Appendix A, Sec. A-9) (4) becomes

$$\nabla(\nabla \cdot \mathbf{A}) - \nabla^2 \mathbf{A} = \mu \mathbf{J} \tag{5}$$

Introducing the condition of (3), this reduces to

$$\nabla^2 \mathbf{A} = -\mu \mathbf{J} \tag{6}$$

or in terms of the three rectangular components of **A** and **J**

$$\hat{\mathbf{x}} \nabla^2 A_x + \hat{\mathbf{y}} \nabla^2 A_y + \hat{\mathbf{z}} \nabla^2 A_z = -\mu(\hat{\mathbf{x}} J_x + \hat{\mathbf{y}} J_y + \hat{\mathbf{z}} J_z) \tag{7}$$

Equation (7) is the vector sum of three scalar equations. Hence,

$$\nabla^2 A_x = -\mu J_x$$
$$\nabla^2 A_y = -\mu J_y \tag{8}$$
$$\nabla^2 A_z = -\mu J_z$$

Each of these relations has the same form as Poisson's equation. Therefore solutions to the three equations of (8) are

$$A_x = \frac{\mu}{4\pi} \iiint \frac{J_x}{r} \, dv$$

$$A_y = \frac{\mu}{4\pi} \iiint \frac{J_y}{r} \, dv \tag{9}$$

$$A_z = \frac{\mu}{4\pi} \iiint \frac{J_z}{r} \, dv$$

---

† The potential function V from which the electric field $\mathbf{E}_c$ can be derived (by the relation $\mathbf{E}_c = -\nabla V$) is a scalar quantity, and hence V is a *scalar potential*.

Taking the vector sum of the components for **A** in (9) gives

$$A = \frac{\mu}{4\pi} \iiint \frac{J}{r} \, dv \qquad (10)$$

According to (10), the vector potential **A** at a point due to a current distribution is equal to the ratio **J**/r integrated over the volume occupied by the current distribution, where **J** is the current density at each volume element dv and r is the distance from each volume element to the point P, where **A** is being evaluated (see Fig. 5-36). If the current distribution is known, **A** can be found. Knowing **A** at a point, the flux density **B** at that point is then obtained by taking the curl of **A** as in (2).

From (2) we note that **A** has the dimensions of

$$\text{Magnetic flux density} \times \text{distance} = \frac{\text{magnetic flux}}{\text{distance}} = \frac{\text{force}}{\text{current}}$$

Hence, the vector potential **A** can be expressed in webers per meter or newtons per ampere. The dimensional symbols for **A** are ML/IT^2.

Whereas the scalar electric potential V (Sec. 2-7) and the scalar magnetic potential U (Sec. 5-16) have simple physical significance, the vector potential **A** does not. Its value here lies in its usefulness as an intermediate mathematical step for calculating the magnetic field **B**.

Example Consider a short copper wire of length l and a cross-sectional area a situated in air coincident with the z axis at the origin, as shown in Fig. 5-37. The current density **J** is in the positive z direction. Assume the hypothetical situation that **J** is uniform throughout the wire and constant with respect to time. Find the magnetic flux density **B** everywhere at a large distance from the wire, using the vector potential to obtain the solution.

SOLUTION The vector potential **A** at any point P produced by the wire is given by (10), where the ratio **J**/r is integrated throughout the volume occupied by the wire. Since we wish to find **B** only at a large distance r from the wire, it suffices to find **A** at a large distance. Specifically the distance r should be large compared with the length of the wire ($r \gg l$). Then, at any point P the distance r to different parts of the wire can be considered constant and (10) written as

$$A = \frac{\mu_0}{4\pi r} \iiint J \, dv \qquad (11)$$

Now **J** is everywhere in the z direction and also is uniform. Thus $J = \hat{z}J_z$, and

$$\iiint J \, dv = \hat{z} \int_{-l/2}^{l/2} \iint_a J_z \, ds \, dl = \hat{z} \int_{-l/2}^{l/2} I \, dl \qquad (12)$$

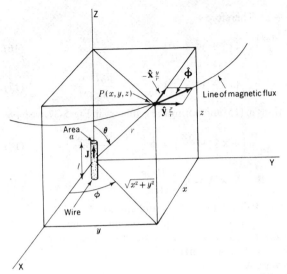

Figure 5-37 Construction for finding the vector potential **A** and flux density **B** due to a short current-carrying wire.

where $I = J_z a$ is the current in the wire. Completing the integration in (12) and substituting this result in (11), we obtain

$$\mathbf{A} = \hat{\mathbf{z}}\,\frac{\mu_0 I l}{4\pi r} = \hat{\mathbf{z}} A_z \tag{13}$$

where $\mathbf{A}$ = vector potential at distance r from wire, Wb m^{-1}
  $\hat{\mathbf{z}}$ = unit vector in positive z direction, dimensionless
  μ_0 = permeability of air = 400π nH m^{-1}
  I = current in wire, A
  l = length of wire, m
  r = distance from wire, m

Equation (13) gives the vector potential $\mathbf{A}$ at a large distance from the wire. It is everywhere in the positive z direction as indicated by the unit vector $\hat{\mathbf{z}}$ and is inversely proportional to the distance r from the wire. It is not a function of angle (ϕ or θ in Fig. 5-37).

Having found the vector potential $\mathbf{A}$, we obtain the flux density $\mathbf{B}$ by taking the curl of $\mathbf{A}$. In rectangular components the curl of $\mathbf{A}$ is given by

$$\nabla \times \mathbf{A} = \hat{\mathbf{x}}\left(\frac{\partial A_z}{\partial y} - \frac{\partial A_y}{\partial z}\right) + \hat{\mathbf{y}}\left(\frac{\partial A_x}{\partial z} - \frac{\partial A_z}{\partial x}\right) + \hat{\mathbf{z}}\left(\frac{\partial A_y}{\partial x} - \frac{\partial A_x}{\partial y}\right) \tag{14}$$

Since $\mathbf{A}$ has only a z component, (14) reduces to

$$\nabla \times \mathbf{A} = \hat{\mathbf{x}}\,\frac{\partial A_z}{\partial y} - \hat{\mathbf{y}}\,\frac{\partial A_z}{\partial x} \tag{15}$$

Now $r = \sqrt{x^2 + y^2 + z^2}$. Therefore

$$\frac{\partial A_z}{\partial y} = \frac{\mu_0 Il}{4\pi} \frac{\partial}{\partial y} (x^2 + y^2 + z^2)^{-1/2} = -\frac{\mu_0 Il}{4\pi} \frac{y}{r^3} \tag{16}$$

and

$$\frac{\partial A_z}{\partial x} = \frac{\mu_0 Il}{4\pi} \frac{\partial}{\partial x} (x^2 + y^2 + z^2)^{-1/2} = -\frac{\mu_0 Il}{4\pi} \frac{x}{r^3} \tag{17}$$

Introducing these relations in (15) and noting the geometry in Fig. 5-37, we have

$$\mathbf{\nabla} \times \mathbf{A} = \frac{\mu_0 Il}{4\pi r^2} \left(-\hat{\mathbf{x}} \frac{y}{r} + \hat{\mathbf{y}} \frac{x}{r} \right) = \hat{\boldsymbol{\phi}} \frac{\mu_0 Il}{4\pi r^2} \frac{\sqrt{x^2 + y^2}}{r} \tag{18}$$

or

$$\mathbf{B} = \mathbf{\nabla} \times \mathbf{A} = \hat{\boldsymbol{\phi}} \frac{\mu_0 Il \sin \theta}{4\pi r^2} \tag{19}$$

where $\mathbf{B}$ = magnetic flux density at distance r and angle θ, T
$\hat{\boldsymbol{\phi}}$ = unit vector in ϕ direction (see Fig. 5-37), dimensionless
θ = angle between axis of wire and radius vector r, dimensionless
μ_0 = permeability of air = 400π nH m^{-1}
I = current in wire, A
l = length of wire, m
r = distance from wire to point where $\mathbf{B}$ is being evaluated, m

According to (19), the flux density $\mathbf{B}$ produced by the wire is everywhere in the ϕ direction. That is, the lines of magnetic flux form closed circles concentric with the z axis. The planes of the circles are parallel to the xy plane. One such line of magnetic flux at a distance r from the origin is indicated in Fig. 5-37. According to (19), $\mathbf{B}$ is also proportional to $\sin \theta$ and inversely proportional to r^2.

Although the result of (19) could have been written down almost directly from (1), without using the vector potential explicitly, the above example serves to illustrate the manner in which the vector potential can be applied. Employing the vector potential in the above example is analogous to using a 10-ton steam hammer to crack a walnut. However, on many problems of a more difficult nature the vector potential is indispensable.

5-23 CHARGED PARTICLES IN ELECTRIC AND MAGNETIC FIELDS

Electric Field

Let a particle of charge e† be placed in a uniform electric field $\mathbf{E}$. Since $\mathbf{E}$ is the force per unit charge (newtons per coulomb), the force $\mathbf{F}$ on the particle is

$$\boxed{\mathbf{F} = e\mathbf{E}} \tag{1}$$

---

† The symbol e is used here to designate the charge of a particle (instead of q) since e is usually employed for the charge of a particle, commonly an electron.

The force is in the same direction as the field if the charge is positive and opposite to the field if the charge is negative. If the particle is at rest and the field is applied, the particle is accelerated uniformly in the direction of the field. According to Newton's second law, the force on a particle is related to its mass and acceleration by

$$\mathbf{F} = m\mathbf{a} \tag{2}$$

where $\mathbf{F}$ = force, N
   m = mass, kg
   $\mathbf{a}$ = acceleration, m s^{-2}

Therefore the acceleration of the particle is

$$\mathbf{a} = \frac{e}{m}\mathbf{E} \tag{3}$$

The velocity $\mathbf{v}$ of the particle after a time t is then

$$\mathbf{v} = \mathbf{a}t = \frac{e}{m}\mathbf{E}t \tag{4}$$

where $\mathbf{v}$ = velocity of particle, m s^{-1}
   e = charge of particle, C
   m = mass of particle, kg
   $\mathbf{E}$ = electric field intensity, V m^{-1} or N C^{-1}
   t = time, s

The field imparts energy to the charged particle. If (4) is reexpressed as

$$m\mathbf{a} = e\mathbf{E} \tag{5}$$

it has the dimensions of force. Integrating this force over the distance moved yields the energy W acquired. Thus

$$W = m\int_1^2 \mathbf{a} \cdot d\mathbf{l} = -e\int_1^2 \mathbf{E} \cdot d\mathbf{l} \tag{6}$$

The line integral of $\mathbf{E}$ between two points, 1 and 2, may be recognized as the potential difference V between the points. When $\mathbf{a} = d\mathbf{v}/dt$ and $d\mathbf{l} = \mathbf{v}\,dt$ are substituted, (6) becomes

$$W = m\int_1^2 \mathbf{v} \cdot d\mathbf{v} = eV \tag{7}$$

or

$$W = \tfrac{1}{2}m(v_2^2 - v_1^2) = eV \tag{8}$$

where W = energy acquired, J
   v_2 = velocity at point 2, or final velocity, m s^{-1}
   v_1 = velocity at point 1, or initial velocity, m s^{-1}
   e = charge on particle, C
   V = potential difference between points 1 and 2, V or J C^{-1}

If the particle starts from rest, the initial velocity is zero; so

$$W = eV = \tfrac{1}{2}mv^2 \qquad (9)$$

where v is the final velocity. Equation (9) has the dimensions of energy. The dimensional relation in SI units is

$$\text{Joules} = \text{coulombs} \times \text{volts} = \text{kilograms} \frac{\text{meters}^2}{\text{seconds}^2}$$

Thus the energy acquired by a particle of charge e starting from rest and passing through a potential drop V is given either by the product of the charge and the potential difference or by one-half the product of the mass of the particle and the square of the final velocity.

Solving (9) for the velocity gives

$$v = \sqrt{\frac{2eV}{m}} \qquad (\text{m s}^{-1}) \qquad (10)$$

The energy acquired by an electron ($e = 1.6 \times 10^{-19}$ C) in falling through a potential difference of 1 V is 1.6×10^{-19} J. This amount of energy is a convenient unit for designating the energies of particles and is called one *electron volt* (eV).

For an electron $e = 1.6 \times 10^{-19}$ C and $m = 0.91 \times 10^{-30}$ kg, so that (10) becomes

$$v = 5.9 \times 10^5 \sqrt{V} \qquad (\text{m s}^{-1}) \qquad (11)$$

Thus, if $V = 1$ V, the velocity of the electron is 0.59 Mm s^{-1}, or 590 km s^{-1}. It is apparent that a relatively small voltage imparts a very large velocity to an electron. If $V = 2.5$ kV, the velocity is 30 Mm s^{-1}, or about one-tenth the velocity of light. The above relations are based on the assumption that the particle velocity is small compared with that of light. The mass of a particle approaches an infinite value as the velocity approaches that of light (relativistic effect), whereas the above relations are based on a constant mass. Actually, however, the mass increase is of negligible consequence for most applications unless the velocity is at least 10 percent that of light. The relation between the mass m of the particle and its mass m_0 at low velocities (rest mass) is given by

$$m = \frac{m_0}{\sqrt{1 - (v/c)^2}} \qquad (12)$$

where v = velocity of particle, m s^{-1}
c = velocity of light = 300 Mm s^{-1}

If the velocity is one-tenth that of light, the mass is only one-half of 1 percent greater than the rest mass.

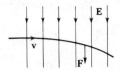

Figure 5-38 Path of positively charged particle in electric field.

If the particle has an initial velocity which is not parallel to the field direction, as assumed above, the particle describes a parabolic path (Fig. 5-38).

Magnetic Field

From (5-9-3) the force **F** on a current element of length dl in a magnetic field is

$$d\mathbf{F} = (\mathbf{I} \times \mathbf{B})\,dl \qquad (N) \qquad (13)$$

where **I** = current (vector indicating magnitude and direction of current), A
      B = magnetic flux density, T
      dl = element length, m

This is the fundamental motor equation of electrical machinery. It also applies to moving charged particles in the absence of any metallic conductor.

The current **I** in a conductor or in a beam of ions or electrons can be expressed in terms of the current density **J**, the charge (volume) density ρ, the beam area A, and the velocity **v** by

$$\mathbf{I} = \mathbf{J}A = \rho \mathbf{v}A \qquad (14)$$

Substituting (14) for **I** in (13) gives

$$d\mathbf{F} = \rho A\,dl(\mathbf{v} \times \mathbf{B}) \qquad (15)$$

But $\rho A\,dl = dq$, the charge in a length dl of the beam. Thus

$$d\mathbf{F} = dq(\mathbf{v} \times \mathbf{B}) \qquad (16)$$

For a single particle of charge e, we have for the (Lorentz) force

$$\boxed{\mathbf{F} = e(\mathbf{v} \times \mathbf{B})} \qquad (17)$$

Consider now the motion of a particle of charge e in a uniform magnetic field of flux density **B**. The velocity of the particle is **v**. From Newton's second law the force on the particle is equal to the product of its mass m and its acceleration **a** ($= d\mathbf{v}/dt$). Thus

$$m\mathbf{a} = e(\mathbf{v} \times \mathbf{B}) \qquad (18)$$

or

$$\mathbf{a} = \frac{e}{m}(\mathbf{v} \times \mathbf{B}) \qquad (19)$$

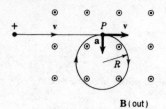

B (out)   **Figure 5-39** Path of positively charged particle in magnetic field.

According to (19), the acceleration is normal to the plane containing the particle path and **B**. If the direction of the particle path (indicated by **v**) is normal to **B**, the acceleration is a maximum. If the particle is at rest, the field has no effect. Likewise, if the particle path is in the same direction as **B**, there is no effect, the particle continuing undeflected. Only when the path or the velocity **v** has a component normal to **B** does the field have an effect.

If a magnetic field of large extent is at right angles to the direction of motion of a charged particle, the particle is deflected into a circular path. Suppose that in a field-free region a positively charged particle is moving to the right, as indicated in Fig. 5-39, and that when it reaches the point P a magnetic field is applied. The direction of **B** is normal to the page (outward). According to the cross product of **v** into **B** in (19), the acceleration **a** is downward, so that the particle describes a circle in the clockwise direction in the plane of the page.

Radius Let us determine the radius R of the circle. The magnitude of the force **F** (radially inward) on the particle is, by (19),

$$F = ma = evB \qquad (20)$$

This force is also given by

$$F = \frac{mv^2}{R} \qquad (21)$$

Equating (20) and (21) yields

$$\frac{mv^2}{R} = evB \qquad (22)$$

or

$$\boxed{R = \frac{mv}{eB}} \qquad (23)$$

where R = radius of particle path, m
m = mass of particle, kg
v = velocity of particle, m s^{-1}
e = charge of particle, C
B = flux density, T

Table 5-2 Charge and mass of particles

Particle	Charge e, C	Rest mass m, kg	Ratio e/m, C kg^{-1}
Electron	-1.602×10^{-19}	9.11×10^{-31}	-1.76×10^{11}
Positron	$+1.602 \times 10^{-19}$	9.11×10^{-31}	$+1.76 \times 10^{11}$
Neutron	0	1.6747×10^{-27}	0
Proton (hydrogen nucleus)	$+1.602 \times 10^{-19}$	1.6725×10^{-27}	$+9.6 \times 10^{7}$
Deuteron (heavy-hydrogen nucleus)	$+1.6 \times 10^{-19}$	3.34×10^{-27}	$+4.8 \times 10^{7}$
Alpha particle (helium nucleus)	$+3.2 \times 10^{-19}$	6.644×10^{-27}	$+4.81 \times 10^{7}$

Thus, the larger the velocity of the particle or the larger its mass, the greater the radius. On the other hand, the larger the charge or the flux density, the smaller the radius.

Frequency The number of revolutions per second of the particle in the circular path is called the frequency f† of the particle. This frequency is

$$f = \frac{v}{2\pi R} = \frac{eB}{2\pi m} \qquad (\text{r s}^{-1}) \tag{24}$$

Example An electron has a velocity of 10 km s^{-1} normal to a magnetic field of 0.1 T flux density. Find the radius of the electron path and also its frequency. (See Table 5-2 or Appendix, Sec. A-2, for charge and mass of electron.)

SOLUTION From (23) the radius is

$$R = \frac{9.1 \times 10^{-31} \times 10^4}{1.6 \times 10^{-19} \times 10^{-1}} = 569 \text{ nm}$$

This is a very small circle. The frequency is

$$f = \frac{10^4}{2\pi \times 5.7 \times 10^{-7}} = 2.8 \times 10^9 \text{ r s}^{-1}$$

If the particle in the above example had an initial velocity component parallel to **B** as well as perpendicular to **B**, the particle would move in a helical path with the axis of the helix parallel to **B**.

For a particle of charge Q in *both* an electric and magnetic field the total force on the particle is as given by (5-9-6).

† This is the same as the *gyro*, or *cyclotron*, frequency (see Prob. 5-4-4).

PROBLEMS†

$\mu = \mu_0$ everywhere.

Group 5-1: Secs. 5-1 through 5-10. Magnetic field of a current element, Biot-Savart law, forces, magnetic flux, Gauss' law, cross product, torque, and magnetic moment

★**5-1-1. Wire and loop** An infinitely long straight wire carries a current of 1 kA. A loop 20 mm in diameter is situated with its center 1 m from the wire. How must the loop be turned and what current is required in it to reduce the field from the long wire to zero at the center of the loop?

★**5-1-2. Two loops** Two identical circular loops of 1-m radius are situated side by side on a common axis. The distance between the loops is 1 m. If both loops carry a current of 1 A in the same direction, find B (a) at the center of one loop and (b) at a point midway between the loops on their common axis.

5-1-3. Parallel conductors. Force Two long parallel linear conductors carry 100 A. If the conductors are separated by 10 mm, what is the force per meter of length on a conductor if the currents flow (a) in opposite directions and (b) in the same direction?

5-1-4. Linear conductor. Force A long linear conductor with current of 10 A is coincident with the z axis. The current flows in the $+z$ direction. If $\mathbf{B} = \hat{x}4 + \hat{y}3$ T, find the vector value of the force $\mathbf{F}$ per meter length of conductor.

★**5-1-5. Tubular conductor. Torque** A long thin-walled tubular conductor of radius r_0 has its axis coincident with the z axis. The tube carries a current $\hat{z}I$ A. (a) What is the torque on the tubular conductor (per unit length) in the presence of a magnetic field $\mathbf{B} = \hat{r}kr^{-1}$ T, where k = constant? (b) Is the torque clockwise or counterclockwise as viewed from the positive z axis? (c) What is the torque if $\mathbf{B} = \hat{\phi}kr^{-1}$ T?

5-1-6. B from parallel conductors Two long straight conductors parallel to the z axis are situated at $(x = 0, y = 1)$ m and $(x = 0, y = -1)$ m. If $I_z = 4$ A for the conductor at $y = 1$ m and $I_z = 2$ A for the conductor at $y = -1$ m, find $\mathbf{B}$ at $(x = 1, y = 0)$ m.

5-1-7. B from loop Show that the flux density from a single-turn wire loop is given by

$$B_\theta = \frac{\mu_0 IA \sin \theta}{4\pi r^3} \quad \text{and} \quad B_r = \frac{\mu_0 IA \cos \theta}{2\pi r^3}$$

where I = current
  A = area
  IA = magnetic moment
  θ = angle between axis of loop and radius vector of length r

It is assumed that r is much greater than the diameter of the loop.

★**5-1-8. Two loops. Torque** A small 10-turn loop is situated at the origin with the loop axis coincident with the z axis. A second similar loop is situated at a radial distance of 3 m from the origin and in a direction which makes an angle of 60° from the z axis. The second loop is oriented so that its axis makes an angle of 30° with the z axis. If each loop has an area of 1 cm² and a current of 3 A, find the torque on the second loop.

5-1-9. Loop and wire A single-turn horizontal loop 2 m in diameter carries a current of $5/\pi$ MA. A long straight horizontal wire situated 1 m below the center of the loop carries a current of 5 MA. Find the magnitude of B at the center of the loop.

★**5-1-10. Field direction from loop** A 1-cm² loop in the xy plane at the origin has 3 turns and carries 2 A. Find the angle with respect to the z direction which a short free-to-rotate bar magnet will assume at a distance of 6 m from the loop and in a direction at an angle of 45° with respect to the z axis (also loop axis).

† Answers to starred problems are given in Appendix C.

5-1-11. Loop torque A 1-cm² loop at the origin has 3 turns and carries 2 A. Find the torque on a second identical loop (same size, current, and number of turns) situated at a distance of 6 m and in a direction at an angle of 45° with respect to the z axis. Both loops are oriented with their axes in the z direction.

5-1-12. Loop and solenoid torque (a) Show that the torque on a pair of single-turn square loops (symmetrical with respect to a central rotation axis as in Fig. P5-1-12) is given by $T = 2IAB \sin \theta$, where θ is the angle between the axis of the loops (as given by $\hat{n}$) and the applied uniform field $\mathbf{B}$. I = loop current, A = area of one loop. (b) Show further that a solenoid of N turns [or $N/2$ pairs of part (a)] has a torque of $T = NIAB \sin \theta$.

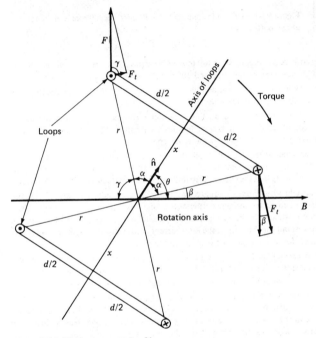

Figure P5-1-12 Torque on pair of loops.

5-1-13. Deflection of indicating meter. Torque The curved pole pieces and cylindrical iron core, as in Fig. P5-1-13, provide a substantially uniform radial magnetic field $B = 0.8T$ for the moving coil of an indicating meter. The coil is restrained by a spring with a torque $T = 2 \mu N$ m per degree. If the coil has 50 turns and measures 16 mm across by 25 mm perpendicular to page, find the current required to deflect the coil through an angle of 45°.

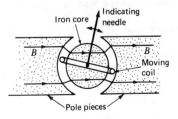

Figure P5-1-13 Moving coil and magnet of indicating meter.

5-1-14. Circular-square loop A loop that is half circular and half square, as in Fig. P5-1-14, carries a current of 20 A. Find B at the two points in the plane of loop (a) P_1 and (b) P_2.

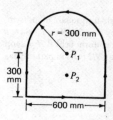

Figure P5-1-14 Circular-square loop. Point P_1 is 300 mm and P_2 150 mm above bottom wire.

5-1-15. B near conducting strip A conducting strip 100 mm wide carries a uniform sheet current density $K = 200$ A m^{-1}. Find B in the plane of the strip and 100 mm from one side of it.

5-1-16. Flux linking wires Two parallel straight conductors carrying 50 A in opposite directions are separated by 300 mm. Find the magnetic flux linking the wires per meter of length.

5-1-17. Flux linking coaxial conductors A coaxial line with outer conductor of 20 mm radius and hollow thin-walled inner conductor of 5 mm radius supplies a current of 125 A to a resistive load. Find the magnetic flux linking the conductors per meter of length.

5-1-18. Force between conducting tubes Two parallel thin-walled tubular conductors of 100-mm diameter are separated 150 mm between centers. If the tubes carry 400 A in opposite directions, find the force **F** between them. Note that the current distribution on the conductors will not be uniform.

5-1-19. Square loop and straight wire. Force A square wire loop 2 m on a side is situated with edges coincident with the positive x and y axes and one corner at the origin. A long straight wire with 100 A flowing in the $+x$ direction is situated in the xy plane and is parallel to the x axis at a distance of 1 m (straight wire crosses loop without touching it). If the loop current is 70 A flowing clockwise, as viewed from the $+z$ direction, find the vector value of the force on the loop.

★**5-1-20. Square and circular areas. Flux** If **B** = $\hat{z}6 \sin(\pi x/2) \sin(\pi y/2)$ T, find the total magnetic flux over a square area 2 m on a side with edges coincident with the positive x and y axes and one corner at the origin. (b) If **B** = $\hat{z}k/r$ T, what is the magnetic flux through a circle of radius r_0?

5-1-21. B from linear conductor A thin linear conductor of length l carrying a current I is coincident with the y axis. The medium surrounding the conductor is air. One end of the conductor is at a distance y_1 from the origin and the other end at a distance y_2. Show that the flux density due to the conductor at a point on the x axis at a distance x_1 from the origin is

$$B = \frac{\mu_0 I}{4\pi x_1}\left(\frac{y_2}{\sqrt{x_1^2 + y_2^2}} - \frac{y_1}{\sqrt{x_1^2 + y_1^2}}\right)$$

Note that if the center of the conductor coincides with the origin $(-y_1 = y_2)$ and if $x_1 \gg l$, the expression reduces to $B = \mu_0 Il/4\pi x_1^2$.

5-1-22. Parabolic B An infinite wire of parabolic shape carries a current I. Find the magnetic flux density B at the focus.

★**5-1-23. Hyperbolic B** An infinite wire of hyperbolic shape carries a current I. Find the magnetic flux density B at the foci.

5-1-24. Current elements. Force Referring to Fig. P5-1-24, show that the force between the two current elements situated in air is given by

$$d\mathbf{F} = \frac{\mu_0 I_2 dl_2 I_1 dl_1}{4\pi r^2}\hat{a}_1 \times (\hat{a}_2 \times \hat{r}) \tag{1}$$

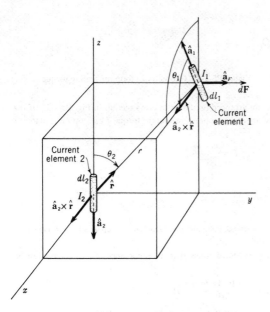

Figure P5-1-24 Geometry of short current-carrying elements for finding force between them.

where $d\mathbf{F}$ = force on element 1 due to current I_2 in element 2, N
  μ_0 = permeability of air, H m^{-1}
  dl_1, dl_2 = lengths of current elements 1 and 2, respectively, m
  I_1, I_2 = currents in elements 1 and 2, respectively, A
  r = distance between elements, m
  $\hat{\mathbf{a}}_1$ = unit vector in direction of current I_1 in element 1, dimensionless
  $\hat{\mathbf{a}}_2$ = unit vector in direction of current I_2 in element 2, dimensionless
  $\hat{\mathbf{r}}$ = unit vector in radial direction (from element 2 to 1), dimensionless

Show further that $d\mathbf{F} = \hat{\mathbf{a}}_F \, dF$, where dF is given by

$$dF = |d\mathbf{F}| = \frac{\mu I_2 \, dl_2 \, I_1 dl_1 \sin \theta_2 \sin \theta_1}{4\pi r^2}$$

where θ_2 = angle between $\hat{\mathbf{a}}_2$ and $\hat{\mathbf{r}}$ (see Fig. P5-1-24)
  θ_1 = angle between $\hat{\mathbf{a}}_1$ and $\hat{\mathbf{a}}_2 \times \hat{\mathbf{r}}$

and $$\hat{\mathbf{a}}_F = \frac{\hat{\mathbf{a}}_1 \times (\hat{\mathbf{a}}_2 \times \hat{\mathbf{r}})}{\sin \theta_2 \sin \theta_1}$$

where $\hat{\mathbf{a}}_F$ is the unit vector in the direction of force $d\mathbf{F}$. *Note*: These equations give the force on element 1 due to the presence of element 2 but not vice versa. That is, they are not symmetrical with respect to elements 1 and 2. However, with two closed circuits the force, as given by an integral of (1), is the same for both circuits. Thus, in the case of actual circuits Newton's third law, that to every action there is an equal (and opposite) reaction, is satisfied.

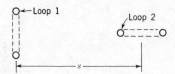

Figure P5-1-25 Geometry of loop for finding torque.

5-1-25. Two loops. Torque Two loops are arranged as shown in cross section in Fig. P5-1-25. If the separation s is large compared with the size of the loops, show that the torque T on loop 2 due to loop 1 is given by

$$T = \frac{\mu_0 mm'}{2\pi s^3}$$

where m = magnetic moment of loop 1
m' = magnetic moment of loop 2

Group 5-2: Secs. 5-11 through 5-17. The solenoid, inductors and inductance, Ampère's law, H, magneto-static potential, field cells and permeability

5-2-1. Inductance defined in terms of energy stored Show that the inductance of an inductor can be expressed as

$$L = \frac{\int \mathbf{B} \cdot \mathbf{H} \, dv}{I^2} \quad \text{(H)}$$

where the numerator involves the energy stored in the inductor and I equals the current flowing in the inductor winding. Compare this energy relation with the energy relation of Prob. 3-2-21 for capacitance and power relation of Prob. 4-1-1 for resistance.

★**5-2-2. Solenoid** A solenoid with 2,000 turns is 300 mm long and 20 mm in diameter. If the current is 600 mA, find (a) inductance of solenoid, (b) energy density at center of solenoid, and (c) energy stored in solenoid.

5-2-3. Conductor parallel to sheet Referring to Fig. P4-3-22, a current is flowing into the page on the cylindrical conductor and an equal return current out of the page on the flat sheet. The medium in this case is nonconducting ($\sigma = 0$) with $\epsilon = \epsilon_0$ and $\mu = \mu_0$. If the sheet current density K at D is $5\,\text{A m}^{-1}$ (out of the page), find (a) H at C; (b) K at A, B, and E; (c) the total current I on the cylindrical conductor; (d) the inductance $L\,\text{km}^{-1}$; and (e) capacitance $C\,\text{km}^{-1}$.

★**5-2-4. H, J, and B** (a) If $\mathbf{H} = \hat{\mathbf{r}}0 + \hat{\boldsymbol{\phi}}r + \hat{\mathbf{z}}0\,\text{A m}^{-1}$, find $\mathbf{J}$. (b) If $\mathbf{J} = \hat{\mathbf{z}}xy^2\,\text{A m}^{-2}$, find I. (c) If $\mathbf{B} = (\hat{\mathbf{x}} + \hat{\mathbf{z}})x^2y$ T, find ψ_m. In parts (b) and (c) find I and ψ_m through the 9 m² area defined by the xy points (0, 0), (3, 0), (3, 3), and (0, 3) m.

5-2-5. Flat coil A flat coil of 100 turns has an inner radius of 20 mm and outer radius of 120 mm. If $I = 1$ A, find (a) B at the center and (b) the inductance.

5-2-6. Split washer The thin flat washer of Fig. P5-2-6 has an inner radius of 20 mm and outer radius of 120 mm. If a uniform voltage V is applied between the edges of the radial slot resulting in an angular

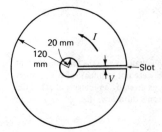

Figure P5-2-6 Split washer.

current of 100 A, find B at the center. Note that the current density will be a function of the radius. Neglect the effect of the slot width.

★5-2-7. B on solenoid axis A uniform cylindrical coil, or solenoid, of 2,000 turns is 600 mm long and 60 mm in diameter. If the coil carries a current of 15 mA, find the flux density (a) at the center of the coil, (b) on the axis at one end of the coil, and (c) on the axis halfway between the center and end of the coil. (d) Calculate and plot a graph of B as function of position along the axis of the solenoid from the center of the solenoid to a distance of 100 mm beyond one end.

5-2-8. B on solenoid axis Show that the flux density at a point P on the axis of a uniform solenoid is given by

$$ B = \mu_0 \frac{NI}{l} \left(1 - \frac{\Omega_1 + \Omega_2}{4\pi} \right) $$

where Ω_1 = solid angle subtended from point P by left end of solenoid ($= 2\pi$ if P is at left end of solenoid)

Ω_2 = solid angle subtended from point P by right end of solenoid ($= 2\pi$ if P is at right end of solenoid)

Note that at the center of a long slender solenoid $\Omega_1 = \Omega_2 \ll 4\pi$ so that $B = \mu_0 NI/l = \mu_0 K$.

5-2-9. Solenoid. Force A solenoid 300 mm long and 15 mm in diameter has a uniform winding of 3,000 turns. If the solenoid is placed in a uniform field of 5 T flux density and a current of 1 A is passed through the solenoid winding, what is (a) maximum force on solenoid, (b) maximum torque on solenoid, and (c) magnetic moment of solenoid?

5-2-10. Solenoid. Energy. Inductance A uniformly wound solenoid of 6,000 turns is 1 m long by 15 mm in diameter. If the current $I = 3$ A, what is (a) the flux density B, (b) magnetic field H, and (c) magnetic energy density w_m at the center of the solenoid? (d) What is the equivalent sheet current density K, (e) the total magnetic energy W_m, and (f) the inductance L?

★5-2-11. Asymmetric coaxial line. H, K, etc. The field map for a coaxial transmission line with asymmetrically located inner conductor is shown in Fig. P5-2-11. If there is 10 V between the conductors and the sheet-current density K at the point P_1 is 2 A m^{-1}, use the map to find (a) the electric field E at P_2, (b) E at P_3, (c) the magnetic field H at P_2, (d) H at P_3, (e) capacitance per unit length of line, (f) inductance per unit length of line, (g) electric flux density D at P_2, (h) D at P_3, (i) magnetic

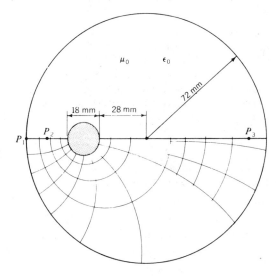

Figure P5-2-11 Coaxial line with asymmetric inner conductor.

flux density B at P_2, (j) B at P_3, (k) electric energy density w_e at P_2, (l) w_e at P_3, (m) magnetic energy density w_m at P_2, (n) w_m at P_3, (o) charge Q per unit length of line, (p) magnetic flux ψ_m per unit length of line, and (q) total current I.

5-2-12. Asymmetric coaxial line. J, R, G If the transmission line of Fig. P5-2-11 is filled with a conductor of uniform conductivity σ and 1 A flows between the inner and outer conductor per unit length of line, find (a) current density J at P_2, (b) J at P_3, (c) shunt resistance R per unit length of line, and (d) shunt conductance G per unit length of line. It is assumed that the transmission-line conductors have infinite conductivity.

5-2-13. Transmission line. Magnetic potential A transmission line consists of two long thin parallel conductors that carry currents of 10 A in opposite directions. The conductors are spaced a distance $2s$ apart. Draw a field map for a plane normal to the wires. Show both H lines and lines of equal magnetic potential. Indicate the value of potential for each equipotential line. Let the line joining the wires be arbitrarily taken to have zero potential. (Compare this map with Fig. 3-16 for two parallel lines of charge spaced a distance $2s$.)

★5-2-14. Tubular conductor. H A long, straight tubular conductor of circular cross section with an outside diameter of 50 mm and wall thickness of 5 mm carries a direct current of 50 A. Find H (a) just inside the wall of the tube, (b) just outside the wall of the tube, and (c) at a point in the tube wall halfway between the inner and outer surfaces.

5-2-15. Coaxial line. Internal inductance A coaxial transmission line has a solid inner conductor of radius a and a thin-walled outer conductor of radius b. The medium between the conductors has permeability μ_0. Find the inductance per meter of length of the line (a) if all current of the inner conductor is on its surface (at radius a) (actually a thin layer of negligible thickness) and (b) if the current of the inner conductor is uniformly distributed throughout its cross-sectional area, that is, if the *internal* inductance of the inner conductor is taken into account. (c) How much is the internal inductance? (d) For what radius ratio b/a will the internal and external inductances be equal? At high frequencies the *skin effect* tends to confine currents to the surface [current not uniformly distributed as in (a)] and the internal inductance may be neglected. Note that the internal inductance is independent of the conductor radius and, hence, always equal to the result in (c). Unless the external inductance is much greater than this, the internal inductance should be taken into account.

5-2-16. Coaxial line. Skin effect An air-filled coaxial transmission line has a solid inner conductor of radius a and an outer conductor of inner radius b and outer radius c. At high frequencies the current is confined by the *skin effect* to a thin layer of thickness δ on the outside of the inner conductor and inside of the outer conductor. Under such conditions show that the total inductance per unit length of line is given by

$$\frac{L}{d} = \frac{\mu_0}{2\pi}\left(\ln\frac{b}{a} + \frac{\delta}{2a} + \frac{\delta}{2b}\right) \quad (\text{H m}^{-1})$$

The skin depth δ is a function of the frequency and conductivity as discussed in Chap. 10.

5-2-17. Hollow conductor. Internal inductance A hollow cylindrical conductor has an inner radius a and outer radius b. If the current I is uniformly distributed, find the internal inductance L per unit length.

Group 5-3: Secs. 5-18 through 5-22. Curl, Maxwell's curl equation, and vector potential

★5-3-1. J in conductor The magnetic field of a long cylindrical conductor is given by

$$H_\phi = kr^2 \quad (\text{A m}^{-1})$$

Find **J**.

5-3-2. Vector potential If the vector potential $\mathbf{A} = \hat{\mathbf{r}}3 + \hat{\boldsymbol{\phi}}r + \hat{\mathbf{z}}6$ Wb m^{-1}, find the magnetic flux density **B**.

⋆5-3-3. B from wire A 2-m-long straight wire extends from $x = 0$, $y = -1$ m through the origin to $(x = 0, y = 1)$ m. If the wire current $I = 50$ A, find B at $(x = 1, y = 0)$ m.

5-3-4. Current element. B. (a) Using the vector potential and its curl, deduce the magnetic flux density B at a distance of 2 m normal to a short current element of length L with current of 10 mA. (b) Find B for the situation in (a) using the Biot-Savart law.

5-3-5. Curl If $F = \hat{x}y + \hat{y}x$, find $\nabla \times F$.

⋆5-3-6. Curl If $F = \hat{x}x^2 + \hat{y}2yz - \hat{z}x^2$, find $\nabla \times F$ and the path of $\nabla \times F$.

5-3-7. Curl If $F = \hat{x}2x + \hat{y}4xy^2z$, find (a) $\nabla \times F$ and (b) $\nabla \times \nabla \times F$.

5-3-8. $\nabla \times H = J$ (a) Find the current I through a square area 2 m on a side with edges coinciding with the $+x$ and $+y$ axes and one corner at the origin if $H = \hat{x}2y^2$ A m^{-1}. (b) Repeat for $H = \hat{z}3x^2y$ A m^{-1}.

Group 5-4: Sec. 5-23. Moving particles in electric and magnetic fields

⋆5-4-1. Particle acceleration A particle with a negative charge of 10^{-17} C and a mass of 10^{-26} kg is at rest in a field-free space. If a uniform electric field $E = 1$ kV m^{-1} is applied for 1 μs, find (a) the velocity of the particle and (b) the radius of curvature of the particle path if the particle enters a magnetic field $B = 2$ mT with this velocity moving normal to **B**.

5-4-2. Betatron The average flux density inside the electron orbit of a betatron is given by $B = 60$ mT, where $\omega = 2\pi \times 60$ rad s^{-1}. If the radius of the electron orbit is 120 mm, find (a) velocity of electrons at $t = 1$ ms if $v \approx 0$ at $t = 0$, (b) energy per unit charge imparted to the electron per revolution, (c) number of revolutions in 1 ms, and (d) electron energy (in electron volts) after 1 ms. Neglect relativistic effects.

5-4-3. Deuteron, proton, and alpha particle (a) A deuteron accelerated through a potential difference V enters a region of uniform magnetic flux density B moving normal to **B**. If the radius of the deuteron path is 120 mm, what is the required relation between B and V? (b) If V and B are the same as in (a) but the particle is a proton, find the radius of its path. (c) Repeat for an alpha particle. (d) Compare the kinetic energies of the three particles.

⋆5-4-4. Cyclotron (a) Find the maximum energy (in MeV) for alpha particles in a cyclotron with maximum usable radius of 450 mm. The flux density $B = 1.2$ T in the air gap. (b) Repeat (a) if protons are used. (c) Repeat (a) if deuterons are used. (d) How many revolutions does each particle make if the peak potential between the dees is 9 kV?

5-4-5. Radius ratio (a) What is the ratio of the path radius of a proton to that of a deuteron if both enter a magnetic field **B** normally with the same kinetic energy? (b) What is the ratio of the path radius of an alpha particle to that of a deuteron?

⋆5-4-6. Radius of curvature Find the radius of curvature for the path of a proton with velocity $v = 10$ Mm s^{-1} moving perpendicular to a magnetic field with (a) $B = 5$ mT (solar-corona magnetic field), (b) $B = 100$ μT (earth's magnetic field), and (c) $B = 10$ nT (interstellar magnetic field).

5-4-7. Helical path A 5-keV proton is injected into a magnetic field $B = 500$ mT at an angle of 85° with respect to **B**. Find the following parameters of the resulting helical path: (a) radius, (b) distance between turns of helix, (c) circular velocity in radians per second and meters per second, (d) axial velocity in meters per second, and (e) period.

⋆5-4-8. Mirror point If an electron is injected into a magnetic field traveling nearly parallel to **B**, it will move in a helical path with large distance between turns. If **B** gradually increases, the distance between turns will decrease until a point is reached where the distance between turns becomes zero and the electron then reverses its direction with respect to **B** and returns toward its starting point (see Fig. P5-4-8). It is as though the electron were reflected by a mirror. What relation must be satisfied between the electron energy and the magnetic field at such a *mirror point*?

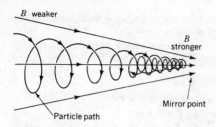

B weaker

B stronger

Mirror point

Particle path **Figure P5-4-8** Magnetic mirror.

5-4-9. Magnetic bottle A charged particle can be contained between two mirror points forming a magnetic "bottle." The magnetosphere of the earth forms such a bottle for particles which travel between northern and southern hemispheres (see Fig. P5-4-9). The maximum value of the earth's magnetic field $B \approx 100 \ \mu T$. What energy-field conditions are required for particles to be contained if the particles are (a) electrons and (b) protons? Higher-energy particles will be dumped, producing an auroral display.

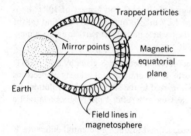

Trapped particles

Mirror points

Magnetic equatorial plane

Earth

Field lines in magnetosphere **Figure P5-4-9** Magnetic bottle in magnetosphere.

5-4-10. Proton curvature. Cosmic ray Find the radius of curvature of a proton with velocity $v = 2 \times 10^8 \ \text{m s}^{-1}$ traveling normally to the following magnetic fields: (a) synchrotron, $B = 1 \ \text{T}$; (b) earth's magnetosphere, $B = 50 \ \mu T$; (c) solar corona, $B = 5 \ \text{mT}$; and (d) interstellar medium, $B = 10 \ \text{nT}$. The particle is relativistic; i.e., its mass is significantly larger than its rest mass [see (5-23-12)]. Such a particle in space could be called a *cosmic ray*.

5-4-11. Synchrotron A *synchrotron* is a particle accelerator similar to a cyclotron but with the following modifications. The magnetic field B and gyro frequency f are both varied in such a way that the particle radius is a constant ($= R_0$), so that an annular, or ring-shaped, magnet can be used, at a substantial cost reduction. Show that the two conditions which must be satisfied in a synchrotron are that $B = 2\pi f m/e$ and $f = v/2\pi R_0$. Thus, increasing the frequency increases the particle velocity (and energy) while B must be increased with the frequency (and particle mass) to keep the particle radius constant.

5-4-12. Cathode-ray tube. Electric field deflection Show that the deflection of an electron at the screen of a cathode-ray tube is given by

$$y = \frac{V_d l x}{2 V_a d} \quad (\text{m})$$

where y = deflection distance at screen, m
 V_d = deflecting potential, V
 l = length of deflecting plates, m
 x = distance from deflecting plates to screen, m
 V_a = accelerating potential, V
 d = spacing of deflecting plates, m

5-4-13. Cathode-ray tube. Magnetic field deflection Show that the deflection of an electron at the screen of a cathode-ray tube is given by

$$y = xBl \sqrt{\frac{e}{2mV_a}} \quad \text{(m)}$$

where y = deflection distance at screen, m
 x = distance from magnetic deflecting field to screen, m
 B = flux density of deflecting field, T
 e = charge on particle, C
 m = mass of particle, kg
 V_a = accelerating voltage, V
 l = axial length of deflecting field, m

★**5-4-14. Particle radius** A particle of charge 10^{-25} C and mass 10^{-20} kg is at rest in a magnetic field $B = 1\ \mu$T. If an electric field of 2 V m^{-1} is applied perpendicular to the direction of B for 1 ns, what is the radius of curvature of the resulting particle path?

5-4-15. Particle radius An electric field of 1 MV m^{-1} is applied to a particle at rest of 1 aC charge and mass of 1 pg for 1 ms, after which the particle enters a magnetic field of 1 mT at right angles to **B**. Find the radius of curvature R of the particle path.

★**5-4-16. Particle radius and braking field** A particle with a charge 10^{-24} C and mass 10^{-24} kg has a linear velocity **v** of 1 km s^{-1}. (a) If the particle enters a magnetic field **B** perpendicular to **v**, find the radius of curvature R. (b) Find R if **v** is parallel to **B**. $B = 2$ T. (c) Find the braking electric field E required to bring the particle to rest in 100 μs. v (linear) = 1 km s^{-1}.

5-4-17. Particle radius An electric field of 10^4 V m^{-1} is applied to a particle at rest having a charge of 10^{-20} C and mass of 10^{-20} kg. After 10 ms the particle enters a magnetic field of 1 T at right angles to **B**. Find the radius of curvature R of the particle path.

Group 5-5: Practical applications

★**5-5-1. Cathode-ray tube** A cathode-ray tube has electrostatic-deflection plates as in Fig. P5-5-1 plus a magnetic field between these plates. The magnetic field is oriented parallel to the electrostatic-deflection field E_d. The tube has an accelerating voltage $V_a = 7$ kV, a plate spacing $d = 20$ mm, and a plate length (and magnetic field length) $l = 20$ mm. Find (a) the deflection-plate voltage and (b) the magnetic flux density B required to deflect the electron beam 50 mm in the y direction and 100 mm in the z direction (perpendicular to page) at a screen placed at a distance $x = 500$ mm from the deflecting plates.

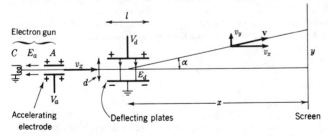

Figure P5-5-1 Cathode-ray tube with electron gun, electrostatic deflection plates, and screen.

5-5-2. TV screen The cathode-ray-tube screen of a television receiver measures 400 mm (horizontally) by 300 mm (vertically). The electron-beam spot scans 525 lines per frame and 30 frames per second. Find (a) the horizontal spot velocity in millimeters per microsecond and (b) the total distance traveled by the spot per second. (*c*) What is the rate of change of magnetic flux density *B* required if the tube parameters are as in Prob. 5-5-1.

★**5-5-3. Smoke precipitator. Rectangular flue** The chimney flue of an industrial factory has a rectangular (interior) cross section of 1 m by 0.5 m. Flat electrodes 1 m² are placed against the long sides of the flue to precipitate dust particles in the smoke electrostatically. If a typical particle has a mass of 1 μg and a charge of 10^{-14} C, what voltage between the electrodes is required to precipitate the particle against an electrode? Take the smoke velocity $v = 1$ m s^{-1} and assume that when the particle enters the electrode space it is at the greatest possible distance (0.25 m) from the electrode to which it will move.

5-5-4. Smoke precipitator. Cylindrical chimney Design a precipitator for a 1-m-diameter cylindrical steel chimney flue capable of precipitating solid particulate wastes being exhausted at the rate of 10 kg min^{-1} with a velocity of 3 m s^{-1}. Assume that the particles have masses of 1 μg each. The flue is 20 m tall. *Hint*: Suspend a charged vertical wire down the center of the flue.

5-5-5. Mine sweeper A trawler used as a magnetic-mine sweeper tows a floating "tail" 300 m long consisting of a single insulated conductor with an electrode at its end which makes contact with the water. Current flows through the conductor and back through the water to an electrode on the hull of the trawler. In sea water 20 m deep how much current is required in the tail to produce a magnetic field *B* of 10 gauss (10 G) at a mine on the sea bottom directly under the middle of the tail? (10 G is an order of magnitude greater than the earth's magnetic field.)

5-5-6. B under transmission line A two-wire transmission line (wires side by side) is suspended 12 m above the ground. If the wires are 6 m apart and carry 2 kA in opposite directions, calculate and graph the magnitude of **B** along the ground to a distance of about 50 m either side of the line.

5-5-7. Helmholtz pair. Field uniformity A simple way of producing a nearly uniform magnetic field over a considerable region of space is by means of two identical coils placed side by side with axes coincident and spaced a distance equal to their radius. This configuration constitutes a *Helmholtz pair* after the German scientist Hermann von Helmholtz. Consider that the 2 coils are 1 m in radius and separated by 1 m. If each coil has 100 turns and carries 10 A in the same sense, (*a*) calculate and plot the variation of *H* along the axis of the coils from the center of one coil to the center of the other. (*b*) Calculate and plot the variation of *H* along the axis of a single 100-turn coil 1 m in radius due to a current *I* in the coil. Let the single coil be situated halfway between the coils of the Helmholtz pair and with its axis coincident with the axis of the pair. Also let *I* have such a value that *H* at the center of the single coil is the same as *H* from the Helmholtz pair at this point. Assume that the coils have negligible cross-sectional area so that each may be represented by a thin single-turn loop. (*c*) What is the ratio of *H* at the midpoint of the Helmholtz pair to *H* at the center of one of the coils? (*d*) What is the ratio of *H* at the center of the single coil to *H* on its axis at a distance equal to its radius? (*e*) What is the percentage improvement in uniformity of field of the Helmholtz pair over the single coil along the axes of the coils? (*f*) Over what axial distance is the Helmholtz pair field constant to within 1, 2, and 5 percent? (*g*) Over what axial distance is the single-coil field constant to within these percentages?

5-5-8. Hall effect. Force relations (*a*) When a current *I* flows as shown (Fig. P5-5-8) in a flat conducting strip with magnetic field **B** normal to the strip, there is a force per unit charge to the right. Thus,

$$\mathbf{F} = (\mathbf{I} \times \mathbf{B})l = I(\mathbf{l} \times \mathbf{B}) = \frac{Q}{t}\mathbf{l} \times \mathbf{B} = Q(\mathbf{v} \times \mathbf{B})$$

or

$$\frac{\mathbf{F}}{Q} = \mathbf{v} \times \mathbf{B}$$

If there are equal numbers of positive and negative charge carriers, the points P_1 and P_2 develop no potential difference. In metals the carriers are negative charges (electrons), and P_2 becomes negative with respect to P_1. This is the *Hall effect*, after Edwin H. Hall, American physicist. Show that the emf developed between P_1 and P_2 is given by $\mathscr{V} = IB/nqd$, where *n* is the number of electrons per cubic meter in the metal, *q* is the electron charge, and *d* is the thickness of the strip. (*b*) If $I = 50$ A, $B = 2$ T,

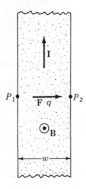

Figure P5-5-8 Conducting strip in magnetic field for determination of Hall effect voltage.

$w = 25$ mm, and $d = 0.5$ mm, find the Hall voltage between P_1 and P_2 for a copper strip. Take $n = 8.4 \times 10^{28}$ m^{-3} for copper. (c) Why does the width w of the strip not appear in the above equation for the Hall voltage?

★**5-5-9. Transmission line. Balance of forces** A transmission line consists of two parallel 10-mm-diameter conductors separated (on centers) by 500 mm. If there is a potential difference $V = 100$ kV between the conductors, is there a line current I which results in a balance between the electric and magnetic forces acting on the wires? If so, what is its value?

5-5-10. Electron propulsion (a) Design an electron gun for the propulsion of a space vehicle. Specify electron beam current and energy required to produce a thrust of 1 N. (b) Design a proton gun for the same purpose. Specify proton beam current and energy required to produce a thrust of 1 N.

5-5-11. Mercury pump with no moving parts Design a pump for liquid mercury using no moving parts which can move 1 liter of mercury per second against a head of 2 m of mercury.

SIX

THE STATIC MAGNETIC FIELD OF FERROMAGNETIC MATERIALS

6-1 INTRODUCTION

Magnetic fields are present around a current-carrying conductor. They also exist around a magnetized object such as an iron bar magnet. The field of the iron bar is not produced by current circuits of the type considered in Chap. 5, but currents are the cause. In the bar the currents flow in circuits of atomically small dimensions. In contrast to these *microscopic* circuits, the circuits considered in Chap. 5 are of *macroscopic* size.

An electron circulating in an orbit around the nucleus of an atom behaves like a tiny current loop. Furthermore, the electron itself may behave like an infinitesimal current loop because of its inherent spin. The nucleus may also act similarly because of its spin, but the effect is usually not significant. Each current loop produces a magnetic field so that an atom with a number of electrons has many possible sources of magnetic field. For most atoms, however, the magnetic effects produced by both the spin and orbit motion of the electrons tend to balance out so that the atom shows very little magnetic effect.

In other atoms, however, the spin and orbit fields do not cancel and the material shows magnetic effects. In iron, nickel, and cobalt these effects are especially strong, and these metals are called *ferromagnetic* materials. Although the effect of each atomic current loop is very small, the combined effect of billions of them in an iron bar results in a strong magnetic field around the bar.

6-2 MAGNETIC DIPOLES, LOOPS, AND SOLENOIDS

If an iron bar is freely suspended, as in a compass, it will turn in the earth's magnetic field so that one end points north. This end is called the *north-seeking pole* of the magnet or simply its *north pole*. The other end of the magnet has a pole of opposite

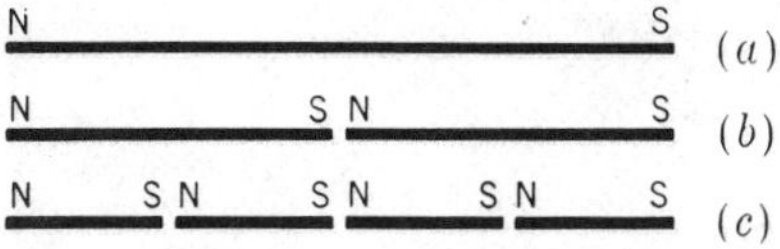

N ————————————————— S　(a)

N ——————— S N ——————— S　(b)

N —— S N —— S N —— S N —— S　(c)

Figure 6-1 New poles appear at each point of division of a bar magnet.

polarity called a *south pole*. It is often convenient to call a north pole a *positive pole* and a south pole a *negative pole*.

All magnetized bodies have both a north and a south pole. They cannot be isolated. For example, consider the long magnetized iron rod of Fig. 6-1a. This rod has a north pole at one end and a south pole at the other. If the rod is cut in half, new poles appear, as in Fig. 6-1b, so that there are two magnets. If each of these is cut in half, we obtain four magnets, as in Fig. 6-1c, each with a north and a south pole. The reason for this is that the ultimate source of ferromagnetism is an atomic current circuit which acts like a tiny magnet with a north and a south pole. Therefore, even if the cutting process could be continued to atomic dimensions and a single iron atom isolated, it would still have a north and a south pole.

The fact that magnetic poles cannot be isolated, whereas electric charges can, is an important point of difference between electric charges and magnetic poles.

If an iron bar is placed on a wooden table and covered with a sheet of paper, iron filings sprinkled on the sheet align themselves along the magnetic field lines of the magnet, as suggested in Fig. 6-2. The field configuration is that of a dipole with magnetic poles separated by a distance approximately equal to the length of the magnet. (Compare with the electric dipole field of Fig. 2-13.)

Figure 6-2 Iron filings align themselves with magnetic field around bar magnet. Near magnet friction was inadequate to keep filings from being pulled in, causing open areas.

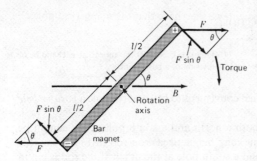

Figure 6-3 Bar magnet experiences torque tending to align it with magnetic field **B**.

Consider now a bar magnet (magnetic dipole) of pole strength Q_m and length l placed in a uniform magnetic field B, as suggested in Fig. 6-3. The north $(+)$ pole experiences a force $F = Q_m B$ to the right and the south $(-)$ pole an equal force to the left. The torque T, or turning moment (force $\times$ distance), on the dipole is

$$T = 2F \frac{l}{2} \sin \theta \qquad \text{(N m)} \tag{1}$$

where $F = Q_m B$

   l = length of dipole

   θ = angle between dipole axis and B

Introducing the value of F, we have

$$T = Q_m l B \sin \theta \qquad \text{(N m)} \tag{1a}$$

where $Q_m l$ = magnetic dipole moment, A m^2

Now the torque on a single-turn current loop in a uniform magnetic field B is, from (5-10-4),

$$T = IAB \sin \theta \tag{2}$$

where IA = magnetic moment of the loop, A m^2

   I = loop current, A

   A = area of loop, m^2

Comparing (1) and (2), it is evident that if $Q_m l = IA$, the dipole and the loop are equivalent.

Furthermore from (5-11-8), the magnetic moment of a solenoid and that of a bar magnet are equivalent if

$$Q_m l = NI'A \qquad \text{(A m}^2\text{)} \tag{3}$$

where N = number of turns of solenoid, dimensionless

   I' = solenoid current, A

   A = area of solenoid cross section, m^2

Letting m equal the magnetic moment, we then have that the magnetic dipole, the loop, and the solenoid will all be equivalent if

$$Q_m l = IA = NI'A = m \qquad (\text{A m}^2) \tag{4}$$

and, in general, we can write the torque on any one as

$$T = mB \sin \theta \qquad (\text{N m}) \tag{5}$$

Regarding the magnetic moment as a vector **m** with direction $\hat{n}$ parallel to the axis of the dipole and solenoid and perpendicular to the plane of the loop, as in Fig. 6-4, (5) can be expressed by the cross product as

$$\mathbf{T} = \mathbf{m} \times \mathbf{B} \qquad (\text{N m}) \tag{6}$$

where **T** = torque on bar, loop, or solenoid, N m

m = $\hat{n}m$ = magnetic moment = $\hat{n}Q_m l = \hat{n}IA = \hat{n}NI'A$, A m^2

B = flux density of applied field, T

A bar, loop, and solenoid are compared in Fig. 6-4. Not only will the torques be the same provided $Q_m l = IA = NI'A$ but, at a sufficient distance and with no applied field, the magnetic fields of all three will be the same. In fact, if the solenoid has the same dimensions as the bar magnet, the external fields of the two may be identical. Note that the positive sense of the unit vector $\hat{n}$ for the dipole is from the negative to positive poles while for the loop and solenoid it is determined by the right-hand rule (fingers in the direction of the current, thumb in the direction of $\hat{n}$).

In all cases the torque tends to align the unit vector $\hat{n}$ with **B**, the torque vector **T** being coincident with the axis of rotation (perpendicular to page) and directed so that **m**, **B**, and **T** form a right-handed set (as for x, y, and z in Fig. 2-3). In other words, turning **m** or $\hat{n}$ into **B**, **T** is then in the direction of motion of a right-handed screw.

An important aspect of (5) or (6) is that although the structure of a magnetic object may not be known, its strength can be described in terms of its *magnetic moment* **m**, which can be determined by measuring its torque **T** in a known magnetic field **B**.

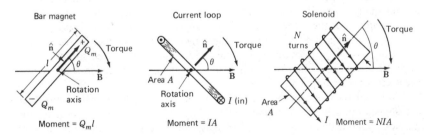

Figure 6-4 Bar magnet (magnetic dipole), current loop, and solenoid in a uniform field **B**. In all three cases the torque is clockwise and tends to align the unit vector $\hat{n}$ with **B**. If $Q_m = IA = NI'A$, the torque is the same in all three cases. The loop is shown in cross section.

6-3 MAGNETIC MATERIALS

All materials show some magnetic effects. In many substances the effects are so weak that the materials are often considered to be *nonmagnetic*. However, a vacuum is the only truly nonmagnetic medium.

In general, materials can be classified according to their magnetic behavior into *diamagnetic, paramagnetic, ferromagnetic, antiferromagnetic, ferrimagnetic,* and *superparamagnetic*. In *diamagnetic* materials magnetic effects are weak. Although the orbit and spin magnetic moments in such materials cancel (net magnetic moment is zero) in the absence of an external magnetic field, an applied field causes the spin moment to slightly exceed the orbital moment, resulting in a small net magnetic moment which opposes the applied field **B**. Thus, if a diamagnetic specimen is brought near either pole of a strong bar magnet, it will be *repelled*, an effect discovered by Michael Faraday in 1846. Faraday's specimen was a piece of bismuth, a substance which shows diamagnetism more strongly than most materials.

Although the diamagnetic effect is present in all substances, it is so weak that it is overshadowed in materials in which the orbit and spin magnetic moments are unequal, resulting in a net magnetic moment for the atom even with no applied field. Random orientation of the atoms may result in little net magnetic moment for a sample of the material, but when an external field is applied, the atomic dipoles tend to line up with the field so that the magnetic moment of the sample is increased in proportion to the number of atoms in the sample (provided perfect alignment is achieved). Internal interactions and thermal agitation tend, however, to inhibit the process so that only partial alignment may actually be achieved. Nevertheless, magnetic effects may be significant, and such substances are called *paramagnetic*. When a paramagnetic substance is brought near the pole of a strong bar magnet, it will be *attracted* to it.

In a few materials, especially iron, nickel, and cobalt, a special phenomenon occurs which greatly facilitates the alignment process. In these substances, called *ferromagnetic*, there is a quantum effect known as "exchange coupling" between adjacent atoms in the crystal lattice of the material which locks their magnetic moments into a rigid parallel configuration over regions, called *domains*, which contain many atoms. However, at temperatures above a critical value, known as the *Curie temperature*, the exchange coupling disappears and the material reverts to an ordinary paramagnetic type. Ferromagnetism is discussed further in Sec. 6-11.

In *antiferromagnetic* materials the magnetic moments of adjacent atoms align in opposite directions so that the net magnetic moment of a specimen is nil even in the presence of an applied field.

Although the magnetic moments of adjacent atoms in *ferrimagnetic* substances are aligned opposite, the moments are not equal so that there is a net magnetic moment, but one that is less than in ferromagnetic materials. In spite of the weaker magnetic effects, some of these materials, known as *ferrites*, have a smaller electrical conductivity, which makes them useful in the cores of alternating-current inductors and transformers since induced (eddy) currents are less and ohmic (heat) losses are reduced.

A *superparamagnetic* material consists of ferrromagnetic particles suspended in a dielectric (plastic) binder or matrix. Each particle may contain many magnetic domains, but exchange forces cannot penetrate to adjacent particles. With the particles suspended in a thin plastic tape it is possible to change the state of magnetization abruptly in a very small tape distance so that a tape can store large amounts of information in magnetic form in convenient lengths. Such tapes are widely used in audio, video, and data recording systems.

A number of substances are classified in Table 6-1 according to their magnetic behavior.

6-4 RELATIVE PERMEABILITY

An important characteristic of a magnetic material is its permeability μ. It is often convenient to speak of the ratio of the permeability of the material to that of vacuum (μ_0). This ratio, or *relative permeability*, μ_r, is thus given by

$$\mu_r = \frac{\mu}{\mu_0} \qquad (1)$$

where μ_r = relative permeability, dimensionless

μ = permeability, H m^{-1}

μ_0 = permeability of vacuum = 400π nH m^{-1}

It is to be noted that the relative permeability is a dimensionless ratio.

The relative permeability of vacuum or free space is unity by definition. The relative permeability of diamagnetic substances is slightly less than 1, while for paramagnetic substances it is slightly greater than 1. The relative permeability of the ferromagnetic materials is generally much greater than 1 and in some special alloys may be as large as 1 million.

The relative permeability of diamagnetic and paramagnetic substances is relatively constant and independent of the applied field, much as the relative permittivity of dielectric substances is independent of the applied electric field intensity. However, the relative permeability of ferromagnetic materials varies over a wide range for different applied fields. It also depends on the previous history of the specimen (see Hysteresis, Sec. 6-13). However, the *maximum* relative permeability is a relatively definite quantity for a particular ferromagnetic material although in different materials the maximum may occur at different values of the applied field. This subject is considered in more detail in Sec. 6-12.

In Table 6-1, the relative permeabilities μ_r are listed for a number of substances. The substances are arranged in order of increasing permeability, and they are also classified as to group type. The value for the ferromagnetic materials is the maximum relative permeability.

Table 6-1 Relative permeabilities

Substance	Group type	Relative permeability μ_r
Bismuth	Diamagnetic	0.99983
Silver	Diamagnetic	0.99998
Lead	Diamagnetic	0.999983
Copper	Diamagnetic	0.999991
Water	Diamagnetic	0.999991
Vacuum	Nonmagnetic	1†
Air	Paramagnetic	1.0000004
Aluminum	Paramagnetic	1.00002
Palladium	Paramagnetic	1.0008
2-81 Permalloy powder (2 Mo, 81 Ni)‡	Ferromagnetic	130
Cobalt	Ferromagnetic	250
Nickel	Ferromagnetic	600
Ferroxcube 3 (Mn-Zn-ferrite powder)	Ferrimagnetic	1,500
Mild steel (0.2 C)	Ferromagnetic	2,000
Iron (0.2 impurity)	Ferromagnetic	5,000
Silicon iron (4 Si)	Ferromagnetic	7,000
78 Permalloy (78.5 Ni)	Ferromagnetic	100,000
Mumetal (75 Ni, 5 Cu, 2 Cr)	Ferromagnetic	100,000
Purified iron (0.05 impurity)	Ferromagnetic	200,000
Supermalloy (5 Mo, 79 Ni)§	Ferromagnetic	1,000,000

† By definition.
‡ Percentage composition. Remainder is iron and impurities.
§ Used in transformer applications with continuous tape-wound (gapless) cores.

6-5 MAGNETIC DIPOLES AND MAGNETIZATION

A loop of area A with current I has a magnetic moment of IA. As already discussed, the fields at a large distance from this loop are identical with those of a bar magnet of dipole moment $Q_m l$, where Q_m is the magnetic pole strength and l is the pole separation, provided the magnetic moment of the bar is equal to that of the loop (see Fig. 6-5). Thus, †

$$Q_m l = IA \tag{1}$$

It was Ampère's theory that the pronounced magnetic effects of an iron bar occur when large numbers of atomic-sized magnets associated with the iron atoms are oriented in the same direction so that their effects are additive. The precise nature of the tiny magnets is not important if we confine our attention to regions containing large numbers of them. Thus, they may be regarded as tiny magnets or

---

† This equivalence may be shown from the fact that a current loop in a field B has a maximum torque $T = IAB$ [from (5-10-4)] while for a bar magnet the maximum torque $T = Q_m lB$. It follows that for equal torques the moments must be equal.

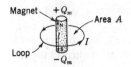

Figure 6-5 Bar magnet of moment $Q_m l$ and equivalent current loop of moment IA.

as miniature current loops. In either case, it is sufficient to describe them by their magnetic moment, which can be expressed either as $Q_m l$ or as IA.

Consider a uniformly magnetized iron rod as shown in cross section in Fig. 6-6. The atomic magnets are all oriented the same way and are uniformly distributed with the north pole of one magnet so close to the south pole of the next magnet that the poles of adjacent magnets annul each other's effects everywhere except at the ends of the rod, leaving a surface layer of north poles on one end and south poles on the other. The situation is analogous to that for a uniformly polarized dielectric slab with the effects of its atomic electric dipoles canceling throughout the interior but resulting in layers of opposite electric (bound) charges on the two surfaces of the slab.

The effect of the atomic magnets can be described by a quantity called the *magnetization M*, defined as the magnetic dipole moment per unit volume, analogous to the polarization P (for the electric dipole moment per unit volume in the dielectric case). Thus, the magnetization is

$$M = \frac{m}{v} = \frac{Q_m l}{v} \tag{2}$$

where $m = Q_m l$ is the magnetic dipole moment in volume v.

Magnetization has the dimensions of both magnetic dipole moment per volume and of magnetic pole strength per area ($IL^2/L^3 = I/L$). It is expressed in amperes per meter (A m^{-1}).

The value of M in (2) is an average for the volume v. To define M at a point, it is convenient to assume that the iron rod has a continuous distribution of infinitesimal magnetic dipoles, i.e., a continuous magnetization, whereas the dipoles actually are of discrete, finite size. Nevertheless, the assumption of continuous

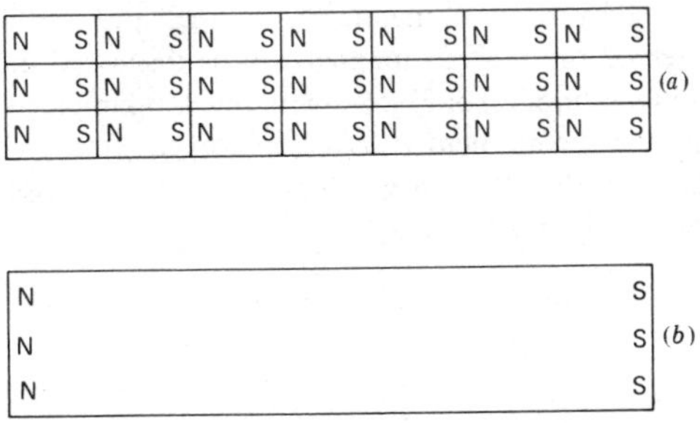

N	S	N	S	N	S	N	S	N	S	N	S	N	S	
N	S	N	S	N	S	N	S	N	S	N	S	N	S	(a)
N	S	N	S	N	S	N	S	N	S	N	S	N	S	

N													S	
N													S	(b)
N													S	

Figure 6-6 (a) The effects of the atomic dipoles of a uniformly magnetized rod annul each other except at the ends of the rod, leaving a layer of north poles at one end and south poles at the other end, as in (b). (Pole model.)

magnetization leads to no appreciable error provided that we restrict our attention to volumes containing many magnetic dipoles. Then, assuming continuous magnetization, the value of M at a point can be defined as the net dipole moment m of a small volume Δv divided by the volume with the limit taken as Δv shrinks to zero around the point. Thus

$$M = \lim_{\Delta v \to 0} \frac{m}{\Delta v} \quad (\text{A m}^{-1}) \tag{3}$$

If M is known as a function of position in a nonuniformly magnetized rod, the total magnetic moment of the rod is given by

$$m = \int_v M \, dv \quad (\text{A m}^2) \tag{4}$$

where the integration is carried out over the volume of the rod.

Example If the long uniformly magnetized rod of Fig. 6-6 has N' elemental magnetic dipoles of moment Δm, find the magnetization of the bar.

SOLUTION From (2) the magnetization is

$$M = \frac{N'}{v} \Delta m = N'' \Delta m$$

where M = magnetization, A m^{-1}

$N'' = N'/v$ = elemental dipoles per unit volume, number m^{-3}

In this case the magnetization M is both an average value and also the value anywhere in the rod since the magnetization is assumed uniform.

6-6 UNIFORMLY MAGNETIZED ROD AND EQUIVALENT AIR-FILLED SOLENOID

In Sec. 6-5 magnetization was explained by means of magnets with north and south poles (pole model). Now the magnetization of a bar magnet will be discussed in terms of the atomic current loops (current model) of Fig. 6-7 with one loop in place of each tiny magnet of Fig. 6-6 and with the moment IA of each loop equal to the moment $Q_m l$ of each magnet. Assuming that there are n loops in a single cross section of the rod (as in the end view in Fig. 6-7), we have

$$nA' = A \tag{1}$$

where A' = area of elemental loop
A = cross-sectional area of rod

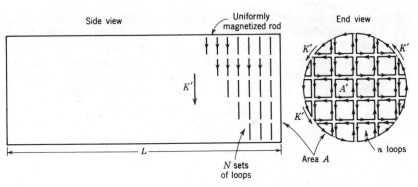

Figure 6-7 Effects of atomic current loops of a uniformly magnetized rod annul each other except on the cylindrical surface of the rod leaving a current sheet around the rod. (Current model.)

Further, let us assume that there are N such sets of loops in the length of the rod (see the side view in Fig. 6-7). Then

$$nN = N' \tag{2}$$

where n = number of loops in a cross section of rod
$\quad N$ = number of such sets of loops
$\quad N'$ = total number of loops in rod

It follows that the magnetization M of the rod is given by

$$M = \frac{m}{v} = \frac{N'IA'}{LA} = \frac{NI}{L}\frac{nA'}{A} = \frac{NI}{L} = K' \tag{3}$$

where K' = equivalent sheet current density on outside surface of rod, A m^{-1}
$\quad L$ = length of rod, m

Referring to the end view of the rod in Fig. 6-7, it is to be noted that there are equal and oppositely directed currents wherever loops are adjacent, so that the currents have no net effect with the exception of the currents at the periphery of the rod. As a result there is the equivalent of a current sheet flowing around the rod, as suggested in Figs, 6-7 and 6-8a. This sheet has a linear current density K' (A m^{-1}).

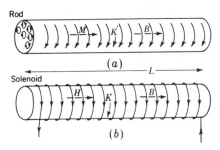

Figure 6-8 (a) Uniformly magnetized rod and (b) equivalent solenoid.

Although the sets of current loops are shown for clarity in Fig. 6-7 with a large spacing, the actual spacing is of atomic dimensions, so that macroscopically we can assume that the current sheet is continuous.

This type of a current sheet is effectively what we also have in the case of a solenoid with many turns of fine wire, as in Fig. 6-8b (see also Fig. 5-15 and in particular Fig. 5-16). The actual sheet-current density K for the solenoid is

$$K = \frac{NI}{L} \quad (\text{A m}^{-1}) \tag{4}$$

where N = number of turns in solenoid, dimensionless
$\quad I$ = current through each turn, A
$\quad L$ = length of solenoid, m

The sheet-current density K is expressed in amperes per meter.

If the solenoid of Fig. 6-8b is the same length and diameter as the rod of Fig. 6-8a, and if $K = K'$, the solenoid is the magnetic equivalent of the rod. In air $B = \mu_0 H$; and noting (5-11-5), we have at the center of the solenoid that

$$B = \mu_0 \frac{NI}{L} = \mu_0 K = \mu_0 H \quad (\text{T}) \tag{5}$$

At the center of the rod

$$B = \mu_0 K' = \mu_0 M \quad (\text{T}) \tag{6}$$

From (5) we have $H = K$, and from (6) we have $M = K'$. These are scalar relations. Vectorially we note that $\mathbf{H}$ is perpendicular to $\mathbf{K}$ and $\mathbf{M}$ is perpendicular to $\mathbf{K}'$. The more general vector relations are

$$\mathbf{K} = \hat{\mathbf{n}} \times \mathbf{H} \quad \text{and} \quad \mathbf{K}' = \mathbf{M} \times \hat{\mathbf{n}} \tag{7}$$

where $\hat{\mathbf{n}}$ is the unit vector normal to the plane containing the field vectors.

From Sec. 6-5 on the pole model and Sec. 6-6 on the current loop model, it is apparent that there are two different, but equivalent, approaches that can be used to explain and calculate the effects of magnetization. Both approaches will be considered further and compared in the next section.

6-7 THE MAGNETIC VECTORS B, H, AND M

Consider an air-filled toroid of area A and radius R with N_0 turns, as in Fig. 6-9. From (5-13-3) the flux density in the toroid is

$$B_0 = \mu_0 \frac{N_0 I}{2\pi R} \tag{1}$$

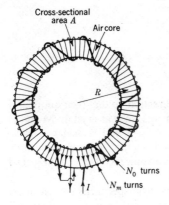

Figure 6-9 Toroid with N_0 turns of coarse winding (producing B_0) and N_m turns of fine winding (producing B_m).

But from (6-6-5) $N_0 I/2\pi R = K$, the sheet-current density. Hence,

$$B_0 = \mu_0 K = \mu_0 H \qquad (2)$$

If the same winding were placed on an iron ring of the same area and radius, the value of B would increase. Imagine now that in place of an iron ring we have the same air-filled toroid of Fig. 6-9 but with the ring's effect simulated by another toroidal winding of N_m turns. The magnetic field from this winding is

$$B_m = \mu_0 \frac{N_m I}{2\pi R} \qquad (3)$$

But $N_m I/2\pi R = K'$, the equivalent sheet-current density. Hence, from (6-6-6)

$$B_m = \mu_0 K' = \mu_0 M \qquad (4)$$

It follows that the total B is given by

$$B = B_0 + B_m = \mu_0(K + K') = \mu_0(H + M) \qquad (5)$$

or

$$\boxed{\mathbf{B} = \mu_0(\mathbf{H} + \mathbf{M})} \qquad (6)$$

where $\mathbf{B}$ = magnetic flux density, T
$\mathbf{H}$ = magnetic field, A m^{-1}
$\mathbf{M}$ = magnetization, A m^{-1}

Although developed for the special case of a toroid, (6) is a (vector) relation which applies in general. Dividing by H gives

$$\mu = \mu_0\left(1 + \frac{M}{H}\right) \qquad (7)$$

Dividing by the permeability μ_0 gives†

$$\mu_r = 1 + \frac{M}{H} \tag{8}$$

or

$$\frac{M}{H} = \mu_r - 1 \tag{9}$$

In isotropic media **M** and **H** are in the same direction, so that their quotient is a scalar and hence μ is a scalar. In nonisotropic media, such as crystals, **M** and **H** are, in general, not in the same direction, and μ is not a scalar. Hence, $\mathbf{B} = \mu_0(\mathbf{H} + \mathbf{M})$ is a general relation, while $\mathbf{B} = \mu\mathbf{H}$ is a more concise expression, which, however, has a simple significance only for isotropic media or certain special cases in nonisotropic media.

A single iron crystal is nonisotropic, but most iron specimens consist of an aggregate of numerous crystals oriented at random, so that macroscopically such specimens may be treated as though they were isotropic. In such cases $\mathbf{B} = \mu\mathbf{H}$ can be applied as a strictly macroscopic, or large-scale, relation.

Since $\nabla \cdot \mathbf{B} = 0$, we have, on taking the divergence of (6) for the pole model, that

$$\boxed{\nabla \cdot \mathbf{H} = -\nabla \cdot \mathbf{M} \qquad (\text{A m}^{-2})} \tag{11}$$

If the divergence of a vector field is not zero, the field has a source, or place of origin. We recall from the polarized dielectric case (Sec. 3-24) that $\nabla \cdot \mathbf{P} = -\rho_p$, which indicates that the polarization field originates on the polarization charge (of apparent volume density ρ_p) at the dielectric surface. In an analogous manner, (11) indicates that the **H** field originates where the magnetization field **M** ends and that the **H** field ends where the **M** field originates. This occurs at the ends of the rod in Fig. 6-6.

The locations where $\nabla \cdot \mathbf{H}$ (or $\nabla \cdot \mathbf{M}$) is not zero may be regarded as the locations of the magnetic poles of a magnetized object. Thus the poles of a uniformly magnetized rod, as in Fig. 6-6, are at the end faces of the rod.‡

Taking the curl of (6), we have in the current model that

$$\nabla \times \mathbf{B} = \mu_0(\nabla \times \mathbf{H}) + \mu_0(\nabla \times \mathbf{M}) \tag{12}$$

or

$$\nabla \times \mathbf{B} = \mu_0\mathbf{J} + \mu_0(\nabla \times \mathbf{M}) \tag{13}$$

---

† The ratio M/H is also sometimes called the *magnetic susceptibility* χ_m, a dimensionless quantity. Introducing it in (8), we have

$$\chi_m = \mu_r - 1 \tag{10}$$

Thus, the magnetic susceptibility $\chi_m = 0$ for vacuum, for which $\mu_r = 1$.

‡ In ordinary magnets with flat ends the magnetization tends to be nonuniform near the edges. Entirely uniform magnetization is possible in spherically or elliptically shaped magnetic objects. However, the assumption of uniform magnetization is a good approximation for a long homogeneous rod magnet, since the magnetization is nearly uniform over all of the rod except near the edges at the ends of the rod. In actual magnets with flat ends the effective separation between the pole centers is slightly less than the physical length of the magnet.

Where there is no magnetization, (13) reduces to $\nabla \times \mathbf{B} = \mu_0 \mathbf{J}$. The curl of $\mathbf{M}$ has the dimensions of current density (amperes per square meter) and represents the equivalent current of density $\mathbf{J}'$ (A m^{-2}) flowing, for example, in a very thin layer around the cylindrical surface of a uniformly magnetized rod. The linear current density for this sheet is $K' = J' \Delta x$ (A m^{-1}), where Δx is the thickness of the layer of current of average density J'. Thus (13) becomes

$$\nabla \times \mathbf{B} = \mu_0(\mathbf{J} + \mathbf{J}') \tag{14}$$

where $\mathbf{J}$ = actual current density, as in current-carrying wire, A m^{-2}

$\mathbf{J}'$ = equivalent current density, as at the surface of magnetized bar, A m^{-2}

The flux density $\mathbf{B}$ is always the result of a current or its equivalent. For example, the magnitude of $\mathbf{B}$ at the center of a long slender iron rod surrounded by a long solenoid is, from (6-6-5) and (6-6-6),

$$B = \mu_0(K + K') \quad \text{(T)} \tag{15}$$

where K = sheet-current density due to solenoid current, A m^{-1}

K' = equivalent sheet-current density due to magnetization of rod, A m^{-1}

In many cases we can conveniently express $\mathbf{B}$ directly in terms of the currents producing it, as in (15). In general, we can also express $\mathbf{B}$ in terms of the vector potential $\mathbf{A}$, which in turn is related to the currents. Thus

$$\mathbf{B} = \nabla \times \mathbf{A} \tag{16}$$

If both conduction currents and magnetization are present,

$$\mathbf{A} = \frac{\mu_0}{4\pi} \int \frac{\mathbf{J} + \mathbf{J}'}{r} \, dv \tag{17}$$

where

$$\mathbf{J} = \nabla \times \mathbf{H} \quad \text{and} \quad \mathbf{J}' = \nabla \times \mathbf{M} \quad \text{(A m}^{-2}\text{)}$$

Example 1 Referring to Fig. 6-10a, an air-core toroidal coil has a radius R and a cross-sectional area $A = \pi r^2$. The coil has a very narrow gap, as shown in the gap detail in Fig. 6-11a. The coil is made of many turns N of fine insulated wire with a current I. Draw graphs showing the variation of $\mathbf{B}$, $\mathbf{M}$, $\mathbf{H}$, and μ along the line of radius R at the gap (centerline of coil).

SOLUTION Neglecting the small effect of the narrow gap, $\mathbf{B}$ is substantially uniform around the inside of the entire toroid. Since $R \gg r$, its magnitude is, from (5-11-5), given approximately by

$$B = \frac{\mu_0 NI}{2\pi R} = \mu_0 K \quad \text{(T)} \tag{18}$$

where K is the magnitude of the linear sheet-current density in amperes per meter. A graph of the magnitude $\mathbf{B}$ along the centerline of the coil at the gap is shown in Fig. 6-11b.

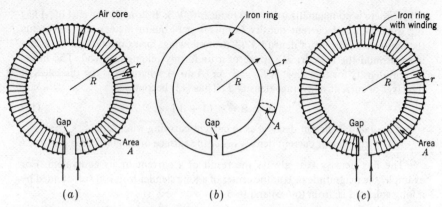

Figure 6-10 (*a*) Toroidal coil with gap. (*b*) Permanently magnetized iron ring with gap. (*c*) Iron-cored toroidal coil with gap.

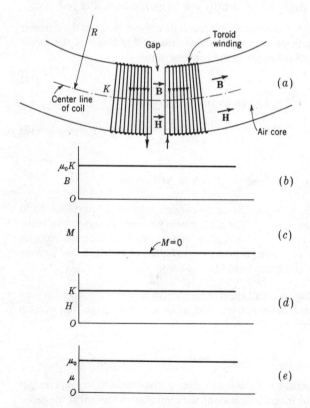

Figure 6-11 Magnitudes of magnetic quantities along the coil centerline at the gap in a toroid (see Fig. 6-10*a*) (Example 1).

No ferromagnetic material is present, so that the magnetization is negligible and $M = 0$, as indicated in Fig. 6-11c. It follows that $\mathbf{V} \cdot \mathbf{M} = 0$ and also $\mathbf{V} \cdot \mathbf{H} = 0$.

Since $M = 0$, we have

$$H = \frac{B}{\mu_0} = \frac{\mu_0 K}{\mu_0} = K = \frac{NI}{2\pi R} \qquad (\text{A m}^{-1}) \tag{19}$$

Therefore, the magnitude of **H** is constant and equal to the sheet-current density K of the coil winding, as indicated in Fig. 6-11d. The permeability everywhere is μ_0 (Fig. 6-11e).

It is to be noted that **B** is continuous and that in this case **H** is also continuous since there is no ferromagnetic material present. Both **B** and **H** have the same direction everywhere in this case.

Example 2 Consider now that the toroidal coil of Example 1 is replaced by an iron ring of the same size and also with a gap of the same dimensions, as suggested in Figs. 6-10b and 6-12a. Assume that the ring has a uniform permanent magnetization **M** that is equal in magnitude to K for the toroid in Example 1. Draw graphs showing the variation of **B**, **M**, **H**, μ, and $\mathbf{V} \cdot \mathbf{H}$ along the centerline of the ring at the gap.

SOLUTION The ring has a north pole at the left side of the gap and a south pole at the right side. Neglecting the small effect of the narrow gap, **B** is substantially uniform around the interior of the entire ring and also across the gap. In the *current model* it is due entirely to the equivalent sheet-current density K' on the surface of the ring. From (6-6-3), $K' = M$. Thus

$$B = \mu_0 M = \mu_0 K' \qquad (\text{T}) \tag{20}$$

where M and K' are, according to the stated conditions, equal to K for the solenoid in Example 1. Hence, B is the same in both examples. Its value at the gap is illustrated in Fig. 6-12b.

In the ring, $M = K'$, but outside the ring and in the gap $M = 0$. Suppose that the change in M from zero to K' at the gap occurs over a short distance Δx rather than as a square step function. The graph for M is then as shown in Fig. 6-12c.

Outside the ring and in the gap $M = 0$; so

$$H = \frac{B}{\mu_0} = K' \qquad (\text{A m}^{-1})$$

Inside the ring

$$H = \frac{B}{\mu_0} - M \qquad (\text{A m}^{-1}) \tag{21}$$

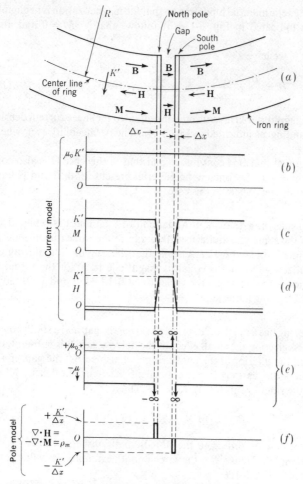

Figure 6-12 Variation of magnetic quantities along the centerline at the gap in a permanently magnetized iron ring (see Fig. 6-10b) (Example 2).

or approximately $H = K' - K' = 0$. The exact value of H is not zero† but is small and negative. The variation of H across the gap is illustrated in Fig. 6-12d.

From (6-7-7) the permeability in the ring is large and negative because H is small compared with M and is negative. In the air gap $\mu = \mu_0$. The variation of μ across the gap is suggested in Fig. 6-12e.

Since K' in this example equals K in Example 1, **B** and **H** in the gap have identical values in both examples. In the gap, the directions of **B** and **H** are the

† The above analysis is approximate since it neglects the effect of the gap. See Sec. 6-24.

same. In the iron ring, **B** is the same as in the toroid of Example 1, but **H** is smaller and in the opposite direction. An **H** direction opposite to that of **B** is characteristic of conditions *inside* a permanent magnet.

In the *pole model*, according to (6-7-11), the divergence of **H** equals the negative divergence of **M**, and this equals the apparent pole volume density ρ_m in the ring on both sides of the gap. Thus

$$\nabla \cdot \mathbf{H} = -\nabla \cdot \mathbf{M} = \rho_m \quad (\text{A m m}^{-3}) \tag{22}$$

This is zero everywhere except at the layers of assumed thickness Δx at the gap. Assuming that **M** changes linearly in magnitude over this thickness, and assuming also that Δx is very small compared with the cross-sectional diameter $2r$ of the ring, we have on the centerline

$$\nabla \cdot \mathbf{M} = \frac{dM_x}{dx} = \frac{\mp K'}{\Delta x} = -\rho_m \tag{23}$$

or

$$\nabla \cdot \mathbf{H} = \frac{\pm K'}{\Delta x} = \rho_m \tag{24}$$

where the upper sign in front of K' applies if M decreases and H increases in proceeding across Δx in a positive direction (from left to right). The variation of $\nabla \cdot \mathbf{H}$ along the centerline is illustrated in Fig. 6-12f. Hence the pole volume density ρ_m has a value only in the layers of assumed thickness Δx at the sides of the gap. This locates the poles of the ring magnet at the sides of the gap, and for this reason the iron surfaces of the gap are called *pole faces*.

Example 3 Suppose now that the iron ring of the previous example has wound over it the toroidal coil of Example 1 with the gap in the toroid coinciding with the gap in the ring, as shown in Fig. 6-10c and in the gap detail of Fig. 6-13a. The combination constitutes an iron-cored toroid as contrasted with the air-cored toroid of Example 1. Let the sheet current density for the toroid winding be K as in the first example. Further, let the *induced* magnetization added to the *permanent* magnetization in the ring yield a *total* uniform magnetization (permanent and induced) that is equal in magnitude to $4K$. Draw graphs showing the variation of **B**, **M**, **H**, μ, and $\nabla \cdot \mathbf{H}$ along the centerline of the ring at the gap.

SOLUTION In this case the total magnetization in the *current model* is $M = K' = 4K$. Neglecting the small effect of the narrow gap, the flux density is substantially uniform around the inside of the ring and across the gap. It is given (see Fig. 6-13b) by

$$B = \mu_0(K + K') = 5\mu_0 K \quad (\text{T}) \tag{25}$$

In the ring $M = 4K$, and in the gap $M = 0$, as shown in Fig. 6-13c. It is again assumed that M changes linearly over a short distance Δx at the pole faces.

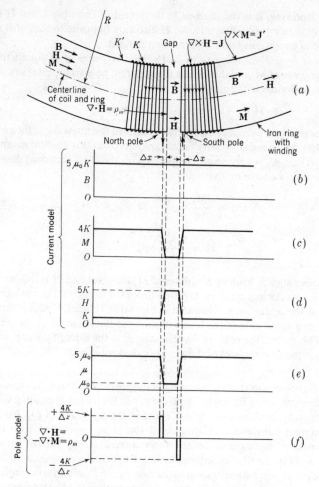

Figure 6-13 Variation of magnetic quantities along coil centerline at the gap in an iron-cored toroid (see Fig. 6-10c) (Example 3).

In the gap

$$H = \frac{B}{\mu_0} = 5K \tag{26}$$

In the ring

$$H = \frac{B}{\mu_0} - M \tag{27}$$

and so we have very nearly that $H = 5K - 4K = K$. The variation of H across the gap is depicted in Fig. 6-13d. In the gap $\mu = \mu_0$. In the ring (see Fig. 6-13e)

$$\mu = \mu_0\left(1 + \frac{M}{H}\right) = \mu_0\left(1 + \frac{4K}{K}\right) = 5\mu_0 \tag{28}$$

In this example, **B** and **H** have the same direction both in the gap and in the ring. In the ring, however, **H** is weaker than in the gap. The toroid has a sheet-current density of K (A m^{-1}), and the ring has an equivalent sheet-current density around its curved surface of $K' = 4K$ (A m^{-1}). Inside a wire of the toroidal coil $\nabla \times \mathbf{H} = \mathbf{J}$ (A m^{-2}) as suggested in Fig. 6-13a. Elsewhere $\nabla \times \mathbf{H} = 0$. At the curved surface of the ring $\nabla \times \mathbf{M} = \mathbf{J'}$ (A m^{-2}). Elsewhere $\nabla \times \mathbf{M} = 0$.

In the *pole model* the divergence of **H** or pole volume density ρ_m is given by the negative of the divergence of **M**. This has a value of $\pm 4K/\Delta x$ over the assumed pole thickness Δx at the pole faces. This is illustrated in Fig. 6-13f. The fact that $\nabla \cdot \mathbf{H} = \rho_m$ at the pole faces is also indicated in Fig. 6-13a. Elsewhere $\nabla \cdot \mathbf{H} = 0$.

In the last two examples involving ferromagnetic material it is to be noted that the magnetization, or **M**, lines originate, or have their source, on a south (negative) pole and end on, or have as a sink, a north (positive) pole. The **H** lines, on the other hand, originate, as in Example 2, on a north pole and end on a south pole. Thus, $\nabla \cdot \mathbf{H}$ has a positive value at a north pole, while $\nabla \cdot \mathbf{M}$ has a positive value at a south pole.

As a final example let us compare the fields around a solenoid and the equivalent permanently magnetized rod.

Example 4 A long uniform solenoid, as in Fig. 6-14a, is situated in air and has NI ampere-turns and a length l. A permanently magnetized iron rod in the pole and current models (Fig. 6-14e) has the same dimensions as the solenoid and has a uniform magnetization M equal to NI/l for the solenoid. Draw graphs showing the variation of **B**, **M**, and **H** along the axes of the solenoid and the rod. Also sketch the configuration of the fields for the two cases.

SOLUTION Since the rod and solenoid have the same dimensions and $M = K' = K = NI/l$, the two are magnetically equivalent. The **B** fields for both are the same everywhere, and the **H** fields for both are the same outside the solenoid and rod. Assuming that the solenoid is long compared with its diameter, the flux density at the center is nearly given by

$$B = \mu_0 \frac{NI}{l} = \mu_0 K \tag{29}$$

At the ends of the solenoid

$$B = \tfrac{1}{2}\mu_0 K \tag{30}$$

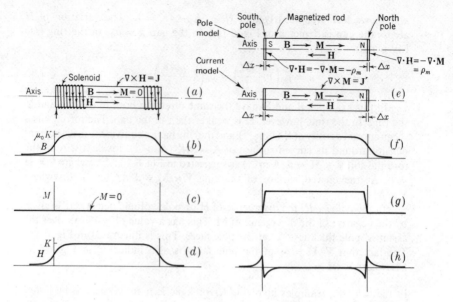

Figure 6-14 Solenoid and equivalent magnetized rod in pole and current models showing fields along axis (Example 4).

The magnitude of **B** at other locations along the solenoid axis can be obtained from (5-11-3) with a suitable change in limits. The variation of **B** along the solenoid axis is shown graphically in Fig. 6-14b. The variation along the rod axis is the same (Fig. 6-14f).

For the solenoid case, $M = 0$ everywhere (Fig. 6-14c). In the rod the magnetization M is assumed to be uniform, as in Fig. 6-14g.

For the solenoid case, $H = B/\mu_0$ everywhere, so that $H = K$ at the center and $H = \frac{1}{2}K$ at the ends. The variation of H along the solenoid axis is shown in Fig. 6-14d. Outside the rod, H is the same as for the solenoid. Inside the rod $H = (B/\mu_0) - M$, so that the variation is as suggested in Fig. 6-14h. It is assumed that M changes from 0 to K over a short distance Δx at the ends of the rod. The direction of **H** in the rod is opposite to that for **B**.

Inside the wires of the solenoid winding, $\nabla \times \mathbf{H} = \mathbf{J}$, as indicated in Fig. 6-14a. On the cylindrical surface of the rod, $\nabla \times \mathbf{M} = \mathbf{J}'$, as suggested in the *current model* in Fig. 6-14e. In the solenoid case, $\nabla \cdot \mathbf{B} = 0$ and $\nabla \cdot \mathbf{H} = 0$ everywhere. In the rod case, $\nabla \cdot \mathbf{B} = 0$ everywhere, but $\nabla \cdot \mathbf{H} = -\nabla \cdot \mathbf{M} = \rho_m$ at the end faces of the rod, as suggested in the *pole model* in Fig. 6-14e.

The **B**, **M**, and **H** fields for the solenoid and rod are sketched in Fig. 6-15. It is to be noted that inside the rod **H** is directed from the north pole to the south pole. Since **M** and **B** have, in general, different directions in the rod, μ loses its simple scalar significance in this case. Here **H** can be obtained by

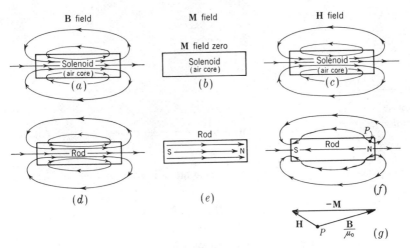

Figure 6-15 Fields of solenoid and equivalent permanently magnetized rod. The **B** fields are the same for both solenoid and rod [see (*a*) and (*d*)]. The **M** field is zero everywhere except inside the rod [see (*b*) and (*e*)]. The **H** fields are the same outside both solenoid and rod but are different inside [see (*c*) and (*f*)]. (Example 4.)

vector addition, using (6-7-6). As an example, **H** at the point P in Fig. 6-15*f* is obtained by the vector addition of $\mathbf{B}/\mu_0$ and $-\mathbf{M}$ as in Fig. 6-15*g*.

Although the magnetization is based on the actual magnetization phenomenon, it is often simpler and more convenient to ignore the mechanism of the phenomenon and use the permeability μ to describe the characteristics of the magnetic medium. This is particularly true where μ can be treated as a scalar. In this case μ is determined experimentally from a sample of the material. However, since μ is not a constant for ferromagnetic materials but a function of **H** and also the previous history of the sample, the methods for dealing with ferromagnetic materials require special consideration (see Sec. 6-11 and following sections).

6-8 ENERGY IN AN INDUCTOR AND ENERGY DENSITY

An inductor stores energy, as may be demonstrated with the aid of the circuit of Fig. 6-16. With the switch S closed the lamp is lighted. When the switch is opened, the lamp increases momentarily in brilliance because the magnetic energy stored in the magnetic field of the inductance induces a current as the field collapses. The (magnetic) energy delivered to the lamp is

$$W_m = \int (\text{power})\, dt = \int VI\, dt \qquad \text{(J)} \qquad (1)$$

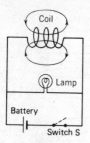

Figure 6-16 Circuit for demonstrating energy storage in a magnetic field.

From circuit theory the voltage across the inductor is given by $V = L\,dI/dt$; so

$$W_m = L \int_0^I I\,dI = \tfrac{1}{2}LI^2 \qquad \text{(J)} \qquad (2)$$

Thus, the magnetic energy has the dimensions of inductance times current squared. Since $L = \Lambda/I$, the energy stored can be variously expressed as

$$W_m = \tfrac{1}{2}LI^2 = \tfrac{1}{2}\Lambda I = \frac{1}{2}\frac{\Lambda^2}{L} \qquad (3)$$

where W_m = energy stored, J
L = inductance of inductor, H
I = current through inductor, A
Λ = flux linkage, Wb turns

The energy possessed by an inductor is stored in its magnetic field. Let us find the density of this energy as a function of the flux density **B**. Consider a small unit cube of side length Δl and volume $\Delta v = \Delta l^3$ situated in a magnetic field as in Fig. 6-17a. Let thin metal sheets be placed on the top and bottom surfaces of the cube, each with a current ΔI as indicated. Also let all the surrounding space be filled with such cubes, as suggested by the cross section of Fig. 6-17b. The directions of the current flow on the sheets are indicated by the circles with a dot (current out of page) and circles with a cross (current into page).

Each cube can be regarded as a magnetic field-cell transmission line of length Δl (into page). Each cell has an inductance $\Delta L = \mu\,\Delta l$. The field H is related to the current ΔI by $H\,\Delta l = \Delta I$. The energy stored in each cell is, from (3),

$$\Delta W_m = \tfrac{1}{2}\,\Delta L\,\Delta I^2 \qquad \text{(J)}$$

It follows that

$$\Delta W_m = \tfrac{1}{2}\mu H^2\,\Delta l^3 = \tfrac{1}{2}\mu H^2\,\Delta v \qquad (4)$$

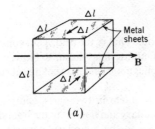

(a)

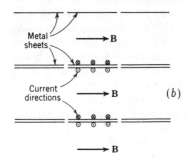

(b)

Figure 6-17 (a) Small cubical volume in a magnetic field. (b) Cross section through region filled with many such cubes.

Dividing (4) by Δv and taking the limit of the ratio $\Delta W_m/\Delta v$ as Δv approaches zero, we obtain the energy per volume, or *energy density*, w_m of the magnetic field at the point around which the cell of volume Δv shrinks to zero. Thus

$$w_m = \lim_{\Delta v \to 0} \frac{\Delta W}{\Delta v} = \tfrac{1}{2}\mu H^2 \qquad (\text{J m}^{-3}) \tag{5}$$

No material medium need be present for energy to be stored by a magnetic field. Thus, energy can be present in the magnetic field of an inductor situated in a vacuum. However, if there is ferrous material (iron) in the inductor, the energy density is increased in proportion to the permeability μ. From (6-7-6) or (6-7-7), we can express the magnetic energy density as

$$w_m = \tfrac{1}{2}(\mu_0 H^2 + \mu_0 MH) = \tfrac{1}{2}(\mu_0 H^2 + MB) = \tfrac{1}{2}\mu H^2 \qquad (\text{J m}^{-3})$$

where $\mu_0 H^2$ = energy density in vacuum, J m^{-3}

$\mu_0 MH = MB$ = energy density in ferrous medium, J m^{-3}

The additional energy in the iron is due to its magnetization. With no iron, there is no magnetization ($M = 0$) and the energy density is simply $\tfrac{1}{2}\mu_0 H^2$. With the iron, it is increased by $\tfrac{1}{2}\mu_0 MH$ to a total energy density of $\tfrac{1}{2}\mu H^2$.

These relations are very similar in form to those for the electric energy density in a capacitor, as discussed in Sec. 3-10.

6-9 BOUNDARY RELATIONS

In a single medium the magnetic field is continuous. That is, the field, if not constant, changes only by an infinitesimal amount in an infinitesimal distance. However, at the boundary between two different media, the magnetic field may change abruptly both in magnitude and direction.

It is convenient to analyze the boundary problem in two parts, considering separately the relation of fields *normal* to the boundary and *tangent* to the boundary.

Taking up first the relation of fields normal to the boundary, consider two media of permeabilities μ_1 and μ_2 separated by the xy plane, as shown in Fig. 6-18. Suppose that an imaginary box is constructed, half in each medium, of area $\Delta x \, \Delta y$ and height Δz. Let B_{n1} be the average component of **B** normal to the top of the box in medium 1 and B_{n2} the average component of **B** normal to the bottom of the box in medium 2. B_{n1} is an outward normal (positive), while B_{n2} is an inward normal (negative). By Gauss' law for magnetic fields (5-8-1), the total magnetic flux over a closed surface is zero. In other words, the integral of the outward normal components of **B** over a closed surface is zero. By making the height Δz of the box approach zero, the contribution of the sides of the box to the surface integral becomes zero even though there may be finite components of **B** normal to the sides. Therefore the surface integral reduces to $B_{n1} \, \Delta x \, \Delta y - B_{n2} \, \Delta x \, \Delta y = 0$ or

$$\boxed{B_{n1} = B_{n2}} \tag{1}$$

According to (1), *the normal component of the flux density* **B** *is continuous across the boundary between two media.*

Turning now to the relation for magnetic fields tangent to the boundary, let two media of permeabilities μ_1 and μ_2 be separated by a plane boundary, as in Fig. 6-19. Consider a rectangular path, half in each medium, of length Δx parallel to the boundary and of length Δy normal to the boundary. Let the average value of **H** tangent to the boundary in medium 1 be H_{t1} and the average value of **H** tangent to the boundary in medium 2 be H_{t2}. According to (5-14-5), the integral of **H** around

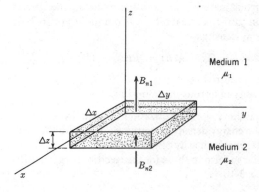

Figure 6-18 Construction for developing continuity relation for normal component of **B**.

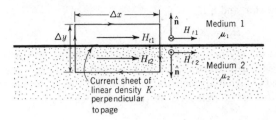

Figure 6-19 Construction for developing continuity relation for tangential component of **H**.

a closed path equals the current I enclosed. By making the path length Δy approach zero, the contribution of these segments of the path becomes zero even though a finite field may exist normal to the boundary. The line integral then reduces to $H_{t1} \Delta x - H_{t2} \Delta x = I$, or

$$H_{t1} - H_{t2} = \frac{I}{\Delta x} = K \qquad (\text{A m}^{-1}) \tag{2}$$

where K is the linear density of any current flowing in an infinitesimally thin sheet at the surface.†

According to (2), *the change in the tangential component of* **H** *across a boundary is equal in magnitude to the sheet-current density K on the boundary.* It is to be noted that **K** is normal to **H**; that is, the direction of the current sheet in Fig. 6-19 is normal to the page. This is expressed by the vector relation

$$\mathbf{K} = \hat{\mathbf{n}} \times (\mathbf{H}_{t1} - \mathbf{H}_{t2}) \tag{3}$$

where $\hat{\mathbf{n}}$ = unit vector normal to boundary, dimensionless
  $\mathbf{H}_{t1}$ = magnetic field tangent to boundary on side 1, A m^{-1}
  $\mathbf{H}_{t2}$ = magnetic field tangent to boundary on side 2, A m^{-1}
  $\mathbf{K}$ = sheet-current density at boundary, A m^{-1}

If the field below the boundary is zero ($H_{t2} = 0$), (3) indicates that the current **K** related to H_{t1} will be into the page in Fig. 6-19, while if the field above the boundary is zero ($H_{t1} = 0$), the current **K** related to H_{t2} will be out of the page.

If $K = 0$, then

$$H_{t1} = H_{t2} \tag{4}$$

According to (4), *the tangential components of* **H** *are continuous across the boundary between two media provided the boundary has no current sheet.*

---

† If J is the current density in amperes per square meter in a thin sheet of thickness $\Delta y'$, then K is defined by

$$K = J \Delta y' \qquad (\text{A m}^{-1})$$

where $J \to \infty$ as $\Delta y' \to 0$.

If medium 1 is a nonconductor, and if $H_{t2} = 0$,

$$H_{t1} = K_2 \tag{5}$$

where K_2 is the sheet-current density in amperes per meter in medium 2 at the boundary (into the page in Fig. 6-19). When medium 1 is air and medium 2 is a conductor, (5) is approximated at high frequencies because the skin effect restricts the current in the conductor to a very thin layer at its surface (see Chap. 10).

Example 1 Consider a plane boundary between two media of permeability μ_1 and μ_2, as in Fig. 6-20. Find the relation between the angles α_1 and α_2. Assume that the media are isotropic with **B** and **H** in the same direction.

SOLUTION From the boundary relations,

$$B_{n1} = B_{n2} \qquad \text{and} \qquad H_{t1} = H_{t2} \tag{6}$$

From Fig. 6-20,

$$B_{n1} = B_1 \cos \alpha_1 \qquad \text{and} \qquad B_{n2} = B_2 \cos \alpha_2 \tag{7}$$

$$H_{t1} = H_1 \sin \alpha_1 \qquad \text{and} \qquad H_{t2} = H_2 \sin \alpha_2 \tag{8}$$

where B_1 = magnitude of **B** in medium 1
B_2 = magnitude of **B** in medium 2
H_1 = magnitude of **H** in medium 1
H_2 = magnitude of **H** in medium 2

Substituting (7) and (8) into (6) and dividing yields

$$\frac{\tan \alpha_1}{\tan \alpha_2} = \frac{\mu_1}{\mu_2} = \frac{\mu_{r1}}{\mu_{r2}} \tag{9}$$

where μ_{r1} = relative permeability of medium 1, dimensionless
μ_{r2} = relative permeability of medium 2, dimensionless

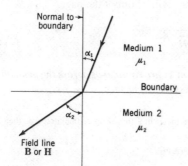

Normal to boundary

α_1

Medium 1
μ_1

Boundary

α_2

Medium 2
μ_2

Field line
B or H

Figure 6-20 Boundary between two media of different permeability showing change in direction of magnetic field line.

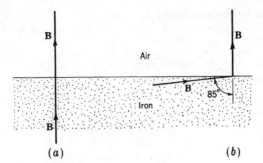

Figure 6-21 B lines at air-iron boundary.

(a)   (b)

Equation (9) gives the relation between the angles α_1 and α_2 for **B** and **H** lines at the boundary between two media.†

Example 2 Referring to Fig. 6-21, let medium 1 be air ($\mu_r = 1$) and medium 2 be soft iron with a relative permeability of 7,000. (a) If **B** in the iron is incident normal on the boundary ($\alpha_2 = 0$), find α_1. (b) If **B** in the iron is nearly tangent to the surface at an angle $\alpha_2 = 85°$, find α_1.

SOLUTION (a) From (9)

$$\tan \alpha_1 = \frac{\mu_{r1}}{\mu_{r2}} \tan \alpha_2 = \frac{1}{7,000} \tan \alpha_2 \tag{10}$$

When $\alpha_2 = 0$, $\alpha_1 = 0$, so that the **B** line in air is also normal to the boundary (see Fig. 6-21a).

(b) When $\alpha_2 = 85°$, we have, from (10), that $\tan \alpha_1 = 0.0016$, or $\alpha_1 = 0.1°$. Thus, the direction of **B** in air is almost normal to the boundary (within 0.1°) even though its direction in the iron is nearly tangent to the boundary (within 5°) (see Fig. 6-21b). Accordingly, for many practical purposes the *direction of* **B** *or* **H** *in air or other medium of low relative permeability may be taken as normal to the boundary of a medium having a high relative permeability.* This property is reminiscent of the one for **E** or **D** at the boundary of a conductor.

6-10 TABLE OF BOUNDARY RELATIONS FOR MAGNETIC FIELDS

Table 6-2 summarizes the boundary relations for magnetic fields developed in Sec. 6-9.

---

† This relation applies only if **B** and **H** have the same direction (μ a scalar). In the absence of magnetization, as in air, **B** and **H** have the same direction. When magnetization is present, as in a soft-iron electromagnet, **B** and **H** also tend to have the same direction. However, this is *not* the situation in a permanent magnet.

Table 6-2 Boundary relations for magnetic fields†

Field component	Boundary relation		Condition
Normal	$B_{n1} = B_{n2}$	(1)	Any two media.
Normal	$\mu_{r1}H_{n1} = \mu_{r2}H_{n2}$	(2)	Any two media.
Tangential	$H_{t1} - H_{t2} = K$	(3)	Any two media with current sheet of infinitesimal
	$\hat{\mathbf{n}} \times (\mathbf{H}_{t1} - \mathbf{H}_{t2}) = \mathbf{K}$	(3a)	thickness at boundary.
Tangential	$H_{t1} = H_{t2}$	(4)	Any two media with no current sheet at boundary.
Tangential	$H_{t1} = K_2$	(5)	$H_{t2} = 0$; also medium 2 has a current sheet of infinitesimal thickness at boundary; H_{t1} and K_2 are normal to each other.

† These relations apply for both static and time-varying fields (see Chap. 8).

6-11 FERROMAGNETISM

Ferromagnetic materials exhibit strong magnetic effects and are the most important magnetic substances. The permeability of these materials is not a constant but is a function both of the applied field and of the previous magnetic history of the specimen. In view of the variable nature of the permeability of ferromagnetic materials, special consideration of their properties is needed.

In ferromagnetic substances, as already explained, the atomic dipoles tend to align in the same direction over regions, or *domains*, containing many atoms. The size of a domain varies but usually contains millions of atoms. In some substances the shape appears to be like a long, slender rod with a transverse dimension of microscopic size but lengths of the order of a millimeter or so. Thus, a domain acts like a small, but not atomically small, bar magnet.

In an unmagnetized iron crystal the domains are parallel to a direction of easy magnetization, but equal numbers have north poles pointing one way as the other, so the external field of the crystal is zero. In an iron crystal there are six directions of easy magnetization. That is, there is a positive and negative direction along each of the three mutually perpendicular crystal axes (Fig. 6-22). Therefore the

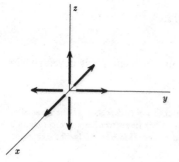

Figure 6-22 Six directions of easy magnetization in an iron crystal.

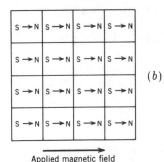

(a)

(b)

Applied magnetic field

Figure 6-23 (a) Domain polarity in an unmagnetized iron crystal. Arrows indicate direction of magnetization. A single N represents a domain with a north pole pointing out of the page; a single S represents a domain with a south pole pointing out of the page. (b) Condition after crystal is saturated by a magnetic field directed to the right.

polarity of the domains in an unmagnetized iron crystal may be as suggested by the highly schematic diagram of Fig. 6-23a. A single N represents a domain with a north pole pointing out of the page and a single S a domain with a south pole pointing out of the page. If the crystal is placed in a magnetic field parallel to one of the directions of easy magnetization, the domains with polarity opposing or perpendicular to the field become unstable and a few of them may rotate so that they have the same direction as the field. With further increase of the field more domains change over, each as an individual unit, until when all the domains are in the same direction, *magnetic saturation* is reached, as suggested by Fig. 6-23b. The crystal is then magnetized to a maximum extent. If the majority of the domains retain their directions after the applied field is removed, the specimen is said to be *permanently magnetized.* Heat and mechanical shock tend to return the crystal to the original unmagnetized state, and if the temperature is raised sufficiently high, the domains themselves are demagnetized (exchange coupling disappears) and the substance changes from ferromagnetic to paramagnetic. For iron this transition temperature, or Curie point, is 770°C. The residual magnetism is so weak compared to the ferromagnetic case that the material is usually considered to be unmagnetized.

Magnetization which appears only in the presence of an applied field may be spoken of as *induced magnetization,* as distinguished from *permanent magnetization,* which is present in the absence of an applied field.

6-12 MAGNETIZATION CURVES

The permeability μ of a substance is given by

$$\mu = \frac{B}{H} = \mu_0 \mu_r$$

where B = magnitude of flux density, T
H = magnitude of field **H**, A m^{-1}
μ_0 = permeability of vacuum = 400π nH m^{-1}
μ_r = relative permeability of substance, dimensionless

Therefore, to illustrate the relation of B to H, a graph showing B (ordinate) as a function of H (abscissa) is used. The line or curve showing B as a function of H on such a BH chart is called a *magnetization curve*. It is noted that μ is not the slope of the curve, which is given by dB/dH, but is equal to the ratio B/H.

To measure a magnetization curve for an iron sample, a thin, closed ring may be cut from the sample. A uniform winding is placed over the ring, forming an iron-cored toroid, as in Fig. 6-24. If the number of ampere-turns in the toroid is NI, the value of H applied to the ring is

$$H = \frac{NI}{l} \quad \text{(A turns m}^{-1})$$

where $l = 2\pi R$ and R is the mean radius of the ring or toroid. This value of H applied to the ring may be called the *magnetizing force*. Hence, in general, H is sometimes called by this name. The flux density B in the ring may be regarded as the result of the applied field H and is measured by placing another (secondary) coil over the ring, as in Fig. 6-24, and connecting it to a fluxmeter.† For a given change in H, produced by changing the toroid current I, there is a change in magnetic flux ψ_m through the ring. Both H and B are substantially uniform in the ring and negligible outside. Therefore the change in flux $\psi_m = BA$, where A is the cross-sectional area of the ring, and the resulting change in the flux density B in the ring is given by $B = \psi_m/A$ where ψ_m is measured by the fluxmeter. This ring method of measuring magnetization curves was used by Rowland in 1873.

A typical magnetization curve for a ferromagnetic material is shown by the solid curve in Fig. 6-25a. The specimen in this case was initially unmagnetized, and the change was noted in B as H was increased from 0. By way of comparison, four dashed lines are also shown in Fig. 6-25a, corresponding to constant relative permeabilities μ_r of 1, 10, 100, and 1,000. The relative permeability at any point on the magnetization curve is given by

$$\mu_r = \frac{B}{\mu_0 H} = 7.9 \times 10^5 \frac{B}{H} \quad \text{(dimensionless)}$$

where B = ordinate of the point, T
H = abscissa of the point, A m^{-1}

---

† The *fluxmeter* operates on the emf induced in the secondary when the magnetic flux through it changes.

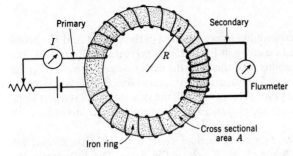

Figure 6-24 Rowland ring method of obtaining magnetization curve.

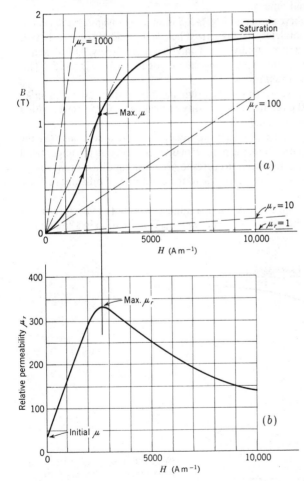

Figure 6-25 (a) Typical magnetization curve and (b) corresponding relation of relative permeability to applied field **H**.

A graph of the relative permeability μ_r as a function of the applied field H, corresponding to the magnetization curve in Fig. 6-25a, is presented in Fig. 6-25b. The maximum relative permeability, and therefore the *maximum permeability*, is at the point on the magnetization curve with the largest ratio of B to H. This is designated "Max. μ"; it occurs at the point of tangency with the straight line of steepest slope that passes through the origin and also intersects the magnetization curve (dash-dot line in Fig. 6-25a).

The magnetization curve for air or vacuum would be given by the dashed line for $\mu_r = 1$ (almost coincident with the H axis) in Fig. 6-25a. The difference in the ordinate B between the magnetization curve of the ferromagnetic sample and the ordinate at the same H value on the $\mu_r = 1$ line is equal to the magnetization M of the ferromagnetic material times μ_0.

The magnetization curve shown in Fig. 6-25a is an *initial-magnetization curve*. That is, the material is completely demagnetized before the field H is applied. As H is increased, the value of B rises rapidly at first and then more slowly. At sufficiently high values of H the curve tends to become flat, as suggested by Fig. 6-25a. This condition is called *magnetic saturation*.

The magnetization curve starting at the origin has a finite slope giving an *initial permeability*. Therefore the relative-permeability curve in Fig. 6-25b starts with a finite permeability for infinitesimal fields.

The initial-magnetization curve may be divided into two sections: (1) the steep section and (2) the flat section, the point P of division being on the upper bend of the curve (Fig. 6-26). The steep section corresponds to the condition of *easy magnetization*, while the flat section corresponds to the condition of difficult, or *hard*, *magnetization*.

Ordinarily a piece of iron consists not of a single crystal but of an aggregate of small crystal fragments with axes oriented at random. The situation in a small piece

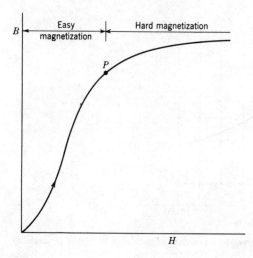

Figure 6-26 Regions of easy and hard magnetization of initial-magnetization curve.

of iron may be represented schematically as in Fig. 6-27. Here a number of crystal fragments are shown, each with a number of magnetic domains, represented in most cases by a small square. The boundaries between crystal fragments are indicated by the heavy lines, and domain boundaries by the light lines, which also indicate the direction of the crystal axes. In Fig. 6-27a, not only is the piece of iron unmagnetized, but also the individual crystal fragments are unmagnetized. The domains in each crystal are magnetized along the directions of easy magnetization,

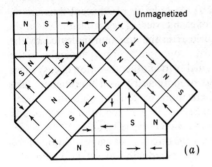

(a)

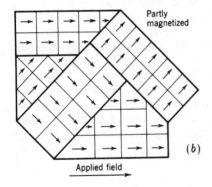

(b)

Applied field

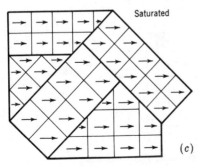

(c)

Figure 6-27 Successive stages of magnetization of a polycrystalline specimen with increasing field. Arrows indicate direction of magnetization of domains. An N represents a domain with a north pole pointing out of the page; an S represents a domain with a south pole pointing out of the page.

(a)

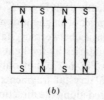

(b)

Figure 6-28 Total energy is reduced when domains go from aligned condition (a) to oppositely oriented condition (b).

i.e., along the three crystal axes. However, the polarity of adjacent domains is opposite, so that the total magnetization of each crystal is negligible.

With the application of a magnetic field H in the direction indicated by the arrow (Fig. 6-27) some domains with polarities opposed to or perpendicular to the applied field become unstable and rotate quickly to another direction of easy magnetization in the same direction as the field, or more nearly so. These changes take place on the steep (easy) part of the magnetization curve. The result, after all domains have changed over, is as suggested in Fig. 6-27b. This condition corresponds roughly to that at the point P on the magnetization curve (Fig. 6-26).

With further increase in the applied field, the direction of magnetization of the domains not already parallel to the field is rotated gradually toward the direction of H. This increase in magnetization is more difficult, and very high fields may be required to reach saturation, where all domains are magnetized parallel to the field, as indicated in Fig. 6-27c. This accounts for the flatness of the upper (hard) part of the magnetization curve.

The tendency of adjacent magnetic domains to be oppositely magnetized can be understood from energy considerations. Thus, when adjacent domains are oriented the same, as in Fig. 6-28a, the total energy is increased. When all domains are oppositely oriented, as in Fig. 6-28b, the energy is decreased. The situation can be illustrated by performing an experiment with two bar magnets arranged to slide easily on a rod. Let both magnets be placed side by side and oriented the same, as in Fig. 6-29a. If the left magnet is held but the right magnet is released, it will move to

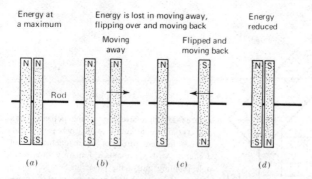

Figure 6-29 Experiment with two bar magnets illustrates the decrease in total energy.

the right, as in Fig. 6-29b, since the adjacent like poles repel. As the right magnet moves farther away, it will rotate on the rod to the position shown in Fig. 6-29c. The opposite poles now attract, and the right magnet moves back to the left until it comes to rest against the left magnet, as in Fig. 6-29d. The pair of magnets now has less energy than at the start (Fig. 6-29a). The decrease in total energy accounts for the work done by the right magnet in moving away, rotating, and moving back.

6-13 HYSTERESIS

If the field applied to a specimen is increased to saturation and is then decreased, the flux density B decreases, but not as rapidly as it increased along the initial-magnetization curve. Thus, when H reaches zero, there is a *residual flux density*, or *remanence*, B_r (Fig. 6-30).

In order to reduce B to zero, a negative field $-H_c$ must be applied (Fig. 6-30).†
This is called the *coercive force*. As H is further increased in the negative direction, the specimen becomes magnetized with the opposite polarity, the magnetization at first being easy and then hard as saturation is approached. Bringing the field to zero again leaves a residual magnetization or flux density $-B_r$, and to reduce B to zero a coercive force $+H_c$ must be applied. With further increase in field, the specimen again becomes saturated with the original polarity.

The phenomenon which causes B to lag behind H, so that the magnetization curve for increasing and decreasing fields is not the same, is called *hysteresis*, and the loop traced out by the magnetization curve is called a *hysteresis loop* (Fig. 6-30). If the substance is carried to saturation at both ends of the magnetization curve, the loop is called the *saturation*, or *major*, *hysteresis loop*. The residual flux density B_r on the saturation loop is called the *retentivity*,‡ and the coercive force H_c on this loop is called the *coercivity*. Thus, the retentivity of a substance is the maximum value which the residual flux density can attain and the coercivity the maximum value which the coercive force can attain. For a given specimen no points can be reached on the BH diagram outside the saturation hysteresis loop, but any point inside can.

In soft, or easily magnetized, materials the hysteresis loop is thin, as suggested in Fig. 6-31, with a small area enclosed. By way of comparison, the hysteresis loop of a hard magnetic material is also shown, the area enclosed in this case being greater.

Turning our attention to the permeability μ, consider the hysteresis loop of Fig. 6-32a. The corresponding graph of μ as a function of H is as shown in Fig. 6-32b. At $H = 0$, it is apparent that μ becomes infinite. On the other hand, when $B = 0$, $\mu = 0$. Under such conditions, the permeability μ becomes meaningless.

---

† By reversing the battery polarity (Fig. 6-24).

‡ The term *retentivity* is also sometimes used to mean the ratio of the residual flux density B_r to the maximum flux density B_m.

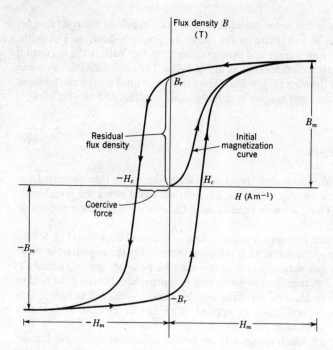

Figure 6-30 Hysteresis loop.

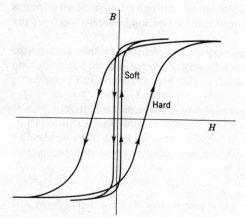

Figure 6-31 Hysteresis loops for soft and hard magnetic materials.

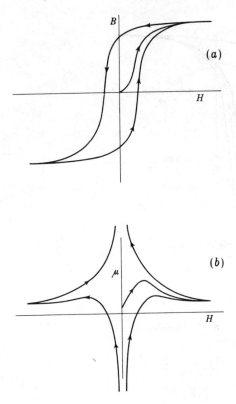

Figure 6-32 (*a*) Hysteresis loop. (*b*) Corresponding permeability curve.

Therefore the use of μ must be confined to situations where it has significance, e.g., the initial magnetization curve. It is to be noted that the term *maximum permeability* signifies specifically the maximum permeability for an initial-magnetization curve and not for a hysteresis loop or other type of magnetization curve.

Another type of magnetization curve for which μ has a definite meaning is the *normal magnetization curve*. This curve is the locus of the tips of a series of hysteresis loops, obtained by cycling the field H over successively smaller ranges. Thus, as shown in Fig. 6-33 the field is changed slowly over the range $\pm H_1$, obtaining the saturation hysteresis loop. The field is next cycled slowly several times over a range $\pm H_2$, obtaining after a few reversals a repeatable hysteresis loop of smaller size. This process is repeated for successively smaller ranges in H, obtaining a series of loops of decreasing size. The curve passing through the tips of these loops is the *normal magnetization curve* (Fig. 6-33). This curve is useful since it is reproducible and is characteristic of the particular type of ferromagnetic material. The normal magnetization curve is actually very similar in shape to the initial-magnetization curve.

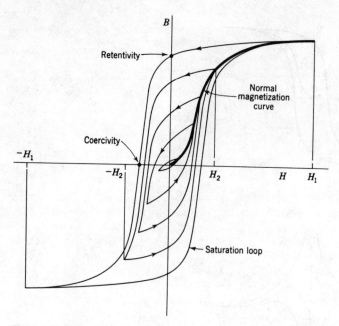

Figure 6-33 Normal magnetization curve with relation to hysteresis loops.

6-14 ENERGY IN A MAGNET

A specimen of iron with residual magnetization contains energy since work has been performed in magnetizing it. The magnetic energy w_m per unit volume of a specimen brought to saturation from an originally unmagnetized condition is given by the integral of the initial-magnetization curve expressed by†

$$w_m = \int_0^B H \, dB \quad (\text{J m}^{-3}) \tag{1}$$

The dimensional relation for (1) is

$$\frac{I}{L} \frac{M}{IT^2} = \frac{M}{LT^2}$$

where M/LT^2 has the dimensions of energy density, which is expressed in joules per cubic meter. Thus, the area between the curve and the B axis is a measure of the

---

† Integrating yields $w_m = \frac{1}{2}\mu H^2$, as in Table 6-5.

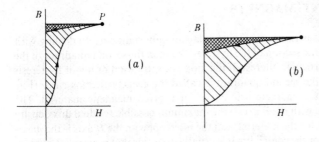

Figure 6-34 Energy density areas for (a) soft and (b) hard magnetic materials.

energy density. This is indicated in Fig. 6-34a for an easily magnetized (magnetically soft) substance which has been carried to the point P in the magnetization process. A magnetically hard substance takes more work to magnetize, as indicated by the larger shaded area in Fig. 6-34b. On bringing H to zero some energy is released, as indicated by the crosshatched areas in Fig. 6-34.

If H is increased and decreased, so that the magnetization of a specimen repeatedly traces out a hysteresis loop, as in Fig. 6-35a, the area enclosed by this loop represents the energy per unit volume expended in the magnetization-demagnetization process in one complete cycle. In general the specimen retains some energy in stored magnetic form at any point in the cycle. However, in going once around the hysteresis loop and back to this point, at which the energy will again be the same, energy proportional to the area of the loop is lost. This energy is expended in stressing the crystal fragments of the specimen and appears as heat. If no hysteresis were present and the initial-magnetization curve were retraced, the area of the loop would be zero (Fig. 6-35b). The magnetization-demagnetization process could then be accomplished with no loss of energy as heat in the specimen, assuming that eddy currents (see Sec. 8-10) are negligible.

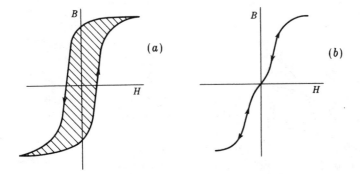

Figure 6-35 Energy lost in magnetization cycle is proportional to area enclosed by hysteresis loop.

6-15 PERMANENT MAGNETS

In many applications permanent magnets play an important part. In dealing with permanent magnets the section of the hysteresis loop in the second quadrant of the *BH* diagram is of particular interest. If the loop is a saturation or major hysteresis loop, the section in the second quadrant is called the *demagnetization curve* (Fig. 6-36a). This curve is a characteristic curve for a given magnetic material. The intercept of the curve with the *B* axis is the maximum possible residual flux density B_r, or the retentivity, for the material, and the intercept with the *H* axis is the maximum coercive force, or the coercivity. It is usually desirable that permanent-magnet materials have a high retentivity, but it is also important that the coercivity be large so that the magnet will not be easily demagnetized.

In Fig. 6-36b, three demagnetization curves are shown. Curve 1 represents a material having a high retentivity but low coercivity, while curve 2 represents a material which is just the reverse; i.e., it has a low retentivity and high coercivity.

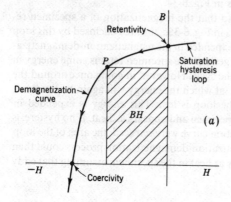

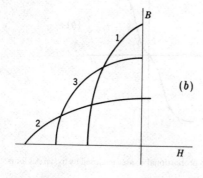

Figure 6-36 Demagnetization curves. (*B* is positive and *H* is negative.)

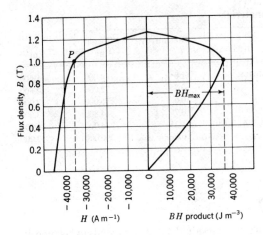

Figure 6-37 Demagnetization and BH product curves for Alnico 5.

Curve 3 represents a material which is a compromise between the other two, having relatively high retentivity and coercivity.

The maximum BH product, abbreviated BH_{max}, is also a quantity of importance for a permanent magnet. In fact, it is probably the best single figure of merit, or criterion, for judging the quantity of a permanent magnet material. Referring to Fig. 6-36b, it is apparent that BH_{max} is greater for curve 3 than for either curves 1 or 2. The maximum BH product for a substance indicates the maximum energy density (in joules per cubic meter) stored in the magnet. A magnet at BH_{max} delivers a given flux with a minimum of magnetic material.

Since the product BH has the dimensions of energy density, it is sometimes called the *energy product* and its maximum value the *maximum energy product*. The product BH for any point P on the demagnetization curve is proportional to the area of the shaded rectangle, as shown in Fig. 6-36a.

Figure 6-37 shows the demagnetization curve for Alnico 5, one of the best permanent-magnet materials, which is an alloy containing iron, cobalt, nickel, aluminum, and copper. A curve showing the BH product is also presented. The maximum BH product is about 36,000 J m^{-3} and occurs at a flux density of about 1 T (see point P).

A discussion of the operating point of permanent magnets is given in Sec. 6-24.

6-16 TABLE OF PERMANENT MAGNETIC MATERIALS

Representative materials for permanent magnets are given in Table 6-3. The materials are listed in the order of increasing maximum BH product. Magnets of cobalt, copper, iron, and either cerium or samarium have been cast with coercivities of over 2 MA m^{-1}.

Table 6-3 Permanent magnetic materials

Material†	Retentivity, T	Coercivity, A m⁻¹	BH_{max}, J m⁻³
Chrome steel (98 Fe, 0.9 Cr, 0.6 C, 0.4 Mn)	1.0	4,000	1,600
Oxide (57 Fe, 28 O, 15 Co)	0.2	72,000	4,800
Alnico 12 (33 Fe, 35 Co, 18 Ni, 8 Ti, 6 Al)	0.6	76,000	12,000
Alnico 2 (55 Fe, 12 Co, 17 Ni, 10 Al, 6 Cu)	0.7	44,800	13,600
Alnico 5 (Alcomax) (51 Fe, 24 Co, 14 Ni, 8 Al, 3 Cu)	1.25	44,000	36,000
Platinum cobalt (77 Pt, 23 Co)	0.6	290,000	52,000

† Compositions in percent.

6-17 DEMAGNETIZATION

A bar of ferromagnetic material that has a residual flux density tends to become demagnetized spontaneously. The phenomenon is illustrated by Fig. 6-38, which shows a bar magnetized so that a north pole is at the left and a south pole at the right. The orientation of a single domain is indicated, and it is evident that the external field of the bar magnet opposes this domain and, hence, will tend to turn it, or reverse its polarity, thereby partially demagnetizing the bar. The tendency for this demagnetization is reduced if the magnet is in the form of a U, as in Fig. 6-39a, since in this case there is but little demagnetizing field along the side of the magnet. The demagnetizing effect can be still further reduced by means of a soft-iron *keeper* placed across the poles, as in Fig. 6-39b.

The process of removing the permanent magnetization of a specimen so that the residual flux density is zero under conditions of zero **H** field is called *demagnetization* or *deperming*. It is evident that B can be reduced to zero by the application of the coercive force H_c, but on removing this field the residual flux density will rise to some value B_0, as suggested in Fig. 6-40. Although it might be possible to end up at $B = 0$ and $H = 0$ by increasing $-H$ to slightly more than the coercive force and then decreasing it to zero, as suggested by the dashed lines, the process requires an accurate knowledge of B and H and the hysteresis loop.

A longer but more simply applied method is called *demagnetization* or *deperming by reversals*. In this method, $\pm H$ is brought to a smaller maximum amplitude on each reversal so that eventually the specimen is left in a demagnetized state at zero field, as suggested by Fig. 6-41. Although such a demagnetization procedure can be completely carried out in a matter of seconds with a small magnetic specimen such as a watch (using ac fields), many seconds or even minutes

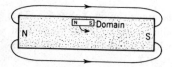

Figure 6-38 Demagnetization effect of bar-magnet field.

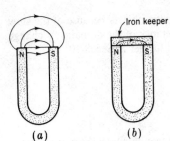

(a)   (b)   **Figure 6-39** U-shaped magnet with and without keeper.

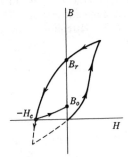

Figure 6-40 Partial hysteresis loop.

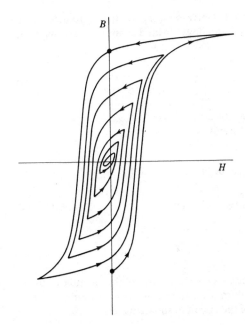

Figure 6-41 Demagnetization by reversals.

may be required *for each reversal* for large magnetic objects because of the slow decay of the induced eddy currents and the reluctance of the domains to change polarity. The matter of eddy currents is discussed further in Sec. 8-10.

6-18 THE MAGNETIC CIRCUIT; RELUCTANCE AND PERMEANCE

An electric circuit forms a closed path or circuit through which the current flows. Magnetic flux tubes are continuous and form closed paths. Hence, by analogy, we may consider that a single flux tube is a *magnetic circuit*. Or all the flux tubes of a magnetic circuit, taken in parallel, may be considered as a magnetic circuit.

Consider first an electric circuit carrying a current I. By Kirchhoff's law the total emf in the circuit is equal to the total IR drop. Thus

$$\mathscr{V}_T = IR_T \tag{1}$$

where $\mathscr{V}_T$ = total emf, V
  R_T = total resistance, Ω

From (1) the total resistance is

$$R_T = \frac{\mathscr{V}_T}{I} \tag{2}$$

Consider now a magnetic circuit. Corresponding to the resistance of an electric circuit as given by (2), we may, by analogy, define a quantity for the magnetic circuit called the *reluctance* $\mathscr{R}$. Thus

$$\mathscr{R}_T = \frac{F_T}{\psi_m} \tag{3}$$

where $\mathscr{R}_T$ = total reluctance of magnetic circuit
  F_T = total mmf of magnetic circuit, A
  ψ_m = flux through magnetic circuit, Wb

In general, the total flux ψ_m in a magnetic circuit is given by

$$\psi_m = \iint \mathbf{B} \cdot d\mathbf{s} \quad \text{(Wb)} \tag{4}$$

where $\mathbf{B}$ = flux density, T
  $d\mathbf{s}$ = element of surface, m^2

The integration is carried out over the cross-sectional area of the flux tube or tubes that constitute the circuit. If $\mathbf{B}$ is uniform over the entire cross section, $\psi_m = BA$, where A is the cross-sectional area of the circuit in square meters.

Reluctance has the dimensions of current per magnetic flux, or in dimensional symbols

$$I \frac{T^2 I}{ML^2} = \frac{T^2 I^2}{ML^2}$$

The relation $T^2 I^2 / ML^2$ has the dimensions of the reciprocal of inductance. Thus the unit for reluctance is the *reciprocal henry* (H^{-1}).

The reciprocal of reluctance $\mathscr{R}$ is called the *permeance* $\mathscr{P}$, which is expressed in henrys. Hence, from (3),

$$\mathscr{P}_T = \frac{1}{\mathscr{R}_T} = \frac{\psi_m}{F_T} \tag{5}$$

where $\mathscr{P}_T$ is the total permeance of the circuit in henrys.

The total mmf of a magnetic circuit is equal to the line integral of **H** around the complete circuit, and this in turn is equal to the ampere-turns enclosed. Therefore, (3) becomes

$$\mathscr{R}_T = \frac{1}{\mathscr{P}_T} = \frac{F_T}{\psi_m} = \frac{\oint \mathbf{H} \cdot dl}{\psi_m} = \frac{NI}{\psi_m} \quad (H^{-1}) \tag{6}$$

where NI is the ampere-turns enclosed.

The above discussion concerns the total reluctance of a circuit. Let us consider next the reluctance of a portion of a magnetic circuit. In an electric circuit, the resistance R between two points having no emfs between them is given by

$$R = \frac{V}{I} \quad (\Omega) \tag{7}$$

where V = potential difference between the points, V
I = current in circuit, A

In the analogous magnetic case, the reluctance $\mathscr{R}$ between two points in a magnetic circuit is given by

$$\mathscr{R} = \frac{U}{\psi_m} \quad (H^{-1}) \tag{8}$$

where U is the magnetic potential difference between the points in amperes. From (5-16-6) for U and (4) for ψ_m we have

$$\mathscr{R} = \frac{\int_1^2 \mathbf{H} \cdot dl}{\iint \mathbf{B} \cdot d\mathbf{s}} \tag{9}$$

where **H** is integrated between the two points (1 and 2) between which we wish to find the magnetic potential difference U.

When the circuit has a uniform cross section of area A and the field is uniform, (9) reduces to

$$\mathscr{R} = \frac{HI}{BA} = \frac{l}{\mu A} \quad (\text{H}^{-1}) \tag{10}$$

where $\mathscr{R}$ = reluctance between points 1 and 2, H^{-1}
$\quad l$ = distance between points 1 and 2, m
$\quad A$ = cross-sectional area of magnetic circuit, m^2
$\quad \mu$ = permeability of medium comprising the circuit, H m^{-1}

The permeance $\mathscr{P}$ between the points 1 and 2 is given by

$$\mathscr{P} = \frac{1}{\mathscr{R}} = \frac{\mu A}{l} \quad (\text{H}) \tag{11}$$

Reluctances in series are additive in the same way that resistances in series are additive. For reluctances in parallel the reciprocal of the total reluctance is equal to the sum of the reciprocals of the individual reluctances. For reluctances in parallel it is usually more convenient to use permeance, the total permeance being equal to the sum of the individual permeances.

Example 1 Find the reluctance and permeance between the ends of the rectangular block of iron shown in Fig. 6-42a, assuming that B is uniform throughout the block and normal to the ends. The permeability of the block is uniform and has a value $\mu_1 = 500\mu_0$, where μ_0 is the permeability of vacuum.

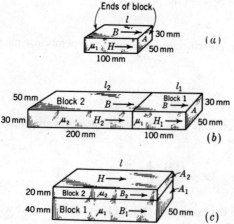

Figure 6-42 Rectangular iron blocks.

SOLUTION The reluctance of the block is, from (10),

$$\mathscr{R} = \frac{l}{\mu A} = \frac{0.1}{500 \times 4\pi \times 10^{-7} \times 15 \times 10^{-4}} = 1.06 \times 10^5 \text{ H}^{-1}$$

The permeance $\mathscr{P}$ is the reciprocal of $\mathscr{R}$; so

$$\mathscr{P} = \frac{1}{1.06 \times 10^5} = 9.4 \ \mu\text{H}$$

Example 2 Find the total reluctance and permeance between the ends of the *series-connected* rectangular iron blocks shown in Fig. 6-42b, assuming that B is uniform throughout the blocks and normal to the ends. The permeability of each block is uniform, the value in block 1 being $\mu_1 = 500\mu_0$ and in block 2, $\mu_2 = 2,000\mu_0$.

SOLUTION The reluctance $\mathscr{R}_1$ of block 1 is given in Example 1. The reluctance of block 2 is

$$\mathscr{R} = \frac{l_2}{\mu_2 A} = \frac{0.2}{2,000 \times 4\pi \times 10^{-7} \times 15 \times 10^{-4}} = 0.53 \times 10^5 \text{ H}^{-1}$$

The total reluctance $\mathscr{R}_T$ equals the sum of the individual reluctances; so

$$\mathscr{R}_T = \mathscr{R}_1 + \mathscr{R}_2 = (1.06 + 0.53) \times 10^5 = 1.59 \times 10^5 \text{ H}^{-1}$$

The total permeance

$$\mathscr{P}_T = \frac{1}{\mathscr{R}_T} = \frac{1}{1.59 \times 10^5} = 6.3 \ \mu\text{H}$$

Example 3 Find the total reluctance and permeance between the ends of the *parallel-connected* rectangular iron blocks shown in Fig. 6-42c, assuming that B is uniform in each block and normal to the ends. The permeability of each block is uniform, the value in block 1 being $\mu_1 = 500\mu_0$ and in block 2 being $\mu_2 = 2,000\mu_0$.

SOLUTION Since the blocks are in parallel, it is more convenient to calculate the total permeance first. The permeance $\mathscr{P}_1$ of block 1 is

$$\mathscr{P}_1 = \frac{\mu_1 A_1}{l} = \frac{500 \times 4\pi \times 10^{-7} \times 20 \times 10^{-4}}{0.2} = 6.28 \ \mu\text{H}$$

The permeance of block 2 is

$$\mathscr{P}_2 = \frac{\mu_2 A_2}{l} = \frac{2,000 \times 4\pi \times 10^{-7} \times 10 \times 10^{-4}}{0.2} = 12.6 \ \mu\text{H}$$

The total permeance equals the sum of the individual permeances; so

$$\mathscr{P}_T = \mathscr{P}_1 + \mathscr{P}_2 = (6.28 + 12.6) \times 10^{-6} = 18.9 \ \mu\text{H}$$

Table 6-4 Comparison of electric and magnetic circuits

Electric circuit	Magnetic circuit

Ohm's law

$\mathscr{V} = IR$ (V)	$F = \psi_m \mathscr{R}$ (A)
where $\mathscr{V}$ = emf (V)	where $F = NI$ = mmf (A)
$I = \iint \mathbf{J} \cdot d\mathbf{s}$ = current (A)	$\psi_m = \iint \mathbf{B} \cdot d\mathbf{s}$ = flux (Wb)
R = resistance (Ω)	$\mathscr{R}$ = reluctance (H^{-1})

Potentials

$\mathscr{V} = \int \mathbf{E} \cdot d\mathbf{l}$	$F = \int \mathbf{H} \cdot d\mathbf{l}$
$E_c = -\nabla V$ (no emfs) (V m^{-1})	$H = -\nabla U$ (no mmfs) (A m^{-1})
$\Delta V = V_1 - V_2$ = potential difference	$\Delta U = U_1 - U_2$ = potential difference
where V = electrical potential	U = magnetic potential

Kirchhoff's law (series circuit)

$\mathscr{V} = I(R_1 + R_2)$	$F = \psi_m(\mathscr{R}_1 + \mathscr{R}_2)$

or, in general

$\sum \mathscr{V} = I \sum R$ (V)	$F = \psi_m \sum \mathscr{R}$ (A)
where $R = \dfrac{l}{\sigma A}$ = resistance (Ω) $= \dfrac{1}{G}$	$\mathscr{R} = \dfrac{l}{\mu A}$ = reluctance (H^{-1}) $= \dfrac{1}{\mathscr{P}}$
$G = \dfrac{\sigma A}{l}$ = conductance ($\mho$)	$\mathscr{P} = \dfrac{\mu A}{l}$ = permeance (H)

Note that whereas σ and R are commonly independent of the current (density), μ and $\mathscr{R}$ of ferromagnetic materials are not independent of B.

The total reluctance is then given by

$$\mathscr{R}_T = \frac{1}{\mathscr{P}_T} = \frac{1}{1.89 \times 10^{-5}} = 5.3 \times 10^4 \text{ H}^{-1}$$

Circuit quantities for electric and magnetic fields are compared in Table 6-4.

6-19 MAGNETIC FIELD MAPPING; MAGNETIC FIELD CELLS

The examples in the preceding section illustrate how the reluctance or permeance may be found for sections of a magnetic circuit that have a uniform cross section and uniform field. In two-dimensional problems where the field and cross section are nonuniform the magnetic field configuration, and consequently the reluctance or permeance, can also be found provided the permeability may be considered constant. Graphical field-mapping techniques such as are employed in Secs. 3-19 and 4-16 are applicable to such situations.

The following basic properties are useful in magnetic field mapping:

1. The field (**H** or **B**) lines and the magnetic potential (U) lines intersect at right angles.

2. At the boundary between air and iron (or other high-permeability medium) the field lines on the air side of the boundary are substantially perpendicular to the boundary surface.
3. The boundary between air and iron (or other high-permeability medium) may be regarded as an equipotential with respect to the air side of the boundary but not, in general, with respect to the iron side.
4. In a uniform field the potential varies linearly with distance.
5. A magnetic flux tube is parallel to the field, and the magnetic flux over any cross section of the tube is a constant.
6. Magnetic flux tubes are continuous.

With these properties in mind a two-dimensional magnetic field can be divided into magnetic flux tubes and then by equipotentials into *magnetic field cells* with sides that are squares or curvilinear squares, using the trial-and-error method described in Sec. 3-19 in connection with electric field mapping.

A *magnetic field cell* is bounded on two sides by equipotential surfaces and on two others by the sidewalls of a flux tube. For instance, the sides of the magnetic field cell in Fig. 6-43 are the walls of a flux tube, while the top and bottom surfaces are equipotentials. The field is parallel to the sides and normal to the top and bottom surfaces. The permeance of a magnetic field cell, as measured between the equipotential surfaces, is, from (6-18-11),

$$\mathscr{P}_0 = \frac{\mu A}{l} = \frac{\mu l d}{l} = \mu d \qquad \text{(H)} \tag{1}$$

and the permeance per unit depth is

$$\boxed{\frac{\mathscr{P}_0}{d} = \mu \qquad (\text{H m}^{-1})} \tag{2}$$

where μ is the permeability of the cell medium in henrys per meter. Thus, the value of μ for a medium is equal to the permeance per unit depth of a magnetic field cell of that medium. For example, a magnetic field cell in air has a permeance per unit

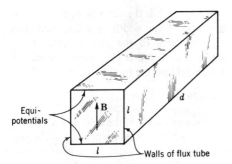

Equi-
potentials

B l d

l

Walls of flux tube

Figure 6-43 Magnetic field cell.

depth of 400π nH m^{-1}, or $1.26\,\mu$H m^{-1}. Thus, if d in Fig. 6-43 equals 1 m and the medium is air, the permeance of the cell is $1.26\,\mu$H.

Any field cell can be subdivided into smaller square-ended cells with as many cells in parallel as in series. Hence the permeance per unit depth of *any* field cell, large or small, exactly square or curvilinear, is equal to μ.

All cells with the same flux through them may be defined as *magnetic field cells of the same kind*. It follows that the magnetic potential difference across all cells of the same kind is the same.

To illustrate some of the principles of magnetic field mapping, let us consider three examples involving three variations of a two-dimensional problem.

Example 1 A magnetic circuit has an air gap of nonuniform separation, as suggested in Fig. 6-44a. The iron has a uniform depth d into the page of 1 m. The geometry of the gap is identical with the region between ff and gg in the capacitor of Fig. 3-19. Find the permeance of the air gap, neglecting fringing of the field.

SOLUTION It may be assumed that the iron permeability is much greater than μ_0 so that the field lines in the gap will be perpendicular to the air-iron boundary and this boundary can be treated as a magnetic equipotential. Since the geometry of the gap is the same as that for the capacitor in Fig. 3-19, the field map in Fig. 3-19 may also serve in the present case, noting that the field lines here are **B** or **H** lines and the equipotentials are surfaces of equal magnetic potential U, as shown in Fig. 6-44b.

With the exception of the cells in the remainder flux tube, all the field cells are of the same kind, and the permeance of the air gap is given in terms of cells of the same kind by

$$\mathscr{P} = \frac{N}{n}\mathscr{P}_0 \qquad (3)$$

where N = number of field cells (or flux tubes) in parallel, dimensionless
n = number of field cells in series, dimensionless
$\mathscr{P}_0$ = permeance of one cell, H

The remainder flux tube has $9\frac{1}{4}$ cells in series, while the other flux tubes have 4. Hence the remainder tube is $4/9\frac{1}{4} = 0.43$ of the width of a full tube, and $N = 15 + 0.43 = 15.43$. The total permeance of the gap is then $\mathscr{P}_T = (15.43/4)\mathscr{P}_0 = 3.86\mathscr{P}_0$. Since the depth of each cell is 1 m, the permeance of one cell is $\mathscr{P}_0 = \mu_0 d = 1.26 \times 1 = 1.26\,\mu$H and the total permeance is

$$\mathscr{P}_T = 3.86 \times 1.26 = 4.86\,\mu\text{H}$$

It is assumed in this example that there is no fringing of the field. For an actual gap there would be fringing at the edges, and the actual permeance of the gap would be somewhat larger than given above.

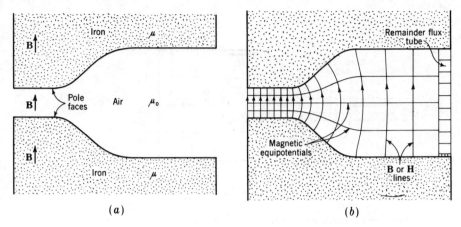

Figure 6-44 Magnetic field in air gap (Example 1).

Example 2 Let the problem of the above example be modified to that shown in Fig. 6-45. Here the gap of the first example is replaced by iron and the iron poles by air. The iron may be regarded as part of a magnetic circuit of iron extending further to the left and to the right as suggested by the dashed lines in Fig. 6-45. The iron extends to a depth of 1 m normal to the page, with the cross section at any depth identical to that in Fig. 6-45. Assume that the iron has a uniform permeability μ which is much larger than μ_0. Find the permeance between the surfaces indicated by the dash-dot lines ff and gg.

SOLUTION The field map for this problem is the same as for Example 1 (Fig. 6-44b) except that the field and equipotential lines are interchanged, as shown

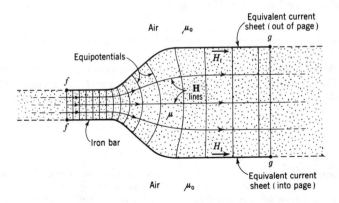

Figure 6-45 Iron bar of nonuniform cross section with internal field (Example 2).

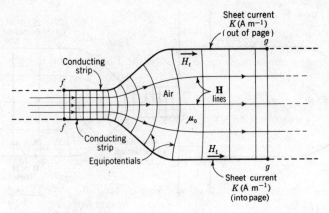

Figure 6-46 Cross section of strip transmission line (Example 3).

in Fig. 6-45. It is assumed that μ is so much greater than μ_0 that in the iron the H field at the air-iron boundary is substantially parallel to the boundary as indicated by the map. The total permeance between ff and gg is, from (3), $\mathscr{P} = (4/15.43)\mathscr{P}_0 = 0.259 \times \mu \times 1 = 0.259 \times 1.26\mu_r = 0.326\mu_r \; \mu H$, where μ_r is the relative permeability of iron.

Example 3 Let the problem of the preceding example be modified to that of a two-strip transmission line having the same cross section as the gap of Example 1 and the iron circuit of Example 2. As shown in Fig. 6-46, the two conducting strips extend normal to the page with a sheet of steady current flowing outward on the upper strip and an equal current flowing inward on the lower strip. The medium in which the strips are located is air. Neglect edge effects. Find the inductance of a 1-m length of the line.

SOLUTION Neglecting edge effects,† the field map between the strips is identical with that for the iron circuit in Fig. 6-45.

If each cell in the map is regarded as a strip transmission line with sheet currents along its upper and lower surfaces, the inductance L_0 for a length d of 1 m of the single-cell line (normal to the page in Fig. 6-46) is, from (5-17-2), given by

$$L_0 = \mu_0 d = 1.26 \; \mu H$$

---

† If the conducting strips are extended an infinite distance to the left and right, as suggested by the dashed lines in Fig. 6-46, the field configuration is precisely as indicated. The field between the strips is produced by the currents on the strips. In Example 2 the field in the iron may be regarded as due to an equivalent current sheet at the surfaces of the iron bar normal to the page (Fig. 6-45).

The total inductance L_T of a 1-m length of the line is then

$$L_T = \frac{4}{15.43} \times 1.26 = 0.326 \, \mu H$$

Note that the permeance of Example 2 is μ_r (of the iron) times this result.

6-20 COMPARISON OF FIELD MAPS IN ELECTRIC, MAGNETIC, AND CURRENT CASES

Graphical field mapping was discussed in Sec. 3-19 for electric fields, in Sec. 4-16 for currents in conductors, and in Sec. 6-19 for magnetic fields. The technique is similar in all these cases. Of particular significance is the fact that a field map for a certain two-dimensional geometry can be applied to numerous problems having this geometry. An illustration of this was provided by the three examples in Sec. 6-19, in which the field map of Fig. 3-19 for a capacitor yielded the solution for the permeance of the volume with the field applied both transversely and longitudinally. The map also gave the inductance of a conducting-strip transmission line.

The same map can, in addition, supply the value of the conductance of a conducting bar with the current flowing transversely and with the current flowing longitudinally. The same map can also be applied to heat- and fluid-flow problems.

To summarize, sketches are given in Fig. 6-47, showing six different problems of the same geometry for which solutions are supplied by one field map. The actual map is shown in Fig. 6-47a, being omitted in the other sketches. The geometry of the problems is that of the capacitor of Fig. 3-19, which was also used in the problems of Figs. 6-44 to 6-46.

In Fig. 6-47a the map represents the electric field in a capacitor with the field transverse. In Fig. 6-47b the map represents the electric field in a conducting bar with current flowing transversely, while in Fig. 6-47c the current flows longitudinally. In Fig. 6-47d the map represents the magnetic field in the air gap between two iron pole faces, while in Fig. 6-47e it represents the magnetic field in an iron bar with the field applied longitudinally. In Fig. 6-47f the map represents the field between two conducting strips acting as a transmission line with current flowing normal to the page. For each case the capacitance, conductance, permeance, or inductance per unit depth (normal to the page) is given as appropriate for the particular problem. Fringing of the field is neglected in all cases.

It is also of interest to compare the significance of the cells (square or curvilinear) of the field maps for the different problems we have considered. Thus, *the capacitance per unit depth of an electric field cell equals the permittivity ϵ of the medium; the conductance per unit depth of a conductor cell equals the conductivity σ of the medium; the permeance per unit depth of a magnetic field cell equals the permeability μ of the medium; and the inductance per unit length of a transmission-line cell equals the permeability μ of the medium.* These relationships are summarized in the last column of Table 6-5. This table also has columns headed Flow lines, Flow tubes, and Equipotentials. Under Flow lines are listed the quantities having the

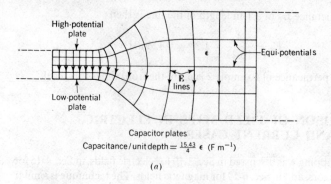

High-potential plate

Low-potential plate

Equi-potentials

ϵ

E lines

Capacitor plates

Capacitance/unit depth $= \frac{15.43}{4} \epsilon$ (F m^{-1})

(a)

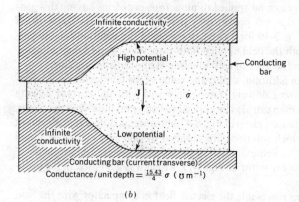

Infinite conductivity

High potential

Conducting bar

J

σ

Infinite conductivity

Low potential

Conducting bar (current transverse)

Conductance/unit depth $= \frac{15.43}{4} \sigma$ (℧ m^{-1})

(b)

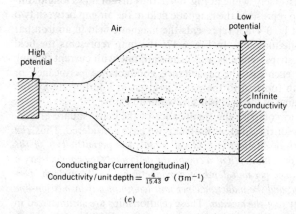

Air

Low potential

High potential

J

σ

Infinite conductivity

Conducting bar (current longitudinal)

Conductivity/unit depth $= \frac{4}{15.43} \sigma$ (℧ m^{-1})

(c)

Figure 6-47 Application of one field map to six situations: (a) capacitor plates, (b) conducting bar with current transverse, (c) conducting bar with current longitudinal.

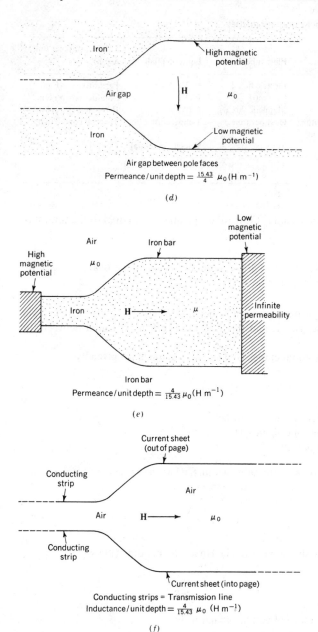

(d)

(e)

(f)

Figure 6-47 Application of one field map to six situations: (d) air gap between magnetic pole pieces, (e) iron bar, and (f) conducting strips as on a transmission line.

Table 6-5 Important field-map quantities

Field	Flow lines	Flow tubes	Equipotentials	Value of cell (per unit depth)
Electric	**D** or **E**	Electric flux ψ	V (V)	Permittivity ϵ (F m^{-1})
Current	**J** or **E**	Current I	V (V)	Conductivity σ (℧ m^{-1})
Magnetic	**B** or **H**	Magnetic flux ψ_m	U (A)	Permeability μ (H m^{-1})
Heat	Temperature gradient	Heat per time	Temperature	Thermal conductivity
Fluid flow (non-turbulent; incompressible)	Velocity	Mass per time	Velocity potential	Density

direction of flow lines and under Flow tubes the quantities equal to the total flux through a tube.

Example Apply the above analogies to find the capacitance of an air capacitor by a resistance measurement.

SOLUTION The capacitor plates are immersed in a large tank filled with a liquid of uniform conductivity σ, and the dc resistance R is measured between the plates.

In general, the conductance G of a certain geometry is given by

$$G = \sigma d \frac{N}{n} \tag{1}$$

where N = number of cells in parallel
n = number of cells in series
d = depth of cells

An actual capacitor with the same geometry has the same field configuration (compare Fig. 6-47a and b); so the capacitance

$$C = \epsilon_0 d \frac{N}{n} \tag{2}$$

where N and n are the same as in (1). Hence, on dividing (2) by (1),

$$C = \frac{\epsilon_0}{\sigma} G = \frac{\epsilon_0}{\sigma R} \tag{3}$$

where C = capacitance of actual capacitor, F
ϵ_0 = permittivity of air = 8.85 pF m^{-1}
σ = conductivity of liquid, ℧ m^{-1}
$R = 1/G$ = measured resistance, Ω

Thus when σ is known (it can be measured with a rectangular volume), the capacitance of an air capacitor can be obtained from (3) by a resistance measurement.

6-21 GAPLESS CIRCUIT

Consider the magnetic circuit of a closed ring of iron of uniform cross section A and mean length l. Suppose that a coil of insulated wire is wound uniformly around the ring and that we wish to know how large the product NI of the number of turns and the current must be to produce a flux density B in the ring.

The coil on the ring in Fig. 6-48a forms a toroid. In the toroid we have, from (5-13-3),

$$B = \frac{\mu NI}{l} = \frac{\mu NI}{2\pi R} \quad \text{(T)} \tag{1}$$

where μ = permeability (assumed uniform) of medium inside of toroid, H m^{-1}
$\quad N$ = number of turns, dimensionless
$\quad I$ = current, A
$\quad l$ = mean length of toroid, m
$\quad R$ = mean radius of toroid, m

Dividing by μ, we have $NI = Hl$ ampere-turns. If a certain flux density B is desired in the ring, the corresponding H value is taken from a BH curve for the ring material and the required number of ampere-turns calculated from $NI = Hl$.

Example 1 An iron ring has a cross-sectional area $A = 1{,}000$ mm^2 and a mean length $l = 600$ mm. Find the number of ampere-turns required to produce a flux density $B = 1$ T. From a BH curve for the iron, $H = 1{,}000$ A m^{-1} at $B = 1$ T.

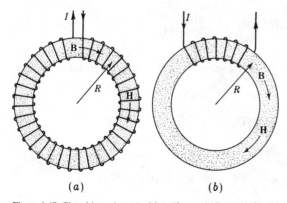

(a)   (b)

Figure 6-48 Closed iron ring (a) with uniform winding and (b) with concentrated winding.

SOLUTION From $NI = Hl$

$$NI = 1,000 \times 0.6 = 600 \text{ A turns}$$

The coil could be 100 turns with a current of 6 A or 1,000 turns with 600 mA. The coil may be uniformly distributed around the ring, as in Fig. 6-48a, or concentrated in a small sector, as in Fig. 6-48b.

The required number of ampere-turns can also be found by calculating the reluctance of the ring circuit, as illustrated by the following example.

Example 2 Find the number of ampere-turns required for the ring of Example 1 for $B = 1$ T by first evaluating the reluctance of the ring.

SOLUTION From (6-18-6)

$$\mathscr{R} = \frac{\oint \mathbf{H} \cdot d\mathbf{l}}{BA} \tag{2}$$

We also have $\oint \mathbf{H} \cdot d\mathbf{l} = Hl = NI$. It follows that $NI = \mathscr{R}BA$, where

$$\mathscr{R} = \frac{l}{\mu A} \tag{3}$$

Since $H = 1,000 \text{ A m}^{-1}$ when $B = 1$ T,

$$\mu = \frac{B}{H} = \frac{1}{1,000} = 1 \text{ mH m}^{-1} \tag{4}$$

It is noted that the relative permeability for this case is

$$\mu_r = \frac{B}{\mu_0 H} = \frac{10^{-3}}{4\pi \times 10^{-7}} = 795 \tag{5}$$

Introducing (4) in (3) and also the value of l and A gives the reluctance of the ring as

$$\mathscr{R} = \frac{0.6}{10^{-3} \times 10^{-3}} = 6 \times 10^5 \text{ H}^{-1} \tag{6}$$

Hence, the required number of ampere-turns is

$$NI = \mathscr{R}BA = 6 \times 10^5 \times 1 \times 10^{-3} = 600$$

as obtained in Example 1.

6-22 MAGNETIC CIRCUIT WITH AIR GAP

Let a narrow air gap of thickness g be cut in the iron ring of Sec. 6-21, as shown in Fig. 6-49a. The gap detail is presented in Fig. 6-49b. By the continuity of the normal component of B the flux density in the gap is the same as in the iron if fringing is neglected. Neglecting the fringing involves but little error where the gap is narrow, as assumed here. The field H_g in the gap is then $H_g = B/\mu_0$, while the field H_i in the iron is

$$H_i = \frac{B}{\mu} = \frac{B}{\mu_r \mu_0} = \frac{H_g}{\mu_r} \tag{1}$$

from which

$$\frac{H_g}{H_i} = \mu_r \tag{2}$$

The number of ampere-turns required to produce a certain flux density B in a magnetic circuit with gap, as in Fig. 6-49a, is a problem for which the solution can be obtained directly. For instance, according to (5-16-9), the line integral of H

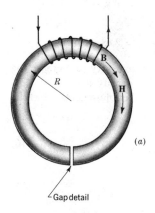

(a)

Gap detail

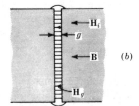

(b)

Figure 6-49 Iron ring with air gap.

once around the magnetic circuit equals the total mmf F, or ampere-turns enclosed. That is,

$$\oint \mathbf{H} \cdot d\mathbf{l} = F = NI \tag{3}$$

Example Let the iron ring of Fig. 6-49 have a cross-sectional area $A = 1,000$ mm², an air gap of width $g = 2$ mm, and a mean length $l = 2\pi R = 600$ mm, including the air gap. Find the number of ampere-turns required to produce a flux density $B = 1$ T.

SOLUTION We have

$$NI = \oint \mathbf{H} \cdot d\mathbf{l} = H_i(l - g) + H_g g \tag{4}$$

where $H_i = H$ field in iron
$H_g = H$ field in gap

From a BH curve for the iron, $H_i = 1,000$ A m^{-1}, and from (2) we know H_g in terms of H_i. Hence (4) becomes

$$NI = H_i[(l - g) + \mu_r g] \tag{5}$$

where $\mu_r = 795$ is the relative permeability of the iron ring at $B = 1$ T. Therefore,

$$NI = 1,000[(0.6 - 0.002) + 795 \times 0.002] = 2,188 \text{ A turns}$$

The introduction of the narrow air gap makes it necessary to increase the ampere-turns from 600 to 2,188 to maintain the flux density at 1 T.

The problem can also be solved by calculating the total reluctance of the magnetic circuit. Thus, from (4) we have

$$NI = \frac{\mu A}{\mu A} H_i(l - g) + \frac{\mu_0 A}{\mu_0 A} H_g g \tag{6}$$

and

$$NI = BA(\mathcal{R}_i + \mathcal{R}_g) \tag{7}$$

where $\mathcal{R}_i = (l - g)/\mu A$ = reluctance of iron part of circuit
$\mathcal{R}_g = g/\mu_0 A$ = reluctance of air gap

6-23 MAGNETIC GAP FORCE

Referring to Fig. 6-49, the effect of the magnetic field is to exert forces which tend to close the air gap. That is, the magnetic poles of opposite polarity at the sides of the

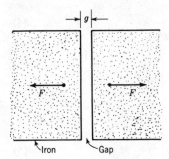

Figure 6-50 Forces at air gap.

gap are attracted to each other. Forces produced by magnetic fields find application in numerous electromechanical devices. In this section an expression for the force between magnetic pole pieces is developed.

The density of energy stored in a magnetic field is

$$w_m = \frac{1}{2}\frac{B^2}{\mu} \quad (\text{J m}^{-3}) \tag{1}$$

If the gap is small, we may assume a uniform field in the air gap. The total energy W_m stored in the gap is then

$$W_m = w_m A g = \frac{B^2 A g}{2\mu_0} \quad (\text{J}) \tag{2}$$

where A = area of gap, m^2
  g = width of gap, m

Suppose now that the iron ring in Fig. 6-49 is perfectly flexible, so that the gap must be held open by a force F as in Fig. 6-50. If the force is increased so as to increase the gap by an infinitesimal amount dg while at the same time the current through the coil is increased to maintain the flux density B constant, the energy stored in the gap is increased by the infinitesimal amount

$$dW_m = \frac{B^2 A}{2\mu_0} dg \tag{3}$$

Equation (3) has the dimensions of energy. But energy may also be expressed as force times distance, which in this case is $F\,dg$, where F is the attractive force between the poles. It is equal in magnitude to the force required to hold them apart. Thus

$$F\,dg = \frac{B^2 A}{2\mu_0} dg$$

or
$$F = \frac{B^2 A}{2\mu_0} \tag{4}$$

where F = attractive force, N
B = flux density, T
A = area of gap, m^2
μ_0 = permeability of air = 400π nH m^{-1}

Dividing by the gap area A yields the pressure P. That is,

$$P = \frac{F}{A} = \frac{B^2}{2\mu_0} \quad (\text{N m}^{-2}) \tag{5}$$

6-24 PERMANENT MAGNET WITH GAP

Suppose first that a closed iron ring is magnetized to saturation with a uniform toroidal coil wound on the ring. When the coil is removed, the flux density in the iron is equal to the retentivity (see Fig. 6-52). If, however, the system has an air gap, as in Fig. 6-51, the flux density has a smaller value as given by a point P which lies somewhere on the demagnetization curve (Fig. 6-52) (see also Sec. 6-15). Further information is needed to locate this point. This may be obtained as follows. The line integral of $\mathbf{H}$ once around a magnetic circuit is $\oint \mathbf{H} \cdot d\mathbf{l} = NI$. Since $NI = 0$,

$$\oint \mathbf{H} \cdot d\mathbf{l} = H_i(l - g) + H_g g = 0$$

or
$$H_i(l - g) = -H_g g \tag{1}$$

where H_i = H field in the iron
$l = 2\pi R$ = total length of magnetic circuit (including gap)
g = width of gap
H_g = H field in gap

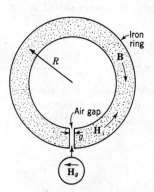

Figure 6-51 Permanently magnetized ring with air gap.

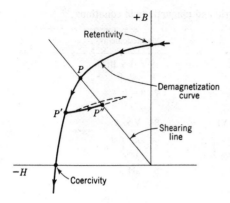

Figure 6-52 Demagnetization curves for permanent magnet.

Thus H_i and H_g are in opposite directions, as indicated in Fig. 6-51. If leakage is neglected, B is uniform around the circuit. Multiplying (1) by μ_0 and solving for the ratio B/H_i, or the permeability of the iron, we obtain

$$\frac{B}{H_i} = -\mu_0 \frac{l-g}{g} \tag{2}$$

This ratio of the flux density B to the field H_i in the iron gives the slope of a line called the *shearing line*, as shown in Fig. 6-52. The intersection of this line with the demagnetization curve determines the position of the iron on the magnetization curve (point P). This location is a function of the ratio of the iron path length $(l-g)$ to the gap length g.

In most permanent-magnet applications, where it desired that B remain relatively constant, a moderate demagnetizing field is applied to the iron, moving the position of the iron to P' (Fig. 6-52). On removing the field, the iron moves to the point P'' on the shearing line. The ring magnet is now said to be *stabilized*, and when fields less than about the difference of H between points P' and P'' are applied to the ring and then removed, the iron will always return to approximately point P''. Under these conditions the iron moves along a minor hysteresis loop, as suggested by the dashed lines in Fig. 6-52.

6-25 A COMPARISON OF STATIC ELECTRIC AND MAGNETIC FIELDS

It is instructive to compare electric and magnetic fields and to note both their differences and their similarities. A partial comparison is given in Table 6-6 involving relations developed in the preceding chapters for static fields. A comparison of relations for nonstatic fields is given in Sec. 8-17.

Table 6-6 A comparison of static electric and magnetic field equations

Description of equation	Electric fields	Magnetic fields
Force	$\mathbf{F} = Q\mathbf{E}$	$d\mathbf{F} = (\mathbf{I} \times \mathbf{B})\, dl$ $\mathbf{F} = Q_m \mathbf{B}$
Basic relations for lamellar and solenoidal fields	$\nabla \times \mathbf{E}_c = 0$†	$\nabla \cdot \mathbf{B} = 0$
Derivation from scalar or vector potential	$\mathbf{E}_c = -\nabla V$ $V = \dfrac{1}{4\pi\epsilon_0} \displaystyle\int_v \dfrac{\rho}{r}\, dv$	$\mathbf{B} = \nabla \times \mathbf{A}$ $\mathbf{A} = \dfrac{\mu_0}{4\pi} \displaystyle\int_v \dfrac{\mathbf{J}}{r}\, dv$
Constitutive relations	$\mathbf{D} = \epsilon\mathbf{E}$	$\mathbf{B} = \mu\mathbf{H}$
Source of electric and magnetic fields	$\nabla \cdot \mathbf{D} = \rho$	$\nabla \times \mathbf{H} = \mathbf{J}$
Energy density	$w_e = \frac{1}{2}\epsilon E^2 = \frac{1}{2}ED$	$w_m = \frac{1}{2}\mu H^2 = \frac{1}{2}BH$
Capacitance and inductance	$C = \dfrac{Q}{V}$	$L = \dfrac{\Lambda}{I}$
Capacitance and inductance per unit length of a cell	$\dfrac{C}{d} = \epsilon$	$\dfrac{L}{d} = \mu$
Closed path of integration	$\displaystyle\oint \mathbf{E} \cdot d\mathbf{l} = \mathscr{V}$ $\displaystyle\oint \mathbf{E}_c \cdot d\mathbf{l} = 0$	$\displaystyle\oint \mathbf{H} \cdot d\mathbf{l} = F = NI$ $\displaystyle\oint \mathbf{H} \cdot d\mathbf{l} = 0 \qquad$ no current enclosed
Derivation from scalar potentials	$\mathbf{E}_c = -\nabla V$	$\mathbf{H} = -\nabla U \qquad$ in current-free region
Dipole fields	$E_r = \dfrac{Ql\cos\theta}{2\pi\epsilon r^3}$ $E_\theta = \dfrac{Ql\sin\theta}{4\pi\epsilon r^3}$	$B_r = \dfrac{\mu_0 Q_m l\cos\theta}{2\pi r^3}$ $B_\theta = \dfrac{\mu_0 Q_m l\sin\theta}{4\pi r^3}$
Loop fields		$B_r = \dfrac{\mu_0 IA\cos\theta}{2\pi r^3}$ $B_\theta = \dfrac{\mu_0 IA\sin\theta}{4\pi r^3}$
Dipole torque	$T = QlE\sin\theta$	$T = Q_m lB\sin\theta$
Loop torque		$T = IAB\sin\theta$
Solenoid torque		$T = NIAB\sin\theta$
Torque (in general)	$\mathbf{T} = \mathbf{m} \times \mathbf{E}$	$\mathbf{T} = \mathbf{m} \times \mathbf{B}$

† $\mathbf{E}_c$ is the static electric field intensity (due to charges). $\mathbf{E}$ (without subscript) implies that emf-producing fields (not due to charges) may also be present.

6-26 COMPARISON OF ELECTRIC AND MAGNETIC RELATIONS INVOLVING POLARIZATION AND MAGNETIZATION

It is interesting to compare the magnetic relations where magnetization M is present with the corresponding electric relations where polarization P is present (see Chap. 3). This is done in the two parts of Table 6-7.

Table 6-7
Comparison of equations involving polarization P and magnetization M

Description of equation	Electric case	Magnetic case
Dipole-moment relations	$P = \dfrac{p}{v} = \dfrac{Ql}{v}$	$M = \dfrac{m}{v} = \dfrac{Q_m l}{v}$
Flux density	$\mathbf{D} = \epsilon_0 \mathbf{E} + \mathbf{P}$	$\mathbf{B} = \mu_0(\mathbf{H} + \mathbf{M})$
Permittivity and permeability	$\epsilon = \epsilon_0 + \dfrac{P}{E}$	$\mu = \mu_0\left(1 + \dfrac{M}{H}\right)$
Relative permittivity and permeability	$\epsilon_r = 1 + \dfrac{P}{\epsilon_0 E}$	$\mu_r = 1 + \dfrac{M}{H}$
Susceptibilities	$\chi = \epsilon_r - 1$	$\chi_m = \mu_r - 1$
Relation to polarization charge density and to pole density and equivalent current density	$\nabla \cdot \mathbf{P} = -\rho_p$ $\nabla \cdot \mathbf{D} = \rho$	$\nabla \cdot \mathbf{M} = -\rho_m$ $\nabla \times \mathbf{M} = \mathbf{J}'$
Relation to surface charges and poles and equivalent current sheet	$P = \rho_{sp}$ $D = \rho_s$	$M = \rho_{sm}$ $M = K'$
Poisson's equations	$\nabla^2 V = -\dfrac{\rho}{\epsilon}$	$\nabla^2 U = \nabla \cdot \mathbf{M} = -\nabla \cdot \mathbf{H}$
Scalar and vector potentials	$V = \dfrac{1}{4\pi\epsilon_0} \displaystyle\int_v \dfrac{\rho - \nabla \cdot \mathbf{P}}{r}\, dv$	$\mathbf{A} = \dfrac{\mu_0}{4\pi} \displaystyle\int_v \dfrac{\mathbf{J} + \nabla \times \mathbf{M}}{r}\, dv$

Comparison of polarization and magnetizations of two models

Electric polarization	Magnetization	
	Current model	Pole model
$P = \dfrac{QL}{v}$	$M = \dfrac{IA}{v}$	$M = \dfrac{Q_m l}{v}$
$P = \rho_{sp}$	$M = K'$	$M = \rho_{sm}$
$\mathbf{P} \cdot \hat{\mathbf{n}} = \rho_{sp}$	$\mathbf{M} \times \hat{\mathbf{n}} = \mathbf{K}'$	$\mathbf{M} \cdot \hat{\mathbf{n}} = \rho_{sm}$
$\nabla \cdot \mathbf{P} = -\rho_p$	$\nabla \times \mathbf{M} = \mathbf{J}'$	$\nabla \cdot \mathbf{M} = -\rho_m = -\nabla \cdot \mathbf{H}$
$\mathbf{D} = \epsilon_0 \mathbf{E} + \mathbf{P}$	$\mathbf{B} = \mu_0(\mathbf{K} + \mathbf{K}')$	$\mathbf{B} = \mu_0(\mathbf{H} + \mathbf{M})$

6-27 DO MAGNETIC MONOPOLES EXIST?

Since electrons (negatively charged) and protons (positively charged) can be separated, one might expect by analogy that magnetic poles should be separable. However, as discussed earlier, the poles of a magnet always come in pairs with a north and south pole, that is, in dipole form. Nevertheless, the idea of separate poles, or monopoles (both north and south), has persisted and, in fact, such particles have been postulated in some Grand Unified Theories. The theories suggest that monopoles were created in large numbers during the early stages of the formation of the universe in the Big Bang explosion and it is thought that some should still be around and observable as part of the steady rain of cosmic rays. Although still of atomic size, a magnetic monopole would be billions of times heavier than a proton.

On a couple of occasions in recent years it has been reported that massive atomic particles have been detected in cosmic ray apparatus, but it could not be established for certain whether they were actually magnetic monopoles or some other massive object. The search for magnetic monopoles continues, and even though one may eventually be detected for certain, they are so rare that they will probably have little effect on our established techniques for dealing with magnetization.

PROBLEMS†

Group 6-1: Secs. 6-1 through 6-7. Magnets, magnetic materials, permeability, magnetization, B, H, and M

6-1-1. Force between conductors in ferrous medium Two long, straight parallel conductors separated by a distance d carry a current I in opposite directions. Find the force per unit length of conductor if (a) the medium is air and (b) the medium is ferrous with $\mu_r = 350$.

6-1-2. Compass deflection by bar magnet A bar magnet with 10 A m² magnetic moment is brought 300 mm from a compass needle. What is the maximum angular deviation of the needle which the magnet can produce? Take the horizontal component of the earth's field as 30 μT.

★6-1-3. Coaxial line filled with ferrous-titanate medium A coaxial transmission line has a solid inner conductor of radius a and a thin-walled outer conductor of radius b. The constants for the ferrous-titanate medium are $\sigma = 0$, $\epsilon_r = 10$, and $\mu_r = 200$. (a) Find the inductance per meter of length of line if the radius b/a is the same as for part (d) of Prob. 5-2-15. (b) Assuming a uniform current distribution over the cross-sectional area of the inner conductor, what is its percentage effect on the total inductance?

★6-1-4. Coaxial line with helical inner conductor (a) To raise the inductance of a coaxial line the inner conductor may be wound as a helix. If the inner conductor is a fine-wire helix of radius a and the outer conductor a thin-walled tube of radius b, find the inductance per unit length. The helix has N turns per meter. (b) If the helical inner conductor is wound on a nonconducting ferrite core of radius a with $\mu_r = 100$, find the line inductance.

6-1-5. Cavities. B and H Two cavities are cut in a ferromagnetic medium ($\mu_r = 300$) of large extent. Cavity 1 is a thin disk-shaped cavity with flat faces perpendicular to the direction of **B** in the ferromagnetic medium. Cavity 2 is a long needle-shaped cavity with its axis parallel to **B**. The cavities are

† Answers to starred problems are given in Appendix C.

filled with air. If $B = 1$ T, what is the magnitude of **H** at (a) the center of cavity 1 and (b) the center of cavity 2?

6-1-6. Bar magnet. Translational force A bar magnet in a uniform magnetic field is acted on only by a torque, there being no translational force on the magnet. In a nonuniform field, there is a net translational force. Find the maximum value of this force on a uniformly magnetized bar magnet 6 mm long with a magnetic moment of 2 A m² situated 100 mm from one pole of a very long slender bar magnet having a pole strength of 600 A m.

6-1-7. Fields of bar magnet Show that the flux density from a bar magnet is given by

$$B_\theta = \frac{\mu_0 Q_m l \sin \theta}{4\pi r^3} \quad \text{and} \quad B_r = \frac{\mu_0 Q_m l \cos \theta}{2\pi r^3}$$

where $Q_m l$ = dipole moment
θ = angle between bar axis and radius vector of length r

It is assumed that $r \gg l$.

★6-1-8. Bar magnet near interface A small bar magnet is situated in air at a distance d from the plane surface of a medium of large extent and permeability μ. The bar axis makes an angle θ with respect to a perpendicular to the surface. If the dipole moment of the bar magnet is $Q_m l$, find the force and torque on the bar magnet due to the magnetization induced in the medium of permeability μ. Make a graph of torque versus θ.

6-1-9. Bar magnet. Magnetization Show that at the center of a long uniformly magnetized bar of uniform cross-sectional area A that $|\mathbf{H}| \approx 2MA/\pi l^2$, where M = magnetization and l = length of bar.

6-1-10. Bar magnet. Magnetization Show that the permeability at the center of a long permanently magnetized rod of uniform magnetization is given by $\mu_0[1 - (\pi l^2/2A)]$, where l = length of rod and A = cross-sectional area of rod.

6-1-11. 4th power effect between magnets Demonstrate that the force between two small bar magnets varies as the inverse fourth power of their separation and is independent of their relative orientation.

Group 6-2: Secs. 6-8 through 6-10. Energy, energy density, and boundary relations

6-2-1. Energy in solenoid A solenoid of 1,000 turns has a length of 600 mm and a diameter of 40 mm. If the solenoid current is 2 A, find (a) the energy density of the center of the solenoid and (b) the total energy in the solenoid.

6-2-2. Energy in bar magnet A bar magnet of 800-A m² magnetic moment has a length of 100 mm and a diameter of 15 mm. Find (a) the energy density at the center of the bar and (b) the total energy in the bar.

6-2-3. Energy between current sheets Two large, parallel flat conducting sheets with 20-mm separation carry a sheet current density of 1,000 A m⁻¹ in opposite directions. Find (a) the energy density between the sheets and (b) the total energy between the sheets over an area of 2 m².

6-2-4. Field angles at interface Two ferromagnetic media are separated by a plane boundary. Medium 1 has a relative permeability $\mu_r = 300$ and medium 2 a relative permeability $\mu_r = 1,800$. If the magnetic field direction in medium 2 is at an angle of 10° with respect to the surface, find the angle between the field direction and the surface in medium 1.

6-2-5. Current sheet at interface A flat sheet of linear current density K separates two media with magnetic flux densities $B_1 = 5$ mT and $B_2 = 25$ mT in opposite directions parallel to the sheet. The relative permeabilities are $\mu_{r1} = 50$ and $\mu_{r2} = 90$. What is K?

6-2-6. Conductor with shield A copper conductor of 50-mm radius is enclosed (magnetically shielded) by a concentric iron tube of inner radius 100 mm and outer radius 150 mm. (a) If the current in the conductor is 300 A, find B and H at radii of 25, 50, 100, 125, 150, and 200 mm. Assume that the current density in the conductor is uniform and that $\mu_r = 500$ in the iron tube. (b) Sketch the variation of B and H as a function of radius from 0 to 250 mm.

Group 6-3: Secs. 6-11 through 6-17. Ferromagnetism, magnetization curves, hysteresis, permanent magnets, and demagnetization

★6-3-1. **Demagnetization curve of permanent magnet** (a) Assuming that the demagnetization curve of a certain ferromagnetic material is a straight line, what is the maximum BH product if the retentivity is 1 T and the coercivity is 20 kA m^{-1}? (b) Prove that this is the maximum value.

★6-3-2. **Rayleigh relation** According to Lord Rayleigh, the bottom part of the normal magnetization curve is given by $B = \mu_i H + aH^2$, where μ_i is the initial permeability (at $H = 0$, $B = 0$) and a is a constant. Assume that this relation applies to the initial-magnetization curve of an iron specimen. What is the expression for the energy density in the iron after the field is raised from $H = 0$ to $H = H_1$? Assume that the specimen is initially unmagnetized.

Group 6-4: Secs. 6-18 through 6-20. Magnetic circuit, reluctance and permeance, and magnetic field mapping

6-4-1. **Permeance of bar** A rectangular iron bar has a length x_1 and a cross-sectional area A. The permeability is a function of x as given by

$$\mu = \mu_0 + \frac{\mu_1 - \mu_0}{x_1} x$$

Find the permeance of the bar.

6-4-2. **Reluctance and permeance of bar with notches. Field map** For an iron bar having the dimensions shown in Fig. 4-23 find (a) reluctance and (b) permeance. $\mu_r = 2,000$.

6-4-3. **Reluctance and permeance of L-shaped bar. Field map** For the L-shaped iron bar having the dimensions shown in Fig. P4-3-3, find (a) reluctance and (b) permeance. $\mu_r = 4,000$. The bar thickness is 20 mm.

★6-4-4. **Finding buried pipeline. Field map** A steel pipeline runs east-west at a constant depth below a flat ground. The undisturbed earth's magnetic field in the region has a dip of 70° (angle of **B** from horizontal) and a declination of 0° (angle of **B** from north). Locate the pipe and its depth if the following dip measurements are obtained with the dip needle 1 m above the ground:

Meters south of reference mark	Dip angle, deg
0	71
2	75
4	77
6	74
8	70
10	65
12	62
14	66
16	69

6-4-5. **Bar with and without notches** How much greater is the reluctance of the bar of Fig. 4-23 as compared to a uniform rectangular block without notches assuming μ_r is the same in both cases?

★6-4-6. **L-shaped vs. straight bar** How much greater is the reluctance of the L-shaped bar of Fig. P4-3-3 as compared to a straight uniform rectangular bar obtained by cutting the one in Fig. P4-3-3 along the diagonal at the bend, turning one half over, and joining both halves to form a gapless straight bar? Assume μ_r to be the same in both cases.

Group 6-5: Secs. 6-21 through 6-27. Gapless circuit, circuit with air gap, gap force, permanent magnet with gap, and magnetic monopoles

6-5-1. Gapless iron ring An iron ring of mean radius 200 mm and 150-mm^2 cross-sectional area is wound with 100 turns of wire. If $B = 0.5$ T and $\mu_r = 250$, find the wire current.

⋆**6-5-2. Iron ring with gap** An iron ring has a uniform cross-sectional area of 150 mm^2 and a mean radius of 200 mm. The ring is continuous except for an air gap 1 mm wide. Find the number of ampere-turns required on the ring to produce a flux density $B = 0.5$ T in the air gap. Neglect fringing. When $B = 0.5$ T in the iron, $\mu_r = 250$.

6-5-3. Composite ferrite-steel ring A ring has a cross-sectional area of 200 mm^2 and a mean path length of 350 mm. Two-thirds of the length of the ring is composed of ferrite ($\mu_r = 1,000$) and the remaining one-third of mild steel ($\mu_r = 2,000$). Find the total magnetic flux in the ring if it is wound with a coil of 750 turns carrying 100 mA. Consider that the two butt joints between the ferrite and steel sections are equivalent to an air gap of 100 μm.

⋆**6-5-4. Iron ring. B in gap** An iron ring of 10-cm^2 cross-sectional area and 150-mm radius has a 3-mm air gap. If 200 turns of wire carrying 10 A is wound on the ring, find B in the gap. Neglect fringing. Assume the iron characteristics are the same as given in Fig. 6-25. This problem requires a trial-and-error or a graphical solution. It is a good example of a practical engineering problem.

6-5-5. Iron ring. B in gap An iron ring of 20 cm^2 cross-sectional area and 200 mm radius has a 4 mm air gap. If 300 turns of wire carrying 12 A is wound on the ring, find B in the gap. Assume that, due to fringing, the effective cross-sectional area of magnetic flux in the gap is increased by 15 percent. Assume the iron characteristics are the same as given in Fig. 6-25.

⋆**6-5-6. Iron ring. Gap force** An iron ring magnet of 0.02-m^2 cross-sectional area and 300-mm radius has a 1-mm air gap and a winding of 1,200 turns. If the current through the coil is 6 A, what is the force tending to close the gap? Take $\mu_r = 1,000$ for the iron and neglect fringing.

6-5-7. Magnet with gap Show that the permeability of a permanent magnet with air gap can be expressed by $\mu = -\mu_0(\mathscr{P}_g/\mathscr{P})$, where $\mu_0 = $ permeability of air, $\mathscr{P}_g = $ permeance of air gap, and $\mathscr{P} = $ permeance of empty space occupied by the magnet.

6-5-8. Closure current for relay An electromagnetic relay has a magnetic circuit like that of an iron ring with gap. If the relay has a 500-turn coil with a 1-mm air gap and 0.5-cm^2 cross section at the gap, find the coil current required for closure. The required closure force is equivalent to the weight of 200 g. Assume that the magnetic material of the relay, including the movable bar at the gap, has a constant cross section of 0.5 cm^2 and a relative permeability $\mu_r = 1,000$.

⋆**6-5-9. Iron ring. Flux** An iron ring has a mean radius of 300 mm and a uniform cross section of 10 cm^2. A wire coil of 1,000 turns carrying 5 A is situated on the ring. Find the magnetic flux ψ_m in the ring. The iron characteristics are as given in Fig. 6-25.

6-5-10. Iron ring with spoke A circular iron ring has an iron spoke bridging the ring across a diameter. Each half of the ring has an identical coil of N turns carrying 3 A, with current directions such that the mmfs of both coils add. The flux in the ring at one of the coils is 100 μWb. The ring and spoke have a uniform cross section of 12 cm^2. The mean radius of the ring is 350 mm. The iron characteristics are as in Fig. 6-25. (a) Find the number of turns N and (b) the flux ψ_m in the spoke.

⋆**6-5-11. Magnetic monopole force** (a) If magnetic monopoles exist and a sufficient quantity of one polarity collected, could they be used to propel a ship along the earth's magnetic field lines? (b) If so, where does the energy come from? (c) How large a monopole would be required to produce a horizontal force equivalent to the weight of 1 tonne (1 metric ton) if the horizontal component of the earth's magnetic field is 50 μT?

Group 6-6: Practical applications

6-6-1. Magnetic levitation Magnetic levitation may be employed for frictionless horizontal movement of objects. An experimental Japanese very high-speed railroad uses magnetic levitation. Design a magnetic levitation system to raise 1,000 kg at the earth's surface by 5 mm using (a) permanent magnets and (b) electromagnets.

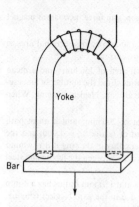

Yoke

Bar

Figure P6-6-2 U-shaped electromagnet.

6-6-2. Electromagnet for lifting An electromagnet consists of a U-shaped iron yoke as shown in Fig. P6-6-2. A thin copper sheet attached to the yoke prevents iron-to-iron contact with an iron load. If the magnetic flux through the circuit is 20 mWb and the yoke area is 0.015 m² per pole, what is the maximum iron-bar weight which can be lifted? Neglect fringing.

6-6-3. Electromagnet for lifting. Tapered poles (a) If the contact area of the electromagnet of Prob. 6-6-2 and Fig. P6-6-2 is reduced to 0.005 m² by means of tapered sections (poles) on the yoke, how much bar weight can be lifted? Assume ψ_m to be the same as before and neglect fringing. (b) In practice what prevents the attactive force from increasing without limit as the contact area is reduced?

6-6-4. Contact pressure What is the contact pressure (a) in Prob. 6-6-2 and (b) in Prob. 6-6-3?

★6-6-5. Cyclotron magnet A cyclotron magnet has the dimensions shown in Fig. P6-6-5. The pole pieces are cylindrical with tapered ends. The diameter of the gap is 1 m, and the gap spacing is 150 mm. How many ampere-turns are required in each of the two windings shown to produce a flux density of 1 T in the air gap? Assume that the magnet is made of iron with $\mu_r = 2,000$ and that there is no fringing at the gap. Neglect leakage along the magnet structure. As a further simplification take the effective

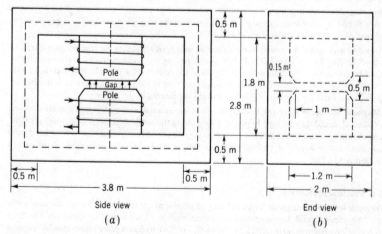

Side view
(a)

End view
(b)

Figure P6-6-5 Cyclotron magnet.

length of sections as the length measured along the centerline (dashed line in Fig. P6-6-5a). Take the diameter of the tapered section of the poles as the average diameter.

*6-6-6. Magnetic field of earth For many purposes the magnetic field of the earth at ionospheric heights or above may be regarded as if it were produced by a short magnetic dipole (or small current loop) situated at the center of the earth with $\mu = \mu_0$ everywhere (except within the dipole). If the horizontal component of **B** at the earth's surface is 20 μT at a point where the dip angle of the field (angle from horizontal) is 72°, (a) find the magnetic moment required for the dipole (or loop) at the center of the earth. Take the earth's radius as 6.37 Mm. (b) Using this model, calculate B at a distance of 40 Mm above the earth's surface in the equatorial and polar directions.

*6-6-7. Degaussing a ship To protect ships against mines triggered by changes in the earth's magnetic field produced by a ship's proximity, the ship's magnetic moment may be reduced by a suitable current-carrying coil of wire wound around it, a process called *degaussing*. Consider two idealized situations. *Case 1*: Let the ship be modeled by a vertically oriented bar magnet of moment 10^5 A m^2. (a) At a depth of 10 m below the ship what is B from the ship? Assume that the magnet is short compared to 10 m and that its middle point is at the water surface. (b) If the earth's field is vertical and equal to 60 μT, how accurately must the coil current be set so that the ship produces less than a 3 percent change in total field? *Case 2*: Let the ship be modeled by a horizontally oriented bar magnet of moment 10^5 A m^2. (c) At a depth of 10 m below the ship what is B from the ship? Assume that the middle of the magnet is at the water surface. (d) If the earth's field is horizontal and equal to 40 μT, how accurately must the coil current be set so that the ship produces less than a 3 percent change in total field? *Note*: An actual ship is much more complex than is suggested by these idealized examples and may require many coils wound vertically and horizontally.

6-6-8. Degaussing the earth If a single turn of wire were laid around the earth at its magnetic equator, a current through this loop could cancel the earth's magnetic field at a distance from the earth. (a) How much current would be required in the wire? The earth's magnetic dipole moment is about 6.4×10^{21} A m^2. (b) What is the radial distance R beyond which the earth's field would be 1 percent or less of its normal value, assuming the current in the wire was adjusted so that the wire-loop magnetic moment was exactly equal to the earth's magnetic moment? (c) With wire-loop current turned on, what would the magnetic field be like at distances of less than R and in particular on the earth's surface? (d) With wire-loop current turned on, what would the magnetic flux density be 1 km from the wire?

SEVEN

BOUNDED FIELDS AND LAPLACE'S EQUATION

7-1 INTRODUCTION

All of electrical science and engineering is based on the action of electric fields (E and D) and magnetic fields (B and H). These fields, in turn, are produced basically by electric charges (Q) and currents (I) in electromagnetic devices of all kinds. To understand how the devices work we must be able to evaluate the fields in and around them. Thus, fields require a space, or three-dimensional, concept, inviting us to make maps or pictures of them showing field lines and equipotential surfaces. The maps give us information about the field intensities, potential differences, and energy, charge, and current densities everywhere on the map. We learn to think of the field as an aggregation of field cells connected in series and parallel, whose totals give us integrated (or circuit) values of capacitance, resistance, and inductance. Furthermore, a map for a given configuration can be applied to many situations. Thus, as illustrated in Fig. 6-47, one map provides the solution for a capacitor, a conducting bar with transverse and longitudinal currents, a magnetic bar and gap, and a strip transmission line.

All of these field maps are solutions of Laplace's equation. In this chapter we go further into more formal, rigorous mathematical methods of finding field configurations for a number of electrostatic problems. In the first examples space is free of charge ($\rho = 0$), and we seek solutions to Laplace's equation† ($\nabla^2 V = 0$) which satisfy the boundary conditions. In a later example space charge is present, and a solution to Poisson's equation ($\Delta^2 V = -\rho/\epsilon$) is obtained. Because of the dependence of the field configuration on its boundaries, these problems are often called *boundary-value problems*.

---

† See Sec. 3-25.

The solution of a boundary-value problem is usually facilitated if it is set up in a coordinate system in which the boundaries can be specified in a simple manner. For instance, a problem involving a rectangular object may be most readily handled with rectangular coordinates, a cylindrical object by cylindrical coordinates, an elliptical object by elliptical-hyperbolic coordinates, etc. A restriction on the formal mathematical method is that the boundaries in many practical problems cannot be simply expressed in any coordinate system, and often in such cases resort must be made to such other methods as graphical, experimental, analog-computer, and iterative (digital-computer) techniques. These various methods will be discussed and compared in the sections which follow.

7-2 LAPLACE'S EQUATION IN RECTANGULAR COORDINATES; SEPARATION OF VARIABLES

In rectangular coordinates Laplace's equation ($\nabla^2 V = 0$) becomes

$$\frac{\partial^2 V}{\partial x^2} + \frac{\partial^2 V}{\partial y^2} + \frac{\partial^2 V}{\partial z^2} = 0 \tag{1}$$

Let us assume that V can be expressed as the product of three functions X, Y, and Z, or

$$V = XYZ \tag{2}$$

where X = function of x only
Y = function of y only
Z = function of z only

Substituting (2) into (1) we get

$$YZ \frac{d^2 X}{dx^2} + XZ \frac{d^2 Y}{dy^2} + XY \frac{d^2 Z}{dz^2} = 0 \tag{3}$$

Dividing by XYZ separates the variables, and so we have

$$\frac{1}{X} \frac{d^2 X}{dx^2} + \frac{1}{Y} \frac{d^2 Y}{dy^2} + \frac{1}{Z} \frac{d^2 Z}{dz^2} = 0 \tag{4}$$

Since the sum of the three terms on the left-hand side is a constant ($=0$) and each variable is independent, each term must equal a constant. That is, we may write

$$\frac{1}{X} \frac{d^2 X}{dx^2} = a_1^2 \tag{5}$$

or

$$\frac{d^2 X}{dx^2} = a_1^2 X \tag{6}$$

and similarly

$$\frac{d^2 Y}{dy^2} = a_2^2 Y \tag{7}$$

and

$$\frac{d^2Z}{dz^2} = a_3^2 Z \tag{8}$$

where

$$a_1^2 + a_2^2 + a_3^2 = 0 \tag{9}$$

The problem now is to find a solution for each of the three variables separately. Hence, the procedure we have been following is often referred to as the *method of separation of variables*.

A solution of (6) is

$$X = C_1 e^{a_1 x} + C_2 e^{-a_1 x} \tag{10}$$

where C_1 and C_2 are arbitrary constants that must be evaluated from the boundary conditions. Either term in (10) is a solution, or the sum is a solution, as may be verified by substituting the solution in (6).

It follows that a general solution of (1) is

$$V = (C_1 e^{a_1 x} + C_2 e^{-a_1 x})(C_3 e^{a_2 y} + C_4 e^{-a_2 y})(C_5 e^{a_3 z} + C_6 e^{-a_3 z})$$

where C_1, C_2, etc., are constants. Thus, solutions may take the form of exponential, trigonometric, or hyperbolic functions.

7-3 EXAMPLE 1: THE PARALLEL-PLATE CAPACITOR

Consider a parallel-plate capacitor as shown in cross section in Fig. 7-1a. The plates are infinite in extent and are separated by a distance x_1. The left-hand plate is at zero potential and the right-hand plate at potential V_1. Let us apply Laplace's equation to find the potential distribution between the plates.

There is no variation in potential in the y and z directions, so that the problem is one-dimensional and Laplace's equation ($\nabla^2 V = 0$) reduces to

$$\boxed{\frac{d^2 V}{dx^2} = 0} \tag{1}$$

For the second derivative of V with respect to x to be zero the first derivative must be equal to a constant. Thus, we have

$$\frac{dV}{dx} = C_1 \tag{2}$$

or

$$dV = C_1 \, dx \tag{3}$$

Integrating (3), we write

$$\int dV = C_1 \int dx$$

or

$$V = C_1 x + C_2 \tag{4}$$

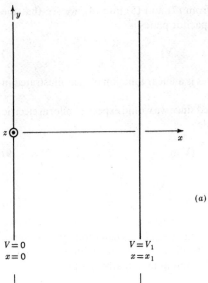

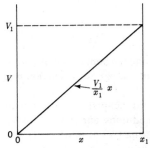

(a)

(b)

Figure 7-1 (*a*) Parallel-plate capacitor and (*b*) potential variation.

The boundary conditions (B.C.) are

B.C. 1: $V = 0$ at $x = 0$

B.C. 2: $V = V_1$ at $x = x_1$

From B.C. 1 (4) becomes

$$0 = 0 + C_2 \tag{5}$$

Hence, $C_2 = 0$. From B.C. 2 (4) becomes

$$V_1 = C_1 x_1 \tag{6}$$

so that

$$C_1 = \frac{V_1}{x_1} \tag{7}$$

Introducing the values for C_1 and C_2 from (7) and (5) into (4), we see that the solution for the potential between the capacitor plates is

$$V = \frac{V_1}{x_1} x \quad \text{(V)} \tag{8}$$

Thus, the variation of V between the plates is a linear function of x, as illustrated in Fig. 7-1b.

This result could have been anticipated since we would expect a uniform electric field **E** between the plates with magnitude

$$E = \frac{V_1}{x_1} \quad (\text{V m}^{-1}) \tag{9}$$

with the potential at any point given by

$$V = Ex \tag{10}$$

The constant C_1 is thus equal to the electric field between the plates. Ordinarily we would not have resorted to a formal solution of Laplace's equation for this simple problem, but its application here serves to illustrate the procedures involved in applying Laplace's equation while using a minimum of mathematics.

7-4 UNIQUENESS

If a function V satisfies Laplace's equation ($\nabla^2 V = 0$) and the boundary conditions, it is a *unique solution*; i.e., it is the only possible solution. No other function will satisfy these requirements.†

Let us reexamine the solution (7-3-8) for the parallel-plate capacitor with respect to its uniqueness. It satisfies the boundary conditions, and it also satisfies Laplace's equation, as can be confirmed by substituting (7-3-8) into (7-3-1) and performing the indicated differentiation. The solution is $V_1 = C_1 x$, where $C_1 = V_1/x_1$. A function of the form $V = C_1 x + C_2$ will also satisfy Laplace's equation but will require different boundary conditions; that is, $V = C_2$ at $x = 0$ and $V = V_1 + C_2$ at $x = x_1$.

Let us try a function of the form $V = C_1 x^2$. This function can satisfy the same boundary conditions as $V = C_1 x$, but it does not satisfy Laplace's equation since substituting it in (7-3-1) gives a nonzero result ($d^2V/dx^2 = 2C_1$). It is apparent that any power of x other than unity will not be a solution. The various functions we have discussed are shown graphically in Fig. 7-2. Many different functions might be tried as solutions, but none except $V = (V_1/x_1)x$ satisfies Laplace's equation *and* the boundary conditions. *It is the unique solution*, and once you have it, you have solved the problem for all time.

More elegant and rigorous proofs of uniqueness are available, but the above discussion illustrates the general approach, which involves a demonstration that

---

† A graphically obtained or computer-generated field map of a two-dimensional problem may likewise be a unique solution.

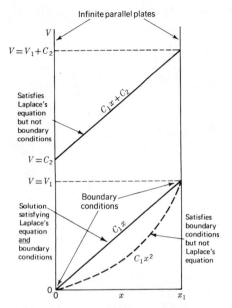

Figure 7-2 The potential variation C_1x satisfies Laplace's equation and the boundary conditions; $C_1x + C_2$ satisfies Laplace's equation but not the boundary conditions (unless $C_2 = 0$); while C_1x^2 satisfies the boundary conditions but not Laplace's equation.

solutions other than the unique one do not satisfy both Laplace's equation and the boundary conditions.

7-5 POINT-BY-POINT, OR ITERATIVE, METHOD

In this section a quantized, or point-by-point, method of solving Laplace's equation is discussed. The method is approximate, but by using sufficient points and by iteration, or repetitive application, the solution can be made as accurate as desired.†

The basic step is to find a solution of Laplace's equation at a point by noting the slope of the second derivative of the potential V in orthogonal directions around the point. In rectangular coordinates Laplace's equation is given by

$$\frac{\partial^2 V}{\partial x^2} + \frac{\partial^2 V}{\partial y^2} + \frac{\partial^2 V}{\partial z^2} = 0 \tag{1}$$

If the potential variation is independent of z, the problem reduces to a two-dimensional one and (1) simplifies to

$$\frac{\partial^2 V}{\partial x^2} + \frac{\partial^2 V}{\partial y^2} = 0 \tag{2}$$

† J. B. Scarborough, "Numerical Mathematical Analysis," 6th ed., pp. 391–422, Johns Hopkins Press, Baltimore, Md., 1966.

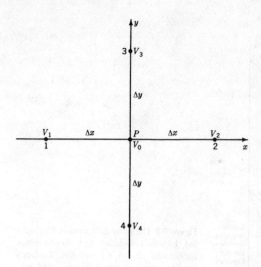

Figure 7-3 Construction to find potential at point P with respect to four surrounding points.

The first term in (2) is the second partial derivative of V with respect to x, that is, the rate of change of the rate of change of V with respect to x.† Similarly, the second term is the rate of change of the rate of change of V with respect to y. The sum of these two terms must be zero.

Consider a two-dimensional potential distribution around a point P, as in Fig. 7-3. Let the potential at P be equal to V_0 and at the four surrounding points be V_1, V_2, V_3, and V_4, as shown. Now $(V_2 - V_0)/\Delta x$ is the slope of V between points 2 and P. This is approximately equal to $\partial V/\partial x$ (becomes exact as $\Delta x \to 0$). Also $(V_0 - V_1)/\Delta x$ is the slope of V between points P and 1. The difference of these slopes (per distance increment Δx) is approximately equal to $\partial^2 V/\partial x^2$. Hence, as discussed in the preceding paragraph, Laplace's equation requires that the difference in slopes of V in the x direction and the difference of the slopes in the y direction must be equal and opposite in sign. Thus, we have

$$\frac{\partial(\partial V/\partial x)}{\partial x} = - \frac{\partial(\partial V/\partial y)}{\partial y} \tag{3}$$

or

$$\frac{[(V_2 - V_0)/\Delta x] - [(V_0 - V_1)/\Delta x]}{\Delta x} \simeq - \frac{[(V_3 - V_0)/\Delta y] - [(V_0 - V_4)/\Delta y]}{\Delta y} \tag{4}$$

Letting $\Delta x = \Delta y$, we have

$$V_1 + V_2 + V_3 + V_4 - 4V_0 \approx 0 \tag{5}$$

and

$$V_0 \approx \tfrac{1}{4}(V_1 + V_2 + V_3 + V_4) \tag{6}$$

---

† Or since $\partial V/\partial x = -E$, $\partial^2 V/\partial x^2$ is equivalent to the negative of the rate of change of E with respect to x.

If we know the potential at points 1, 2, 3, and 4, then according to Laplace's equation, the potential at the point P is as given by (6). In other words, the physical significance of Laplace's equation is simply that *the potential at a point is the average of the potential at four surrounding points.*† What could be simpler than that?

7-6 EXAMPLE 2: THE INFINITE SQUARE TROUGH

To illustrate the point-by-point method consider the infinitely long square trough of sheet metal shown in cross section in Fig. 7-4a. The sides and bottom of the trough are at zero potential. A cover, separated by small gaps from the trough, is at a potential of 40 V. Let us use the point-by-point method to obtain the potential distribution. First, we find the potential at the center of the trough from (7-5-6) as

$$\frac{40 + 0 + 0 + 0}{4} = 10 \text{ V}$$

Next, we find the potential at the center of the four quadrants of the trough. For this we rotate the xy axes 45° and take the potential at the cover-trough gap as 20 V (average of 40 V and 0). Again from (7-5-6) we have

$$\frac{20 + 40 + 0 + 10}{4} = 17.5 \text{ V}$$

for the upper left and right quadrants, as in Fig. 7-4b, and

$$\frac{0 + 10 + 0 + 0}{4} = 2.5 \text{ V}$$

for the lower left and right quadrants. Next, we can find V at four more points, with voltages as shown in Fig. 7-4c (xy axes returned to usual orientation).

The procedure is now repeated starting at the upper left, obtaining

$$\frac{40 + 21.2 + 7.5 + 0}{4} = 17.2 \text{ V}$$

This new value of V is now used to recalculate the potential at the adjacent point to the right as

$$\frac{40 + 17.2 + 10 + 17.2}{4} = 21.1 \text{ V}$$

---

† In a three-dimensional problem we would find the average of the potential for six surrounding points (four as shown in Fig. 7-3 plus one a distance Δz from P into the page and one a distance Δz from P out of the page; $\Delta z = \Delta x = \Delta y$) and the potential at P would be given by

$$V_0 = \tfrac{1}{6}(V_1 + V_2 + V_3 + V_4 + V_5 + V_6)$$

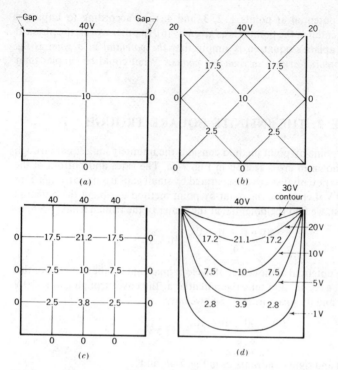

Figure 7-4 Application of point-by-point method to determine potential at nine points inside an infinite square trough. The map at (d) is a solution of Laplace's equation which satisfies the boundary conditions.

See Fig. 7-4d.† All points are recalculated in this way proceeding left to right and top to bottom over and over again until the values no longer change. This gives the most accurate solution to Laplace's equation we can get by this method with the number of points used. A still more accurate solution can be obtained by using a larger number of points (subdividing the area of the trough into a finer grid). As the distance between points (Δx or Δy) approaches zero, the solution can be made to approach an exact one. This method is well adapted for very accurate calculations using digital computers. However, manual calculations with relatively coarse grids can often provide solutions of sufficient accuracy for many purposes. Thus, a determination, as above, of the potential at nine points in this trough example is sufficient for drawing the potential contours as in Fig. 7-4d.

Referring to the first step (determination of V at the center of the trough), let us reexamine the problem. Using the center point and four neighboring points,

† In practice it is convenient to tabulate the voltage obtained in each iteration in column format at each grid point so that a running record is kept. When the voltages stabilize, the iteration can stop.

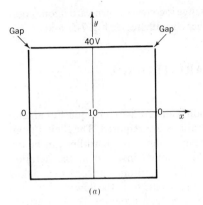

(a)

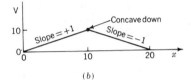

(b)

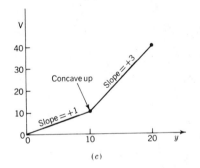

(c)

Figure 7-5 (a) Infinite square trough. (b) Slope or gradient of potential in x direction. (c) Slope or gradient of potential in y direction.

as in Fig. 7-5a, we get the slopes of the potential in the x and y directions as shown in Fig. 7-5b and c. (We assume $\Delta x = \Delta y = 10$ units.) Thus, $\Delta V/\Delta x = 10/10 = \pm 1$, and $\Delta V/\Delta y = 10/10 = 1$ or $30/10 = 3$. Laplace's equation is satisfied since the difference of the slopes in the x direction equals the negative of the difference of the slopes in the y direction.† Thus,

$$+1 - (-1) = 2$$

and

$$+1 - (+3) = -2$$

---

† Note that in the one-dimensional case the difference in slopes in one coordinate direction must be zero; that is, V must vary linearly with distance as in the parallel-plate-capacitor example (Sec. 7-3). In the three-dimensional case all one can say is that the *sum* of the slope differences in the three orthogonal coordinate directions must be zero.

We note in this problem that if the slope change is concave downward in one coordinate, it must be concave upward in the other coordinate (see Fig. 7-5b and c).

7-7 EXAMPLE 3: INFINITE SQUARE TROUGH, DIGITAL-COMPUTER SOLUTION

The method discussed in the preceding section is an iterative technique. For greater accuracy many more points and more iterations are required. The digital computer is ideally suited for making such computations. An algorithm can be developed for systematic computation of the averages involved in the iteration method [see (7-5-6)]. Having developed a suitable algorithm, one constructs a flow chart, which is a map of the strategy to be pursued, and from the flow chart a series of program statements is written. This is the computer program. To demonstrate the method, a flow chart and a Fortran IV program† are given for a square trough with different potentials (0, 20, 60, and 100 V) on each side. A grid of 7 × 7 interior points is used and the results are shown with contours in Fig. 7-6.

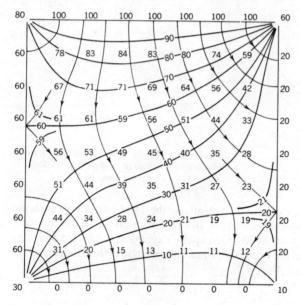

Figure 7-6 Infinite square trough with different potential (in volts) on each side with resulting potentials (in volts) at 49 interior points as obtained by Fortran IV computer program. Potential contours (heavy) and field lines (light and with arrows) are also shown.

† This flow chart and program were written by Dr. Robert S. Dixon, Ohio State University Radio Observatory.

FLOW CHART

```
( Start )
```

Fill in the known boundary values

Calculate the center point value

Calculate the values at the centers of the four 5 X 5 squares, 45° angle

Calculate the four values between the previous four on the cardinal axes

Calculate the values at the centers of the 3 X 3 squares. 45° angle

Calculate the values of all remaining points in the even rows. 2 X 2 squares. cardinal axes

Same, but for odd rows

Choose the 2nd column

Choose the 2nd row

Calculate a new value at this point

Measure the change between the new and old values

Is this change greater than the largest change found previously? — Yes → Retain this change as the greatest found so far

No

Increment the row by one

Have we reached the 9th row? — No

Yes

Increment the column by one

Have we reached the 9th column? — No

Yes

Increment the number of iterations

Is the largest change less than 0.1 volt? — No

Yes

Print out the number of iterations

Print out the final array of values obtained

```
( Stop )
```

Fortran IV Program

```
C  PROGRAM TO CALCULATE ELECTRIC POTENTIAL BY THE RELAXATION METHOD.
      COMMON Z(9,9)
      DIMENSION Z2(9,9)
C  THE DESIRED POINTS LIE ON A 9 X 9 SQUARE GRID IN THE FIRST QUADRANT.
C  I.E. - Z(1,1) IS IN THE LOWER LEFT CORNER.
      DATA ITERAT/0/
C  FILL IN THE KNOWN VALUES ON THE BOUNDARIES.
      DO 1 J=2,8
      Z(1,J)=60.
1     Z(9,J)=20.
      DO 2 I=2,8
      Z(I,1)=0.
2     Z(I,9)=100.
      Z(1,1)=30.
      Z(1,9)=80.
      Z(9,9)=60.
      Z(9,1)=10.
C  CALCULATE THE CENTER VALUE, CASE 0, 9 X 9 SQUARE.
      Z(5,5)=AVERAG(5,5,0,4)
C  CALCULATE THE 4 SURROUNDING POINTS, CASE 1, 5 X 5 SQUARES.
      Z(3,7)=AVERAG(3,7,1,2)
      Z(7,7)=AVERAG(7,7,1,2)
      Z(7,3)=AVERAG(7,3,1,2)
      Z(3,3)=AVERAG(3,3,1,2)
C  CALCULATE THE 4 SURROUNDING POINTS, CASE 0, 3 X 3 SQUARES.
      Z(3,5)=AVERAG(3,5,0,2)
      Z(5,7)=AVERAG(5,7,0,2)
      Z(7,5)=AVERAG(7,5,0,2)
      Z(5,3)=AVERAG(5,3,0,2)
C  NOW GO TO THE 3 X 3 SQUARES, CASE 1.
      DO 3 IROW=2,8,2
      DO 3 ICOL=2,8,2
3     Z(ICOL,IROW)=AVERAG(ICOL,IROW,1,1)
C  NEXT THE IN-BETWEEN 2 X 2 SQUARES IN THE EVEN NUMBERED ROWS, CASE 0.
      DO 4 IROW=2,8,2
      DO 4 ICOL=3,7,2
4     Z(ICOL,IROW)=AVERAG(ICOL,IROW,0,1)
C  THEN THE IN-BETWEEN 2 X 2 SQUARES IN THE ODD NUMBERED ROWS, CASE 0.
      DO 5 IROW=3,7,2
      DO 5 ICOL=2,8,2
5     Z(ICOL,IROW)=AVERAG(ICOL,IROW,0,1)
C  AT THIS POINT, ALL POINTS HAVE BEEN GIVEN AN INITIAL VALUE.
C  SO WE START ITERATING ALONG UNTIL THE VALUES STOP CHANGING.
C  WE DO ONE COMPLETE ITERATION AT A TIME, THEN CHECK TO SEE IF THE LARGEST
C  CHANGE IS LESS THEN 0.1 VOLT. CASE 0.
8     DO 6 I=2,8
      DO 6 J=2,8
6     Z2(I,J)=AVERAG(I,J,0,1)
C  MOVE THE NEW VALUES BACK TO THE ORIGINAL ARRAY, AND FIND THE LARGEST
C  DIFFERENCE.
      DIFF2=0.
      DO 7 I=2,8
      DO 7 J=2,8
      DIFF1=Z2(I,J)-Z(I,J)
      IF(DIFF1.GT.DIFF2)DIFF2=DIFF1
7     Z(I,J)=Z2(I,J)
C  TEST FOR DIFFERENCE, REITERATE AS NECESSARY.
C  KEEP COUNT OF THE NUMBER OF ITERATIONS.
      ITERAT=ITERAT+1
      IF(DIFF2.GT.0.1)GO TO 8
C  WRITE OUT THE Z ARRAY AND THE NUMBER OF ITERATIONS.
      WRITE(6,11)
      WRITE(6,12)ITERAT
12    FORMAT(' SOLUTION OBTAINED AFTER ',I5,' ITERATIONS.',/////)
11    FORMAT(1H1)
      J=10
      DO 10 K=1,9
      J=J-1
10    WRITE(6,9) (Z(I,J),I=1,9)
9     FORMAT(1X,9(3X,F5.1)///)
C  GO HOME.
      CALL EXIT
      STOP
      END

C  A FUNCTION TO COMPUTE THE AVERAGE OF FOUR POINTS, CENTERED AT
C  SUBSCRIPT (I,J), SPACED "ISPACE" POINTS AWAY.
C  IF ICASE=0, THE 4 POINTS ARE IN THE CARDINAL DIRECTIONS FROM THE
C  CENTER.
C  IF ICASE=1, THE 4 POINTS ARE IN THE 45 DEGREE DIRECTIONS FROM THE
C  CENTER.
      FUNCTION AVERAG(I,J,ICASE,ISPACE)
      COMMON Z(9,9)
      AVERAG=0.25*(Z(I-ISPACE,J-ICASE*ISPACE)
     X             +Z(I-ISPACE*ICASE,J+ISPACE)
     X             +Z(I+ISPACE,J+ISPACE*ICASE)
     X             +Z(I+ISPACE*ICASE,J-ISPACE))
      RETURN
      END
```

It is instructive to compare the times required using the manual and computer techniques. Thus, a manually constructed map can be produced in about an hour, although this time can be reduced with the aid of a vest-pocket or desk-top micro- or minicomputer. The programmable digital computer will do the computations required in seconds, but automatic plotting may require minutes. However, many hours may be required to construct the flow chart and write the program statements, and it takes additional time before the program is ready to run on the computer. The advantages of the programmable computer is that once a program has been successfully executed, large numbers of similar problems can be solved very quickly.

7-8 ANALOG-COMPUTER SOLUTION WITH LAPLACE'S AND KIRCHHOFF'S LAWS

In an analog computer the problem is simulated in analog form. Thus, in the solution of an electromagnetic field problem there is a useful analogy between Laplace's equation and Kirchhoff's current law. Hence, to determine the potential V of the point P at the center of the trough of Example 2, we connect four resistors of resistance R as shown in Fig. 7-7a. According to Kirchhoff's current law at a point (Sec. 4-12), the sum of the currents at a junction is zero, or

$$I_1 + I_2 + I_3 + I_4 = 0 \tag{1}$$

From Ohm's law $I_1 = (V_1 - V)/R, I_2 = (V_2 - V)/R$, etc.; so we have

$$\frac{V_1 - V}{R} + \frac{V_2 - V}{R} + \frac{V_3 - V}{R} + \frac{V_4 - V}{R} = 0 \tag{2}$$

or

$$V_1 + V_2 + V_3 + V_4 - 4V = 0 \tag{3}$$

and

$$V = \tfrac{1}{4}(V_1 + V_2 + V_3 + V_4) \tag{4}$$

which is the same result we obtained by applying Laplace's equation. Hence, *Kirchhoff's current law at a point and Laplace's equation are equivalent.* This is readily seen since Kirchoff's current law from (4-13-1) can be expressed as

$$\mathbf{\nabla \cdot J} = 0 \tag{5}$$

But $\mathbf{J} = \sigma\mathbf{E}$ and $\mathbf{E} = -\nabla V$, and so we have

$$\mathbf{\nabla \cdot J} = \sigma\mathbf{\nabla \cdot E} = \sigma\nabla^2 V = 0 \tag{6}$$

or

$$\nabla^2 V = 0 \tag{7}$$

which is Laplace's equation.

If the grid is made finer by using more resistors and junction points, the potential distribution can be determined more accurately. An experimental arrangement for doing this using a high-resistance voltmeter is shown in Fig. 7-7b. An even more practical and very accurate method is to use the equivalent of an

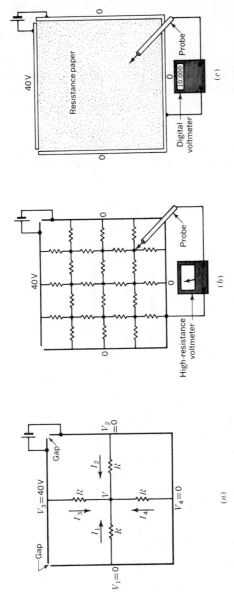

Figure 7-7 (a) Application of circuit analysis to determination of potential at center of infinite square trough. (b) Experimental measurement of potential at nine points inside trough and (c) at any point using resistance paper and digital voltmeter (DVM).

infinite grid of points, namely, a sheet of uniform-resistance paper (such as Teledeltos paper). The outline (or boundaries) of the trough are drawn on the resistance paper with silver (conducting) paint. A battery is then connected, as in Fig. 7-7c, causing current to flow through the paper between the boundaries. The potential contours are then traced out with a high-resistance voltmeter and probe as suggested.

To facilitate the measurements a digital voltmeter (DVM) can be employed. For example, with 40 V applied as in Fig. 7-7c and equipotential contours desired at 5-V intervals, the probe is moved until 5.000 appears on the readout, the point is marked, and then more points at this value are found, next progressing to 10.000 on the readout, etc.

Another field-mapping arrangement using resistance paper is shown in Fig. 7-8. A 10-turn potentiometer and galvanometer null-indicator are used to locate equipotential points. Thus, for each turn on the potentiometer one can trace out

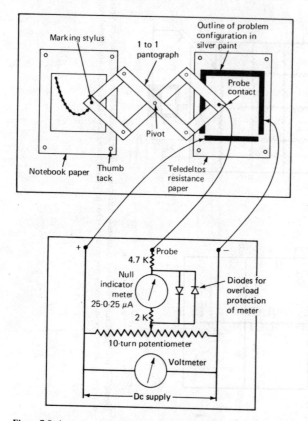

Figure 7-8 Arrangement and circuit of analog field plotter. Configuration shown is a square trough with insulated cover (Examples 2 and 3, Secs. 7-6 and 7-7).

Figure 7-9 Analog field plotter with L-shaped conductor similar to Probs. P4-3-3, P7-1-9, and P7-1-10. Null indicator and 10-turn potentiometer are on left side of control panel. Meter at right reads dc supply voltage (not critical).

equipotentials at 10 percent intervals of the total voltage difference. The arrangement of Fig. 7-8, shown also in the photograph of Fig. 7-9, has a pantograph for transferring the equipotential points from the resistance paper to a separate sheet of ordinary paper on which the map will appear. The probe is moved over the resistance paper until a balance, or null, is obtained. Then the pantograph-marking stylus is pressed to imprint a dot at the corresponding point on the map sheet. The pantograph in Fig. 7-8 and 7-9 has a 1:1 ratio, so that the map is of the same size as its resistance-paper analog. With other pantograph ratios, maps can be enlarged or reduced as desired.†

A variety of other experimental analog field-mapping techniques have been devised. In some, stretched rubber membranes have been used. In others, as developed by Moore,‡ controlled fluid flow is used as a hydraulic analogy to the electrostatic problem in which fluid flow replaces current flow.

We have discussed analytical (formal mathematical), graphical, digital-computer, analog-computer, and experimental analog procedures for solving Laplace's equation. Each method has its advantages and disadvantages. In solving a particular problem one should consider all facets of the various techniques and then select the one best suited for the particular problem.

7-9 EXAMPLE 4: CONDUCTING SHEET BETWEEN TWO CONDUCTING PLATES

Referring to Fig. 7-10, two infinite parallel conducting plates are spaced a distance a. An infinitely long conducting strip is placed between the plates and normal to them, as shown in the figure.

---

† Commerical analog field plotters are available from Sunshine Scientific Co., Philadelphia, PA 19115.

‡ A. D. Moore, Mapping Techniques Applied to Fluid-mapper Patterns, *Proc. A.I.E.E.*, **71**, 1952.

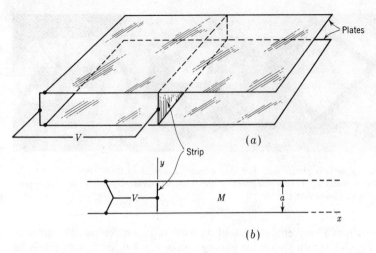

Figure 7-10 Infinite conducting strip between two infinite parallel conducting plates (*a*) in perspective view and (*b*) in cross section.

The width of the strip is only very slightly less than the spacing between the plates. The strip is insulated from the plates. Let the two plates be connected together and a constant potential V be applied between the plates and the conducting strip. The medium between the plates is air. Suppose that the plates are at zero potential and that the strip is at a positive potential of 1 V ($V = 1$ V). The problem is to find the potential distribution in the region M between the plates to the right of the strip, as indicated in the cross section of Fig. 7-10*b*.†

It is possible to find the potential distribution by solving Laplace's equation ($\nabla^2 V = 0$) subject to the boundary conditions. It is most convenient to handle this problem in rectangular coordinates, the relation of the conductor boundaries to the coordinate axes being as in Fig. 7-11*a*. Expanding Laplace's equation in the two rectangular coordinates of the problem (x and y), we have

$$\frac{\partial^2 V}{\partial x^2} + \frac{\partial^2 V}{\partial y^2} = 0 \tag{1}$$

This differential equation is the most general way of expressing the variation of potential with respect to x and y. It is a partial differential equation of the second order and first degree. However, this equation does not tell us anything about the particular potential distribution in the problem. For this we must obtain a solution

---

† This problem is similar to that for the square trough of Example 2 (Sec. 7-6) except that here the trough is not closed at the bottom (sides infinitely long). Also in the present problem the trough has been turned on its side.

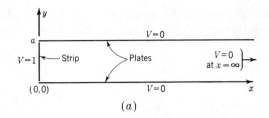

(a)

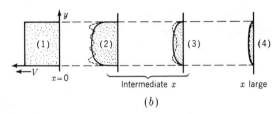

(b)

Figure 7-11 (a) Boundary conditions for potential-distribution problem of Fig. 7-10. (b) Potential variation between plates as obtained from solution of Laplace's equation at different distances (x) from the strip.

of the differential equation which is appropriate to the boundary conditions of the problem. These *boundary conditions* (B.C.) are

B.C. 1: $V = 0$ at $\begin{array}{l} y = 0 \\ 0 \le x \le \infty \end{array}$

B.C. 2: $V = 0$ at $\begin{array}{l} y = a \\ 0 \le x \le \infty \end{array}$

B.C. 3: $V = 1$ at $\begin{array}{l} x = 0 \\ 0 \le y \le a \end{array}$

B.C. 4: $V = 0$ at $\begin{array}{l} x = \infty \\ 0 \le y \le a \end{array}$

Proceeding now to find a solution of (1) by the method of separation of variables, let us assume that a solution of V can be expressed as

$$V = XY \tag{2}$$

where X = function of x alone
Y = function of y alone

Substituting (2) into (1) and dividing by XY, we have

$$\frac{1}{X}\frac{d^2X}{dx^2} + \frac{1}{Y}\frac{d^2Y}{dy^2} = 0 \tag{3}$$

In (3) the variables are separated. Since X and Y are independent and the sum of the two terms is a constant (zero), each term alone must equal a constant. Thus, we can write

$$\frac{1}{X}\frac{d^2 X}{dx^2} = k^2 \tag{4}$$

and

$$\frac{1}{Y}\frac{d^2 Y}{dy^2} = -k^2 \tag{5}$$

where k equals a constant and $k^2 - k^2 = 0$.

These equations may be rearranged to the form

$$\frac{d^2 X}{dx^2} - k^2 X = 0 \tag{6}$$

and

$$\frac{d^2 Y}{dy^2} + k^2 Y = 0 \tag{7}$$

Thus, the second-order partial differential equation of (1) has been reduced to two second-order ordinary differential equations, each involving but one variable. These equations, (6) and (7), have the solutions

$$X = C_1 e^{kx} + C_2 e^{-kx} \tag{8}$$

and

$$Y = C_3 e^{jky} + C_4 e^{-jky} \tag{9}$$

One may readily confirm that these are solutions by substituting them into (6) and (7), in each case obtaining an identity. Introducing (8) and (9) into (2) yields the general solution

$$V = C_1 C_3 e^{kx} e^{jky} + C_2 C_3 e^{-kx} e^{jky} + C_1 C_4 e^{kx} e^{-jky} + C_2 C_4 e^{-kx} e^{-jky} \tag{10}$$

which reduces to

$$V = C_1' e^{k(x \pm jy)} + C_2' e^{-k(x \pm jy)} \tag{11}$$

where $C_1' = C_1 C_3$ or $C_1 C_4$ and $C_2' = C_2 C_3$ or $C_2 C_4$, depending on which sign is chosen in $x \pm jy$.

Because of B.C. 4 ($V = 0$ at $x = \infty$), $C_1' = 0$, so that only the second term of (11) applies in our problem.† Also, using de Moivre's theorem, (11) then becomes

$$V = C_2' e^{-kx}(\cos ky \pm j \sin ky) \tag{12}$$

To satisfy B.C. 1 ($V = 0$ at $y = 0$) we should retain only the imaginary part of (12). That is,

$$V = C_2' e^{-kx} \sin ky \tag{13}$$

---

† If we were interested in the potential distribution to the *left* instead of to the right of the strip (Fig. 7-10), the boundary condition is $V = 0$ at $x = -\infty$ so that $C_2' = 0$.

This is a particular solution of Laplace's equation appropriate to our problem. It indicates that the potential V falls off exponentially with x and also that it varies as a sine function of y. To satisfy B.C. 2 ($V = 0$ at $y = a$) requires that

$$k = \frac{n\pi}{a} \tag{14}$$

where n is a positive integer $(1, 2, 3, \dots)$. Introducing (14) into (13) yields

$$V = C_2' e^{-n\pi x/a} \sin \frac{n\pi y}{a} \tag{15}$$

All the boundary conditions are now satisfied except for B.C. 3 ($V = 1$ at $x = 0$), that is, $V = 1$ at $x = 0$ for all values of y between 0 and a. Obviously (15) does not satisfy this requirement, and hence a more general solution is required. A more general solution can be obtained by taking the sum of expressions like (15) for different integral values of n. We then have

$$V = C_1 e^{-\pi x/a} \sin \frac{\pi y}{a} + C_2 e^{-2\pi x/a} \sin \frac{2\pi y}{a} + C_3 e^{-3\pi x/a} \sin \frac{3\pi y}{a} + \cdots \tag{16}$$

where C_1, C_2, etc., are new constants.

Equation (16) can be expressed more concisely by

$$V = \sum_{n=1}^{n=\infty} C_n e^{-n\pi x/a} \sin \frac{n\pi y}{a} \tag{17}$$

The solution for V given by (16) or (17) is still incomplete since the coefficients C_1, C_2, etc., are not evaluated. To find their values we impose B.C. 3 ($V = 1$ at $x = 0$), so that (16) reduces to

$$1 = C_1 \sin \frac{\pi y}{a} + C_2 \sin \frac{2\pi y}{a} + C_3 \sin \frac{3\pi y}{a} + \cdots \tag{18}$$

Now by the Fourier sine expansion,

$$f(by) = a_1 \sin by + a_2 \sin 2by + a_3 \sin 3by + \cdots + a_n \sin nby \tag{19}$$

where

$$a_n = \frac{2}{\pi} \int_0^\pi f(by) \sin nby \, d(by)$$

It follows that for $f(by) = 1$

$$a_n = \begin{cases} \dfrac{4}{n\pi} & \text{for } n \text{ odd} \\[2ex] 0 & \text{for } n \text{ even} \end{cases}$$

Therefore, for $f(by) = 1$, (19) reduces to

$$1 = \frac{4}{\pi}(\sin by + \tfrac{1}{3}\sin 3by + \tfrac{1}{5}\sin 5by + \cdots) \qquad (20)$$

Comparing (20) with (18), it follows that

$$C_1 = \frac{4}{\pi} \qquad C_2 = 0 \qquad C_3 = \frac{4}{3\pi} \qquad C_4 = 0 \qquad C_5 = \frac{4}{5\pi} \text{ etc.}$$

and

$$b = \frac{\pi}{a}$$

Introducing these constants into (16), we have

$$V = \frac{4}{\pi}e^{-\pi x/a}\sin\frac{\pi y}{a} + \frac{4}{3\pi}e^{-3\pi x/a}\sin\frac{3\pi y}{a} + \frac{4}{5\pi}e^{-5\pi x/a}\sin\frac{5\pi y}{a} + \cdots \qquad (21)$$

This is the complete solution of Laplace's equation for the potential V appropriate to the boundary conditions of the problem. It gives the potential V as a function of position between the plates and to the right of the strip (Fig. 7-10).

Although an infinite number of terms is required for an exact representation of the potential distribution as a function of x and y, an approximate solution of practical value can be obtained with a finite number of terms. Each term attenuates at a different rate. Since the higher terms fall off more rapidly with x, only a few terms are needed to give a good approximation except where x is small. That is, convergence is rapid at large x. Thus, when x is very large, the distribution is nearly equal to $\sin(\pi y/a)$, since the contribution of the harmonics higher than the first may be neglected. At $x = 0$, $V = 1$ for all values of y between 0 and a. The variation of V as a function of y at $x = 0$ and at a large value of x is shown at (1) and (4) in Fig. 7-11b. The distributions at two intermediate values of x are presented at (2) and (3). The actual distribution is shown by the solid curves, the dashed curves giving the

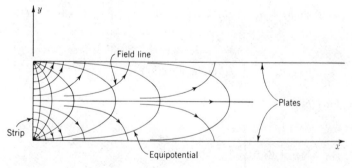

Figure 7-12 Electric field lines (with arrows) and equipotentials in space between conducting plates.

approximate distribution as obtained by four terms ($n = 1, 3, 5, 7$) of the series at (2) and by two terms ($n = 1$ and 3) at (3). It is apparent that as x decreases, the effect of the higher terms becomes more important.

The potential distribution can also be presented by means of equipotential contours with orthogonal field lines, as suggested in Fig. 7-12. It is apparent that graphical, digital-computer, and analog-computer methods could also have been used instead of the above analytical method to obtain an engineering solution for the potential distribution.

7-10 SOLUTION OF LAPLACE'S EQUATION IN CYLINDRICAL AND SPHERICAL COORDINATES

Laplace's equation in cylindrical coordinates (r, ϕ, z) (see Fig. A-1 in Appendix A and inside back cover) is

$$\frac{\partial^2 V}{\partial r^2} + \frac{1}{r}\frac{\partial V}{\partial r} + \frac{1}{r^2}\frac{\partial^2 V}{\partial \phi^2} + \frac{\partial^2 V}{\partial z^2} = 0 \tag{1}$$

Using the method of separation of variables (as illustrated in Sec. 7-2 for rectangular coordinates), a solution is

$$V = V_0(C_1 e^{kz} + C_2 e^{-kz})(C_3 \cos n\phi + C_4 \sin n\phi)[C_5 J_n(kr) + C_6 N_n(kr)] \tag{2}$$

where n = integer, dimensionless

$J_n(kr)$ = Bessel function of first kind and of order n, with argument kr, dimensionless

$N_n(kr)$ = Bessel function of second kind and of order n (also called a Neumann function), with argument kr, dimensionless

k = constant, m^{-1}

V_0 = constant, V

C_1, C_2, etc. = constants, dimensionless

Laplace's equation in spherical coordinates (r, θ, ϕ) (see Fig. A-1 in Appendix A and inside back cover) is

$$\frac{\partial^2 V}{\partial r^2} + \frac{1}{r^2}\frac{\partial^2 V}{\partial \theta^2} + \frac{1}{r^2 \sin^2 \theta}\frac{\partial^2 V}{\partial \phi^2} + \frac{2}{r}\frac{\partial V}{\partial r} + \frac{\cot \theta}{r^2}\frac{\partial V}{\partial \theta} = 0 \tag{3}$$

Using the method of separation of variables, we find that a solution is

$$V = V_0(C_1 r^n + C_2 r^{-(n+1)})(C_3 \cos m\phi + C_4 \sin m\phi)[C_5 P_n^m(\cos \theta) + C_6 Q_n^m(\cos \theta)] \tag{4}$$

where n, m = integers, dimensionless

$P_n^m(\cos \theta)$ = associated Legendre function (solid zonal harmonic) of the first kind, dimensionless

$Q_n^m(\cos \theta)$ = associated Legendre function (solid zonal harmonic) of the second kind, dimensionless

C_1, C_2, etc. = constants, dimensionless

V_0 = constant, V

The numerical calculations involved in solutions like (2) and (4) are conveniently carried out with a digital computer especially if subroutines are stored for needed functions such as $J_n(kr)$ or $P_n^m(\cos \theta)$.

7-11 EXAMPLE 5: COAXIAL LINE

Consider the section of coaxial transmission line shown in Fig. 7-13a. One end (at the origin) is short-circuited, and the other end is open. The radius of the inner conductor is a, and the inside radius of the outer conductor is b. The conductivity of the inner conductor is finite, but the conductivity of the outer conductor and of the short-circuiting disk is assumed to be infinite. A constant voltage V_1 is applied between the inner and outer conductor at the open end of the line. The length z_1 of the line is long compared with its radius ($z_1 \gg b$). The problem is to find the potential V everywhere inside the line, except near the open end.

The *boundary conditions* (B.C.) for this problem are

B.C. 1: $\qquad\qquad V = 0 \qquad$ at $\begin{aligned} z &= 0 \\ a &\leq r \leq b \end{aligned}$

B.C. 2: $\qquad\qquad V = 0 \qquad$ at $\begin{aligned} r &= b \\ 0 &\leq z \leq z_1 \end{aligned}$

B.C. 3: $\qquad\qquad V = V_1 \qquad$ at $\begin{aligned} z &= z_1 \\ r &= a \end{aligned}$

There is also the condition that at any distance z from the origin the electric field inside the inner conductor and also along its surface ($r = a$) is given by

$$\frac{\partial V}{\partial z} = \frac{V_1}{z_1} \tag{1}$$

from which we have on integration that

$$V = \frac{V_1}{z_1} z \tag{2}$$

At $z = z_1$, (2) reduces to B.C. 3 above.

We wish to find a solution of Laplace's equation ($\nabla^2 V = 0$) that satisfies the boundary conditions. By symmetry V is independent of ϕ; so expanding Laplace's equation in the other two cylindrical coordinates of the problem (r and z), we have†

$$\frac{\partial^2 V}{\partial \rho^2} + \frac{1}{\rho} \frac{\partial V}{\partial \rho} + \frac{\partial^2 V}{\partial z^2} = 0 \tag{3}$$

---

† Although V is dependent only on ρ and z, this problem is not two-dimensional in the sense that the problem of Example 5 is two-dimensional. Here the potential distribution for a longitudinal plane through the axis differs from the distribution for all planes parallel to it.

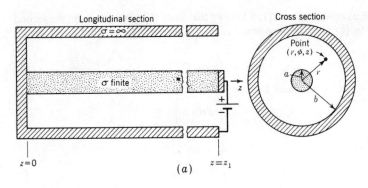

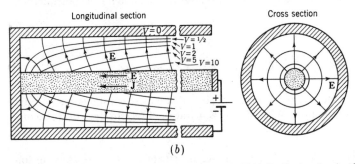

Figure 7-13 (*a*) Section of coaxial transmission line. (*b*) Field distribution showing field lines (with arrows) and equipotentials.

Using the method of separation of variables, let

$$\frac{V}{V_0} = RZ \tag{4}$$

where R = function only of r
$\quad\ \ Z$ = function only of z
$\quad\ \ V_0$ = constant, V

Introducing (4) into (3) and dividing by RZ yields

$$\frac{1}{R}\frac{d^2R}{dr^2} + \frac{1}{rR}\frac{dR}{dr} + \frac{1}{Z}\frac{d^2Z}{dz^2} = 0 \tag{5}$$

The last term is a function only of z. Thus we can write

$$\frac{d^2Z}{dz^2} = a_z^2 Z \tag{6}$$

where a_z = a constant. From (1) it follows that since V_1/z_1 is a constant, the second derivative in (6) must be zero and hence a_z must be zero. A solution of (6) for $a_z = 0$ is

$$Z = C_1 z + C_2 \tag{7}$$

The last term of (5) may now be set equal to zero, so that the equation reduces to

$$r \frac{d^2 R}{dr^2} + \frac{dR}{dr} = 0 \tag{8}$$

A solution is

$$R = C_3 \ln r + C_4 \tag{9}$$

Introducing (9) for R and (7) for Z in (4) gives the solution for the potential

$$\frac{V}{V_0} = (C_1 z + C_2)(C_3 \ln r + C_4) \tag{10}$$

or

$$\frac{V}{V_0} = C_5 z \ln r + C_6 z + C_7 \ln r + C_8 \tag{11}$$

where C_5, C_6, etc., are new constants. To evaluate these constants, we introduce the boundary conditions. When B.C. 1 ($V = 0$ at $z = 0$) is introduced, (11) becomes

$$0 = C_7 \ln r + C_8 \tag{12}$$

For (12) to be satisfied for all values of r requires that $C_7 = C_8 = 0$. Thus our solution reduces to

$$\frac{V}{V_0} = C_5 z \ln r + C_6 z \tag{13}$$

Introducing now B.C. 2 ($V = 0$ at $r = b$), we have

$$0 = C_5 z \ln b + C_6 z \tag{14}$$

From B.C. 3 ($V = V_1$ at $z = z_1$ and $r = a$) we get

$$\frac{V_1}{V_0} = C_5 z_1 \ln a + C_6 z_1 \tag{15}$$

From (14) and (15) we find that

$$C_5 = \frac{V_1}{V_0 z_1 \ln (a/b)} \quad \text{and} \quad C_6 = -\frac{V_1 \ln b}{V_0 z_1 \ln (a/b)}$$

When the values for these constants are introduced into (13) the complete solution for the problem is

$$V = V_1 \frac{z}{z_1} \frac{\ln (b/r)}{\ln (b/a)} \tag{16}$$

This solution satisfies Laplace's equation and the boundary conditions and hence must represent the potential distribution at all points between the inner and outer conductors ($a \leq r \leq b$) except near the open end. This potential distribution is portrayed in Fig. 7-13b, the relative potential being indicated for the equipotential lines. Electric field lines (with arrows) are also shown, being normal to the equipotentials.

It is interesting to note in Fig. 7-13b that although the field lines are normal to the perfectly conducting surfaces (outer conductor and short-circuiting disk), they are not normal to the finitely conducting inner conductor. The current and field direction in the inner conductor is to the left ($-z$ direction). Comparison of Fig. 7-13b should be made with Figs. 4-18 and 4-20. In Figs. 4-18 and 4-20 both inner and outer conductors are assumed to have finite conductivity.

7-12 POISSON'S EQUATION

In the preceding sections, the volume charge density ρ was assumed to be zero. We now consider Poisson's equation (see Sec. 3-25), which applies to problems involving space charge ($\rho \neq 0$). Poisson's equation is

$$\nabla^2 V = -\frac{\rho}{\epsilon} \tag{1}$$

where V = potential, V

ρ = volume free-charge density, C m^{-3}

ϵ = permittivity of medium, F m^{-1}

Poisson's equation is a second-order inhomogeneous differential equation which applies at a point. From the theory of differential equations, it must have a solution of the following form:

$$V = \text{particular solution} + \text{complementary solution}$$

The particular solution must satisfy the inhomogeneous equation (1) whereas the complementary solution (which contains two linearly independent solution terms) must satisfy the homogeneous equation, i.e., Laplace's equation. In the first five examples of this chapter the volume charge density was zero, and the solution of Laplace's equation was the complementary solution. In Example 6 of Sec. 7-13, the volume charge density is not zero, and a particular solution is found.

In rectangular coordinates, (1) becomes

$$\frac{\partial^2 V}{\partial x^2} + \frac{\partial^2 V}{\partial y^2} + \frac{\partial^2 V}{\partial z^2} = -\frac{\rho}{\epsilon} \tag{2}$$

If V is independent of y and z, (2) becomes

$$\frac{d^2 V}{dx^2} = -\frac{\rho}{\epsilon} \tag{3}$$

7-13 EXAMPLE 6: PARALLEL-PLATE CAPACITOR WITH SPACE CHARGE†

Consider the parallel-plate capacitor shown in Fig. 7-14a with potential difference V_1 applied between the plates as indicated. If the left-hand plate is coated with suitable material and both plates enclosed in a glass envelope and evacuated, an electron current passes between the plates when the left-hand plate is heated. This arrangement is called a *diode* or *two-element vacuum tube*. The left-hand plate or emitter is the *cathode*, and the right-hand plate or collector is the *anode*. Let us solve for the potential variation between the plates subject to the applied boundary conditions. Since there is electric charge between the plates, we must use Poisson's equation.

Let the cathode be held at potential $V = 0$ and a positive voltage $V = V_1$ be applied to the anode. When the cathode is heated, electrons are emitted, so that there is a volume charge density $\rho(x)$ between the plates.

From (7-12-3), Poisson's equation for this case is

$$\frac{d^2V}{dx^2} = -\frac{\rho}{\epsilon_0} \tag{1}$$

The solution to this differential equation must satisfy two *boundary conditions* (B.C.) as follows:

B.C. 1: $\qquad\qquad\qquad V = 0 \qquad \text{at } x = 0$

B.C. 2: $\qquad\qquad\qquad V = V_1 \qquad \text{at } x = x_1$

In addition it is assumed that emitted electrons experience a repulsive force from other electrons already between the plates and are barely able to escape from the cathode (initial velocity zero). The result of this space-charge limitation is that the electric field $E(= dV/dx)$ is zero at $x = 0$. This can be expressed as a third *boundary condition*

B.C. 3: $\qquad\qquad\qquad E = -\frac{dV}{dx} = 0 \qquad \text{at } x = 0$

Equating the kinetic energy of an electron to the energy it acquires from the field, we have

$$\tfrac{1}{2}mv^2 = eV \tag{2}$$

where m and e are the mass and charge of the electron, respectively (see Appendix A). It is assumed in (2) that the initial velocity of the electron is zero.

The electrons in motion constitute a steady current of density

$$J = -\rho v \qquad (\text{A m}^{-2}) \tag{3}$$

† K. R. Spangenberg, "Vacuum Tubes," p. 170, McGraw-Hill Book Company, New York, 1948.

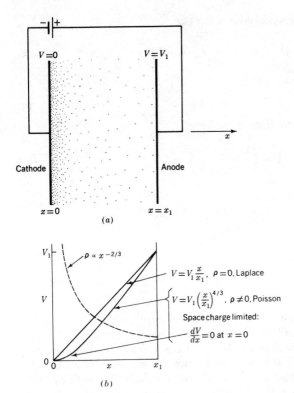

Figure 7-14 (*a*) Parallel-plate diode with space charge. (*b*) Potential and charge distribution.

Solving for v from (2) and substituting into (3) gives

$$\rho = -J \sqrt{\frac{m}{2eV}} \tag{4}$$

Substituting (4) into Poisson's equation (1) yields

$$\frac{d^2V}{dx^2} = \frac{J}{\epsilon_0} \sqrt{\frac{m}{2eV}} \tag{5}$$

Multiplying by $2\,dV/dx$ and integrating, we have

$$\left(\frac{dV}{dx}\right)^2 = \frac{4J}{\epsilon_0} \sqrt{\frac{mV}{2e}} + C_1 \tag{6}$$

where C_1 is a constant. Applying B.C. 1 and B.C. 3, we get

$$0 = 0 + C_1 \tag{7}$$

Hence, $C_1 = 0$. Integrating again gives

$$\tfrac{4}{3}V^{3/4} = 2\sqrt{\frac{J}{\epsilon_0}}\sqrt{\frac{m}{2e}}\,x + C_2 \tag{8}$$

where C_2 is a constant. From B.C. 1

$$0 = 0 + C_2 \tag{9}$$

Hence, $C_2 = 0$. Solving for V from (8) we have†

$$V = \left(\frac{3}{2}\sqrt{\frac{J}{\epsilon_0}}\sqrt{\frac{m}{2e}}\right)^{4/3}x^{4/3} \tag{10}$$

From B.C. 2 this becomes

$$V_1 = \left(\frac{3}{2}\sqrt{\frac{J}{\epsilon_0}}\sqrt{\frac{m}{2e}}\right)^{4/3}x_1^{4/3} \tag{11}$$

Dividing (10) by (11) yields

$$V = V_1\left(\frac{x}{x_1}\right)^{4/3} \tag{12}$$

which is the unique solution to Poisson's equation satisfying the three boundary conditions. This variation of V between the plates is shown by the lower solid curve in Fig. 7-14b. Note that the slope dV/dx is zero at $x = 0$ as a result of the space-charge limitation (B.C. 3). For comparison the upper solid (straight) line gives the potential variation in the absence of space charge (solution of Laplace's equation, see Fig. 7-1). The difference between solutions is in the exponent for the ratio x/x_1, which is unity for the solution to Laplace's equation and 4/3 for the solution to Poisson's equation. Solving (11) for J gives

$$J = \tfrac{4}{9}\epsilon_0\sqrt{\frac{2e}{m}}\frac{1}{x_1^2}V_1^{3/2} \tag{13}$$

This three-halves power-voltage relation for the diode current is the *Child–Langmuir* space-charge law.‡ Substituting (10) into (4), we can solve for the charge-density variation, obtaining

$$\rho \propto x^{-2/3} \tag{14}$$

This variation is as suggested by the dashed curve in Fig. 7-14b. (See Prob. 7-2-12 for a semiconductor junction example.)

---

† Equation (10) is the particular solution to Poisson's equation. The complementary solution is zero for this problem.

‡ D. C. Child, Discharge from Hot CaO, *Phys. Rev.*, **32**: 492–511 (May 1911); I. Langmuir, The Effect of Space Charge and Residual Gases on Thermionic Currents in High Vacuums, *Phys. Rev.* **2**(2): 450–486 (December 1913).

7-14 THE THEORY OF IMAGES

Consider an arbitrary distribution of charge in space, with a known equipotential surface such as the $V = 10$ V surface of the dipole shown in Fig. 7-15a (same as Fig. 2-13). Confining our attention to the volume v bounded by the $V = 10$ surface, it follows by superposition that the potential V inside v is a result of charges both interior to and exterior to the volume. If we now remove the charge exterior to v, V inside will remain unchanged if an equivalent surface charge distribution is placed on the surface, as suggested in Fig. 7-15b. In effect, the lower charge of the dipole ($-Q$) has been spread over the $V = 10$ contour surface.

It follows that the converse of this argument holds. If we start out with some known potential at any point in a volume v bounded by a charged surface, that potential can also be produced by an equivalent charge distribution exterior to v, the charge distribution on the surface now being removed. The charges exterior to v are called *image charges*, as previously discussed in Sec. 3-18. The examples of the following sections illustrate the theory of images.

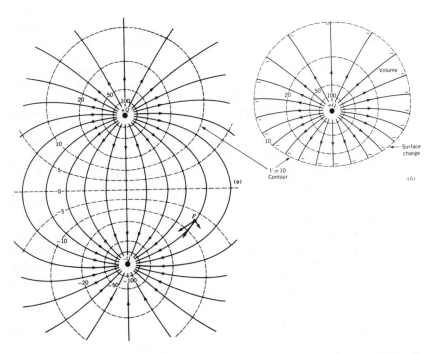

Figure 7-15 (*a*) Field and potential distribution of an electric dipole. (*b*) Distribution inside $V = 10$ contour remains unchanged when an equivalent surface charge is placed on the contour and the lower charge of the dipole is removed. Field inside contour is same as in (*a*), but field outside is zero.

In Fig. 7-15 the $-Q$ charge is the image of the $+Q$ charge with respect to the $V = 10$ surface, or for that matter to any of the equivalent potential surfaces. Thus, for example, if the $V = 0$ surface is chosen, we have the case of a charge over a flat plane. A charged cylindrical conductor over a flat plane is discussed in the next section.

7-15 EXAMPLE 7: CHARGED CYLINDRICAL CONDUCTOR OVER AN INFINITE METAL GROUND PLANE

Consider a long thin uniformly charged cylindrical conductor at a distance s from an infinite metal ground plane, as in Fig. 7-16a. From the theory of images this configuration is equivalent to the upper half of Fig. 7-16b, where we have two identical parallel conductors of opposite charge separated by a distance $2s$.

It is instructive to examine first the two-conductor system. The behavior of the electric field of these two conductors can be found by superimposing the electric fields due to either conductor alone. This procedure gives the electric field as the vector superposition of the two components. (An alternative method, discussed in Sec. 2-13, is to find the scalar electrostatic potential V as the scalar superposition of the potential due to each conductor alone. Then the electric field is found as the negative gradient of V.) The electric field lines due to either conductor alone would point radially outward from the conductor. Thus, on the equipotential surface equidistant from the two conductors, a simple graphical addition at any point shows that the resultant electric field is directed downward, as suggested in Fig. 7-16b. Thus, since $\mathbf{E}$ is orthogonal to the surface, it must be an equipotential surface.†

According to the previously developed image theory, the fields above the surface will remain unchanged if we remove the lower conductor and place a surface charge distribution on the surface. This surface charge is

$$\rho_s = D_n = -\epsilon_0 E_n \tag{1}$$

Consider first the case where both conductors are thin lines of charge. The electric field from the positively charged conductor is $\mathbf{E}_+$, and the electric field from the negatively charged conductor is $\mathbf{E}_-$. If each line charge is now replaced by a cylindrical conductor of radius r, positioned so that its surface (shown dashed in Fig. 7-16b) coincides with an equipotential surface, the field outside the cylindrical conductors will be undisturbed. From (3-13-1) and Fig. 7-16,

$$|\mathbf{E}_+| = \frac{\rho_L}{2\pi\epsilon_0 \sqrt{s^2 + x^2}} = |\mathbf{E}_-| \tag{2}$$

† Therefore, without formally solving for all equipotential surfaces, we already know from symmetry considerations where one such surface is.

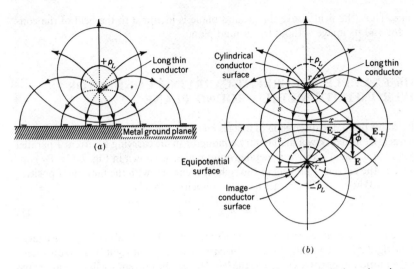

Figure 7-16 (a) Field of conductor over a ground plane. (b) Field of two oppositely charged conductors. Field above flat equipotential surface is same as in (a).

As shown in Fig. 7-16b, the normal component of the electric field can be found from superposition as

$$E_n = 2 \cos \phi \, |\mathbf{E}_+| = 2 \cos \phi \, \frac{\rho_L}{2\pi\epsilon_0 \sqrt{s^2 + x^2}} \qquad (3)$$

But since $\cos \phi = s/\sqrt{s^2 + x^2}$, (3) becomes

$$E_n = \frac{\rho_L}{\pi\epsilon_0} \frac{s}{s^2 + x^2} \qquad (4)$$

Thus, substituting (4) into (1) gives

$$\rho_s = -\frac{\rho_L}{\pi} \frac{s}{s^2 + x^2} \qquad (\text{C m}^{-2}) \qquad (5)$$

Equation (5) indicates that the equivalent surface density of charge on the equipotential surface is strongest at the point of closest approach to the upper conductor, that is, at $x = 0$.

If we now imagine that a large, thin, perfectly conducting flat plane is slipped in along the equipotential surface, the field lines will be undisturbed since they are already normal to this metallic sheet, thus satisfying the boundary conditions. Since the field lines have not been disturbed, it follows from the preceding discussion that a charge distribution according to (5) will be set up on this conducting ground plane. Thus, the field above the ground plane is identical to the field of the conductor and its image without the ground plane.

7-16 EXAMPLE 8: CURRENT-CARRYING CONDUCTOR OVER INFINITE METAL GROUND PLANE

Up to this point, the discussion of images has been limited to distributions of static charges. Consider now a filamentary conductor in air carrying a current I parallel to, and at a height s above, a metal ground plane, as shown in Fig. 7-17a. By convention, the current I flowing to the right is identified with the motion of positive charges to the right, the charges having a velocity v, that is,

$$I = \rho_L v \quad \text{(A)} \tag{1}$$

where ρ_L is the charge per unit length of conductor. The motion of *positive* charge to the right is imaged by the motion of *negative* charge to the right. This is equivalent to the motion of positive charge to the left. Hence, the current I will have an image current I' below the ground plane, equal in magnitude but flowing in the opposite direction, as suggested in Fig. 7-17a, i.e.,

$$I' = -\rho_L v = -I \quad \text{(A)} \tag{2}$$

Next consider a filamentary conductor carrying a current I perpendicular to a metal ground plane, as shown in Fig. 7-17b. Using similar reasoning, it can be seen that this filamentary conductor has a vertical image conductor carrying current in the same direction, as shown. The image of a conductor making an arbitrary angle with a metal ground plane can be found by resolving it into vertical and horizontal components, as shown in Fig. 7-17c.

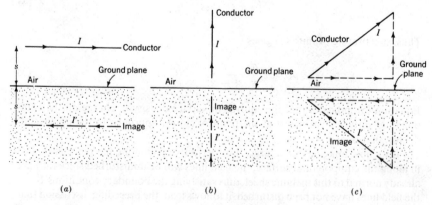

Figure 7-17 Image of a horizontal, vertical, and slant current-carrying conductor over a ground plane.

PROBLEMS†

Group 7-1: Secs. 7-1 through 7-9. Laplace's equation in rectangular coordinates and analytical, iterative, and analog solutions

7-1-1. Laplace's equation. Mean value significance The significance of Laplace's equation is simply that the potential at a point in a charge-free region is the *average*, or *mean value*, of the potential at the surrounding points. Show, as a corollary, that the potential at a point cannot be higher than the maximum value or lower than the minimum value of the 4, 6, or more surrounding potentials.

★**7-1-2. Potential in cubical box** A cubical metal box is provided with a cover separated from it by small gaps like the cover of the trough of Example 2 in Sec. 7-6 is separated or insulated from the trough. The box (sides and bottom) is at zero potential and the cover is at 40 V. (a) Find the potential at the center of the box. (b) Compare this value with the potential at the center of the trough (Example 2). Note that this is a three-dimensional problem. See footnotes to Sec. 7-5.

7-1-3. Potential in cylindrical can A cylindrical metal can of height equal to its diameter has a cover separated from it by small gaps like the cover of the trough of Example 2 in Sec. 7-6 is separated or insulated from the trough. The can (sides and bottom) is at zero potential and the cover is at 40 V. (a) Find the potential at the center of the can. (b) Compare this value with the potential at the center of the trough (Example 2). Note that this is a three-dimensional problem. See footnotes to Sec. 7-5.

7-1-4. Strip transmission line (a) A transmission line consists of four flat strips of equal width. In cross section these form the four sides of a square box with small gaps at the four corners. The top and left-hand strips are at +100 V while the bottom and right-hand strips are at zero potential. Determine the interior potentials for a 9-point grid similar to that for Example 2 in Sec. 7-6. (b) Repeat the problem for the case when the top and bottom strips are at zero potential while the left and right strips are at 100 V.

7-1-5. Laplace's equation. Mean value when $\Delta x \neq \Delta y$ Show that if $\Delta x \neq \Delta y$ in Fig. 7-3, the potential V_0 at the center point is given by

$$V_0 = \frac{V_1}{2\left[1 + \left(\frac{\Delta x}{\Delta y}\right)^2\right]} + \frac{V_2}{2\left[1 + \left(\frac{\Delta x}{\Delta y}\right)^2\right]} + \frac{V_3}{2\left[1 + \left(\frac{\Delta y}{\Delta x}\right)^2\right]} + \frac{V_4}{2\left[1 + \left(\frac{\Delta y}{\Delta x}\right)^2\right]}$$

7-1-6. Laplace's equation. Mean value when all lengths unequal Show that if all length values in Fig. 7-3 are unequal, the potential V_0 at 0 (point P) in Fig. 7-3 is given by

$$V_0 = \frac{V_1}{(1 + a)\left(1 + \frac{b}{c}\right)} + \frac{V_2}{(1 + a^{-1})\left(1 + \frac{b}{c}\right)} + \frac{V_3}{(1 + c)\left(1 + \frac{d}{b}\right)} + \frac{V_4}{(1 + c^{-1})\left(1 + \frac{d}{b}\right)}$$

where $a = \Delta x_1/\Delta x_2$
$b = \Delta x_1/\Delta x_2$
$c = \Delta y_3/\Delta y_4$
$d = \Delta y_3/\Delta y_4$
Δx_1 = distance from point P at 0 to 1
Δx_2 = distance 0 to 2
Δy_3 = distance 0 to 3 and so forth

7-1-7. Transmission line with conductors of square cross section A transmission line has an inner conductor of square cross section of 100 mm by 100 mm symmetrically situated inside an outer conductor of square cross section of 200 mm by 200 mm. The medium between the conductors is air. Determine (a) the electric potential and field configuration, (b) the capacitance per meter of length, and (c) the inductance per meter of length. Use mean-value method or graphical solution.

† Answers to starred problems are given in Appendix C.

★7-1-8. Square enclosure with sides at different potentials Four long plates of equal width form a square enclosure as shown in cross section in Fig. P7-1-8. If the plates are insulated and have voltages as indicated, what is the potential of the point *P* one-half of the distance from the center of the square to the center of the lower plate? If iteration is used, list the successive potentials obtained.

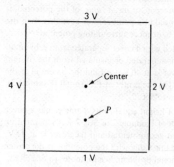

Figure P7-1-8 Square enclosure with different potential on four sides.

★7-1-9. L-shaped conductor. Potential in corner A uniform conducting sheet has potentials applied as shown in Fig. P7-1-9. Find *V* at *P*. Note that the grid does *not* represent a field map.

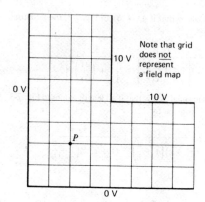

Figure P7-1-9 L-shaped conductor.

7-1-10. L-shaped conductor. Resistance (*a*) Prepare a curvilinear square map of a conductor of uniform thickness with right-angle bend connected as in Fig. P7-1-10. The map may be constructed by experimental (Teledeltos paper), iterative, or graphical methods. Take $a = 3b$. (*b*) From the map find the resistance of the configuration in terms of the resistance per square. (*c*) Compare the result with that given by the following formula:

$$\frac{R}{R_s} = \frac{a + 2b}{a - b} + 0.65e^{(a + 2b)/2} + \frac{0.04}{a - b} - 0.441$$

[see B. R. Chawla, *Proc. IEEE*, **60**:151 (January 1972).] *Word of caution*: Although the above formula and other such handbook formulas may be useful in special cases, they tend to be somewhat arbitrary

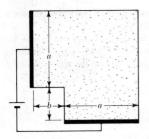

Figure P7-1-10 Conductor with right-angle bend.

and to mask the basic physical relationships and should be checked where possible by simpler, more direct methods, such as graphical field maps, to guard against gross errors.

7-1-11. Deep and shallow troughs Referring to the trough with insulated cover in Example 2 in Sec. 7-6, find the potential at the center of the trough (a) if the trough is twice as high as it is wide and (b) if it is twice as wide as it is high.

7-1-12. Tall and short cans Referring to the cylindrical metal can of Prob. 7-1-3, find the potential at the center of the can (a) if the height of the can is twice its diameter and (b) if the height is half the diameter.

7-1-13. Rectangular trough A long rectangular metal trough with insulated cover is shown in cross section in Fig. P7-1-13. The trough is at zero potential, and the cover is at potential V_1. Show that the potential at any point (x, y) inside is given by

$$V = \sum_{n \text{ odd}} \frac{4V_1 \sin \dfrac{n\pi}{x_1} x \; \sinh \dfrac{n\pi}{x_1} y}{n\pi \sinh \dfrac{n\pi y_1}{x_1}}$$

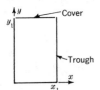

Figure P7-1-13 Trough with flat cover.

7-1-14. Rectangular trough with curved electrode A conducting rectangular trough with electrode is shown in cross section in Fig. P7-1-14. If the trough is at zero potential, find the required shape of the electrode at potential V_1 such that the potential in the trough below the electrode is of the form $V = C \sin kx \sinh ky$. Reduce the equation for the electrode to its simplest form and construct an accurate graph of the trough and electrode for the case $x_1 = y_1$.

Figure P7-1-14 Trough with curved cover.

Group 7-2: Secs. 7-10 through 7-16. Laplace's equation in cylindrical and spherical coordinates, Poisson's equation, and images

★**7-2-1. Capacitor plates at an angle** Determine the potential distribution V between two infinite flat metal sheets meeting at an angle θ, as shown in cross section in Fig. P7-2-1. The left edges of the sheets are separated by an infinitesimal gap. The lower sheet is at zero potential, and the upper sheet is at potential V_1.

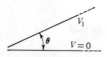

Figure P7-2-1 Plates at an angle.

7-2-2. Concentric cones (a) Find the potential distribution between two infinite metal cones as shown in cutaway view in Fig. P7-2-2. The potential of the outer cone is zero and of the inner cone is V_1. The half-angle of the inner cone is θ_i and of the outer cone is θ_0. (b) Find the potential distribution between a single cone at potential V_1 and a flat metal sheet at zero potential. This is a special case of (a) for which the outer cone half-angle $\theta_0 = 90°$.

Figure P7-2-2 Concentric cones.

★**7-2-3. Uniform B inside sphere** (a) What sheet-current distribution on the surface of a sphere will provide a uniform magnetic field inside? $\mu = \mu_0$ everywhere. (b) How should the turns of a winding be arranged on the sphere to produce this result?

7-2-4. Spherical sheet current. Boundary relations A steady electric surface current given by $\mathbf{K} = \hat{\phi} k_0 \sin \theta$ exists on the surface of a sphere of radius R. k_0 is a constant and $\mu = \mu_0$ everywhere. The magnetic flux density set up by the current is given by $\mathbf{B} = \hat{z} C_1$ for $r < R$ and $\mathbf{B} = (C_2/r^3)(\hat{z} - \hat{r}3 \cos \theta)$ for $r > R$. Find C_1 and C_2.

★**7-2-5. Electric field for insulated wire** A conducting wire of radius a is situated in air in an originally uniform field E_0 perpendicular to the wire. The wire has an insulating coating of radius b and permittivity ϵ, as shown in Fig. P7-2-5. Find the potential V_0 everywhere outside the insulating coating and the potential V_i everywhere inside the insulating coating.

Figure P7-2-5 Insulated wire.

7-2-6. Two hemispheres A thin conducting spherical shell of radius r_1 is cut into two hemispheres separated by a very small air gap. If one hemisphere is at a potential V_1 and the other at a potential V_2, find the potential everywhere (a) outside the sphere and (b) inside the sphere.

7-2-7. Leakage field through slot The electric field is uniform and equal to E_0 and is normal to one side of an infinite plane conducting sheet. The field is zero on the other side. If a slot of width s is cut in the sheet, show that the surface charge density on the E_0 side is

$$\rho_s = \frac{\epsilon_0 E_0}{2}\left\{1 + \frac{u}{[u^2 - (s/2)^2]^{1/2}}\right\}$$

and on the side where the field was originally zero is given by

$$\rho_s = \frac{\epsilon_0 E_0}{2}\left\{1 - \frac{u}{[u^2 - (s/2)^2]^{1/2}}\right\}$$

where u is the distance from the center of the slot.

7-2-8. Leakage field through slot at various distances A slot of width s is cut in an infinite horizontal flat conducting sheet. Before the slot was cut, the electric field above the sheet was everywhere vertical and equal to E_0, while below the sheet the field was zero. At what distance below the center of the slot is the potential V equal to (a) $0.1E_0s$, (b) $0.01E_0s$, and (c) $0.001E_0s$?

★7-2-9. Hemisphere on conducting sheet A conducting hemisphere of radius a is placed on a flat conducting sheet of infinite extent, as shown in Fig. P7-2-9. Before the hemisphere was introduced, the electric field everywhere above the sheet was normal to it and equal to E_0. The medium is air, and the sheet and hemisphere are at zero potential. (a) Find V as a function of r and θ everywhere above the sheet and hemisphere. (b) Find the surface charge density at all points on the sheet and hemisphere. (c) Plot a graph of the surface charge density on the hemisphere and along the flat sheet to a distance of $5a$.

Figure P7-2-9 Hemisphere on conducting sheet.

7-2-10. Half-cylindrical tube A half-cylindrical metal tube of radius a, shown in cross section in Fig. P7-2-10, is at zero potential. If a metal plate at a potential V_1 is placed across one end of the tube but not in contact with it, show that the potential inside the tube at a large distance z from the plate is of the form

$$V = Ce^{-(R_{11}/a)z}\sin\phi\ \ J_1\left(\frac{R_{11}}{a}r\right)$$

where C = constant
 J_1 = first-order Bessel function
 R_{11} = first root of $J_1 (= 3.832)$

Assume that the tube is infinitely long.

Figure P7-2-10 Half-cylindrical tube.

★7-2-11. One-dimensional device. Poisson's equation In a one-dimensional device the charge density $\rho = \rho_0(x/x_1)$. If $E = 0$ at $x = 0$ and $V = 0$ at $x = x_1$, find V.

★7-2-12. Semiconductor junction. Poisson's equation A semiconductor of constant cross section A has a junction at $x = 0$. If the charge density on both sides of the junction is given by $\rho = \rho_L x(x^2 + x_1^2)^{-3/2}$, find (a) the voltage V as a function of x and (b) the capacitance of the junction between the points $x = \pm 100x_1$. Take $E = 0$ at large x and $V = 0$ at $x = 0$. (c) Plot graphs of ρ, E, and V vs. x to $x = \pm 10x_1$.

★**7-2-13. Semiconductor junction. Capacitance** A semiconductor junction diode has a charge density distribution as given in Prob. 7-2-12. If the distance between the charge peaks is 1 μm and the voltage is applied between points 50 μm either side of the junction, find the junction capacitance. $\epsilon_r = 10$ and the junction cross section is 10^{-2} mm^2.

7-2-14. Conductor at interface. Images A long, thin linear conductor carrying a current I extends along the plane boundary between two media, air and iron. Assuming that the permeability of the iron is uniform, show that the magnetic field H in air at a radius r from the conductor is $H_0 = \mu I/[(\mu + \mu_0)\pi r]$, while in the iron at a radius r from the conductor it is given by $H_i = \mu_0 I/[(\mu + \mu_0)\pi r]$.

Group 7-3: Practical applications

7-3-1. Coaxial transmission line A cylindrical coaxial transmission line is shown in cross section in Fig. P7-3-1. The medium filling the line is air. If 50 V is applied between the conductors and the total current is 15 mA, find (a) electric field E at P_1 and P_2, (b) electric flux density D at P_1 and P_2, (c) magnetic flux density B at P_1 and P_2, (d) magnetic field H at P_1 and P_2, (e) exact capacitance per unit length (C/l), (f) exact inductance per unit length (L/l), (g) total electric flux per unit length (Q/l), and (h) total magnetic flux per unit length (ψ_m/l). (i) Draw a curvilinear square map (both V and E lines) with equipotential contours at 10, 20, 30, and 40 V; and (j) use a count of the series and parallel squares of the map of (i) for comparison with the exact capacitance and inductance values of (e) and (f). Give units for all results.

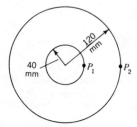

Figure P7-3-1 Coaxial transmission line.

7-3-2. Fringing field Although in most problems the effect of the *fringing field* (extending beyond the edges of the configuration under study) is neglected, fringing may be important in practical applications. In this problem the fringing field is the main consideration. Consider the fringing field of a strip transmission line or of two long capacitor plates. The fringing field may be mapped by an analog method or by a conformal transformation in which a uniform field in the xy plane (Fig. P7-3-2a) is transformed into the fringing field in the uv plane (Fig. P7-3-2b). Thus, if $u = (h/\pi)(e^x \cos y - x)$ and $v = (h/\pi)(e^x \sin y - y - \pi)$, we have for $y = 0$, $u = (h/\pi)(e^x - x)$ and $v = h$, while for $y = \pi$, $u = (h/\pi)(-e^x - x)$ and $v = 0$. Hence, the $y = 0$ line, when mapped in the uv plane, is folded back on itself and turned around so as to lie at $v = h$ and at u values equal to or greater than h/π. The $y = \pi$ line is transformed so as to lie along the u axis ($v = 0$). It undergoes also a scale change and reversal of direction (Fig. P7-3-2b). The rectangle $ABCDEFA$ in the uniform field (xy plane) is transformed to the figure $ABCDEFA$ when mapped as the fringing field (uv plane). By these methods the fringing field map of Fig. P7-3-2c is obtained. (a) If the strip transmission line or capacitor plates are 4 times the width shown in Fig. P7-3-2c and the fringing field is identical on both the left and right sides, find the percentage increase in the capacitance per unit depth due to fringing over the capacitance if fringing is neglected. (b) If the strips or plates have the dimensions of the heavy dashed contour, find the breakdown voltage in air (at atmospheric pressure) if the plate separation is 30 mm.

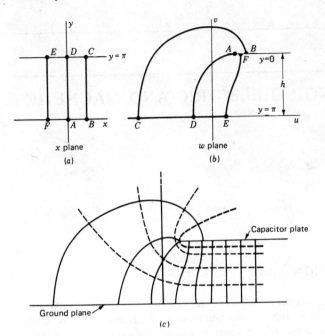

Figure P7-3-2 Coordinate transformation for fringing field at edge of plate.

CHAPTER
EIGHT

TIME-CHANGING ELECTRIC AND MAGNETIC FIELDS

8-1 INTRODUCTION

In the preceding chapters the principles of static electric and magnetic fields have been considered. In this chapter electric and magnetic fields that change with time are discussed, and a number of new relations and concepts are introduced. Some of the more important of these are (1) *Faraday's law*, which gives the emf induced in a closed circuit due to a change of magnetic flux linking it; (2) a relation giving the emf induced in a conductor moving in a magnetic field; (3) *Maxwell's displacement current*, which represents an extension of the current concept to include charge-free space; (4) the equivalent conductivity of a dielectric; and (5) an extension of the boundary relations developed in earlier chapters to include time-varying situations.

8-2 FARADAY'S LAW

In Chap. 5 we observed that a current-carrying conductor produces a magnetic field. About 1831 Michael Faraday in London and Joseph Henry in Albany found independently that the reverse effect is also possible. That is, a magnetic field can produce a current in a closed circuit but with the important qualification that the magnetic flux linking the circuit must be changing.

Consider the wire loop shown in cross section in Fig. 8-1a with a bar magnet moving up toward the loop, so that the magnet's *B* flux through the loop is increasing. This results in an induced current in the loop flowing in a direction such that the loop's magnetic field *opposes* the motion (like poles repel).

In Fig. 8-1b with the magnet moving down and away from the loop, the magnet's *B* flux through the loop is decreasing, resulting in an induced current flowing in a direction such that the loop's magnetic field again opposes the motion

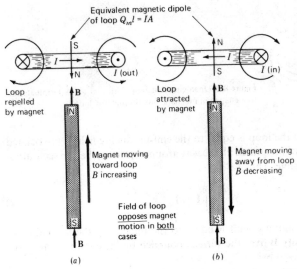

Figure 8-1 Changing the magnetic flux through the wire loop by moving the magnet induces a current in the loop.

(unlike poles attract). Thus, the loop current direction in Fig. 8-1b is reversed from its direction in Fig. 8-1a.

By moving the magnet alternately up and down, an alternating current (ac) is induced in the loop, the arrangement constituting a simple ac generator.

The fact that the induced current in the loop is always in such a direction as to oppose the change that produced it is a statement of *Lenz's law*, deduced in 1834 by Heinrich Friedrich Lenz.

If the loop is open-circuited, as in Fig. 8-2, an emf appears at its terminals which is equal to the time rate of decrease of the magnetic flux linking the loop. Thus,†

$$\mathscr{V} = -\frac{d\psi_m}{dt} \tag{1}$$

where $\mathscr{V}$ = total emf, V
ψ_m = total flux, Wb
t = time, s

Equation (1) is a statement of *Faraday's law.*‡

---

† With loops of N turns where each turn links the same amount of flux one can also write $\Lambda = N\psi_m$, where Λ = total flux linkage and ψ_m = flux linked by each turn. Thus, (1) becomes

$$\mathscr{V} = -N\frac{d\psi_m}{dt} = -\frac{d\Lambda}{dt}$$

‡ Michael Faraday, "Experimental Researches In Electricity," B. Quaritch, London, 1839.

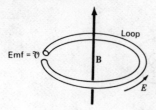

Figure 8-2 Open-circuited loop with emf at its terminals due to a change in magnetic flux through the loop.

The emf induced in the loop is equal to the emf-producing field **E** (associated with the induced current) integrated all the way around the loop, the gap separation being considered negligible. Thus,†

$$\mathscr{V} = \oint \mathbf{E} \cdot d\mathbf{l} \qquad (2)$$

The total flux through a circuit is equal to the integral of the normal component of the flux density **B** over the surface bounded by the circuit. That is, the total magnetic flux is given by

$$\psi_m = \iint \mathbf{B} \cdot d\mathbf{s} \qquad (3)$$

The surface over which the integration is carried out is the surface bounded by the periphery of the circuit, as in Fig. 8-3. Equation (3) applies to a closed single-conductor circuit of any number of turns or loops. It is important to note that *any closed circuit with any number of turns forms the boundary of a single surface* (see Fig. 8-3) and ∫∫ **B** · d**s** over this surface yields the total flux. Thus, integrating **B** over the surface in Fig. 8-3 yields all the flux. Lines of flux passing through the surface only once are integrated once, but those linking all 4 turns are integrated 4 times since they pass through the surface 4 times. Substituting (3) in (1) leads to‡

$$\mathscr{V} = -\frac{d}{dt} \int_s \mathbf{B} \cdot d\mathbf{s} \qquad (4)$$

where $\mathscr{V}$ = induced emf, V
  B = flux density, T
  d**s** = surface element, m²
  t = time, s

When the loop or closed circuit is stationary or fixed, (4) reduces to

$$\mathscr{V} = -\int_s \frac{\partial \mathbf{B}}{\partial t} \cdot d\mathbf{s} \qquad (5)$$

---

† To be more explicit, **E** in (2) could be written as $\mathbf{E}_e$ to indicate that it is an emf-producing field (see Sec. 4-10).

‡ The symbol $\int_s$ indicates a double or surface integral ($\iint$) over a surface s.

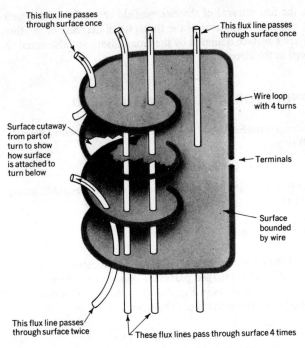

This flux line passes through surface once

This flux line passes through surface once

Wire loop with 4 turns

Surface cutaway from part of turn to show how surface is attached to turn below

Terminals

Surface bounded by wire

This flux line passes through surface twice

These flux lines pass through surface 4 times

Figure 8-3 Circuit with four-turn coil. The wire forms the boundary of a single continuous surface. Part of the surface of one turn has been cut away to show how the surface is bounded by the turn below.

This form of Faraday's law gives the induced emf due specifically to a time rate of change of **B** for a loop or circuit that is fixed with respect to the observer. This is sometimes called the *transformer induction equation*.

Combining (2) and (5)

$$\mathscr{V} = \oint \mathbf{E} \cdot d\mathbf{l} = -\int_s \frac{\partial \mathbf{B}}{\partial t} \cdot d\mathbf{s} \qquad (6)$$

where $\mathscr{V}$ = induced emf, V
$\mathbf{E} = \mathbf{E}_e$ = emf-producing electric field, V m^{-1}
$d\mathbf{l}$ = element of path, m
$\mathbf{B}$ = flux density, T
$d\mathbf{s}$ = element of area, m^2
t = time, s

This relation is referred to as *Maxwell's equation as derived from Faraday's law*. It appears in (6) in its integral form. The corresponding differential relation is given in Sec. 8-7.

According to (6), the line integral of the electric field around a fixed closed circuit is equal to the normal component of the time rate of decrease of the flux density **B** integrated over a surface bounded by the circuit. Both are also equal to the total emf $\mathscr{V}$ induced in the circuit.

8-3 MOVING CONDUCTOR IN A MAGNETIC FIELD

From (5-23-17) the Lorentz force **F** on a particle of charge Q moving with a velocity **v** in a magnetic field **B** is

$$\mathbf{F} = Q(\mathbf{v} \times \mathbf{B}) \quad \text{(N)} \tag{1}$$

Dividing by the charge Q, we obtain the force per charge, or electric field **E**, acting on the charged particle as

$$\mathbf{E} = \frac{\mathbf{F}}{Q} = \mathbf{v} \times \mathbf{B} \quad (\text{N C}^{-1} \text{ or V m}^{-1}) \tag{2}$$

This electric field **E** is of the emf-producing type and is in a direction normal to the plane containing **v** and **B**. If the charged particle is one of many electrons in a wire moving across a magnetic field, as in Fig. 8-4, then the emf $\mathscr{V}$ induced between its end points is given by the integral of **E** between 1 and 2, or

$$\mathscr{V} = \int_{1}^{2} \mathbf{E} \cdot d\mathbf{l} = \int (\mathbf{v} \times \mathbf{B}) \cdot d\mathbf{l} \quad \text{(V)} \tag{3}$$

where $\mathscr{V}$ = emf induced over a length l of wire, V
   E = electric field, V m^{-1}
   $d\mathbf{l}$ = element of length of wire, m
   v = velocity of wire, m s^{-1}
   B = flux density of magnetic field, T

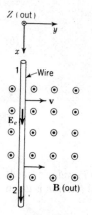

Figure 8-4 An emf is induced in a wire moving across a magnetic field.

For a straight wire where **v**, **B**, and the wire are mutually perpendicular, **B** is uniform, and **v** is the same for all parts of the wire, (3) reduces to

$$\mathscr{V} = El = vBl \quad (V) \tag{4}$$

where l is the length of the wire in meters.

Equations (3) and (4) are *motional-induction, or flux-cutting, laws* giving the emf induced in a conductor moving with respect to the observer. Equation (3) is the more general form, while (4) applies to the special case where the directions of the wire, its motion, and the magnetic field are all mutually perpendicular.

The relations may be used to find the emf induced in any part of a circuit due to its motion through a magnetic field. They also can be applied to find the total emf induced in a closed circuit that is moved or deformed in a magnetic field that does not change with time. For a closed circuit (3) becomes

$$\mathscr{V} = \oint \mathbf{E} \cdot d\mathbf{l} = \oint (\mathbf{v} \times \mathbf{B}) \cdot d\mathbf{l} \tag{5}$$

where $\mathscr{V}$ is the total emf induced in the circuit.

8-4 GENERAL CASE OF INDUCTION

Equation (8-2-6) gives the emf induced in a closed circuit due to the time rate of change of **B** (transformer induction). Equation (8-3-5) gives the emf induced in a closed circuit due to its motion. When both kinds of changes are occurring simultaneously, i.e., when **B** changes with time and the circuit is also in motion, the total emf induced is equal to the sum of the emfs given by (8-2-6) and (8-3-5), or

$$\boxed{\mathscr{V} = \oint (\mathbf{v} \times \mathbf{B}) \cdot d\mathbf{l} - \int_s \frac{\partial \mathbf{B}}{\partial t} \cdot d\mathbf{s}} \tag{1}$$

The first term of the right-hand member gives the emf induced by the motion, while the second term gives the emf induced by the time change in **B**. The line integral in the first term is taken around the entire circuit, while the surface integral in the second term is taken over the entire surface bounded by the circuit.

Equation (1) is a general relation and gives the correct value of total induced emf in all cases. For the special case of motion only, $\partial B/\partial t = 0$, and (1) reduces to

$$\mathscr{V} = \oint (\mathbf{v} \times \mathbf{B}) \cdot d\mathbf{l} \tag{2}$$

For the special case of time change of flux density only, $\mathbf{v} = 0$, and (1) reduces to

$$\mathscr{V} = - \int_s \frac{\partial \mathbf{B}}{\partial t} \cdot d\mathbf{s} \tag{3}$$

To summarize, the induction relations are

(I) $\quad \mathscr{V} = \oint (\mathbf{v} \times \mathbf{B}) \cdot d\mathbf{l} - \int_s \dfrac{\partial \mathbf{B}}{\partial t} \cdot d\mathbf{s} \quad$ general case

(II) $\quad \mathscr{V} = \oint (\mathbf{v} \times \mathbf{B}) \cdot d\mathbf{l} \quad$ motion only (motional induction)

(III) $\quad \mathscr{V} = - \int_s \dfrac{\partial \mathbf{B}}{\partial t} \cdot d\mathbf{s} \quad$ B change only (transformer induction)

8-5 EXAMPLES OF INDUCTION

In this section seven examples are given in which the total emf induced in a closed circuit (total induction) is calculated. The general relation (I) gives the correct result in all cases. In some cases (motion only or time change only) (II) or (III) alone is sufficient.

Example 1: Loop: B change with no motion Consider the fixed rectangular loop of area A shown in Fig. 8-5. The flux density **B** is normal to the plane of the loop (outward in Fig. 8-5) and is uniform over the area of the loop. However, the magnitude of **B** varies harmonically with respect to time as given by

$$B = B_0 \cos \omega t \qquad (1)$$

where B_0 = maximum amplitude of B, T

$\qquad \omega$ = radian frequency ($= 2\pi f$, where f = frequency), rad s^{-1}

$\qquad t$ = time, s

Find the total emf induced in the loop.

SOLUTION This is a pure case of B change only, there being no motion. Hence, from (III), the total emf induced in the loop is

$$\mathscr{V} = - \int_s \dfrac{\partial \mathbf{B}}{\partial t} \cdot d\mathbf{s} = A\omega B_0 \sin \omega t \qquad (V) \qquad (2)$$

This emf appears at the terminals of the loop (Fig. 8-5). Since the velocity **v** = 0, the emf calculated by (II) is zero and by (I) is identical with that in (2).

Example 2: Loop: Motion only: No B change Consider the rectangular loop shown in Fig. 8-6. The width l of the loop is constant, but its length x is increased uniformly with time by moving the sliding conductor at a uniform

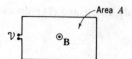

Figure 8-5 Fixed loop of area A (for Example 1).

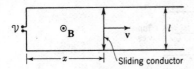

Figure 8-6 Sliding conductor for increasing loop area (for Examples 2 and 3).

velocity **v**. The flux density **B** is everywhere the same (normal to the plane of the loop) and is constant with respect to time. Find the total emf induced in the loop.

SOLUTION This is a pure case of motion only, the flux density **B** being constant. Hence, from (II),

$$\mathscr{V} = \oint (\mathbf{v} \times \mathbf{B}) \cdot d\mathbf{l} = vBl \qquad \text{(V)} \qquad (3)$$

The entire emf in this case is induced in the moving conductor of length l. Since $\partial B/\partial t = 0$, the emf by (III) is zero and by (I) is identical with (3).

Example 3: Loop: Motion and B change Consider again the same loop with sliding conductor discussed in the preceding example (Fig. 8-6). The flux density **B** is normal to the plane of the loop and is uniform everywhere. The sliding conductor moves with a uniform velocity **v**. These conditions are the same as in the preceding example. However, in this case let the magnitude of the flux density **B** vary harmonically with time as given by

$$B = B_0 \cos \omega t \qquad (4)$$

Find the total emf induced in the loop.

SOLUTION This is a case involving both motion and a time-changing **B**. The emf $\mathscr{V}_m$ due to the motion is given, from (II), by

$$\mathscr{V}_m = \int (\mathbf{v} \times \mathbf{B}) \cdot d\mathbf{l} = vBl = vlB_0 \cos \omega t \qquad (5)$$

The emf $\mathscr{V}_t$ due to a time-changing **B** is, from (III),

$$\mathscr{V}_t = - \int_s \frac{\partial \mathbf{B}}{\partial t} \cdot d\mathbf{s} = \omega x l B_0 \sin \omega t \qquad (6)$$

According to (I) the total emf $\mathscr{V}$ is the sum of the emfs of (5) and (6), or

$$\mathscr{V} = \mathscr{V}_m + \mathscr{V}_t = \oint (\mathbf{v} \times \mathbf{B}) \cdot d\mathbf{l} - \int_s \frac{\partial \mathbf{B}}{\partial t} \cdot d\mathbf{s}$$

$$= vB_0 l \cos \omega t + \omega x B_0 l \sin \omega t$$

$$= B_0 l \sqrt{v^2 + (\omega x)^2} \, \sin (\omega t + \delta) \qquad (7)$$

where $\delta = \tan^{-1} (v/\omega x)$

x = instantaneous length of loop

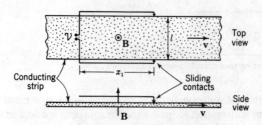

Figure 8-7 Fixed loop with sliding strip (for Examples 4 and 5).

Example 4: Moving strip: No B change The circuit for a rectangular loop of width l and length x_1 is completed by sliding contacts through a thin conducting strip, as suggested in Fig. 8-7. The loop is stationary, but the strip moves longitudinally with a uniform velocity $\mathbf{v}$. The magnetic flux density $\mathbf{B}$ is normal to the strip and the plane of the loop. It is constant with respect to time and is uniform everywhere. The width of the loop is l, the same as for the strip, although for clarity the loop is shown with a slightly greater width in Fig. 8-7. Find the total emf induced in the circuit.

SOLUTION This is another case of motion only. Therefore from (II) the total emf is given by

$$\mathscr{V} = \oint (\mathbf{v} \times \mathbf{B}) \cdot d\mathbf{l} = vBl \qquad (8)$$

The entire emf in this case is induced in the moving strip and appears at the terminals. Since $\partial B/\partial t = 0$, the emf by (III) is zero and by (I) is identical with (8). A variation of the arrangement of Example 4 is provided by the Faraday disk generator (see Prob. 8-4-2).

Example 5: Moving strip with B change Consider now the same loop and strip as in the preceding example (Fig. 8-7), but let the magnitude of the flux density vary harmonically with time as given by

$$B = B_0 \cos \omega t \qquad (9)$$

Find the total emf induced in the circuit.

SOLUTION This case involves both motion and a time-changing $\mathbf{B}$. From (II) the emf $\mathscr{V}_m$ due to the motion is

$$\mathscr{V}_m = \oint (\mathbf{v} \times \mathbf{B}) \cdot d\mathbf{l} = vBl = vB_0 l \cos \omega t \qquad (10)$$

From (III) the emf $\mathscr{V}_i$ due to a time-changing $\mathbf{B}$ is†

$$\mathscr{V}_i = - \int_s \frac{\partial \mathbf{B}}{\partial t} \cdot d\mathbf{s} = \omega x_1 B_0 l \sin \omega t \qquad (11)$$

† At low frequencies the effect of eddy currents in the strip can be neglected. The effect of eddy currents will be even less if the strip is very thin and its conductivity poor.

According to (I), the total emf $\mathscr{V}$ is the sum of (10) and (11), or

$$\mathscr{V} = \mathscr{V}_m + \mathscr{V}_t = vB_0 l \cos \omega t + \omega x_1 B_0 l \sin \omega t$$

$$= B_0 l \sqrt{v^2 + (\omega x_1)^2} \sin (\omega t + \delta) \tag{12}$$

where $\delta = \tan^{-1}(v/\omega x_1)$.

Example 6: Rotating loop: No B change Consider next a rotating rectangular loop in a steady magnetic field as in Fig. 8-8a. The loop rotates with a uniform angular velocity ω radians per second. This arrangement represents a simple ac generator, the induced emf appearing at terminals connected to the slip rings. If the radius of the loop is R and its length l, find the total emf induced.

SOLUTION Since this is a case of motion only, the total emf can be obtained from (II). Referring to Fig. 8-8b, it is given by

$$\mathscr{V} = \oint (\mathbf{v} \times \mathbf{B}) \cdot d\mathbf{l} = 2vBl \sin \theta \tag{13}$$

Since $\theta = \omega t$, we have

$$\mathscr{V} = 2\omega R l B \sin \omega t \tag{14}$$

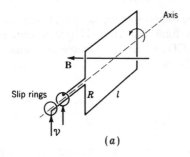

(a)

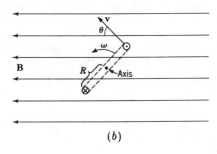

(b)

Figure 8-8 AC generator (for Examples 6 and 7): (a) perspective view; (b) cross section perpendicular to axis.

The factor 2 is necessary because there are two conductors of length l moving through the field, the emfs in both aiding. Since $2Rl = A$, the area of the loop, (14) reduces to

$$\mathscr{V} = \omega BA \sin \omega t \tag{15}$$

Since $\partial B/\partial t = 0$, the emf calculated by (III) is zero. Hence from (I) the emf is as given by (14) or (15).

Example 7: Rotating loop with B change Consider finally the same rotating loop as in the preceding example with the modification that B varies with time as given by $B = B_0 \sin \omega t$ (ω same as rotation angular velocity). When $t = 0$, $B = 0$ and $\theta = 0$ (Fig. 8-8b). Find the total emf induced.

SOLUTION This case involves both motion and a time-changing B. From (II) the emf $\mathscr{V}_m$ due to the motion is

$$\mathscr{V}_m = 2\omega Rl B_0 \sin^2 \omega t = \omega Rl B_0 - \omega Rl B_0 \cos 2\omega t \tag{16}$$

From (III) the emf $\mathscr{V}_t$ due to a time-changing B is

$$\mathscr{V}_t = -2\omega Rl B_0 \cos^2 \omega t = -\omega Rl B_0 - \omega Rl B_0 \cos 2\omega t \tag{17}$$

From (I) the total emf $\mathscr{V}$ is given by the sum of (16) and (17), or

$$\mathscr{V} = \mathscr{V}_m + \mathscr{V}_t = -2\omega Rl B_0 \cos 2\omega t \tag{18}$$

The emf in this example is at twice the rotation, or magnetic field, frequency. It is to be noted that the emf calculated from either (II) or (III) alone contains a dc component. In adding the emfs of (II) and (III) the dc component cancels, yielding the correct total emf given by (18).

8-6 STOKES' THEOREM

In Sec. 8-2 Maxwell's equation from Faraday's law is stated in integral form. This equation may be transformed from an integral to a differential form by means of Stokes' theorem, which is developed in this section and applied to Maxwell's equation in Sec. 8-7.

Consider a square of area Δs in the xy plane as in Fig. 8-9. Let the electric field $\mathbf{E}$ have components E_x and E_y as shown. The work per coulomb required to move a charge around the perimeter of the square is given by the line integral of $\mathbf{E}$ around the perimeter. This work equals the total emf around the perimeter. That is,

$$\mathscr{V} = \oint \mathbf{E} \cdot d\mathbf{l} \tag{1}$$

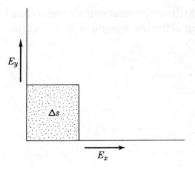

Figure 8-9 Small rectangular area.

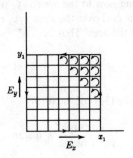

Figure 8-10 Illustration for Stokes' theorem.

Dividing by the area Δs and taking the limit of this ratio as Δs approaches zero yields the curl of **E** normal to Δs at the point around which Δs shrinks to zero (see Sec. 5-18). Thus

$$\lim_{\Delta s \to 0} \frac{\oint \mathbf{E} \cdot d\mathbf{l}}{\Delta s} = \nabla \times \mathbf{E} \tag{2}$$

Consider now a surface of area $x_1 y_1$ as shown in Fig. 8-10. Let the area be divided into infinitesimal areas as suggested. From (2) the work per coulomb to carry a charge around an infinitesimal loop divided by its area is equal to the curl of **E** at the point. If the curl of **E** is integrated over the entire area $x_1 y_1$, all contributions to the total work cancel except for the work along the periphery of the area $x_1 y_1$.

The situation here is analogous to that of a single current loop with current I, Fig. 8-11a, whose effect is the same as a mesh of current loops, each with a current I, as suggested in Fig. 8-11b. It is assumed that the adjacent sides of the small loops are very close together. Since the currents in adjacent sides are oppositely directed, their fields cancel. The only currents whose effects are not canceled are those along the periphery of the area of radius R.

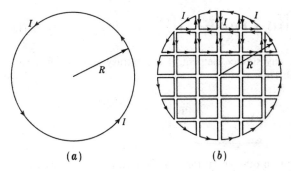

Figure 8-11 (a) Single current loop and (b) equivalent mesh of small current loops.

Returning now to the area x_1y_1 in Fig. 8-10, the integral of the normal component of the curl over the area x_1y_1 must equal the line integral of $\mathbf{E}$ around the periphery of the area. That is,

$$\oint \mathbf{E} \cdot d\mathbf{l} = \int_s (\nabla \times \mathbf{E}) \cdot d\mathbf{s} \tag{3}$$

Dimensionally (3) is of the form

$$\frac{\text{Force}}{\text{Charge}} \times \text{distance} = \frac{\text{work/charge}}{\text{area}} \times \text{area}$$

Since force $\times$ distance = work, (3) is balanced dimensionally. In (3) it is understood that if the curl of $\mathbf{E}$ is integrated over an area s, the line integral of $\mathbf{E}$ is taken around the periphery of the same area s. That is,

$$\oint_{\substack{\text{periphery} \\ \text{of } s}} \mathbf{E} \cdot d\mathbf{l} = \int_{\substack{\text{surface} \\ \text{of } s}} (\nabla \times \mathbf{E}) \cdot d\mathbf{s} \tag{4}$$

The relation expressed by (3) or (4) is called *Stokes' theorem* as applied to electric fields. In general, Stokes' theorem states that *the line integral of a vector function around a closed contour C is equal to the integral of the normal component of the curl of that vector function over any surface having the contour C as its bounding edge.*

8-7 MAXWELL'S EQUATION FROM FARADAY'S LAW: DIFFERENTIAL FORM

By means of Stokes' theorem (8-6-4) let us substitute the surface integral of the curl of $\mathbf{E}$ for the line integral of $\mathbf{E}$ in (8-2-6). That is,

$$\iint (\nabla \times \mathbf{E}) \cdot d\mathbf{s} = -\iint \frac{\partial \mathbf{B}}{\partial t} \cdot d\mathbf{s} \tag{1}$$

Since $d\mathbf{s}$ in (1) applies to any surface element, it is arbitrary and therefore the integrands in (1) are equal. Thus

$$\nabla \times \mathbf{E} = -\frac{\partial \mathbf{B}}{\partial t} \tag{2}$$

This is Maxwell's equation, in differential form, as derived from Faraday's law. The integral form of the equation was given in (8-2-6).

8-8 INDUCTANCE, SELF AND MUTUAL

Case 1 Consider that two uniform toroidal coils are interwound as in Fig. 8-12a. Coil 1 of N_1 turns is indicated by a heavy wire and coil 2 of N_2 turns by a fine wire. There is no electrical connection between the coils. The ring-shaped form on which the coils are wound is assumed to have a constant permeability μ. Coil 1 will be called the *primary winding* and coil 2 the *secondary*. A schematic diagram of the arrangement is shown in Fig. 8-12b.

If the primary current I_1 is constant in value, the emf $\mathscr{V}_2$ appearing at the terminals of the secondary coil is zero, since the flux ψ_{m1} produced by the primary coil is not changing. It is assumed that all of the magnetic field produced by I_1 is confined to the region inside the toroidal windings.

Suppose now that the resistance R is decreased so that I_1 increases. This increases the magnetic flux ψ_{m1}. Disregarding the negative sign, we have from Faraday's law that the magnitude of the emf $\mathscr{V}_2$ induced in coil 2 and appearing at its terminals is

$$\mathscr{V}_2 = N_2 \frac{d\psi_{m1}}{dt} \tag{1}$$

where ψ_{m1} is the magnetic flux produced by the primary coil. Assuming that the radius r of the toroid is large compared with the radius s of the winding (Fig. 8-12a), the flux density **B** may be considered constant over the interior of the winding.

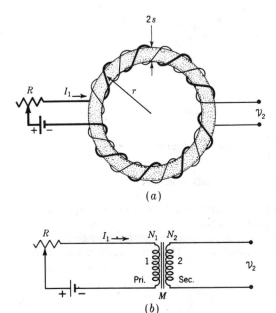

(a)

(b)

Figure 8-12 Toroidal coil with two windings.

Obtaining the magnitude of **B** from (5-13-3), we see that the total flux ψ_{m1} through the toroid is

$$\psi_{m1} = \frac{\mu N_1 I_1 A}{l} \quad \text{(Wb)} \tag{2}$$

where A = area of winding cross section, m^2
l = mean length of the toroidal coil = $2\pi r$, m

Substituting (2) in (1) gives

$$\mathcal{V}_2 = N_1 N_2 \frac{\mu A}{l} \frac{dI_1}{dt} \tag{3}$$

According to (3), the secondary emf $\mathcal{V}_2$ is proportional to the number of primary turns N_1, the number of secondary turns N_2, the permeability μ of the medium inside the winding, the cross-sectional area A of the winding, and the time rate of change of the primary current I_1 and is inversely proportional to the average length l of the winding. When we put

$$M = N_1 N_2 \frac{\mu A}{l} \tag{4}$$

(3) reduces to

$$\mathcal{V}_2 = M \frac{dI_1}{dt} \tag{5}$$

Dimensionally (5) is

$$\text{Emf} = M \frac{\text{current}}{\text{time}}$$

or

$$M = \frac{\text{emf}}{\text{current}} \times \text{time} = \text{resistance} \times \text{time}$$

or

$$M = \text{ohm-seconds} = \text{henrys}$$

Thus, M has the dimensions of inductance, and since M involves two coils, it is called the *mutual inductance* of the two coils.

The inductance L discussed in previous sections involves a single coil. Therefore, in contrast, L is called the *self-inductance*. The emf $\mathcal{V}_1$ developed for a coil of self-inductance L_1 is

$$\mathcal{V}_1 = L_1 \frac{dI_1}{dt} \tag{6}$$

where I_1 is the current in the coil. This relation involving the self-inductance of a coil is similar in form to (5), which involves the mutual inductance of two coils.

From (5-13-4) the self-inductance of a toroid is

$$L = \frac{N^2}{l/\mu A} = \frac{N^2}{\mathscr{R}} \quad \text{(H)} \tag{7}$$

where N = number of turns of toroid, dimensionless

$\mathscr{R} = l/\mu A$ = reluctance of region enclosed by toroid winding, H^{-1}

From (4) the mutual inductance M of two coils (as in Fig. 8-12) is

$$M = \frac{N_1 N_2}{l/\mu A} = \frac{N_1 N_2}{\mathscr{R}} \quad \text{(H)} \tag{8}$$

where N_1 = number of primary turns, dimensionless

N_2 = number of secondary turns, dimensionless

$\mathscr{R}$ = reluctance of magnetic circuit linking primary and secondary windings, H^{-1}

Case 2 Consider next the converse of the situation described above. That is, let the battery and resistance be connected across coil 2 (Fig. 8-12), and let the terminals of coil 1 be left open. Then the emf $\mathscr{V}_1$ at the terminals of coil 1 is

$$\mathscr{V}_1 = N_1 \frac{d\psi_{m2}}{dt} \tag{9}$$

where ψ_{m2} is the magnetic flux produced by the secondary coil. But

$$\psi_{m2} = \frac{\mu N_2 I_2 A}{l}$$

and so (9) becomes

$$\mathscr{V}_1 = N_1 N_2 \frac{\mu A}{l} \frac{dI_2}{dt} \tag{10}$$

and from (4) for the mutual inductance

$$\mathscr{V}_1 = M \frac{dI_2}{dt} \tag{11}$$

Thus, from (5) and (11),

$$M = \frac{\mathscr{V}_1}{dI_2/dt} = \frac{\mathscr{V}_2}{dI_1/dt} \tag{12}$$

Therefore, if a given time rate of change of current in the primary induces a certain voltage in the secondary, the same time rate of change of current applied to the secondary will induce the same voltage in the primary. In effect this is a statement of the *reciprocity theorem* as applied to a special case.†

8-9 ALTERNATING-CURRENT BEHAVIOR OF FERROMAGNETIC MATERIALS

We noted in Chap. 6 that the permeability of iron is not a constant. In spite of this, the permeability of the iron in an iron-cored coil carrying alternating current may be taken as a constant for certain applications, but its value, in this case, needs further explanation.

Where μ is not a constant, we have from (5-12-2) that the inductance L of a coil is given by

$$L = \frac{d\Lambda}{dI} \quad \text{(H)} \tag{1}$$

If there is no flux leakage, $\Lambda = N\psi_m$; so

$$L = N \frac{d\psi_m}{dI} \tag{2}$$

For a toroidal type of coil, $d\psi_m = A\, dB$, and $dI = l\, dH/N$, where A equals the area and l equals the mean length of the coil. Therefore (2) becomes

$$L = \frac{N^2 A}{l} \frac{dB}{dH} \tag{3}$$

In (3) dB/dH has the dimensions of permeability. It is equal to the slope of the hysteresis curve. At some point P, as in Fig. 8-13, this is different from the ordinary permeability, B_1/H_1, which is equal to the slope of the line from the origin to the point P. Since dB/dH involves infinitesimals, it is sometimes called the *infinitesimal* or *differential permeability*.

If alternating current is applied to an iron-cored coil so that the condition of the iron moves around a hysteresis loop (Fig. 8-13) once per cycle, the slope dB/dH varies over a wide range and the instantaneous value of the inductance will, from (3), vary over a corresponding range. Under these conditions it is often convenient to consider the average inductance (over one cycle) as obtained from (3), using the average value of the slope dB/dH. This is equal to the ordinary permeability at the maximum value of B attained in the cycle (see Fig. 8-13). Thus

$$L_{av} = \frac{N^2 A}{l} \left(\frac{dB}{dH}\right)_{av} = \frac{N^2 A}{l} \mu \tag{4}$$

where $\mu = B_{max}/H_{max}$ is the ordinary permeability at B_{max}.

---

† If the currents vary harmonically (alternating current), (12) can be written (see Sec. 8-13) as

$$M = \frac{\mathscr{V}_1}{j\omega I_2} = \frac{\mathscr{V}_2}{j\omega I_1} \quad \text{and} \quad \frac{\mathscr{V}_1}{I_2} = \frac{\mathscr{V}_2}{I_1} = j\omega M = Z_m \quad (\Omega)$$

where Z_m is the *mutual impedance*, which may be complex.

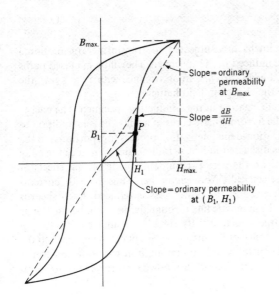

Figure 8-13 Hysteresis loop illustrating ordinary and differential permeabilities.

The above discussion is for alternating current only through the coil. If a small alternating current is superimposed on a relatively large steady, or direct, current through the coil, the situation is as suggested in Fig. 8-14. The magnetic condition of the iron then follows a minor hysteresis loop as indicated. In this case the average value of the slope dB/dH is given by the line passing through the tips of the minor hysteresis loop and is called the *incremental permeability* μ_{inc}. Referring to Fig. 8-14,

$$\mu_{inc} = \left(\frac{dB}{dH}\right)_{av} = \frac{\Delta B}{\Delta H} \qquad (5)$$

The incremental permeability is much less than the ordinary permeability $B_1 H_1$ for a point at the center of the minor loop in Fig. 8-14.

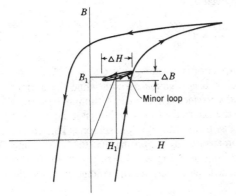

Figure 8-14 Minor hysteresis loop illustrating incremental permeability.

8-10 EDDY CURRENTS

When large conducting specimens are subjected to transformer or motional induction, currents tend to be induced in the specimen. They flow in closed paths in the specimen and are called *eddy currents*. In accordance with Lenz's law, the eddy current tends to oppose the change in field inducing it.

Eddy currents result in joule heating in the conducting specimen. The energy loss due to eddy currents in the ferromagnetic cores of ac devices is in addition to the energy lost in the magnetization-demagnetization process (proportional to the area of the hysteresis loop), as discussed in Sec. 6-15. In order to reduce the eddy currents in iron-cored ac devices, the core is commonly made of thin sheets or laminations of iron insulated electrically from each other. Thus the eddy currents are confined to individual sheets, and the power loss is reduced. Each sheet is continuous in the direction of the magnetic flux through the core, but because of its thinness it has a relatively large reluctance. By stacking a sufficient number of sheets in parallel the total reluctance of the magnetic circuit can be reduced to the desired value. To reduce eddy currents to a minimum, iron wires are sometimes used in place of sheets, while at radio frequencies powdered-iron cores are commonly employed.

8-11 DISPLACEMENT CURRENT

In this section a new concept is introduced, that of *displacement current*. Consider a voltage applied to a resistor and a capacitor in parallel, as in Fig. 8-15a. The nature of the current flow through the resistor is different from that through the capacitor. Thus, a constant voltage across a resistor produces a continuous flow of current of constant value. On the other hand, there will be a current through the capacitor only while the voltage is changing.

For a voltage V across a resistor of resistance R and capacitor of capacitance C in parallel, as in Fig. 8-15a, we have a current through the resistor given by $i_1 = V/R$ and a current through the capacitor given by

$$i_2 = \frac{dQ}{dt} = C \frac{dV}{dt} \tag{1}$$

The instantaneous charge Q in the capacitor is given by $Q = CV$.

The current through the resistor is a *conduction current*, while the current "through" the capacitor may be called a *displacement current*. Although the current does not flow through the capacitor, the external effect is as though it did, since as much current flows out of one plate as flows into the opposite one. This circuit concept may be extended to three dimensions by supposing that the resistor and capacitor elements each occupies a volume as in Fig. 8-15b. Fringing of the field is neglected. Inside each element the electric field E equals the voltage V across the

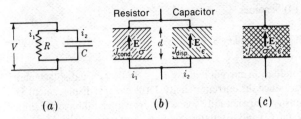

Figure 8-15 (*a*) and (*b*) Conduction current through resistor and displacement current through capacitor. (*c*) Conduction and displacement currents through conducting dielectric medium.

element divided by its length d. That is, $E = V/d$. From (4-7-2) the current density J_1 inside the resistor equals the product of the electric field E and the conductivity σ of the medium inside the resistor element; it is also equal to i_1 divided by the cross-sectional area A. Or

$$J_1 = E\sigma = \frac{i_1}{A} \tag{2}$$

The dimensional form of (2) in SI units is

$$\frac{\text{Amperes}}{\text{Meter}^2} = \frac{\text{volts}}{\text{meter}} \times \frac{\text{mhos}}{\text{meter}}$$

The capacitance of a parallel-plate capacitor is $C = \epsilon A/d$, where A is the area of the plates and d is the spacing between them. Subsituting this value for C, and $V = Ed$, into (1) yields

$$i_2 = \frac{\epsilon A d}{d}\frac{dE}{dt} = \epsilon A \frac{dE}{dt} \tag{3}$$

Dividing (3) by the area A gives the relation that the current density J_2 inside the capacitor equals the permittivity of the nonconducting medium filling the capacitor element multiplied by the time rate of change of the electric field. Thus

$$\frac{i_2}{A} = J_2 = \epsilon \frac{dE}{dt} \tag{4}$$

The dimensions of (4) in SI units are

$$\frac{\text{Amperes}}{\text{Meter}^2} = \frac{\text{farads}}{\text{meter}} \times \frac{\text{volts/meter}}{\text{second}}$$

Recalling that $D = \epsilon E$, (4) becomes

$$J_2 = \frac{dD}{dt} \tag{5}$$

In this example J_1 is a conduction current density J_{cond}, while J_2 is a displacement current density J_{disp}. Also, since the current density $\mathbf{J}$, the electric displacement $\mathbf{D}$, and the electric field intensity $\mathbf{E}$ are actually space vectors, which all have the same direction in isotropic media, (2) and (5) may be expressed in more general form as

$$\mathbf{J}_{cond} = \sigma\mathbf{E} \tag{6}$$

and

$$\mathbf{J}_{disp} = \epsilon \frac{d\mathbf{E}}{dt} = \frac{d\mathbf{D}}{dt} \tag{7}$$

As a final step suppose that instead of having two separate elements in parallel, one of which acts like a pure resistance and the other like a pure capacitance, there is only one, which has both capacitance and resistance. Thus, as in Fig. 8-15c, there is a capacitor filled with a conducting dielectric so that both conduction and displacement currents are present. Then the total current density $\mathbf{J}_{total}$ is

$$\mathbf{J}_{total} = \mathbf{J}_{cond} + \mathbf{J}_{disp} \tag{8}$$

The concept of displacement current, or displacement-current density, was introduced by James Clerk Maxwell to account for the production of magnetic fields in empty space. In empty space the conduction current is zero, and the magnetic fields are due entirely to displacement currents.

8-12 MAXWELL'S EQUATION FROM AMPERE'S LAW: COMPLETE EXPRESSION

According to Ampère's law, the line integral of $\mathbf{H}$ around a closed contour is equal to the current enclosed. Where both conduction and displacement currents are present, this current is the *total current*. Thus (5-15-1), which applies only to conduction currents, may be extended as follows when both conduction and displacement currents are present:

$$\oint \mathbf{H} \cdot d\mathbf{l} = \int_s (\mathbf{J}_{cond} + \mathbf{J}_{disp}) \cdot d\mathbf{s} \tag{1}$$

or

$$\oint \mathbf{H} \cdot d\mathbf{l} = \int_s \left(\sigma\mathbf{E} + \epsilon \frac{\partial \mathbf{E}}{\partial t} \right) \cdot d\mathbf{s} \tag{2}$$

The line integral of $\mathbf{H}$ on the left side of (2) is around the boundary of the surface s, over which the surface integral is taken on the right side of (2). Each term in (2) has the dimensions of current. The conduction current through the surface s is given by

$$\int_s \sigma \mathbf{E} \cdot d\mathbf{s}$$

while the displacement current through the surface s is given by

$$\int_s \epsilon \frac{\partial \mathbf{E}}{\partial t} \cdot d\mathbf{s}$$

Equation (2) is the complete expression in integral form of Maxwell's equation derived from Ampère's law. It is also often written

$$\oint \mathbf{H} \cdot d\mathbf{l} = \int_s \left(\mathbf{J} + \frac{\partial \mathbf{D}}{\partial t} \right) \cdot d\mathbf{s} \tag{3}$$

where $\mathbf{J}$ without a subscript is understood to refer only to conduction current density.

By an application of Stokes' theorem to (3) or by an extension of (5-18-12) to include displacement currents, the complete expression in *differential form of Maxwell's equation derived from Ampère's law* is

$$\nabla \times \mathbf{H} = \mathbf{J} + \frac{\partial \mathbf{D}}{\partial t} \tag{4}$$

or

$$\nabla \times \mathbf{H} = \sigma \mathbf{E} + \epsilon \frac{\partial \mathbf{E}}{\partial t} \tag{5}$$

It should be noted that when the electric field varies harmonically with time $(\mathbf{E} = \mathbf{E}_0 \sin \omega t)$, the conduction and displacement currents are in time-phase quadrature. That is,

$$\mathbf{J} = \sigma \mathbf{E} = \sigma \mathbf{E}_0 \sin \omega t \tag{6}$$

and

$$\frac{\partial \mathbf{D}}{\partial t} = \epsilon \frac{\partial \mathbf{E}}{\partial t} = \omega \epsilon \mathbf{E}_0 \cos \omega t \tag{7}$$

Thus, when $\omega t = 0$, the displacement current is a maximum and the conduction current is zero. On the other hand, when $\omega t = \pi/2$, the conduction current is a maximum and the displacement current is zero. Since the displacement current is a maximum one-quarter cycle $(\omega t = \pi/2)$ before the conduction current, it is said that the displacement current leads the conduction current by 90°. This is similar to the situation in a circuit having a resistor and a capacitor in parallel (Fig. 8-15a) in which the current "through" the capacitor leads the current through the resistor by 90°.

8-13 PHASORS

In general, electric and magnetic fields are a function of position (x, y, z) and, in time-varying situations, also a function of time t. Thus, one may write that an electric field

$$E = \overset{\text{Position}}{|E|} \quad \overset{\text{Time}}{\cos(\omega t + \phi)} \tag{1}$$

where $|E|$ = magnitude or maximum value of E at x, y, z

$\omega = 2\pi f$ = radian frequency, rad s^{-1}

$f = \dfrac{1}{T}$ = frequency or cycles per second, Hz

T = period or time of one cycle, s

ϕ = phase angle (to be determined), rad or deg

Quantities varying sinusoidally with time (harmonically varying) can be represented by complex quantities in the sense that only the real (or imaginary) part has physical significance. Thus, from Euler's identity the complex quantity $e^{j(\omega t + \phi)}$ can be written in terms of its real and imaginary parts as

$$e^{j(\omega t + \phi)} = \cos(\omega t + \phi) + j \sin(\omega t + \phi) \tag{2}$$

where $j = \sqrt{-1}$. Taking the real part (Re) of (2),

$$\text{Re } e^{j(\omega t + \phi)} = \cos(\omega t + \phi) \tag{3}$$

Thus, (1) can be re-expressed as

$$E = \text{Re} |E| e^{j(\omega t + \phi)} \tag{4}$$

Dropping Re and suppressing $e^{j\omega t}$ gives

$$E = |E| e^{j\phi} \tag{5}$$

where $E(= |E| e^{j\phi})$ is called a *phasor*.†

To convert a phasor (not a function of time) to an instantaneous time-varying quantity, it is understood that it is multiplied by $e^{j\omega t}$ and the real part (Re) taken. We could equally well take the imaginary part of (2) but take the real part in accord with the usual convention.

The phasor E may be a scalar or a (space) vector and, in either case, a function of position (x, y, z) with time implicit. Referring to Fig. 8-16, let E_z (a phasor) be the z component of a (space) vector **E**. Thus,

$$E_z = |E_z| e^{j\phi} \tag{6}$$

where $|E_z|$ = magnitude or maximum value of E_z at x, y, z.

---

† Note that $e^{j\phi} = \cos \phi + j \sin \phi = \sqrt{\cos^2 \phi + \sin^2 \phi} \underline{/\phi} = \underline{/\phi}$, where the first member is the exponential (complex) form, the second member the rectangular (complex) form, and the third and fourth members polar (complex) forms.

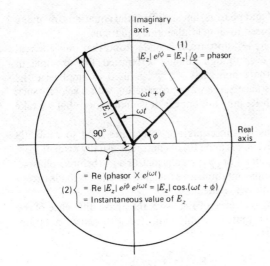

Figure 8-16 Time-phase diagram showing the relation between (1) a phasor representation of a field component E_z and (2) its instantaneous value.

Multiplying by $e^{j\omega t}$ rotates $|E_z|$ counterclockwise by ωt, and taking the real part gives the instantaneous value of E_z as the projection of $|E_z|e^{j(\omega t + \phi)}$ on the real axis. As ωt increases with time, the instantaneous value of E_z varies as a cosine function of time.

Consider the time-varying field

$$E = |E| \cos (\omega t + \phi) = \operatorname{Re} [E_{ph} e^{j\omega t}] \tag{7}$$

where $E_{ph} = |E|e^{j\phi}$ = phasor (ph) field. Now consider the partial time derivative

$$\frac{\partial E}{\partial t} = -\omega |E| \sin (\omega t + \phi) \tag{8}$$

Next consider the phasor representation of the field:

$$\frac{\partial E}{\partial t} = \frac{\partial}{\partial t} \operatorname{Re} [E_{ph} e^{j\omega t}] = \operatorname{Re} \left[\frac{\partial}{\partial t} (E_{ph} e^{j\omega t}) \right]$$

$$= \operatorname{Re} [j\omega E_{ph} e^{j\omega t}]$$

$$= j\omega E_{ph} \tag{9}$$

Similarly,

$$\frac{\partial^2 E}{\partial t^2} = -\omega^2 |E| \cos (\omega t + \phi) = \operatorname{Re} (j\omega)^2 E_{ph} e^{j\omega t}$$

$$= -\omega^2 E_{ph} \tag{10}$$

Accordingly, the time derivative of a time-varying quantity is equivalent to the corresponding phasor multiplied by $j\omega$, and the second derivative is equivalent to multiplying by $(j\omega)^2$. With $e^{j\omega t}$ and phasors, linear differential equations involving harmonic variation can be reduced to simpler algebraic equations.

In some later sections where $e^{j\omega t}$ appears, it is understood that only the real part has physical significance even though Re may be omitted. For example, in wave-propagation equations the exponential $e^{j(\omega t - \beta x)}$ is used to emphasize the relation between time t and distance x, the wave propagating with a velocity such that $\omega t = \beta x$, and even though Re may not appear, it is understood that we take the real part.

Although special notations or subscripts are sometimes used to explicitly indicate a phasor, as done above in (7) and (10), its nature is usually evident from the context and the presence of an $e^{j\phi}$ or $\underline{/\phi}$ (in a phasor), or a $j\omega$ (before a phasor). Accordingly, to avoid cumbersome notation, no explicit distinction is made except in cases where the meaning is not self-evident.

Referring to the conducting dielectric of Fig. 8-15, the phase difference of the conduction and displacement currents is readily indicated with phasor notation. Thus, (8-12-5) in phasor form is

$$\nabla \times \mathbf{H} = \sigma\mathbf{E} + j\omega\epsilon\mathbf{E} = (\sigma + j\omega\epsilon)\mathbf{E} \tag{11}$$

Here the operator j with the displacement current density indicates that it leads the conduction current density by 90°.

8-14 DIELECTRIC HYSTERESIS

In dielectrics that are good insulators the dc conduction current may be negligible. However, an appreciable ac current in phase with the applied field may be present because of *dielectric hysteresis*. This phenomenon is analogous to magnetic hysteresis in ferromagnetic materials. Materials, such as glass or plastics, which are good insulators under static conditions, may consume considerable energy in alternating fields. The heat generated in this way is sometimes applied in industrial radio-frequency heating processes.

As explained in Sec. 3-5, the electron cloud of an atom in a dielectric becomes slightly displaced or asymmetrical when an electric field is applied. This produces an electric dipole (moment Ql), and the atom is said to be polarized. When the electric field is removed, the atom returns to its normal, or unpolarized, state. If the electric field is again applied but in the opposite direction, the dipole will be reversed. Thus, when an alternating field is applied to a dielectric atom, the dipole goes through the successive stages suggested in Fig. 8-17a. An equivalent mechanical system is shown in Fig. 8-17b. The large sphere represents the large mass of the nucleus. The small sphere (attached by a spring and moving through a tunnel in the large sphere) represents the small mass of the electron cloud.

The atom constitutes an electromechanical system with mass m, damping (or friction) coefficient d, and tension (or spring) constant s. The behavior of the system

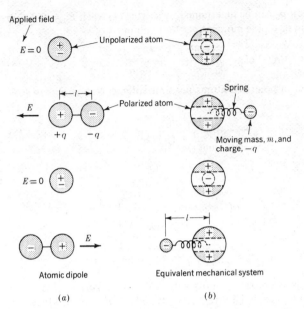

Figure 8-17 (*a*) Behavior of atomic dipole in alternating electric field and (*b*) equivalent mechanical system.

is described by†

$$m\frac{d^2l}{dt^2} + d\frac{dl}{dt} + sl = qE_0 e^{j\omega t} \tag{1}$$

where l = dipole length, or separation
q = dipole charge

The first term involves mass times acceleration, the second term damping coefficient times velocity, and the third term spring constant times displacement. The right-hand side is the driving force (qE_0 = peak force) resulting from the harmonically varying applied electric field. Equation (1) is a second-order differential equation (of standard form) for a damped harmonic oscillation with a driving function. A solution is

$$l = l_0 e^{j\omega t} \tag{2}$$

where

$$l_0 = \frac{(q/m)E_0}{\omega_0^2 - \omega^2 + (j\omega d/m)} \tag{3}$$

where ω = driving radian frequency
ω_0 = resonant or natural radian frequency = $\sqrt{s/m}$ (for $d = 0$).

† P. Debye, "Polar Molecules," chap. 5, Chemical Catalog Company, Inc., New York, 1929.

The resonant frequency ω_0 can be determined by solving (1) when $E_0 = 0$. From (3-5-1) the polarization or dipole moment per unit volume is

$$P = Nql = \frac{(Nq^2/m)E_0 \, e^{j\omega t}}{\omega_0^2 - \omega^2 + (j\omega d/m)} \tag{4}$$

where N is the number of polarized atoms per unit volume. Now $\epsilon = \epsilon_0 + P/E$. But $E = E_0 \, e^{j\omega t}$; so

$$\epsilon = \epsilon_0 + \frac{Nq^2/m}{\omega_0^2 - \omega^2 + (j\omega d/m)} \tag{5}$$

But writing

$$\epsilon = \epsilon' - j\epsilon''$$

we have

$$\epsilon' = \epsilon_0 \left[1 + \frac{(Nq^2/\epsilon_0 m)(\omega_0^2 - \omega^2)}{(\omega_0^2 - \omega^2)^2 + (\omega d/m)^2} \right] \tag{6}$$

$$\epsilon'' = \epsilon_0 \frac{(Nq^2/\epsilon_0 m)(\omega d/m)}{(\omega_0^2 - \omega^2)^2 + (\omega d/m)^2} \tag{7}$$

We observe that ϵ is a complex quantity with real and imaginary parts which are both dependent on frequency,† Putting $\mathbf{J} = \sigma \mathbf{E}$ and $\epsilon = \epsilon' - j\epsilon''$ in Maxwell's equation $\nabla \times \mathbf{H} = \mathbf{J} + j\omega \epsilon \mathbf{E}$, we obtain

$$\nabla \times \mathbf{H} = j\omega \epsilon' \mathbf{E} + (\sigma + \omega \epsilon'')\mathbf{E} \tag{8}$$

It is apparent that ϵ'' (imaginary part of ϵ) is involved in a frequency-dependent term with the dimensions of conductance. At dc ($\omega = 0$ and therefore $\omega \epsilon'' = 0$) power loss is small in a good dielectric for which σ is small. However, at high frequencies (ω large) losses can become larger as $\omega \epsilon''$ becomes significant. The sum of σ and $\omega \epsilon''$ constitutes what may be called the *equivalent conductivity* σ'. Thus,

$$\sigma' = \sigma + \omega \epsilon'' \tag{9}$$

Now (8) can be expressed

$$\nabla \times \mathbf{H} = \mathbf{J}_{\text{total}} \tag{10}$$

It follows that

$$J_{\text{total}} = \sigma' E + j\omega \epsilon' E \tag{11}$$

Referring to Fig. 8-18, we see that the total-current density J_{total} is the sum of a conduction-current density ($\sigma' E$) and a displacement-current density $\omega \epsilon' E$ in time-phase quadrature. From Fig. 8-18 we have

$$\tan \delta = \frac{\sigma'}{\omega \epsilon'} \tag{12}$$

---

† Probs. 8-3-7 and 8-3-8 illustrate how ϵ' and ϵ'' change with frequency. Over very wide frequency ranges (radio to ultraviolet) materials may exhibit several such resonances and associated permittivity changes due to various vibrational modes.

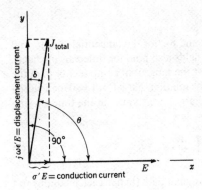

Figure 8-18 Time-phase diagram for dielectric with losses.

The quantity $\tan \delta$ is called the *loss tangent*. Also the cosine of the angle $\theta \ (= 90° - \delta)$ is the *power factor* (PF). Thus (for small δ)

$$\text{PF} = \cos \theta \simeq \tan \delta = \frac{\sigma'}{\omega \epsilon'} \tag{13}$$

The power dissipated per unit volume is

$$p = \frac{\text{power}}{\text{volume}} = \frac{\text{current}}{\text{area}} \frac{\text{voltage}}{\text{length}} = JE = \sigma' E^2 \quad (\text{W m}^{-3}) \tag{14}$$

Example Find the average power dissipated per cubic meter in a nonconducting dielectric medium with relative permittivity of 4 and a loss tangent of 0.001 if $E = 1 \text{ kV m}^{-1}$ rms and the frequency is 10 MHz.

SOLUTION Since $\sigma = 0$, $\sigma' = \omega \epsilon''$ and $\tan \delta = \sigma'/\omega \epsilon' = \epsilon''/\epsilon'$, or $\epsilon'' = \epsilon' \tan \delta$, and since $\tan \delta$ is small, $\epsilon'' \ll \epsilon'$ and $\epsilon' \approx \epsilon$; so $\sigma' = \omega \epsilon' \tan \delta \approx \omega \epsilon \tan \delta$, or

$$\sigma' \approx 2\pi \times 10^7 \times 4 \times 8.85 \times 10^{-12} \times 10^{-3} = 2.22 \ \mu\mho \text{ m}^{-1}$$

The power p dissipated per unit volume is then

$$p = E^2 \sigma' = 10^6 \times 2.22 \times 10^{-6} = 2.22 \text{ W m}^{-3}$$

It is apparent that the real part (ϵ') of the complex permittivity is associated with the displacement-current density and, hence, with the stored-energy density ($= \epsilon' E^2$). The imaginary part (ϵ'') is associated with the conduction-current density and, hence, with the power dissipated per unit volume as heat $[= (\sigma + \omega \epsilon'')E^2 = \sigma' E^2]$.

Referring to Fig. 8-18, the current density J and field intensity E both have the same space direction but different time phase. Thus, the scalar magnitude J_{total} leads the scalar magnitude E by the time phase angle θ. With increasing time both J_{total} and E rotate counterclockwise on the time-phase diagram but the phase difference θ remains fixed.

8-15 BOUNDARY RELATIONS

The boundary relations given in Tables 3-2 and 6-2 for the tangential and normal components of static electric and magnetic fields also hold for time-varying fields. This can be shown as follows. Consider first the tangential components E_t of the electric field (see Fig. 3-3). Instead of using the relation that $\oint \mathbf{E} \cdot d\mathbf{l} = 0$ for a closed path, which is true for static fields due to charges, we should, in the time-varying case, use Maxwell's equation from Faradays' law

$$\oint \mathbf{E} \cdot d\mathbf{l} = - \int_s \frac{\partial \mathbf{B}}{\partial t} \cdot d\mathbf{s} \tag{1}$$

If there is a flux density $\mathbf{B}$ normal to the rectangular path (half in each medium) and $\mathbf{B}$ changes with time, then $\oint \mathbf{E} \cdot d\mathbf{l}$ is not zero if the path encloses a finite area. However, it is assumed that the dimension Δy approaches zero so that E_{t1} and E_{t2} are separated by only an infinitesimal distance. Therefore the area of the rectangle approaches zero, and the surface integral of $\partial B/\partial t$ vanishes. Thus the work around the path is given by $E_{t1} \Delta x - E_{t2} \Delta x = 0$, as before, and it follows that $E_{t1} = E_{t2}$ holds for both static and time-varying situations.

Consider next the tangential components of the $\mathbf{H}$ field (see Fig. 6-19). Instead of using the relation $\int \mathbf{H} \cdot d\mathbf{l} = \iint \mathbf{J} \cdot d\mathbf{s}$ for steady fields, we should, in the time-varying case, use Maxwell's equation from Ampère's law in its complete form,

$$\oint \mathbf{H} \cdot d\mathbf{l} = \int_s \left(\mathbf{J} + \frac{\partial \mathbf{D}}{\partial t} \right) \cdot d\mathbf{s} \tag{2}$$

If there is a time-changing $\mathbf{D}$ normal to the rectangular path (half in each medium), there will be a contribution due to $\mathbf{D}$. However, it is assumed that the dimension Δy approaches zero, so that the surface integral of $\partial D/\partial t$ vanishes. In (2) the conduction-current density $\mathbf{J}$ may also change with time. However, its surface integral also vanishes as Δy approaches zero unless the conduction current is assumed to exist in an infinitesimally thin layer at the conductor surface. Thus, for a sheet current of linear density $\mathbf{K}$ at the surface

$$\mathbf{K} = \hat{\mathbf{n}} \times (\mathbf{H}_{t1} - \mathbf{H}_{t2}) \tag{3}$$

as before, while in the absence of such a sheet

$$H_{t1} = H_{t2} \tag{4}$$

as before. Thus, the relations for the tangential $\mathbf{H}$ field of Table 6-2 hold for both static and time-changing situations.

The formal approach in obtaining the continuity relations for the normal components of $\mathbf{D}$ and $\mathbf{P}$ is the same under time-varying conditions as for static conditions, and the relations given in Tables 3-2 and 6-2 apply under both conditions.

Table 8-1 summarizes the boundary relations developed for electric and magnetic fields. These relations apply under all situations, except as noted.

Table 8-1 Boundary relations for electric and magnetic fields

Field component	Boundary relation		Condition
Tangential	$E_{t1} = E_{t2}$	(1)	Any two media.
Tangential	$E_{t1} = 0$	(2)	Medium 2 is a perfect conductor ($\sigma_2 = \infty$).†
Tangential	$H_{t1} = H_{t2}$	(3)	Any two media.
Tangential	$H_{t1} - H_{t2} = K$	(4)‡	Current sheet at boundary.
Tangential	$H_{t1} = K$	(5)‡	Medium 2 is a perfect conductor ($\sigma_2 = \infty$) with current sheet at surface.
Tangential	$H_{t1} = 0$	(6)	Medium 2 has infinite permeability ($\mu_2 = \infty$) (no currents).
Normal	$D_{n1} - D_{n2} = \rho_s$	(7)	Any two media with charge at boundary.
Normal	$D_{n1} = D_{n2}$	(8)	Any two media with no charge at boundary.
Normal	$D_{n1} = \rho_s$	(9)	Medium 2 is a perfect conductor with charge at surface.
Normal	$B_{n1} = B_{n2}$	(10)	Any two media.

† Under static conditions it suffices for medium 2 to be a conductor (σ_2 finite). However, for E_{t2} to be zero under time-varying conditions requires that $\sigma_2 = \infty$ (see Chap. 10).

‡ Note that although **K** and the components of **H** are measured parallel to the boundary, they are normal to each other. Thus, in vector notation (5) is expressed by $\mathbf{K} = \hat{\mathbf{n}} \times (\mathbf{H}_{t1} - \mathbf{H}_{t2})$, where $\hat{\mathbf{n}}$ = unit vector normal to the boundary.

8-16 GENERAL FIELD RELATIONS

In Chap. 5 it was shown that the divergence of the curl of a vector function **F** is zero. Thus,

$$\mathbf{V} \cdot (\mathbf{V} \times \mathbf{F}) = 0 \tag{1}$$

As a corollary, any vector function with no divergence must be the curl of some other vector function. Thus, if $\mathbf{V} \cdot \mathbf{G} = 0$, we can write $\mathbf{G} = \mathbf{V} \times \mathbf{F}$, where **F** is some other vector function. As an example, $\mathbf{V} \cdot \mathbf{B} = 0$, so that **B** can be expressed as the curl of a vector potential ($\mathbf{B} = \mathbf{V} \times \mathbf{A}$).

It was also shown in Chap. 5 that the curl of the gradient of a scalar function f is zero. Thus $\mathbf{V} \times (\mathbf{V}f) = 0$. As a corollary, any vector function with no curl is the gradient of some scalar function. Thus, if $\mathbf{V} \times \mathbf{F} = 0$, we can write $\mathbf{F} = \mathbf{V}g$, where g is a scalar function. As an example, the curl of the static electric field due to electric charges is zero ($\mathbf{V} \times \mathbf{E} = 0$). It follows that a static electric field due to charges may be expressed as the gradient of a scalar function. That is, $\mathbf{E} = -\mathbf{V}V$, where V is the electric scalar potential.

According to Maxwell's equation derived from Faraday's law, we note, however, that in time-changing situations the curl of the electric field is not zero but is equal to the time rate of decrease of **B**. Thus

$$\mathbf{V} \times \mathbf{E} = -\frac{\partial \mathbf{B}}{\partial t} \tag{2}$$

Since $\nabla \times \mathbf{E}$ is not zero, the relation $\mathbf{E} = -\nabla V$ is not sufficient for time-varying fields. An additional term is required. This may be found as follows: Since $\mathbf{B} = \nabla \times \mathbf{A}$, (2) becomes

$$\nabla \times \mathbf{E} = -\frac{\partial(\nabla \times \mathbf{A})}{\partial t} \tag{3}$$

from which

$$\nabla \times \left(\mathbf{E} + \frac{\partial \mathbf{A}}{\partial t}\right) = 0 \tag{4}$$

Since the curl of the expression in parentheses in (4) equals zero, it must be equal to the gradient of a scalar function. Thus we can write

$$\mathbf{E} + \frac{\partial \mathbf{A}}{\partial t} = \nabla f \tag{5}$$

where f is a scalar function. If the electric scalar potential V is taken to be this scalar function, a relation is obtained that satisfies the requirements for both static and time-varying situations. Thus let $f = -V$, so that from (5) we have

$$\mathbf{E} = -\nabla V - \frac{\partial \mathbf{A}}{\partial t} \tag{6}$$

For static fields this reduces to $\mathbf{E} = -\nabla V$, as it should. In the general case, where the field may vary with time, $\mathbf{E}$ is given by both a scalar potential V and a vector potential $\mathbf{A}$, as in (6). If the time variation is harmonic, (6) becomes

$$\mathbf{E} = -\nabla V - j\omega \mathbf{A} \tag{7}$$

When the vector potential $\mathbf{A}$ and the scalar potential V are known, the electric and magnetic fields can be obtained under static or time-varying situations from the relations

$$\mathbf{E} = -\nabla V - \frac{\partial \mathbf{A}}{\partial t} \qquad (\text{V m}^{-1}) \tag{8}$$

and

$$\mathbf{B} = \nabla \times \mathbf{A} \qquad (\text{T}) \tag{9}$$

where

$$V = \frac{1}{4\pi\epsilon_0} \int_v \frac{\rho}{r}\, dv \qquad (\text{V}) \tag{10}$$

and

$$\mathbf{A} = \frac{\mu}{4\pi} \int_v \frac{\mathbf{J}}{r}\, dv \qquad (\text{Wb m}^{-1}) \tag{11}$$

It is assumed that the distance r in the expressions for V and $\mathbf{A}$ is small compared with a wavelength, so that propagation-time effects can be neglected. If this is not the case, the propagation time must be considered and the more general retarded form used for ρ and $\mathbf{J}$, as explained in Chap. 14.

8-17 COMPARISON OF ELECTRIC AND MAGNETIC FIELD RELATIONS

In Table 6-6 a comparison is made of electric and magnetic field equations. All of these apply in static or slowly time-varying situations. Certain of the relations can be extended so as to apply under rapidly time-varying conditions. These relations are listed in Table 8-2. It is to be noted that under static conditions the time de-

Table 8-2 Comparison of the electric and magnetic field relations for time-changing situations

Description of equation	Electric field	Magnetic field
Curl equations (point relations)	$\nabla \times \mathbf{E} = -\dfrac{\partial \mathbf{B}}{\partial t}$	$\nabla \times \mathbf{H} = \mathbf{J} + \dfrac{\partial \mathbf{D}}{\partial t}$
Closed path of integration	$\mathscr{V} = \oint \mathbf{E} \cdot d\mathbf{l} = -\int_s \dfrac{\partial \mathbf{B}}{\partial t} \cdot d\mathbf{s}$	$F = \oint \mathbf{H} \cdot d\mathbf{l} = \int_s \left(\mathbf{J} + \dfrac{\partial \mathbf{D}}{\partial t} \right) \cdot d\mathbf{s}$
Derivation of fields from scalar and vector potentials†	$\mathbf{E} = -\nabla V - \dfrac{\partial \mathbf{A}}{\partial t}$	$\mathbf{B} = \nabla \times \mathbf{A}$

† V and $\mathbf{A}$ are as indicated in (8-16-10) and (8-16-11).

rivatives are zero, and these relations reduce to the corresponding special cases given in Table 6-6. These static relations (Table 6-6) are also applicable in time-changing situations provided the variations are slow enough to permit the time derivatives to be neglected. In more rapidly time-varying situations, where the time derivatives cannot be neglected, the expressions of Table 8-2 must be employed.

PROBLEMS†

Group 8-1: Secs. 8-1 through 8-7. Faraday's law and induction

8-1-1. Induction for stationary loop A fixed, square, 6-turn wire loop has corners at (x, y) values of $(0, 0)$, $(2, 0)$, $(0, 3)$, and $(2, 3)$ m. If a magnetic field normal to the loop varies as a function of position, as given by $B = 8 \sin (\pi x/2) \cos (\pi y/3)$, find the rms emf induced in the loop if B also varies harmonically with time at 500 Hz.

∗8-1-2. Induction for moving conductor (single spoke) A conducting wheel of rim and single spoke rotates perpendicular to a uniform magnetic field B (Fig. P8-1-2). The magnetic field is confined to a radius R of the pole pieces of a magnet. An external circuit makes contact with the axle and rim through brushes. (a) If the wheel is rotated N r s^{-1}, find the emf induced in the circuit. (b) If a current I is sent through the circuit, find the torque on the wheel. (c) If the current flows as shown, is the torque clockwise (cw) or counterclockwise (ccw)?

† Answers to starred problems are given in Appendix C.

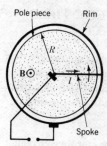

Pole piece Rim

Spoke

Figure P8-1-2 Moving spoke.

8-1-3. Induction for rotating disk If the one-spoked wheel of Prob. 8-1-2 is replaced by a solid metal disk, what difference (if any) occurs in both generator and motor cases?

8-1-4. Moving spoke and varying field If the magnetic field for the conducting rim and spoke of Prob. 8-1-2 is given by $B = B_0 \cos \omega t$, find the induced emf.

8-1-5. Swinging plumb bob over mercury pool A brass plumb bob swings in a circle of 100-mm radius over a pool of mercury, its tip just making contact with the liquid, as suggested in Fig. P8-1-5. The 4-m wire suspending the plumb bob sweeps out a cone as the bob makes one revolution in 4 s. The hook holding the swinging wire also supports a stationary vertical wire along the axis of the cone, which makes contact with the mercury in the middle of the circle, thus completing an electrical circuit. If there is a uniform horizontal magnetic field of 50 μT, as for example near the earth's equator, find the emf induced in the circuit.

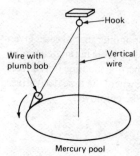

Hook

Wire with plumb bob

Vertical wire

Mercury pool

Figure P8-1-5 Plumb bob over mercury pool.

★8-1-6. Generator-motor If a rectangular loop rotating in a uniform magnetic field B is equipped with a two-segment commutator, as in Fig. P8-1-6, the arrangement can function as either a dc generator or a motor. (a) If the loop rotates $N \, \text{r s}^{-1}$, find the average dc voltage generated. (b) If a current I flows in the loop, find the average torque. (c) If the current flows as shown, find the rotation direction (cw or ccw). Assume commutator gap is negligible.

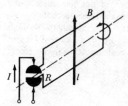

B

I R l

Figure P8-1-6 DC machine.

8-1-7. Generator-motor interchangeability Is it true that any machine which functions satisfactorily as a generator will also work as a motor when a voltage is applied which is of the same type as that generated?

8-1-8. Rotatable bar magnet A bar magnet is attached perpendicular to a vertical axle shaft which extends along the centerline of a cylindrical container of conducting fluid, as in Fig. P8-1-8. (a) If a current I is passed axially through the conducting fluid, will the device operate as a motor? (b) If the shaft with magnet is rotated, will the device operate as a generator?

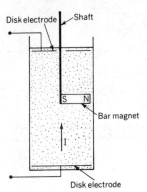

Figure P8-1-8 Magnet in conducting fluid.

8-1-9. Magnetic levitation A straight conducting bar with a weight attached is suspended by wire springs in a uniform magnetic field B as in Fig. P8-1-9. The length of the bar is 500 mm. Find the current I (magnitude and direction) required to "float," or balance, the bar and weight if $B = 3$ T and the mass of the bar and weight is 8 kg.

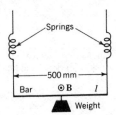

Figure P8-1-9 Floating bar.

★8-1-10. Pendulum A wire pendulum with brush swings normal to a uniform magnetic field of 250 mT, as shown in Fig. P8-1-10. The velocity of any point on the pendulum at a distance r from its support point is given by $v = \omega d(r/R) \cos \omega t$, where d is the maximum horizontal displacement, or half-amplitude. If the length R of the pendulum is 4 m, its period T at the earth's surface will be about 4 s $[T(s) = 2\pi\sqrt{R(m)/9.8(m\,s^{-2})}]$. Using this value for the period, determine the peak emf appearing at the terminals if $d = 100$ mm.

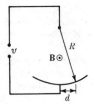

Figure P8-1-10 Pendulum.

8-1-11. Moving loop A uniform magnetic field $B = 250$ mT extends over a square area 100 mm on a side, as in Fig. P8-1-11, with zero field outside. A rectangular wire loop 40 mm by 80 mm is moved through the field with a velocity $v = 125$ mm s^{-1}. Find the emf induced in the loop and plot the results on a graph of emf vs. the distance x for $-20 \le x \le 120$ mm.

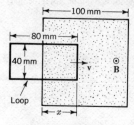

Figure P8-1-11 Moving loop.

*8-1-12. Moving wire Find the emf induced in a straight wire moving perpendicularly to a uniform magnetic field B with velocity v as in Fig. P8-1-12. The magnetic field is confined to the radius R of the pole pieces of a magnet.

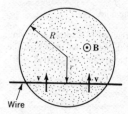

Figure P8-1-12 Moving wire.

*8-1-13. Induction in loop. Current direction Is the direction of the current I in the closed loop at the right of Fig. P8-1-13 clockwise (cw) or counterclockwise (ccw) when the switch is (a) closed and (b) opened?

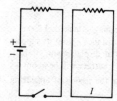

Figure P8-1-13 Stationary loops.

8-1-14. Induction in loop from rotating bar magnet A short bar magnet of 8 A m^2 magnetic moment rotates around its center point at 3 r s^{-1}. A 50-turn loop of 1-m^2 area is located 6 m from the bar magnet. Find the maximum rms emf induced in the loop by the magnet.

8-1-15. Induction in loop (a) A 5-turn loop with 0.5-m^2 area situated in air has a uniform magnetic field normal to the plane of the loop. If the flux density changes 8 mT s^{-1}, what is the emf appearing at the terminals of the loop? (b) If the emf at the loop terminals is 150 mV, what is the rate of change of the magnetic field?

8-1-16. Loop and inducing field A 15-turn wire loop of 0.5-m^2 area is situated in air in the presence of a 3-MHz radio wave. If the loop develops an emf of 10 mV rms when oriented for maximum response, find the rms magnetic field H of the wave.

8-1-17. Flip coil A coil of 50 turns and 1,000-mm^2 area has its plane normal to a magnetic field. If the charge flowing through the coil is measured to be 2 mC when the coil is flipped 180°, find B. The coil and attached measuring circuit have a resistance of 80 Ω.

8-1-18. Loop at equator of contracting balloon A conducting loop is painted around the equator of a spherical rubber balloon. A magnetic field $B = B_0 \cos \omega t$ is applied perpendicularly to the plane of the equator. If the balloon contracts inward with a constant velocity v, find the emf induced in the loop.

8-1-19. Force on loop Find the maximum force (not torque) on a loop 10 mm in diameter at a distance of 100 mm from a small bar magnet of 8 A m^2 moment if the magnet is approaching the loop at 20 mm s^{-1}.

8-1-20. Force on loop If the bar magnet in Fig. 8-1 is turned around so that the south pole is up, will the loop still be repelled on its approach and attracted on its retreat?

★8-1-21. Loop in two situations (*a*) A square loop 1 m on a side is normal to a uniform magnetic field $B_0 = 1$ mT. Find the peak emf induced in the loop if the loop is stationary but the field varies as $B_0 \cos \omega t$, where $\omega = 159$ Hz. (*b*) Find the peak emf induced in the loop if the field is fixed at B_0 but the loop rotates 159 r s^{-1}.

★8-1-22. Moving arm. Induction A moving radial arm contacts a circular wire as shown in Fig. P8-1-22. The arm velocity is 3 rad s^{-1}. B is perpendicular to plane of the figure and is everywhere uniform but varies with time as given by $B = 3t + 4$ (T), where t = time (s). If $R = 4$ m, $\theta = 120°$, and $t = 3$ s, find V at the terminals (*a*) due to motion only and (*b*) due to time change only and find (*c*) the total due to both motion and time change. Note that the sign of the voltage is important.

Figure P8-1-22 Moving arm.

8-1-23. Expanding circular loop A circular conducting rubber loop expands at a uniform velocity v. There is a uniform magnetic field perpendicular to the loop given by $B = B_0 t$, where t = time. The radius of the loop is given by $r = vt$. (*a*) Find the emf induced in the loop (in symbols). (*b*) Evaluate this result if $v = 2$ m s^{-1}, $t = 9$ s, and $B_0 = 2$ T.

8-1-24. Expanding square loop Four straight conductors form a square with a magnetic field B perpendicular to the square. If all conductors move outward with the same velocity v while contacting each other at the corners, find V (rms) induced in the square loop at the instant when its area is 4 m^2, $v = 8$ m s^{-1}, and $B = 3 \cos (2\pi ft)$, where $f = 5$ kHz.

8-1-25. Spoke and rim A conducting spoke makes a sliding contact with a stationary conducting rim of radius R, forming a closed circuit, as in Fig. P8-1-25. A uniform magnetic field perpendicular to the

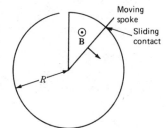

Figure P8-1-25 Moving spoke.

loop is given by $B = B_0\sqrt{t} \cos \omega t$. The spoke sweeps 360° in 2 s. Find the emf induced in the loop at time $t = \frac{1}{2}$ s if $B_0 = 4\,\mathrm{T}$ and $R = 0.4\,\mathrm{m}$. The spoke is at the "12 o'clock," or up, position at $t = 0$.

★8-1-26. Pie-shaped loop A pie-shaped closed loop, as in Fig. P8-1-26, rotates in the plane of the page with an angular velocity $v = \omega r$. A uniform magnetic field perpendicular to the loop is given by $B = B_0 \cos 4\omega t$. (a) Find the emf induced in the loop in symbol form and (b) evaluate numerically if $R = 2\,\mathrm{m}$, $B_0 = 2\,\mathrm{T}$, and $f = 100\,\mathrm{Hz}$. Disregard the brush at P_2.

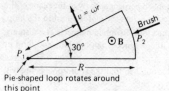

Pie-shaped loop rotates around this point

Figure P8-1-26 Rotating pie-shaped loop.

8-1-27. Pie-shaped loop with brush If the pie-shaped loop of Prob. 8-1-26 has a brush in contact with its edge as in Fig. P8-1-26, (a) find the emf induced between points P_1 and P_2 in symbol form and (b) evaluate numerically the emf using data of Prob. 8-1-26.

8-1-28. Parallel wires with sliders Two parallel wires with separation l have two sliding conductors bridging the wires as in Fig. P8-1-28. The left-hand slider moves to the left with a velocity v while the right-hand slider moves to the right with a velocity $v/2$. A uniform magnetic field perpendicular to the plane of the wires is given by $B = B_0 t^2 \cos 2\omega t$. (a) Find the emf induced in the closed circuit in symbol form and (b) evaluate numerically if $B_0 = 3\,\mathrm{T}$, $l = 500\,\mathrm{mm}$, $t = 2\,\mathrm{s}$, and $v = 0.5\,\mathrm{m\,s^{-1}}$. Both sliders are together at $t = 0$.

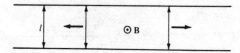

Figure P8-1-28 Parallel wires with sliders.

Group 8-2: Secs. 8-8 through 8-13. Self-inductance, mutual inductance, ac behavior of ferrous materials, displacement current, Maxwell's equation, and phasors

★8-2-1. Mutual inductance of wire and loop Find the mutual inductance between a long straight wire and a rectangular wire loop as in Fig. P8-2-1.

Long wire

Figure P8-2-1 Wire and loop.

8-2-2. Mutual inductance of concentric loops Find the mutual inductance between two concentric circular wire loops of radius r_1 and r_2 (a) when the radius of the inner loop is much smaller than the outer loop and (b) without this restriction.

★**8-2-3. Mutual inductance of parallel wires** (a) Find the mutual inductance per unit length for two long straight parallel wires with separation s. (b) If the current in one wire is I, find the emf induced per unit length in the other wire if it is at a separation s and approaching at a velocity v.

★**8-2-4. Coil and long wire. Mutual inductance and induction** (a) A long straight wire of radius a carries a current $I = I_0 \sin \omega t$. A soft-iron sleeve of permeability μ, length l, inner radius b, and outer radius c surrounds the wire. The sleeve has a winding of N turns as shown in Fig. P8-2-4. Find the mutual inductance between the wire and the winding. (b) Find the open circuit voltage induced in the winding.

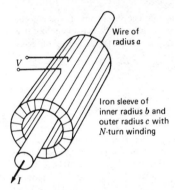

Wire of radius a

V

Iron sleeve of inner radius b and outer radius c with N-turn winding

I

Figure P8-2-4 Coil around wire.

8-2-5. Toroid. Inductance A toroid of 400 turns has a mean radius of 200 mm and a radius for the winding of 15 mm. Find the average inductance for (a) an air core and (b) an iron core with average relative incremental permeability of 700.

8-2-6. Displacement-current density Find the displacement-current density of a magnetic field in air given by (a) $H_y = H_0 \cos (\omega t - \beta x)$ and (b) $\mathbf{H} = \hat{x} H_x \cos 2x \cos (\omega t - \beta y) + \hat{z} H_z \cos 2x \sin (\omega t - \beta y)$.

8-2-7. Capacitor. Conduction and displacement currents A parallel-plate air capacitor has a 1,000-Ω resistor connected between the centers of the plates. The plates are 100 mm square and separated by 1 mm. If 10 V rms is applied to the capacitor, find (a) rms displacement current through the capacitor, (b) rms conduction current, (c) total current, and (d) power dissipated in the resistor at each of the following frequencies: (1) 1 Hz, (2) 1 kHz, (3) 1 MHz, and (4) 1 GHz. Make a table with four columns for the quantities (a), (b), (c), and (d) and four rows for the four different frequencies. Neglect fringing. Discuss the significance of the results.

Group 8-3: Secs. 8-14 through 8-17. Dielectric hysteresis, boundary relations, and general field relations

★**8-3-1. Capacitor. Dielectric heating** A parallel-plate capacitor has a plate area of $0.2 \, m^2$ and a plate separation of 6 mm. If 50 V (rms) is applied at 100 MHz, find the power lost as heat in the dielectric between the plates if its constants are $\sigma = 5 \times 10^{-3} \, \mho \, m^{-1}$, $\epsilon_r' = 20$, $\epsilon_r'' = 2$, and $\mu_r = 1$. Neglect fringing.

★**8-3-2. Capacitor. Dielectric heating** A parallel-plate capacitor with 10-mm plate separation is filled with a medium for which $\sigma = 10^{-3} \, \mho \, m^{-1}$, $\mu_r = 1$, and $\epsilon_r = 20 - j0.2$. Find the heat loss in the capacitor if 100 V (rms) is applied at 300 MHz. The volume of the capacitor is 200 ml. Neglect fringing.

8-3-3. Capacitor. Dielectric heating A capacitor consists of two parallel square plates 500 mm on a side and separated by 2 mm. If the medium filling the capacitor has constants $\sigma = 0$, $\mu_r = 1$, and $\epsilon_r = 15 - j2$, find the power lost as heat in the capacitor at 1 GHz if 100 V (rms) is applied to the plates. Neglect fringing.

★8-3-4. Capacitor current and heating A parallel-plate capacitor 1 m square with 100-mm plate separation is filled with a medium for which $\sigma = 0.005$ $\mho m^{-1}$, $\epsilon_r = 20 - j10$, and $\mu_r = 1$. If 100 V (rms) at 10 MHz is applied, find (a) total current and (b) power lost as heat. Assume a uniform field in the capacitor. Neglect fringing.

8-3-5. Capacitor. Dielectric heating Given that the average power dissipated per unit volume of a medium is $E(dD/dt) \cos \theta$ W m^{-3}, where θ is the time-phase angle between E and D, show that the total average power dissipated in a parallel-plate capacitor is $VI \cos \theta$ (W), where V = voltage across capacitor and I = current through capacitor. Neglect fringing. E, D, V, and I are rms values.

★8-3-6. Capacitor. Power factor A parallel-plate capacitor is filled with a dielectric of 0.003 power factor and $\epsilon_r = 10$. The plates are 250 mm square, and the distance between is 10 mm. If 500 V rms at 2 MHz is applied to the capacitor, find the power dissipated as heat.

8-3-7. Ice. Permittivity as a function of temperature and frequency (a) The permittivity of ice is a function of both temperature and frequency. In E. R. Pounder, "Physics of Ice," p. 129, Pergamon Press, New York, 1965, the complex relative permittivity is given by $\epsilon_r = \epsilon_r' - j\epsilon_r''$ and the Drude-Debye relations by

$$\epsilon_r' = \frac{\epsilon_1 + \epsilon_2 \alpha^2 f^2}{1 + \alpha^2 f^2} \quad \text{and} \quad \epsilon_r'' = \frac{(\epsilon_1 - \epsilon_2)\alpha f}{1 + \alpha^2 f^2}$$

where T = temperature, °C
$\quad\quad f$ = frequency, Hz

and where for pure water ice $\epsilon_1 = 75$, $\epsilon_2 = 3$, and $\alpha = 1.2 \times 10^{-4} e^{-0.1T}$ (s). The relations are applicable from 0 to 10 GHz and from 0 to $-70°C$. The effect of impurities is not included, but they would probably increase the magnitude of the imaginary part (ϵ_r''). Calculate the complex relative permittivity of pure water ice at 0°C at 0, 1, 10, and 100 Hz; 1, 10, and 100 kHz; and 1, 10, and 100 MHz. (b) Calculate the equivalent conductivity σ' as a function of frequency (0 to 100 MHz). Take $\sigma = 0$. (c) Make a log-log graph of ϵ_r', ϵ_r'', and σ' as a function of frequency; also a linear graph of ϵ_r', ϵ_r'', and σ' versus log frequency.

8-3-8. ϵ_r vs. frequency. Setting the constants $Nq^2/\epsilon_0 m$ and d/m equal to unity, calculate the variation of ϵ_r' and ϵ_r'' from (8-14-6) and (8-14-7) as a function of ω for $\omega_0 = 1$. Display results graphically in both log-log and linear-log form, as in Prob. 8-3-7.

Group 8-4: Practical applications

★8-4-1. Earth inductor (a) How many turns are required for a circular loop of 100-mm radius to develop a peak emf of 10 mV if the loop rotates 30 r s^{-1} in the earth's magnetic field? Take $B = 60$ μT. (b) How must the axis of rotation be oriented to reduce the voltage to zero, that is, to null the observation? This arrangement consitutues an *earth* inductor, useful for measuring the magnitude and direction of the earth's field.

8-4-2. Faraday disk generator (a) A thin copper disk 200 mm in diameter is situated with its plane normal to a constant, uniform magnetic field $B = 600$ mT. If the disk rotates 30 r s^{-1}, find the emf developed at the terminals connected to brushes as shown in Fig. P8-4-2. One brush contacts the periphery

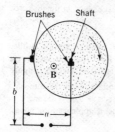

Figure P8-4-2 Faraday disk generator.

of the disk and the other contacts the axle or shaft. This arrangement is called a *Faraday disk generator*. (b) If the magnetic field varies with time, as given by $B = B_0 \sin \omega t$, where $B_0 = 600$ mT and $\omega = 2\pi \times 5$ rad s^{-1}, find the emf developed at the terminals.

★8-4-3. Power-line spillover. A typical 765-kV, 60-Hz 3-phase power transmission line has 3 conductors at equal heights spaced 20 m apart, as in Fig. P8-4-3, with 4,000 A per phase. The power-line easement extends 20 m to either side. If a 1-turn loop of number 6-gauge aluminum wire with dimensions 20 m by 100 m is situated outside the power-line easement area as shown, how much power will be induced from the line? Assume that the loop is in the same plane as the power-line conductors. (Number 6 gauge is 4.1 mm in diameter.)

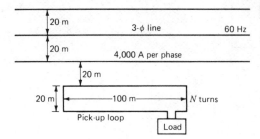

Figure P8-4-3 Power line and induction loop.

8-4-4. Power-line spillover. Design economics This is an extension of Prob. 8-4-3 to determine the optimum number of turns of the loop, the best metal for the loop, and whether the arrangement is cost effective. Assume that copper wire costs $3.30/kg, aluminum $2.65/kg, and steel $0.88/kg. Also that electric energy is 5¢/kWh. Assuming a total investment of $2,000 ($1,000 for loop wire and $1,000 for installation with load), find: (a) best metal for loop and (b) optimum number of turns; and for these optimum conditions find (c) load power and (d) time to pay off investment. (e) If the investment were increased, would it take longer to recover the cost? Note that this is a hypothetical problem which, like many engineering problems, also has legal ramifications. Although the power is being induced on property outside the power-line easement, the power company might claim that this spillover energy belongs to it. On the other hand, the owner of the property adjacent to the power-line easement could claim that the spillover energy is an unauthorized, uncompensated intrusion into private property.

★8-4-5. Lightning conductor impedance (a) A typical thunderstorm cloud may build up a negative charge of 100 C, inducing an equal positive charge on the ground..If the charges are neutralized by a lightning stroke of 2-ms duration, what is the average current of the stroke? (b) Typically the lightning stroke has a fast rise with a gradual decay. If the rise time is 2 μs, what is the maximum impedance presented by a 50-m vertical grounding conductor of 0.1-Ω resistance? The wire diameter is 4 mm. Take the peak current as twice the value calculated in (a).

8-4-6. Rowland disk. Convection current A plastic disk of 200-mm radius carries a uniform total electric charge Q and rocks back and forth through an arc of 90° at the rate of 10 times per second. The disk shaft is of soft iron and is mounted in a soft-iron yoke having a 50,000-turn coil as indicated in Fig. 8-4-6. The yoke has an average area of 1 cm^2 and a length of 500 mm. Find the rms voltage induced in the coil. Assume that the disk is charged to 1 percent of the breakdown field for air. *Note*: An experiment with a rotating charged disk was performed by Henry Rowland in 1876 to demonstrate for the first time that a moving electric charge, or *convection current*, was capable of generating a magnetic field in the same way as a *conduction current* on a wire. Rowland used a continuously rotating disk and may have detected its magnetic field with a compass-type magnetometer. The reciprocating disk of the above problem generates an ac output which can be amplified to facilitate detection of the effect.

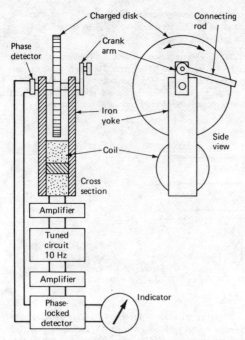

Figure P8-4-6 Rowland disk.

★8-4-7. Faraday disk generator in a superconducting solenoid A Faraday disk generator with a superconducting solenoid is shown in cross section in Fig. P8-4-7. The disk has a radius of 100 mm, and it rotates at 3,600 r min^{-1}. If $B = 3$ T, find (a) the required solenoid sheet-current density and (b) the emf generated. (See *Machine Design*, Nov. 11, 1971, p. 18.)

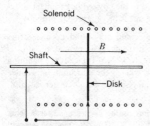

Figure P8-4-7 Faraday disk in superconducting solenoid.

8-4-8. Baked Idaho potato. Dielectric hysteresis A homogeneous 200-cm^3, 200-g Idaho potato has a relative permittivity $\epsilon_r = 65 - j15$. If $E(\text{rms}) = 30$ kV m^{-1} at 2.45 GHz (alongside potato), how long will it take to bake the potato? Take $\sigma = 0$. [See N. E. Bengtsson and P. O. Risman, Dielectric Properties of Foods at 3 GHz, *J. Microwave Power*, **6**(2):107–123 (1972).]

★8-4-9. RF loop A copper conductor of diameter $d = 10$ mm is formed into a circular loop of radius $R = 215$ mm and connected to a parallel-plate capacitor, as in Fig. P8-4-9. The capacitor plates are square (side length $l = 100$ mm) and spaced a distance $s = 5$ mm apart. The medium is air. If the loop is placed in the field of a 40-MHz radio wave with $H = 4$ mA m^{-1}, and if the equivalent (series)

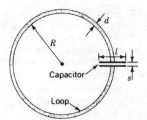

Figure P8-4-9 Radio loop.

resistance of the loop-capacitor circuit is $3\,\Omega$, find (a) the rms displacement current density in the capacitor and (b) the rms voltage across the capacitor. (c) Repeat (a) and (b) if the frequency of the wave is adjusted to the resonant frequency of the circuit. Assume H is unchanged. *Note*: The dc resistance of the loop is only about $0.0002\,\Omega$, but its radiation resistance (see Chap. 14) is about $3\,\Omega$.

8-4-10. Linear-motor vehicle design (a) Design a linear dc motor for propelling and braking a track-guided vehicle. The wheels are not to be used for traction or braking except in emergency. The vehicle contains no power source other than that for magnetic field generation. and no power is supplied to the vehicle from any external source. Currents in the track region are controlled from the vehicle by radio command. Figure P8-4-10 shows a suggested scheme. (b) What current and magnetic field are required to accelerate a 50×10^3 kg vehicle from 0 to 100 km h^{-1} in 60 s? The track width is 1.5 m. Neglect friction. [For other arrangements see E. R. Laithwaite, Linear Motion Electrical Machines, *Proc. IEEE*, **58**:531–542 (April 1970).]

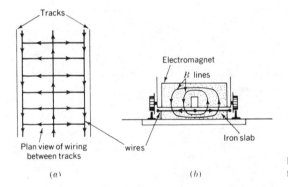

Plan view of wiring between tracks

(a)   (b)

Figure P8-4-10 Linear motor for track-guided vehicle.

★**8-4-11. Water generator.** (a) A rectangular nonconducting trough 3 m wide carries a uniform flow of brackish water 250 mm deep with a velocity $v = 3$ m s^{-1}. The water has a conductivity $\sigma = 1\,\mho$ m^{-1}. If conducting plates 500 mm wide by 250 mm deep are placed at the sides of the trough and a magnetic field $B = 1$ T is impressed, as shown in Fig. P8-4-11, find the current I flowing in a load resistor

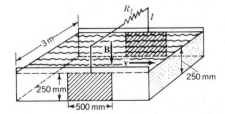

Figure P8-4-11 Water generator.

$R_l = 10\,\Omega$. Neglect resistance of plates and connecting wires. (*b*) Find the power delivered to the load. (*c*) Is this the maximum power which can be delivered to a load if R_l is variable? (*d*) Discuss the practical aspects of using a similar system to generate electric power directly from the flow of water in a large river using the earth's magnetic field. (*e*) Could the system be used to measure the velocity of flow of a river? If so, explain how such variables as depth and conductivity could be accommodated. (*f*) Can the arrangement of Fig. P8-4-11 be modified so that it can be used to pump a conducting liquid?

8-4-12. Metal detector design Metal detectors operate on several principles. In one design a coil and capacitor form the resonant circuit of an oscillator at, say, 100 kHz. The introduction of a metal object into the field of the coil changes, or detunes, the oscillator frequency, and this change is noted. Another design operates on the null principle. Thus, two flat exciter coils are placed side by side in the same plane. Both are connected to the oscillator ($\sim$100 kHz modulated at $\sim$500 Hz), but the coils are so connected that the field between them is zero (balanced out). A third detector, or search coil, is situated at this point. Its rectified output is connected to an audio amplifier and transducer. When a metal object is introduced into the field, balance is upset and a voltage is induced in the search coil, producing an audible 500-Hz tone. (*a*) Design a device for detecting automobiles for traffic-light control at a highway intersection. (*b*) Design a device for locating buried pipes or coins. (*c*) Discuss the pros and cons of frequency-shift vs. null-principle designs for the above applications. (*d*) Can these detectors be used to locate dielectric bodies or anomalies? [See L. C. Chan, D. L. Moffatt, and L. Peters, Jr., "Subsurface Radar Target Imaging Effects," *IEEE Trans. on Ant. and Prop.*, **AP-29**:413–418 (March 1981).]

THE RELATION BETWEEN FIELD AND CIRCUIT THEORY; MAXWELL'S EQUATIONS

9-1 INTRODUCTION

In circuit theory we deal with circuit elements, the voltage V across them, and the total current I through them. In field theory we deal with the field vectors ($\mathbf{E}$, $\mathbf{D}$, $\mathbf{B}$, $\mathbf{H}$, and $\mathbf{J}$) and their values as a function of position.

Consider, for instance, a short rod of length l and cross-sectional area A in Fig. 9-1. In low-frequency-circuit theory it is convenient to describe the rod in terms of one quantity, its resistance R. Its length, area, and shape are of secondary importance. Thus the voltage difference between the ends of the rod is, from Ohm's law,

$$V = IR \tag{1}$$

where I is the current through the rod.

From the field-theory point of view we consider the value of the electric field $\mathbf{E}$ at a point in the rod. From Ohm's law at a point (see Sec. 4-7)

$$\mathbf{E} = \frac{\mathbf{J}}{\sigma} \quad (\mathrm{V\ m^{-1}}) \tag{2}$$

where $\mathbf{J}$ = conduction-current density, $\mathrm{A\ m^{-2}}$
  σ = conductivity, $\mho\ \mathrm{m^{-1}}$

Now, integrating (2) over the length of the rod, we obtain the voltage difference V between the ends. That is,

$$V = \int \mathbf{E} \cdot d\mathbf{l} = \int \frac{\mathbf{J}}{\sigma} \cdot d\mathbf{l} \tag{3}$$

Figure 9-1 Conducting rod.

For a uniform rod with uniform current density this becomes

$$V = \frac{Jl}{\sigma} = JA \frac{l}{\sigma A} \quad \text{(V)} \tag{4}$$

where $JA = I$ = current through rod, A

$l/\sigma A = R$ = resistance of rod, Ω

A = cross-sectional area of rod, m^2

Thus, from (4) we have

$$V = IR \tag{5}$$

Starting with field theory, we have arrived at the circuit relation known as *Ohm's law*.

Historically, this and other circuit relations were postulated and verified first. Then, as a generalization, they were extended to apply to the more general field situation. It follows, therefore, that circuit relations are simply special cases of field equations and may be deduced from them. Although field relations are more general, it is usually much simpler to use circuit equations wherever they are applicable.

Equation (1) is a pure circuit relation. On the other hand, (2) is a pure field relation. Many equations are not purely one or the other but are a combination or mixture. Such mixed relations are necessary, for example, in order to provide a connection between field and circuit theory. Two important equations that provide such connecting links are

$$V = \int \mathbf{E} \cdot d\mathbf{l} \quad \text{(V)} \tag{6}$$

and

$$I = \oint \mathbf{H} \cdot d\mathbf{l} \quad \text{(A)} \tag{7}$$

Equation (6) relates V (a circuit quantity) between two points to the line integral of $\mathbf{E}$ (a field quantity) between those points. Likewise (7), which is Ampère's law, relates I (a circuit quantity) to the line integral of $\mathbf{H}$ (a field quantity) around a closed path.

9-2 APPLICATIONS OF CIRCUIT AND FIELD THEORY

While field relations are applicable in general, circuit relations are usually more convenient wherever V and I have a simple, well-defined significance.

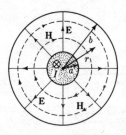

Figure 9-2 Coaxial transmission line with Transverse Electro-Magnetic (TEM) mode. E lines are radial and solid. H lines are circles and dashed.

Thus, in determining the capacitance of a capacitor of irregular shape with the aid of a graphical field map (see Fig. 3-19) we are in effect directing our attention to the field and its value as a function of position in the capacitor. However, once we have determined the capacitance, we may at low frequencies consider it as simply a two-terminal circuit element of capacitance C with a voltage difference V. The physical size and shape of the capacitor and the field configuration within it are then relegated to positions of secondary importance.

As another illustration let us consider the coaxial transmission line shown in cross section in Fig. 9-2 under two conditions, one where V and I are useful quantities and one where they are not. The coaxial line has an inner conductor of radius a and an outer conductor of inside radius b. With a steady potential difference between the conductors the electric field lines are radial, as shown. If a current I is flowing, the magnetic field lines **H** are circles, as indicated. Now by (9-1-6) the potential difference between the inner and outer conductors is

$$V = \int_a^b \mathbf{E} \cdot d\mathbf{r} \quad (V) \tag{1}$$

Likewise from (9-1-7) the current I in the inner conductor is

$$I = \oint \mathbf{H} \cdot d\mathbf{l} = \int_0^{2\pi} Hr\, d\theta \quad (A) \tag{2}$$

In (1) V is independent of the path between the conductors, while in (2) the value of I, obtained by integrating **H**, is independent of the radius r provided it is between a and b. Hence, V and I have a simple, definite significance in this case and are useful quantities.

The field configuration shown in Fig. 9-2 is called a *Transverse Electro Magnetic* (TEM) *field* because the electric and magnetic fields are entirely transverse (no component in the axial direction). This type of field is the only configuration or field mode possible under steady conditions and for time-varying situations where the wavelength is of the order of $4b$ or greater.† At higher frequencies (shorter

---

† That is, the frequency is so high that a disturbance traveling with the velocity of light can travel only about a distance equal to the diameter $(2b)$ in one-half period. In free space a wave has a wavelength λ in meters that is related to the frequency f in hertz and the velocity of light c as follows: $\lambda = c/f$, where $c = 300$ Mm s^{-1}. For a further discussion of field modes in lines and guides see Chaps. 10 and 13.

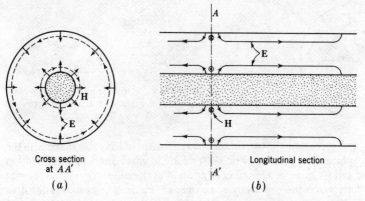

Figure 9-3 Coaxial transmission line with higher-order Transverse Magnetic (TM) mode.

wavelengths) more complex field configurations known as higher-order modes become possible. These modes are characterized by having some field components in the axial direction. Although coaxial lines are seldom used under such conditions, suppose that the frequency is sufficiently high for the mode or configuration shown in Fig. 9-3 to exist. Both a cross (or transverse) section and a longitudinal (or axial) section are needed to show the field configuration. This field is called a *Transverse Magnetic* (TM) *mode* because the magnetic field is entirely transverse, while the electric field has a longitudinal component. For this mode, the voltage V between the conductors as obtained by (1) may become negligible, while the current I obtained by (2) depends on the radius r at which **H** is integrated. Hence V and I no longer have a simple significance and are not as useful as the field quantities themselves. The breakdown of the circuit concept occurs here when the transverse dimensions become comparable with the wavelength.

9-3 THE SERIES CIRCUIT; COMPARISON OF FIELD AND CIRCUIT THEORY

From (4-10-1) we may express the total electric field $\mathbf{E}_{\text{total}}$ as the sum of a field $\mathbf{E}_e$ related to emfs and a field **E** induced by charges and currents. Thus,

$$\mathbf{E}_{\text{total}} = \mathbf{E}_e + \mathbf{E} \qquad \text{or} \qquad \mathbf{E}_e = \mathbf{E}_{\text{total}} - \mathbf{E} \tag{1}$$

From (4-7-2)

$$\mathbf{E}_{\text{total}} = \frac{\mathbf{J}}{\sigma} \tag{2}$$

and from (8-16-8)

$$\mathbf{E} = -\boldsymbol{\nabla}V - \frac{\partial \mathbf{A}}{\partial t} \tag{3}$$

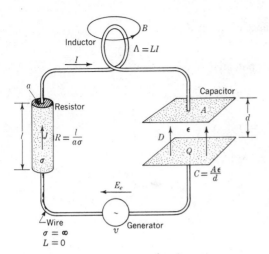

Figure 9-4 Series circuit with resistance, inductance, capacitance, and ac generator used to illustrate equivalence of field and circuit theory.

Consider now a series circuit of resistance, inductance, and capacitance connected to a generator as shown in Fig. 9-4. It is assumed that all resistance in the circuit is confined to the resistor, all inductance to the inductor, all capacitance to the capacitor, and all emf to the generator. Let us apply the above field relations to this circuit. First substitute (2) and (3) in (1). Then to convert field into circuit quantities we integrate all terms completely around the circuit in a clockwise direction. This gives

$$\oint \mathbf{E}_e \cdot d\mathbf{l} = \oint \frac{\mathbf{J}}{\sigma} \cdot d\mathbf{l} + \oint \nabla V \cdot d\mathbf{l} + \oint \frac{\partial \mathbf{A}}{\partial t} \cdot d\mathbf{l} \tag{4}$$

Integrating the left side yields the emf $\mathscr{V}$ of the generator. Also integrating the first two terms on the right and noting that $\nabla V = -\mathbf{E}$, we have

$$\mathscr{V} = \frac{Jl}{\sigma} + Ed + \frac{d}{dt} \oint \mathbf{A} \cdot d\mathbf{l} \tag{5}$$

The last term may be reexpressed as†

$$\frac{d}{dt} \oint \mathbf{A} \cdot d\mathbf{l} = \frac{dI}{dt} \oint \frac{\mathbf{A}}{I} \cdot d\mathbf{l} = L \frac{dI}{dt} \tag{6}$$

---

† The transformation of (6) may be also made with the aid of Stokes' theorem, recalling that $\mathbf{B} = \nabla \times \mathbf{A}$ and $\Lambda = LI$, as follows:

$$\frac{d}{dt} \oint \mathbf{A} \cdot d\mathbf{l} = \frac{d}{dt} \int_s (\nabla \times \mathbf{A}) \cdot d\mathbf{s} = \frac{d}{dt} \int_s \mathbf{B} \cdot d\mathbf{s} = \frac{d\Lambda}{dt} = L \frac{dI}{dt}$$

where L = inductance of loop
I = current through loop

where $L = \oint \dfrac{\mathbf{A}}{I} \cdot d\mathbf{l} = $ inductance of circuit

$\mathbf{A} = \dfrac{\mu_0}{4\pi} \displaystyle\int_v \dfrac{\mathbf{J}}{r}\, dv = $ vector potential

Noting also that $J = I/a$ and $E = D/\epsilon$, (5) becomes

$$\mathscr{V} = I\,\frac{l}{a\sigma} + \frac{Dd}{\epsilon} + L\,\frac{dI}{dt} \tag{7}$$

Now $l/a\sigma = R$ and $D = Q/A$; so we have

$$\mathscr{V} = IR + \frac{Q}{A\epsilon/d} + L\,\frac{dI}{dt} \tag{8}$$

But $A\epsilon/d = C$ and the charge $Q = \int I\, dt$; so (8) takes the form

$$\mathscr{V} = IR + \frac{1}{C}\int I\, dt + L\,\frac{dI}{dt} \tag{9}$$

Thus, beginning with the field relations (1), (2), and (3) as applied to a series circuit, we have arrived at the familiar circuit relation (9) for a series circuit. For harmonic variation of the current the phasor form of (9) is

$$\mathscr{V} = IR + \frac{I}{j\omega C} + j\omega LI \tag{10}$$

or
$$\mathscr{V} = IR + jI\left(\omega L - \frac{1}{\omega C}\right) \tag{11}$$

In deriving (11) the assumption was made that at any instant the current is the same at all parts of the circuit. This implies that a disturbance is propagated around the circuit instantaneously. If the circuit length is small compared with the wavelength, this is a satisfactory assumption. However, if the circuit length is appreciable compared with the wavelength (say at least $\lambda/8$), the variation in current and phase around the circuit may become significant. Under these circumstances the simple circuit concepts tend to become inadequate and inaccurate. It is also to be noted that the above circuit treatment ignores the phenomenon of radiation, which is so important at high frequencies (see Chap. 14).

There are certain exceptions to the above statement that circuit concepts become inadequate when the circuit length is comparable with the wavelength. For example, circuit concepts are successfully applied to the long transmission line. In this case, the distributed inductance and capacitance are represented by suitable lumped elements (see Chap. 10). Although the length of the line can be many wavelengths, it is significant that even in this extension of circuit theory the treatment is adequate only for lines with transverse dimensions that are very small compared with the wavelength.

9-4 MAXWELL'S EQUATIONS AS GENERALIZATIONS OF CIRCUIT EQUATIONS†

In the remainder of this chapter a number of relations developed in the preceding chapters are brought together and considered as a group. These relations, known as *Maxwell's equations*, consist of four expressions: one derived from Ampère's law, one derived from Faraday's law, and two derived from Gauss' law. These equations are of profound importance and, together with boundary, continuity, and other auxiliary relations, form the basic tools for the analysis of most electromagnetic problems.

In Chap. 5 Ampère's law relating the line integral of **H** around a closed path to the current I enclosed was given as

$$\oint \mathbf{H} \cdot d\mathbf{l} = I \tag{1}$$

Replacing the current I by the surface integral of the conduction-current density **J** over an area bounded by the path of integration of **H**, we have the more general relation

$$\oint \mathbf{H} \cdot d\mathbf{l} = \int_s \mathbf{J} \cdot d\mathbf{s} \tag{2}$$

In Chap. 8 this relation was made even more general by adding a displacement-current density to the conduction-current density so that (2) becomes

$$\oint \mathbf{H} \cdot d\mathbf{l} = \int_s \left(\mathbf{J} + \frac{\partial \mathbf{D}}{dt} \right) \cdot d\mathbf{s} \tag{3}$$

This relation is called *Maxwell's equation as derived from Ampères' law*. In (3) it is given in its integral form, the line integral of **H** being taken over a closed path bounding the surface s. In circuit parlance a closed path or loop is often called a *mesh*. Hence, (3) is a *mesh relation*. Applying Stokes' theorem to (3), we obtain the corresponding *point relation*

$$\nabla \times \mathbf{H} = \mathbf{J} + \frac{\partial \mathbf{D}}{\partial t} \tag{4}$$

Equation (4) is a differential relation and relates the field quantities at a point. It is the differential form of Maxwell's equation as derived from Ampère's law.

In Chap. 8 Faraday's law relating the emf $\mathscr{V}$ induced in a circuit to the time rate of decrease of the total magnetic flux linking the circuit was given as

$$\mathscr{V} = -\frac{d\Lambda}{dt} \tag{5}$$

† James Clerk Maxwell, "A Treatise on Electricity and Magnetism," 2 vols., Oxford University Press, London, 1873; 3d ed., 1904.

Replacing the flux linkage Λ by the surface integral of $\mathbf{B}$ over the area bounded by the circuit, we have the more general equation

$$\mathscr{V} = -\frac{d}{dt}\int_s \mathbf{B} \cdot d\mathbf{s} \tag{6}$$

Replacing $\mathscr{V}$ in (6) by the line integral of $\mathbf{E}$ around the circuit, we have the still more general relation (for stationary circuits) that

$$\oint \mathbf{E} \cdot d\mathbf{l} = -\int_s \frac{\partial \mathbf{B}}{\partial t} \cdot d\mathbf{s} \tag{7}$$

This field relation is a generalization of Faraday's circuit law (5). Equation (7) is called *Maxwell's equation as derived from Faraday's law*. It is given in (7) in integral form; i.e., it is a mesh equation. The corresponding point relation can be obtained from (7) by an application of Stokes' theorem, yielding

$$\nabla \times \mathbf{E} = -\frac{\partial \mathbf{B}}{\partial t} \tag{8}$$

Equation (8) is a differential relation and relates the field quantities at a point. It is the differential form of Maxwell's equation as derived from Faraday's law.

In Chap. 2 Gauss' law relating the surface integral of the electric flux density $\mathbf{D}$ to the charge Q enclosed was given as

$$\oint_s \mathbf{D} \cdot d\mathbf{s} = Q \tag{9}$$

Replacing Q in (9) by the volume integral of the charge density ρ throughout the volume enclosed by the surface s, we can write (9) in a more general form as

$$\oint_s \mathbf{D} \cdot d\mathbf{s} = \int \rho \, dv \tag{10}$$

This field relation is a generalization of Gauss' law and is called *Maxwell's electric field equation as derived from Gauss' law*. In (10) it appears in integral form and applies to a finite volume v. Applying (10) to an infinitesimal volume, we can obtain the corresponding differential relation that relates the field quantities at a point as given by

$$\nabla \cdot \mathbf{D} = \rho \tag{11}$$

Equation (11) is Maxwell's electric field equation as derived from Gauss' law in differential form.

For magnetic fields the surface integral of $\mathbf{B}$ over a closed surface s yields zero. Thus the magnetic counterpart of Gauss' electric field relation (9) is

$$\oint_s \mathbf{B} \cdot d\mathbf{s} = 0 \tag{12}$$

The corresponding differential or point relation is

$$\mathbf{V} \cdot \mathbf{B} = 0 \tag{13}$$

Equations (12) and (13) may be referred to as *Maxwell's magnetic field equations as derived from Gauss' law*, (12) being the integral and (13) the differential form.

The development of Maxwell's equations as generalizations of circuit relations involves both inductive and physical reasoning. In developing his equations, Maxwell drew heavily on Faraday's experimental work. Maxwell also used analogies with hydraulic and other mechanical systems to assist in formulating his ideas.

It is not implied that the "derivation" of Maxwell's equations is rigorous. His equations are justified by the fact that conclusions based on them have been found in innumerable cases to be in excellent agreement with experiment, in the same way that the earlier circuit relations are justified within their more restricted domain by the excellent agreement of conclusions based on them with experiment. It is perhaps worth recalling that Maxwell's equations were *not* generally accepted for many years after they were postulated (1873). His curl equations (involving $\mathbf{V} \times \mathbf{E}$ and $\mathbf{V} \times \mathbf{H}$) implied that time-varying electric and magnetic fields in empty space were interdependent, a changing electric field being able to generate a magnetic field, and vice versa. The inference from this is that a time-changing electromagnetic field would propagate energy through empty space with the velocity of light (see Chap. 10) and, further, that light is electromagnetic in nature. Radio waves were unknown at the time, and it was 15 years (1888) before Hertz demonstrated that electromagnetic (or radio) waves were possible, as predicted by Maxwell.

There is no guarantee that Maxwell's equations are exact. However, insofar as the precision of experimental measurements allows, they appear to be, and therefore we may regard them as exact.

Along with Maxwell's equations certain other fundamental relations are of importance in dealing with electromagnetic problems. Among these may be mentioned *Ohm's law at a point* (4-7-2)

$$\mathbf{J} = \sigma \mathbf{E} \tag{14}$$

the *continuity relation* (4-13-3)

$$\mathbf{V} \cdot \mathbf{J} = -\frac{\partial \rho}{\partial t} \tag{15}$$

the *force relations*

$$\mathbf{F} = q\mathbf{E}$$
$$d\mathbf{F} = (\mathbf{I} \times \mathbf{B})dl \tag{16}$$

and the *constitutive relations* between $\mathbf{E}$ and $\mathbf{D}$ and between $\mathbf{B}$ and $\mathbf{H}$ as given by

$$\mathbf{D} = \epsilon\mathbf{E} = \epsilon_0 \mathbf{E} + \mathbf{P} \tag{17}$$

$$\mathbf{B} = \mu\mathbf{H} = \mu_0(\mathbf{H} + \mathbf{M}) \tag{18}$$

9-5 MAXWELL'S EQUATIONS IN FREE SPACE

In the preceding section, Maxwell's equations are stated in their general form. For the special case of free space, where the current density **J** and the charge density ρ are zero, the equations reduce to a simpler form. In integral form the equations are

$$\oint \mathbf{H} \cdot d\mathbf{l} = \int_s \frac{\partial \mathbf{D}}{\partial t} \cdot d\mathbf{s} \tag{1}$$

$$\oint \mathbf{E} \cdot d\mathbf{l} = -\int_s \frac{\partial \mathbf{B}}{\partial t} \cdot d\mathbf{s} \tag{2}$$

$$\oint_s \mathbf{D} \cdot d\mathbf{s} = 0 \tag{3}$$

$$\oint_s \mathbf{B} \cdot d\mathbf{s} = 0 \tag{4}$$

In differential form the equations are

$$\nabla \times \mathbf{H} = \frac{\partial \mathbf{D}}{\partial t} \tag{5}$$

$$\nabla \times \mathbf{E} = -\frac{\partial \mathbf{B}}{\partial t} \tag{6}$$

$$\nabla \cdot \mathbf{D} = 0 \tag{7}$$

$$\nabla \cdot \mathbf{B} = 0 \tag{8}$$

9-6 MAXWELL'S EQUATIONS FOR HARMONICALLY VARYING FIELDS

For harmonic variation, the phasor forms of Maxwell's integral and differential equations are

$$\oint \mathbf{H} \cdot d\mathbf{l} = (\sigma + j\omega\epsilon) \int_s \mathbf{E} \cdot d\mathbf{s} \quad (1) \qquad \nabla \times \mathbf{H} = (\sigma + j\omega\epsilon)\mathbf{E} \quad (5)$$

$$\oint \mathbf{E} \cdot d\mathbf{l} = -j\omega\mu \int_s \mathbf{H} \cdot d\mathbf{s} \quad (2) \qquad \nabla \times \mathbf{E} = -j\omega\mu\mathbf{H} \quad (6)$$

$$\oint_s \mathbf{D} \cdot d\mathbf{s} = \int_s \rho \, dv \quad (3) \qquad \nabla \cdot \mathbf{D} = \rho \quad (7)$$

$$\oint_s \mathbf{B} \cdot d\mathbf{s} = 0 \quad (4) \qquad \nabla \cdot \mathbf{B} = 0 \quad (8)$$

9-7 TABLES OF MAXWELL'S EQUATIONS

Maxwell's equations are summarized in Tables 9-1 and 9-2. Table 9-1 gives Maxwell's equations in integral form and Table 9-2 in differential form. The equations are stated for the general case, free-space case, harmonic-variation case,

Table 9-1 Maxwell's equations in integral form

Dimensions and SI units / Case	From Ampère — mmf, A	From Faraday — emf, V	From Gauss — Electric flux, C	Magnetic flux, Wb
General	$F = \oint_l \mathbf{H} \cdot d\mathbf{l} = \int_s \left(\mathbf{J} + \dfrac{\partial \mathbf{D}}{\partial t}\right) \cdot d\mathbf{s} = I_{\text{total}}$	$\mathscr{V} = \oint_l \mathbf{E} \cdot d\mathbf{l} = -\int_s \dfrac{\partial \mathbf{B}}{\partial t} \cdot d\mathbf{s}$	$\psi = \oint_s \mathbf{D} \cdot d\mathbf{s} = \int_v \rho\, dv$	$\psi_m = \oint_s \mathbf{B} \cdot d\mathbf{s} = 0$
Free space	$F = \oint_l \mathbf{H} \cdot d\mathbf{l} = \int_s \dfrac{\partial \mathbf{D}}{\partial t} \cdot d\mathbf{s} = I_{\text{disp}}$	$\mathscr{V} = \oint_l \mathbf{E} \cdot d\mathbf{l} = -\int_s \dfrac{\partial \mathbf{B}}{\partial t} \cdot d\mathbf{s}$	$\psi = \oint_s \mathbf{D} \cdot d\mathbf{s} = 0$	$\psi_m = \oint_s \mathbf{B} \cdot d\mathbf{s} = 0$
Harmonic variation	$F = \oint_l \mathbf{H} \cdot d\mathbf{l} = (\sigma + j\omega\epsilon) \int_s \mathbf{E} \cdot d\mathbf{s} = I_{\text{total}}$	$\mathscr{V} = \oint_l \mathbf{E} \cdot d\mathbf{l} = -j\omega\mu \int_s \mathbf{H} \cdot d\mathbf{s}$	$\psi = \oint_s \mathbf{D} \cdot d\mathbf{s} = \int_v \rho\, dv$	$\psi_m = \oint_s \mathbf{B} \cdot d\mathbf{s} = 0$
Steady	$F = \oint_l \mathbf{H} \cdot d\mathbf{l} = \int_s \mathbf{J} \cdot d\mathbf{s} = I_{\text{cond}}$	$V = \oint_l \mathbf{E} \cdot d\mathbf{l} = 0$	$\psi = \oint_s \mathbf{D} \cdot d\mathbf{s} = \int_v \rho\, dv$	$\psi_m = \oint_s \mathbf{B} \cdot d\mathbf{s} = 0$
Static	$U = \oint_l \mathbf{H} \cdot d\mathbf{l} = 0$	$V = \oint_l \mathbf{E} \cdot d\mathbf{l} = 0$	$\psi = \oint_s \mathbf{D} \cdot d\mathbf{s} = \int_v \rho\, dv$	$\psi_m = \oint_s \mathbf{B} \cdot d\mathbf{s} = 0$

Table 9-2 Maxwell's equations in differential form

Case / Dimensions	From Ampère Electric current / area	From Faraday Electric potential / area	From Gauss Electric flux / volume	From Gauss Magnetic flux / volume
General	$\nabla \times \mathbf{H} = \mathbf{J} + \dfrac{\partial \mathbf{D}}{\partial t}$	$\nabla \times \mathbf{E} = -\dfrac{\partial \mathbf{B}}{\partial t}$	$\nabla \cdot \mathbf{D} = \rho$	$\nabla \cdot \mathbf{B} = 0$
Free space	$\nabla \times \mathbf{H} = \dfrac{\partial \mathbf{D}}{\partial t}$	$\nabla \times \mathbf{E} = -\dfrac{\partial \mathbf{B}}{\partial t}$	$\nabla \cdot \mathbf{D} = 0$	$\nabla \cdot \mathbf{B} = 0$
Harmonic variation	$\nabla \times \mathbf{H} = (\sigma + j\omega\epsilon)\mathbf{E}$	$\nabla \times \mathbf{E} = -j\omega\mu\mathbf{H}$	$\nabla \cdot \mathbf{D} = \rho$	$\nabla \cdot \mathbf{B} = 0$
Steady	$\nabla \times \mathbf{H} = \mathbf{J}$	$\nabla \times \mathbf{E} = 0$	$\nabla \cdot \mathbf{D} = \rho$	$\nabla \cdot \mathbf{B} = 0$
Static	$\nabla \times \mathbf{H} = 0$	$\nabla \times \mathbf{E} = 0$	$\nabla \cdot \mathbf{D} = \rho$	$\nabla \cdot \mathbf{B} = 0$

Table 9-3 Resistance, capacitance, and inductance in terms of circuit and field quantities

	Circuit definition	Physical definition	Field cell	Energy definition
Resistance R	$\dfrac{\int \mathbf{E} \cdot d\mathbf{l}}{\oint \mathbf{H} \cdot d\mathbf{l}}$	$\dfrac{l}{\sigma A}$	$\dfrac{1}{\sigma d}$	$\dfrac{\iiint \dfrac{J^2}{\sigma}\, dv}{I^2}$
Capacitance C	$\dfrac{\iint \mathbf{D} \cdot d\mathbf{s}}{\oint \mathbf{E} \cdot d\mathbf{l}}$	$\dfrac{\epsilon A}{l}$	ϵd	$\dfrac{\iiint \epsilon E^2\, dv}{V^2}$
Inductance L	$\dfrac{\iint \mathbf{B} \cdot d\mathbf{s}}{\oint \mathbf{H} \cdot d\mathbf{l}}$	$\dfrac{\mu A}{l}$	μd	$\dfrac{\iiint \mu H^2\, dv}{I^2}$

steady case (static fields but with steady conduction currents), and static case (static fields with no currents). In Table 9-1 the equivalence is also indicated between the various field quantities and the electric potential V, the emf $\mathscr{V}$, the magnetic potential U, the mmf F, the total current I_{total}, the displacement current I_{disp}, the conduction current I_{cond}, the electric flux ψ, and the magnetic flux ψ_m. It should be noted that Maxwell's equations as tabulated here apply specifically to stationary systems or bodies at rest.

PROBLEMS†

Group 9-1: All sections. Interrelation of field and circuit theory and Maxwell's equations

9-1-1. Maxwell's equation from Ampère's law Give the step-by-step development of Maxwell's equation from Ampère's law in (*a*) integral form and (*b*) using Stokes' theorem in differential, or point, form. (*c*) Modify both integral and differential forms for harmonically varying fields.

† Answers to starred problems are given in Appendix C.

9-1-2. Maxwell's equation from Faraday's law Give the step-by-step development of Maxwell's equation from Faraday's law in (*a*) integral form and (*b*) using Stokes' theorem in differential, or point, form. (*c*) Modify both integral and differential forms for harmonically varying fields.

9-1-3. Maxwell's equation from Gauss' law. Electric fields Give the step-by-step development of Maxwell's equation from Gauss' law in (*a*) integral form and (*b*) in differential, or point, form.

9-1-4. Maxwell's equation from Gauss' law. Magnetic fields Give the step-by-step development of Maxwell's equation from Gauss' law in (*a*) integral form and (*b*) in differential, or point, form.

★**9-1-5. Maxwell's equations. Symmetry** Why are Maxwell's equations not completely symmetrical?

9-1-6. Objection to Maxwell's theory An objection to Maxwell's theory raised 100 years ago was that the relative permittivity or "dielectric constant" of water was 81 according to Maxwell's theory, whereas the established value was 1.77, an order-of-magnitude difference. Is this still regarded as an objection to Maxwell's theory? Explain.

9-1-7. Parallel circuit. Field and circuit theory Apply field theory to a circuit consisting of a resistor, inductor, and capacitor connected in parallel with an alternating-current source *I*. Proceeding in an analogous manner to that used for the series circuit in Sec. 9-3, show that

$$I = \frac{V}{R} + C\frac{dV}{dt} + \frac{1}{L}\int V\,dt$$

where V = voltage across parallel combination
R = resistance of resistor
C = capacitance of capacitor
L = inductance of inductor

★**9-1-8. Pole pieces. Displacement current** The magnetic field between two pole pieces of radius R varies as $B = B_0[1 - \frac{1}{2}(\beta r)^2]\sin \omega t$. Assume $R \ll \lambda$. Find the displacement-current density as a function of radius r (for $r < R$).

9-1-9. Capacitor. Displacement current A parallel-plate capacitor with plates of radius R and separation d has a voltage applied at the center as given by $V = V_0 \sin \omega t$. As a function of radius r (for $r < R$), find (*a*) the displacement-current density $J_d(r)$ and (*b*) the magnetic field $H(r)$. Take $d \ll R$.

9-1-10. Neumann's inductance formula Show that the expression for the low-frequency inductance $L = \oint (A/I) \cdot d\mathbf{l}$ reduces for a conducting circuit to Neumann's low-frequency inductance formula

$$L = \frac{\mu_0}{4\pi} \oint\!\!\oint \frac{d\mathbf{l}'}{r} \cdot d\mathbf{l}$$

9-1-11. Transmission-line inductance A two-conductor transmission line of length *l* has a conductor (center-to-center) separation D and conductor radius a. The conductors are thin-walled tubes. Referring to Fig. P9-1-11, apply Neumann's low-frequency inductance formula (Prob. 9-1-10) to show that the inductance of the line is

$$L = \frac{\mu_0 l}{\pi} \ln \frac{D}{a} \qquad \text{(H)}$$

Compare with (5-13-12). Assume $l \gg D$, and neglect end effects.

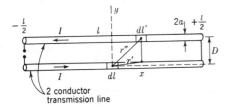

2 conductor
transmission line

Figure P9-1-11 Two-conductor transmission line.

TEN

WAVES AND TRANSMISSION LINES

10-1 INTRODUCTION

In an electromagnetic wave a changing electric field produces a changing magnetic field, which in turn generates an electric field, and so on with a resulting propagation of energy. In this chapter the propagation of waves in space and on transmission lines will be discussed.

A transmission line may be defined as a device for transmitting or guiding energy from one point to another. The energy may be for lighting, heating, or performing work, or it may be in the form of signal information (speech, pictures, data, music). Basically a transmission line has two terminals into which power (or information) is fed and two terminals from which power (or information) is received. Thus, a transmission line may be regarded as a four-terminal device for connecting any and all electrical devices.

The power cord on a lamp or appliance is a transmission line and so are the wires from a generating station to a factory or home. Telephone and telegraph wires; audio, video, and radio cables; and the myriad of nerve fibers in our bodies are all transmission lines. The interconnections of all electrical circuits are transmission lines and in a broad sense waveguides and optical fibers, and even radio links may be regarded as transmission lines. A few examples are shown in Fig. 10-1.

Transmission lines are everywhere and are of infinite variety but regardless of type, length, or construction, all operate according to the same basic principles which will be discussed in the following pages.

It is convenient to classify transmission lines into three main groups: (1) those with transverse electromagnetic (TEM) modes, (2) those with higher-order modes, and (3) those with transverse electromagnetic space waves (as in a radio link). In a TEM mode both the electric and magnetic fields are entirely transverse to the direction of propagation. There is no component of either $\mathbf{E}$ or $\mathbf{H}$ in the direction of transmission. Higher-order modes, on the other hand, always have at least one

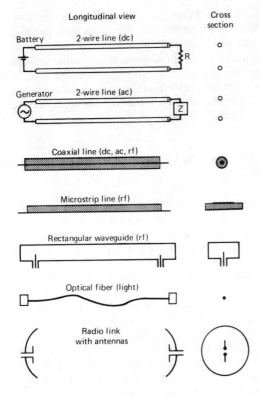

Figure 10-1 A few examples of transmission lines.

field component in the direction of transmission. All two-conductor lines such as coaxial or two-wire transmission lines are examples of TEM-mode types, while hollow single-conductor waveguides or dielectric rods are examples of higher mode types.

To summarize, transmission lines may be classified as follows:

1. TEM-mode type: **E** and **H** entirely transverse. All two-conductor types, including coaxial lines. Power flows along and between conductors.
2. Higher-mode type: **E** and **H** or both have components in the direction of transmission. Hollow single-conductor waveguides, dielectric rods, and optical fibers. Power flows in space inside conductor or in or close to dielectric rod or fiber.
3. TEM space waves between antennas of a radio link. Power radiates through space.

The propagation of transverse electromagnetic waves is considered first and it is shown how these lead to the equations for waves guided along a two-conductor transmission line. The velocity of propagation and attenuation on the two-conductor line are discussed, followed by treatments of impedance matching and bandwidth. Then unguided (space) waves are considered and their behavior in dielectric and conducting media is discussed. The analogies and similarities between unguided waves in space and guided waves on transmission lines are used to relate the circuit theory of transmission lines to the field theory of space waves and vice versa. A key consideration in the interrelations is the fact that space may be regarded as an array of field-cell transmission lines.

10-2 THE WAVE EQUATION FOR WAVES IN SPACE AND ON TRANSMISSION LINES

The interdependence of electric and magnetic fields is demonstrated in a striking manner by an electromagnetic wave propagating through space. In such a wave the time-changing magnetic field may be regarded as generating a time-varying electric field, which in turn generates a magnetic field, and as the process repeats, energy is propagated through empty space at the velocity of light.

The field lines for a wave propagating toward the reader (out of the page) are indicated in Fig. 10-2. The directions of **E** and **H** are everywhere perpendicular. In a uniform plane wave **E** and **H** lie in a plane and have the same value everywhere in that plane. A wave of this type with both **E** and **H** transverse to the direction of propagation is called a *Transverse ElectroMagnetic* (TEM) wave.

Referring to Fig. 10-3, assume that a plane wave is traveling in the direction of the x axis. The electric field **E** has only a component E_y in the y direction and the magnetic field **H** only a component H_z in the z direction. It is said that this wave is polarized in the y direction (vertically polarized) (see Chap. 11).

For a nonconducting medium, the conduction-current density **J** is zero. Thus Maxwell's equation from Ampère's law reduces to

$$\nabla \times \mathbf{H} = \frac{\partial \mathbf{D}}{\partial t} \tag{1}$$

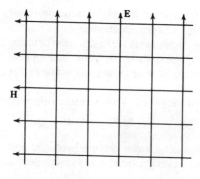

Figure 10-2 Plane traveling wave approaching reader (out of page).

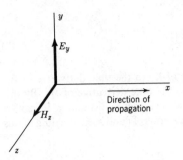

Figure 10-3 Field components of plane wave with relation to coordinate system.

or in rectangular coordinates

$$\hat{\mathbf{x}}\left(\frac{\partial H_z}{\partial y} - \frac{\partial H_y}{\partial z}\right) + \hat{\mathbf{y}}\left(\frac{\partial H_x}{\partial z} - \frac{\partial H_z}{\partial x}\right) + \hat{\mathbf{z}}\left(\frac{\partial H_y}{\partial x} - \frac{\partial H_x}{\partial y}\right) = \frac{\partial}{\partial t}(\hat{\mathbf{x}}D_x + \hat{\mathbf{y}}D_y + \hat{\mathbf{z}}D_z)$$

(2)

For a plane wave traveling in the x direction the only components of (2) that contribute are

$$-\hat{\mathbf{y}}\frac{\partial H_z}{\partial x} = \hat{\mathbf{y}}\frac{\partial D_y}{\partial t}$$

(3)

Therefore

$$\frac{\partial H_z}{\partial x} = -\epsilon\frac{\partial E_y}{\partial t}$$

(4)

Maxwell's equation from Faraday's law is

$$\mathbf{\nabla} \times \mathbf{E} = -\frac{\partial \mathbf{B}}{\partial t}$$

(5)

or in rectangular coordinates

$$\hat{\mathbf{x}}\left(\frac{\partial E_z}{\partial y} - \frac{\partial E_y}{\partial z}\right) + \hat{\mathbf{y}}\left(\frac{\partial E_x}{\partial z} - \frac{\partial E_z}{\partial x}\right) + \hat{\mathbf{z}}\left(\frac{\partial E_y}{\partial x} - \frac{\partial E_x}{\partial y}\right) = -\frac{\partial}{\partial t}(\hat{\mathbf{x}}B_x + \hat{\mathbf{y}}B_y + \hat{\mathbf{z}}B_z)$$

(6)

For a plane wave traveling in the x direction the only components of (6) that contribute are†

$$\hat{\mathbf{z}}\frac{\partial E_y}{\partial x} = -\hat{\mathbf{z}}\frac{\partial B_z}{\partial t}$$

(7)

---

† If we had specified only that the wave is linearly polarized with **E** in the y direction and that the wave travels in the x direction, it follows from (6) and also from (7) that **B** and hence **H** must be in the z direction.

Therefore

$$\frac{\partial E_y}{\partial x} = -\mu \frac{\partial H_z}{\partial t} \tag{8}$$

Equation (4) relates the space derivative of H_z to the time derivative of E_y, while (8) relates the space derivative of E_y to the time derivative of H_z. By differentiating (4) with respect to time t and (8) with respect to distance x, H_z can be eliminated and an expression obtained for E_y in terms of t and x. Proceeding in this way, we obtain, from (4),

$$\frac{\partial}{\partial t}\left(\frac{\partial H_z}{\partial x}\right) = -\epsilon \frac{\partial^2 E_y}{\partial t^2} \tag{9}$$

and, from (8),

$$\frac{\partial^2 E_y}{\partial x^2} = -\mu \frac{\partial}{\partial x}\left(\frac{\partial H_z}{\partial t}\right) \tag{10}$$

Dividing (10) by $-\mu$ yields

$$-\frac{1}{\mu}\frac{\partial^2 E_y}{\partial x^2} = \frac{\partial}{\partial x}\left(\frac{\partial H_z}{\partial t}\right) \tag{11}$$

Since in (9) it does not matter whether we differentiate first with respect to x and then with respect to t or vice versa, the left-hand side of (9) is equal to the right-hand side of (11) and it follows that

$$\boxed{\frac{\partial^2 E_y}{\partial t^2} = \frac{1}{\mu\epsilon}\frac{\partial^2 E_y}{\partial x^2}} \tag{12}$$

Equation (12) relates the space and time variation of the scalar magnitude E_y of the electric field intensity and is called a *wave equation* in E_y. It is, in fact, a scalar wave equation of the simplest form.

Differentiating (4) and (8) in the reverse order, i.e., (4) with respect to x and (8) with respect to t, we can eliminate E_y and obtain a wave equation in H_z as

$$\frac{\partial^2 H_z}{\partial t^2} = \frac{1}{\mu\epsilon}\frac{\partial^2 H_z}{\partial x^2} \tag{13}$$

Both (12) and (13) are of the same form. A wave equation as given by (12) and (13) has many important physical applications and is sometimes called *D'Alembert's equation*, having been integrated by him in 1747. If E_y in (12) is a transverse displacement, the equation can represent the motion of a disturbance on a stretched string. This was D'Alembert's problem. If E_y is a mechanical compression, the equation can describe the motion of small oscillations of air in a narrow pipe. In our case E_y represents the scalar magnitude of the electric field intensity of a plane electromagnetic wave progressing in the x direction, and the equation is the most general way of describing the motion of this field as a function of time and space.

Let us now introduce a quantity v in (12) such that

$$v^2 = \frac{1}{\mu\epsilon} \tag{14}$$

Equation (12) then becomes

$$\frac{\partial^2 E_y}{\partial t^2} = v^2 \frac{\partial^2 E_y}{\partial x^2} \tag{15}$$

Dimensionally (15) is

$$\frac{\text{Volts}}{\text{Meters seconds}^2} = v^2 \frac{\text{volts}}{\text{meters}^3}$$

so that

$$v = \frac{\text{meters}}{\text{second}}$$

Thus, it appears that v has the dimensions of velocity. This velocity is a characteristic of the medium, being dependent on the constants μ and ϵ for the medium. For free space (vacuum) v is approximately equal to 300 Mm s^{-1}.

The wave equation (15) is a linear partial differential equation of the second order. To apply the equation, a solution must be found for E_y. Methods of solving this type of equation are discussed in texts on differential equations. It will suffice here to say that if we take the following trial solution

$$E_y = \sin \beta(x + mt) \tag{16}$$

where $\beta = 2\pi/\lambda$
λ = wavelength
m = a constant (to be determined)
t = time

we find on substitution in (15) that (16) is a solution provided that

$$m = \pm v \tag{17}$$

where v is the velocity. Hence a general solution for (15) is

$$E_y = \sin \beta(x + vt) + \sin \beta(x - vt) \tag{18}$$

Either term alone is a solution, or the sum, as in (18), is a solution. This can be verified by taking the second derivatives of the solution in terms of t and x and substituting them in (15).

Since $v = f\lambda$, it follows that

$$\beta v = \frac{2\pi}{\lambda} f\lambda = 2\pi f = \omega \tag{19}$$

Thus, (18) can also be expressed as

$$E_y = \sin (\beta x + \omega t) + \sin (\beta x - \omega t) \tag{20}$$

Suppose that the first term of (18) is considered by itself as a solution. That is,

$$E_y = \sin \beta(x + vt) \qquad (21)$$

The significance of (21) can be illustrated by evaluating E_y as a function of x for several values of the time t. First let us take $t = 0$. Then $E_y = \sin \beta x$. The curve for this instant of time is shown by Fig. 10-4a. Next consider the situation one-quarter period later, i.e., when $t = T/4$, where T is the time of one period. Then

$$\beta vt = \omega t = (2\pi f)t = \frac{2\pi}{T} t = \frac{2\pi}{T} \frac{T}{4} = \frac{\pi}{2} \qquad (22)$$

The curve for $t = T/4$ or $\omega t = \pi/2$ rad is shown in Fig. 10-4b. One-half period later, $t = T/2$, and $\omega t = \pi$, yielding the curve of Fig. 10-4c. Focusing our attention on the crest of one of the waves, as indicated by the point P, we note that as time progresses, P moves to the left. From Fig. 10-4 we can thus interpret (21) as representing a wave traveling to the left, or in the negative x direction. The maximum value of E_y for this wave is unity.

The point P is a point of constant phase and is characterized by the condition that

$$x + vt = \text{constant} \qquad (23)$$

Taking the time derivative of (23) gives

$$\frac{dx}{dt} + v = 0 \qquad (24)$$

and

$$\frac{dx}{dt} = -v \qquad (25)$$

In (25), dx/dt is the rate of change of distance with respect to time, or velocity, of a constant-phase point. Hence, v is the velocity of a constant-phase point and is called the *phase velocity*. We note also that v is negative, which means that the wave is traveling in the negative x direction.

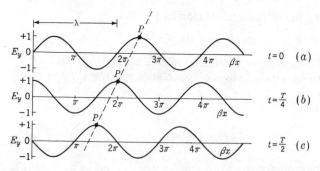

Figure 10-4 Curves for $E_y = \sin(\beta x + \omega t)$ at three instants of time: $t = 0$, $t = T/4$, and $t = T/2$. A constant-phase point P moves to the left as time progresses.

Next consider the last term of (18) as a solution by itself. Then

$$E_y = \sin \beta(x - vt) = \sin (\beta x - \omega t) \qquad (26)$$

Putting in values for $t = 0$, $T/4$, and $T/2$, we obtain from (26) the curves of Fig. 10-5. Here a constant-phase point P moves to the right as time progresses. Hence (26) represents a wave traveling in the positive x direction.

If we set $x - vt$ equal to a constant and proceed in the same manner as for (24) and (25), we find in this case that

$$\frac{dx}{dt} = +v \qquad (27)$$

Thus, the wave travels with a velocity v in the positive x direction.

To summarize, a negative sign in $x \pm vt$ or in $\beta x \pm \omega t$ is associated with a wave to the right, while a positive sign is associated with a wave to the left. Accordingly, when solutions with both positive and negative signs are given, as in (18), two waves are represented, one to the left and one to the right, and the complete solution is equal to the sum of both waves.

Let us now treat in somewhat more detail the wave traveling in the positive x direction. A number of forms can be used which are equivalent except for a phase displacement. Four such forms are

$$\begin{aligned} E_y &= \sin (\beta x - \omega t) \\ E_y &= \sin (\omega t - \beta x) \\ E_y &= \cos (\beta x - \omega t) \\ E_y &= \cos (\omega t - \beta x) \end{aligned} \qquad (28)$$

If the phase displacement is disregarded, any of the four forms given by (28) can be selected to represent a wave traveling in the positive x direction. Suppose we choose the form

$$E_y = \cos (\omega t - \beta x) \qquad (29)$$

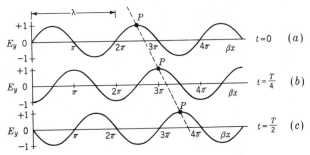

Figure 10-5 Curves for $E_y = \sin (\beta x - \omega t)$ at three instants of time: $t = 0$, $t = T/4$, and $t = T/2$. A constant-phase point P moves to the right as time progresses.

Thus far, it has been assumed that the maximum amplitude of E_y is unity. If now we specify the maximum amplitude as E_0, we have

$$E_y = E_0 \cos(\omega t - \beta x) \tag{30}$$

Since $f = 1/T$, (30) can be expressed in a form in which the period T appears explicitly. For the sake of symmetry, let us also put $\beta = 2\pi/\lambda$, obtaining

$$E_y = E_0 \cos\left(2\pi\frac{t}{T} - 2\pi\frac{x}{\lambda}\right) \tag{31}$$

These expressions, (30) and (31), represent a wave traveling in the positive x direction. The corresponding expressions for a wave traveling in the negative x direction are

$$E_y = E_0 \cos(\omega t + \beta x) \tag{32}$$

and

$$E_y = E_0 \cos\left(2\pi\frac{t}{T} + 2\pi\frac{x}{\lambda}\right) \tag{33}$$

The solutions of the wave equation given by (30) and (32) are *trigonometric* solutions. We can also express the solution in *exponential* form. Thus

$$E_y = E_0\, e^{j(\omega t \pm \beta x)} \tag{34}$$

where it is understood that the instantaneous value of the field is given by the real (or imaginary) part of the exponential function. Thus, taking the real part (Re), we have

$$E_y = E_0 \operatorname{Re} e^{j(\omega t - \beta x)} = E_0 \cos(\omega t - \beta x) \tag{35}$$

We have seen that $x - vt$ is a constant for a point of constant phase in a traveling wave. It follows that $\omega t - \beta x$ is a constant. That is, t and x must vary together, so that

$$\omega t - \beta x = \text{constant} \tag{36}$$

Differentiating (36) with respect to time to find the velocity of the constant-phase point, as done in (24), yields

$$\omega - \beta\frac{dx}{dt} = 0 \tag{37}$$

or

$$\frac{dx}{dt} = \frac{\omega}{\beta} \tag{38}$$

Thus, the *phase velocity*, or velocity of a constant-phase point, is given by ω/β.

That ω/β has the dimensions of velocity is more apparent if it is reexpressed as

$$\frac{\omega}{\beta} = \frac{2\pi f}{2\pi/\lambda} = \lambda f = \text{velocity} \tag{39}$$

where ω = radian frequency = $2\pi f$, rad or deg
$\quad \beta$ = phase constant = $2\pi/\lambda$, rad m^{-1}
$\quad f$ = frequency, Hz (cycles s^{-1})
$\quad \lambda$ = wavelength, m

Thus, the product λf has the dimensions of wavelength (distance) times frequency (reciprocal of time) which is distance per time, or velocity.

From (14) we have that this (phase) velocity v is

$$\frac{\omega}{\beta} = v = \frac{1}{\sqrt{\mu\epsilon}} \quad (\text{m s}^{-1}) \tag{40}$$

Equation (40) gives the phase velocity of a wave in an unbounded medium of permeability μ and permittivity ϵ. For free space (vacuum) the velocity is a well-known constant (usually designated by c) equal to the *velocity of light*. Thus

$$c = \frac{1}{\sqrt{\mu_0\epsilon_0}} = 299.79 \text{ Mm s}^{-1} \quad \text{(measured)} \tag{41}$$

The SI unit for the permeability of vacuum is

$$\mu_0 = 400\pi \text{ nH m}^{-1} \quad \text{(exactly by definition)}$$

Therefore, from the definition value of μ_0 and the measured value of c, the permittivity ϵ_0 of vacuum is

$$\epsilon_0 = \frac{1}{\mu_0 c^2} = 8.85 \text{ pF m}^{-1} \tag{42}$$

The wave equation (12) is for a *lossless medium* ($\sigma = 0$). In the more general situation, for a medium that is not lossless (σ finite), Maxwell's curl equations are

$$-\frac{\partial H_z}{\partial x} = \sigma E_y + \epsilon \frac{\partial E_y}{\partial t} \tag{43}$$

and
$$\frac{\partial E_y}{\partial x} = -\mu \frac{\partial H_z}{\partial t} \tag{44}$$

or in phasor form

$$\frac{\partial H_z}{\partial x} = -(\sigma + j\omega\epsilon)E_y \tag{45}$$

and
$$\frac{\partial E_y}{\partial x} = -j\omega\mu H_z \tag{46}$$

Differentiating (46) with respect to x and substituting (45) yields

$$\frac{\partial^2 E_y}{\partial x^2} = (j\omega\mu\sigma - \omega^2\mu\epsilon)E_y \tag{47}$$

This is the wave equation in E_y for a plane wave in a *conducting medium*. As before, these equations are for a linearly polarized wave (**E** in the y direction) traveling in the x direction.

Space may be regarded as an array of field-cell transmission lines, as in Fig. 10-6. Directing our attention to a single transmission line cell, the upper and lower surfaces of the cell can be regarded as consisting of conducting strips of width w and of infinite length in the direction of wave propagation (x direction, out of page). Recalling from earlier chapters that for field cells the inductance L per unit length (in x direction) is equal to the permeability μ of the medium, the capacitance C per unit length is equal to the permittivity ϵ of the medium, and the conductance G per unit length is equal to the conductivity σ of the medium, we can write

$$L = \mu = \text{inductance per unit length, H m}^{-1}$$

$$C = \epsilon = \text{capacitance per unit length, F m}^{-1}$$

$$G = \sigma = \text{conductance per unit length, } \mho \text{ m}^{-1}$$

where the symbols L, C, and G are now understood to be *distributed quantities*, that is, per unit length.

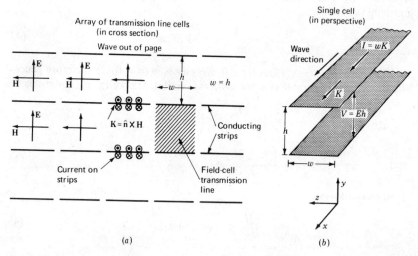

(a)

(b)

Figure 10-6 Space may be regarded as an array of field-cell transmission lines as in (a). One field cell is shown in perspective in (b). **E** is in the y direction and **H** is in the z direction.

Introducing L, C, and G into (47), we obtain

$$\frac{\partial^2 E_y}{\partial x^2} - (j\omega LG - \omega^2 LC)E_y = 0 \tag{48}$$

or

$$\frac{\partial^2 E_y}{\partial x^2} - j\omega L(G + j\omega C)E_y = 0 \tag{49}$$

Integrating $E_y\ (= |\mathbf{E}|)$ between the upper and lower strips yields the potential difference $V(= E_y h$, or $E_y = V/h)$, so that (49) can be expressed as

$$\frac{\partial^2 V}{\partial x^2} - j\omega L(G + j\omega C)V = 0 \tag{50}$$

This differential equation is the wave equation for a field-cell transmission line in terms of the voltage V between the conducting strips.

It has been assumed that the conducting strips are lossless (resistance zero). For finite resistance R per unit length, its effect can be included by writing (50) as

$$\frac{\partial^2 V}{\partial x^2} - (G + j\omega C)(R + j\omega L)V = 0 \tag{51}$$

This can be simplified to

$$\frac{\partial^2 V}{\partial x^2} - YZV = 0 \tag{52}$$

where $Y = G + j\omega C = G + jB,\ \mho\ \mathrm{m}^{-1}$
$Z = R + j\omega L = R + jX,\ \Omega\ \mathrm{m}^{-1}$

where Y = shunt admittance, $\mho\ \mathrm{m}^{-1}$
Z = series impedance, $\Omega\ \mathrm{m}^{-1}$
G = shunt conductance, $\mho\ \mathrm{m}^{-1}$
C = shunt capacitance, $\mathrm{F}\ \mathrm{m}^{-1}$
$B = \omega C$ = shunt susceptance, $\mho\ \mathrm{m}^{-1}$
R = series resistance, $\Omega\ \mathrm{m}^{-1}$
L = series inductance, $\mathrm{H}\ \mathrm{m}^{-1}$
$X = \omega L$ = series reactance, $\Omega\ \mathrm{m}^{-1}$

Note that all of these parameters are *distributed quantities*, that is, per unit length.

Thus, starting with Maxwell's equations for a space wave, we have developed the wave equation of a two-conductor transmission line in terms of its circuit parameters of series resistance and inductance and shunt conductance and capacitance per unit length of line. An equivalent circuit for a section of the transmission line is shown in Fig. 10-7 with the parameters R, L, G, and C as indicated.

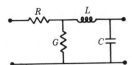

Figure 10-7 Equivalent circuit for a section of transmission line.

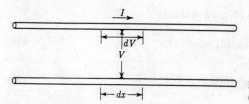

Figure 10-8 Section of an infinite two-wire (or two-conductor) transmission line.

If we had obtained a wave equation in H_z instead of E_y, the corresponding transmission line equation would have involved the current I in each conductor as given by

$$\frac{\partial^2 I}{\partial x^2} - ZYI = 0 \tag{53}$$

Turning from a *field approach*, as above, to a strictly *circuit approach*, let us develop the transmission-line wave equation using circuit theory. Thus, from Ohm's law the voltage change dV over a length of line dx equals the IZ change per unit length multiplied by dx, where I is the line current (see Fig. 10-8). In symbols

$$\frac{dV}{dx} = IZ \tag{54}$$

Likewise, the change in current in length dx of line is equal to the shunt current VY between conductors multiplied by dx, or

$$\frac{dI}{dx} = VY \tag{55}$$

Differentiating (54) and (55) with respect to x yields

$$\frac{d^2V}{dx^2} = I\frac{dZ}{dx} + Z\frac{dI}{dx} = I\frac{dZ}{dx} + ZVY \tag{56}$$

$$\frac{d^2I}{dx^2} = V\frac{dY}{dx} + Y\frac{dV}{dx} = V\frac{dY}{dx} + YIZ \tag{57}$$

For a uniform line (no variation of Z or Y with x), (56) and (57) reduce to

$$\frac{d^2V}{dx^2} - ZYV = 0 \tag{58}$$

$$\frac{d^2I}{dx^2} - ZYI = 0 \tag{59}$$

which are identical with (52) and (53) above as obtained with Maxwell's equations applied to a field-cell transmission line.

10-3 COAXIAL, TWO-WIRE, AND FIELD-CELL TRANSMISSION LINES

Before proceeding with a development of transmission-line theory, it is instructive to consider how the field-cell transmission lines of space relate to actual two-conductor transmission lines.

As shown in Fig. 10-9a, an array of field-cell transmission lines can represent a space wave propagating out of the page with **E** (solid line) and **H** (dashed line) as indicated. The presence of the conducting strips does not affect the wave since the strips are perpendicular to **E** and parallel to **H**. Consider now a group of many side-by-side cells, as in Fig. 10-9b, connected by continuous upper and lower conducting sheets. The wave between the sheets is a plane wave with **E** and **H** identical to the

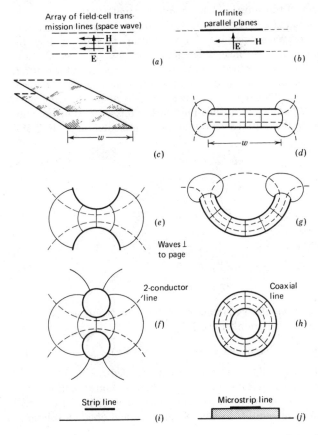

Figure 10-9 Evolution of two-wire, coaxial and microstrip transmission lines from field-cell transmission lines.

space wave but, if the sheets are infinite in extent, we now have an infinite parallel-plane transmission line. Taking a section of this line of width w, as shown in perspective in (c) and in cross section in (d), and bending the sheets outward from each other, as in (e), or bending them both in the same direction as in (g), we end up with two-conductor transmission line of (f) and coaxial line of (h). However, if the upper sheet of the line in (c) and (d) is reduced in width as suggested in (i) and dielectric material introduced between it and the lower sheet, as in (j), we obtain a microstrip transmission line, a type used extensively in integrated circuitry.

10-4 THE INFINITE UNIFORM TRANSMISSION LINE: CHARACTERISTIC IMPEDANCE

Continuing the analysis of transmission lines, consider a uniform two-conductor line of infinite extent, as in Fig. 10-8. Equations (10-2-58) and (10-2-59) are the basic differential equations, or wave equations, for a uniform transmission line. In mathematical terminology they are linear differential equations of the second order with constant coefficients. They are the most general way of expressing the natural law relating the variation of voltage and current with distance along a uniform transmission line. However, they tell us nothing specifically about the voltage or current distribution on a particular transmission line. For this we must first obtain a solution appropriate to the imposed conditions. As a trial solution of (10-2-58) let us substitute

$$V = e^{\gamma x} \tag{1}$$

from which

$$\frac{d^2 V}{dx^2} = \gamma^2 e^{\gamma x} = \gamma^2 V \tag{2}$$

Thus, (10-2-58) becomes

$$(\gamma^2 - ZY)e^{\gamma x} = 0 \tag{3}$$

and

$$\gamma^2 - ZY = 0 \tag{4}$$

Equation (4), known as the *auxiliary equation*, has two unequal roots $+ \sqrt{ZY}$ and $- \sqrt{ZY}$, so that the general solution for (10-2-58) is

$$V = C_1 \exp (\sqrt{ZY}\, x) + C_2 \exp (- \sqrt{ZY}\, x) \tag{5}†$$

where C_1 and C_2 are constants. Thus (1) is a solution provided $\gamma = \pm \sqrt{ZY}$.

If (10-2-59) is solved in the same fashion as (10-2-58), we obtain a solution for I similar in form to (5) but having two more constants. Instead of solving for I in this manner, let us proceed along another avenue of approach and obtain a solution

---

† Read $\exp (\sqrt{zy}\, x)$ as $e^{\sqrt{zy}\, x}$.

for I from (5). To do this, let (5) be differentiated with respect to x. Recalling also (10-2-54), we obtain

$$\frac{dV}{dx} = C_1\sqrt{ZY}\exp(\sqrt{ZY}\,x) - C_2\sqrt{ZY}\exp(-\sqrt{ZY}\,x) = IZ \qquad (6)$$

from which it follows that

$$I = \frac{C_1}{\sqrt{Z/Y}}\exp(\sqrt{ZY}\,x) - \frac{C_2}{\sqrt{Z/Y}}\exp(-\sqrt{ZY}\,x) \qquad (7)$$

This is a solution for the current. To evaluate the constants, we note from (5) that when $x = 0$,

$$V = C_1 + C_2 \qquad (8)$$

where V is the instantaneous voltage at the point $x = 0$ on the line. We may regard this voltage as the sum of two voltages which, in general, are unequal in amplitude and vary harmonically with time. Let V_1 and V_2 be the amplitudes of the voltages. The quantities C_1 and C_2 are constants with respect to x but may be regarded as variables with respect to time. Thus we can put $C_1 = V_1 e^{j\omega t}$ and $C_2 = V_2 e^{j\omega t}$. Substituting these into (5) and (7) yields

$$V = V_1 e^{j\omega t}\exp(\sqrt{ZY}\,x) + V_2 e^{j\omega t}\exp(-\sqrt{ZY}\,x) \qquad (9)$$

and

$$I = \frac{V_1 e^{j\omega t}}{\sqrt{Z/Y}}\exp(\sqrt{ZY}\,x) - \frac{V_2 e^{j\omega t}}{\sqrt{Z/Y}}\exp(-\sqrt{ZY}\,x) \qquad (10)$$

The quantity $\sqrt{ZY} = \gamma$ is called the *propagation constant*. In general it is complex, with a real part α called the *attenuation constant* and an imaginary part β called the *phase constant*. Thus

$$\gamma = \sqrt{ZY} = \alpha + j\beta \qquad (11)$$

or

$$\alpha = \operatorname{Re}\sqrt{ZY} \qquad \text{Np m}^{-1} \qquad (12)$$

and

$$\beta = \operatorname{Im}\sqrt{ZY} \qquad \text{rad m}^{-1} \qquad (13)$$

Introducing (11) into (9) and (10) and rearranging, we obtain

$$V = V_1 e^{\alpha x} e^{j(\omega t + \beta x)} + V_2 e^{-\alpha x} e^{j(\omega t - \beta x)} \qquad (14)$$

and

$$I = \frac{V_1}{\sqrt{Z/Y}} e^{\alpha x} e^{j(\omega t + \beta x)} - \frac{V_2}{\sqrt{Z/Y}} e^{-\alpha x} e^{j(\omega t - \beta x)} \qquad (15)$$

Equation (14) is the solution for the voltage on the transmission line. The solution has two terms. The first term, involving $\omega t + \beta x$, represents a wave traveling in the *negative* x direction along the line. The magnitude of this wave at

$x = 0$ and $t = 0$ is V_1, and the factor $e^{\alpha x}$ indicates that this wave decreases in magnitude as it proceeds in the negative x direction. The second term, involving $\omega t - \beta x$, represents a wave traveling in the *positive* x direction along the line. The magnitude of this wave at $x = 0$ and $t = 0$ is V_2, and the factor $e^{-\alpha x}$ indicates that this wave decreases in magnitude as it proceeds in the positive x direction. The factors $e^{\alpha x}$ and $e^{-\alpha x}$ are *attenuation factors*, α being the *attenuation constant*. The factors $e^{j(\omega t + \beta x)}$ and $e^{j(\omega t - \beta x)}$ are *phase factors*, β being the *phase constant*.

The solution for the current in (15) also has two terms, the first term representing a current wave traveling in the negative x direction and the second term a current wave traveling in the positive x direction.

Thus, the total voltage or the total current at any point is the resultant of two traveling-wave components.

Confining our attention now to a single wave traveling in the negative x direction as represented by the first terms of (14) and (15), we note that V and I are identical functions of x and t. The amplitudes differ. Taking the ratio of the voltage V across the line to the current I through the line for a single traveling wave, we obtain an impedance Z_0, which is called the *characteristic impedance* of the line. That is,

$$\frac{V}{I} = \sqrt{\frac{Z}{Y}} = Z_0 \qquad (\Omega) \qquad (16)$$

This impedance is a function of the line parameters of series impedance Z per unit length and shunt admittance Y per unit length. Expanding Z and Y as in (10-2-52), we obtain, from (16),

$$\boxed{Z_0 = \sqrt{\frac{R + j\omega L}{G + j\omega C}} \qquad (\Omega)} \qquad (17)$$

When R and G are zero (line lossless) or when the frequency is large, so that $\omega L \gg R$ and $\omega C \gg G$, (17) reduces to

$$Z_0 = \sqrt{\frac{L}{C}} \qquad (\Omega) \qquad (18)$$

where $Z_0 = $ characteristic impedance, Ω
$\quad L = $ series inductance, H m^{-1}
$\quad C = $ shunt capacitance, F m^{-1}

In (18) Z_0 is entirely real, or resistive, so that in this case we may, to be explicit, speak of the *characteristic resistance* R_0 of the line. That is, for this case

$$\boxed{Z_0 = \sqrt{\frac{L}{C}} = R_0 \qquad (\Omega)} \qquad (19)$$

In general, when R and G cannot be neglected, Z_0 is complex and the term *characteristic impedance* should be used.

Table 10-1 Characteristic impedance of transmission lines

Condition	Characteristic impedance, Ω
General case	$Z_0 = \sqrt{\dfrac{Z}{Y}} = \sqrt{\dfrac{R + j\omega L}{G + j\omega C}}$
Small losses	$Z_0 = \sqrt{\dfrac{L}{C}}\left[1 + j\left(\dfrac{G}{2\omega C} - \dfrac{R}{2\omega L}\right)\right]$
Lossless case, $R = 0$, $G = 0$	$Z_0 = \sqrt{\dfrac{L}{C}} = R_0$

When R and G are small, but not small enough to be neglected, (17) may be reexpressed approximately in the following form:

$$Z_0 = \sqrt{\frac{L}{C}}\left[1 + j\left(\frac{G}{2\omega C} - \frac{R}{2\omega L}\right)\right] \tag{20}$$

Thus Z_0 for this case is, in general, complex. However, if

$$\frac{G}{C} = \frac{R}{L} \tag{21}$$

Z_0 is real. This situation is *Heaviside's condition* for a distortionless line.

The relations developed above for the characteristic impedance of a uniform transmission line are summarized in Table 10-1.

The phase velocity v of a wave traveling on the line is given by ω/β. That is,

$$v = \frac{\omega}{\beta} = \frac{\omega}{\mathrm{Im}\,\gamma} = \frac{\omega}{\mathrm{Im}\,\sqrt{ZY}} \tag{22}$$

If the line is lossless ($R = 0$ and $G = 0$) or $R \ll \omega L$ and $G \ll \omega C$,

$$\boxed{v = \frac{\omega}{\omega\sqrt{LC}} = \frac{1}{\sqrt{LC}} \quad (\mathrm{m\ s^{-1}})} \tag{23}$$

where L = series inductance, $\mathrm{H\ m^{-1}}$
      C = shunt capacitance, $\mathrm{F\ m^{-1}}$

10-5 IMPEDANCE OF TRANSMISSION LINES AND OF MEDIA

As discussed in Sec. 10-4, the characteristic impedance of a transmission line is equal to the ratio of the voltage V across the line to the current I through the line for a single traveling wave, or

$$Z_0 = \frac{V}{I} \quad (\Omega) \tag{1}$$

For a field-cell transmission line (see Fig. 10-6), $V = Eh$ and $I = Hw$, where E is the electric field intensity and H the magnetic field. Since the height h and width w of a field cell are equal, the characteristic impedance of a field-cell transmission line is

$$Z_0 = \frac{V}{I} = \frac{Eh}{Hw} = \frac{E}{H} \quad (\Omega) \qquad (2)$$

From (10-4-19) the characteristic impedance Z_0 of lossless transmission lines (including field-cell transmission lines) is

$$Z_0 = \sqrt{\frac{L}{C}} \quad (\Omega) \qquad (3)$$

In (3) the impedance Z_0 is expressed in terms of *circuit quantities*.

For a field-cell transmission line, $L = \mu$ and $C = \epsilon$, so that its impedance can be expressed in terms of *field quantities* as

$$Z_0 = \sqrt{\frac{\mu}{\epsilon}} \quad (\Omega) \qquad (4)$$

and from (2) as

$$Z_0 = \frac{E}{H} = \sqrt{\frac{\mu}{\epsilon}} \quad (\Omega) \qquad (5)$$

Here Z_0 is called the *intrinsic impedance* of the medium to distinguish it from the *characteristic impedance* of a transmission line. Thus, *the characteristic impedance of a field-cell transmission line is equal to the intrinsic impedance of the medium.* If the medium is vacuum, we have

$$Z_0 = \sqrt{\frac{\mu_0}{\epsilon_0}} = \sqrt{\frac{4\pi \times 10^{-7} \text{ H m}^{-1}}{8.85 \times 10^{-12} \text{ F m}^{-1}}} = 376.731 \simeq 120\pi \quad (\Omega) \qquad (6)$$

This value (376.7 Ω) is the *intrinsic impedance of vacuum* or empty space.

In the more general situation where the medium is conducting, we have, by comparison to the equivalent transmission-line case, that the intrinsic impedance is given by

$$Z_0 = \sqrt{\frac{j\omega\mu}{\sigma + j\omega\epsilon}} \quad (\Omega) \qquad (7)$$

In (6) Z_0 is real (resistive), but in (7) it is complex.†

---

† If the medium has both conductivity and dielectric hysteresis losses, then the denominator in (7) becomes $\sigma + j\omega(\epsilon' - j\epsilon'')$. If we go further and define a general permittivity $\epsilon = \epsilon' - j(\epsilon'' + \sigma/\omega)$, then the simple expression for impedance $Z_0 = \sqrt{\mu/\epsilon}$ can apply in general.

The permeability may also be complex; that is, $\mu = \mu' - j\mu''$. Introducing this complex permeability into (7), the intrinsic impedance is

$$Z_0 = \sqrt{\frac{\omega\mu'' + j\omega\mu'}{\sigma + j\omega\epsilon}} \quad (\Omega) \qquad (8)$$

By comparison with (10-4-17) for the transmission line, it is apparent that $\omega\mu''$ is analogous to a series resistance.

The concept of intrinsic impedance and field-cell transmission lines is useful in connection with the determination of the characteristic impedance of lossless transmission lines operating in the TEM mode. With a field map, each square (or curvilinear square) represents the cross section of a field-cell transmission line of characteristic impedance $\sqrt{\mu/\epsilon}$. The characteristic impedance of a transmission line is then

$$Z_0 = \frac{N_s}{N_p} \sqrt{\frac{\mu}{\epsilon}} \quad (\Omega) \tag{9}$$

where N_s = number of cells in series

N_p = number of cells in parallel

For air, $\sqrt{\mu/\epsilon} = 376.7\ \Omega$.

Consider the *strip line* in Fig. 10-10a with three cells in parallel directly under the strip ($w/h = 3$) and additional cells in parallel due to fringing field. Taking two cells as representing the fringing field, the characteristic impedance of this line is given approximately by

$$Z_0 \simeq \frac{377}{3 + 2} = 75\ \Omega \tag{10}$$

In general, for strip lines with w greater than the height h ($w > h$), the characteristic impedance is approximately

$$Z_0 \simeq \frac{377}{(w/h) + 2}\ \Omega \tag{10a}$$

where w = strip width

h = height above ground plane (in same units as w)

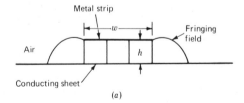

(a)

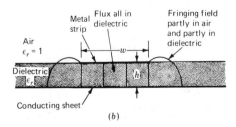

(b)

Figure 10-10 (a) Strip transmission line. (b) Microstrip transmission line with conducting strip of width w mounted on dielectric substrate of height (or thickness) h over a flat conducting sheet.

For strips of width w less than the height h, the formula for a single conductor above a ground plane (with image below) can be used, taking the impedance as half that of the corresponding two-conductor line, or

$$Z_0 \simeq 138 \log \frac{8}{w/h} \; \Omega \tag{11}$$

the flat strip being considered equivalent to a circular conductor with diameter d half the width w of the strip ($w = 2d$). [Compare (11) with (26).]

In many transmission line applications the metal strips are placed on a dielectric substrate, as in Fig. 10-10b. For this arrangement, called a *microstrip line*, the characteristic impedance will be less than with an air dielectric and given approximately by

$$Z_0 \simeq \frac{377}{\sqrt{\epsilon_r}[(w/h) + 2]} \tag{11a}$$

where $\epsilon_r =$ relative permittivity of substrate material. Since the fringing field is only partially in the dielectric, (11a) is good to within a few percent only if $w \geq 2h$. Both (10a) and (11a) approach exactness as the ratio w/h becomes very large.

By dividing the cross section of a transmission line into curvilinear squares by field-mapping techniques, the characteristic impedance of any shape of lossless TEM mode line can be determined by this method. From the field map the inductance per unit length and the capacitance per unit length can also be determined from the number of cells in series and in parallel. Thus,

$$\text{Inductance per unit length} = L = \frac{N_s}{N_p} \mu \quad (\text{H m}^{-1}) \tag{12}$$

$$\text{Capacitance per unit length} = C = \frac{N_p}{N_s} \epsilon \quad (\text{F m}^{-1}) \tag{12a}$$

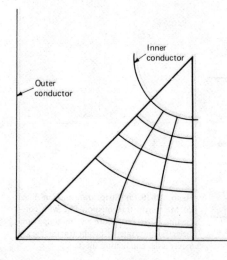

Figure 10-11 Coaxial transmission line with square outer conductor showing field map for one octant. The field maps for all other octants are the same (or mirror images). There are 4.4 field cells in series and $3 \times 8 = 24$ in parallel.

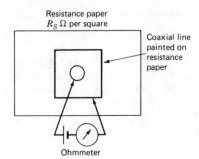

Resistance paper
R_s Ω per square

Coaxial line
painted on
resistance
paper

Ohmmeter

Figure 10-12 Determination of characteristic impedance of transmission line by simple dc measurement.

The characteristic impedance of lossless transmission lines of any shape can
also be obtained by a simple direct-current (dc) measurement. For example, if we
wish to find the parameters of the coaxial line with square outer conductor shown
in cross section in Fig. 10-11 by this method, the conductor cross section is drawn
to scale with conducting paint (such as silver paint) on a sheet of resistance (Teledeltos) paper of uniform resistance R_s per square, as suggested in Fig. 10-12.†
Then by connecting the terminals of an ohmmeter to the inner and outer conductors,
as indicated, a dc resistance R_m is measured.

Since $R_m = (N_s/N_p)R_s$ and $R_m/R_s = N_s/N_p$, it follows that

$$\text{Inductance per unit length} = L = \mu \frac{R_m}{R_s} \quad (\text{H m}^{-1}) \qquad (13)$$

$$\text{Capacitance per unit length} = C = \epsilon \frac{R_s}{R_m} \quad (\text{F m}^{-1}) \qquad (14)$$

$$\text{Characteristic impedance} = Z_0 = \sqrt{\frac{\mu}{\epsilon}} \frac{R_m}{R_s} \quad (\Omega) \qquad (15)$$

where R_s is the resistance of a square piece of the resistance paper measured
between opposite edges.

Thus, if the line is air-filled, $\sqrt{\mu/\epsilon} = \sqrt{\mu_0/\epsilon_0} = 376.7$ Ω and (15) becomes

$$Z_0 = \frac{376.7}{R_s} R_m \quad (\Omega) \qquad (16)$$

Hence, if "space paper" ($R_s = 376.7$ Ω per square) is used,

$$Z_0 = R_m \quad (\Omega) \qquad (17)$$

and the *ohmmeter reads directly the characteristic impedance of the line.*

---

† Since only the shape is important, the cross section may be scaled to any convenient size.

Example 1 Find L, C, and Z for the coaxial line with square outer conductor (Fig. 10-11).

SOLUTION The symmetry is such that a field map needs to cover only one octant. There are three squares in parallel and 4.4 in series in the octant so that for the entire line $N_p = 24$ and $N_s = 4.4$. Thus, from (9) for an air-filled line

$$Z = \frac{N_s}{N_p} Z_0 = \frac{4.4}{24} 376.7 = 69 \ \Omega \tag{18}$$

From (12) and (12a)

$$L = \frac{N_s}{N_p} \mu_0 = \frac{4.4}{24} 4\pi \times 10^{-7} = 230 \text{ nH m}^{-1} \tag{19}$$

$$C = \frac{N_p}{N_s} \epsilon_0 = \frac{24}{4.4} 8.85 \times 10^{-12} = 48 \text{ pF m}^{-1} \tag{20}$$

If a cross section of the coaxial line is drawn to scale with conducting paint on a sheet of space paper, 69 Ω would be measured directly on an ohmmeter connected between the inner and outer conductors.

When this resistance-measurement method is applied to open types of line such as a two-wire line, the sheet of resistance material should extend out to a distance that is large compared with the line cross section if accurate results are to be obtained.

Although graphical, analog-computer, and digital-computer methods can be applied to two-conductor lines of any shape, some configurations yield to a simple calculation. Thus, for the coaxial line of Fig. 10-13 we have from (5-13-8) for L and (3-15-2) for C that its characteristic impedance

$$Z_0 = \sqrt{\frac{L}{C}} = \frac{1}{2\pi} \sqrt{\frac{\mu}{\epsilon}} \ln \frac{b}{a} = 0.367 \sqrt{\frac{\mu}{\epsilon}} \log \frac{b}{a} \quad (\Omega) \tag{21}$$

If there is no ferromagnetic material present, $\mu = \mu_0$ and (21) reduces to

$$Z_0 = \frac{138}{\sqrt{\epsilon_r}} \log \frac{b}{a} \quad (\Omega) \tag{22}$$

where ϵ_r = relative permittivity of medium filling line
a = outside radius of inner conductor
b = inside radius of outer conductor

For an air-filled line $\epsilon_r = 1$, and (22) becomes

$$Z_0 = 138 \log \frac{b}{a} \quad (\Omega) \tag{23}$$

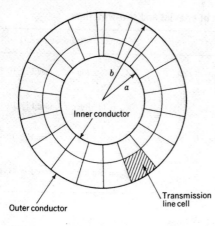

Figure 10-13 Coaxial transmission line with 18.3 transmission-line cells in parallel and 2 in series.

Example 2 The air-filled coaxial line in Fig. 10-13 has a radius ratio $b/a = 2$. Find its characteristic impedance.

SOLUTION From (23)

$$Z_0 = 138 \log 2 = 41.4 \ \Omega$$

From the graphical field map in Fig. 10-13

$$Z_0 = \frac{N_s}{N_p} Z_0' = \frac{2}{18.3} \, 376.7 = 41.2 \ \Omega$$

This value agrees well with the exact value (41.4 Ω).

In a similar way, the characteristic impedance can be obtained for a *two-wire line*, as in Fig. 10-14. Thus, if $D \gg a$, we have

$$Z_0 = \frac{1}{\pi} \sqrt{\frac{\mu}{\epsilon}} \ln \frac{D}{a} = 0.73 \sqrt{\frac{\mu}{\epsilon}} \log \frac{D}{a} \qquad (\Omega) \qquad (24)$$

If there is no ferromagnetic material present, $\mu = \mu_0$ and (24) reduces to

$$Z_0 = \frac{276}{\sqrt{\epsilon_r}} \log \frac{D}{a} \qquad (\Omega) \qquad (25)$$

where ϵ_r = relative permittivity of medium
D = center-to-center spacing (see Fig. 10-14)
a = radius of conductor (in same units as D)

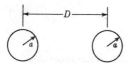

Figure 10-14 Two-wire transmission line.

Table 10-2 Characteristic impedance of coaxial and two-wire lines

Type of line	Characteristic impedance, Ω	
Coaxial (filled with medium of relative permittivity ϵ_r)	$Z_0 = \dfrac{138}{\sqrt{\epsilon_r}} \log \dfrac{b}{a}$	(see Fig. 10-13)
Coaxial (air-filled)	$Z_0 = 138 \log \dfrac{b}{a}$	(see Fig. 10-13)
Two-wire (in medium of relative permittivity ϵ_r) ($D \gg a$)	$Z_0 = \dfrac{276}{\sqrt{\epsilon_r}} \log \dfrac{D}{a}$	(see Fig. 10-14)
Two-wire (in air) ($D \gg a$)	$Z_0 = 276 \log \dfrac{D}{a}$	(see Fig. 10-14)

If the medium is air, $\epsilon_r = 1$ and (25) becomes

$$Z_0 = 276 \log \frac{D}{a} \quad (\Omega) \qquad (26)$$

The characteristic impedances obtained above are summarized in Table 10-2.

It is assumed throughout this section that the line is lossless (or $R \ll \omega L$ and $G \ll \omega C$) and also that the currents are confined to the conductor surfaces to which the radii refer. This condition is approximated at high frequencies owing to the small depth of penetration. This condition may also be approximated at low frequencies by the use of thin-walled tubes. It is also assumed that the lines are operating in the TEM mode.

10-6 THE TERMINATED UNIFORM TRANSMISSION LINE

Thus far we have considered only lines of infinite length. Let us now analyze the situation where a line of characteristic impedance Z_0 is terminated in a load impedance Z_L, as in Fig. 10-15. The load is at $x = 0$, and positive distance x is measured to the left along the line. The total voltage and total current are expressed as the resultant of two traveling waves moving in opposite directions as on an

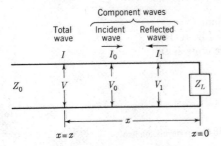

Figure 10-15 Terminated transmission line.

infinite transmission line. However, on the terminated line the wave to the right may be regarded as the incident wave and the wave to the left as the reflected wave, with the reflected wave related to the incident wave by the load impedance Z_L.

At a point on the line at a distance x from the load let the voltage between the wires and the current through one wire due to the incident wave traveling to the right be designated V_0 and I_0, respectively. Let V_1 and I_1 be the voltage and current due to the wave traveling to the left that is reflected from the load. The resultant voltage V at a point on the line is equal to the sum of the voltages V_0 and V_1 at the point. That is, in phasor notation

$$V = V_0 + V_1 \tag{1}$$

where $V_0 = |V_0|e^{\gamma x}$
$V_1 = |V_1|e^{-\gamma x + j\xi}$
γ = propagation constant = $\alpha + j\beta$
ξ = phase shift at load

At the load ($x = 0$) we have $V_0 = |V_0|$ and $V_1 = |V_1|e^{j\xi} = |V_1| \angle \xi$, so that *at the load*

$$\frac{V_1}{V_0} = \frac{|V_1|}{|V_0|} \angle \xi = \rho_v \tag{2}$$

where ρ_v is the *reflection coefficient for voltage* (dimensionless). It follows that

$$V = |V_0|(e^{\gamma x} + \rho_v e^{-\gamma x}) \tag{3}$$

The resultant current I at a point on the line is equal to the sum of the currents I_0 and I_1 at the point. That is,

$$I = I_0 + I_1 \tag{4}$$

where $I_0 = |I_0|e^{\gamma x - j\delta}$
$I_1 = |I_1|e^{-\gamma x + j(\xi - \delta)}$
δ = phase difference between current and voltage

At the load

$$\frac{I_1}{I_0} = \frac{|I_1|}{|I_0|} \angle \xi = \rho_i \tag{5}$$

where ρ_i is the *reflection coefficient for current* (dimensionless). It follows that

$$I = |I_0|e^{-j\delta}(e^{\gamma x} + \rho_i e^{-\gamma x}) \tag{6}$$

Now ρ_v and ρ_i may be expressed in terms of the characteristic impedance Z_0 and the load impedance Z_L. Thus we note that at any point on the line

$$Z_0 = \frac{V_0}{I_0} = \frac{|V_0|}{|I_0|} \angle \delta = -\frac{V_1}{I_1} = -\frac{|V_1|}{|I_1|} \angle \delta \tag{7}$$

while at the load ($x = 0$)

$$Z_L = \frac{V}{I} \tag{8}$$

It follows from (4) that at the load

$$\frac{V}{Z_L} = \frac{V_0}{Z_0} - \frac{V_1}{Z_0} = \frac{V_0 - V_1}{Z_0} \tag{9}$$

But $V = V_0 + V_1$; so we have

$$\frac{V_0 + V_1}{Z_L} = \frac{V_0 - V_1}{Z_0} \tag{10}$$

Solving for V_1/V_0 yields

$$\boxed{\frac{V_1}{V_0} = \frac{Z_L - Z_0}{Z_L + Z_0} = \rho_v} \tag{11}$$

For real load impedances Z_L ranging from 0 to ∞, ρ_v ranges from -1 to $+1$ in value.

In a similar way it can be shown that

$$\rho_i = -\frac{Z_L - Z_0}{Z_L + Z_0} = -\rho_v \tag{12}$$

The ratio V/I at any point x on the line gives the impedance Z_x at the point looking toward the load. Taking this ratio and introducing the relation (12) in (6) for I, we obtain

$$Z_x = \frac{V}{I} = \frac{|V_0|}{|I_0|} \underline{/\delta} \; \frac{e^{\gamma x} + \rho_v e^{-\gamma x}}{e^{\gamma x} - \rho_v e^{-\gamma x}} \tag{13}$$

Noting (7) and (11), we can reexpress this as

$$\boxed{Z_x = Z_0 \frac{Z_L + Z_0 \tanh \gamma x}{Z_0 + Z_L \tanh \gamma x}} \quad (\Omega) \tag{14}$$

where Z_x = impedance at distance x looking toward load, Ω
  Z_0 = characteristic impedance of line, Ω
  Z_L = load impedance, Ω
  γ = propagation constant = $\alpha + j\beta$, m^{-1}
  x = distance from load, m

This is the *general expression* for the line impedance Z_x as a function of the distance x from the load.

If the line is *open-circuited*, $Z_L = \infty$ and (14) reduces to

$$Z_x = \frac{Z_0}{\tanh \gamma x} = Z_0 \coth \gamma x \tag{15}$$

If the line is *short-circuited*, $Z_L = 0$ and (14) reduces to

$$Z_x = Z_0 \tanh \gamma x \tag{16}$$

It is to be noted that, in general, γ is complex $(= \alpha + j\beta)$. Thus

$$\tanh \gamma x = \frac{\sinh \alpha x \cos \beta x + j \cosh \alpha x \sin \beta x}{\cosh \alpha x \cos \beta x + j \sinh \alpha x \sin \beta x} \tag{17}$$

or

$$\tanh \gamma x = \frac{\tanh \alpha x + j \tan \beta x}{1 + j \tanh \alpha x \tan \beta x} \tag{18}$$

Note that the product of the impedance of the line when it is open-circuited and when it is short-circuited equals the square of the characteristic impedance; i.e.,

$$\boxed{Z_0^2 = Z_{oc} Z_{sc}} \tag{19}$$

where $Z_{oc} = Z_x$ for open-circuited line $(Z_L = \infty)$
$Z_{sc} = Z_x$ for short-circuited line $(Z_L = 0)$

If the line is *lossless* ($\alpha = 0$), the above relations reduce to the following: *In general,*

$$\boxed{Z_x = Z_0 \frac{Z_L + jZ_0 \tan \beta x}{Z_0 + jZ_L \tan \beta x}} \tag{20}$$

When the line is *open-circuited* ($Z_L = \infty$),

$$\boxed{Z_x = \frac{Z_0}{j \tan \beta x} = -jZ_0 \cot \beta x} \tag{21}$$

When the line is *short-circuited* ($Z_L = 0$),

$$\boxed{Z_x = jZ_0 \tan \beta x} \tag{22}$$

We note that (19) is also fulfilled on the lossless line. Furthermore, the impedance for an open- or short-circuited lossless line is a pure reactance.

The impedance relations developed above apply to all uniform two-conductor lines, such as coaxial and two-wire lines. They give the input impedance Z_x of a uniform transmission line of length x and characteristic impedance Z_0 terminated in a load Z_L (see Fig. 10-16). These relations are summarized in Table 10-3.

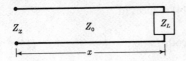

Figure 10-16 Terminated transmission line.

On a lossless line the *Voltage Standing-Wave Ratio* (VSWR) is given by

$$\text{VSWR} = \frac{V_{\text{max}}}{V_{\text{min}}} = \frac{I_{\text{max}}}{I_{\text{min}}} \qquad (23)$$

It follows that

$$\text{VSWR} = \frac{|V_0| + |V_1|}{|V_0| - |V_1|} = \frac{1 + (|V_1|/|V_0|)}{1 - (|V_1|/|V_0|)} \qquad (24)$$

But

$$\frac{|V_1|}{|V_0|} = |\rho_v| \qquad (25)$$

and so

$$\text{VSWR} = \frac{1 + |\rho_v|}{1 - |\rho_v|} \qquad (26)$$

where ρ_v is the reflection coefficient for voltage. This relation is identical with that given by (10-13-16) for the VSWR of plane waves. Solving (26) for the magnitude of the reflection coefficient gives

$$|\rho_v| = \frac{\text{VSWR} - 1}{\text{VSWR} + 1} \qquad (27)$$

It is often of interest to know the voltage V at the load in terms of the voltage V_0 of the incident wave. This is given by the *transmission coefficient for voltage* τ_v. That is, at the load

$$V = \tau_v V_0 \qquad \text{or} \qquad \tau_v = \frac{V}{V_0} \qquad (28)$$

Table 10-3 Input impedance of terminated transmission line†

Load condition	General case ($\alpha \neq 0$)	Lossless case ($\alpha = 0$)
Any value of load Z_L	$Z_x = Z_0 \dfrac{Z_L + Z_0 \tanh \gamma x}{Z_0 + Z_L \tanh \gamma x}$	$Z_x = Z_0 \dfrac{Z_L + jZ_0 \tan \beta x}{Z_0 + jZ_L \tan \beta x}$
Open-circuited line ($Z_L = \infty$)	$Z_x = Z_0 \coth \gamma x$	$Z_x = -jZ_0 \cot \beta x$
Short-circuited line ($Z_L = 0$)	$Z_z = Z_0 \tanh \gamma x$	$Z_x = jZ_0 \tan \beta x$

† $\gamma = \alpha + j\beta$, where α = attenuation constant in nepers per meter and $\beta = 2\pi/\lambda$ = phase constant in radians per meter, where λ is the wavelength.

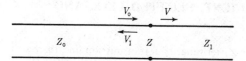

Figure **10-17** Junction of transmission lines of different characteristic impedance.

The load impedance may be a lumped element, as suggested in Fig. 10-15 or Fig. 10-16, or it may be the impedance presented by another line of characteristic impedance Z_1, as suggested in Fig. 10-17. In the latter case (28) gives the voltage V of the wave transmitted beyond the junction.

It may be shown that the coefficient τ_v is related to Z_L and Z_0 by

$$\tau_v = \frac{2Z_L}{Z_L + Z_0} = 1 + \rho_v \qquad (29)$$

where Z_L = load impedance presented to line of characteristic impedance Z_0
   ρ_v = reflection coefficient for voltage

As Z_L ranges from 0 to ∞, τ_v ranges from 0 to 2.

It also follows that the *transmission coefficient for current* τ_i is given by

$$\tau_i = \frac{I}{I_0} = \frac{2Z_0}{Z_0 + Z_L} = 1 + \rho_i \qquad (30)$$

As Z_L ranges from 0 to ∞, τ_i varies from 2 to 0.

The relations for reflection and transmission coefficients developed in this section are summarized in Table 10-4.

Table 10-4 Relations for reflection and transmission coefficients

Reflection coefficient for voltage	$\rho_v = \dfrac{Z_L - Z_0}{Z_L + Z_0}$
Reflection coefficient for current	$\rho_I = \dfrac{Z_0 - Z_L}{Z_0 + Z_L} = -\rho_v$
Transmission coefficient for voltage	$\tau_v = \dfrac{2Z_L}{Z_0 + Z_L} = 1 + \rho_v$
Transmission coefficient for current	$\tau_i = \dfrac{2Z_0}{Z_0 + Z_L} = 1 + \rho_i$
Voltage standing-wave ratio (VSWR)	$\dfrac{1 + |\rho_v|}{1 - |\rho_v|} = \dfrac{1 + |\rho_i|}{1 - |\rho_i|}$
Magnitude of reflection coefficient	$|\rho_v| = |\rho_i| = \dfrac{\text{VSWR} - 1}{\text{VSWR} + 1}$

10-7 REFLECTION COEFFICIENT, SLOTTED LINE, AND SMITH CHART

In general, for any load impedance Z_L terminating a transmission line, there will be a reflected wave with reflection coefficient ρ_v and a *Voltage Standing Wave Ratio* (VSWR) related as follows:

$$|\rho_v| = \frac{|\text{reflected voltage}|}{|\text{incident voltage}|} = \frac{\text{VSWR} - 1}{\text{VSWR} + 1} \qquad (1)$$

where the VSWR is the ratio of the maximum to minimum voltage on the line.

The reflection coefficient is a complex quantity with magnitude $|\rho_v|$ and phase angle θ_v. Thus,

$$\rho_v = |\rho_v|\underline{/\theta_v} \qquad (2)$$

The VSWR can be measured by moving a voltage probe along the line, as will be discussed in detail in connection with the slotted (measuring) line. The value of $|\rho_v|$ is then given from (1). Using the probe also to locate the voltage minimum point on the line, the reflection coefficient phase angle θ_v is found from

$$\theta_v = 720° \left(\frac{x_{vm}}{\lambda} - \frac{1}{4} \right) \qquad (3)$$

where x_{vm} = distance of voltage minimum from load, m
λ = wavelength, m

Knowing both the magnitude and phase angle of the reflection coefficient, the load impedance Z_L is given by

$$Z_L = Z_0 \frac{1 + |\rho_v|\underline{/\theta_v}}{1 - |\rho_v|\underline{/\theta_v}} \qquad (4)$$

Thus, an unknown load impedance can be determined from measurements of the standing wave pattern on a transmission line.

The relation of the VSWR, x_{vm}, ρ_v, and Z_L for several special cases are illustrated in Fig. 10-18 with the voltage variation along the line shown for each case.

Case 1 is a *matched line* with load impedance (100 Ω) equal to the characteristic impedance of the line (100-Ω resistive). There is no reflected wave, and so $\rho_v = 0$; and there are no standing waves (VSWR = 1).

Case 2 is an *open-circuited line* ($Z_L = \infty$) so that the reflected wave is equal in magnitude to the incident wave ($|\rho_v| = 1$) and the VSWR = ∞. At the load the reflected voltage wave is in phase with the incident voltage wave ($\theta_v = 0$) so that there is a voltage maximum at the load with a voltage minimum $\lambda/4$ away ($x_{vm} = \lambda/4$).

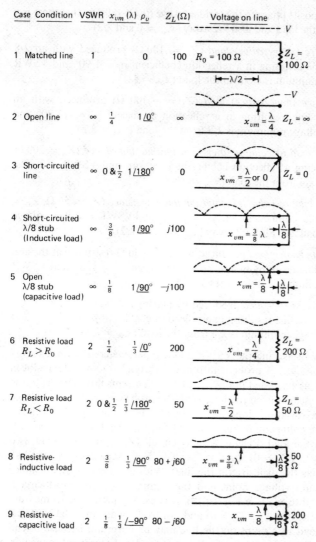

Case	Condition	VSWR	x_{vm} (λ)	ρ_v	Z_L (Ω)	Voltage on line
1	Matched line	1		0	100	$R_0 = 100\ \Omega$, $Z_L = 100\ \Omega$
2	Open line	∞	$\frac{1}{4}$	$1\underline{/0°}$	∞	$x_{vm} = \frac{\lambda}{4}$, $Z_L = \infty$
3	Short-circuited line	∞	$0\ \&\ \frac{1}{2}$	$1\underline{/180°}$	0	$x_{vm} = \frac{\lambda}{2}$ or 0, $Z_L = 0$
4	Short-circuited λ/8 stub (Inductive load)	∞	$\frac{3}{8}$	$1\underline{/90°}$	$j100$	$x_{vm} = \frac{3}{8}\lambda$, $\frac{\lambda}{8}$
5	Open λ/8 stub (capacitive load)	∞	$\frac{1}{8}$	$1\underline{/90°}$	$-j100$	$x_{vm} = \frac{\lambda}{8}$, $\frac{\lambda}{8}$
6	Resistive load $R_L > R_0$	2	$\frac{1}{4}$	$\frac{1}{3}\underline{/0°}$	200	$x_{vm} = \frac{\lambda}{4}$, $Z_L = 200\ \Omega$
7	Resistive load $R_L < R_0$	2	$0\ \&\ \frac{1}{2}$	$\frac{1}{3}\underline{/180°}$	50	$x_{vm} = \frac{\lambda}{2}$, $Z_L = 50\ \Omega$
8	Resistive-inductive load	2	$\frac{3}{8}$	$\frac{1}{3}\underline{/90°}$	$80 + j60$	$x_{vm} = \frac{3}{8}\lambda$, $\frac{\lambda}{8}$ 50 Ω
9	Resistive-capacitive load	2	$\frac{1}{8}$	$\frac{1}{3}\underline{/-90°}$	$80 - j60$	$x_{vm} = \frac{\lambda}{8}$, $\frac{\lambda}{8}$ 200 Ω

Figure 10-18 Transmission line of 100-Ω characteristic resistance for nine different terminations with columns giving VSWR, distance of voltage minimum from terminals (x_{vm}), reflection coefficient (ρ_v), and load impedance (Z_L) at the terminals. In the sketch of the line (at right) the dashed lines show the relative voltage on the line.

Case 3 is a *short-circuited line* ($Z_L = 0$) with the reflected wave equal in magnitude to the incident wave ($|\rho_v| = 1$) and the VSWR $= \infty$. At the load the reflected voltage wave is in opposite phase to the incident wave ($\theta_v = 180°$) so that there is a voltage minimum at the load and also $\lambda/2$ away ($x_{v,n} = 0$ and $\lambda/2$).

Case 4 is an *inductive* (*nonresistive*) *load* ($Z_L = j100 \, \Omega$) produced with an $\lambda/8$ section of shorted line resulting in a reflection coefficient $\rho_v = 1\underline{/90°}$ and VSWR $= \infty$. The voltage minimum is $\frac{3}{8}\lambda$ from the load ($x_{vm} = \frac{3}{8}\lambda$).

Case 5 is a *capacitive* (*nonresistive*) *load* ($Z_L = -j100 \, \Omega$) produced with an $\lambda/8$ section of open line, resulting in a reflection coefficient $\rho_v = 1\underline{/-90°}$ and VSWR $= \infty$. The voltage minimum is $\lambda/8$ from the load ($x_{vm} = \lambda/8$).

Case 6 is a *resistive load of value greater than the line impedance* ($Z_L = 200 \, \Omega$, $Z_0 = 100 \, \Omega$), producing a reflection coefficient $\rho_v = \frac{1}{3}\underline{/0°}$ and VSWR $= 2$. The voltage minimum is $\lambda/4$ from the load ($x_{vm} = \lambda/4$).

Case 7 is a *resistive load of value less than the line impedance* ($Z_L = 50 \, \Omega$, $Z_0 = 100 \, \Omega$) producing a reflection coefficient $\rho_v = \frac{1}{3}\underline{/180°}$ and VSWR $= 2$. The voltage minimum is at the load and also $\lambda/2$ away ($x_{vm} = 0$ and $\lambda/2$).

Case 8 is a *resistive-inductive load* produced by placing 50-Ω resistance at the end of an $\lambda/8$ section of line resulting in a reflection coefficient $\rho_v = \frac{1}{3}\underline{/90°}$ and VSWR $= 2$. The voltage minimum is $\frac{3}{8}\lambda$ from the load ($x_{vm} = \frac{3}{8}\lambda$).

Case 9 is a *resistive-capacitive load* produced by placing a 200-Ω resistance at the end of an $\lambda/8$ section of line resulting in a reflection coefficient $\rho_v = \frac{1}{3}\underline{/-90°}$ and VSWR $= 2$. The voltage minimum is $\lambda/8$ from the load ($x_{vm} = \lambda/8$).

One method for measuring the VSWR and x_{vm} parameters is with a *slotted (coaxial) transmission line*.† A probe is introduced through a longitudinal slot in the outer conductor, as suggested in Fig. 10-19. Since currents flow parallel to the slot, very little field escapes from the line. With the probe connected to a voltage indicator the voltage variation along the line can be determined, giving both the VSWR (= maximum voltage/minimum voltage) and x_{vm}, the distance of the voltage minimum from the load. Placing a short circuit ($Z_L = 0$) at the load, two successive minima, $V_{\min}(Z_L = 0)$, can be located. Their separation is one-half wavelength (for air as the dielectric in the line). With any load Z_L the distance between the minimum voltage point and the minimum for the short circuit, $V_{\min}(Z_L = 0)$, is equal to x_{vm} (see Fig. 10-19). It is usually preferable to measure x_{vm} with respect to the first $V_{\min}(Z_L = 0)$ and not to the load since the electrical distance from the load to a point on the line may be uncertain because of end effects caused, for example, by the use of a dielectric bead in the terminal connector, which modifies its electrical length.

An overall picture of what is happening on a transmission line can be obtained by constructing a polar coordinate chart, as in Fig. 10-20, in which radial distance gives the magnitude of the reflection coefficient $|\rho_v|$ [and also the VSWR, related

---

† The slotted line is valuable as an instructional aid since it illustrates nicely the relation between the magnitude and location of a standing wave and the corresponding point on a Smith chart.

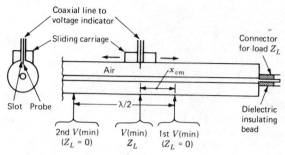

Figure 10-19 Slotted line in longitudinal and cross sections with movable probe for measuring voltages along the line.

to $|\rho_v|$ as in (1)] while the angle represents the reflection coefficient angle θ_v and also the distance x_{vm} of the voltage minimum from the load (or first voltage minimum for a short circuit at the load).

Referring to the nine situations discussed above, Case 1 is represented on the chart by a point at the origin (center of chart) with $\rho_v = 0$ and VSWR = 1, while Cases 2, 3, 4, and 5 are represented by points on the periphery of the chart ($|\rho_v| = 1$

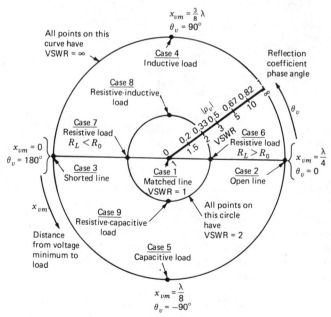

Figure 10-20 Chart of reflection coefficient ρ_v with magnitude and VSWR as radial distance. The reflection coefficient phase and the distance from the load appear as an angle. Points are marked on the chart corresponding to the nine different cases of Fig. 10-18.

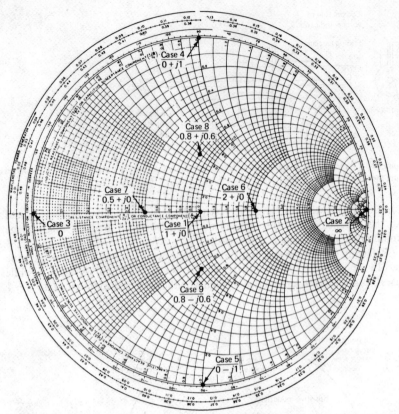

Figure 10-21 A Smith chart (above) is like a reflection coefficient diagram, as in Fig. 10-20, but with normalized impedance (R_n and X_n) or normalized admittance (G_n and B_n) values superposed. The nine cases of Fig. 10-20 are indicated with corresponding normalized impedance values. Note the same positions of the points for the nine cases in Figs. 10-20 and 10-21. See Appendix for full-size Smith chart.

and VSWR $= \infty$). Cases 6, 7, 8, and 9 are represented by points on the VSWR $= 2$ circle ($|\rho_v| = \frac{1}{3}$).

Each point on the chart also represents a load impedance Z_L as given by (4). By a suitable coordinate transformation (see Prob. 10-2-27) the real and imaginary parts of Z_L may be superposed forming a *Smith chart*† (see Fig. 10-21). The positions of each of the nine cases are shown on the chart in terms of a *normalized impedance* Z_n (dimensionless). Since a lossless line is assumed ($Z_0 = R_0$),

$$Z_n = \frac{Z_L}{R_0} = \frac{R_L}{R_0} + j\frac{X_L}{R_0} = R_n + jX_n$$

† P. H. Smith, Transmission Line Calculator, *Electronics*, **12**:29–31 (January 1939).

Since $Z_0 = R_0 = 100\ \Omega$, we have for the nine cases

Case	$Z_n = R_n + jX_n$ (dimensionless)
1	$1 + j0$
2	∞
3	0
4	$0 + j1$
5	$0 - j1$
6	$2 + j0$
7	$0.5 + j0$
8	$0.8 + j0.6$
9	$0.8 - j0.6$

The Smith chart can also be used for admittances, the normalized admittance being given by

$$Y_n = \frac{1}{Z_n} = G_n + jB_n$$

where G_n = normalized conductance $= \dfrac{G}{G_0}$

B_n = normalized susceptance $= \dfrac{B}{G_0}$

and where

$$G_0 = \text{characteristic conductance of line} = \frac{1}{R_0}, \mho$$

Moving on a constant VSWR circle halfway around the chart converts from impedance to admittance or vice versa. For example, to convert a normalized impedance $Z_n = 1 + j1$ to a normalized admittance, we move halfway around the chart, obtaining $Y_n = 0.5 - j0.5$. Moving halfway around the chart is equivalent to moving $\lambda/4$ along the transmission line.

The Smith chart is extremely useful in dealing with transmission lines. It not only shows the relation of ρ_v, VSWR, and the impedance of the load Z_L, but it also gives the impedance Z_x anywhere along the line as points on the appropriate VSWR circle. The VSWR circles are usually not shown on the chart but can be constructed as needed with a compass.

The utility of the Smith chart will be illustrated by several worked examples.

Example 1 Referring to Fig. 10-22, a transmission line of characteristic resistance $R_0 = 100\ \Omega$ is terminated in a load of $50 + j40\ \Omega$. What is (a) the VSWR, (b) the impedance Z_x at $x = 0.2\lambda$, and (c) the two shortest distances for which Z_x is nonreactive?

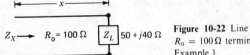

Figure 10-22 Line with characteristic resistance $R_0 = 100\,\Omega$ terminated in load $Z_L = 50 + j40\,\Omega$ for Example 1.

SOLUTION The normalized load impedance

$$Z_n = \frac{50 + j40\ \Omega}{100\ \Omega} = 0.5 + j0.4 \qquad \text{(dimensionless)}$$

Locating this Z_n value (point P_1) on the Smith chart, a compass is used to draw the VSWR circle through P_1, as in Fig. 10-23. This circle crosses the real axis at 2.4 so the VSWR = 2.4 [answer to (a)].

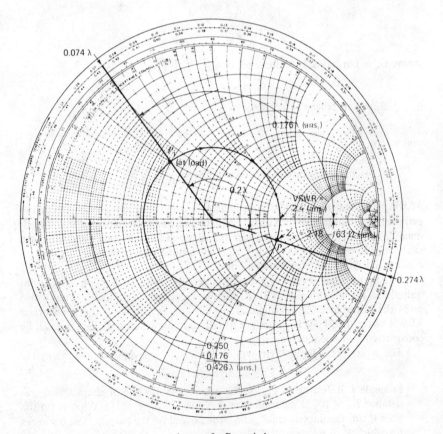

Figure 10-23 Smith chart positions and moves for Example 1.

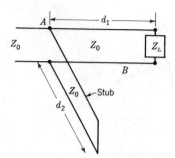

Figure 10-24 Terminated transmission line with single-stub tuner. Both the stub position d_1 and its length d_2 are adjustable. See Example 2.

Now drawing a radial line from the center of the chart through P_1, we note that its distance x is 0.074λ (at chart periphery). To move 0.2λ toward the generator (away from the load) we move clockwise on the VSWR = 2.4 circle 0.2λ to the line intersecting the periphery at $0.074 + 0.2 = 0.274\lambda$ (point P_2). The normalized impedance at this point is $2.18 - j0.63$. The impedance Z_x at 0.2λ from the load is this value times 100 Ω, or

$$Z_x = 218 - j63 \ \Omega \qquad [\text{answer to } (b)]$$

Finally, we note that proceeding clockwise on the VSWR = 2.4 circle from the load toward the generator, the first crossing of the real axis (horizontal bisector of chart) occurs at a distance $x = 0.25 - 0.074 = 0.176\lambda$ and a second crossing at an additional 0.25λ distance or $x = 0.176 + 0.25 = 0.426\lambda$ [answer to (c)]. Added crossings occur at intervals of 0.25λ. The impedance Z_x on the line at $x = 0.176\lambda$ is $240 + j0 \ \Omega$ and at $x = 0.426\lambda$ is $42 + j0 \ \Omega$.

Example 2 Referring to the terminated transmission line with short-circuited stub shown in Fig. 10-24 the load $Z_L = 150 + j50 \ \Omega$. The line and stubs have a characteristic impedance $Z_0 = R_0 = 100 \ \Omega$. Find the shortest values of d_1 and d_2 for which there is no reflected wave at A (VSWR = 1).

SOLUTION The normalized impedance of the load is

$$Z_n = \frac{Z_L}{R_0} = \frac{150 + j50 \ \Omega}{100 \ \Omega} = 1.5 + j0.5 \qquad \text{(dimensionless)}$$

The Smith chart is entered at the point $1.5 + j0.5$, as indicated by P_1 in Fig. 10-25. For clarity only a few coordinate lines are shown in this figure. Point P_1 is on the VSWR circle intersecting the real axis at 1.77, so that the VSWR = 1.77 at B. Since the stub is connected in parallel with the line, it will be *advantageous to work in admittances*; so moving from P_1 $\lambda/4$ (90 electrical degrees) to P_2 converts Z_n to a normalized admittance $Y_n = 0.6 - j0.2$ at the load.† [Matching a line using a Smith chart is like playing a game in which

---

† Moving $\lambda/4$ on the chart transforms an impedance to an admittance (or vice versa) at the same location on the line.

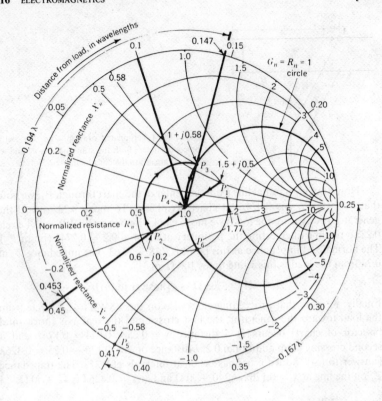

Figure 10-25 Smith chart positions and moves for single-stub tuner of Example 2.

the object is to reach the center of the chart (home) with a minimum of moves and shortest line distances.] Now to get home, we proceed on a constant VSWR circle until we reach a constant admittance curve which leads home ($G_n = R_n = 1$ circle). Thus, we move a distance $d_1 = 0.194\lambda$ from P_2 to P_3 on the chart to the normalized admittance $Y_n = 1.0 + j0.58$. We are now at point A and ready to connect the stub. If the stub length is now adjusted so that the stub presents a normalized admittance of $-j0.58$ at A, the total admittance at A is $1.0 + j0.58 - j0.58 = 1.0 + j0$ and the line is matched (VSWR = 1). This corresponds to a move from P_3 to P_4 on the chart. A normalized admittance $Y_n = j0.58$ (pure susceptance) is indicated at P_5, and we note that the distance required from a short circuit ($Y_n = \pm j\infty$) to obtain this value is 0.167λ. A short-circuited stub has an infinite VSWR so that the admittance at points along the stub are on the circle at the edge of the chart. Thus the required stub length is

$$d_2 = 0.417 - 0.25 = 0.167\lambda$$

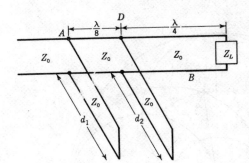

Figure 10-26 Double-stub tuner with short-circuited stubs for Example 3.

The required distance of the stub from the load as obtained above is

$$d_1 = 0.147 + (0.500 - 0.453) = 0.194\lambda$$

We note that matching with the stub can be accomplished from any point on the $R_n = 1$ circle. Thus, we might have moved from P_2 beyond P_3 all the way to P_6. However, both d_1 and d_2 would then be larger.

Example 3 Consider the terminated line with the two short-circuited stubs portrayed in Fig. 10-26. The position at which the stubs connect to the line is fixed, as shown, but the stub lengths, d_1 and d_2, are adjustable. This kind of arrangement is called a *double-stub tuner*. The load $Z_L = 50 + j100\,\Omega$. The line and stubs have a characteristic impedance $Z_0 = R_0 = 100\,\Omega$. Find the shortest values of d_1 and d_2 such that there is no reflected wave at A (VSWR $= 1$).

SOLUTION The normalized value of the load impedance is

$$Z_n = \frac{50 + j100}{100} = 0.5 + j1.0$$

The chart (Fig. 10-27) is entered at this normalized impedance, as indicated by the point P_1. Constructing a constant VSWR curve through P_1, we note that the VSWR at B (Fig. 10-26) is 4.6. Next, constructing the diametric line through P_1, we locate P_2 halfway around the constant VSWR circle from P_1. Thus, the normalized load admittance is $0.4 - j0.8$. Now, moving clockwise along the constant VSWR circle from P_2 a distance $\lambda/4$ away from the load (toward the generator), we arrive back at P_1.† Thus at the point D on the line the normalized admittance of the main line (looking toward the load) is $0.5 + j1.0$. Since the reflection at A must be zero, we need to look ahead and note the

---

† We arrive back at P_1 because the point D on the line is $\lambda/4$ from the load. For other line lengths, we would arrive at some other point on the chart.

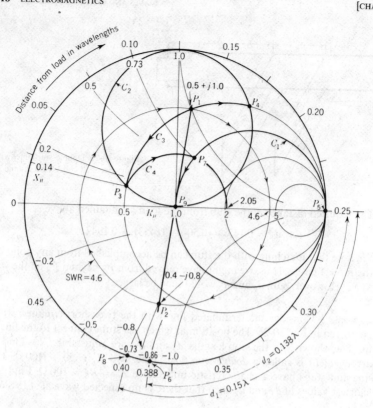

Figure 10-27 Smith chart positions and moves for double-stub tuner of Example 3.

fact that the admittance of the main line at A (without the stub of length d_1 connected) must fall on the $G_n = 1$ circle marked C_1 (Fig. 10-27). Therefore, at the junction of the stub of length d_2 the admittance should fall on circle C_2 since proceeding $\lambda/8$ further from the load, circle C_2 will be rotated clockwise 90° and coincide with circle C_1 ($G_n = 1$ circle), and by a suitable adjustment of the stub of length d_1 a match can be achieved.

The admittance added by the stub of length d_2 will cause the total admittance to move from P_1 along a *constant-conductance* line C_3. In order to reach the circle C_2, we can move either to the left, arriving at P_3, or to the right, arriving at P_4. Moving to P_3 results in shorter stubs; so we will make the stub of such length as to bring the total admittance to P_3. This requires a stub admittance (pure susceptance) of

$$Y_n = -j(1.0 - 0.14) = -j0.86$$

A short-circuited stub has an infinite VSWR so that the admittance at points along the stub are on the circle at the periphery of the chart. At the short circuit the admittance is infinite (point P_5). Therefore, in order to present a value

$$Y_n = -j0.86 \text{ (point } P_6)$$

(noting that the outside scale on the chart reads 0.25 at P_5) the stub length must be given by

$$d_2 = 0.388 - 0.25 = 0.138\lambda$$

Next, moving along the constant VSWR curve C_4 from P_3 to P_7, we find that the line admittance at A is $Y_n = 1.0 + j0.73$. Hence a stub admittance of $Y_n = -j0.73$ is required in order to make the total normalized admittance at A equal to $1.0 + j0$, and therefore the actual impedance at A equal to $100 + j0 \, \Omega$. A value $Y_n = -j0.73$ falls at point P_8. Therefore the length of the stub is given by

$$d_1 = 0.40 - 0.25 = 0.15\lambda$$

Connecting this stub brings the total admittance (or impedance) to the center of the chart (point P_9) and the line is matched (VSWR = 1).

To summarize, the required stub lengths are

$$d_1 = 0.15\lambda$$

$$d_2 = 0.138\lambda$$

We note from Figs. 10-26 and 10-27 that

$$\text{VSWR} = \begin{cases} 4.6 & \text{at } B \text{ (points } P_1 \text{ and } P_2) \\ 2.05 & \text{between } D \text{ and } A \text{ (points } P_3 \text{ and } P_7) \\ 1 & \text{at } A \text{ (point } P_9) \\ \infty & \text{on stubs (points } P_6 \text{ and } P_8) \end{cases}$$

If we had moved to P_4 instead of to P_3, we would have ended up with longer stubs, namely,

$$d_1 = 0.443\lambda \quad \text{and} \quad d_2 = 0.364\lambda$$

and also a larger VSWR between D and A.

10-8 DIRECTIONAL COUPLER

A directional coupler is a device which permits a determination of the reflection coefficient $|\rho_v|$ and the VSWR without using a slotted line. Two short auxiliary conductors are installed in a coaxial line, as indicated in Fig. 10-28. Each is connected at one end to a resistor matching its characteristic resistance R_1 and at the other end to an output terminal which connects to an indicating device.

The electric field of the coaxial line induces a field component E_x in the auxiliary conductor developing a voltage over the length l of the conductor proportional to

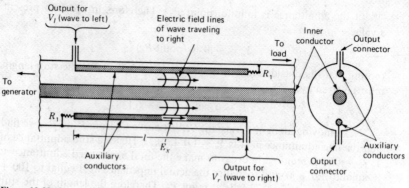

Figure 10-28. Longitudinal and cross-sectional views of coaxial transmission line showing auxiliary conductors acting as directional couplers. Electric field line curvature is exaggerated.

$E_x l$. Thus, a wave traveling to the right in Fig. 10-28 will induce a voltage V_r at the output of the lower auxiliary conductor. In like manner, a wave traveling to the left in the coaxial line produces a voltage V_l at the output of the upper conductor. The matched terminations on the auxiliary conductors prevent voltages from the wave to the right appearing at the upper terminal or from the wave to the left at the lower terminal. If the auxiliary conductors are symmetrical, the ratio of V_l to V_r gives both the reflection coefficient of the load and the VSWR. Thus,

$$|\rho_v| = \frac{V_l}{V_r}$$

and

$$\text{VSWR} = \frac{1 + (V_l/V_r)}{1 - (V_l/V_r)} = \frac{1 + |\rho_v|}{1 - |\rho_v|}$$

With a directional coupler, $|\rho_v|$ and the VSWR can be monitored continuously while adjustments are made on the load impedance Z_L.†

10-9 $\lambda/4$ TRANSFORMERS AND BANDWIDTH

There are many situations where a $\lambda/4$ section of transmission line of suitable impedance may be useful. Suppose that we wish to connect a transmission line of 100-Ω characteristic impedance to a load of $400 + j0\ \Omega$, as in Fig. 10-29. This can be done with a $\lambda/4$ transformer of proper characteristic impedance Z_1. From (10-6-20) we note that when $x = \lambda/4$ ($\beta x = \pi/2$) (line assumed lossless),

$$Z_x = \frac{(Z_1)^2}{Z_L} \quad \text{or} \quad Z_1 = \sqrt{Z_L Z_x} \tag{1}$$

† With directional couplers and a sweep-frequency generator a VSWR or impedance vs. frequency curve can be presented in real time on a cathode-ray-tube screen in either Smith chart form (like the dashed line in Fig. 10-30) or in rectangular form (as in Fig. 10-31).

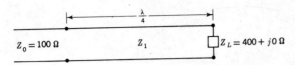

Figure 10-29 $\lambda/4$ transformer.

Here Z_1 is the *geometric mean* of Z_L and Z'_x. In the present example $Z_L = 400 + j0\ \Omega$, while Z_x must be equal to $100 + j0\ \Omega$. It follows that

$$Z_1 = \sqrt{400(100)} = 200\ \Omega$$

Therefore, a $\lambda/4$ section of line of characteristic impedance $Z_1 = R_1 = 200\ \Omega$ provides the desired transformation, eliminating a reflected wave on the 100-Ω line at the design frequency (or wavelength). However, there will be reflection at adjacent frequencies. In other words, the transformer is a frequency-sensitive device. Referring to Fig. 10-29, we have $Z_L = 400 + j0\ \Omega$, $Z_1 = 200\ \Omega$, and $Z_0 = 100\ \Omega$.

Suppose that this system is operating at ultrahigh frequency (UHF) with 300 MHz as the design frequency for which reflection is to be zero (line assumed lossless). At 300 MHz, $\lambda = 1$ m, and a $\lambda/4$ section is 250 mm long. Using a Smith chart, as in Fig. 10-30, we enter at the normalized impedance of the load (i.e., normalized with respect to the characteristic impedance of the $\lambda/4$ section) which is equal to $2 + j0$ [$= (400 + j0)/200$]. This is the point P on the chart. At 300 MHz the $\lambda/4$ section is 0.25λ long, so that to find the impedance at the input (or left) end of the transformer section we move clockwise from P along the VSWR $= 2$ circle a distance of $\lambda/4$, which is halfway around. This point, marked 300 MHz in Fig. 10-30, gives the normalized impedance at the input, which is $0.5 + j0$. The actual impedance is this value multiplied by the characteristic impedance of the transformer section ($= 200\ \Omega$) or $100 + j0\ \Omega$. On the Smith chart the normalized impedance with respect to the 100-Ω line is then $1 + j0$, which is at the center of the chart. Thus, the system is matched and the VSWR $= 1$ on the 100-Ω line.

Suppose we wish to find the VSWR of this $\lambda/4$ transformer at adjacent frequencies in the range between 200 and 400 MHz. Consider first the situation at 200 MHz. At this frequency the wavelength is longer, and the transformer section is now less than $\lambda/4$ long. Its length at 200 MHz is 0.167λ [$= 0.25/(300/200)$], so that again entering the Smith chart at P, we move only 0.167λ along the VSWR $= 2$ circle to the point marked 200 in Fig. 10-30. To refer this normalized impedance to the 100-Ω line we multiply by 2 (as before), which moves us to the point P'. The resulting VSWR on the 100-Ω line is thus about 2.1. Calculating the VSWR at other frequencies in the range 200 to 400 MHz, as shown on the Smith chart of Fig. 10-30, we can draw a VSWR-vs.-frequency curve as in Fig. 10-31 (solid line). We note that the VSWR $= 1$ (zero reflection) at 300 MHz but that the VSWR rises to more than 2 at 200 and 400 MHz. As indicated in the figure, the bandwidth for

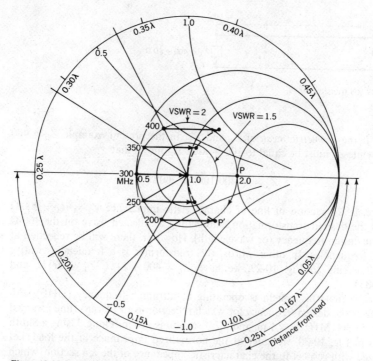

Figure 10-30 Smith chart analysis of performance of $\lambda/4$ transformer over a band of frequencies.

which the VSWR ≤ 1.2 is about 50 MHz ($= 325 - 275$). This yields a bandwidth with respect to the design frequency of

$$\tfrac{50}{300} \times 100 = 16.7\%$$

One method of increasing the bandwidth of the transformer is to connect two $\lambda/4$ transformers in series, as in Fig. 10-32a. The impedances chosen satisfy the requirement of the geometric mean. Thus, $\sqrt{400(100)} = 200\,\Omega$, which should be the impedance seen looking either way at the midpoint A. Then, $141.4\,\Omega$ $[= \sqrt{100(200)}]$ and $283\,\Omega$ $[= \sqrt{200(400)}]$ are the impedances required for the two sections. The VSWR for this double (or two-section) $\lambda/4$ transformer is shown by the dashed line in Fig. 10-31.† It is evident that this combination possesses considerably greater bandwidth. The bandwidth for which the VSWR ≤ 1.2 is about 150 MHz ($= 375 - 225$), or

$$\tfrac{150}{300} \times 100 = 50\%$$

---

† The VSWR can be found with a Smith chart by entering at $1.41 + j0$ $[= (400 + j0)/283]$ and moving clockwise the length of the transformer on a constant-VSWR circle to get the impedance at A (Fig. 10-32a), as above for the single $\lambda/4$ section. The process is then repeated for the second $\lambda/4$ section to get the impedance at the input.

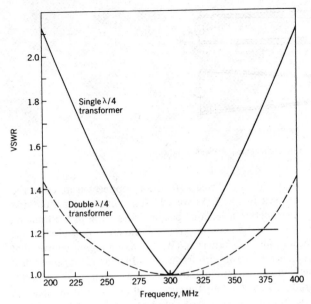

10-31 Comparison of VSWR with single and double $\lambda/4$ transformers over a 2 to 1 bandwidth. The double $\lambda/4$ transformer has a much wider bandwidth.

compared to 16.7 percent for the single section. Over the 200- to 400-MHz range the VSWR does not exceed about 1.5 for the two-section type. With a two-wire transmission line, shown schematically in Fig. 10-32a, the different impedances might be achieved in practice with a constant spacing but different conductor diameter, as in Fig. 10-32b, or with a constant conductor diameter and different spacing, as in Fig. 10-32c, or by using dielectrics of different permittivity between the conductors, or some combination of these methods.

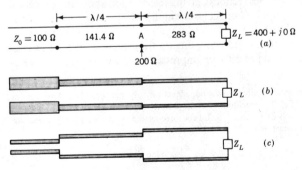

Figure 10-32 (a) Double $\lambda/4$ transformer dimensions and impedances, (b) constant conductor spacing but different conductor diameter, and (c) constant conductor diameter but different spacing.

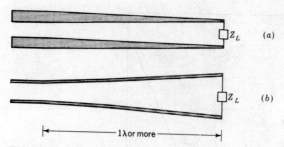

Figure 10-33 Transformers with gradual taper for wide bandwidth.

With a transformer consisting of three $\lambda/4$ sections connected in series the bandwidth could be increased further.† As we add still more sections, the discontinuity (impedance step) between sections becomes less, until in the limit we would approach an infinitely long, smooth, gradually tapered transmission line with no reflection over an infinite bandwidth. (We are assuming, of course, that the load impedance remains constant with frequency.) In practice we must deal with lines of finite length and hence finite bandwidth. However, if the transition is made smoothly and gradually by changing the conductor diameter or the spacing, as suggested in Fig. 10-33a and Fig. 10-33b, it may be possible to achieve a very wide bandwidth in practice. The characteristic impedance of the line can be made to follow an exponential taper. It is necessary, however, that the transition be at least 1λ long at the lowest frequency of operation (see Prob. 10-2-22). We may conclude that spreading out a discontinuity into a long gradual transition will provide broadband matching.

10-10.1 PULSES AND TRANSIENTS

If a pulse of electromagnetic waves is sent along a transmission line, discontinuities on the line will reflect back waves (echoes), giving information about their location (from delay time) and magnitude (from echo strength). The principle is the same as with radar.

As an introduction to pulses and transients on transmission lines, let us consider what happens when a dc impulse is impressed on a two-conductor line of

---

†To justify the geometric-mean requirement the impedances Z_1, Z_2, and Z_3, of the three sections would be 119, 200, and 336 Ω; i.e., the logarithms of the impedance ratios are related as the coefficients of the binomial series (1, 2, 1 for two sections; 1, 3, 3, 1 for three sections; 1, 4, 6, 4, 1 for four sections, etc.). Thus, we have the requirement that

$$3 \log \frac{Z_1}{100} = \log \frac{Z_2}{Z_1} = \log \frac{Z_3}{Z_2} = 3 \log \frac{400}{Z_3}$$

from which $Z_1 = 119$, $Z_2 = 200$, and $Z_3 = 336\ \Omega$. See J. C. Slater, "Microwave Transmission," pp. 57–61, McGraw-Hill Book Company, New York.

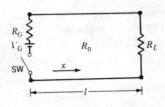

Figure 10-34 Transmission line of characteristic resistance R_0 and length l with load resistance R_L, generator (battery) internal resistance R_G, and generator (battery) open-circuit voltage V_G.

characteristic resistance R_0 with dc generator (battery) and load as in Fig. 10-34.†
The load resistance is R_L, the generator (battery) internal resistance R_G, and the generator (battery) open-circuit voltage V_G. If the switch (sw) is closed at time $t = 0$, a voltage

$$V_0 = \frac{R_0}{R_G + R_0} V_G \qquad \text{(V)} \qquad (1)$$

is applied to the line. The load has no effect until the voltage pulse V_0 arrives after a time

$$t_1 = \frac{l}{v} \qquad \text{(s)} \qquad (2)$$

where l = line length, m
    v = velocity of propagation on line, m s^{-1}

At t_1 the voltage at the load is the sum of the incident and reflected voltages, or

$$V_{t1} = V_0 + \rho_L V_0 = (1 + \rho_L)V_0 \qquad (3)$$

where ρ_L = voltage reflection coefficient of load = $(R_L - R_0)/(R_L + R_0)$

The reflected voltage $\rho_L V_0$ bounces back from the load toward the generator and is in addition to the voltage already on the line.

At time $t = 2t_1$ the impulse arrives back at the generator, resulting in a total voltage

$$\begin{aligned} V_{2t1} &= V_0 + \rho_L V_0 + \rho_L \rho_G V_0 \\ &= (1 + \rho_L + \rho_L \rho_G)V_0 \end{aligned} \qquad (4)$$

where ρ_G = voltage reflection coefficient of generator = $(R_G - R_0)/(R_G + R_0)$

An impulse of magnitude $\rho_L \rho_G V_0$ now travels from the generator to the load, and the cycle repeats with a period $2t_1$.

† R. K. Moore, "Traveling Wave Engineering," p. 99, McGraw-Hill Book Company, New York, 1960.

The zigzag line in Fig. 10-35a indicates the position of the voltage impulse step with respect to distance from the generator (up on the page) and time (to right). With each reflection the voltage changes in steps as follows:

Status	Voltage
No reflection	V_0
After 1st reflection from load	$V_0(1 + \rho_L)$
After 1st reflection from generator	$V_0(1 + \rho_L + \rho_L\rho_G)$
After 2d reflection from load	$V_0(1 + \rho_L + \rho_L\rho_G + \rho_L^2\rho_G)$
After 2d reflection from generator, etc.	$V_0(1 + \rho_L + \rho_L\rho_G + \rho_L^2\rho_G + \rho_L^2\rho_G^2)$

(5)

If the transmission line parameters are

$$\text{Length} = 300 \text{ m}$$
$$\text{Characteristic resistance } R_0 = 100 \ \Omega$$
$$\text{Load resistance } R_L = 200 \ \Omega$$
$$\text{Generator internal resistance } R_G = 20 \ \Omega$$
$$\text{Generator open-circuit voltage } V_G = 10 \text{ V}$$

we have from (1) that the incident voltage

$$V_0 = \frac{R_0 V_G}{R_G + R_0} = \frac{100 \times 10}{20 + 100} = 8.333 \text{ V} \tag{6}$$

Assuming that the wave velocity on the line equals the velocity of light,

$$t_1 = \frac{300}{3 \times 10^8} = 10^{-6} \text{ s} = 1 \ \mu\text{s} \tag{7}$$

For the reflection coefficients we have

$$\rho_L = \frac{R_L - R_0}{R_L + R_0} = \frac{200 - 100}{200 + 100} = \frac{1}{3} \tag{8}$$

and

$$\rho_G = \frac{R_G - R_0}{R_G + R_0} = \frac{20 - 100}{20 + 100} = -\frac{2}{3} \tag{9}$$

Thus, after the first reflection from the load we have from (5)

$$V = V_0(1 + \rho_L) = V_0(1 + \tfrac{1}{3}) = 8.33(\tfrac{4}{3}) = 11.1 \text{ V} \tag{10}$$

and after the first reflection from the generator we have

$$V = V_0(1 + \rho_L + \rho_L\rho_G) = 8.33(\tfrac{10}{9}) = 9.26 \text{ V} \tag{11}$$

In like manner, the voltage at later times after more reflections can be determined with values vs. time as shown by the solid line in Fig. 10-35b. After 6 or 7 μs (6 or 7 bounces) the voltage settles to a value of 9.091 V, which is substantially the

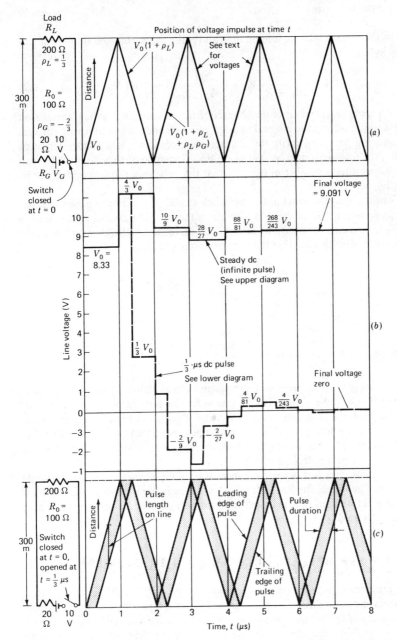

Figure 10-35 (a) DC transmission line showing position of the voltage step with respect to distance (up) and time (to right) after switch is closed at $t = 0$. The variation of voltage is shown by the solid curve in (b). (c) Same transmission line showing position of $\frac{1}{3}$-μs voltage pulse with respect to distance (up) and time (to right). The variation of voltage is shown by the dashed curve in (b).

same as the voltage after an infinite time and is the value obtained from dc circuit theory as

$$V = \frac{R_L}{R_L + R_G} V_G = \frac{200}{200 + 20} 10 = 9.091 \text{ V} \tag{12}$$

This voltage is independent of the characteristic resistance R_0 of the transmission line. However, during the interval of a few microseconds, while transient voltages are bouncing back and forth, the characteristic resistance is a factor. Although (ideally) the transients may be said to continue indefinitely, they decrease to such small values after a number of bounces that the voltage approaches closely to its final (dc) value.

Consider next the same line with switch closed at $t = 0$ but opened again at $t = \frac{1}{3} \mu s$ so that a $\frac{1}{3}$-μs dc pulse is sent along the line. The double zigzag lines in Fig. 10-35c indicate the position of the pulse with respect to distance (up) and time (to right). With each reflection the voltage changes in steps as follows:

Status	Voltage	
No reflection	V_0	
After 1st reflection from load	$V_0 \rho_L$	
After 1st reflection from generator	$V_0 \rho_L \rho_G$	(13)
After 2d reflection from load	$V_0 \rho_L^2 \rho_G$	
After 2d reflection from generator, etc.	$V_0 \rho_L^2 \rho_G^2$	

Evaluating (13), the pulse voltages vs. time are as shown by the dashed line in Fig. 10-35b. The pulse voltage is the same as for the earlier case until after the 1st reflection from the load, after which the voltage decreases rapidly, settling finally to zero, as we would expect with the switch open. Note that for $\frac{1}{3} \mu s$ at each reflection, incident and reflected voltages add, resulting in $\frac{1}{3}$-μs steps prior to the $\frac{2}{3}$-μs step before the next reflection.

Transients or pulses occur on transmission lines whenever a line is suddenly energized. The longer the line (if lossless), the longer the transients will persist. If the line is terminated in a matched load the transient or pulse disappears after reaching the load.

10-10.2 WAVE REFLECTIONS ON A $\lambda/4$ TRANSFORMER

In the previous section the generator was a dc type. Let us consider now ac or rf generators and how reflections are eliminated with a $\lambda/4$ transformer. The reflectionless property of the $\lambda/4$ transformer is achieved by adjusting reflections from two points (one at each end of the transformer) so that they balance out at the design frequency. We will investigate this in more detail using the wave-reflection method.

Consider the $\lambda/4$ transformer of Fig. 10-29 but with a short section of 400-Ω line inserted between the $\lambda/4$ transformer and the load, as suggested in Fig. 10-36.

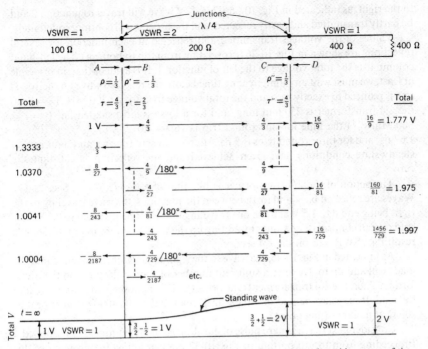

Figure 10-36 Reflections of a continuous wave on a $\lambda/4$ transformer. Note that with passage of time the voltage approaches 1.0 V on the 100-Ω line and 2.0 V on the 400-Ω line with a transition from 1 to 2 V on the $\lambda/4$ transformer (VSWR = 2).

We need to consider three waves: A approaching junction 1 from the left, B approaching junction 1 from the right, and C approaching junction 2 from the left. In a general situation we would also need to consider a wave D approaching junction 2 from the right, but here this wave is absent since the load and line are matched (no reflection at load). The reflection and transmission coefficients (ρ and τ) for each wave are as indicated in Fig. 10-36.

Assume that the initial incoming wave from the left is equal to 1 V and that the lines are lossless. Then at junction 1, $\frac{4}{3}$-V will be transmitted and $\frac{1}{3}$-V reflected. When the $\frac{4}{3}$-V wave arrives at junction 2, four-thirds of it will be transmitted and one-third reflected, giving rise to a $\frac{16}{9}$-V wave proceeding to the load and a $\frac{4}{9}$-V wave traveling back to the left from junction 2, as in Fig. 10-36. When this $\frac{4}{9}$-V wave arrives back at junction 1, it will be reversed in phase (having traveled $\lambda/2$, or 180°)† and two-thirds of it will be transmitted and one-third reflected (with sign change), resulting in a $-\frac{8}{27}$-V wave traveling to the left from junction 1 and a $+\frac{4}{27}$-V wave

---

† In ac or rf situations phase change must be considered. In the dc case of the previous section the phase change along the line is zero.

to the right, as indicated in Fig. 10-36. This $\frac{4}{27}$-V wave will travel to junction 2 and be partly transmitted and partly reflected, and the process will continue indefinitely.

Values of the voltages transmitted and reflected at each junction for several reflections are shown in Fig. 10-36 (proceeding from top to bottom). The left-hand column lists the total voltage to the left of junction 1. It is evident that, in principle at least, we must wait an infinitely long time before the reflected wave on the 100-Ω line is reduced to exactly zero and the total voltage to unity. At 300 MHz ($\lambda = 1$ m) the $\lambda/4$ transformer is 250 mm long, and for a lossless air-insulated line ($v = c = 300$ Mm s^{-1}) the time for one round trip is 16.7×10^{-10} s ($= 0.5/3 \times 10^8$), or 1.67 ns; and for three round trips is 5 ns. After a sufficient time we may assume that steady-state conditions have been achieved and the reflection is substantially zero.†

The action of the $\lambda/4$ transformer is like that of a *resonator* (see Sec. 13-17). Waves are reflected back and forth between the junctions, the wave traveling to the right being equal to 1.5 V and the one traveling to the left being equal to 0.5 V. The student can confirm these values by adding up the wave fractions in Fig. 10-36. The resulting VSWR = 2 on the $\lambda/4$ section.

As indicated in Fig. 10-36, the wave to the right of junction 2 (traveling into the load) will add up to 2-V after a sufficient number of wave reflections on the transformer. Thus, the $\lambda/4$ transformer transforms the 1-V input wave to a 2-V wave into the load. It may also be said that the transformer makes the 400-Ω load appear as 100 Ω at its input (junction 1).

Consider next that a square pulse of duration t_1 and magnitude 1 V is applied.‡ Proceeding from top to bottom in Fig. 10-37, we can follow the progress of this pulse at successive instants of time until it has disappeared. It is clear that the transformer fails to function as a reflectionless device for a pulse as short as indicated in Fig. 10-37 ($t_1 \ll l/v$, where l is the length of the transformer and v is the wave velocity). There is no way for the pulse reflected from junction 2 to catch up with the one reflected from junction 1 and reduce its magnitude.§ Only for the steady-state condition or for a very long pulse ($t_1 \gg l/v$) can the multiple reflections reduce the wave reflected to the left from junction 1 to its desired low value. To avoid reflections with short pulses one must construct lines without discontinuities. Conversely, one can use short pulses to locate and measure the magnitude of discontinuities on transmission lines. To obtain short enough pulses for such diagnostic measurements, it may be desirable to use frequencies which are much higher than the design frequency.

For a single (short) incident pulse, the reflected pulses on a $\lambda/4$ transformer are shown versus time in Fig. 10-37c. The pulses are spaced the length of time it

---

† The time for two round trips is the same as for one cycle. Thus, the steady-state condition is substantially achieved within a few cycles.

‡ The pulse is given by

$$V(t) = \begin{cases} 1 & \text{for } 0 \leq t \leq t_1 \\ 0 & \text{for } t < 0 \text{ and } t > t_1 \end{cases}$$

§ Hence, a Smith chart analysis of a line is of no value if the line is to be used with very short pulses.

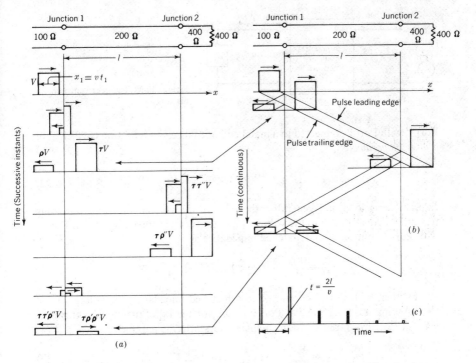

Figure 10-37 Progress of short pulse along transmission line with resulting pulses reflected and transmitted at junctions. (*a*) Situation at successive instants. (*b*) Situation on continuous basis. Note that an incident pulse splits into a reflected and transmitted pulse at each junction. A CRT display connected to the line at the left would show a series of echo pulses of decreasing magnitude as in (*c*). The magnitude of the pulses after each encounter with a junction is indicated in (*a*) by the τ, ρ values. For example, the pulse traveling through junction 1, reflected from junction 2 and then reflected from junction 1 has a magnitude $\tau\rho'\rho''V$, where τ is the transmission coefficient through junction 1 (left to right), ρ' is the reflection coefficient at junction 2, and ρ'' is the reflection coefficient at junction 1 (right to left) [see bottom of (*a*) and (*b*)]. In (*c*) the pulse spacing $= 2l/v$ s.

takes for the pulse to travel the $\lambda/4$ section down and back ($= 2l/v$). Although the reflected pulses continue indefinitely, their magnitude becomes very small after a few round-trip periods.

10-10.3 *S* OR SCATTERING PARAMETERS†

Consider the two-port junction of Fig. 10-38. The ratio of the voltage V_{1o} of the *outgoing* traveling wave at Port 1 to the voltage V_{1i} of the *incident* traveling wave at

† C. G. Montgomery, R. H. Dicke, and E. M. Purcell, "Principles of Microwave Circuits," p. 146, McGraw-Hill Book Company, New York, 1948.

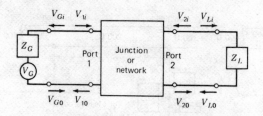

Figure 10-38 Two-port junction or network.

Port 1, with Port 2 connected to a matched impedance, is a reflection coefficient or *scattering parameter*

$$s_{11} = \frac{V_{1o}}{V_{1i}} \qquad \text{(Port 2 matched, } V_{2i} = 0) \qquad (1)$$

Interchanging load and generator, the ratio of the voltage V_{2o} of the outgoing traveling wave at Port 2 to the voltage V_{2i} of the incident traveling wave at Port 2, with Port 1 connected to a matched impedance, is

$$s_{22} = \frac{V_{2o}}{V_{2i}} \qquad \text{(Port 1 matched, } V_{1i} = 0) \qquad (2)$$

Further, with Port 2 connected to a matched impedance, the voltage V_{2o} of the outgoing traveling wave at Port 2 to the voltage V_{1i} of the incident traveling wave at Port 1 is a transmission coefficient or scattering parameter

$$s_{21} = \frac{V_{2o}}{V_{1i}} \qquad \text{(Port 2 matched, } V_{2i} = 0) \qquad (3)$$

Finally, with Port 1 connected to a matched impedance, the voltage V_{1o} of the outgoing traveling wave at Port 1 to the voltage V_{2i} of the incident traveling wave at Port 2 is

$$s_{12} = \frac{V_{1o}}{V_{2i}} \qquad \text{(Port 1 matched, } V_{1i} = 0) \qquad (4)$$

If reciprocity applies, the four scattering parameters constitute a symmetrical scattering matrix

$$\begin{bmatrix} s_{11} & s_{12} \\ s_{21} & s_{22} \end{bmatrix}$$

with $s_{21} = s_{12}$.

Extending the above concept to junctions or networks of any number of ports, n, yields an $n \times n$ scattering matrix in which the diagonal elements are reflection coefficients ($s_{11}, s_{22}, s_{33}, \ldots$) and the off-diagonal elements are transmission coefficients ($s_{12}, s_{21}, s_{13}, s_{31}, \ldots$). Connecting a matched generator successively to each port of an n-port junction, all other ports being connected to matched loads, the scattering parameters are measured as the reflection and transmission coefficients at the different ports.

The following worked example illustrates the application of scattering parameters to the calculation of a *Field-Effect Transistor* (FET) amplifier.

Example Calculate the voltage gain of a *Gallium-Arsenide Field-Effect Transistor* (GaAsFET) amplifier at 2 GHz if the GaAsFET scattering parameters are $s_{11} = 0.9 \underline{/-45°}$, $s_{12} = 0.03 \underline{/45°}$, $s_{21} = 3.9 \underline{/135°}$, and $s_{22} = 0.7 \underline{/-30°}$. The amplifier is connected as in Fig. 10-39. Connections for gate and drain voltages are high impedance and are not shown.

SOLUTION In general, with ports not necessarily matched we have (see Fig. 10-38)

$$V_{1o} = s_{11}V_{1i} + s_{12}V_{2i} \tag{5}$$

$$V_{2o} = s_{22}V_{2i} + s_{21}V_{1i} \tag{6}$$

Also the reflection coefficient at the load connected to Port 2 is

$$\rho_L = \frac{V_{Lo}}{V_{Li}} = \frac{V_{2i}}{V_{2o}} \tag{7}$$

while the reflection coefficient of the generator connected to Port 1 is

$$\rho_G = \frac{V_{Go}}{V_{Gi}} = \frac{V_{1i}}{V_{1o}} \tag{8}$$

From the above equations the reflection coefficient at Port 1 is

$$\frac{V_{1o}}{V_{1i}} = \rho_{11} = s_{11} + \frac{s_{12}s_{21}\rho_L}{1 - s_{22}\rho_L} = s'_{11} \tag{9}$$

For a matched load $\rho_L = 0$ and $s'_{11} = s_{11}$.
The transmission coefficient between Port 2 and Port 1 is

$$\frac{V_{2o}}{V_{1i}} = \tau_{21} = \frac{s_{21}}{1 - s_{22}\rho_L} \tag{10}$$

For a matched load $\rho_L = 0$ and $\tau_{21} = s_{21}$.

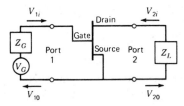

Figure 10-39 Two-port device with GaAsFET as active network for gain calculation in worked example.

The reflection coefficient at Port 2 is

$$\frac{V_{2o}}{V_{2i}} = \rho_{22} = s_{22} + \frac{s_{12}s_{21}\rho_G}{1 - s_{11}\rho_G} = s'_{22} \tag{11}$$

For a matched generator $\rho_G = 0$ and $s'_{22} = s_{22}$.

Finally, the transmission coefficient between Port 1 and Port 2 is

$$\frac{V_{1o}}{V_{2i}} = \tau_{12} = \frac{s_{12}}{1 - s_{11}\rho_G} \tag{12}$$

For a matched generator $\rho_G = 0$ and $\tau_{12} = s_{12}$.

The amplifier voltage gain G_V is given by the ratio of the total output voltage to the total input voltage, or

$$G_V = \frac{V_{2i} + V_{2o}}{V_{1i} + V_{1o}} = \frac{V_{2o}(1 + \rho_L)}{V_{1i}(1 + s'_{11})} = \tau_{21}\frac{1 + \rho_L}{1 + s'_{11}} \tag{13}$$

One extreme case is when the load matches Port 2, so $\rho_L = 0$. The voltage gain

$$G_V = \frac{s_{21}}{1 + s_{11}} = \frac{3.9\underline{/135°}}{1 + 0.9\underline{/-45°}}$$

and $\qquad\qquad G_V = 2.22$ or 6.9 dB   (answer)

Another extreme case is when the load is a very high resistance, so that $\rho_L \approx 1\underline{/0°}$. Then

$$G_V = 9.68 \text{ or } 19.7 \text{ dB (answer)}$$

Note that for amplifiers the network is not reciprocal (the amplifier is an active device), so $s_{21} \neq s_{12}$.

10-11 RELATIVE PHASE VELOCITY AND INDEX OF REFRACTION

The phase velocity relative to the velocity of light, or the *relative phase velocity*, is

$$\rho = \frac{v}{c} = \frac{\sqrt{\mu_0\epsilon_0}}{\sqrt{\mu\epsilon}} = \frac{1}{\sqrt{\mu_r\epsilon_r}} \qquad \text{(dimensionless)} \tag{1}$$

where μ_r = relative permeability of medium
ϵ_r = relative permittivity of medium

The phase velocity of a plane wave in an unbounded lossless medium is equal to or less than the velocity of light ($p \leq 1$). In general, however, the phase velocity may have values both greater and less than the velocity of light. For example, in a hollow metal waveguide v is always equal to or greater than c (see Chap. 13).

In optics the *index of refraction* η is defined as the reciprocal of the relative phase velocity p. That is,

$$\eta = \frac{1}{p} = \frac{1}{v/c} = \frac{c}{v} = \sqrt{\mu_r \epsilon_r} \qquad (2)$$

For nonferrous media μ_r is very nearly unit so that

$$\eta = \sqrt{\epsilon_r} \qquad (3)$$

Example 1 Paraffin has a relative permittivity $\epsilon_r = 2.1$. Find the index of refraction for paraffin and also the phase velocity of a wave in an unbounded medium of paraffin.

SOLUTION The index of refraction

$$\eta = \sqrt{2.1} = 1.45$$

The phase velocity

$$v = \frac{c}{\sqrt{2.1}} = 207 \text{ Mm s}^{-1}$$

Example 2 Distilled water has the constants $\sigma \approx 0$, $\epsilon_r = 81$, $\mu_r = 1$. Find η and v.

SOLUTION

$$\eta = \sqrt{81} = 9$$

$$v = \frac{c}{\sqrt{81}} = 0.111c = 33.3 \text{ Mm s}^{-1}$$

The index of refraction given for water in the above example is the value at low frequencies ($f \to 0$). At light frequencies, say for sodium light ($\lambda = 589 \text{ nm}$), the index of refraction is observed to be about 1.33 instead of 9 as calculated on the basis of the relative permittivity. This difference was at one time cited as invalidating Maxwell's theory. The explanation for the difference is that the permittivity ϵ is not a constant but is a function of frequency. At zero frequency $\epsilon_r = 81$, but at light frequencies $\epsilon_r = 1.33^2 = 1.77$. The index of refraction and permittivity of many other substances also vary as a function of the frequency.

10-12 GROUP VELOCITY†

Consider a plane wave traveling in the positive x direction, as in Fig. 10-3. Let the total electric field be given by

$$E_y = E_0 \cos (\omega t - \beta x) \tag{1}$$

Suppose now that the wave has not one but two frequencies of equal amplitude expressed by

$$\omega_0 + \Delta\omega$$

and

$$\omega_0 - \Delta\omega$$

It follows that the β values corresponding to these two frequencies are

$$\beta_0 + \Delta\beta \qquad \text{corresponding to } \omega_0 + \Delta\omega$$

and

$$\beta_0 - \Delta\beta \qquad \text{corresponding to } \omega_0 - \Delta\omega$$

For frequency 1

$$E_y' = E_0 \cos [(\omega_0 + \Delta\omega)t - (\beta_0 + \Delta\beta)x] \tag{2}$$

and for frequency 2

$$E_y'' = E_0 \cos [(\omega_0 - \Delta\omega)t - (\beta_0 - \Delta\beta)x] \tag{3}$$

Adding gives the total field

$$E_y = E_y' + E_y'' \tag{4}$$

or

$$E_y = E_0 \{\cos [(\omega_0 + \Delta\omega)t - (\beta_0 + \Delta\beta)x] + \cos [(\omega_0 - \Delta\omega)t - (\beta_0 - \Delta\beta)x]\} \tag{5}$$

Multiplying out (5) and by trigonometric transformation, we get

$$E_y = 2E_0 \cos (\omega_0 t - \beta_0 x) \cos (\Delta\omega t - \Delta\beta x) \tag{6}$$

The two cosine factors indicate the presence of beats, i.e., a slow variation superimposed on a more rapid one.

For a *constant-phase* point

$$\omega_0 t - \beta_0 x = \text{constant}$$

and

$$\frac{dx}{dt} = \frac{\omega_0}{\beta_0} = v = f_0 \lambda_0 \tag{7}$$

---

† Leon Brillouin, "Wave Propagation in Periodic Structures," chap. 5, McGraw-Hill Book Company, New York, 1946; J. A. Stratton, "Electromagnetic Theory," p. 330, McGraw-Hill Book Company, New York, 1941.

where v is the *phase velocity*. Setting the argument of the second cosine factor equal to a constant, we have

$$\Delta\omega\, t - \Delta\beta\, x = \text{constant}$$

and
$$\frac{dx}{dt} = \frac{\Delta\omega}{\Delta\beta} = u = \Delta f \, \Delta\lambda \tag{8}$$

where u is the phase velocity of the wave envelope, which is usually called the *group velocity*. In the above development we can consider $\omega_0 + \Delta\omega$ and $\omega_0 - \Delta\omega$ as the two sideband frequencies due to the modulation of a carrier frequency ω_0 by a frequency $\Delta\omega$, the carrier frequency being suppressed.

In nondispersive media the group velocity is the same as the phase velocity. Free space is an example of a lossless, nondispersive medium, and in free space $u = v = c$. However, in dispersive media the phase and group velocities differ.

A *dispersive medium* is one in which the phase velocity is a function of the frequency (and hence of the free-space wavelength). Dispersive media are of two types:

1. *Normally dispersive.* In these media the change in phase velocity with wavelength is positive; that is, $dv/d\lambda > 0$. For these media $u < v$.
2. *Anomalously dispersive.* In these media the change in phase velocity with wavelength is negative; that is, $dv/d\lambda < 0$. For these media $u > v$.

The terms *normal* and *anomalous* are arbitrary, the significance being simply that anomalous dispersion is different from the type of dispersion described as normal.

For a particular frequency (bandwidth vanishingly small)

$$u = \lim_{\Delta\omega \to 0} \frac{\Delta\omega}{\Delta\beta} = \frac{d\omega}{d\beta} \tag{9}$$

But $\omega = 2\pi f = 2\pi f \lambda/\lambda = \beta v$; so

$$u = \frac{d\omega}{d\beta} = \frac{d(\beta v)}{d\beta} = \beta\frac{dv}{d\beta} + v \tag{10}$$

or
$$u = v + \beta\frac{dv}{d\beta} \tag{11}$$

It may also be shown that

$$u = v - \lambda\frac{dv}{d\lambda} \tag{12}$$

Equations (11) and (12) are useful in finding the group velocity for a given phase-velocity function.

Example A 1-MHz (300-m-wavelength) plane wave traveling in a normally dispersive, lossless medium has a phase velocity at this frequency of 300 Mm s^{-1}. The phase velocity as a function of wavelength is given by

$$v = k\sqrt{\lambda}$$

where k is a constant. Find the group velocity.

SOLUTION From (12) the group velocity is

$$u = v - \lambda \frac{dv}{d\lambda} = v - \frac{k}{2}\sqrt{\lambda}$$

or $$u = v(1 - \tfrac{1}{2})$$

Hence

$$u = \frac{v}{2} = 150 \text{ Mm s}^{-1}$$

To illustrate graphically the difference between phase and group velocity, let us consider a wave of the same phase-velocity characteristics as in the above example and assume, further, that the wave has two frequencies, $f_0 + \Delta f$ and $f_0 - \Delta f$, of equal amplitude, where $f_0 = 1$ MHz and $\Delta f = 100$ kHz. This is equivalent to a 1-MHz carrier modulated at 100 kHz with the carrier suppressed. From (6) graphs of the instantaneous magnitude of E_y as a function of distance (plotted in meters) are presented in Fig. 10-40 for three instants of time, $t = 0$, $t = T/4$, and $t = T/2$. The point P is a point of constant phase of the wave proper and moves with the phase velocity v. The point P' is a point of constant phase of the envelope enclosing the wave and moves with the group velocity u. It is apparent that in one-half period ($T/2$) the point P' has moved a distance d', which is one-half the distance d which the point P has moved. That is to say, the group velocity u is one-half the phase velocity v. The intelligence conveyed by the modulation moves with the velocity of the envelope, i.e., at the group velocity.†

The difference between phase and group velocities is also illustrated by a crawling caterpillar. The humps on its back move forward with phase velocity, while the caterpillar as a whole progresses with group velocity.

For a single-frequency constant-amplitude (steady-state) wave the group velocity is not apparent. However, if the wave consists of two or more frequencies, or a frequency group, as in a modulated wave, the group velocity may be observed because the wave amplitude is nonuniform and the individual waves appear to form groups that may be enclosed by an envelope, as in Fig. 10-40.

---

† In a lossless medium the energy is also conveyed at the group velocity.

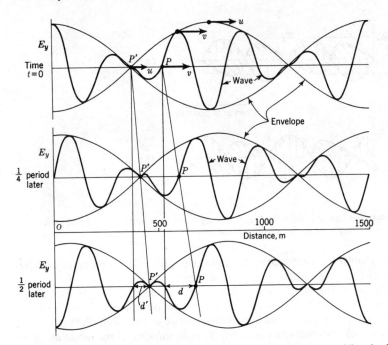

Figure 10-40 Constant-phase point P of the wave moves with phase velocity v, while point P' on the envelope moves with group velocity u. In this example the group velocity is one-half the phase velocity.

10-13 TRAVELING WAVES AND STANDING WAVES

The instantaneous values of E_y and H_z for a plane traveling wave are illustrated in Fig. 10-41 with the wave progressing in the positive x direction. Figure 10-41a shows the condition at the time $t = 0$, while Fig. 10-41b shows the condition one-quarter period later ($t = T/4$). The maximum values of E_y and H_z (E_0 and H_0) are shown to be equal. Hence, if the medium is free space, the scale in volts per meter along the y axis should be 377 times the scale in amperes per meter along the z axis. The scales would be equal, however, if the medium had an intrinsic impedance of 1 Ω. In Fig. 10-41 both the magnitudes and directions of E_y and H_z are shown for points along the x axis. Since we are considering a plane wave traveling in the direction of the x axis, the relations of E_y and H_z along all lines parallel to the x axis are the same as those shown.

Thus far, we have considered only a single traveling wave, such as a wave moving in the positive *or* the negative x direction. Let us now examine the situation which exists when there are two waves traveling in opposite directions, such as the negative *and* positive x directions. Assume that the two waves are of the same frequency and of sinusoidal form. The condition that the waves be of the same

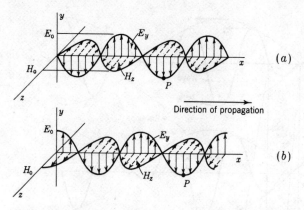

Figure 10-41 Instantaneous values of E_y and H_z along x axis (a) at time $t = 0$ and (b) $\frac{1}{4}$ period later. In this interval the point P has advanced $\lambda/4$ to the right.

frequency and form is automatically fulfilled if one wave is a reflection of the other since both then originate from the same source.

Referring to Fig. 10-42, assume that space is divided into two media, 1 and 2, with a plane boundary between as shown. A wave originating in medium 1 and incident on the boundary is said to be the *incident wave*. The wave reflected from the boundary back into medium 1 is called the *reflected wave*. If the reflection of the incident wave at the boundary is not complete, some of the wave continues on into medium 2 and this wave is referred to as the *transmitted wave*.

In the solution of the wave equation for E_y as given by (10-2-20) there are two terms, the first representing a wave in the negative x direction (to the left) and the second a wave in the positive x direction (to the right). Referring now to (10-2-34), let the incident wave (traveling to the left) be given by

$$E_{y0} = E_0 e^{j(\omega t + \beta x)} \tag{1}$$

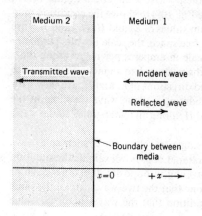

Figure 10-42 Relation of incident, reflected, and transmitted waves.

and the reflected wave (traveling to the right) by

$$E_{y1} = E_1 e^{j(\omega t - \beta x + \delta)} \tag{2}$$

where δ = time-phase lead of E_{y1} with respect to E_{y0} at $x = 0$, that is, δ = phase
    shift at point of reflection
  E_0 = amplitude of incident wave
  E_1 = amplitude of reflected wave

The total electric field E_y is

$$E_y = E_{y0} + E_{y1} \tag{3}$$

The instantaneous magnitude of the fields is obtained by taking either the real
(Re) or imaginary (Im) parts of (1) and (2). Thus, taking the imaginary parts, the
total instantaneous electric field is

$$E_y = E_0 \sin(\omega t + \beta x) + E_1 \sin(\omega t - \beta x + \delta) \tag{4}$$

If $\delta = 0$ or $180°$, (4) can be expanded as follows:†

$$E_y = E_0 \sin \omega t \cos \beta x + E_0 \cos \omega t \sin \beta x + E_1 \sin \omega t \cos \beta x - E_1 \cos \omega t \sin \beta x \tag{5}$$

Collecting terms, we have

$$E_y = (E_0 + E_1) \sin \omega t \cos \beta x + (E_0 - E_1) \cos \omega t \sin \beta x \tag{6}$$

If medium 2 is a perfect conductor, the reflected wave is equal in magnitude to the
incident wave. If $x = 0$ is taken to be at the boundary between media 1 and 2, the
boundary relation for the tangential component of **E** requires that $E_y = 0$, so that
$E_1 = -E_0$ at the boundary ($\delta = 180°$). Thus (6) becomes

$$E_y = 2E_0 \cos \omega t \sin \beta x \tag{7}$$

This represents a wave which is stationary in space. The values of E_y at a
particular instant are a sine function of x. The instantaneous values at a particular
point are a cosine function of t. The peak value of the wave is the sum of the incident
and reflected peak values or $2E_0$. A stationary wave of this type for which $|E_1| = |E_0|$ is a *pure standing wave*. This type of wave is associated with resonators.

The space and time variations of E_y for a pure standing wave are shown by the
curves of Fig. 10-43. It is to be noted that a constant-phase point, such as P, does
not move in the x direction but remains at a fixed position as time passes.

Now let us examine the conditions when the reflected wave is smaller than the
incident wave, say one-half as large. Then, $E_1 = -0.5E_0$. (In the analogous transmission line case, $Z_L = 3Z_0$ and $\rho_v = \frac{1}{2}$.) Evaluating (6) for this case at four
instants of time gives the curves of Fig. 10-44. The curves show the values of E_y as a
function of βx at times equal to 0, $\frac{1}{8}$, $\frac{1}{4}$, and $\frac{3}{8}$ period. The peak values of E_y range

† $\sin(a \pm b) = \sin a \cos b \pm \cos a \sin b$

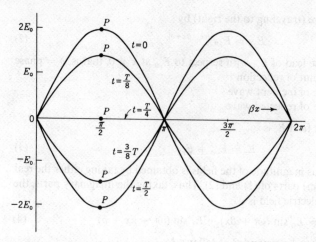

Figure 10-43 Pure standing wave showing E_y at various instants of time.

from $1.5E_0$ at $t = 0$ to $0.5E_0$ at $t = \frac{1}{4}$ period. The peak values as a function of x as observed over an interval of time greater than one cycle correspond to the envelope as indicated. This envelope remains stationary, but focusing our attention on a constant-phase point P of the wave, we note that the total instantaneous wave travels to the left. It will also be noted that the velocity with which P moves is not constant. Between time 0 and $\frac{1}{8}$ period P moves about 0.05λ (0.1π), while in the next $\frac{1}{8}$ period P moves about 4 times as far, or about 0.2λ (0.4π). Although the average velocity of the constant-phase point is the same as for a pure traveling wave, its instantaneous magnitude varies between values which are greater and less.

To summarize, there are two E_y waves, one traveling in the negative x direction and another one-half as large traveling in the positive x direction. The waves

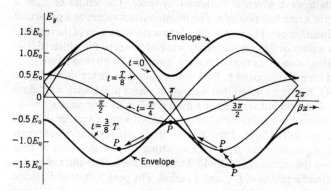

Figure 10-44 Standing-wave envelope for $E_1 = -0.5E_0$ with associated (traveling) wave at four instants of time: $t = 0$, $t = T/8$, $t = T/4$, and $t = 3T/8$.

reinforce each other at some points and subtract from each other at other points. The resultant wave travels in the negative x direction.

The envelope of the instantaneous curves in Fig. 10-44 can be called a *standing-wave curve* or *envelope*. At any position βx the maximum value of the field at some time during the cycle is equal to the ordinate value of the envelope.

To calculate the value of the standing-wave envelope, we may proceed as follows. In (6) put

$$A = (E_0 + E_1) \cos \beta x \tag{8}$$

$$B = (E_0 - E_1) \sin \beta x \tag{9}$$

Expanding $\sin \omega t$ and $\cos \omega t$ in terms of exponentials, we can show that

$$A \sin \omega t + B \cos \omega t = \sqrt{A^2 + B^2} \sin (\omega t + \gamma) \tag{10}$$

Equation (6) can then be written

$$E_y = \sqrt{A^2 + B^2} \sin (\omega t + \beta x) \tag{11}$$

Expanding (11) by means of (8) and (9) yields

$$E_y = \sqrt{(E_0 + E_1)^2 \cos^2 \beta x + (E_0 - E_1)^2 \sin^2 \beta x} \, \sin (\omega t + \beta x) \tag{12}$$

The maximum value of E_y at some position βx as observed over an interval of at least one period occurs when $\sin (\omega t + \beta x) = 1$. Thus for the shape of the standing-wave envelope of E_y we have

$$E_y = \sqrt{(E_0 + E_1)^2 \cos^2 \beta x + (E_0 - E_1)^2 \sin^2 \beta x} \tag{13}$$

Ordinarily we are not so much interested in the shape of the standing-wave envelope as given by (13) as in the ratio of the maximum to minimum values for the envelope, which is called the *Standing-Wave Ratio* (SWR). The potential or voltage at any distance x will be proportional to the field,† and so the SWR in this case may be referred to as the *Voltage Standing-Wave Ratio* (VSWR). The maximum value of the envelope corresponds to the sum of the amplitudes of the incident and reflected waves $(E_0 + E_1)$, while the minimum corresponds to the difference between the two $(E_0 - E_1)$. With this information we can determine the fraction of the incident E_y wave which is reflected, forming the reflected wave, and also that which is transmitted (see Fig. 10-42). As will be noted later, this knowledge is of value in determining the nature of the conditions at the point of reflection.

Thus, for the VSWR we can write

$$\boxed{\text{VSWR} = \frac{E_{\max}}{E_{\min}} = \frac{E_0 + E_1}{E_0 - E_1}} \tag{14}$$

---

† Provided E is integrated over equal paths l ($V = \int \mathbf{E} \cdot d\mathbf{l} = El$).

When the reflected wave is zero ($E_1 = 0$), the VSWR is unity. When the reflected wave is equal to the incident wave ($E_1 = E_0$), the VSWR is infinite. Hence for all intermediate values of the reflected wave, the VSWR lies between 1 and infinity.

The ratio of the reflected wave to the incident wave is the *reflection coefficient*. Thus, at the point of reflection ($x = 0$) and at the time $t = 0$, the ratio of (2) to (1) is

$$\rho = \frac{E_{y1}}{E_{y0}} = \frac{E_1 e^{j\delta}}{E_0} = \frac{E_1 \underline{/\delta}}{E_0} = \rho \underline{/\delta} \tag{15}$$

The magnitude of ρ can range between 0 and 1 with phase angles between 0 and $\pm 180°$.

Rewriting (14) and substituting (15) gives

$$\boxed{\text{VSWR} = \frac{1 + (E_1/E_0)}{1 - (E_1/E_0)} = \frac{1 + |\rho|}{1 - |\rho|}} \tag{16}$$

Solving for $|\rho|$ gives an expression for the magnitude of the reflection coefficient in terms of the VSWR:

$$|\rho| = \frac{\text{VSWR} - 1}{\text{VSWR} + 1} \tag{17}$$

In Fig. 10-45, standing-wave envelopes are presented for three magnitudes of the reflected wave as given by reflection coefficients $\rho = 0, 0.5,$ and 1. The amplitude of the incident wave is taken as unity. The curves show E_y as a function of position in terms of both βx and wavelength. For complete reflection ($\rho = 1$) we have a pure standing wave with a VSWR of infinity. For zero reflection ($\rho = 0$), the VSWR is unity, and E_y is constant as a function of position. For a reflection coefficient of 0.5, the curve varies between 1.5 and 0.5 so that VSWR = 3. *In general*, the standing-wave envelope is *not* a sine curve. This is illustrated by the curve for $\rho = 0.5$.

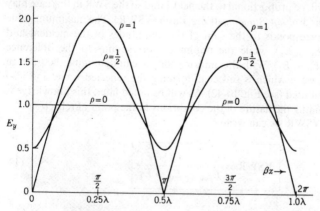

Figure 10-45 Standing-wave envelopes for three magnitudes of reflection coefficient, $\rho = 0, 0.5,$ and 1.

However, in the limiting condition of $\rho = 1$ the curve does have the form of a rectified sine function ($|\sin \beta x|$). Also, as the condition $\rho = 0$ is approached, the curve approximates a sinusoidal variation.

This entire section applies not only to waves in media but also to waves on transmission lines, requiring only that the field E be replaced by the transmission line voltage V. Note also that the VSWR and reflection coefficient equations are identical with those in Sec. 10-6 for transmission lines provided that we set $\rho = \rho_v$.

10-14 CONDUCTORS AND DIELECTRICS

According to Maxwell's curl equation from Ampère's law,

$$\nabla \times \mathbf{H} = \mathbf{J} + \frac{\partial \mathbf{D}}{\partial t} \tag{1}$$

since $\mathbf{J} = \sigma\mathbf{E}$, (1) becomes

$$\nabla \times \mathbf{H} = \sigma\mathbf{E} + \frac{\partial \mathbf{D}}{\partial t} \tag{2}$$

For a linearly polarized plane wave traveling in the x direction with $\mathbf{E}$ in the y direction, the vector equation (2) reduces to the scalar phasor equation

$$-\frac{\partial H_z}{\partial x} = \sigma E_y + j\omega\epsilon E_y \tag{3}$$

The terms in (3) each have the dimensions of current density, which is expressed in amperes per square meter. The term σE_y represents the *conduction-current density*, while the term $j\omega\epsilon E_y$ represents the *displacement-current density*. Thus, according to (3) the space rate of change of H_z equals the sum of the conduction- and displacement-current densities. If the conductivity is zero, the conduction-current term vanishes and we have the condition considered in previous sections. If σ is not equal to zero, one may arbitrarily define three conditions as follows:†

(1) $\omega\epsilon \gg \sigma$

(2) $\omega\epsilon \sim \sigma$

(3) $\omega\epsilon \ll \sigma$

When the displacement current is much greater than the conduction current, as in condition (1), the medium behaves like a dielectric. If $\sigma = 0$, the medium is a perfect, or lossless, dielectric. For σ not equal to zero the medium is a lossy, or imperfect, dielectric. However, if $\omega\epsilon \gg \sigma$, it behaves more like a dielectric than anything else and may, for practical purposes, be classified as a *dielectric*. On the other

---

† Referring to Sec. 8-14, condition (1) would be modified in the case of a lossy dielectric to $\omega\epsilon' \gg \sigma'$ and condition (2) to $\omega\epsilon' \sim \sigma'$, but for condition (3) we have $\omega\epsilon' \approx \omega\epsilon$ and $\sigma' = \sigma$, and hence we can write $\omega\epsilon \ll \sigma$, as indicated.

hand, when the conduction current is much greater than the displacement current, as in condition (3), the medium may be classified as a *conductor*. Under conditions midway between these two, when the conduction current is of the same order of magnitude as the displacement current, the medium may be classified as a *quasiconductor*.

We can be even more specific and arbitrarily classify media as belonging to one of three types according to the value of the ratio $\sigma/\omega\epsilon$ as follows:

Dielectrics:
$$\frac{\sigma}{\omega\epsilon} < \frac{1}{100}$$

Quasiconductors:
$$\frac{1}{100} < \frac{\sigma}{\omega\epsilon} < 100$$

Conductors:
$$100 < \frac{\sigma}{\omega\epsilon}$$

where σ = conductivity of medium, $\mho \ m^{-1}$
  ϵ = permittivity of medium, $F \ m^{-1}$
  ω = radian frequency = $2\pi f$, where f is the frequency, Hz

The ratio $\sigma/\omega\epsilon$ is dimensionless.

It is to be noted that frequency is an important factor in determining whether a medium acts like a dielectric or a conductor. For example, take the case of average rural ground (Ohio) for which $\epsilon_r = 14$ (at low frequencies) and $\sigma = 10^{-2} \ \mho \ m^{-1}$.

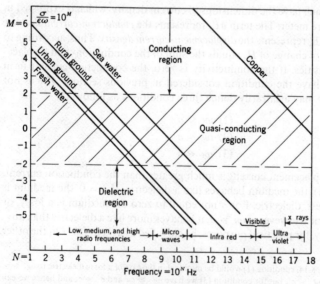

Figure 10-46 Ratio $\sigma/\omega\epsilon$ as a function of frequency for some common media (log-log plot).

Table 10-5 Table of constants for some common media

Medium	Relative permittivity ϵ_r, dimensionless	Conductivity σ, $\mho\, m^{-1}$
Copper	1	5.8×10^7
Seawater	80	4
Rural ground (Ohio)	14	10^{-2}
Urban ground	3	10^{-4}
Fresh water	80	10^{-3}

Assuming no change in these values as a function of frequency, the ratio $\sigma/\omega\epsilon$ at three different frequencies is as tabulated.

Frequency, Hz	Ratio $\sigma/\omega\epsilon$
10^3	1.3×10^4
10^7	1.3
$3 \times 10^{10} (\lambda = 10 \text{ mm})$	4.3×10^{-4}

At 1 kHz rural ground behaves like a conductor, while at the microwave frequency of 30 GHz it acts like a dielectric. At 10 MHz its behavior is that of a quasiconductor.

In Fig. 10-46 the ratio $\sigma/\omega\epsilon$ is plotted as a function of frequency for a number of common media. In preparing Fig. 10-46 the constants were assumed to maintain their low-frequency values at all frequencies. The curves in Fig. 10-46 should therefore not be regarded as accurate above the microwave region since the constants of media may vary with frequency, particularly at frequencies of the order of 1 GHz and higher. A list of the low-frequency constants for the media of Fig. 10-46 is presented in Table 10-5.

Referring to Fig. 10-46, we note that copper behaves like a conductor at frequencies far above the microwave region. On the other hand, fresh water acts like a dielectric at frequencies above about 10 MHz. The $\sigma/\omega\epsilon$ ratios for seawater, rural ground, and urban ground are between the extremes of copper and fresh water.

10-15 CONDUCTING MEDIA AND LOSSY LINES

Referring to Fig. 10-47, consider a wave that is transmitted into the conducting medium. Let $x = 0$ at the boundary of the conducting medium with x increasing positively into the conducting medium.

As given by (10-2-47), the wave equation for a conducting medium is

$$\frac{\partial^2 E_y}{\partial x^2} - \gamma^2 E_y = 0 \tag{1}$$

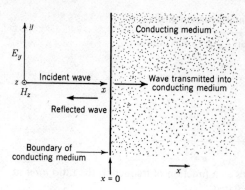

Figure 10-47 Plane wave entering conducting medium at normal incidence.

where

$$\gamma^2 = j\omega\mu\sigma - \omega^2\mu\epsilon \tag{2}$$

A solution of (1) for a wave traveling in the positive x direction is

$$E_y = E_0 e^{-\gamma x} \tag{3}$$

For conductors, $\sigma \gg \omega\epsilon$, so that (2) reduces to

$$\gamma^2 \approx j\omega\mu\sigma \tag{4}$$

and†

$$\gamma \approx \sqrt{j\omega\mu\sigma} = (1 + j)\sqrt{\frac{\omega\mu\sigma}{2}} \tag{5}$$

Thus, γ has a real and imaginary part. Putting $\gamma = \alpha + j\beta$, we see that α, the real part, is associated with attenuation and β, the imaginary part, is associated with phase. Hence,

$$E_y = E_0 e^{-\alpha x} e^{-j\beta x} \tag{6}$$

where $\alpha = \mathrm{Re}\,\gamma = \sqrt{\dfrac{\omega\mu\sigma}{2}} =$ attenuation constant, Np m^{-1}

$\beta = \mathrm{Im}\,\gamma = \sqrt{\dfrac{\omega\mu\sigma}{2}} =$ phase constant, rad m^{-1}

$\omega =$ radian frequency $= 2\pi f$, rad s^{-1}

$\mu =$ permeability of medium, H m^{-1}

$\sigma =$ conductivity of medium, $\mho$ m^{-1}

$x =$ distance, m

$j =$ complex operator $= \sqrt{-1}$, dimensionless

† *Note:*
$$\sqrt{j} = \sqrt{\frac{2j}{2}} = \sqrt{\frac{1 + 2j - 1}{2}} = \sqrt{\frac{(1 + j)^2}{2}}$$
$$= \frac{1 + j}{\sqrt{2}} = 1\underline{/45°}$$

Equation (6) is a solution of the wave equation for a plane wave traveling in the positive x direction in a conducting medium. It gives the variation of E_y in both magnitude and phase as a function of x. The field attenuates exponentially and is retarded linearly in phase with increasing x.

Let us now obtain a quantitative measure of the penetration of a wave into a conducting medium. Referring to Fig. 10-47, let (6) be written in the form

$$E_y = E_0 e^{-x/\delta} e^{-j(x/\delta)} \tag{7}$$

where $\delta = \sqrt{2/\omega\mu\sigma}$. At $x = 0$, $E_y = E_0$. This is the amplitude of the field at the surface on the conducting medium. Now δ in (7) has the dimension of distance. At a distance $x = \delta$ the amplitude of the field is

$$|E_y| = E_0 e^{-1} = E_0 \frac{1}{e} \tag{8}$$

Thus, E_y decreases to $1/e$ (36.8 percent) of its initial value, while the wave penetrates to a distance δ. Hence δ is called the $1/e$ *depth of penetration*. See Fig. 10-48.

As an example, consider the depth of penetration of a plane electromagnetic wave incident normally on a good conductor, such as copper. Since $\omega = 2\pi f$, the $1/e$ depth becomes

$$\delta = \frac{1}{\sqrt{f\pi\mu\sigma}} \tag{9}$$

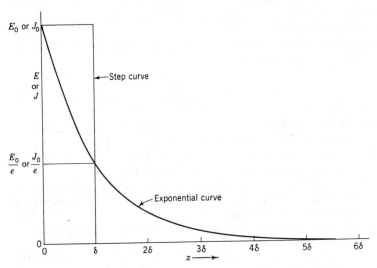

Figure 10-48 Relative magnitude of electric field **E** or current density **J** ($=\sigma$**E**) as a function of depth of penetration δ for a plane wave traveling in x direction into conducting medium. The abscissa gives the penetration distance x and is expressed in $1/e$ depths (δ). The wavelength in the conductor equals $2\pi\delta$.

For copper $\mu_r = 1$, so that $\mu = 1.26 \, \mu H \, m^{-1}$, and the conductivity $\sigma = 58 \, M\mho \, m^{-1}$. Putting these values in (9), we obtain for copper

$$\delta = \frac{6.6 \times 10^{-2}}{\sqrt{f}} \quad (10)$$

where $\delta = 1/e$ depth of penetration, m
$\quad f$ = frequency, Hz

Evaluating (10) at specific frequencies, we find that

At 60 Hz: $\qquad\qquad \delta = 8.5 \times 10^{-3} \, m$

At 1 MHz: $\qquad\qquad \delta = 6.6 \times 10^{-5} \, m$

At 30 GHz: $\qquad\qquad \delta = 3.8 \times 10^{-7} \, m$

Thus, while at 60 Hz the $1/e$ depth of penetration is 8.5 mm, the penetration decreases in inverse proportion to the square root of the frequency. At 10 mm wavelength (30 GHz) the penetration is only 0.00038 mm, or less than $\frac{1}{2} \, \mu m$. This phenomenon is often called *skin effect*.

Thus, a high-frequency field is damped out as it penetrates a conductor in a shorter distance than a low-frequency field.†

In addition to the $1/e$ depth of penetration, we can speak of other depths for which the electric field decreases to an arbitrary fraction of its original value. For example, consider the depth at which the field is 0.01 (1 percent) of its original value. This depth is obtained by multiplying the $1/e$ depth by 4.6 and may be called the 1 *percent depth of penetration*.

Phase velocity is given by the ratio ω/β. In the present case, $\beta = 1/\delta$ so that the phase velocity in the conductor is

$$v_c = \omega\delta = \sqrt{\frac{2\omega}{\sigma\mu}} \quad (11)$$

Since the $1/e$ depth is small, the phase velocity in conductors is small. It is apparent from (11) that the velocity is a function of the frequency and hence of the wavelength. In this case, $dv/d\lambda$ is negative, where λ is the free-space wavelength. Hence, conductors are anomalously dispersive media (Sec. 10-12).

The ratio of the velocity of a wave in free space to that in the conducting medium is the index of refraction for the conducting medium. At low frequencies the index for conductors is very large.

To find the wavelength λ_c in the conductor, we have from (11) that $f\lambda_c = \omega\delta$, or

$$\lambda_c = 2\pi\delta \quad (12)$$

---

† This is analogous to the way in which a rapid-temperature variation at the surface of a thermal conductor penetrates a shorter distance than a slow-temperature variation.

Table 10-6 Penetration depths, wavelength, velocity, and refractive index for copper

	Frequency		
	60 Hz	1 MHz	30 GHz
Wavelength in free space λ	5 Mm	300 m	10 mm
$1/e$ depth, m	8.5×10^{-3}	6.6×10^{-5}	3.8×10^{-7}
1 percent depth, m	3.9×10^{-2}	3×10^{-4}	1.7×10^{-6}
Wavelength in conductor λ_c, m	5.3×10^{-2}	4.1×10^{-4}	2.4×10^{-6}
Velocity in conductor v_c, m s^{-1}	3.2	4.1×10^{2}	7.1×10^{4}
Index of refraction, dimensionless	9.5×10^{7}	7.3×10^{5}	4.2×10^{3}

In (12), both λ_c and δ are in the same units of length. Hence the wavelength in the conductor is 2π times the $1/e$ depth. Since the $1/e$ depth is small for conductors, the wavelength in conductors is small.

Values of the $1/e$ depth, 1 percent depth, wavelength, velocity, and refractive index for a medium of copper are given in Table 10-6 for three frequencies.

It is interesting to note that the electric field is damped to 1 percent of its initial amplitude in about $\frac{3}{4}\lambda$ in the metal.

Since the penetration depth is inversely proportional to the square root of the frequency, a thin sheet of conducting material can act as a low-pass filter for electromagnetic waves.

For the case of a conducting medium where $\sigma \gg \omega\epsilon$, we have from (10-5-7) that the characteristic impedance

$$Z_0 = \sqrt{\frac{j\omega\mu}{\sigma}} = \sqrt{\frac{\omega\mu}{\sigma}}\ \underline{/45^\circ} \quad (\Omega) \tag{13}$$

Example 1 Calculate the ocean depths at which a 1 μV m^{-1} field will be obtained with E at the surface equal to 1 V m^{-1} at frequencies of 1, 10, 100, and 1,000 kHz. What is the most suitable frequency for communication with submerged submarines?

SOLUTION From Table 10-5, $\sigma = 4\,\mho$ m^{-1} and $\epsilon_r = 80$ for seawater. At the highest frequency (1,000 kHz), $\sigma \gg \omega\epsilon$, so that $\alpha = \sqrt{\omega\mu\sigma/2}$ can be used at all four frequencies. At 1 kHz

$$\alpha = \sqrt{\frac{2\pi 10^3 4\pi \times 10^{-7} 4}{2}} = 0.13 \text{ Np m}^{-1}$$

Since

$$\frac{E}{E_0} = 10^{-6} = e^{-\alpha x}, \quad x = \frac{6}{\alpha}\log e = \frac{13.8}{\alpha}$$

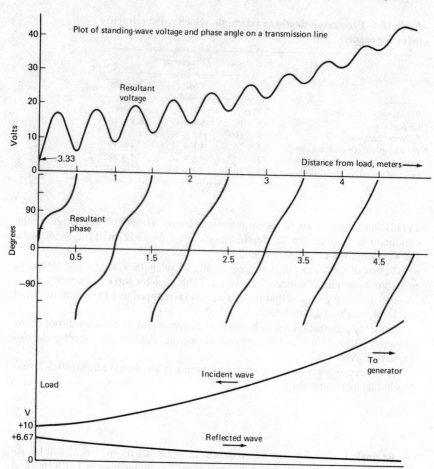

Figure 10-49 Lossy transmission line with load at left. Two-thirds of incident voltage is reflected at load (with sign reversed). Near load VSWR is large (nearly pure standing wave). Remote from load VSWR is small (nearly pure traveling wave). Phase changes stepwise near load but changes linearly with distance remote from load.

```
DIMENSION V(500),X(500),P(500)
                          ;arrays for plot-point data
OPEN 12,'PLOTFILE',ATT='SOL'
                          ;file for plot information to be
                          ;used by the printer.
BETA=2.*3.14159261        ;two times pi (value of beta here)
XP=0.0                    ;x-position
DO 10 I=1,500             ;calculate the 500 points
BX=BETA*XP                ;beta times x-position
A=10.*EXP(.3*XP)          ;coefficient for V incident
B=COS(BX)                 ;real part of exp(j beta x)
C=SIN(BX)                 ;imaginary part of exp(j beta x)
D=-C                      ;for sin of minus beta x
E=-20./3.*EXP(-0.3*XP)    ;coefficient for V reflected
RE=A*B+E*B                ;real part of Vi+Vr
RI=A*C+E*D                ;imaginary part of Vi+Vr
V(I)=(RE**2.+RI**2.)**.5  ;magnitude of V in array for plot
P(I)=ATAN2(RI,RE)*180./3.14159261
                          ;phase angle in degrees between
                          ;plus 180 and minus 180, in array
X(I)=XP                   ;x position for this plot point
XP=XP+.01                 ;next x position
CALL PLOTS(0,0,1)         ;intialize plotting routine
XAX=10.0                  ;x-axis in inches
CALL PLOT(0.5,0.5,-3)     ;set cursor at (.5,.5)
YAX=4.0                   ;y-axis in inches
FX=0.0                    ;lowest x-value
DX=0.5                    ;user x-units per inch
FY=0.0                    ;smallest y-value
DY=40./4.                 ;y-values per inch
CALL AXIS(0.0,0.0,'DISTANCE, METERS',-16,XAX,0.0,FX,DX)
                          ;draws and labels the x-axis
CALL AXIS(10.0,0.0,'MAGNITUDE, VOLTS',-16,YAX,90.0,FY,DY)
                          ;draws and labels the right y-axis
CALL ALINE(X,V,500,1,0,0,FX,DX,FY,DY)
                          ;arranges the printfile to plot the
                          ;magnitude function
FY1=-180.                 ;lowest y-axis value
DY1=360./YAX              ;number of y-units per inch
CALL AXIS(0.,4.,'PHASE ANGLE, DEGREES',20,YAX,90.,FY1,DY1)
                          ;draws and labels the left y-axis
FY1=-540                  ;to put the phase graph above
CALL ALINE(X,P,500,1,0,1,FX,DX,FY1,DY1)
                          ;arranges printfile to plot the
                          ;phase angle function
CALL PLOT(0.,8.,3)        ;       to
CALL PLOT(10.,8.,2)       ;       complete
CALL PLOT(10.,4.,2)       ;         the margin
CALL PLOT(0.,4.,3)        ;           around the
CALL PLOT(0.,0.,2)        ;             four sides
CALL SYMBOL(0.,8.5,.25,
'UNIFORM TRANSMISSION LINE AT 300 MEGAHERTZ',0.,42)
                          ;to print the label at the top
CALL PLOT(0.,0.,999)      ;to terminate the plot routine
CALL EXIT                 ;fortran housekeeping
END
PROGRAM BY PAUL HERMAN
```

and at 1 kHz, $x = 13.8/0.13 = 106$ m. At 10 kHz, $x = 35$ m; at 100 kHz, $x = 11$ m; and at 1,000 kHz, $x = 3.5$ m, where $x = $ depth.

Although 1 kHz would appear to be the best of the above four frequencies, an even lower frequency might be desirable depending on other factors including the antennas used for transmitting and receiving. This problem will be discussed further in Chap. 14.

Example 2 A uniform lossy transmission line is connected to a load. Let the load be at $x = 0$ and the line extend to positive values of x. An incident wave has voltage V_i and the wave reflected from the load has voltage V_r. Use a computer to calculate and plot the magnitude and phase of the total voltage $V = V_i + V_r$ at 100-mm intervals to a distance of 5 m from the load (50 points) for

$$V_i = V_0 e^{\alpha x} e^{j(\omega t + \beta x)}$$

$$V_r = -\tfrac{2}{3} V_0 e^{-\alpha x} e^{j(\omega t - \beta x)}$$

$$V_0 = 10 \text{ V}$$

$$f = 300 \text{ MHz } (\lambda = 1 \text{ m})$$

$$\alpha = 0.3 \text{ Np m}^{-1}$$

SOLUTION The results are shown in Fig. 10-49. Because of the attenuation, or lossiness, of the line, the effect of the reflected wave rapidly becomes small compared to the incident wave as the distance x increases. Thus, at a distance of 5 m from the load the VSWR is small and we have a nearly pure traveling wave. However, close to the load, the VSWR is large and we have a nearly pure standing wave. Near the load the phase changes stepwise, but remote from the load the phase changes smoothly in a linear manner with distance.

10-16 PLANE WAVES AT INTERFACES

Consider a linearly polarized wave traveling in the positive x direction with **E** in the y direction and **H** in the z direction. The wave is incident normally on the boundary between two media, as in Fig. 10-50a. Assume that the incident traveling wave has field components E_i and H_i at the boundary. Part of the incident wave is, in general, reflected while another part is transmitted into the second medium. The reflected traveling wave has field components E_r and H_r at the boundary. The transmitted wave has field components E_t and H_t at the boundary.

The situation here is analogous to that of a terminated transmission line as discussed in Sec. 10-6 using circuit theory. Now analyzing the wave interaction at the interface between two media with field theory, it will be shown that we arrive at identical expressions for the reflection and transmission coefficients.

From the continuity of the tangential field components at a boundary

$$E_i + E_r = E_t \tag{1}$$

and

$$H_i + H_r = H_t \tag{2}$$

The electric and magnetic fields of a plane wave are related by the intrinsic impedance of the medium. Thus

$$\frac{E_i}{H_i} = Z_1 \qquad \frac{E_r}{H_r} = -Z_1 \qquad \frac{E_t}{H_t} = Z_2 \tag{3}$$

The impedance of the reflected wave (traveling in the negative x direction) is taken to be negative Z_1 and of the incident wave, positive Z_1. From (2) and (3)

$$H_t = \frac{E_t}{Z_2} = \frac{E_i}{Z_1} - \frac{E_r}{Z_1} \tag{4}$$

or

$$E_t = \frac{Z_2}{Z_1} E_i - \frac{Z_2}{Z_1} E_r \tag{5}$$

Multiplying (1) by Z_2/Z_1 gives

$$\frac{Z_2}{Z_1} E_t = \frac{Z_2}{Z_1} E_i + \frac{Z_2}{Z_1} E_r \tag{6}$$

Adding (5) and (6), we get

$$E_t \left(1 + \frac{Z_2}{Z_1}\right) = \frac{2Z_2}{Z_1} E_i \tag{7}$$

or

$$E_t = \frac{2Z_2}{Z_2 + Z_1} E_i = \tau E_i \tag{8}$$

where τ is called the *transmission coefficient*. It follows that

$$\tau = \frac{E_t}{E_i} = \frac{2Z_2}{Z_2 + Z_1} \tag{9}$$

Subtracting (5) from (6) gives

$$E_t \left(\frac{Z_2}{Z_1} - 1\right) = \frac{2Z_2}{Z_1} E_r \tag{10}$$

Substituting E_t from (8) into (10) and solving for E_r, we have

$$E_r = \frac{Z_2 - Z_1}{Z_2 + Z_1} E_i = \rho E_i \tag{11}$$

where ρ is called the *reflection coefficient*. It follows that

$$\rho = \frac{E_r}{E_i} = \frac{Z_2 - Z_1}{Z_2 + Z_1} \tag{12}$$

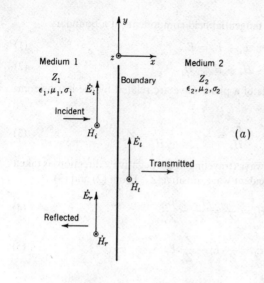

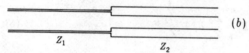

Figure 10-50 (a) Plane wave incident (b) normally on boundary between two media and (b) analogous transmission line.

From (9) and (12)

$$\tau = \rho + 1 \tag{12a}$$

The situation (Fig. 10-50) of a plane wave incident normally on a boundary between two different media of infinite extent, with intrinsic impedances Z_1 and Z_2, is analogous to the situation of a wave on an infinite transmission line having an abrupt change in impedance from Z_1 to Z_2 (Fig. 10-50b). The transmission and reflection coefficients for voltage across the transmission line are identical to those given in (9) and (12) if the intrinsic impedance Z_1 of medium 1 is taken to be the characteristic impedance of the line to the left of the junction (Fig. 10-50b) and the intrinsic impedance Z_2 of medium 2 is taken to be the characteristic impedance of the line to the right of the junction.

Returning now to the case of a plane wave incident normally on the boundary between two media of infinite extent as in Fig. 10-50a, let us consider several special cases.

Case 1 Assume that medium 1 is air and medium 2 is a conductor, so that $Z_1 \gg Z_2$. Then, from (8) we have the approximate relation

$$E_t \approx \frac{2Z_2}{Z_1} E_i \tag{13}$$

But from (3) this becomes

$$H_t Z_2 \approx \frac{2Z_2}{Z_1} H_i Z_1 \tag{14}$$

from which

$$H_t \approx 2H_i \tag{15}$$

Thus, for a plane wave in air incident normally on a conducting medium, the magnetic field is, to a good approximation, doubled in intensity at the boundary. It also follows that $H_r \approx H_i$, so that there is a nearly pure standing wave to the left of the boundary (in medium 1).

Case 2 Consider now the opposite situation, where medium 1 is a conductor and medium 2 is air so that $Z_1 \ll Z_2$. Then, from (8) we have approximately

$$E_t \approx 2E_i \tag{16}$$

Thus, for a wave leaving a conducting medium, the electric field is nearly doubled at the boundary. It follows that $E_r \approx E_i$, so that there is a nearly pure standing wave (VSWR $= \infty$) immediately to the left of the boundary (in medium 1). However, owing to the attenuation of waves in medium 1, the VSWR decreases rapidly as one moves away from the boundary (to the left).

Case 3 In case 1 it is assumed that $Z_1 \gg Z_2$. Consider now that $Z_2 = 0$ (medium 2 a perfect conductor). Then from (12) the reflection coefficient $\rho = -1$, and from (12a) the transmission coefficient $\tau = 0$. Thus, the wave is completely reflected, and no field is transmitted into medium 2. Further $E_r = -E_i$, and $H_r = H_i$, so that the magnetic field intensity exactly doubles at the boundary. This situation is analogous to a short-circuited transmission line.

Case 4 In case 2 it is assumed that $Z_1 \ll Z_2$. Consider now the hypothetical situation where Z_2 is infinite.† Then from (12) $\rho = +1$ and from (12a) $\tau = 2$. Thus the wave is completely reflected, but $E_r = +E_i$, so that the electric field intensity at the boundary is exactly doubled. The situation is analogous to an open-circuited transmission line.

Case 5 Assume that both medium 1 *and* medium 2 are lossless nonferromagnetic dielectrics ($\mu_1 = \mu_2 = \mu_0$). Then it follows from (12) that

$$\rho = \frac{\sqrt{\epsilon_1 \epsilon_2} - 1}{\sqrt{\epsilon_1/\epsilon_2} + 1} \tag{17}$$

---

† It is to be noted that for free space the intrinsic impedance is only 376.7 Ω. To obtain a higher impedance would require that $\mu_r > 1$ such as in ferromagnetic media.

and from (9) that

$$\tau = \frac{2}{1 + \sqrt{\epsilon_2/\epsilon_1}}$$ (18)

Case 6 Take now the case where $Z_2 = Z_1$. Then $\rho = 0$ and $\tau = 1$, so that the wave propagates into medium 2 without any reflection. This situation is similar to that on a continuous transmission line of uniform characteristic impedance.

$\lambda/4$ Plate

A $\lambda/4$ transformer for impedance matching was discussed in Sec. 10-9. An important application of the $\lambda/4$ transformer principle is to the $\lambda/4$ plate, used to eliminate reflections in many optical devices in which $\lambda/4$ coatings are deposited on the surfaces of lenses and prisms in cameras, binoculars, and telescopes to improve their efficiency.

Thus, for example, a plane wave in air incident normally on a half-space filled with a lossless dielectric medium of relative permittivity $\epsilon_r = 4$ will be partly reflected and partly transmitted. The reflection can be eliminated, as shown in Fig. 10-51a by placing a $\lambda/4$ plate between the air and the dielectric medium provided the plate has an intrinsic impedance

$$Z_1 = \sqrt{Z_0 Z_2}$$

where Z_0 = intrinsic impedance of air
Z_2 = intrinsic impedance of dielectric medium

In the present example,

$$Z_0 = \sqrt{\frac{\mu_0}{\epsilon_0}} = 376.7 \ \Omega$$

$$Z_2 = \frac{Z_0}{\sqrt{\epsilon_r}} = \frac{376.7}{\sqrt{4}} = 188 \ \Omega$$

Therefore, the intrinsic impedance of the plate must be

$$Z_1 = \sqrt{Z_0 Z_2} = 266 \ \Omega$$

and its relative permittivity must be

$$\epsilon_r = \frac{Z_0}{Z_2} = 2$$

The analogous transmission-line equivalent is also shown in Fig. 10-51a. It is assumed that no ferromagnetic material is present, and so $\mu = \mu_0$. It is to be noted that the plate is $\lambda/4$ thick as measured in terms of the wavelength *in the plate*.

For reflectionless transmission of a plane wave *through* a dielectric medium of finite thickness, a $\lambda/4$ plate is required on both sides to prevent reflection at each

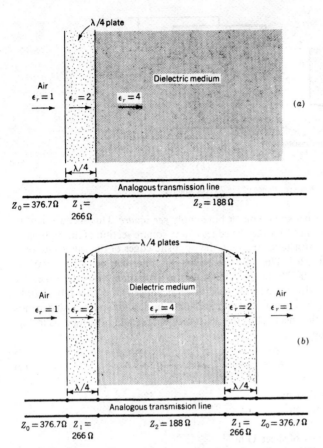

Figure 10-51 (*a*) λ/4 plate at interface of air and dielectric medium and (*b*) two *λ*/4 plates for transmission through a dielectric medium.

dielectric-air interface. The arrangement is suggested in Fig. 10-51*b*, where the analogous transmission-line equivalent is also shown.

Wave Absorption with Space Cloth

Let us consider now the situation where a wave is completely absorbed and there is no reflected wave.

The intrinsic impedance of free space is 376.7 Ω. This concept of an impedance for free space takes on more physical significance if we consider the properties of a resistive sheet having a resistance of 376.7 Ω *per square*. Material so treated is often called *space paper* or *space cloth*. It should be noted that the resistance is not per

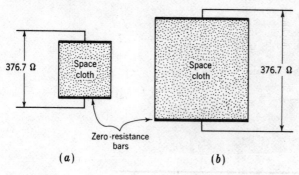

Figure 10-52 Space cloth has a resistance of 376.7 Ω per square.

square centimeter or per square meter but simply *per square*. This is equivalent to saying that the resistance between the edges of any square section of the material is the same. Hence the resistance between the opposite edges of the small square of space cloth in Fig. 10-52a is 376.7 Ω, as is also the resistance between the edges of the large square in Fig. 10-52b. In this illustration it is assumed that the edges are clamped with zero-resistance bars and that the impedance of the leads is negligible.

The conductivity of the material required for a sheet of the space cloth depends on the thickness of the sheet. Thus the resistance R of a square section as in Fig. 10-53 is expressed by

$$R = \frac{l}{\sigma a} = \frac{l}{\sigma h l} = \frac{1}{\sigma h} \quad (\Omega) \qquad (19)$$

where l = length of side, m
  a = area of edge, m^2
  h = thickness of sheet, m
  σ = conductivity of sheet, ℧ m^{-1}

It follows that the required conductivity is

$$\sigma = \frac{1}{Rh} = \frac{1}{376.7h} \quad (\text{℧ m}^{-1}) \qquad (20)$$

Consider now the behavior of a sheet of space cloth placed in the path of a plane wave. Suppose, as shown in Fig. 10-54a, that a plane wave in free space traveling to the right is incident normally on a sheet of space cloth of infinite extent.

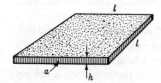

Figure 10-53 Square of space cloth of thickness h.

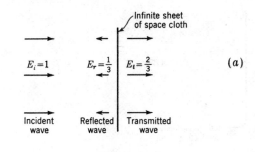

Incident wave   Reflected wave   Transmitted wave

Figure 10-54 (a) Plane wave traveling to right incident normally on sheet of space cloth and (b) analogous transmission-line arrangement.

Taking the amplitude of the incident wave as 1 V m^{-1}, we have from (9) that there is a transmitted wave continuing to the right of the sheet of amplitude

$$E_t = \tau E_i = \frac{2 \times 188.3}{188.3 + 376.7} = \frac{2}{3} \text{ V m}^{-1}$$

and from (12) that there is a reflected wave to the left of the sheet of amplitude

$$E_r = \rho E_i = \frac{188.3 - 376.7}{188.3 + 376.7} = -\frac{1}{3} \text{ V m}^{-1}$$

It is to be noted that the impedance presented to the incident wave at the sheet is the resultant of the space cloth in parallel with the impedance of the space behind it. This is one-half of 376.7, or 188.3 Ω.

It is apparent that a sheet of space cloth by itself is insufficient to terminate a wave. This may also be seen by considering the analogous transmission arrangement shown in Fig. 10-54b.

In order completely to absorb or terminate the incident wave without reflection or transmission, let an infinite, perfectly conducting sheet be placed parallel to the space cloth and $\lambda/4$ behind it, as portrayed in Fig. 10-55a. Now the impedance presented to the incident wave at the sheet of space cloth is 376.7 Ω, being the impedance of the sheet in parallel with an infinite impedance. As a consequence, this arrangement results in the total absorption of the wave by the space cloth, with no reflection to the left of the space cloth. There is, however, a standing wave and energy circulation between the cloth and the conducting sheet. The analogous transmission-line arrangement is illustrated in Fig. 10-55b.

In the case of the plane wave, the perfectly conducting sheet effectively isolates the region of space behind it from the effects of the wave. In a roughly analogous

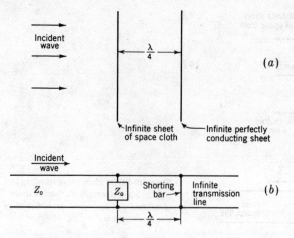

Figure 10-55 (*a*) Plane wave traveling to right incident normally on sheet of space cloth backed by conducting sheet is absorbed without reflection. (*b*) Wave traveling to right on transmission line is absorbed without reflection by analogous arrangement.

manner, the shorting bar on the transmission line reduces the wave beyond it to a small value.

A transmission line may also be terminated by placing an impedance across the line which is equal to the characteristic impedance of the line, as in Fig. 10-54*b*, and disconnecting the line beyond it. Although this provides a practical method of terminating a transmission line, there is no analogous counterpart in the case of a wave in space because it is not possible to "disconnect" the space to the right of the termination. A region of space may only be isolated or shielded, as by a perfectly conducting sheet.

Wave Absorption with Ferrite-Titanate Medium

A lossy mixture of a high-μ (ferrite) material and a high-ϵ (barium titanate) material can be used effectively for wave absorption with both μ and ϵ being complex and with the ratio μ/ϵ equal to that for free space ($\mu_r/\epsilon_r = 1$). Although the mixture constitutes a physical discontinuity, an incident wave enters it without reflection. The velocity of the wave is reduced, and large attenuation can occur in a short distance.

Example: Lossy medium with same impedance as space Let a plane 100-MHz wave be incident normally on a solid ferrite-titanate slab of 10-mm thickness for which $\mu_r = \epsilon_r = 60(2 - j1)$. See Fig. 10-56. The medium is backed by a flat conducting sheet. What is the level of the reflected wave (in decibels) below that of the incident wave? The medium is nonconducting ($\sigma = 0$).

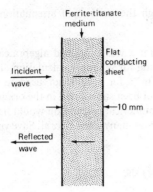

Ferrite-titanate
medium

Flat
conducting
sheet

Incident
wave

←10 mm

Reflected
wave

Figure 10-56 Ferrite-titanate medium with complex μ and ϵ matches the intrinsic impedance of space but has high attenuation.

SOLUTION The wave enters the ferrite-titanate slab without reflection since its intrinsic impedance Z_0 is the same as that of space. Thus,

$$Z_0 = \sqrt{\frac{\mu_0}{\epsilon_0}}\sqrt{\frac{\mu_r}{\epsilon_r}} = \sqrt{\frac{\mu_0}{\epsilon_0}}\sqrt{\frac{60(2-j1)}{60(2-j1)}} = \sqrt{\frac{\mu_0}{\epsilon_0}} = \text{intrinsic impedance of space}$$

From (10-15-2) the attenuation constant, in general, is given by

$$\alpha = \text{Re}\,\sqrt{j\omega\mu\sigma - \omega^2\mu\epsilon}$$

If both μ and ϵ are complex, so that $\mu = \mu' - j\mu''$ and $\epsilon = \epsilon' - j\epsilon''$, then, in general,

$$\alpha = \text{Re}\,\sqrt{(\omega\mu'' + j\omega\mu')[\sigma + j\omega(\epsilon' - j\epsilon'')]}$$

For $\sigma = 0$, we have

$$\alpha = \text{Re}\, j\frac{2\pi}{\lambda_0}\sqrt{(\mu_r' - j\mu_r'')(\epsilon_r' - j\epsilon_r'')} = \text{Re}\, j\frac{2\pi}{\lambda_0}\sqrt{\mu_r\epsilon_r}$$

$$= \text{Re}\, j\frac{2\pi}{3}60(2-j1) = \text{Re}\,40\pi(1+j2)$$

$$= 126\ \text{Np m}^{-1} = 1092\ \text{dB m}^{-1}$$

In traveling 10 mm the attenuation is 11 dB. After reflection from the flat conducting sheet an equal attenuation occurs. This reflected wave passes on out into space again without reflection (intrinsic impedances matched) and is 22 dB down from the incident wave. Thus, the reflected wave power is less than $\frac{1}{100}$ of the incident wave power. The reflected wave is not completely eliminated but it is reduced enough to be satisfactory for many practical purposes, such as applications on moving vehicles, ships, and aircraft to reduce the likelihood of radar detection.

Whereas the space cloth termination is matched only at frequencies for which the spacing is $\lambda/4$ (between it and the conducting sheet), the ferrite-titanate medium matching is independent of frequency, although the amount of attenuation is proportional to the frequency.

For the student: Note that the expression for α was simplified algebraically before numerical values were introduced, making for a simple calculation. The $4\pi \times 10^{-7}\,\mathrm{H\,m^{-1}}$ for μ_0 and $8.85 \times 10^{-12}\,\mathrm{F\,m^{-1}}$ for ϵ_0 did not need to be introduced at all. If the full μ, ϵ, and ω values had been substituted in the original equation for α (in eight places), a much longer numerical calculation would have been required with possible introduction of errors at many steps. Moral: Save your calculator until it's really needed.

10-17 POWER AND ENERGY RELATIONS

Consider a region of space represented by an array of field-cell transmission lines of total width w and total height h, as in Fig. 10-57, with a plane wave traveling from left to right. The electric field E is vertical and the magnetic field H is horizontal. The voltage $V = Eh$ and the current $I = Hw$. By analogy to circuits the power conveyed is

$$P = VI = EhHw = EHhw = EHA \qquad \text{(W)} \qquad (1)$$

where $A = hw$ = area of field cell array.

The power (surface) density is then

$$S = \frac{P}{A} = EH \qquad (\mathrm{W\,m^{-2}}) \qquad (2)$$

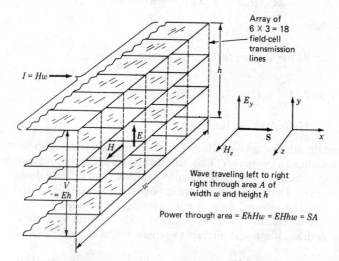

Array of $6 \times 3 = 18$ field-cell transmission lines

$I = Hw$

h

E_y y

S x

H_z z

E

H

$V = Eh$

w

Wave traveling left to right right through area A of width w and height h

Power through area $= EhHw = EHhw = SA$

Figure 10-57 Power flow of wave traveling left to right through area of width w and height h is equal to $EhHw$.

Equation (2) relates the scalar magnitudes. The power flow is perpendicular to **E** and **H** and it can be shown that in vector notation

$$\boxed{\mathbf{S} = \mathbf{E} \times \mathbf{H} \qquad (\text{W m}^{-2})}$$ (3)

Turning **E** into **H** and proceeding as with a right-handed screw gives the direction of **S** perpendicular to both **E** and **H**. **S** is a power surface density called the *Poynting vector*.† Its value in (3) is the *instantaneous Poynting vector*. The *average Poynting vector* is obtained by integrating the instantaneous Poynting vector over one period and dividing by one period. It is also readily obtained in complex notation from

$$\mathbf{S}_{av} = \tfrac{1}{2} \operatorname{Re} \mathbf{E} \times \mathbf{H}^* = \tfrac{1}{2}\hat{\mathbf{x}}|E_y||H_z|\cos\xi \qquad (\text{W m}^{-2})$$ (4)‡

where $\mathbf{S}_{av} = \hat{\mathbf{x}}S =$ average Poynting vector, W m^{-2}

$\mathbf{E} = \hat{\mathbf{y}}E_y = \hat{\mathbf{y}}|E_y|e^{j\omega t}$, V m^{-1}

$\mathbf{H}^* = \hat{\mathbf{z}}H_z^* = \hat{\mathbf{z}}|H_z|e^{-j(\omega t - \xi)}$, A m^{-1}

$\xi =$ time phase angle between E_y and H_z, rad or deg

H* is called the *complex conjugate* of **H**, where

$$\mathbf{H} = \hat{\mathbf{z}}H_z = \hat{\mathbf{z}}|H_z|e^{j(\omega t - \xi)} \qquad (\text{A m}^{-1})$$

The quantities **H** and its complex conjugate **H*** have the same space direction but they differ in sign in their phase factors. Note that if E_y and H_z in (4) are rms values instead of (peak) amplitudes, the factor $\tfrac{1}{2}$ in (4) is omitted.

The magnitude of the *average Poynting vector*

$$S_{av} = \tfrac{1}{2}\operatorname{Re} E_y H_z^* = \tfrac{1}{2}|E_y||H_z|\cos\xi \qquad (\text{W m}^{-2})$$ (5)

The relation corresponding to (5) for the average power of a traveling wave on a transmission line is

$$P_{av} = \tfrac{1}{2}\operatorname{Re} VI^* = \tfrac{1}{2}|V||I|\cos\theta \qquad (\text{W})$$ (6)

where $V =$ voltage between conductors of transmission line, V

$I =$ current through one conductor, A

$I^* =$ complex conjugate of I

$\theta =$ time phase angle between V and I, rad or deg

Since the intrinsic impedance of the medium

$$Z_0 = \frac{E}{H} = \frac{|E|}{|H|}\underline{/\xi} = |Z_0|\underline{/\xi}$$ (6a)

---

† J. H. Poynting, On the Transfer of Energy in the Electromagnetic Field, *Phil. Trans.*, **174**:343 (1883); Oliver Heaviside, "Electromagnetic Theory," vol. 1, p. 78, Ernest Benn, Ltd., London, 1893.

‡ Note that if **E** and **H** are not perpendicular but with a space angle ϕ between them, (4) would need to include a sin ϕ factor.

the magnitude of the average Poynting vector can also be written

$$S_{av} = \tfrac{1}{2} \operatorname{Re} H_z H_z^* Z_0 = \tfrac{1}{2}|H_z|^2 \operatorname{Re} Z_0 \qquad (\text{W m}^{-2}) \tag{7}$$

or
$$S_{av} = \tfrac{1}{2} \operatorname{Re} \frac{E_y E_y^*}{Z_0} = \tfrac{1}{2}|E_y|^2 \operatorname{Re} \frac{1}{Z_0} \qquad (\text{W m}^{-2}) \tag{8}†$$

Equation (7) is very useful since if the intrinsic impedance Z_0 of a conducting medium and also the magnetic field H_0 at the surface are known, it gives the average Poynting vector (or average power per unit area) into the conducting medium.

Example A plane 1-GHz traveling wave in air with peak electric field intensity of 1 V m^{-1} is incident normally on a large copper sheet. Find the average power absorbed by the sheet per square meter of area.

SOLUTION First let the intrinsic impedance of copper be calculated at 1 GHz. From (10-15-13)

$$Z_0 = \sqrt{\frac{\omega\mu}{\sigma}} \; \underline{/45°}$$

For copper $\mu_r = \epsilon_r = 1$ and $\sigma = 58$ M℧ m^{-1}. Hence the real part of Z_0 is

$$\operatorname{Re} Z_0 = \cos 45° \sqrt{\frac{2\pi \times 10^9 \times 4\pi \times 10^{-7}}{5.8 \times 10^7}} = 8.2 \text{ m}\Omega$$

Next we find the value of H_0 at the sheet (tangent to the surface). This is very nearly double H for the incident wave. Thus

$$H_0 = 2\frac{E}{Z} = \frac{2 \times 1}{376.7} \text{ A m}^{-1}$$

From (7) the average power per square meter into the sheet is

$$S_{av} = \frac{1}{2}\left(\frac{2}{376.7}\right)^2 8.2 \times 10^{-3} = 116 \text{ nW m}^{-2}$$

The relations corresponding to (7) and (8) for the average power of a traveling wave on a transmission line are

$$P_{av} = \tfrac{1}{2} \operatorname{Re} II^* Z_0 = \tfrac{1}{2}|I|^2 \operatorname{Re} Z_0 \qquad (\text{W}) \tag{9a}$$

and
$$P_{av} = \tfrac{1}{2} \operatorname{Re} \frac{VV^*}{Z_0} = \tfrac{1}{2}|V|^2 \operatorname{Re} \frac{1}{Z_0} \qquad (\text{W}) \tag{9b}$$

---

† Note that in general

$$\operatorname{Re} \frac{1}{Z_0} \neq \frac{1}{\operatorname{Re} Z_0}$$

When Z_0 is real ($\xi = 0$) and E and H are rms values, we have for the *traveling space wave.*

$$S_{av} = EH = H^2 Z_0 = \frac{E^2}{Z_0} \quad \text{(W m}^{-2}) \tag{10}$$

and for the *traveling wave on a transmission line* ($\theta = 0$ and V and I rms)

$$P_{av} = VI = I^2 Z_0 = \frac{V^2}{Z_0} \quad \text{(W)} \tag{11}$$

From (3-10-12) the energy density w_e at a point in an electric field is

$$w_e = \tfrac{1}{2}\epsilon E^2 \quad \text{(J m}^{-3}) \tag{12}$$

where ϵ = permittivity of medium, F m^{-1}
E = electric field intensity, V m^{-1}

From (3-10-12) the energy density w_e at a point in an electric field is

$$w_m = \tfrac{1}{2}\mu H^2 \quad \text{(J m}^{-3}) \tag{13}$$

where μ = permeability of medium, H m^{-1}
H = magnetic field, A m^{-1}

In a traveling wave in an unbounded, lossless medium

$$\frac{E}{H} = \sqrt{\frac{\mu}{\epsilon}} \tag{14}$$

Substituting for H from (14) in (13), we have

$$w_m = \tfrac{1}{2}\mu H^2 = \tfrac{1}{2}\epsilon E^2 = w_e \tag{15}$$

Thus the electric and magnetic energy densities in a plane traveling wave are equal, and the total energy density w is the sum of the electric and magnetic energies. Thus

$$\boxed{w = w_e + w_m = \tfrac{1}{2}\epsilon E^2 + \tfrac{1}{2}\mu H^2} \tag{16}$$

or

$$w = \epsilon E^2 = \mu H^2 \quad \text{(J m}^{-3}) \tag{17}$$

Next let us consider the energy and power relations for two plane waves traveling in opposite directions. Assume that both waves are polarized with **E** in the y direction. Assume further that one wave travels in the negative x direction and has an amplitude E_0, while the other wave travels in the positive x direction and has an amplitude E_1. In this case the instantaneous value of E_y, resulting from the two waves, is given by

$$E_y = E_0 \sin(\omega t + \beta x) + E_1 \sin(\omega t - \beta x) \tag{18}$$

We may find a corresponding relation for H_z as follows. Let us start with (10-2-8), i.e.,

$$\frac{\partial E_y}{\partial x} = -\mu \frac{\partial H_z}{\partial t} \tag{19}$$

Substituting E_y from (18) into (19), differentiating with respect to x, and integrating with respect to t, we obtain

$$H_z = -\sqrt{\frac{\epsilon}{\mu}} E_0 \sin(\omega t + \beta x) + \sqrt{\frac{\epsilon}{\mu}} E_1 \sin(\omega t - \beta x) \tag{20}$$

The magnitude of the Poynting vector is

$$S_x = E_y H_z \tag{21}$$

Substituting (18) and (20) in (21) yields

$$S_x = -\sqrt{\frac{\epsilon}{\mu}} [E_0^2 \sin^2(\omega t + \beta x) - E_1^2 \sin^2(\omega t - \beta x)] \tag{22}$$

According to (22), the net Poynting vector is in the negative x direction provided $E_0 > E_1$. Furthermore, the net Poynting vector is equal to the difference of the Poynting vectors for the two traveling waves. Suppose that the wave to the left is incident on a plane boundary at $x = 0$ (as in Fig. 10-42). The wave to the right then becomes a reflected wave. If the medium to the left of the boundary is a perfect conductor, we have the condition at the boundary that $E_1 = -E_0$, resulting in a pure standing wave to the right of the boundary. We note that for this condition the net Poynting vector is zero, and hence no power is transmitted.

It is interesting to examine the condition of a pure standing wave ($E_1 = -E_0$) in more detail, particularly from the standpoint of concentrations of energy. Accordingly, let us find the values of the electric and magnetic energy densities separately. Substituting (10-13-7) into (12), we obtain, for the electric energy density of a pure standing wave,

$$w_e = 2\epsilon E_0^2 \cos^2 \omega t \sin^2 \beta x \tag{23}$$

Taking (20), expanding, collecting terms, and putting $E_1 \sqrt{\epsilon/\mu} = H_0$, we get

$$H_z = -2H_0 \sin \omega t \cos \beta x \tag{24}$$

Substituting this in (13) yields the value of the magnetic energy density of a pure standing wave,

$$w_m = 2\mu H_0^2 \sin^2 \omega t \cos^2 \beta x \tag{25}$$

Comparing (23) and (25), we see that the electric energy density is a maximum when the magnetic is zero, and vice versa. Furthermore, the points where they are maximum are $\lambda/4$ apart. In other words, the electric and magnetic energy densities of a pure standing wave are in space and time quadrature. This condition is typical of a pure resonator (see Sec. 13-17). The energy oscillates back and forth from the electric form to the magnetic. Energy in this condition is often spoken of as reactive or stored energy. It is not transmitted but circulates from one form to the other. Simultaneously with the change from the electric to the magnetic form of energy there is a space motion of the energy back and forth over a distance of $\lambda/4$. These relations are shown graphically in Fig. 10-58. Here the energy densities are shown at three instants of time, $t = 0$, $T/8$, and $T/4$. The dashed curves show the instantaneous electric energy density w_e as evaluated from (23), and the solid curves show the instantaneous magnetic energy density w_m as evaluated from (25).

Finally, let us find an expression for the magnitude of the Poynting vector of a pure standing wave. To do this, we substitute (24) and (10-13-7) in (21), obtaining

$$S_x = -4E_0 H_0 \cos \omega t \sin \omega t \cos \beta x \sin \beta x \qquad (26)$$

Putting H_0 in terms of E_0 gives

$$S_x = -4\sqrt{\frac{\epsilon}{\mu}} E_0^2 \cos \omega t \sin \omega t \cos \beta x \sin \beta x \qquad (27)$$

and the peak value of the Poynting vector is

$$\text{Peak } S_x = \sqrt{\frac{\epsilon}{\mu}} E_0^2 \qquad (28)$$

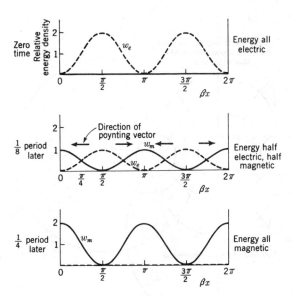

Figure 10-58 Total electric and magnetic energy densities at three instants of time for a pure standing wave. Conditions are shown over a distance of $1\lambda(\beta x = 2\pi)$. There is no net transmission of energy in a pure standing wave. The situation here (pure standing wave) is identical with that in a resonator (Chap. 13).

From (27) it is clear that S_x is a maximum at $\omega t = \pi/4$ ($\frac{1}{8}$ period). At this instant the position of one maximum is at $\beta x = \pi/4$ ($\lambda/8$) and is directed to the left as shown by the arrow in Fig. 10-58. The other arrows indicate other Poynting vector maxima and illustrate that at $t = T/8$ the energy is flowing from the regions of electric energy density to those of magnetic energy density.

The situation here with a pure standing wave is identical with that in a resonator as will be discussed in Chap. 13.

10-18 POWER FLOW ON A TRANSMISSION LINE

The electric and magnetic fields around a two-conductor transmission line are shown in Fig. 10-59a (see also Fig. 3-16). At any point the power flow parallel to the line (perpendicular to page) is given by the Poynting vector. But $E/H = Z_0 (= \sqrt{\mu_0/\epsilon_0} = 376.7\ \Omega$ for air or vacuum), so that the Poynting vector $(= E^2/Z_0)$ is proportional to the square of the electric field E. Calculating the Poynting vector for the transmission line of Fig. 10-59a, we can obtain the power-flow map of Fig. 10-59b. The contours indicate power flow in watts per square meter for the

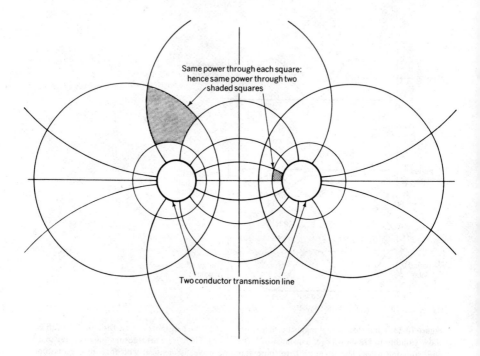

Same power through each square:
hence same power through two
shaded squares

Two conductor transmission line

Figure 10-59 (*a*) Electric field map for infinite two-conductor lossless transmission line.

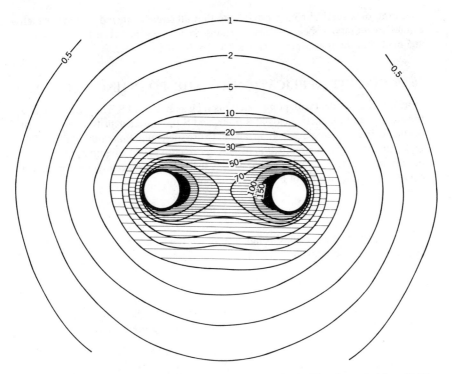

Figure 10-59 (*b*) Power-flow map; the contours give the power density (Poynting vector) in watts per square meter. The same power is transmitted through each curvilinear square in (*a*). The two conductors are 25.4 mm in diameter and are spaced 76.2 mm between centers. The maximum power density is between the conductors and close to each one as shown by the contours in (*b*) and by the fact that the smallest curvilinear squares are so located in (*a*).

case where each conductor is 25.4 mm in diameter with spacing between centers of 76.2 mm and a potential difference between conductors of 10 V rms. It is interesting to note that the maximum power flow is concentrated between the two conductors and close to each one. Integrating the Poynting vector over an infinite plane perpendicular to the conductors yields the total power transmitted. The total power can also be obtained from the interconductor voltage V and the characteristic impedance Z_c of the line, or in this case 465 mW ($= V^2/Z_c = 10^2/215$).

The field map of Fig. 10-59a has been drawn with true curvilinear squares (see Sec. 3-19). An important property of this map is that *the same power is transmitted through each curvilinear square*. Thus, if we know the power through any square and the number of squares, we can obtain the total power transmitted. Or conversely, from a knowledge of the total power and the number of squares we can obtain the power per square. In the present case there are $3 \times 5\frac{1}{4} = 15\frac{3}{4}$ squares in each quadrant, or a total of 63 squares. (Note that all squares, including those which extend to infinity, appear at least partially in Fig. 10-59a.) The total power is 465 mW, and the power per square is 7.4 mW ($= 465/63$).

10-19 CIRCUIT APPLICATION OF THE POYNTING VECTOR

In field theory we deal with point functions such as $\mathbf{E}$, $\mathbf{H}$, and $\mathbf{S}$. Thus, $\mathbf{E}$ and $\mathbf{H}$ give the electric and magnetic fields at a point and $\mathbf{S}$ the power density at a point. In dealing with waves in space it is convenient to use such point functions. On the other hand, in dealing with circuits it is usually more convenient to employ integrated quantities such as V, I, and P. That is, V is the voltage between two points and is equal to the line integral of $\mathbf{E}$ between the points, or

$$V = \int_1^2 \mathbf{E} \cdot d\mathbf{l} \tag{1}$$

The quantity I is the current through a conductor, which is equal to the integral of $\mathbf{H}$ around the conductor or

$$I = \oint \mathbf{H} \cdot d\mathbf{l} \tag{2}$$

The quantity P is, for example, the power dissipated in a load and is equal to the integral of the Poynting vector over a surface enclosing the load, or

$$P = \oint_s \mathbf{S} \cdot d\mathbf{s} \tag{3}$$

Consider as an illustration the simple circuit of Fig. 10-60a, consisting of a battery of voltage V connected by a conductor with current I to a load of resistance R. All the emf of the circuit is confined to the battery and all resistance to the load (conductivity of the conductor is infinite).

There exists an electric field E extending between the conductors ($\int \mathbf{E} \cdot d\mathbf{l} = V$, where $\mathbf{E}$ is integrated from the upper to the lower conductor) and a magnetic field $\mathbf{H}$ around the conductors ($\oint \mathbf{H} \cdot d\mathbf{l} = I$), as suggested in Fig. 10-60a. At any point the Poynting vector $\mathbf{S}$ (in watts per square meter) is given by the cross product of $\mathbf{E}$ and $\mathbf{H}$ ($\mathbf{S} = \mathbf{E} \times \mathbf{H}$). The direction of the Poynting vector is suggested at several points in Fig. 10-60a. Around the battery (power source) the Poynting vector is outward (positive). Around the resistor (load) the Poynting vector is inward (negative). Over an imaginary plane through the middle of the circuit the Poynting vector is to the right (from battery to load). The total power leaving the battery or entering the load is given by the integral of the Poynting vector over a surface enclosing the battery (positive power), over a surface enclosing the load (negative power), or over an infinite plane dividing the circuit as suggested.

The total power into the load is given by

$$P = \oint_s (\mathbf{E} \times \mathbf{H}) \cdot d\mathbf{s} \tag{4}$$

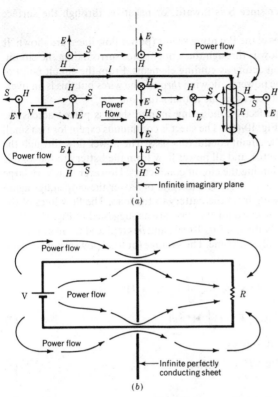

Figure 10-60 Simple circuit showing Poynting vector or power-flow lines from battery (source) to resistor (load) (*a*) without and (*b*) with intervening conducting sheet.

where the integration is over a surface enclosing the load, such as the cylindrical surface shown in Fig. 10-60*a*. The fields **E** and **H** are everywhere normal to each other. They are also both tangent to the cylindrical surface enclosing the resistor or load. Thus, from the geometry in this case (4) reduces to

$$P = -\int \mathbf{E} \cdot d\mathbf{l} \oint \mathbf{H} \cdot d\mathbf{l} \tag{5}$$

or

$$P = -VI \tag{6}$$

where $V = \int \mathbf{E} \cdot d\mathbf{l}$ = voltage between ends of the cylinder (or load), V
$I = \oint \mathbf{H} \cdot d\mathbf{l}$ = current through load, A

Thus, integrating the Poynting vector over the resistor† yields the same power *VI* as given by circuit theory. However, from the field point of view the power entering

---

† We have neglected the contribution of power flowing through the flat end surfaces of the cylinder. However, this can be made negligibly small if the cylinder diameter is very small compared to its length.

the load is negative power since **S** is inward, or negative, through the surface enclosing the load.

In Fig. 10-60a flow lines of the Poynting vector (power flow lines) are shown. It is evident that the power flow is through the empty space surrounding the circuit, the conductors of the circuit acting as guiding elements. From the circuit point of view we usually think of the power as flowing *through* the wires but this is an over-simplification and does not represent the actual situation.

Suppose now that a perfectly conducting infinite sheet is placed across the middle of the circuit, as in Fig. 10-60b. The sheet is continuous except for two small openings through which the circuit conductors pass. The power flow through the sheet is zero (**E** in sheet is zero), and all power flow from the battery to the load is through the two holes containing the circuit conductors. However, **E** is very large in these openings, and **S** is also large. The integral of **S** over these openings again equals the total power flowing from the battery to the load. The flow lines of the Poynting vector (power flow lines) for this case are as suggested in Fig. 10-60b.

Consider now that the battery in Fig. 10-60a and b is replaced by an alternator whose emf varies harmonically with time. The average (or total power) is now given by

$$P_{av} = \oint_s \text{Re } \mathbf{S} \cdot d\mathbf{s} = \frac{1}{2} \oint_s \text{Re } (\mathbf{E} \times \mathbf{H}^*) \cdot d\mathbf{s}$$
$$= \tfrac{1}{2} \text{Re} \int \mathbf{E} \cdot d\mathbf{l} \oint \mathbf{H}^* \cdot d\mathbf{l} \tag{7}$$

Integrating over the alternator gives

$$P_{av} = \tfrac{1}{2} \text{Re } VI^* = \tfrac{1}{2}V_0 I_0 \text{ Re } e^{j\xi} = \tfrac{1}{2}V_0 I_0 \cos \xi \tag{8}$$

where $V = V_0 e^{j\omega t}$
$I^* = I_0 e^{-j(\omega t - \xi)}$
V_0 = amplitude or peak value of $V = |V|$
I_0 = amplitude or peak value of $I = |I|$
ξ = time-phase angle between current and voltage
$\cos \xi$ = power factor

Integrating (7) over the load, we obtain

$$P_{av} = -\tfrac{1}{2}V_0 I_0 \cos \xi \tag{9}$$

which is equal in magnitude but opposite in sign to (8). Thus, from field theory and the Poynting vector we arrive at familiar circuit relations.

Since the load impedance $Z = V/I$, we also have (omitting the negative sign) that

$$P_{av} = \tfrac{1}{2}I_0^2 \text{ Re } Z = \tfrac{1}{2}I_0^2 R \tag{10}$$

where R is the resistance of the load (real part of Z) in ohms.

Although **E** × **H** generally indicates power flow, it is possible to contrive a situation where it does not, as, for example, with an **E** from a static electric charge and an **H** from a nearby permanent magnet. However, when both **E** and **H** are related to each other as with an electric current or in a wave, **E** × **H** does represent power flow and, in all cases, without exception, the integral of **E** × **H** over a *closed* surface always equals the total power flow through the surface.

10-20 THE AXON: AN ACTIVE, LOSSLESS, NOISELESS TRANSMISSION LINE

The nervous systems of animals consist of many *neurons* (nerve cells) each having an active transmission line, or *axon*, with input and output terminals. At the input end, structures called *dendrites* interface with specialized transducers sensitive to heat, pressure, or other stimuli. The dendrites are connected to a central cell body (soma), and when the algebraic sum of the excitations it receives from the dendrites exceeds a certain threshold value, it fires a signal down the axon to the terminal region, activating a motor unit (muscle) or another axon. A neuron with axon 1 m long is shown in Fig. 10-61.

The axon is an *active* transmission line with emf inputs all along it, and this results in zero attenuation of the signal. The other transmission lines we have considered are *passive*, having no energy inputs except at the input terminals.

Many neurons may be connected in series by structures called *synapses*, in which the output dendrites of one neuron connect with the input dendrites of the next neuron. The velocity of signal propagation along a particular axon transmission line is constant, but different axons may have different velocities. Axons of larger dimeter (20 μm) may have signal velocities of 100 m s^{-1}.

The axon is enclosed in a myelin sheath, which is electrically passive and acts as an insulator. At millimeter intervals along the axon the sheath may be interrupted at nodes exposing the axon to the surrounding medium. By diffusion of ions from the surrounding medium through the outer membrane of the axon, emfs are

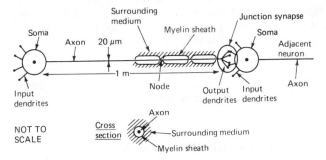

Figure 10-61 Idealized diagram of a typical neuron, as found in the sciatic (leg) nerve of a large mammal, with connection shown to an adjacent neuron. The axon acts as the inner conductor, the myelin sheath as its insulation, and the surrounding medium as the outer conductor of a coaxial transmission line. A bundle (cable) of thousands of such axons or nerve fibers forms the sciatic nerve.

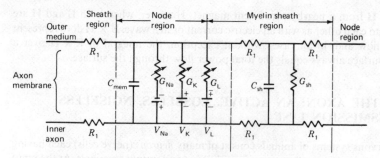

Figure 10-62 Equivalent circuit of an axon transmission line, divided into node and sheath regions. The axon has only a membrane separating it from the surrounding medium in the node region but a myelin enclosure in the sheath region.

applied between the inner axon (as one conductor) and the surrounding medium (as the second conductor) like the voltage across a two-conductor transmission line. This voltage produces a current via the axon and surrounding medium through the next node, triggering emfs there and so on down the line.

The first definitive theory of axon behavior was published by A. L. Hodgkin and A. F. Huxley in 1952. Their work, based on research on the properties of a giant axon of the squid, earned them a Nobel Prize.

The equivalent circuit of an axon transmission line according to Hodgkin and Huxley's theory is shown in Fig. 10-62. The line has series resistance and shunt conductance and capacitance. There is no series inductance, but some models include a shunt inductance. In addition, shunt emfs are applied through variable conductances which act like switching elements. Normally the diffusion of potassium (K) ions and miscellaneous leakage (L) ions keeps the inner axon negative by about 100 mV. But on excitation, the diffusion of sodium (Na) ions swings the potential positive for the period of the impulse, which is typically a few tenths of a millisecond (ms). The recovery of the axon to its normal negative potential after the passage of the impulse is accomplished in less than 1 ms. Since the full impulse voltage is received at the terminals, the axon transmission line has zero attenuation. It is also a "noiseless" line in that it either transmits a full impulse or none at all. There is no intermediate condition.

Although this brief discussion is greatly oversimplified, it provides some insights into the remarkable properties of the active two-conductor coaxial transmission lines present in great numbers in all animals.

10-21 SHIELDING OF TRANSMISSION LINES

The presumed purpose of a transmission line is to convey energy (or information) from one point to another as efficiently as possible. With direct current the only losses involve the series resistance and shunt conductance. With alternating current

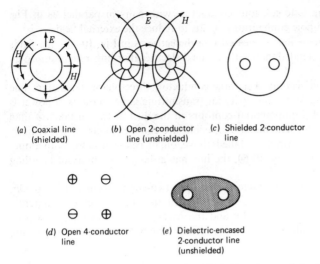

(a) Coaxial line   (b) Open 2-conductor   (c) Shielded 2-conductor
    (shielded)   line (unshielded)   line

(d) Open 4-conductor   (e) Dielectric-encased
    line   2-conductor line
                              (unshielded)

Figure 10-63 Different types of shielded and unshielded (open) transmission lines.

(ac) and especially with high-frequency ac (radio frequency), the transmission line may have mutual coupling (via mutual impedance) to adjacent lines or structures and in addition may act to some extent as an antenna so that power may be radiated into or received from space. If the transmission line conveys information (speech, pictures, data, music), such coupling or antenna action can introduce unwanted noise and interference onto the line, degrading the line's performance and making the line active in the sense that energy enters it at other than its input terminals. The isolation of a transmission line's fields from its surroundings may be referred to as its *shielding*.

With a coaxial transmission line (Fig. 10-63a) having a perfectly conducting solid tubing as outer conductor, its electric and magnetic fields are all confined internally so that ideally the line is perfectly shielded. But, unless operating as superconductors, the conductors will have less than infinite conductivity and the shielding will be imperfect.† In flexible coaxial cables with a wire braid instead of solid outer conductor, a second wire braid outer conductor is sometimes placed over the outer conductor to improve the shielding.

With a two-conductor transmission line (Fig. 10-63b) the electric and magnetic fields are not confined and extend theoretically to infinity. For dc this may be no problem, but with ac it can be troublesome. To improve shielding the two conductors can be enclosed in a metal pipe, as in Fig. 10-63c. Another technique for

---

† Even the best shielded coaxial cable will introduce (thermal) noise to a system unless its temperature is reduced to absolute zero (or until the cable becomes superconducting). This noise contribution to the system temperature is discussed in Sec. 14-24 on radio telescopes and remote sensing with analogies to an emitting-absorbing transmission line.

improving shielding is to use four conductors with pairs in parallel as in Fig. 10-63d. This quadrupole configuration tends to reduce the external fields.

If a two-conductor line is encased in a dielectric as in Fig. 10-63a, the fields are somewhat less extended than without the dielectric, affording a slight improvement in shielding.

The termination section of a three-conductor, three-phase metal duct line designed at the Ohio State University for transmitting large amounts of power is shown in Fig. 10-64. Equipotential contours in the duct at two instants of time are given in Fig. 10-65. The metal duct enclosure confines fields almost entirely to its interior. The duct can be installed above, on, or below ground. With a duct 2.5 m in diameter, as in Fig. 10-64, the line has a design capability of handling more than 10 billion watts.

If the space s between conductors of an open two-conductor line (Fig. 10-63b) becomes an appreciable fraction of a wavelength, the line will act as an antenna for either transmission or reception and the series resistance of the line will be increased by what is called a *radiation resistance* (the power dissipated in it being

Figure 10-64 Termination section of three-phase duct line constructed at the Ohio State University with design capability of handling powers in excess of 10 billion watts.

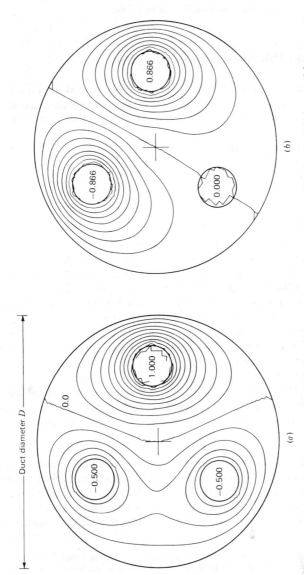

Figure 10-65 Three-conductor three-phase transmission line inside grounded metal duct (shield) with electric potential contours at two instants of time. (*a*) Relative conductor voltages of 1.0, −0.5, and −0.5 and (*b*) −0.866, 0.866, and 0.0. This duct line with diameter $D = 2.5$ m has a power transmission capability in excess of 10 billion watts.

equal to the radiated power) (see Chap. 14). If a transmission line is superconducting, the series (conduction) resistance may be reduced to zero. However, if shielding is inadequate (outer conductor of a coaxial line slotted or not completely closed), some series (radiation) resistance could still be present.

10-22 RADIO-LINK TRANSMISSION LINE

A radio link may be regarded as a transmission line in the sense that it is a circuit with input and output terminals. Consider a link with two antennas separated by a distance r (Fig. 10-66). As developed in Chap. 14, the attenuation is given by

$$\text{Attenuation} = 10 \log \frac{r^2 \lambda^2}{A_{et} A_{er}} \quad \text{(dB)} \tag{1}$$

where r = distance between antennas, m
  λ = wavelength, m
  A_{et} = effective aperture of transmitting antenna, m^2
  A_{er} = effective aperture of receiving antenna, m^2

Equation (1) is the Friis formula for the ratio of the power input to the transmitting antenna to the power output of the receiving antenna, the antennas being assumed to have zero losses (see Chap. 14). The attenuation results from the spherical spreading out of the wave front, E decreasing as $1/r$ and the Poynting vector as $1/r^2$. Thus, the wave is not a plane wave although over small regions of space at large distances from its source it is essentially a plane wave.

For free space, the velocity of propagation is $1/\sqrt{\mu_0 \epsilon_0} = c =$ velocity of light $= 3 \times 10^8$ m s^{-1}. Although the impedance of space $Z_0 = \sqrt{\mu_0/\epsilon_0} = 376.7 \, \Omega$, the terminal impedances of the antennas may be any convenient value according to their design and matching arrangements.

Except for frequency separation and beaming with directional antennas, the radio link is a completely open (unshielded) system. The receiver is highly susceptible to interference and the transmitter can readily interfere with other radio systems.

Example: Radio link between earth and Clarke satellite Calculate the path attenuation and time delay for a radio link between a Clarke orbit† satellite and the earth directly below it at 4 GHz. The antennas are parabolic dishes of 50 percent aperture efficiency with the satellite antenna diameter being 3 m and the earth station 5 m. The Clarke orbit is 36,000 km above the equator, and a

---

† In 1945, while an officer in the Royal Air Force, Arthur C. Clarke proposed that a solution to the world's communication problem could be by means of satellites in synchronous, or geostationary, orbit. This was 12 years before Sputnik, and although the idea then seemed farfetched, the Clarke orbit is now crowded with satellites handling a large and growing fraction of the earth's communications.

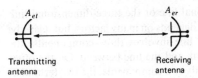

Transmitting
antenna

Receiving
antenna   **Figure 10-66** Radio-link transmission line.

satellite in this orbit revolves in synchronism with the earth as though attached
to it so that the satellite remains fixed in the sky above a particular point on the
earth's equator. Having a TV antenna on a Clarke orbit satellite is like having
it on an invisible TV tower 36,000 km high.

SOLUTION From (1)

$$\text{Attenuation} = 10 \log \frac{(36 \times 10^6)^2(0.075)^2}{\dfrac{\pi 2.5^2}{2} \dfrac{\pi 1.5^2}{2}} = 10 \log (2.1 \times 10^{11}) = 113 \text{ dB}$$

The time is 36×10^6 (m)$/3 \times 10^8$ (m s^{-1}) = 0.12 s. The path attenuation
up and back is 226 dB and the time delay up and back is 0.24 s. If 113-dB
attenuation is just acceptable one way, then a relay amplifier (transponder) in
the satellite will need to introduce 113 dB gain into an up-down circuit.
In spite of such large attenuations, adequate *signal-to-noise ratios* can be
achieved (see Prob. 10-6-1).

A space radio link is three-dimensional, with waves expanding spherically
and power attenuation proportional to $1/r^2$, where r = distance. The other transmission lines we have considered are one-dimensional, with attenuation proportional to $e^{-2\alpha x}$, where x = distance along the line. Thus, as indicated in Fig.
10-67, attenuation along the one-dimensional line in decibels is linear with distance
while that of the three-dimensional radio link amounts to 6 dB for each doubling

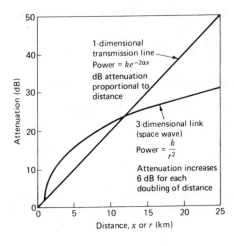

Attenuation (dB)

1-dimensional
transmission line
Power = $ke^{-2\alpha x}$

dB attenuation
proportional to
distance

3-dimensional link
(space wave)
Power = $\dfrac{k}{r^2}$

Attenuation increases
6 dB for each
doubling of distance

Distance, x or r (km)

Figure 10-67 Attenuation of one-dimensional transmission line and three-dimensional radio link. Attenuations are adjusted
equal to the same value at a distance of 1 km.
For large distances space wave transmission
involves less attenuation.

of the distance. Whether the one-dimensional line or the three-dimensional link will be more economical in a given situation depends on many factors, but it is to be noted that the cost of a one-dimensional line involves three components: the transmitting terminal, the receiving terminal, and the line between. On the other hand, the three-dimensional link involves only two components: the two terminals. The connecting medium between the terminals is provided by nature.

10-23 GENERAL DEVELOPMENT OF THE WAVE EQUATION

In this chapter we have dealt with plane waves (traveling in the x direction). The wave equation was developed for this special case, and appropriate solutions were obtained. A more general development of the wave equation will now be given, and it will be shown that for a plane wave it reduces to the expressions obtained previously.

Maxwell's curl equations are

$$\nabla \times \mathbf{H} = \mathbf{J} + \frac{\partial \mathbf{D}}{\partial t} = \sigma \mathbf{E} + \epsilon \frac{\partial \mathbf{E}}{\partial t} \tag{1}$$

and

$$\nabla \times \mathbf{E} = -\frac{\partial \mathbf{B}}{\partial t} = -\mu \frac{\partial \mathbf{H}}{\partial t} \tag{2}$$

Taking the curl of (2) and introducing the value of $\nabla \times \mathbf{H}$ from (1) gives

$$\nabla \times (\nabla \times \mathbf{E}) = -\mu \frac{\partial (\nabla \times \mathbf{H})}{\partial t} = -\mu \frac{\partial}{\partial t} \left(\sigma \mathbf{E} + \epsilon \frac{\partial \mathbf{E}}{\partial t} \right) \tag{3}$$

But by a vector identity, meaningful only in rectangular coordinates, the left-hand side of (3) can be expressed

$$\nabla \times (\nabla \times \mathbf{E}) = \nabla(\nabla \cdot \mathbf{E}) - \nabla^2 \mathbf{E} \tag{4}$$

Equating (4) and (3), and noting that in space having no free charge $\nabla \cdot \mathbf{E} = 0$, we get

$$\nabla^2 \mathbf{E} = \mu\epsilon \frac{\partial^2 \mathbf{E}}{\partial t^2} + \mu\sigma \frac{\partial \mathbf{E}}{\partial t} \tag{5}$$

With the assumption of harmonic variation of the field with time, (5) reduces to

$$\nabla^2 \mathbf{E} = (-\omega^2 \mu\epsilon + j\omega\mu\sigma)\mathbf{E} = \gamma^2 \mathbf{E} \tag{6}$$

or

$$\nabla^2 \mathbf{E} - \gamma^2 \mathbf{E} = 0 \tag{7}$$

From (4) we can also write

$$\nabla \times \nabla \times \mathbf{E} + \gamma^2 \mathbf{E} = 0 \tag{8}$$

Equations (5) to (8) are vector wave equations in $\mathbf{E}$. In (5) the time is explicit, while in the other three it is implicit (phasor form). These wave equations incorporate all four of Maxwell's equations.

10-24 TABLE OF CIRCUIT AND FIELD RELATIONS

Circuit relation for transmission line	SI units	Field relation for space wave
Series inductance $= L$	H m^{-1}	Permeability $= \mu$
Shunt capacitance $= C$	F m^{-1}	Permittivity $= \epsilon$
Shunt conductance $= G$	℧ m^{-1}	Conductivity $= \sigma$
Series impedance $= Z = R + j\omega L$	Ω m^{-1}	$j\omega\mu$
Shunt admittance $= Y = G + j\omega C$	℧ m^{-1}	$\sigma + j\omega\epsilon$
Velocity $= \dfrac{\omega}{\beta} = f\lambda$	m s^{-1}	Velocity $= \dfrac{\omega}{\beta} = f\lambda$
Velocity $= \dfrac{\omega}{\mathrm{Im}\sqrt{(R + j\omega L)(G + j\omega C)}}$	m s^{-1}	Velocity $= \dfrac{\omega}{\mathrm{Im}\sqrt{(\omega\mu'' + j\omega\mu')[\sigma + j\omega(\epsilon' - j\epsilon'')]}}$ †
Velocity $= \dfrac{1}{\sqrt{LC}}$ $R = G = 0$	m s^{-1}	Velocity $= \dfrac{1}{\sqrt{\mu\epsilon}}$ $\sigma = 0$ μ & ϵ real
Characteristic impedance $= Z_0 = \sqrt{\dfrac{R + j\omega L}{G + j\omega C}}$	Ω	Intrinsic impedance $= Z_0 = \sqrt{\dfrac{\omega\mu'' + j\omega\mu'}{\sigma + j\omega(\epsilon' - j\epsilon'')}}$
Characteristic impedance $= Z_0 = \sqrt{\dfrac{L}{C}}$ $R = G = 0$	Ω	Intrinsic impedance $= Z_0 = \sqrt{\dfrac{\mu}{\epsilon}}$ $\sigma = 0$ μ & ϵ real
$\dfrac{\text{Voltage}}{\text{Distance}} = \dfrac{V}{x}$	V m^{-1}	Electric field $= E$
$\dfrac{\text{Current}}{\text{Distance}} = \dfrac{I}{x}$	A m^{-1}	Magnetic field $= H$
Average power P_{av} $= \begin{cases} = V_{\mathrm{rms}} I_{\mathrm{rms}} \cos\theta \\ = I_{\mathrm{rms}}^2 \operatorname{Re} Z_0 \\ = V_{\mathrm{rms}}^2 \operatorname{Re} \dfrac{1}{Z_0} \end{cases}$	W W m^{-2}	Average Poynting vector S_{av} $= \begin{cases} = E_{\mathrm{rms}} H_{\mathrm{rms}} \cos\xi \\ = H_{\mathrm{rms}}^2 \operatorname{Re} Z_0 \\ = E_{\mathrm{rms}}^2 \operatorname{Re} \dfrac{1}{Z_0} \end{cases}$

† Rearranging $\sigma + j\omega(\epsilon' - j\epsilon'')$ to read $(\sigma + \omega\epsilon'') + j\omega\epsilon'$ makes the equivalence to the circuit relation $G + j\omega C$ more obvious.

10-25 TABLE OF IMPEDANCE, VELOCITY, ATTENUATION, PHASE, AND OTHER CONSTANTS FOR TRANSMISSION LINES AND WAVES

Transmission lines		Plane space waves			Spherical space waves (radio link)
General case	**$R = G = 0$**	**General case**	**$\sigma \gg \omega\epsilon$**	**$\sigma = 0$, μ and ϵ real**	

Attenuation constant

Transmission lines, General case:
$$\alpha = \text{Re}\sqrt{ZY}$$
$$= \text{Re}\sqrt{(R + j\omega L)(G + j\omega C)} \ \text{Np m}^{-1}$$

Transmission lines, $R = G = 0$:
$$\alpha = \text{Re } j\omega\sqrt{LC}$$
$$= 0 \ \text{Np m}^{-1}$$

Plane space waves, General case:
$$\alpha = \text{Re}\sqrt{(\omega\mu'' + j\omega\mu')[\sigma + j\omega(\epsilon' - j\epsilon'')]} \ \text{Np m}^{-1}$$
$$= \text{Re}\frac{2\pi}{\lambda_0}\sqrt{(\mu_r' - j\mu_r'')\left[\zeta_r' - j\left(\zeta_r'' + \frac{\sigma}{\omega\epsilon_0}\right)\right]} \ \text{Np m}^{-1}$$

Plane space waves, $\sigma \gg \omega\epsilon$ (Attenuation constant = phase constant):
$$\alpha = \beta = \sqrt{\frac{\omega\mu\sigma}{2}} \ \text{Np m}^{-1} \text{ or rad m}^{-1}$$

Plane space waves, $\sigma = 0$:
$$\alpha = \text{Re } j\omega\sqrt{\mu\epsilon}$$
$$= 0 \ \text{Np m}^{-1}$$

Spherical space waves (Attenuation):
$$\alpha = 10\log\frac{r^2\lambda^2}{A_{et}A_{er}} \ \text{dB}$$

Phase constant

Transmission lines, General case:
$$\beta = \text{Im}\sqrt{ZY}$$
$$= \text{Im}\sqrt{(R + j\omega L)(G + j\omega C)}$$
$$= \frac{2\pi}{\lambda} \ \text{rad m}^{-1}$$

Transmission lines, $R = G = 0$:
$$\beta = \text{Im } j\omega\sqrt{LC}$$
$$= \omega\sqrt{LC} \ \text{rad m}^{-1}$$

Plane space waves, General case:
$$\beta = \text{Im}\sqrt{(\omega\mu'' + j\omega\mu')[\sigma + j\omega(\epsilon' - j\epsilon'')]} \ \text{rad m}^{-1}$$
$$= \text{Im } j\frac{2\pi}{\lambda_0}\sqrt{(\mu_r' - j\mu_r'')\left[\zeta_r' - j\left(\zeta_r'' + \frac{\sigma}{\omega\epsilon_0}\right)\right]} \ \text{rad m}^{-1}$$

Plane space waves, $\sigma \gg \omega\epsilon$ ($1/e$ depth of penetration):
$$\delta = \sqrt{\frac{1}{f\pi\mu\sigma}} \ \text{m}$$

Plane space waves, $\sigma = 0$:
$$\beta = \text{Im } j\omega\sqrt{\mu\epsilon}$$
$$= \omega\sqrt{\mu\epsilon} \ \text{rad m}^{-1}$$

Spherical space waves (Phase constant):
$$\beta = \frac{2\pi}{\lambda_0} \ \text{rad m}^{-1}$$

Velocity

Transmission lines, General case:
$$v = \frac{\omega}{\beta} \ \text{m s}^{-1}$$

Transmission lines, $R = G = 0$:
$$v = \frac{1}{\sqrt{LC}} \ \text{m s}^{-1}$$

Plane space waves, General case:
$$v = \frac{\omega}{\beta} \ \text{m s}^{-1}$$

Plane space waves, $\sigma = 0$:
$$v = 1/\sqrt{\mu\epsilon} \ \text{m s}^{-1}$$

Spherical space waves (Velocity):
$$\frac{\omega}{\beta} = f\lambda_0 = c$$
$$= 3 \times 10^8 \ \text{m s}^{-1}$$

Characteristic impedance / Intrinsic impedance

Transmission lines, General case (Characteristic impedance):
$$Z_0 = \sqrt{\frac{R + j\omega L}{G + j\omega C}} \ \Omega$$

Transmission lines, $R = G = 0$ (Characteristic impedance):
$$Z_0 = \sqrt{\frac{L}{C}} \ \Omega$$

Plane space waves, General case (Intrinsic impedance):
$$Z_0 = \sqrt{\frac{\omega\mu' + j\omega\mu''}{\sigma + j\omega(\epsilon' - j\epsilon'')}}$$
$$= 376.7\sqrt{\frac{\mu_r' - j\mu_r''}{\zeta_r' - [(\zeta_r'' + (\sigma/\omega\epsilon_0)]}} \ \Omega$$

Plane space waves, $\sigma = 0$ (Intrinsic impedance):
$$Z_0 = \sqrt{\frac{\mu}{\epsilon}} \ \Omega$$

Spherical space waves (Intrinsic impedance):
$$Z_0 = \sqrt{\frac{\mu_0}{\epsilon_0}}$$
$$= 376.7 \ \Omega \text{ per square}$$

To convert α in nepers per meter to decibels per meter, multiply by 8.69. Thus, $8.69\,\alpha$ (Np m⁻¹) $= \alpha$ (dB m⁻¹).

PROBLEMS†

Group 10-1: Secs. 10-1 through 10-5. Waves in space and on transmission lines and impedance of lines and media

10-1-1. Coaxial lines A coaxial transmission line has an outer conductor of 10-mm inside diameter and a thin-walled inner conductor of 3-mm outside diameter. Find (a) inductance per meter, (b) capacitance per meter, and (c) characteristic impedance.

10-1-2. Microstrip line A transmission line consists of a thin conducting strip 4 mm wide separated from a flat conducting ground plane by a dielectric substrate 0.5 mm thick. Find (a) characteristic impedance and (b) wave velocity as percentage of velocity of light. $\epsilon_r = 2.6$ for the substrate.

10-1-3. Lossless line Show that a transmission line having no attenuation must also have a nonreactive characteristic impedance.

10-1-4. Perfect vacuum Does the term perfect *vacuum* (or free space) imply the absence of everything?

★10-1-5. Microstrip line For a polystyrene substrate ($\epsilon_r = 2.7$) what w/h ratio results in a 50-Ω microstrip transmission line?

10-1-6. Coaxial line. Helical inner conductor. Delay line If the inner conductor of the coaxial line of Prob. 10-1-1 is replaced by a thin wire wound as a helix of 3-mm outside diameter with 500 turns mm^{-1}, find (a) inductance per meter, (b) capacitance per meter, and (c) characteristic impedance. Helical inner conductors increase line impedance and reduce velocity of propagation. The latter property is useful in *delay lines*.

10-1-7. Line impedance and attenuation A uniform transmission line has constants $R = 12$ mΩ, $G = 0.8$ $\mu\mho$, $L = 1.3$ μH, and $C = 0.7$ nF, all per meter. At 5 kHz find (a) characteristic impedance, (b) v/c velocity ratio, and (c) dB attenuation in 2 km.

Group 10-2: Secs. 10-6 through 10-9. Terminated line, reflection coefficient, slotted line, Smith chart, couplers, transformers, and bandwidth

★10-2-1. Terminated line A lossless 50-Ω transmission line is terminated in $25 + j50$ Ω. Find (a) voltage reflection coefficient, (b) VSWR, (c) impedance 0.3λ from load, (d) shortest length of line for which impedance is purely resistive, and (e) the value of this resistance. Solve using Smith chart.

10-2-2. Terminated line Solve Prob. 10-2-1 by analytical methods. Compare results.

10-2-3. Terminated line A lossless 100-Ω transmission line is terminated in $200 + j200$ Ω. Find (a) voltage reflection coefficient, (b) VSWR, (c) impedance 0.375λ from load, (d) shortest length of line for which impedance is purely resistive, and (e) the value of this resistance. Solve using Smith chart.

10-2-4. Terminated line Solve Prob. 10-2-3 by analytical methods. Compare results.

★10-2-5. Series matching section A 150-Ω transmission line is connected through a section of length d and characteristic impedance Z_1 to a load of $250 + j100$ Ω. Find d and Z_1 which match the load to the 150-Ω line.

10-2-6. Input impedance A uniform transmission line has constants $R = 15$ mΩ, $G = 1.2$ $\mu\mho$, $L = 2$ μH, and $C = 1.1$ nF, all per meter. The line is 1 km long. At 8 kHz find (a) the line impedance, (b) the input impedance, and (c) the dB attenuation. The line is terminated in $80 + j0\Omega$.

10-2-7. Single-stub VSWR A uniform, lossless 50-Ω transmission line is terminated in a load impedance of $80 + j40$ Ω. A short-circuited stub of 50-Ω characteristic impedance and 0.3λ long is connected in parallel 0.2λ from the load. Find the VSWR on the line 0.5λ from the load.

10-2-8. Single-stub matching A uniform, lossless 100-Ω line is connected to a load of $200 - j150$ Ω. A short-circuited stub of the same characteristic impedance is connected in parallel with the line at a

† Answers to starred problems are given in Appendix C.

distance d_1 from the load. (*a*) Find the shortest values of d_1 and the stub length d_2 for a match. (*b*) Find the next shortest values.

★10-2-9. Single-stub matching A uniform, lossless 400-Ω line is connected to a load of $200 + j300\,\Omega$. A short-circuited 400-Ω line (stub) is connected in parallel with the line at a distance d_1 from the load. Find the shortest values of d_1 and the stub length d_2 for a match.

10-2-10. Single-stub matching A uniform, lossless 50-Ω transmission line is terminated in a load impedance of $80 + j40\,\Omega$. A short-circuited stub of length d_1 is connected in parallel at a distance d_2 from the load. Find d_1 and d_2 for a match (VSWR = 1).

★10-2-11. Single-stub matching A uniform, lossless 100-Ω line is connected to a load of $35 - j35\,\Omega$. A short-circuited 100-Ω stub is connected in parallel with the line at a distance d_1 from the load. Find the shortest values of d_1 and the stub length d_2 for a match ($d_1 + d_2$ a minimum).

10-2-12. Transformer matching A line of length d and characteristic resistance R_1 acts as a transformer to match a load impedance of $150 + j0\,\Omega$ to a 300-Ω line at 300 MHz. Find (*a*) d, (*b*) R_1, and (*c*) VSWR on transformer section at 400 MHz.

★10-2-13. Double-stub matching For the double-stub tuner of Fig. P10-2-13, find the shortest values of d_1 and d_2 for a match (VSWR = 1) if $Z_L = 400 + j200\,\Omega$ and $R_0 = 200\,\Omega$.

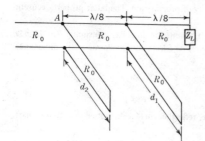

Figure P10-2-13 Double-stub tuner.

10-2-14. Double-stub mismatch For the double-stub tuner of Fig. P10-2-13, find the VSWR on the line to the left of the left-hand stub if $d_1 = 0.18\lambda$ and $d_2 = 0.16\lambda$.

10-2-15. Double-stub matching A uniform, lossless 100-Ω line is connected to a load of $100 - j100\,\Omega$. Stub A is connected in parallel with the line $\frac{3}{16}$-λ from the load and stub B $\lambda/4$ from the load. Find the shortest lengths of the two stubs for a match. The stubs are of 100-Ω characteristic resistance.

★10-2-16. Double-stub matching A 100-Ω line is connected to a load of $20 + j100\,\Omega$. Stub A is connected in parallel with the line $\lambda/8$ from the load and stub B $\lambda/4$ from the load. Find the shortest lengths of the two stubs for a match. The stubs are of 100-Ω characteristic resistance.

10-2-17. $\lambda/4$ transformer A $\lambda/4$ transformer of characteristic resistance R_0 is inserted in a 300-Ω line at a distance d from the load Z_L. Find d and R_0 for a match. $Z_L = 150 + j200\,\Omega$.

★10-2-18. Single-stub matching A uniform, lossless 200-Ω line is connected to a load of $100 - 150\,\Omega$. A single stub of 200-Ω characteristic resistance is connected in parallel with the line at a distance d_1 from the load. Find the shortest values of d_1 and the stub length d_2 for a match.

10-2-19. Double-stub tuner A uniform, lossless 100-Ω line is connected to a load of $40 + j80\,\Omega$. Stub A is connected in parallel with the line 0.24λ from the load and stub B 0.365λ from the load. Find the shortest stub lengths for a match. The stubs are of 100-Ω characteristic resistance.

10-2-20. Power divider A 150-Ω transmission line is connected to two loads as shown in Fig. P10-2-20. Find d_1, d_2, R_1, and R_2 such that the loads receive equal powers but with voltages of opposite phase.

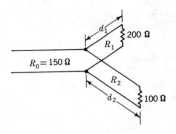

Figure P10-2-20 Line with two loads.

★10-2-21. Series resistors in line Referring to Fig. P10-2-21, find (a) the power in the load and (b) the maximum and minimum voltages on the 2λ section for the following two cases: (1) R_0 (both sections) = 300 Ω, and (2) R_0 (both sections) = 150 Ω.

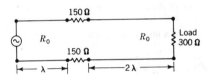

Figure P10-2-21 Line with series resistors.

10-2-22. Exponential line Show that the input impedance to a lossless exponential line of length y connected as in Fig. P10-2-22 to a load Z_L is given by

$$Z(y) = Z_0(0)e^{-ky}\frac{Z_L + jZ_0(0)\tan\beta y}{Z_0(0) + jZ_L \tan\beta y}$$

where $Z_0(0)$ is the characteristic impedance of the line at the load ($y = 0$).

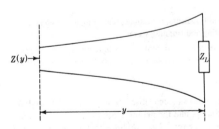

Figure P10-2-22 Exponential line.

10-2-23. Power divider A 100-Ω line (line A) is connected to two 100-Ω lines (B and C) in parallel by a Y junction as in Fig. P10-2-20. Line B is λ/4 long and is terminated in 50 + j0 Ω, while line C is λ/2 long and is terminated in 200 Ω. Find (a) VSWR on line A and (b) relative powers in the two loads.

10-2-24. Auxiliary line A 600-Ω, 2-conductor open-wire line is terminated in a load of 350 − j400 Ω. Find the length l and connection distances d_1 and d_2 for an auxiliary line of the same type in order to produce a match. The arrangement of the auxiliary line is shown in Fig. P10-2-24.

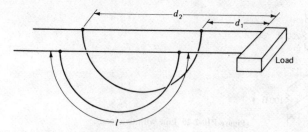

Figure P10-2-24 Auxiliary matching line.

10-2-25. VSWR in terms of Z_L and Z_0 If the load impedance Z_L and the line impedance Z_0 are real and $Z_L > Z_0$, show that VSWR $= Z_L/Z_0$, and if $Z_L < Z_0$, show that VSWR $= Z_0/Z_L$.

10-2-26. VSWR and reflected wave If the incident wave voltage is unity, show that (a) the reflected wave voltage magnitude is $(\text{VSWR} - 1)/(\text{VSWR} + 1)$ and (b) the relative reflected wave power is this quantity squared.

10-2-27. Smith chart The reflection coefficient for voltage on a line of characteristic impedance Z_0 is

$$\rho_v = \frac{Z - Z_0}{Z + Z_0} = \frac{(Z/Z_0) - 1}{(Z/Z_0) + 1} = \frac{Z_n - 1}{Z_n + 1}$$

where Z = load impedance

Z_n = normalized impedance = $R_n + jX_n$, dimensionless

It follows that

$$Z_n = \frac{1 + \rho_v}{1 - \rho_v}$$

Show that if R_n and X_n are obtained as functions of the real and imaginary parts of ρ_v, a map of R_n and X_n in the ρ_v plane yields the Smith impedance chart.

Group 10-3: Secs. 10-10.1 through 10-13. Pulses and transients, reflections, scattering parameters, index of refraction, phase and group velocities, and traveling and standing waves

10-3-1. VSWR vs. index of refraction (a) For a wave in air incident normally onto an infinite half-space of refractive index η, show that the resulting VSWR $= \eta$. (b) Show that the relative reflected wave power is $[(\eta - 1)/(\eta + 1)]^2$.

10-3-2. S parameters Prove Eq. (10-10.3-9).

10-3-3. Line reflections A 50-Ω coaxial transmission line has a small discontinuity at $\lambda/2$ intervals. If the reflection coefficient $\rho_v = 0.01\underline{/0°}$ at each discontinuity, find the net reflection for a 3λ section of line (a) for a pulse which is short compared to the travel time for $\lambda/2$ and (b) for a continuous steady-state wave. The far end of the line is matched. Note that the pulse reflection will consist of a succession of pulses. Assume that the line is lossless.

10-3-4. Line reflections Find the reflections for the line of Prob. 10-3-3 for both short pulse and continuous wave if the discontinuity separation is $\lambda/4$ instead of $\lambda/2$.

10-3-5. Addition of waves of different frequencies Given a plane wave for which $E = E_0 \operatorname{Re} e^{j(\omega t - \beta x)}$. Show that the resultant of two waves of this type of equal amplitude and of two frequencies $\omega_0 + \omega_1$ and $\omega_0 - \omega_1$ (and two corresponding phase constants given by $\beta_0 + \beta_1$ and $\beta_0 - \beta_1$) may be expressed as $E = 2E_0 \cos(\omega_0 t - \beta_0 x) \cos(\omega_1 t - \beta_1 x)$.

10-3-6. Velocities. Phase and group The group velocity $u = d\omega/d\beta$. Show that u can be expressed also in the following forms: $u = v + \beta(dv/d\beta) = v - \lambda(dv/d\lambda) = df/d\lambda^{-1}$, where v = phase velocity.

$\star$**10-3-7. Velocities. Phase and group** Find the group velocity for a 100-MHz wave in a normally dispersive, lossless medium for which the phase velocity $v = 2 \times 10^7 \lambda^{2/3}$ m s^{-1}.

10-3-8. Standing wave Show that when the incident-wave amplitude E_0 is much greater than the reflected-wave amplitude E_1, the standing-wave-envelope expression becomes approximately $E_y = E_0 + E_1 \cos 2\beta x$.

10-3-9. Standing wave Show that in a standing wave the average phase velocity of the resultant field is equal to the geometric mean of the maximum and minimum velocities; that is, $v_{av} = \sqrt{v_{max}v_{min}}$.

Group 10-4: Secs. 10-14 through 10-16. Conductors and dielectrics, lossy lines, interfaces, and wave absorption

10-4-1. Medium. Conductor or dielectric A medium has constants $\sigma = 0.1 \; \mho m^{-1}, \mu_r = 1$, and $\epsilon_r = 40$. Assuming that these values do not change with frequency, does the medium behave like a conductor or a dielectric at (a) 50 kHz and (b) 10 GHz?

10-4-2. Line with small R and G Show that when the series resistance R and the shunt conductance G of a transmission line are small but not negligible, the attenuation constant may be written as

$$\alpha \approx \frac{R}{2}\sqrt{\frac{C}{L}} + \frac{G}{2}\sqrt{\frac{L}{C}}$$

and the phase constant as $\beta \approx \omega\sqrt{LC}$.

$\star$**10-4-3. Phase shift via hole** A plane 3-GHz wave is incident normally on a large sheet of polystyrene ($\epsilon_r = 2.7$). How thick must the sheet be to retard the wave in phase by 180° with respect to a wave which travels through a large hole in the sheet?

$\star$**10-4-4. Matching plate** A plane 1-GHz wave is incident normally on the plane surface of a half-space of material having constants $\mu_r = 1$ and $\epsilon_r = 3.5$. Find (a) the thickness in millimeters and (b) the relative permittivity required for a matching plate which will eliminate reflection of the incident wave.

10-4-5. 1/e depths A medium has constants $\sigma = 10^2 \; \mho m^{-1}$, $\mu_r = 2$, and $\epsilon_r = 3$. If the constants do not change with frequency, find the $1/e$ and 1 percent depths of penetration at (a) 60 Hz, (b) 2 MHz, and (c) 3 GHz.

10-4-6. Lossy medium At 200 MHz a solid ferrite-titanate medium has constants $\sigma = 0, \mu_r = 15(1 - j3)$, and $\epsilon_r = 50(1 - j1)$. Find (a) Z/Z_0, (b) λ/λ_0, (c) v/v_0, (d) $1/e$ depth, (e) dB attenuation for a 5-mm thickness, and (f) reflection coefficient ρ for a wave in air incident normally on the flat surface of the medium. The zero subscripts refer to parameters for air (or vacuum).

$\star$**10-4-7. Lossy medium. Complex constants** A medium has constants $\sigma = 3.34 \; \mho m^{-1}$ and $\mu_r = \epsilon_r = 5 + j2$. Find the $1/e$ depth of penetration at 30 GHz.

10-4-8. Attenuation by lossy slab A nonconducting slab 200 mm thick has constants $\mu_r = \epsilon_r = 2 - j2$. Find the dB attenuation of the slab to a 600-MHz wave.

$\star$**10-4-9. Attenuation of lossy sheet** A nonconducting sheet 6 mm thick has constants $\mu_r = \epsilon_r = 5 - j5$. Find the dB attenuation of the sheet to a 300-MHz wave.

10-4-10. Attenuation in lossy sheet A nonmagnetic conducting sheet 2 mm thick has a conductivity $\sigma = 10^3 \; \mho m^{-1}$. Find the dB attenuation of the sheet to an 800-MHz wave.

$\star$**10-4-11. Attenuation in lossy medium** A medium has constants $\sigma = 1.112 \times 10^{-2} \; \mho m^{-1}, \mu_r = 5 - j4$, and $\epsilon_r = 5 + j2$. At 100 MHz find (a) impedance of the medium and (b) distance required to attenuate a wave by 20 dB after entering the medium.

10-4-12. Attenuation on lossy line. Graphs Use a computer to calculate and plot the magnitude and phase of the total voltage on a lossy transmission line, as done in Example 2 of Sec. 10-15, for line parameters

$$V_i = V_0 e^{\alpha x} e^{j(\omega t + \beta x)}$$

$$V_r = -\tfrac{3}{4} V_0 e^{-\alpha x} e^{j(\omega t - \beta x)}$$

$$V_0 = 10 \text{ V}$$

$$f = 250 \text{ MHz } (\lambda = 1.2 \text{ m})$$

$$\alpha = 0.2 \text{ Np m}^{-1}$$

10-4-13. Power loss at junction Show that at the junction between two infinite transmission lines the relative power loss due to mismatch at the junction is given by

$$10 \log \frac{1}{1 - \rho_v^2} = 10 \log \frac{Z_t}{\tau_v^2 Z_i}$$

where Z_i = impedance of line for incident wave
$\quad\quad Z_t$ = impedance of line for transmitted wave

10-4-14. Absorbing sheet A large flat sheet of nonconducting material is backed by aluminum foil. At 500 MHz the constants of the ferrous dielectric medium are $\mu_r = \epsilon_r = 6 - j6$. How thick must the sheet be for a 500-MHz wave (in air) incident on the sheet to be reduced upon reflection by 30 dB if the wave is incident normally?

Group 10-5: Secs. 10-17 through 10-23. Power and energy relations. Poynting vector, axon, shielding, space links, and general development of wave equation

★**10-5-1. Reflection from dielectric medium** A plane 3-GHz wave is incident normally from air onto a half-space of dielectric with constants $\sigma = 0$, $\mu_r = 1$, and $\epsilon_r = 2 - j2$. Find the dB value of the reflected power.

10-5-2. Power per square Prove that the power through each curvilinear square of Fig. 10-59a is the same.

★**10-5-3. Two-conductor line. Radiation** (a) Find the characteristic impedance Z_0 for a two-conductor transmission line in air with center-to-center conductor spacing of 50 mm and conductor diameter of 6 mm. (b) What is the upper frequency limit at which this line should be used if losses due to radiation are to be minimized? (See Chap 14.)

10-5-4. Energy velocity For a plane wave in a dielectric medium show that the energy velocity can be expressed as $1/\epsilon Z$ or Z/μ, where Z is the intrinsic impedance of the medium.

10-5-5. Traveling wave. Poynting vector A plane traveling wave has a peak electric field $E_0 = 4 \text{ V m}^{-1}$. If the medium is lossless with $\mu_r = 1$ and $\epsilon_r = 3.5$, find (a) velocity of wave, (b) peak Poynting vector, (c) average Poynting vector, (d) impedance of medium, and (e) peak value of the magnetic field H.

10-5-6. Traveling wave. Poynting vector A plane traveling 100-MHz wave has an average Poynting vector of 4 W m^{-2}. If the medium is lossless with $\mu_r = 2$ and $\epsilon_r = 3$, find (a) velocity of wave, (b) wavelength, (c) impedance of medium, (d) rms electric field E, and (e) rms magnetic field H.

10-5-7. Traveling wave. Energies Show that for a traveling wave the peak energy density stored in the electric field equals that stored in the magnetic field.

★10-5-8. Solar power The earth receives $2.0 \, \text{g cal min}^{-1} \, \text{cm}^{-2}$ of sunlight. (a) What is the Poynting vector in watts per square meter? (b) What is the power output of the sun in sunlight assuming that the sun radiates isotropically? (c) What is the rms electric field E at the earth assuming that the sunlight is all at a single frequency? (d) How long does it take the sunlight to reach the earth? Take the earth-sun distance as 150 Gm. ($1 \, \text{W} = 14.3 \, \text{g cal min}^{-1}$.)

10-5-9. Solar power The earth receives about $1.5 \, \text{kW m}^{-2}$ of power from the sun (integrated over all frequencies) but the amount reaching the earth's surface depends on atmospheric conditions. (a) What is the sun's total power output assuming it radiates isotropically? Take the earth-sun distance as 150 Gm. (b) What is the total power received by the earth? Take the earth's radius as 6.4 Mm. (c) If the sun's mass is 2×10^{30} kg and its mass is converted to radiant energy according to Einstein's relation (energy $= mc^2$) at 1 percent efficiency, how long can the sun radiate at its present level?

★10-5-10. Solar waves 3-GHz radio waves from the sun have a flux density of $10^{-20} \, \text{W m}^{-2} \, \text{Hz}^{-1}$. (a) What is the Poynting vector for a 1-GHz bandwidth assuming the flux density is constant over this bandwidth? (b) What is the rms electric field E assuming that the power in the 1-GHz bandwidth is at a single frequency? (c) What is the radio power output of the sun for the 1-GHz bandwidth assuming that the sun radiates isotropically? Take earth-sun distance as 150 Gm.

10-5-11. Extragalactic radio source A 1.4-GHz radio wave received from an extragalactic radio source has an average Poynting vector per unit bandwidth (or flux density) of $10^{-26} \, \text{W m}^{-2} \, \text{Hz}^{-1}$. (a) How much power is intercepted by an area of 1,000 m^2 over a bandwidth of 1 MHz? The area is normal to the wave direction. (b) If the radio source is at a distance of 10^9 light-years and radiates isotropically, what is the power radiated by the source in a 1-MHz bandwidth? One light-year equals the distance light or radio waves travel in 1 year. Assume that the medium is free space. (c) How does this power compare with the earth's *total* power requirements for electricity, heating, industry, transportation, etc.?

10-5-12. Planck's law (a) Referring to Prob. 10-5-8, what is the Poynting vector per unit bandwidth of sunlight at the earth? Take the sunlight band as 400 to 800 nm (4,000 to 8,000 Å). (b) What is the ratio of this Poynting vector to that given in Prob. 10-5-10 for 3-GHz radio waves? (c) According to Planck's radiation law, the radiation brightness $\mathscr{B}$ from a blackbody is given by

$$\mathscr{B} = \frac{2hf^3}{c^2} \frac{1}{e^{hf/kt} - 1} \quad (\text{W m}^{-2} \, \text{Hz}^{-1} \, \text{rad}^{-2})$$

where h = Planck's constant = 6.63×10^{-34} J s
$\quad f$ = frequency, Hz
$\quad c$ = velocity of light or radio waves = 300 Mm s^{-1}
$\quad k$ = Boltzmann's constant = 1.38×10^{-23} J K^{-1}
$\quad T$ = blackbody temperature, K

Taking $T = 6000$ K for the sun, would you say that the sun behaves as a blackbody (or Planck) radiator over the above optical-radio range?

10-5-13. Photons According to quantum theory, electromagnetic waves are transmitted by photons of energy hf, where h = Planck's constant = 6.63×10^{-34} J s and f = frequency in hertz. Find the number of photons per second which are intercepted by an area 1 m^2 for a wave with average Poynting vector of $10^{-20} \, \text{W m}^{-2}$ for the following two cases: (a) frequency = 2 GHz and (b) wavelength = 600 nm. The area is normal to the wave direction. (c) What advantage do radio waves have over optical waves for transmission of information if the wave is so weak that the number of photons is small?

10-5-14. Resistor and power flow A cylindrical resistor of length l, radius a, and conductivity σ carries an rms current I. (a) Show that the Poynting vector is directed normally *inward* with respect to the surface of the resistor. (b) Show that the integral of the Poynting vector over the surface of the resistor equals I^2R, where R is the resistance of the resistor.

Group 10-6: Practical applications

★10-6-1. Satellite TV downlink As discussed in Sec. 14-24, the criterion of detectability of a signal is the *signal-to-noise ratio* (*S/N*). For isotropic (nondirectional) antennas and a transmitter power of 1 W, the TV downlink from a Clarke-orbit satellite to an earth station has a signal-to-noise ratio given by (see Prob. 14-5-14)

$$\frac{S}{N} = \frac{\lambda^2}{16\pi^2 r^2 k T_{sys} B}$$

where λ = wavelength, m
  r = distance of downlink, m
  k = Boltzmann's constant = 1.38×10^{-23} J K^{-1}
  T_{sys} = system temperature, K
  B = bandwidth, Hz

A typical 4-GHz satellite TV transponder may have 5-W output and an antenna gain of 29.4 dB, making its Effective Radiated Power

$$\text{ERP} = 10 \log 5 + 29.4 = 36.4 \text{ dBW (dB over 1 W isotropic)}$$

on the antenna beam axis (boresight)

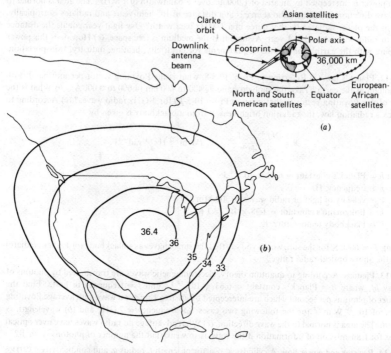

(a)

(b)

Figure P10-6-1 (*a*) Geostationary TV and communication relay satellites in Clarke orbit around earth. (*b*) Effective radiated power contours, or footprint, in decibels above 1 W isotropic for TV transponder downlink. Note that the footprint gives radiation contours over a spherical surface (the earth's) and, in most cases, at an oblique angle. Only when the satellite beams directly down on the earth's equator with a relatively narrow pattern will the footprint contours approximate the true antenna pattern contours.

Figure P10-6-1a shows the Clarke orbit geostationary satellites in relation to the earth. Figure P10-6-1b shows the ERP contours (footprint) of the above typical transponder over North America. (a) Determine the S/N ratio (dB) if the earth station antenna diameter is 3 m, the antenna temperature 25 K, the receiver temperature 75 K, and the bandwidth 30 MHz. The system temperature is the sum of the antenna and receiver temperatures. Take the satellite distance as 36,000 km. The ERP at the earth station location is 35 dBW. Assume the antenna is a parabolic reflector (dish-type) of 50 percent efficiency with gain G, given by $G = 2\pi^2(r/\lambda)^2$, where r = dish radius, m. (b) If a 10-dB S/N ratio is acceptable, what is the required diameter of the earth station antenna? (c) Calculate the diameter for the satellite antenna with a gain = 29.4 dB. *Note*: For FM-modulated video signals, as employed by the Clarke orbit satellites, the S/N ratio as used above is actually a carrier-to-noise (C/N) ratio, the ultimate video signal-to-noise ratio for typical North American domestic system satellites being almost 40 dB higher. This is an advantage of FM modulation. If the C/N exceeds a few decibels, then, in principle, a perfect picture results. However, a $C/N \geq 10$ dB is desirable to allow for misalignment of the earth-station antenna, attenuation due to water or snow in the dish, a decrease in transponder power, etc.

10-6-2. RG-59/U coaxial line The specifications for RG-59/U flexible radio-frequency coaxial cable (6.1 mm O.D.) are: $Z_0 = 73\,\Omega$; $v/c = 0.66$; capacitance = 69 pF m^{-1}; attenuation = 7.9 dB per 100 m at 50 MHz, 11.2 dB at 100 MHz, 16.7 dB at 200 MHz, 26.9 dB at 500 MHz, and 45.6 dB at 900 MHz. For a 50-m length at 300 MHz find (a) attenuation, (b) time delay, (c) reflected signal at input if load impedance is $85 + j20\,\Omega$. Express in decibels relative to input signal.

10-6-3. RF shielding (a) A 1-GHz traveling wave in air has an electric field $E = 100$ MV m^{-1}. If the wave is incident normally on a flat copper sheet, how thick must the sheet be to reduce the wave transmitted through the sheet to the nonhazardous level of 100 W m^{-2}? Frequencies in the 1- to 3-GHz range are used by home microwave ovens. *Hint*: Consider the copper sheet as a short section of lossy transmission line connected between two lossless lines of characteristic impedance $Z_0 = 376.7\,\Omega$. [Regarding hazardous radiation levels, see W. W. Mumford, Some Technical Aspects of Microwave Radiation Hazards, *Proc. IRE*, **49**:427–447 (February 1961).] (b) What power is dissipated per square meter in the sheet? (c) Will dissipation of this power be a problem, and, if so, how can it be accomplished?

10-6-4. Ice caps Antarctica and Greenland are covered with ice caps of great thickness. Radio-equipped surface vehicles are used to measure the thickness of the ice by sending radio waves down through the ice and obtaining a reflection from the ground under the ice. It is assumed that the ice is homogeneous and of uniform thickness over a flat perfectly conducting earth. Take the constants for the ice at 30 MHz as $\mu_r = 1$ and $\epsilon_r = 3 - j0.01$. (a) Find the ratio of the power received per unit area to the power radiated at 30 MHz for a 2-km thickness of ice. Express in decibels. (b) Compare this value with the ratio in decibels if the ice were replaced by air. The difference is the attenuation due to the presence of the ice. In (a) and (b) assume that the vehicle transmitter radiates isotropically and that there is no loss (mismatch) at the air-ice interface. (c) What is the travel time for the wave down and back through the ice? (d) Would it be more accurate to determine the ice thickness by pulsing the transmitter and measuring the delay time of the reflected pulse, as in a radar system, rather than by measuring the wave attenuation? (e) Which scheme would be more practical?

★10-6-5. Spaceship near moon A spaceship at lunar distance from the earth transmits 2-GHz waves. If a power of 10 W is radiated isotropically, find (a) average Poynting vector at the earth, (b) rms electric field E at the earth, and (c) the time it takes for the radio waves to travel from the spaceship to the earth. (Take the earth-moon distance as 380 Mm.) (d) How many photons per unit area per second fall on the earth from the spaceship transmitter?

10-6-6. Superconducting line Overhead long-distance high-power transmission lines radiate electromagnetic interference, generate poisonous ozone, and otherwise disrupt the ecology. Design a transmission line for handling terawatt power levels over megameter distances which avoids the above effects. Assume that a 25 K superconductor is available. *Hint*: Consider a buried thermally insulated liquid-hydrogen-cooled dc line operating at levels of a few volts. Alternatively consider a liquid-hydrogen-cooled cylindrical waveguide operating at 100 GHz in a higher-order mode or modes (see Chap. 13).

10-6-7. Superconducting world grid The designer of the geodesic dome, Buckminster Fuller, has proposed a superconducting power grid over the North Pole to link America and Eurasia so that solar power stations on the sunlit side of the earth could gather solar energy and transmit it as electric energy to the dark side of the earth. Discuss feasibility, costs, etc.

10-6-8. Superconducting cable. Nanosecond response In Prob. 10-6-6 superconductors are considered for low-loss power transmission. Another advantage is that superconductive cables can provide faster response (wider bandwidth) for transmission of information. Design a superconductive two-conductor cable with nanosecond response. See, for example, H. J. Jensen, Nuclear Test Instrumentation with Miniature Superconductive Cables, *IEEE Spectrum*, **5**:91–99 (September 1968).

ELEVEN

WAVE POLARIZATION

11-1 INTRODUCTION†

In this chapter the polarization of electromagnetic waves is discussed. Linear, elliptical, and circular polarization are considered, and their relation on the Poincaré sphere is explained. These concepts are first developed for coherent, completely polarized waves and then extended to the case of partially and completely unpolarized or incoherent radiation. Finally Stokes parameters and coherency matrices are introduced for handling waves of this kind.

11-2 LINEAR, ELLIPTICAL, AND CIRCULAR POLARIZATION

Consider a plane wave traveling out of the page (positive z direction), as in Fig. 11-1a, with the electric field at all times in the y direction. This wave is said to be *linearly polarized* (in the y direction). As a function of time and position the electric field of a linearly polarized wave (as in Fig. 11-1a) traveling in the positive z direction (out of page) is given by

$$E_y = E_2 \sin(\omega t - \beta z) \tag{1}$$

In general the electric field of a wave traveling in the z direction may have both a y component and an x component, as suggested in Fig. 11-1b. In this more general situation the wave is said to be *elliptically polarized*. At a fixed value of z the electric vector $\mathbf{E}$ rotates as a function of time, the tip of the vector describing an ellipse called the *polarization ellipse*. The ratio of the major to minor axes of the polarization ellipse is called the *axial ratio* (AR). Thus, for the wave in Fig. 11-1b, AR $= E_2/E_1$. Two extreme cases of elliptical polarization correspond to *circular polarization*, as in Fig. 11-1c, and *linear polarization*, as in Fig. 11-1a. For circular polarization $E_1 = E_2$ and AR $= 1$, while for linear polarization $E_1 = 0$ and AR $= \infty$.

† A more detailed discussion of wave polarization is given in J. D. Kraus, "Radio Astronomy," Cygnus-Quasar Books, Powell, Ohio, 1982.

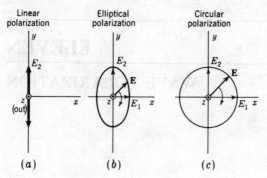

Figure 11-1 Linear, elliptical, and circular polarization.

In the most general case of elliptical polarization the polarization ellipse may have any orientation, as suggested in Fig. 11-2. The elliptically polarized wave may be expressed in terms of two linearly polarized components, one in the x direction and one in the y direction. Thus, if the wave is traveling in the positive z direction (out of the page), the electric field components in the x and y directions are

$$E_x = E_1 \sin(\omega t - \beta z) \tag{2}$$

$$E_y = E_2 \sin(\omega t - \beta z + \delta) \tag{3}$$

where E_1 = amplitude of wave linearly polarized in x direction
E_2 = amplitude of wave linearly polarized in y direction
δ = time-phase angle by which E_y leads E_x

Combining (2) and (3) gives the instantaneous total vector field $\mathbf{E}$:

$$\mathbf{E} = \hat{\mathbf{x}}E_1 \sin(\omega t - \beta z) + \hat{\mathbf{y}}E_2 \sin(\omega t - \beta z + \delta) \tag{4}$$

At $z = 0$, $E_x = E_1 \sin \omega t$ and $E_y = E_2 \sin(\omega t + \delta)$. Expanding E_y yields

$$E_y = E_2(\sin \omega t \cos \delta + \cos \omega t \sin \delta) \tag{5}$$

From the relation for E_x we have $\sin \omega t = E_x/E_1$ and $\cos \omega t = \sqrt{1 - (E_x/E_1)^2}$. Introducing these in (5) eliminates ωt, and on rearranging we obtain

$$\frac{E_x^2}{E_1^2} - \frac{2E_x E_y \cos \delta}{E_1 E_2} + \frac{E_y^2}{E_2^2} = \sin^2 \delta \tag{6}$$

or

$$aE_x^2 - bE_x E_y + cE_y^2 = 1 \tag{7}$$

where $\qquad a = \dfrac{1}{E_1^2 \sin^2 \delta} \qquad b = \dfrac{2 \cos \delta}{E_1 E_2 \sin^2 \delta} \qquad c = \dfrac{1}{E_2^2 \sin^2 \delta}$

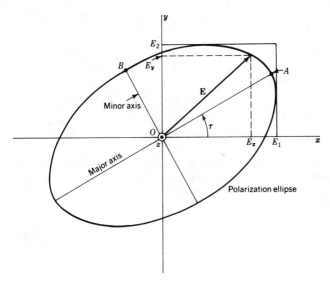

Figure 11-2 Polarization ellipse at tilt angle τ showing instantaneous components E_x and E_y and amplitudes (or peak values) E_1 and E_2.

Equation (7) describes a (polarization) ellipse, as in Fig. 11-2. The line segment OA is the semimajor axis, and the line segment OB is the semiminor axis. The tilt angle of the ellipse is τ. The axial ratio is

$$\mathrm{AR} = \frac{OA}{OB} \qquad (1 \leq \mathrm{AR} \leq \infty) \tag{8}$$

If $E_1 = 0$, the wave is linearly polarized in the y direction. If $E_2 = 0$, the wave is linearly polarized in the x direction. If $\delta = 0$ and $E_1 = E_2$, the wave is also linearly polarized but in a plane at an angle of $45°$ with respect to the x axis ($\tau = 45°$).

If $E_1 = E_2$ and $\delta = \pm 90°$, the wave is circularly polarized. When $\delta = +90°$, the wave is *left circularly polarized*, and when $\delta = -90°$, the wave is *right circularly polarized*. For the case $\delta = +90°$ and for $z = 0$ and $t = 0$ we have from (2) and (3) that $\mathbf{E} = \hat{y}E_2$, as in Fig. 11-3a. One-quarter cycle later ($\omega t = 90°$) $\mathbf{E} = \hat{x}E_1$, as in Fig. 11-3b. Thus, at a fixed position ($z = 0$) the electric field vector rotates clockwise (viewing the wave approaching). According to the IEEE definition, this corresponds to left circular polarization.† The opposite rotation direction ($\delta = -90°$) corresponds to right circular polarization.

If the wave is viewed receding (from negative z axis in Fig. 11-3), the electric vector appears to rotate in the opposite direction. Hence, clockwise rotation of $\mathbf{E}$ with the wave approaching is the same as counterclockwise rotation with the wave

---

† This IEEE definition is opposite to the classical optics definition.

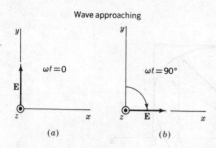

Figure 11-3 Instantaneous orientation of electric field vector **E** at two instants of time for a left circularly polarized wave which is approaching (out of page).

receding. Thus, unless the wave direction is specified, there is a possibility of ambiguity as to whether the wave is left- or right-handed. This can be avoided by defining the polarization with the aid of helical-beam antennas (see Chap. 14). Thus, a right-handed helical-beam antenna radiates (or receives) right circular (IEEE) polarization.† A right-handed helix, like a right-handed screw, is right-handed regardless of the position from which it is viewed. There is no possibility here of ambiguity.

11-3 POYNTING VECTOR FOR ELLIPTICALLY OR CIRCULARLY POLARIZED WAVES

In complex notation the Poynting vector is

$$\mathbf{S} = \tfrac{1}{2}\mathbf{E} \times \mathbf{H}^* \tag{1}$$

The average Poynting vector is the real part of (1), or

$$\mathbf{S}_{av} = \text{Re } \mathbf{S} = \tfrac{1}{2} \text{Re } \mathbf{E} \times \mathbf{H}^* \tag{2}$$

Referring to Fig. 11-2, let the elliptically polarized wave have x and y components with a phase difference δ as given by

$$E_x = E_1 e^{j(\omega t - \beta z)} \tag{3}$$

$$E_y = E_2 e^{j(\omega t - \beta z + \delta)} \tag{4}$$

At $z = 0$ the total electric field (vector) is then

$$\mathbf{E} = \hat{\mathbf{x}}E_x + \hat{\mathbf{y}}E_y = \hat{\mathbf{x}}E_1 e^{j\omega t} + \hat{\mathbf{y}}E_2 e^{j(\omega t + \delta)} \tag{5}$$

where $\hat{\mathbf{x}}$ = unit vector in x direction
$\hat{\mathbf{y}}$ = unit vector in y direction

Note that **E** has two components each involving both a space vector and a time phase factor (phasor, with ωt explicit).

The **H**-field component associated with E_x is

$$H_y = H_1 e^{j(\omega t - \beta z - \xi)} \tag{6}$$

---

† A left-handed helical-beam antenna radiates (or receives) left circular (IEEE) polarization.

where ξ is the phase lag of H_y with respect to E_x. The H-field component associated with E_y is

$$H_x = -H_2 e^{j(\omega t - \beta z + \delta - \xi)} \tag{7}$$

The total **H** field (vector) at $z = 0$ for a wave traveling in the positive z direction is then

$$\mathbf{H} = \hat{\mathbf{y}}H_y - \hat{\mathbf{x}}H_x = \hat{\mathbf{y}}H_1 e^{j(\omega t - \xi)} - \hat{\mathbf{x}}H_2 e^{j(\omega t + \delta - \xi)} \tag{8}$$

The complex conjugate of **H** is equal to (8) except for the sign of the exponents. That is,

$$\mathbf{H}^* = \hat{\mathbf{y}}H_1 e^{-j(\omega t - \xi)} - \hat{\mathbf{x}}H_2 e^{-j(\omega t + \delta - \xi)} \tag{9}$$

Substituting (5) and (9) in (2) gives the average Poynting vector at $z = 0$ as

$$\begin{aligned}\mathbf{S}_{av} &= \tfrac{1}{2} \operatorname{Re} \left[(\hat{\mathbf{x}} \times \hat{\mathbf{y}})E_x H_y^* - (\hat{\mathbf{y}} \times \hat{\mathbf{x}})E_y H_x^* \right] \\ &= \tfrac{1}{2}\hat{\mathbf{z}} \operatorname{Re} (E_x H_y^* + E_y H_x^*) \end{aligned} \tag{10}$$

where $\hat{\mathbf{z}}$ is the unit vector in the z direction (direction of propagation of wave). It follows that

$$\begin{aligned}\mathbf{S}_{av} &= \tfrac{1}{2}\hat{\mathbf{z}}(E_1 H_1 \operatorname{Re} e^{j\xi} + E_2 H_2 \operatorname{Re} e^{j\xi}) \\ &= \tfrac{1}{2}\hat{\mathbf{z}}(E_1 H_1 + E_2 H_2) \cos \xi \end{aligned} \tag{11}$$

It is to be noted that $\mathbf{S}_{av}$ is independent of δ.

In a lossless medium $\xi = 0$ (electric and magnetic fields in time phase) and $E_1/H_1 = E_2/H_2 = Z_0$, where Z_0, the intrinsic impedance of the medium, is real; so

$$\begin{aligned}\mathbf{S}_{av} &= \tfrac{1}{2}\hat{\mathbf{z}}(E_1 H_1 + E_2 H_2) \\ &= \tfrac{1}{2}\hat{\mathbf{z}}(H_1^2 + H_2^2)Z_0 = \tfrac{1}{2}\hat{\mathbf{z}}H^2 Z_0 \end{aligned} \tag{12}$$

where $H = \sqrt{H_1^2 + H_2^2}$ is the amplitude of the total **H** field. We can also write

$$\boxed{\mathbf{S}_{av} = \tfrac{1}{2}\hat{\mathbf{z}}\frac{E_1^2 + E_2^2}{Z_0} = \tfrac{1}{2}\hat{\mathbf{z}}\frac{E^2}{Z_0}} \tag{13}$$

where $E = \sqrt{E_1^2 + E_2^2}$ is the amplitude of the total **E** field.

Example An elliptically polarized wave traveling in the positive z direction in air has x and y components

$$E_x = 3 \sin (\omega t - \beta z) \qquad (\text{V m}^{-1})$$
$$E_y = 6 \sin (\omega t - \beta z + 75°) \qquad (\text{V m}^{-1})$$

Find the average power per unit area conveyed by the wave.

SOLUTION The average power per unit area is equal to the average Poynting vector, which from (13) has a magnitude

$$S_{av} = \frac{1}{2}\frac{E^2}{Z} = \frac{1}{2}\frac{E_1^2 + E_2^2}{Z}$$

From the stated conditions the amplitude $E_1 = 3 \text{ V m}^{-1}$, and the amplitude $E_2 = 6 \text{ V m}^{-1}$. Also for air $Z = 376.7 \ \Omega$. Hence

$$S_{av} = \frac{1}{2} \frac{3^2 + 6^2}{376.7} = \frac{1}{2} \frac{45}{376.7} \approx 60 \text{ mW m}^{-2}$$

11-4 THE POLARIZATION ELLIPSE AND THE POINCARÉ SPHERE

In the Poincaré sphere† representation of wave polarization, the *polarization state* is described by a point on a sphere where the longitude and latitude of the point are related to parameters of the polarization ellipse (see Fig. 11-4) as follows:

$$\text{Longitude} = 2\tau$$
$$\text{Latitude} \ \ = 2\epsilon \tag{1}$$

where τ = tilt angle, $0° \leq \tau \leq 180°$
  $\epsilon = \cot^{-1}(\mp AR)$, $-45° \leq \epsilon \leq +45°$

The axial ratio (AR) is negative for right-handed and positive for left-handed (IEEE) polarization.

The polarization state described by a point on a sphere can also be expressed in terms of the angle subtended by the great circle drawn from a reference point on the equator and the angle between the great circle and the equator (see Fig. 11-4) as follows:

$$\text{Great-circle angle} = 2\gamma$$
$$\text{Equator-to-great-circle angle} = \delta \tag{2}$$

where $\gamma = \tan^{-1}(E_2/E_1)$, $0° \leq \gamma \leq 90°$
  δ = phase difference between E_y and E_x, $-180° \leq \delta \leq +180°$

The geometric relation of τ, ϵ, and γ to the polarization ellipse is illustrated in Fig. 11-5. The trigonometric interrelations of τ, ϵ, γ, and δ are as follows:‡

$$
\boxed{
\begin{aligned}
\cos 2\gamma &= \cos 2\epsilon \cos 2\tau \\
\tan \delta &= \frac{\tan 2\epsilon}{\sin 2\tau} \\
\tan 2\tau &= \tan 2\gamma \cos \delta \\
\sin 2\epsilon &= \sin 2\gamma \sin \delta
\end{aligned}
}
\tag{3}
$$

---

† H. Poincaré, "Théorie mathématique de la lumière," G. Carré, Paris, 1892; G. A. Deschamps, Geometrical Representation of the Polarization of a Plane Electromagnetic Wave, *Proc. IRE*, **39**:540 (May 1951).

‡ These relations involve spherical trigonometry. See M. Born and E. Wolf, "Principles of Optics," pp. 24–27, The Macmillan Company, New York, 1964.

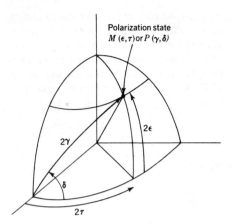

Figure 11-4 Poincaré sphere showing relation of angles ϵ, τ, γ, and δ.

Knowing ϵ and τ, one can determine γ and δ or vice versa. It is convenient to describe the *polarization state* by either of the two sets of angles (ϵ, τ) or (γ, δ) which describe a point on the Poincaré sphere (Fig. 11-4). Let the polarization state as a function of ϵ and τ be designated by $M(\epsilon, \tau)$ or simply M and the polarization state as a function of γ and δ be designated by $P(\gamma, \delta)$ or simply P, as in Fig. 11-4. Two special cases are of interest.

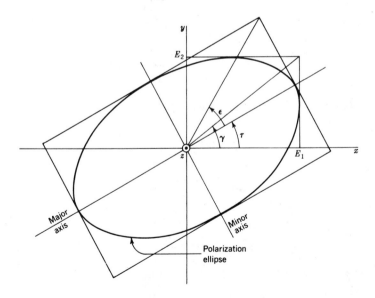

Figure 11-5 Polarization ellipse showing relation of angles ϵ, γ, and τ.

Case 1　For $\delta = 0$ or $\delta = \pm 180°$, E_x and E_y are exactly in phase or out of phase, so that any point on the equator represents a state of linear polarization. At the origin ($\epsilon = \tau = 0$) the polarization is linear and in the x direction ($\tau = 0$), as suggested in Fig. 11-6a. On the equator 90° to the right the polarization is linear with a tilt angle $\tau = 45°$, while 180° from the origin the polarization is linear and in the y direction ($\tau = 90°$). See Fig. 11-6a and b. One octant of the Poincaré sphere is shown in Fig. 11-6a, and the full sphere is shown in Fig. 11-6b in rectangular projection.

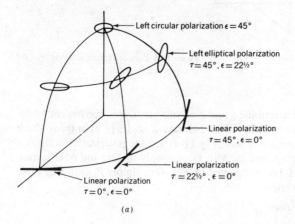

(a)

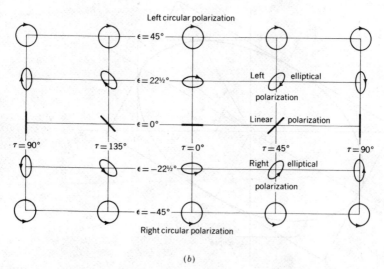

(b)

Figure 11-6 (a) One octant of Poincaré sphere with polarization states. (b) Rectangular projection of Poincaré sphere showing full range of polarization states.

Case 2 For $\delta = \pm 90°$ and $E_2 = E_1$ $(2\gamma = 90°$ and $2\epsilon = \pm 90°)$ E_x and E_y have equal amplitudes but are in phase quadrature, which is the condition for circular polarization. Thus, the poles represent a state of circular polarization, the upper pole representing left circular polarization and the lower pole right circular (IEEE) polarization, as suggested in Fig. 11-6a and b.

Cases 1 and 2 represent limiting conditions. In the general case any point on the upper hemisphere describes a left elliptically polarized wave ranging from pure left circular at the pole to linear at the equator. Likewise, any point on the lower hemisphere describes a right elliptically polarized wave ranging from pure right circular at the pole to linear at the equator. Several elliptical states of polarization are shown by ellipses with appropriate tilt angles τ and axial ratios AR at points on the Poincaré sphere in Fig. 11-6a and b.

As an application of the Poincaré sphere representation it may be shown that the voltage response V of an antenna to a wave of arbitrary polarization is given by†

$$V = k \cos \frac{MM_a}{2} \tag{4}$$

where MM_a = angle subtended by great-circle line from polarization state M to M_a

M = polarization state of wave

M_a = polarization state of antenna

k = constant

The polarization state of the antenna is defined as the polarization state of the wave radiated by the antenna when it is transmitting. The factor k in (4) involves the field strength of the wave and the size of the antenna. An important result to note is that if $MM_a = 0°$, the antenna is matched to the wave (polarization state of wave same as for antenna) and the response is maximized. However, if $MM_a = 180°$, the response is zero. This can occur, for example, if the wave is linearly polarized in the y direction while the antenna is linearly polarized in the x direction; or if the wave is left circularly polarized while the antenna is right circularly polarized. More generally we may say that *an antenna is blind to a wave of opposite (or antipodal) polarization state.*

11-5 PARTIAL POLARIZATION AND THE STOKES PARAMETERS

The previous sections deal with completely polarized waves, i.e., waves for which E_1, E_2, and δ are constants (or at least slowly varying functions of time). The radiation from a single-frequency (monochromatic) radio transmitter is of this

† G. Sinclair, The Transmission and Reception of Elliptically Polarized Waves, *Proc. IRE*, **38**: 151 (1950).

type. However, the radiation from many celestial radio sources extends over a wide frequency range and within any bandwidth Δf consists of the superposition of a large number of statistically independent waves of a variety of polarizations. The resultant wave is said to be *incoherent* or *unpolarized*. For such a wave we can write

$$E_x = E_1(t) \sin \omega t \tag{1}$$

$$E_y = E_2(t) \sin [\omega t + \delta(t)] \tag{2}$$

where all the time functions are independent. The variations of $E_1(t)$, $E_2(t)$, and $\delta(t)$ with time are slow compared with that of the mean frequency f ($\omega = 2\pi f$) and are of the order of the bandwidth Δf. A wave of this type could be generated by connecting one noise generator to an antenna which is linearly polarized in the y direction and another noise generator to an adjacent antenna which is linearly polarized in the x direction. If the waves from both antennas have the same average power at an observing point, the total wave at this point will be completely unpolarized.

The most general situation of wave polarization occurs when a wave is *partially polarized*; i.e., it may be considered to be of two parts, one completely polarized and the other completely unpolarized.

To deal with partial polarization it is convenient to use the *Stokes parameters*, introduced in 1852 by Sir George Stokes.† The Stokes parameters I, Q, U, and V are defined as follows:

$$I = S = S_x + S_y = \frac{\langle E_1^2 \rangle}{Z} + \frac{\langle E_2^2 \rangle}{Z} \tag{3}$$

$$Q = S_x - S_y = \frac{\langle E_1^2 \rangle}{Z} - \frac{\langle E_2^2 \rangle}{Z} \tag{4}$$

$$U = \frac{2}{Z} \langle E_1 E_2 \cos \delta \rangle = S \langle \cos 2\epsilon \sin 2\tau \rangle \tag{5}$$

$$V = \frac{2}{Z} \langle E_1 E_2 \sin \delta \rangle = S \langle \sin 2\epsilon \rangle \tag{6}$$

where S = total Poynting vector magnitude for the wave
S_x = Poynting vector component of wave polarized in x direction
S_y = Poynting vector component of wave polarized in y direction
E_1 = amplitude of electric field component of wave polarized in x direction
E_2 = amplitude of electric field component of wave polarized in y direction
Z = intrinsic impedance of medium

† G. Stokes, On the Composition Resolution of Streams of Polarized Light from Different Sources, *Trans. Camb. Phil. Soc.*, **9**(3):399–416 (1852).

The angles δ, ϵ, and τ are as defined in the previous section. The angle brackets $\langle\ \rangle$ indicate the time average. It is understood that in general E_1, E_2, δ, ϵ, and τ are functions of time. Thus, for example,

$$\langle E_1^2 \rangle = \frac{1}{T} \int_0^T [E_1(t)]^2 \, dt \tag{7}$$

For a *completely unpolarized wave* $S_x = S_y$ and E_1 and E_2 are uncorrelated. It may be shown that under these conditions

$$\langle E_1 E_2 \cos \delta \rangle = \langle E_1 E_2 \sin \delta \rangle = 0$$

and we have

$$I = S$$
$$Q = 0$$
$$U = 0$$
$$V = 0$$

where S is the total Poynting vector of the wave. The condition $Q = U = V = 0$ is a requirement for a completely unpolarized wave.

For a *completely polarized wave* E_1, E_2, δ, ϵ, and τ may be considered to be constant, so that the Stokes parameters as in (3) through (6) do not require time averages. Thus, for example, we would have $\langle E_1^2 \rangle = E_1^2$.

For a *linearly (completely) polarized wave with* **E** *in the x direction* ($E_2 = 0$, and $\tau = \epsilon = 0$) the Stokes parameters are

$$I = S$$
$$Q = S$$
$$U = 0$$
$$V = 0$$

For a *linearly (completely) polarized wave with* **E** *in the y direction* ($E_1 = 0$, $\tau = 90°$, $\epsilon = 0$) the Stokes parameters are

$$I = S$$
$$Q = -S$$
$$U = 0$$
$$V = 0$$

For a *left circularly (completely) polarized wave* ($E_1 = E_2$, $\delta = 90°$)

$$I = S$$
$$Q = 0$$
$$U = 0$$
$$V = S$$

By consideration of other polarization states it is possible to show that the Stokes parameter I is always equal to the total power (density) of the wave, Q is equal to the power in the linearly polarized components (in x or y directions), U is equal to the power in the linearly polarized components at tilt angles $\tau = 45°$ or $135°$, and V is equal to the power in the circularly polarized components (left- or right-handed).

It is often convenient to normalize the Stokes parameters by dividing each parameter by S, obtaining the *normalized Stokes parameters* $s_0, s_1, s_2,$ and s_3, where $s_0 = I/S = 1$, $s_1 = Q/S$, $s_2 = U/S$, and $s_3 = V/S$. The normalized Stokes parameters for the cases discussed above are summarized in Table 11-1.

If any of the parameters Q, U, or V (or s_1, s_2, or s_3) has a nonzero value, it indicates the presence of a polarized component in the wave. The *degree of polarization d* is defined as the ratio of completely polarized power to the total power, or

$$d = \frac{\text{polarized power}}{\text{total power}} \qquad 0 \le d \le 1$$

or

$$d = \frac{\sqrt{Q^2 + U^2 + V^2}}{I} = \frac{\sqrt{s_1^2 + s_2^2 + s_3^2}}{1} \tag{8}$$

The condition $Q^2 + U^2 + V^2 = I^2$ or $s_1^2 + s_2^2 + s_3^2 = 1$ indicates a completely polarized wave. A partially polarized wave may be regarded as the sum of a completely unpolarized wave and a completely polarized wave. Thus, we may write for a partially polarized wave

$$\begin{bmatrix} s_0 \\ s_1 \\ s_2 \\ s_3 \end{bmatrix} = \begin{bmatrix} 1 - d \\ 0 \\ 0 \\ 0 \end{bmatrix} + \begin{bmatrix} d \\ d \cos 2\epsilon \cos 2\tau \\ d \cos 2\epsilon \sin 2\tau \\ d \sin 2\epsilon \end{bmatrix} \tag{9}$$

where the first term on the right of (9) represents the unpolarized part and the second term the polarized part of the wave.

Table 11-1 Normalized Stokes parameters for seven wave-polarization states

| Normalized Stokes parameter | Completely unpolarized wave | Completely polarized waves | | | | | |
| | | Linearly polarized | | | | Circularly polarized | |
		$\tau = 0°$	$\tau = 90°$	$\tau = 45°$	$\tau = 135°$	Left-hand	Right-hand
s_0	1	1	1	1	1	1	1
s_1	0	1	-1	0	0	0	0
s_2	0	0	0	1	-1	0	0
s_3	0	0	0	0	0	1	-1

Consider now a wave with Stokes parameters s_0, s_1, s_2, s_3 incident on an antenna with Stokes parameters a_0, a_1, a_2, a_3.† The parameters a_0, a_1, a_2, a_3 are those for a wave radiated by the antenna when it is transmitting. It follows that the total power P available (to a receiver) from the antenna due to the incident wave is given by

$$P = \tfrac{1}{2} A_e S [a_0 \quad a_1 \quad a_2 \quad a_3] \begin{bmatrix} s_0 \\ s_1 \\ s_2 \\ s_3 \end{bmatrix} \tag{10}$$

or

$$\boxed{P = \tfrac{1}{2} A_e S (a_0 s_0 + a_1 s_1 + a_2 s_2 + a_3 s_3) \qquad \text{(W)}} \tag{11}$$

where A_e = effective aperture of antenna (see Chap. 14), m²
     S = Poynting vector of incident wave, W m⁻²

This result may also be expressed as

$$\boxed{P = \tfrac{1}{2} A_e S (1 + d \cos MM_a)} \tag{12}$$

or

$$P = \tfrac{1}{2} A_e S (1 - d) + dS A_e \cos^2 \frac{MM_a}{2} \tag{13}$$

where MM_a is the angle subtended by the great-circle line between the wave- and antenna-polarization states on the Poincaré sphere. In (13) the first term on the right side represents the unpolarized power and the second term the polarized power. This form emphasizes the fact that only one-half the incident unpolarized power is available to a receiver but all the completely polarized power is available.

By arrangement of (13) we have

$$\boxed{P = \left(\frac{1 - d}{2} + d \cos^2 \frac{MM_a}{2} \right) A_e S = F A_e S \qquad \text{(W)}} \tag{14}$$

where F is the factor in parentheses, which may be called the *wave-to-antenna coupling factor*. This factor is a function of the degree of polarization and the angle MM_a and gives the fraction of the power $A_e S$ which is received. It is dimensionless with values between 0 and 1. Its possible range of values as a function of the degree of polarization is shown by the shaded area in Fig. 11-7. Thus, for a perfect match ($MM_a = 0$) F ranges from $\tfrac{1}{2}$ to 1 as d goes from 0 to 1, while for a complete mismatch ($MM_a = 180°$) F ranges from $\tfrac{1}{2}$ to 0 as d goes from 0 to 1. Between these extremes ($MM_a = 0$ and $MM_a = 180°$) we have conditions of partial mismatching between antenna and wave, for example, $MM_a = 90°$.

† H. C. Ko, On the Reception of Quasi-monochromatic Partially Polarized Radio Waves, *Proc. IRE*, **50**:1950 (September 1962).

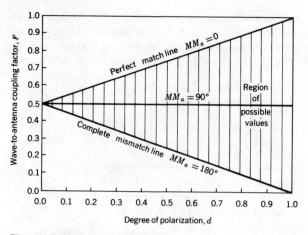

Figure 11-7 Chart showing wave-to-antenna coupling factor F versus the degree of polarization d for various values of the wave-to-antenna angle MM_a. The region of possible values is shaded. These values range from a perfect match ($MM_a = 0$) to a complete mismatch ($MM_a = 180°$). No values are possible outside this region. Note that for completely polarized waves ($d = 1$) matching is important, but for a completely unpolarized wave ($d = 0$) the condition of match makes no difference. (*After H. C. Ko.*)

Equation (14) may be interpreted as signifying that the *response P* of the system is given by the maximum available power $A_e S$ times the factor F which characterizes the receiving antenna. The relation (14) is typical of all observing or measuring systems in that the response (or observed result) may be modified from the optimum or ideal result by a quantity characterizing the observing or measuring system.

If we define four new parameters

$$s_{11} = \tfrac{1}{2}(s_0 + s_1) \qquad s_{12} = \tfrac{1}{2}(s_2 + js_3)$$
$$s_{21} = \tfrac{1}{2}(s_2 - js_3) \qquad s_{22} = \tfrac{1}{2}(s_0 - s_1) \tag{15}$$

and four similar parameters ($a_{11}, a_{12}, a_{21}, a_{22}$) for the antenna, we can express the available power with 2×2 matrices as

$$P = A_e S \operatorname{Tr} \left\{ \begin{bmatrix} a_{11} & a_{12} \\ a_{21} & a_{22} \end{bmatrix} \begin{bmatrix} s_{11} & s_{12} \\ s_{21} & s_{22} \end{bmatrix} \right\} \tag{16}$$

where Tr signifies the *trace*, i.e., the sum of the diagonal elements of a square matrix or

$$P = A_e S(a_{11}s_{11} + a_{12}s_{21} + a_{21}s_{12} + a_{22}s_{22}) \tag{17}$$

More concisely

$$\boxed{P = A_e S \operatorname{Tr} \{[a_{ij}][s_{ij}]\} = A_e SF \qquad \text{(W)}} \tag{18}$$

where F is the wave-to-antenna coupling factor, as in (14). The above matrix relations are similar to ones used in optics, which is appropriate if we regard the antenna receiver as a detector of radio photons. In optics the matrices are called *coherency matrices*.†

PROBLEMS‡

Group 11-1: Secs. 11-1 through 11-4. Linear, elliptical, and circular polarization and Poincaré sphere

⋆**11-1-1. CP waves** A wave traveling normally out of the page is the resultant of two circularly polarized components $E_{\text{right}} = 5e^{j\omega t}$ and $E_{\text{left}} = 2e^{-j(\omega t + 90°)}$ (V m^{-1}). Find (a) axial ratio AR, (b) tilt angle τ, and (c) hand of rotation (left or right).

11-1-2. EP wave A wave traveling normally out of the page (toward the reader) is the resultant of two linearly polarized components $E_x = 3 \cos \omega t$ and $E_y = 2 \cos (\omega t + 90°)$. For the resultant wave find (a) axial ratio AR, (b) tilt angle τ, and (c) hand of rotation (left or right).

⋆**11-1-3. CP waves** Two circularly polarized waves traveling normally out of the page have fields given by $E_{\text{left}} = 2e^{-j\omega t}$ and $E_{\text{right}} = 3e^{j\omega t}$ V m^{-1} (rms). For the resultant wave find (a) AR, (b) hand of rotation, and (c) Poynting vector.

11-1-4. CP wave reflection and transmission A circularly polarized 200-MHz wave in air with $E = 2$ V m^{-1} (rms) is incident normally on a half-space of nonconducting medium with $\mu_r = \epsilon_r = 3 - j3$. Find the Poynting vector (a) for the reflected wave and (b) for the transmitted wave at a depth of 200 mm in the medium.

⋆**11-1-5. EP waves** A wave traveling normally out of the page is the resultant of two elliptically polarized (EP) waves, one with components $E_x = 5 \cos \omega t$ and $E_y = 3 \sin \omega t$ and another with components $E_r = 3e^{j\omega t}$ and $E_l = 4e^{-j\omega t}$. For the resultant wave, find (a) AR, (b) τ, and (c) hand of rotation.

11-1-6. CP waves A wave traveling normally out of the page is the resultant of two circularly polarized components $E_r = 2e^{j\omega t}$ and $E_l = 4e^{-j(\omega t + 45°)}$. For the resultant wave, find (a) AR, (b) τ, and (c) hand of rotation.

11-1-7. More power with CP Show that the average Poynting vector of a circularly polarized wave is twice that of a linearly polarized wave if the maximum electric field E is the same for both waves. This means that a medium can handle twice as much power before breakdown with circular polarization (CP) than with linear polarization (LP).

11-1-8. PV constant for CP Show that the instantaneous Poynting vector (PV) of a plane circularly polarized traveling wave is a constant.

⋆**11-1-9. EP wave power** An elliptically polarized wave in a medium with constants $\sigma = 0$, $\mu_r = 2$, $\epsilon_r = 5$ has H-field components (normal to the direction of propagation and normal to each other) of amplitudes 3 and 4 A m^{-1}. Find the average power conveyed through an area of 5 m^2 normal to the direction of propagation.

Group 11-2: Sec. 11-5. Partial polarization and Stokes parameters

11-2-1. Power for partially polarized wave A partially polarized wave for which $d = \frac{1}{2}$, AR $= +2$, $\tau = 0°$, and $S = 1$ (W m^{-2}) is incident on an antenna for which AR $= +4$, $\tau = 22.5°$, and $A_e = 1$ m^2. Find the received power.

⋆**11-2-2. Stokes parameters** A wave for which $s_1^2 = s_2^2 = s_3^2$ is incident on an antenna for which AR $= -2.5$, $\tau = 90°$, and $A_e = 3$ m^2. If the wave is two-thirds polarized with $S = 2$ W m^{-2}, find the received power.

---

† M. Born and E. Wolf, "Principles of Optics," The Macmillan Company, New York, 1964.

‡ Answers to starred problems are given in Appendix C.

11-2-3. One wave and six antennas (a) A wave for which $d = 0.4$, AR $= 3$, and $\tau = 45°$ is received by six antennas of unit effective aperture with polarization as follows: (1) linear horizontal, (2) linear vertical, (3) linear slant (45°), (4) linear slant (135°), (5) left circular, and (6) right circular. If the wave has unit Poynting vector, find the six power responses. (b) Is there a wave to which all six antennas of part (a) respond equally? If so, what are the wave parameters?

★**11-2-4. Three waves and three antennas** Three waves have the following characteristics: (a) unpolarized; (b) left elliptically polarized with $d = \frac{1}{2}$, AR $= 3$, and $\tau = 135°$; (c) left circularly polarized. Three antennas produce the following types of waves when transmitting: (1) linear horizontal polarization; (2) right elliptical polarization with AR $= 3$ and $\tau = 45°$; (3) right circular polarization. If all the antennas have unit effective aperture and all waves have unit Poynting vector, find the received power for the nine wave-antenna combinations a1, a2, a3, b1, b2, etc. Arrange answers in 3×3 row-column display with top row a1, a2, a3, second row b1, b2, etc.

11-2-5. Ten waves Ten waves have the following characteristics:

(1) $d = 0$ (2) $d = \frac{1}{2}$, AR $= \infty$, $\tau = 0°$

(3) $d = \frac{2}{3}$, AR $= \infty$, $\tau = 45°$ (4) $d = \frac{3}{4}$, AR $= -1$

(5) $d = 1$, AR $= +1$ (6) $d = 1$, AR $= \infty$, $\tau = 90°$

(7) $d = 1$, AR $= \infty$, $\tau = 135°$ (8) $d = \frac{1}{4}$, AR $= 10$, $\tau = 0°$

(9) $d = \frac{1}{3}$, AR $= 4$, $\tau = 30°$ (10) $d = \frac{1}{2}$, AR $= 3$, $\tau = 120°$

Find the normalized Stokes parameters and the coherency matrices for these waves.

11-2-6. Stokes parameters for polarized wave The degree of polarization d of a wave can be resolved into a *linearly polarized* component $d_l = \sqrt{s_1^2 + s_2^2}$ and a *circularly polarized* component $d_c = \sqrt{s_3^2}$, with $d = \sqrt{d_l^2 + d_c^2}$. What are the Stokes parameters for a completely polarized wave for which $d_l = d_c$ and for which $s_1 = s_2$?

11-2-7. Wave with equal Stokes parameters (a) What are the wave characteristics (τ and AR) for a completely polarized wave for which $Q^2 = U^2 = V^2$? (b) Draw the polarization ellipse.

★**11-2-8. Five waves. Power** Give the percentage of polarized wave power in each of five waves characterized by the following Stokes parameters: (a) 1, 0, 0, 0; (b) 1, 1, 0, 0; (c) 1, 0, $-\frac{1}{2}$, 0; (d) 1, $\frac{1}{2}$, $\frac{1}{2}$, 0; (e) 1, 0, $1/\sqrt{2}$, $1/\sqrt{2}$.

11-2-9. Coherency matrix. Power A wave of unit Poynting vector with coherency matrix

$$\begin{bmatrix} \frac{1}{2} & \frac{1}{6} \\ \frac{1}{6} & \frac{1}{2} \end{bmatrix}$$

is incident on a right circularly polarized antenna of unit effective aperture. Find the received power.

11-2-10. Ten waves. Stokes parameters Describe the characteristics (AR, left- or right-handed, tilt angle) of waves having Stokes parameters as follows:

(1) 1, 0, 0, 1   (2) 1, 0, 0, -1

(3) 1, 1, 0, 0   (4) 1, -1, 0, 0

(5) 1, 0, 1, 0   (6) 1, 0, -1, 0

(7) 1, 0.577, 0.577, 0.577   (8) 1, 0, 0, 0

(9) 1, 0, 0, $-\frac{1}{2}$ (10) 1, $\frac{1}{2}$, 0, 0

(11) 1, 0, $\frac{1}{3}$, 0   (12) 1, $-\frac{1}{3}$, 0, $\frac{1}{3}$

REFLECTION, REFRACTION, AND DIFFRACTION

12-1 INTRODUCTION

The case of a plane wave normally incident on a boundary between two media was discussed in Chap. 10. Here the treatment is extended to reflection and refraction of plane waves obliquely incident on boundaries, including cases of total internal reflection with surface waves. Then diffraction is considered with examples of diffraction behind knife-edge obstacles.† Finally, the difference between physical optics (wave-front) and geometrical optics (ray-path) methods is discussed.

Although there is a simple analogy between waves at normal incidence and transmission lines as shown in Chap. 10, there is no simple analogy for oblique incidence and diffraction, and accordingly these topics are dealt with entirely by field theory.

12-2 OBLIQUE INCIDENCE

Consider a linearly polarized plane wave obliquely incident on a boundary between two media, as shown in Fig. 12-1. The incident wave (from medium 1) makes an angle of θ_i with the y axis, the *reflected wave* (medium 1) makes an angle of θ_r with the y axis, and the *transmitted wave*‡ makes an angle of θ_t with the negative y axis.

Consider two cases: (1) the electric field perpendicular to the *plane of incidence* (the xy plane) and (2) the electric field parallel to the plane of incidence. These

---

† Put simply, the difference between *diffraction* and *refraction* may be stated in connection with a transparent prism: Waves bent in traveling through the prism are *refracted* while waves scattered off the edges are *diffracted*.

‡ The transmitted wave is also called the *refracted wave*. Hence, θ_t is called the *angle of refraction*.

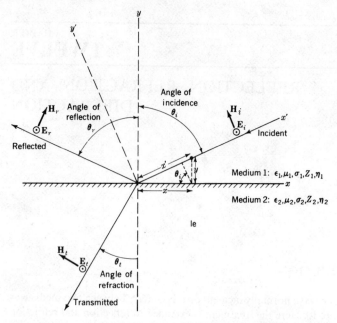

Figure 12-1 Geometry in the plane of incidence (xy plane) for linearly polarized wave at oblique incidence and for perpendicular polarization. The z direction is outward from the page.

waves are said to be *perpendicularly polarized* and *parallel polarized*, respectively. The field vectors shown in Fig. 12-1 are for the case of *perpendicular polarization*. It is clear that any arbitrary plane wave can be resolved into perpendicular and parallel components.

Referring to Fig. 12-1, the $x'y'$ axes are orthogonal, with the x' axis oriented along the direction of the incident wave. It is seen that

$$x' = x \sin \theta_i + y \cos \theta_i \tag{1}$$

and that a unit vector in the y' direction can be expressed as

$$\hat{\mathbf{y}}' = -\hat{\mathbf{x}} \cos \theta_i + \hat{\mathbf{y}} \sin \theta_i \tag{2}$$

Note that the z axis (normally outward from page) is common to both coordinate systems, i.e., the z axis and the z' axis are the same.

Perpendicular Case ($\mathbf{E}_\perp$)

Consider an incident *perpendicularly* ($\perp$) *polarized wave* propagating in the negative x' direction, i.e.,

$$\mathbf{E}_i = \hat{\mathbf{z}} E_0 e^{j\beta_1 x'} \tag{3}$$

$$\mathbf{H}_i = \hat{\mathbf{y}}' \frac{E_0}{Z_1} e^{j\beta_1 x'} \tag{4}$$

Substituting (1) and (2) into (3) and (4) gives

$$\mathbf{E}_i = \hat{\mathbf{z}} E_0 \exp \left[j\beta_1(x \sin \theta_i + y \cos \theta_i) \right] \tag{5}$$

$$\mathbf{H}_i = (-\hat{\mathbf{x}} \cos \theta_i + \hat{\mathbf{y}} \sin \theta_i) \frac{E_0}{Z_1} \exp \left[j\beta_1(x \sin \theta_i + y \cos \theta_i) \right] \tag{6}$$

The reflected fields are denoted by $(\mathbf{E}_r, \mathbf{H}_r)$ and the transmitted fields by $(\mathbf{E}_t, \mathbf{H}_t)$. From (10-16-12) and (10-16-9) the reflection and transmission coefficients for a perpendicularly ($\perp$) polarized wave are (at origin)

$$\rho_\perp = \frac{E_r}{E_i} \tag{7}$$

$$\tau_\perp = \frac{E_t}{E_i} \tag{8}$$

By developing relations similar to (1) and (2) for the reflected and transmitted waves, it follows that

$$\mathbf{E}_r = \hat{\mathbf{z}} \rho_\perp E_0 \exp \left[j\beta_1(x \sin \theta_r - y \cos \theta_r) \right] \tag{9}$$

$$\mathbf{H}_r = (\hat{\mathbf{x}} \cos \theta_r + \hat{\mathbf{y}} \sin \theta_r) \rho_\perp \frac{E_0}{Z_1} \exp \left[j\beta_1(x \sin \theta_r - y \cos \theta_r) \right] \tag{10}$$

$$\mathbf{E}_t = \hat{\mathbf{z}} \tau_\perp E_0 \exp \left[j\beta_2(x \sin \theta_t + y \cos \theta_t) \right] \tag{11}$$

$$\mathbf{H}_t = (-\hat{\mathbf{x}} \cos \theta_t + \hat{\mathbf{y}} \sin \theta_t) \tau_\perp \frac{E_0}{Z_2} \exp \left[j\beta_2(x \sin \theta_t + y \cos \theta_t) \right] \tag{12}$$

The boundary conditions are as follows: (1) the tangential components of the electric fields in both media must be equal at $y = 0$, and (2) the tangential components of the magnetic fields in both media must be equal at $y = 0$. Condition (1) can be written using (5), (9), and (11), to yield

$$\exp \left(j\beta_1 x \sin \theta_i \right) + \rho_\perp \exp \left(j\beta_1 x \sin \theta_r \right) = \tau_\perp \exp \left(j\beta_2 x \sin \theta_t \right) \tag{13}$$

From (10-16-12a), $1 + \rho_\perp = \tau_\perp$. It follows that the exponential arguments are all equal, i.e.,

$$\beta_1 \sin \theta_i = \beta_1 \sin \theta_r = \beta_2 \sin \theta_t \tag{14}$$

From the first equality,

$$\boxed{\theta_r = \theta_i} \tag{15}$$

i.e., *the angle of reflection is equal to the angle of incidence.*† From the second equality and using (15),

$$\boxed{\sin \theta_t = \frac{\eta_1}{\eta_2} \sin \theta_i} \tag{16}$$

---

† This is sometimes called *Snell's law of reflection.*

where η_1 and η_2 are the indexes of refraction of medium 1 and medium 2, respectively. Equation (16) is known as *Snell's law*† and is a relation of fundamental importance in geometrical optics. For a lossless medium the index of refraction η can be written as equal to $\sqrt{\mu_r \epsilon_r}$, and Snell's law becomes

$$\sin \theta_t = \sqrt{\frac{\mu_1 \epsilon_1}{\mu_2 \epsilon_2}} \sin \theta_i \tag{17}$$

Example 1: Polystyrene-air interface Polystyrene has a relative permittivity of 2.7. If a wave is incident at an angle of $\theta_i = 30°$ from air onto polystyrene, calculate the angle of transmission θ_t. Interchange polystyrene and air and repeat the calculation.

SOLUTION From air onto polystyrene, $\epsilon_1 = \epsilon_0$, $\mu_1 = \mu_0$, $\epsilon_2 = 2.7\epsilon_0$, and $\mu_2 = \mu_0$. From (16)

$$\sin \theta_t = \sqrt{\frac{1}{2.7}} (0.5) = 0.304$$

$$\theta_t = 17.7°$$

From polystyrene onto air, $\epsilon_1 = 2.7\epsilon_0$, $\mu_1 = \mu_0$, $\epsilon_2 = \epsilon_0$, $\mu_2 = \mu_0$.

$$\sin \theta_t = \sqrt{2.7} (0.5) = 0.822$$

$$\theta_t = 55.2°$$

Boundary condition (2) can be written using (6), (10), (12), and (14) with $y = 0$ to yield

$$-\cos \theta_i + \rho_\perp \cos \theta_i = -\tau_\perp \frac{Z_1}{Z_2} \cos \theta_t \tag{18}$$

but from (10-16-12a)

$$1 + \rho_\perp = \tau_\perp \tag{19}$$

and on substituting (19) into (18) and solving for the Fresnel reflection coefficient $\rho_\perp$, we have

$$\rho_\perp = \frac{Z_2 \cos \theta_i - Z_1 \cos \theta_t}{Z_2 \cos \theta_i + Z_1 \cos \theta_t} \tag{20}$$

where Z_1 and Z_2 are the impedances of medium 1 and medium 2, respectively. It is seen that the previously derived reflection coefficient for normal incidence, (10-16-12), is obtained as a special case of (20) when $\theta_i = 0$.

---

† This is sometimes called *Snell's law of refraction*.

If medium 2 is a perfect conductor, $Z_2 = 0$ and $\rho_\perp = -1$. If both media are lossless nonmagnetic dielectrics, (20) becomes

$$\rho_\perp = \frac{\cos \theta_i - \sqrt{(\epsilon_2/\epsilon_1) - \sin^2 \theta_i}}{\cos \theta_i + \sqrt{(\epsilon_2/\epsilon_1) - \sin^2 \theta_i}} \tag{21}$$

Provided medium 2 is a more dense dielectric than medium 1 ($\epsilon_2 > \epsilon_1$), the quantity under the square root will be positive and $\rho_\perp$ will be real. If, however, the wave is incident from the more dense medium onto the less dense medium ($\epsilon_1 > \epsilon_2$), and if $\sin^2 \theta_i \geq \epsilon_2/\epsilon_1$, then $\rho_\perp$ becomes complex and $|\rho_\perp| = 1$. Under these conditions, the incident wave is *totally internally reflected* back into the more dense medium.† The incident angle for which $\rho_\perp = 1\underline{/0°}$ is called the *critical angle* θ_{ic}. From (21) it is seen that this happens when the radical is zero; so that

$$\theta_{ic} = \sin^{-1} \sqrt{\frac{\epsilon_2}{\epsilon_1}} \tag{22}$$

defines the critical angle. For all angles greater than the critical angle, $|\rho_\perp| = 1$. Using Snell's law, it is seen that when $\theta_i > \theta_{ic}$, then $\sin \theta_t > 1$,‡ and $\cos \theta_t$ must be imaginary, i.e.,

$$\cos \theta_t = \sqrt{1 - \sin^2 \theta_t} = jA \tag{23}$$

where $A = \sqrt{(\epsilon_1/\epsilon_2) \sin^2 \theta_i - 1}$ is a real number. The electric field in the less dense medium can now be written, from (11), as

$$\mathbf{E}_t = \hat{z} \tau_\perp E_0 \exp(-\alpha y) \exp(j\beta_2 x \sin \theta_t) \tag{24a}$$

where

$$\alpha = \beta_2 A = \omega \sqrt{\mu_2 \epsilon_2} \sqrt{\frac{\epsilon_1}{\epsilon_2} \sin^2 \theta_i - 1} \tag{24b}$$

Thus, $E_\perp$ in the less dense medium has a magnitude $\tau_\perp E_0$, decaying exponentially away from the surface (y direction) and propagating without loss in the $-x$ direction. Waves whose fields are of the form of (24a) are called *surface waves*. These results can be summarized by the *Principle of Total Internal Reflection* as follows. *When a wave is incident from the more dense onto the less dense medium at an angle equal to or exceeding the critical angle, the wave will be totally internally reflected and will also be accompanied by a surface wave in the less dense medium.*

Example 2: Total internal reflection with surface wave Referring to Fig. 12-2, a plane wave is incident from water onto the water-air interface at 45°. Calculate the magnitude of the electric field in air (a) at the interface and

---

† It can be shown that this is true for either perpendicular or parallel polarization.

‡ Although the sine is greater than unity and the cosine is imaginary, there is a simple interpretation of the resulting field in such a case. See Example 2.

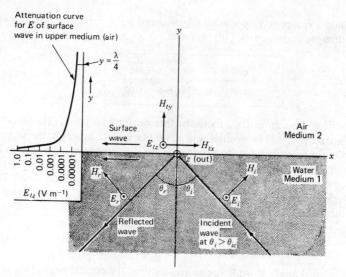

Figure 12-2 Total internal reflection of incident wave with accompanying surface wave which attenuates exponentially above surface (y direction) as shown by graph at left. No power is transmitted in y direction.

(b) $\lambda/4$ above the surface if the incident electric field $E = 1\,\mathrm{V\,m^{-1}}$. Take the water constants to be those of distilled water: $\epsilon_r = 81$, $\mu_r = 1$, $\sigma \simeq 0$.

SOLUTION From (22) the critical angle

$$\theta_{ic} = \sin^{-1}\sqrt{\frac{1}{81}} = 6.38°$$

Thus, the angle of incidence $\theta_i(= 45°)$ exceeds the critical angle and the wave will be totally internally reflected (see Fig. 12-2). From (17)

$$\sin\theta_t = \sqrt{81}\,(0.707) = 6.36$$

From (23)

$$\cos\theta_t = jA = \sqrt{1 - 6.36^2} = j6.28$$

From (24b)

$$\alpha = \beta_2 A = \frac{2\pi}{\lambda_0} 6.28 = \frac{39.49}{\lambda_0}\,\mathrm{Np\,m^{-1}}$$

From (19) and (21)

$$\tau_\perp = 1 + \rho_\perp = 1 + \frac{0.707 - \sqrt{\frac{1}{81} - 0.5}}{0.707 + \sqrt{\frac{1}{81} - 0.5}} = 1.42\underline{/-44.64°}$$

Therefore, the magnitude of the field strength is:

(a)   at the interface: $|E_t| = 1.42$ V m^{-1}

(b) $\lambda/4$ away from the interface:

$$|E_t| = 1.42 \exp\left(-\frac{39.49\, \lambda_0}{\lambda_0}\frac{}{4}\right) = 73.2 \ \mu\text{V m}^{-1}$$

Thus, the field $\lambda/4$ above the surface is

$$20 \log \frac{73.2 \times 10^{-6}}{1.42} = -85.8 \text{ dB}$$

less than the field at the surface. Recalling that a power ratio of 1 billion equals 90 dB, it is evident that the field attenuates very rapidly above the surface (in the y direction) (see Fig. 12-2), meaning that the surface wave is very tightly bound to the water surface. Note that $\sin \theta_t$ is greater than 1 but real, while $\cos \theta_t$ is imaginary. From (24)

$$\mathbf{E}_t = \hat{\mathbf{z}}\tau_\perp E_0 e^{-(\beta_2 A)y}e^{j\beta_2 x \sin \theta_t} \tag{25a}$$

and from (12)

$$\mathbf{H}_t = (-\hat{\mathbf{x}}jA + \hat{\mathbf{y}} \sin \theta_t)\tau_\perp \frac{E_0}{Z_2} e^{-(\beta_2 A)y}e^{j\beta_2 x \sin \theta_t} \tag{25b}$$

where $A = -\sqrt{\sin^2 \theta_t - 1}$.

From (25a) and (25b) the average Poynting vector of the wave in the y direction in air (above the water surface) is

$$S_{y(\text{av})} = \tfrac{1}{2} \text{Re } \mathbf{E} \times \mathbf{H}^* = \hat{\mathbf{y}}\tfrac{1}{2}E_{tz}H_{tx} \sin \phi \cos \theta \tag{25c}$$

where ϕ = space angle between $\mathbf{E}$ and $\mathbf{H}$ ($= 90°$)
θ = time phase angle between $\mathbf{E}$ and $\mathbf{H}$

The exponentials in (25a) and (25b) are identical. However, H_{tx} has a j factor whereas E_{tz} does not, indicating a 90° time-phase difference between $\mathbf{E}$ and $\mathbf{H}$. Hence, $\theta = 90°$ and since $\sin \phi = \sin 90° = 1$

$$S_{y(\text{av})} = \tfrac{1}{2}E_{tz}H_{tx} \cos 90° = 0 \tag{25d}$$

Thus, no power is transmitted in the y direction (wave reactive). Both E_t and H_t decay exponentially with y. Similar waves, called *evanescent waves*, exist in hollow conducting waveguides at wavelengths too long to propagate through the guide (see Chap. 13).

The waves in medium 2 (air) involving E_{tz} and H_{ty} propagate without attenuation as a surface wave in the $-x$ direction with a velocity v_x equal to the wave velocity in the water (medium 1) as observed parallel to the x axis ($v_x = v_\text{water}/\sin \theta_i$). The traveling wave is simply the matching field at the boundary. Total internal reflection with a surface wave can also occur for $E_\parallel$, but the details differ.

Parallel Case ($E_\parallel$)

Consider now the case of *parallel* ($\parallel$) *polarization*. The geometry is the same as in Fig. 12-1 but with $\mathbf{E}_i$, $\mathbf{E}_r$, and $\mathbf{E}_t$ parallel to the plane of incidence as would be obtained by replacing $\mathbf{H}_i$ by $\mathbf{E}_i$, $\mathbf{H}_r$ by $\mathbf{E}_r$, and $\mathbf{H}_t$ by $\mathbf{E}_t$. It can be shown that the fields are given by

$$\mathbf{E}_i = (-\hat{\mathbf{x}} \cos \theta_i + \hat{\mathbf{y}} \sin \theta_i)E_0 \exp \left[j\beta_1(x \sin \theta_i + y \cos \theta_i) \right] \tag{26}$$

$$\mathbf{H}_i = -\hat{\mathbf{z}} \frac{E_0}{Z_1} \exp \left[j\beta_1(x \sin \theta_i + y \cos \theta_i) \right] \tag{27}$$

$$\mathbf{E}_r = (-\hat{\mathbf{x}} \cos \theta_r - \hat{\mathbf{y}} \sin \theta_r)\rho_\parallel E_0 \exp \left[j\beta_1(x \sin \theta_r - y \cos \theta_r) \right] \tag{28}$$

$$\mathbf{H}_r = \hat{\mathbf{z}}\rho_\parallel \frac{E_0}{Z_1} \exp \left[j\beta_1(x \sin \theta_r - y \cos \theta_r) \right] \tag{29}$$

$$\mathbf{E}_t = (-\hat{\mathbf{x}} \cos \theta_t + \hat{\mathbf{y}} \sin \theta_t)\tau_\parallel E_0 \exp \left[j\beta_2(x \sin \theta_t + y \cos \theta_t) \right] \tag{30}$$

$$\mathbf{H}_t = -\hat{\mathbf{z}}\tau_\parallel \frac{E_0}{Z_2} \exp \left[j\beta_2(x \sin \theta_t + y \cos \theta_t) \right] \tag{31}$$

By matching boundary conditions, as before, it is found that the angle of incidence equals the angle of reflection and that Snell's law (16) holds. It can also be shown that

$$1 + \rho_\parallel = \frac{\cos \theta_t}{\cos \theta_i} \tau_\parallel \tag{32}$$

The Fresnel reflection coefficient is found to be

$$\rho_\parallel = \frac{Z_2 \cos \theta_t - Z_1 \cos \theta_i}{Z_1 \cos \theta_i + Z_2 \cos \theta_t} \tag{33}$$

which for lossless nonmagnetic dielectrics becomes

$$\rho_\parallel = \frac{-(\epsilon_2/\epsilon_1) \cos \theta_i + \sqrt{(\epsilon_2/\epsilon_1) - \sin^2 \theta_i}}{(\epsilon_2/\epsilon_1) \cos \theta_i + \sqrt{(\epsilon_2/\epsilon_1) - \sin^2 \theta_i}} \tag{34}$$

and reduces to $\rho_\parallel = -1$ if medium 2 is a perfect conductor.

It is of especial interest that, for parallel polarization, it is possible to find an incident angle so that $\rho_\parallel = 0$ and the wave is *totally transmitted* into medium 2. This angle, called the *Brewster angle* θ_{iB}, can be found by setting the numerator of (34) equal to zero, giving

$$\theta_{iB} = \sin^{-1} \sqrt{\frac{\epsilon_2/\epsilon_1}{1 + (\epsilon_2/\epsilon_1)}} = \tan^{-1} \sqrt{\frac{\epsilon_2}{\epsilon_1}} \tag{35}$$

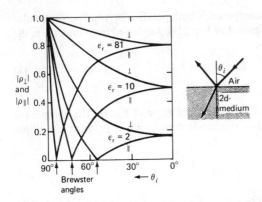

Figure 12-3 Magnitude of reflection coefficient for perpendicular ($\perp$) and parallel ($\parallel$) polarization vs. angle of incidence θ_i for distilled water ($\epsilon_r = 81$), flint glass ($\epsilon_r = 10$), and paraffin ($\epsilon_r = 2$). The coefficient $\rho_{\parallel} = 0$ at the Brewster angle; it also reverses sign at the Brewster angle (see Example 3).

The Brewster angle is also sometimes called the *polarizing angle* since a wave composed of both perpendicular and parallel components and incident at the Brewster angle produces a reflected wave with only a perpendicular component. Thus, a circularly polarized wave incident at the Brewster angle becomes linearly polarized on reflection.

Example 3 A parallel-polarized wave is incident from air onto (*a*) distilled water ($\epsilon_r = 81$), (*b*) flint glass ($\epsilon_r = 10$), and (*c*) paraffin ($\epsilon_r = 2$). Find the Brewster angle for each of these cases.

SOLUTION

$$(a) \qquad \theta_{iB} = \tan^{-1}\sqrt{81} = 83.7°$$

$$(b) \qquad \theta_{iB} = \tan^{-1}\sqrt{10} = 72.4°$$

$$(c) \qquad \theta_{iB} = \tan^{-1}\sqrt{2} = 54.7°$$

The reflection coefficients $\rho_{\perp}$ and ρ are shown in Fig. 12-3 for waves incident from air onto distilled water ($\epsilon_r = 81$), flint glass ($\epsilon_r = 10$), and paraffin ($\epsilon_r = 2$). The phase angle for $\rho_{\perp}$ is always π, whereas for $\rho_{\parallel}$ the phase angle jumps from 0 to π at the Brewster angle. Furthermore, the reflection coefficients have unit magnitude at grazing incidence ($\theta_i = 90°$).

The quantities $\rho_{\parallel}$ and $\rho_{\perp}$ as given in (21) and (34) are sometimes called the *Fresnel* (fray-nel') *reflection coefficients*.

12-3 ELLIPTICALLY POLARIZED PLANE WAVE, OBLIQUE INCIDENCE

We now consider an elliptically polarized plane wave obliquely incident on a boundary. The problem is to find the magnitude and polarization of the reflected and transmitted waves.

The incident electric field is composed of both parallel ($E_{i\parallel}$) and perpendicular ($E_{i\perp}$) components, as shown in inset A of Fig. 12-4. As viewed from the origin, the orthogonal field components are

$$E_z = E_{i\perp} \tag{1}$$

$$E_{y'} = E_{i\parallel} e^{j\delta_i} \tag{2}$$

where δ_i is the angle by which E_y' leads E_z and $E_{i\perp}$ and $E_{i\parallel}$ are the component amplitudes. Thus, as in Sec. 11-4, the incident wave has a polarization ellipse specified by (γ_i, δ_i), where

$$\gamma_i = \tan^{-1} \frac{E_{i\parallel}}{E_{i\perp}} \tag{3}$$

The polarization ellipse can also be specified in terms of its axial ratio AR_i and tilt angle τ_i. From (11-4-3) we can relate (AR_i, τ_i) to (γ_i, δ_i):

$$\tan 2\tau_i = \tan 2\gamma_i \cos \delta_i \tag{4}$$

$$\sin 2\epsilon_i = \sin 2\gamma_i \sin \delta_i \tag{5}$$

$$\cos 2\gamma_i = \cos 2\epsilon_i \cos 2\tau_i \tag{6}$$

$$\tan \delta_i = \frac{\tan 2\epsilon_i}{\sin 2\tau_i} \tag{7}$$

where
$$\epsilon_i = \cot^{-1}(\mp AR_i) \tag{8}$$

If the wave is right elliptically polarized, the minus sign is used; if it is left elliptically polarized, the plus sign is used.

Considering the reflected wave as viewed from point P (see inset B), the orthogonal field components are

$$E_z = E_{r\perp} = |\rho_\perp| E_{i\perp} \exp(j\phi_\perp) \tag{9}$$

$$E_{y''} = -E_{r\parallel} = |\rho_\parallel| E_{i\parallel} \exp[j(\phi_\parallel + \delta_i + \pi)] \tag{10}$$

where $\phi_\parallel$ and $\phi_\perp$ are the phase angles of the parallel and perpendicular reflection coefficients $\rho_\parallel$ and $\rho_\perp$, respectively. The phase angle by which $E_{y''}$ leads E_z is thus given by

$$\delta_r = \delta_i + \pi + (\phi_\perp - \phi_\parallel) \tag{11}$$

and similarly [see (3)]

$$\gamma_r = \tan^{-1}\left(\frac{|\rho_\parallel|}{|\rho_\perp|} \tan \gamma_i\right) \tag{12}$$

The relations (11) and (12) specify the polarization state of the reflected wave, as seen by an observer at point P. To find the axial ratio and tilt angle of the polarization ellipse of the reflected wave, the relations (4), (5), and (8) are used with the subscript i replaced by the subscript r.

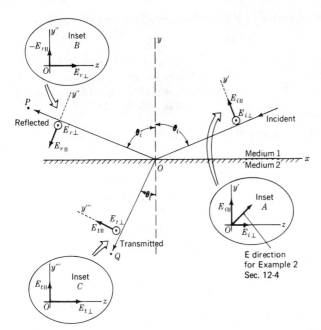

Figure 12-4 Geometry in the plane of incidence (xy plane) for elliptically polarized wave incident obliquely on a plane surface. The incident electric field has parallel ($E_{i\parallel}$) and perpendicular ($E_{i\perp}$) components which appear as in inset A when viewed from the origin 0. The reflected and transmitted fields shown in insets B and C are viewed looking toward the origin from points P and Q, respectively.

In similar fashion, it can be shown that an observer at point Q (see inset C) will see a transmitted wave with polarization ellipse given by

$$\delta_t = \delta_i + (\xi_\perp - \xi_\parallel) \tag{13}$$

$$\gamma_t = \tan^{-1}\left(\frac{|\tau_\parallel|}{|\tau_\perp|}\tan\gamma_i\right) \tag{14}$$

where $\xi_\parallel$ and $\xi_\perp$ are the phase angles of the parallel and perpendicular transmission coefficients $\tau_\parallel$ and $\tau_\perp$, respectively.†

Example 1: Circularly polarized wave reflection A right circularly polarized wave (RCP) is incident at an angle of 45° from air onto (a) a perfect conductor and (b) polystyrene ($\epsilon_r = 2.7$). What is the polarization state of the reflected wave for these two cases?

---

† Note that when τ is used in this section to denote the *tilt angle*, it is associated with the incident, reflected, or transmitted wave (that is, τ_i, τ_r, or τ_t), but when it is used to denote the *transmission coefficient*, it is associated with either a parallel or perpendicular case (that is, $\tau_\parallel$ or $\tau_\perp$).

SOLUTION (a) When medium 1 is air and medium 2 is a perfect conductor,

$$\rho_{\parallel} = 1\underline{/180°} \qquad \rho_{\perp} = 1\underline{/180°}$$

For a right circularly polarized wave, $\gamma_i = 45°$, $\delta_i = -90°$, so that from (11), $\delta_r = -90° + 180° + (180° - 180°) = 90°$ and from (12), $\gamma_r = \tan^{-1}(\frac{1}{1} \times 1) = 45°$. Therefore the reflected wave is *left circularly polarized* (LCP), as shown in Fig. 12-5a. Conversely, if the incident wave had been LCP, the reflected wave would be RCP.

(b) When medium 1 is air and medium 2 is polystyrene, *then* from (12-2-34)

$$\rho_{\parallel} = \frac{-2.7(0.707) + \sqrt{2.7 - 0.5}}{2.7(0.707) + \sqrt{2.7 - 0.5}} = 0.126\underline{/180°}$$

and from (12-2-21)

$$\rho_{\perp} = \frac{0.707 - \sqrt{2.7 - 0.5}}{0.707 + \sqrt{2.7 - 0.5}} = 0.354\underline{/180°}$$

Therefore,

$$\delta_r = -90° + 180° + (180° - 180°) = 90°$$

$$\gamma_r = \tan^{-1}\frac{0.126}{0.354} = 19.6°$$

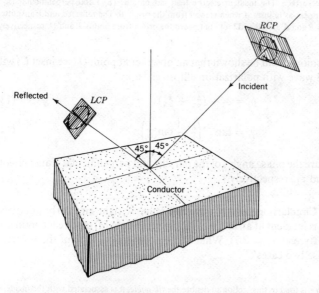

Figure 12-5a (Example 1a) Right circularly polarized (RCP) wave incident on a perfect conductor. The reflected wave is left circularly polarized (LCP). No wave is refracted (transmitted).

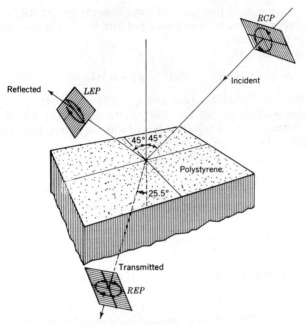

Figure 12-5b (Example 1b) Right circularly polarized (RCP) wave incident on dielectric slab of polystyrene. The reflected wave is left elliptically polarized (LEP) with major axis of the polarization ellipse horizontal and the transmitted wave right elliptically polarized (REP) with major axis of the polarization ellipse in the (vertical) plane of incidence.

If the angle of incidence is increased to 58.7° (= Brewster angle), the angle of reflection increases to 58.7° and the angle of refraction increases from 25.5 to 31.3°. Also the polarization ellipse of the reflected wave collapses to a straight horizontal line (linear $\perp$ polarization).

Substituting these values into (4), (5), and (8) (with subscripts i replaced by r), we find $\tau_r = 0°$ and $AR_r = 2.81$. Thus the reflected wave is left elliptically polarized (LEP), as shown in Fig. 12-5b.

Example 2: Linearly polarized wave with equal $E_\perp$ and $E_\parallel$ components A linearly polarized wave whose electric field vector bisects the zy' axis (see Fig. 12-4, inset A) is incident at an angle of 45° from air onto (a) a perfect conductor, and (b) polystyrene ($\epsilon_r = 2.7$). What is the polarization state of the reflected wave for these two cases? (c) If the angle of incidence onto polystyrene is 58.7° (Brewster angle), what is the polarization state of the reflected wave?

SOLUTION (a) Since $\rho_\parallel = 1\underline{/180°}$ and $\rho_\perp = 1\underline{/180°}$,

$$\delta_r = 0° + 180° + (180° - 180°) = 180°$$

$$\gamma_r = \tan^{-1}(\tfrac{1}{1} \times 1) = 45°$$

From (4), $\tan 2\tau_r = -\infty$, so that $\tau_r = 135°$; and from (5) and (8), $AR_r = \infty$. Therefore the reflected wave is *linearly polarized* with the electric field vector bisecting the $-z$ and y'' axes.

(b) From Example 1,

$$\rho_{\parallel} = 0.126\underline{/180°} \qquad \text{and} \qquad \rho_{\perp} = 0.354\underline{/180°}$$

so that $\gamma_r = \tan^{-1}(0.126/0.354) = 19.6°$ and $\delta_r = 180°$. From (4), $\tan 2\tau_r = \tan 39.2° \cos \pi = -0.815$, so that $\tau_r = 70.4°$. From (5) and (8), $AR_r = \infty$. Therefore the reflected wave is *linearly polarized* with the electric field vector making an angle in the z-y'' plane of $70.4°$ with the z axis.

(c) When $\theta_i = \theta_{iB} = 58.7°$.

$$\rho_{\parallel} = 0$$

$$\rho_{\perp} = \frac{0.520 - 1.40}{0.520 + 1.40} = 0.46\underline{/180°}$$

Therefore the reflected wave is perpendicularly polarized.

12-4 HUYGENS' PRINCIPLE AND PHYSICAL OPTICS

Huygens' principle[†] states that *each point on a primary wavefront can be considered to be a new source of a secondary spherical wave and that a secondary wavefront can be constructed as the envelope of these secondary spherical waves*, as suggested in Fig. 12-6. This fundamental principle of physical optics can be used to explain the apparent bending of radio waves around obstacles, i.e., the *diffraction* of waves. A *diffracted* ray is one that follows a path that cannot be interpreted as either reflection or refraction.

As an example, consider a uniform plane wave incident on a conducting half-plane, as in Fig. 12-7a.[‡] We want to calculate the electric field at point P by using Huygens' principle; i.e.,

$$E = \int_{\substack{\text{over} \\ x \text{ axis}}} dE \tag{1}$$

where dE is the electric field at P due to a point source at a distance x from the origin O, as in Fig. 12-7b; i.e.,

$$dE = \frac{E_0}{r} e^{-j\beta(r+\delta)} dx \tag{2}$$

so that

$$E = \frac{E_0}{r} e^{-j\beta r} \int_a^\infty e^{-j\beta\delta} dx \tag{3}$$

---

† C. Huygens, "Traité de la lumière," Leyden, 1690. English translation by S. P. Thompson, London, 1912, reprinted by The University of Chicago Press.

‡ J. D. Kraus, "Radio Astronomy," pp. 194–198, Cygnus-Quasar Books, Powell, Ohio, 1982.

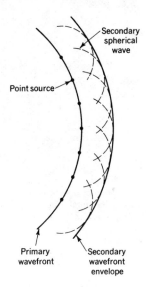

Figure 12-6 Illustrating Huygens' principle of physical optics (point-to-wave correspondence).

If $\delta \ll r$, it follows that

$$\delta = \frac{x^2}{2r} \tag{4}$$

When we let $\kappa^2 = 2/r\lambda$ and $u = \kappa x$, (3) becomes

$$E = \frac{E_0}{\kappa r} e^{-j\beta r} \int_{\kappa a}^{\infty} e^{-j\pi u^2/2}\, du \tag{5}$$

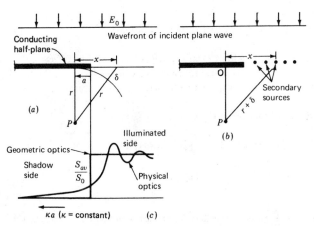

Figure 12-7 Plane wave incident from above onto a conducting half-plane with resultant power-density variation below the plane as obtained by physical optics.

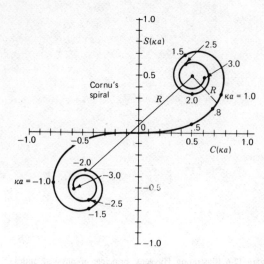

Figure 12-8 Cornu's spiral showing $C(\kappa a)$ and $S(\kappa a)$ as a function of κa values along the spiral. For example, when $\kappa a = 1.0$, $C(\kappa a) = 0.780$ and $S(\kappa a) = 0.338$. When $\kappa a = \infty$, $C(\kappa a) = S(\kappa a) = \frac{1}{2}$.

which can be rewritten as

$$E = \frac{E_0}{\kappa r} e^{-j\beta r} \left(\int_0^\infty e^{-j\pi u^2/2} \, du - \int_0^{\kappa a} e^{-j\pi u^2/2} \, du \right) \tag{6}$$

The integrals in (6) have the form of Fresnel integrals† so that (6) can be written

$$E = \frac{E_0}{\kappa r} e^{-j\beta r} \{ \tfrac{1}{2} + j\tfrac{1}{2} - [C(\kappa a) + jS(\kappa a)] \} \tag{7}$$

where

$$C(\kappa a) = \int_0^{\kappa a} \cos \frac{\pi u^2}{2} \, du = \text{Fresnel cosine integral} \tag{8}$$

$$S(\kappa a) = \int_0^{\kappa a} \sin \frac{\pi u^2}{2} \, du = \text{Fresnel sine integral} \tag{9}$$

A graph of $C(\kappa a)$ and $S(\kappa a)$ yields *Cornu's spiral* (Fig. 12-8). Since $C(-\kappa a) = -C(\kappa a)$, and $S(-\kappa a) = -S(\kappa a)$, the spiral for negative values of κa is in the third quadrant and is symmetrical with respect to the origin for the spiral in the first quadrant.

The power density as a function of κa is then

$$S_{av} = \frac{EE^*}{2Z} = S_0 \tfrac{1}{2} \{ [\tfrac{1}{2} - C(\kappa a)]^2 + [\tfrac{1}{2} - S(\kappa a)]^2 \} \quad (\text{W m}^{-2}) \tag{10}$$

where

$$S_0 = \frac{E_0^2 \lambda}{2Zr} \quad (\text{W m}^{-2}) \tag{11}$$

---

† See M. Abramowitz and I. A. Stegun, "Handbook of Mathematical Functions," pp. 295–329, National Bureau of Standards, U.S. Government Printing Office, 1964. E. Jahnke and F. Emde, "Tables of Functions," pp. 35–38, Dover Publications, New York, 1943.

The power density variation (10) as a function of κa (with r, λ, and κ constant) is shown in Fig. 12-7c. Note that when $\kappa a = -\infty$, which corresponds to no edge present, $S_{av} = S_0$; when $\kappa a = 0$, $S_{av} = S_0/4$; and when $\kappa a = +\infty$, which corresponds to complete obscuration of the plane wave by the half-plane, $S_{av} = 0$. Furthermore, the power density does not abruptly go to zero as the point of observation goes from the illuminated side ($\kappa a < 0$) to the shadow side ($\kappa a > 0$); rather, there are fluctuations followed by a gradual decrease in power density.

From (10) and (11) the relative power density as a function of κa is

$$S_{av}(\text{rel}) = \frac{S_{av}}{S_0} = \tfrac{1}{2}\{[\tfrac{1}{2} - C(\kappa a)]^2 + [\tfrac{1}{2} - S(\kappa a)]^2\} \tag{12}$$

The relative power density (12) is equal to $\tfrac{1}{2}R^2$, where R is the distance from a κa value on the Cornu spiral to the point $(\tfrac{1}{2}, \tfrac{1}{2})$. (See Fig. 12-8.) For large positive values of κa, R approaches $1/\pi\kappa a$, so that (12) reduces approximately to

$$S_{av}(\text{rel}) = \frac{1}{2}\left(\frac{1}{\pi\kappa a}\right)^2 = \frac{r\lambda}{4\pi^2 a^2} \tag{13}$$

where r = distance from obstacle (conducting half-plane), m
   λ = wavelength, m
   a = distance into shadow region, m

Equation (13) gives the relative power density for large κa (>3) (well into the shadow region). For this condition it is apparent that the power flux density (Poynting vector) due to diffraction increases with wavelength and with distance r (below edge) but decreases as the square of the distance a into the shadow region.

Example 1: Radio wave diffraction over a mountain   Referring to Fig. 12-9, the direct path between a transmitting antenna and a receiving antenna is blocked by a mountain. Assuming that the ridge of the mountain acts as a knife edge, calculate the level of the signal diffracted over the mountain with respect to the direct path level at a frequency of 30 MHz. Transmitting and receiving antennas are separated by 10 km. The mountain ridge extends 1 km above the reference baseline (see Fig. 12-9).

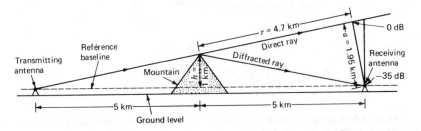

Figure 12-9 Radio-wave diffraction over knife-edge obstacle (mountain). The diffracted ray is 35 dB below the direct ray level.

SOLUTION From geometry the receiving antenna distance r behind the mountain ridge is 4.7 km and the distance a into the shadow region is 1.95 km. From $\lambda = c/f$, the wavelength is 10 m. From (13)

$$S_{av}(rel) = \frac{r\lambda}{4\pi^2 a^2} = \frac{4700 \times 10}{4\pi^2 1950^2} = \frac{1}{3194} \quad \text{or} -35 \text{ dB}$$

Thus, the mountain causes 35-dB attenuation as compared to a direct path signal. Since an actual mountain ridge may not be knife-edge sharp with respect to dimensions of the order of a wavelength ($= 10$ m), the actual attenuation may be 10 or 20 dB more (rounded edges diffract less power into shadow region than sharp edges).

Example 2: Diffraction behind a conducting strip Referring to Fig. 12-10, a conducting strip of width D ($D \gg 2$) acts as an obstacle to a plane wave traveling downward from above with diffraction into the region below around both right and left edges. On the centerline of the strip diffraction fields from both edges are equal in magnitude and in the same phase since the path lengths from the right and left edges are equal ($r_r = r_l$). Thus, the diffraction field has a maximum or central peak on the centerline. To either side, when r_r and r_l differ by one wavelength or a multiple thereof, the fields from both edges will also be in phase and there will be a maximum, but at intermediate points where r_r and r_l differ by an odd multiple of $\lambda/2$, the fields will be 180° out of phase and there will be a minimum. The result is an interference pattern, as suggested in Fig. 12-10. The pattern is symmetrical around the central peak.

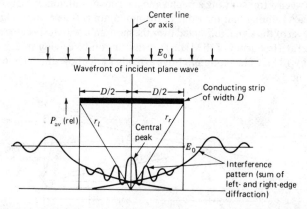

Figure 12-10 Plane wave incident on conducting strip of width D produces diffraction field below the strip with a central peak and interference pattern. The relative power-density scale is not linear and the magnitude of the weak interference pattern has been exaggerated to show the lobe pattern more clearly.

If the strip in Fig. 12-10 is replaced by a disk (diameter $= D$) with axis on the centerline, there will be diffraction around its entire edge and all diffraction fields will arrive in phase on the centerline below the disk. The sum total of these fields will produce a larger central peak. In optics this peak is called the *axial bright spot*. In a similar way the diffracted fields from the feed system at the focus of a parabolic dish antenna can produce a back lobe on the axis of the parabola (see Chap. 14).

Another illustration of Huygens' principle occurs in *holography*. In an ordinary photograph only amplitude information is recorded. In a *hologram* both amplitude information and phase information are recorded. Thus, when the hologram is illuminated with coherent light, it generates waves (in both amplitude and phase) which, in accord with Huygens' principle, produce a three-dimensional picture.†

12-5 GEOMETRICAL-OPTICS CONCEPTS

Imagine that a family of ray paths is drawn from each point on a primary wavefront to corresponding points on a secondary wavefront, as in Fig. 12-11. The ray paths are perpendicular to the wavefronts and are in the direction of the Poynting vector at each point. Geometrical-optics theory uses a *point-to-point ray correspondence* between two successive positions of a wavefront, in contrast to the physical-optics theory of Sec. 12-4, which postulates a point-to-spherical-wave correspondence. Put simply, physical optics involves *wavefronts* while geometrical optics involves *ray paths*.

To illustrate this, let us reconsider the example of Sec. 12-4, where a plane wave is incident on a straight edge, as in Fig. 12-7. According to the theory of geometrical optics, the power density on the left side of the edge will drop abruptly to zero (solid line) as an observer goes from the illuminated region to the shadow region, in contrast to the physical-optics solution (dashed line), which predicts fluctuations followed by a gradual decrease. The geometrical-optics approximation can be considered as the high-frequency limit of the physical optics approximation. To take suitable account of edge diffraction, the simple geometrical-optics concepts have been extended in a *Geometrical Theory of Diffraction* (GTD) by adding a diffraction field E_d to the simple geometrical-optics field E_G. Referring to Fig. 12-12 the diffracted field is given approximately by

$$E_d = -\tfrac{1}{2}E_0 \frac{\sqrt{\lambda\rho}}{\pi x} e^{-j\beta\rho} \qquad (- \text{ for positive } x, + \text{ for negative } x) \qquad (1)\ddagger$$

† D. Gabor, A New Microscopic Principle, *Nature*, **161**:777 (May 15, 1948); E. N. Leith and J. Upatnieks, Progress in Holography, *Phys. Today*, **25**(3):28–34 (March 1972).

‡ For $\sqrt{\dfrac{\pi^2 x^2}{\lambda\rho}} > 6$.

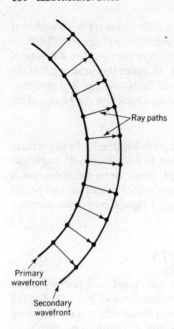

Ray paths

Primary
wavefront

Secondary
wavefront

Figure 12-11 The principle of geometrical optics (ray paths).

The simple geometrical-optics field is given by

$$E_G = E_o \qquad \text{(for } x > 0\text{)} \tag{2}$$

$$E_G = 0 \qquad \text{(for } x < 0\text{)} \tag{3}$$

The total field E_T is then the sum of E_d and E_G, or

$$E_T = E_G + E_d \tag{4}$$

giving a result which agrees with that of physical optics.

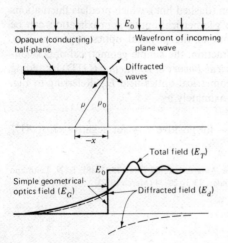

Opaque (conducting)
half-plane

E_0

Wavefront of incoming
plane wave

Diffracted
waves

ρ ρ_0

$-x$

Total field (E_T)

E_0

Simple geometrical-
optics field (E_G)

Diffracted field (E_d)

Figure 12-12 Plane wave incident on conducting half-plane with resultant fields below the plane as obtained by geometrical optics.

An advantage of GTD is that it permits an engineering approach to many diffraction problems that would be difficult or impossible to solve by physical optics or other methods. A further discussion involving GTD is given in connection with Fig. 14-47 (Sec. 14-19.2).

PROBLEMS†

Group 12-1: Secs. 21-1 through 12-3. Oblique incidence for linear and elliptical polarization

12-1-1. LP wave at angle A linearly polarized plane traveling wave in air is incident at an angle θ_i on the flat surface of a dielectric medium of large extent. The constants of the dielectric medium are $\sigma = 0$, $\mu_r = 1$, and $\epsilon_r = 8$. Calculate the magnitude of the reflected field E_r and the transmitted field E_t relative to the incident field E_i as a function of θ_i when $\mathbf{E}_i$ is (a) parallel and (b) perpendicular to the plane of incidence. (c) Prepare graphs showing E_t/E_i and E_r/E_i as ordinates vs. θ_i as abscissa. (d) What value of θ_i is the Brewster angle?

★**12-1-2. CP wave at 45°** A circularly polarized 2-GHz wave in air is incident on a half-space of lossless dielectric medium ($\epsilon_r = 3$) at 45°. Find the axial ratio AR of the reflected wave.

★**12-1-3. CP wave at Brewster angle** A circularly polarized 1-GHz wave in air is incident on a half-space of lossless dielectric medium ($\epsilon_r = 5$) at the Brewster angle. Find the axial ratio AR for (a) transmitted wave and (b) reflected wave. (c) Find angle of transmitted wave.

12-1-4. CP wave at Brewster angle. Power If the incident wave of Prob. 12-1-3 has a Poynting vector of 10 W m⁻², find the Poynting vector of (a) the reflected wave and (b) the transmitted wave.

★**12-1-5. CP wave at Brewster angle** A circularly polarized 3-GHz wave in air is incident on a lossless dielectric medium ($\epsilon_r = 6$) at the Brewster angle. Find the axial ratio AR for the (a) transmitted wave and (b) reflected wave.

★**12-1-6. CP wave and prism** A circularly polarized wave enters a 45° glass prism ($\epsilon_r = 1.5$) normally as shown in Fig. P12-1-6. Find the axial ratio AR of the wave leaving the prism.

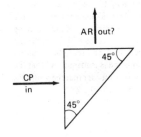

Figure P12-1-6 Circularly polarized wave and prism.

12-1-7. CP wave at 50° A right circularly polarized 4-GHz wave in air of 15 W m⁻² is incident at an angle of 50° (from the normal) on a large flat dielectric medium ($\epsilon_r = 5$). For the reflected wave find (a) AR, (b) τ, and (c) Poynting vector. For the transmitted wave find (d) AR, (e) τ, and (f) Poynting vector.

12-1-8. Wavelength change When an electromagnetic wave of frequency f and wavelength λ traveling in air enters a different medium, why does the wavelength change but not the frequency?

12-1-9. Unpolarized wave at angle (a) An *unpolarized* plane traveling wave in air is incident at an angle θ_i on the flat surface of a dielectric medium of large extent. The constants of the dielectric medium are $\sigma = 0$, $\mu_r = 1$, and $\epsilon_r = 8$. Calculate the ratio of the transmitted Poynting vector S_t and the reflected

† Answers to starred problems are given in Appendix C.

Poynting vector S_r relative to the incident Poynting vector S_i as a function of θ_i. Prepare graphs showing S_t/S_i and S_r/S_i as ordinates vs. θ_i as abscissa. (b) Is there any angle for which the reflected or transmitted wave becomes (linearly) polarized?

12-1-10 Unpolarized wave from dielectric medium (a) Repeat Prob. 12-1-9a with circumstances reversed, i.e., with the unpolarized wave originating in the dielectric medium. (b) Is there an angle or angles for which there is no transmitted wave?

Group 12-2: Secs. 12-4 through 12-5. Physical optics and geometrical optics

12-2-1. Lunar occultation. Diffraction When the moon occults (passes in front of) a distant celestial radio source, a Fresnel type of diffraction pattern is obtained, as in Fig. 12-7. (a) Calculate and plot the occultation pattern if the source subtends an angle of 8 seconds of arc. The wavelength is 1 m. Take the earth-moon distance as 380 Mm. (b) Repeat (a) for the case of a point source (zero subtended angle) and compare with the pattern for (a). Part (a) may be done as a convolution (see Kraus, "Radio Astronomy," pp. 194–198) or by superimposing the patterns of several point sources situated within an angle of 8 seconds of arc.

12-2-2. Submerged transmitter (a) A 60-MHz radio transmitter which radiates isotropically is situated 20 m below the surface of a deep freshwater lake with constants $\mu_r = 1$ and $\epsilon_r = 80 - j4$. If the power radiated is 100 W and the polarization is circular, what is the ratio of the power received per unit area at a radius of 1 km as a function of the zenith angle from 0 to 90°? The radial distance is measured from a point on the water surface directly above the transmitter. Express the power ratio in decibels and plot this ratio as ordinate vs. zenith angle as abscissa. (b) Discuss the practical aspects of radio communication over such paths. (c) Compare the situation of this problem with transmission from the earth to an extraterrestrial point above the ionosphere.

Group 12-3: Practical applications

12-3-1. Overland TV (a) A typical overland microwave communication circuit for AM, FM, or TV between a transmitter on a tall building and a distant receiver involves two paths of transmission, one a direct path (length r_0) and one an indirect path with ground reflection (length $r_1 + r_2$), as suggested in Fig. P12-3-1. Let $h_1 = 300$ m and $d = 5$ km. For a frequency of 100 MHz calculate the ratio of the power received per unit area to the transmitted power as a function of the height h_2 of the receiving antenna. Plot these results in decibels as abscissa vs. h_2 as ordinate for three cases with transmitting and receiving antennas both (a) vertically polarized, (b) horizontally polarized, and (c) right circularly polarized for h_2 values from 0 to 100 m. Assume that the transmitting antenna is isotropic and that the receiving antennas are also isotropic and all have the same effective aperture. Consider that the ground is

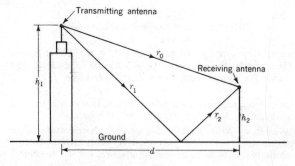

Figure P12-3-1 Microwave circuit.

flat and perfectly conducting. (b) Compare the results for the three types of polarization and show that circular polarization is best from the standpoint of both the noncriticalness of the height h_2 and of the absence of echo or ghost signals. Thus, for horizontal polarization the direct and ground-reflected waves may reinforce at a certain height h_2 (path-length difference equals an odd number of half wavelengths), but if this path-length difference yields delays of the order of microseconds, objectionable ghost images can appear on a TV screen. (c) Extend the comparison of (b) to consider the effect of other buildings or structures which may produce additional paths of transmission.

Note that direct satellite-to-earth TV downlinks are substantially free of these reflection and ghost image effects (see Chaps. 10 and 14).

12-3-2. Over-mountain TV A TV receiving antenna is 25 km from a 60-MHz transmitter. A 2-km-high mountain ridge is situated halfway between. The receiving antenna is at a height of 10 m and the transmitting antenna at a height of 200 m. Find the added attenuation caused by the mountain as compared to the signal level without the mountain.

THIRTEEN

WAVEGUIDES AND RESONATORS

13-1 INTRODUCTION

In our discussion of transmission lines in Chap. 10, the emphasis was on lines with transverse electromagnetic (TEM) waves, that is, waves with electric and magnetic fields entirely transverse to the direction of propagation. In this chapter we continue the discussion of transmission lines with emphasis on waves of higher-order modes, that is, waves having components of $\mathbf{E}$ or $\mathbf{H}$ in the direction of propagation. Transmission lines which can convey electromagnetic waves *only* in higher-order modes are usually called *waveguides* or simply *guides*.

The infinite-parallel-plane transmission line or guide will be discussed first, leading to the hollow rectangular and cylindrical waveguides and their field configurations, cutoff wavelengths, and attenuation. Waves propagating along the outside of single conductors, with and without dielectric coatings, and waves guided in dielectric sheets, rods, and fibers are considered next. Finally, highly resonant systems with waves trapped inside metal enclosures or cavities are treated.

13-2 CIRCUITS, LINES, AND GUIDES: A COMPARISON

At low frequencies a concept of currents, voltages, and lumped circuit elements is practical. Thus, for the simple circuit of Fig. 13-1a, consisting of a generator G and resistor R, circuit theory involving lumped elements can be used.

At somewhat higher frequencies these ideas can be extended satisfactorily to lines of considerable length provided that the velocity of propagation and the distributed constants of the line are considered. Thus, the behavior of the transmission line of Fig. 13-1b can be handled by an extension of circuit theory involving distributed elements, as in Chap. 10.

Consider now another type of transmission system, as shown in Fig. 13-1c, consisting of a hollow cylindrical or rectangular pipe or tube of metal. Suppose we

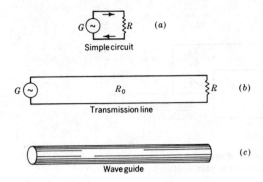

Figure 13-1 Comparison of (*a*) circuit, (*b*) two-conductor transmission line, and (*c*) hollow single-conductor waveguide.

ask: can such a pipe convey electromagnetic energy? If our experience were restricted to simple circuits or transmission lines, as in Fig. 13-1*a* and *b*, our answer would be *no*, since there is only a single conductor and no return circuit for the current. However, if our experience were restricted to optics, our answer would be *yes*, since light will pass through a straight metal pipe and light consists of electromagnetic waves of extremely high frequency (10^{16} Hz).

The complete answer is *yes* and *no*, depending on the frequency. Carrying our reasoning further, we might deduce that if the metal pipe will not transmit low frequencies but will transmit extremely high frequencies, there must be some intermediate frequency at which there is a transition from one condition to the other. In the following sections on waveguides we shall find that this transition, or low-frequency *cutoff*, occurs when the wavelength is of the same order of magnitude as the diameter of the pipe.

In explaining the transmission of electromagnetic energy through the pipe of Fig. 13-1*c* it is found that the circuit theory which worked for lumped circuits and two-conductor transmission lines is inadequate. For the hollow metal pipe or tube it is necessary to direct our attention to the empty space inside the tube and to the electric and magnetic fields **E** and **H** inside the tube. From the field-theory point of view we realize that the energy is actually conveyed through the empty space inside the tube and that the currents or voltages are merely associated effects.

13-3 TE MODE WAVE IN THE INFINITE-PARALLEL-PLANE TRANSMISSION LINE OR GUIDE

As an introduction to waveguides let us consider an infinite-parallel-plane transmission line, as in Fig. 13-2. This is a two-conductor line which is capable of guiding energy in a TEM mode with **E** in the *y* direction. However, at sufficiently high frequencies it can also transmit high-order modes, and this type of transmission between the parallel planes serves as a good starting point for our discussion of higher-order modes.

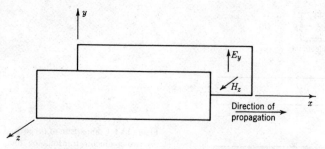

Figure 13-2 Transmission system consisting of two conducting planes parallel to the xy plane. The planes are assumed to be infinite in extent (infinite-parallel-plane transmission line).

Consider the higher-order mode where the electric field is everywhere in the y direction, with transmission in the x direction; i.e., the electric field has only an E_y component. Since E_y is transverse to the direction of transmission, this mode is designated a *transverse electric* (TE) mode. Although **E** is everywhere transverse, **H** has longitudinal, as well as transverse, components. Assuming perfectly conducting sheets, boundary conditions require that E_y vanish *at* the sheets. However, E_y need not be zero at points removed from the sheets. It is possible to determine the properties of a TE wave of the type under discussion by regarding it as being made up of two plane TEM waves reflected obliquely back and forth between the sheets.

First, however, consider the situation that exists when two plane TEM waves of the same frequency traveling in free space intersect at an angle, as suggested in Fig. 13-3. It is assumed that the waves are linearly polarized with **E** normal to the page. Wavefronts, or surfaces of constant phase, are indicated for the two waves.

The solid lines (marked "max") show where the field is a maximum with **E** directed out from the page. These lines may be regarded as representing the crests of the waves. The dashed lines (marked "min") show where the field is a minimum, i.e., where **E** is of maximum absolute magnitude but directed into the page. These lines may be regarded as representing the troughs of the waves. Wherever the crest of one wave coincides with the trough of the other wave there is cancellation, and the resultant **E** at that point is zero. Wherever crest coincides with crest or trough with trough there is reinforcement, and the resultant **E** at that point doubles. Referring to Fig. 13-3, it is therefore apparent that at all points along the dash-dot lines the field is always zero, while along the line indicated by dash and double dots the field will be reinforced and will have its maximum value.

Since **E** is zero along the dash-dot lines, boundary conditions will be satisfied at plane, perfectly conducting sheets placed along these lines normal to the page. The waves, however, will now be reflected at the sheets with an angle of reflection equal to the angle of incidence, and waves incident from the outside will not penetrate to the region between the sheets. But if two plane waves (A and B) are launched between the sheets from the left end, they will travel to the right via multiple reflections between the sheets, as suggested by the wave paths in Fig. 13-4a. The wavefronts (normal to the wave paths) for these waves are as indicated in Fig. 13-4b.

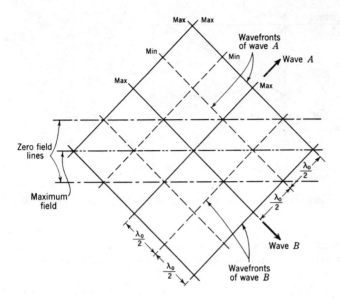

Figure 13-3 Two plane TEM waves traveling in free space in different directions.

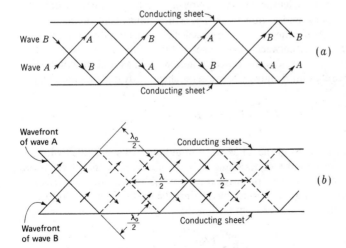

Figure 13-4 (a) Wave paths and (b) wave fronts between infinite parallel conducting sheets acting as a waveguide.

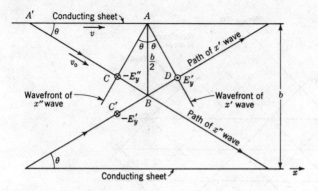

Figure 13-5 Component waves between infinite-parallel-plane conducting sheets acting as a waveguide.

Here the field between the sheets is the same as in Fig. 13-3, with solid lines indicating that **E** is outward (a maximum) and dashed lines that **E** is inward (a minimum). At the sheets the resultant **E** is always zero.

Although the two component waves we have been considering are plane TEM-mode waves, the *resultant wave* belongs to a higher-order TE mode. It is an important property of the TE-mode wave that it will not be transmitted unless the wavelength is sufficiently short. The critical wavelength at which transmission is no longer possible is called the *cutoff wavelength*. It is possible by a very simple analysis, which will now be given, to calculate the cutoff wavelength as a function of the sheet spacing.

Referring to Fig. 13-5, let the TE wave be resolved into two component TEM waves traveling in the x' and x'' directions. These directions make an angle θ with the conducting sheets (and the x axis). The electric field is in the y direction (normal to the page). The spacing between the sheets is b. From Fig. 13-5 we note that E'_y of the x' wave and E''_y of the x'' wave cancel at a point such as A at the conducting sheet and reinforce at point B midway between the sheets provided that the distance

$$CB = BD = C'B = \frac{\lambda_0}{4} \tag{1}$$

where λ_0 is the wavelength of the TEM wave in unbounded space filled with the same medium as between the sheets. Thus, if E''_y is into the page (negative) at the point C and E'_y is out of the page (positive) at the point D, the two waves will cancel at A. They will also reinforce at B since by the time the field $-E''_y$ moves from C to B the field $-E'_y$ will have moved from C' to B. More generally we can write

$$CB = \frac{n\lambda_0}{4} \tag{2}$$

where n is an integer $(1, 2, 3, \ldots)$.† It follows that

$$AB \sin \theta = \frac{b}{2} \sin \theta = \frac{n\lambda_0}{4} \tag{3}$$

or

$$\lambda_0 = \frac{2b}{n} \sin \theta \tag{4}$$

where λ_0 = wavelength, m
  b = spacing of conducting sheets, m
  $n = 1, 2, 3, \ldots$
  θ = angle between component wave direction and conducting sheets

According to (4), we note that for a given sheet separation b, the longest wavelength that can be transmitted in a higher-order mode occurs when $\theta = 90°$. This wavelength is the cutoff wavelength λ_{oc} of the higher-order mode. Thus, for $\theta = 90°$,

$$\lambda_{oc} = \frac{2b}{n} \tag{5}$$

Each value of n corresponds to a particular higher-order mode. When $n = 1$, we find that

$$\boxed{\lambda_{oc} = 2b} \tag{6}$$

This is the longest wavelength which can be transmitted between the sheets in a higher-order mode. That is, the spacing b must be at least $\lambda/2$ for a higher-order mode to be transmitted.

When $n = 1$, the wave is said to be the lowest of the higher-order types. When $n = 2$, we have the next higher-order mode and for this case

$$\lambda_{oc} = b \tag{7}$$

Thus the spacing b must be at least 1λ for the $n = 2$ mode to be transmitted. For $n = 3$, $\lambda_{oc} = \frac{2}{3}b$, etc.

Introducing (5) in (4) yields $\sin \theta = \lambda_0/\lambda_{oc}$, or

$$\theta = \sin^{-1} \frac{\lambda_0}{\lambda_{oc}} \tag{8}$$

Hence, at cutoff for any mode ($\lambda_0 = \lambda_{oc}$) the angle $\theta = 90°$. Under these conditions the component waves for this mode are reflected back and forth between the sheets, as in Fig. 13-6a, and do not progress in the x direction. Hence there is a standing wave between the sheets, and no energy is propagated. If the wavelength λ_0 is slightly less than λ_{oc}, θ is less than $90°$ and the wave progresses in the x direction

† For n even, the field halfway between the sheets is zero, with maximum fields either side of the centerline.

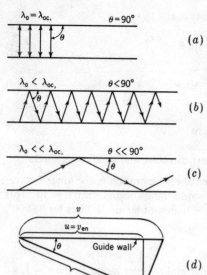

Figure 13-6 (a) to (c) Reflection of waves between walls of waveguide. (d) Triangle showing magnitude of phase velocity v, group velocity u, and energy velocity v_{en} in the guide relative to phase velocity v_0 of the component wave (equal to phase velocity of wave in an unbounded medium).

although making many reflections from the sheets, as in Fig. 13-6b. As the wavelength is further reduced, θ becomes less, as in Fig. 13-6c, until at very short wavelengths the transmission for this mode approaches the conditions in an unbounded medium.

It is apparent from Fig. 13-5 that a constant-phase point of the TE wave moves in the x direction with a velocity v that is greater than that of the component waves. The phase velocity v_0 of the component TEM waves is the same as for a wave in an unbounded medium of the same kind as fills the space between the conducting sheets. That is,

$$v_0 = \frac{1}{\sqrt{\mu\epsilon}} \quad (\text{m s}^{-1}) \tag{9}$$

where μ = permeability of medium, H m^{-1}

ϵ = permittivity of medium, F m^{-1}

From Fig. 13-5 it follows that

$$\frac{v_0}{v} = \frac{A'C}{A'A} = \cos\theta \tag{10}$$

or

$$\boxed{v = \frac{v_0}{\cos\theta} = \frac{1}{\sqrt{\mu\epsilon}\,\cos\theta} \quad (\text{m s}^{-1})} \tag{11}$$

According to (11), the phase velocity v of a TE wave approaches an infinite value as the wavelength is increased toward the cutoff value. On the other hand, v approaches the phase velocity v_0 in an unbounded medium as the wavelength becomes very short. Thus, the phase velocity of a higher-order mode wave in the guide formed by the sheets is always equal to or greater than the velocity in an unbounded medium. The energy, however, is propagated with the velocity of the zigzag component wave. Thus $v_{en} = v_0 \cos \theta$. Accordingly, the energy velocity v_{en} is always equal to or less than the velocity in an unbounded medium.† When, for instance, the wavelength approaches cutoff, the phase velocity becomes infinite while the energy velocity approaches zero. This is another way of saying that the wave degenerates into a standing wave and does not propagate energy at the cutoff wavelength or longer wavelengths. The relative magnitudes of the various velocities are shown by the triangle in Fig. 13-6d.

Since the wavelength is proportional to the phase velocity, the wavelength λ of the higher-order mode in the guide is given in terms of the wavelength λ_0 in an unbounded medium by

$$\lambda = \frac{\lambda_0}{\cos \theta} \tag{12}$$

The phase velocity and group (or energy) velocity in the guide as a function of θ are shown in Fig. 13-7. As θ approaches 90°, the phase velocity becomes infinite while the energy velocity goes to zero. The velocities are expressed in terms of the phase velocity v_0 of the wave in an unbounded medium. The situation here is analogous to the action of water waves at a breakwater. Thus, as suggested in Fig. 13-8, a water plume moves along the breakwater where a wave crest (constant-phase point) strikes the breakwater. The velocity v of the plume is greater than the wave velocity v_0. The plume velocity can become infinite if θ becomes 90°.

The infinite-parallel-plane transmission line we have been considering is an idealization and not a type to be applied in practice. Actual waveguides for higher-order modes usually take the form of a single hollow conductor. The hollow rectangular guide is a common form. The above analysis for the infinite-parallel-plane transmission line is of practical value, however, because the properties of TE-mode waves, such as are discussed above, are the same in a rectangular guide of width b as between two infinite parallel planes separated by a distance b. This follows from the fact that if infinitely conducting sheets are introduced normal to **E** between the parallel planes, the field is not disturbed. Thus, if a TE-mode wave with electric field in the y direction is traveling in the x direction, as indicated in Fig. 13-9a, the introduction of sheets normal to E_y, as in Fig. 13-9b, does not disturb the field. The conducting sheets now form a complete enclosure of rectangular shape.

---

† The waveguide behaves like a lossless dispersive medium. It follows that

$$u = v_{en} = \frac{v_0^2}{v}$$

where u = group velocity
v_{en} = energy velocity
v_0 = phase velocity in unbounded medium
v = phase velocity in guide

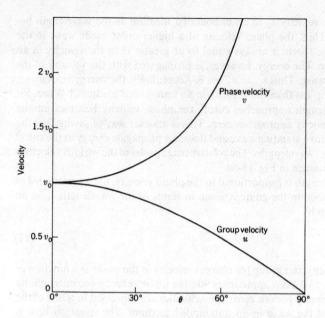

Figure 13-7 Phase and group velocity as a function of wave angle. The ordinate gives v and u in terms of the velocity v_0 for a wave in an unbounded medium of the same type as fills the waveguide.

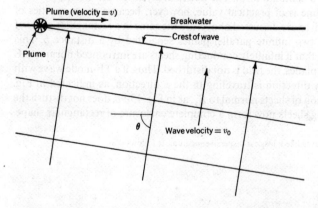

Figure 13-8 A plume of water moves along a breakwater with a phase velocity v that is greater than the wave velocity v_0.

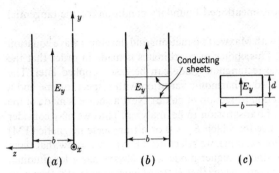

Figure 13-9 (*a*) Infinite-parallel-plane transmission line acting as a waveguide for TE wave. **E** is in *y* direction with wave in *x* direction (out of page). The guide consists of two parallel conducting sheets separated by a distance *b*. (*b*) Additional sheets introduced normal to E_y. (*c*) Hollow rectangular waveguide.

Proceeding a step further, let the sheets beyond the rectangular enclosure be removed, leaving the hollow rectangular waveguide shown in Fig. 13-9*c*. The cutoff wavelengths for the TE modes as given by (5) for the infinite-parallel-plane line also apply for this rectangular guide of width *b*. For the type of TE modes we have thus far considered (E_y component only) the dimension *d* (Fig. 13-9*c*) is not critical.

Although the above simple analysis yields information about cutoff wavelength, phase velocity, etc., it gives little information concerning the field configuration and fails to consider more complex higher-order modes of wave transmission in which, for example, **E** is transverse but with both *y* and *z* components. To obtain complete information concerning the waves in a hollow waveguide, we shall solve the wave equation subject to the boundary conditions for the guide. This is done for the hollow rectangular guide in the next section.

13-4 THE HOLLOW RECTANGULAR WAVEGUIDE†

In Sec. 13-3 certain properties of an infinite-parallel-plane transmission line and of a hollow rectangular guide were obtained by considering that the higher-mode wave consists of two plane TEM component waves and then applying the boundary condition that the tangential component of the resultant **E** must vanish at the perfectly conducting walls of the guide. This method could be extended to provide more complete information about the waves in a hollow waveguide. However, in this section we shall use another approach, which involves the solution of the wave

---

† Lord Rayleigh, On the Passage of Electric Waves through Tubes, *Phil. Mag.*, **43**: 125–132 (February 1897); L. J. Chu and W. L. Barrow, Electromagnetic Waves in Hollow Metal Tubes of Rectangular Cross Section, *Proc. IRE*, **26**: 1520–1555 (December 1938). A general dyadic Green's function method for solving transmission line and waveguide problems is developed by C-T Tai, "Dyadic Green's Functions in Electromagnetic Theory," International Textbook Company, Scranton, Pa., 1971.

equation subject to the above-mentioned boundary condition for the tangential component of **E**.

In this method we start with Maxwell's equations and develop a wave equation in rectangular coordinates. This choice of coordinates is made in order that the boundary conditions for the rectangular guide can be easily applied later. The restrictions are then introduced of harmonic variation with respect to time and a wave traveling in the x direction (direction of guide). Next a choice is made of the type of higher-order mode of transmission to be analyzed. Thus we may consider a transverse electric (TE) wave for which $E_x = 0$ or a transverse magnetic (TM) wave for which $H_x = 0$. If, for example, we select the TE type, we know that there must be an H_x component, since a higher-mode wave always has a longitudinal field component and E_x being zero means that H_x must have a value. It is then convenient to write the remaining field components in terms of H_x. Next a solution of a scalar-wave equation in H_x is obtained that fits the boundary conditions of the rectangular guide. This solution is substituted back into the equations for the other field components (E_y, E_z, H_y, and H_z). In this way we end up with a set of equations giving the variation of each field component with respect to space and time. This method of solution is very general and can be applied to many problems.

We shall develop the method in detail for TE waves in a hollow rectangular waveguide. First, however, the procedure will be outlined in step form as follows:

1. Start with Maxwell's equations.
2. Apply restriction of harmonic variation with respect to time.
3. Apply restriction of harmonic variation and attenuation with respect to x.
4. Select the type or mode of wave transmission (TE in this case; so $E_x = 0$ and $H_x \neq 0$).
5. Find equations for other four field components (E_y, E_z, H_y, and H_z) in terms of H_x.
6. Develop scalar wave equation for H_x.
7. Solve this wave equation for H_x subject to boundary conditions of waveguide.
8. Substitute H_x back into equations of step 5, giving a set of equations expressing each field component as a function of space and time. This constitutes the complete solution of the problem.

Beginning now with **Step 1** of the procedure, we have from Maxwell's curl equations in rectangular coordinates the following set of six scalar equations:

$$\frac{\partial H_z}{\partial y} - \frac{\partial H_y}{\partial z} - \sigma E_x - \epsilon \frac{\partial E_x}{\partial t} = 0 \tag{1}$$

$$\frac{\partial H_x}{\partial z} - \frac{\partial H_z}{\partial x} - \sigma E_y - \epsilon \frac{\partial E_y}{\partial t} = 0 \tag{2}$$

$$\frac{\partial H_y}{\partial x} - \frac{\partial H_x}{\partial y} - \sigma E_z - \epsilon \frac{\partial E_z}{\partial t} = 0 \tag{3}$$

$$\frac{\partial E_z}{\partial y} - \frac{\partial E_y}{\partial z} + \mu \frac{\partial H_x}{\partial t} = 0 \tag{4}$$

$$\frac{\partial E_x}{\partial z} - \frac{\partial E_z}{\partial x} + \mu \frac{\partial H_y}{\partial t} = 0 \tag{5}$$

$$\frac{\partial E_y}{\partial x} - \frac{\partial E_x}{\partial y} + \mu \frac{\partial H_z}{\partial t} = 0 \tag{6}$$

From Maxwell's divergence equations in rectangular coordinates we have in space free from charge ($\rho = 0$) the following two scalar equations:

$$\frac{\partial E_x}{\partial x} + \frac{\partial E_y}{\partial y} + \frac{\partial E_z}{\partial z} = 0 \tag{7}$$

$$\frac{\partial H_x}{\partial x} + \frac{\partial H_y}{\partial y} + \frac{\partial H_z}{\partial z} = 0 \tag{8}$$

Let us assume now that any field component varies harmonically with both time and distance and also may attenuate with distance (**Steps 2** and **3**). Thus, confining our attention to waves traveling in the positive x direction, we have, for instance, that the field component E_y is expressed by

$$E_y = E_1 e^{j\omega t - \gamma x} \tag{9}$$

where γ = propagation constant = $\alpha + j\beta$
α = attenuation constant
β = phase constant

When the restriction of (9) is introduced into the equations, (1) to (8) reduce to

$$\frac{\partial H_z}{\partial y} - \frac{\partial H_y}{\partial z} - (\sigma + j\omega\epsilon)E_x = 0 \tag{10}$$

$$\frac{\partial H_x}{\partial z} + \gamma H_z - (\sigma + j\omega\epsilon)E_y = 0 \tag{11}$$

$$-\gamma H_y - \frac{\partial H_x}{\partial y} - (\sigma + j\omega\epsilon)E_z = 0 \tag{12}$$

$$\frac{\partial E_z}{\partial y} - \frac{\partial E_y}{\partial z} + j\omega\mu H_x = 0 \tag{13}$$

$$\frac{\partial E_x}{\partial z} + \gamma E_z + j\omega\mu H_y = 0 \tag{14}$$

$$-\gamma E_y - \frac{\partial E_x}{\partial y} + j\omega\mu H_z = 0 \tag{15}$$

$$-\gamma E_x + \frac{\partial E_y}{\partial y} + \frac{\partial E_z}{\partial z} = 0 \tag{16}$$

$$-\gamma H_x + \frac{\partial H_y}{\partial y} + \frac{\partial H_z}{\partial z} = 0 \tag{17}$$

The above eight equations can be simplified by introducing a series impedance Z and shunt admittance Y, analogous to a transmission line (see Sec. 10-24), where

$$Z = -j\omega\mu \qquad (\Omega \, \text{m}^{-1}) \qquad (18)$$

$$Y = \sigma + j\omega\epsilon \qquad (\mho \, \text{m}^{-1}) \qquad (19)$$

Substituting these relations in (10) to (17) yields

$$\frac{\partial H_z}{\partial y} - \frac{\partial H_y}{\partial z} - Y E_x = 0 \qquad (20)$$

$$\frac{\partial H_x}{\partial z} + \gamma H_z - Y E_y = 0 \qquad (21)$$

$$-\gamma H_y - \frac{\partial H_x}{\partial y} - Y E_z = 0 \qquad (22)$$

$$\frac{\partial E_z}{\partial y} - \frac{\partial E_y}{\partial z} - Z H_x = 0 \qquad (23)$$

$$\frac{\partial E_x}{\partial z} + \gamma E_z - Z H_y = 0 \qquad (24)$$

$$-\gamma E_y - \frac{\partial E_x}{\partial y} - Z H_z = 0 \qquad (25)$$

$$-\gamma E_x + \frac{\partial E_y}{\partial y} + \frac{\partial E_z}{\partial z} = 0 \qquad (26)$$

$$-\gamma H_x + \frac{\partial H_y}{\partial y} + \frac{\partial H_z}{\partial z} = 0 \qquad (27)$$

These are the general equations for the steady-state field of a wave traveling in the x direction. No restrictions have as yet been made on the mode of the wave or the shape of the guide. We are now ready to proceed with **Step 4** and introduce the condition for a TE wave that $E_x = 0$. The equations then reduce to

$$\frac{\partial H_z}{\partial y} - \frac{\partial H_y}{\partial z} = 0 \qquad (28)$$

$$\frac{\partial H_x}{\partial z} + \gamma H_z - Y E_y = 0 \qquad (29)$$

$$-\gamma H_y - \frac{\partial H_x}{\partial y} - Y E_z = 0 \qquad (30)$$

$$\frac{\partial E_z}{\partial y} - \frac{\partial E_y}{\partial z} - Z H_x = 0 \qquad (31)$$

$$\gamma E_z - Z H_y = 0 \qquad (32)$$

$$-\gamma E_y - Z H_z = 0 \qquad (33)$$

$$\frac{\partial E_y}{\partial y} + \frac{\partial E_z}{\partial z} = 0 \tag{34}$$

$$-\gamma H_x + \frac{\partial H_y}{\partial y} + \frac{\partial H_z}{\partial z} = 0 \tag{35}$$

Proceeding to **Step 5**, let us rewrite these equations so that each field component is expressed in terms of H_x. To do this, we note from (32) and (33) that

$$\frac{E_z}{H_y} = -\frac{E_y}{H_z} = \frac{Z}{\gamma} \quad (\Omega) \tag{36}$$

The ratio E_z/H_y or E_y/H_z is a quantity which corresponds, in the case of a waveguide, to the characteristic impedance of a transmission line. Since (36) involves only transverse field components, it may be called the *transverse-wave impedance* Z_{yz} of the waveguide. Thus

$$Z_{yz} = \frac{E_y}{H_z} = -\frac{E_z}{H_y} = -\frac{Z}{\gamma} = \frac{j\omega\mu}{\gamma} \quad (\Omega) \tag{37}$$

Introducing (37) into (30) and solving for H_y yields

$$H_y = \frac{-1}{\gamma - YZ_{yz}} \frac{\partial H_x}{\partial y} \tag{38}$$

In a like manner we have, from (29),

$$H_z = \frac{-1}{\gamma - YZ_{yz}} \frac{\partial H_x}{\partial z} \tag{39}$$

Now, substituting (39) into (37), we obtain

$$E_y = \frac{-Z_{yz}}{\gamma - YZ_{yz}} \frac{\partial H_x}{\partial z} \tag{40}$$

and substituting (38) into (37) gives

$$E_z = \frac{Z_{yz}}{\gamma - YZ_{yz}} \frac{\partial H_x}{\partial y} \tag{41}$$

Equations (38) to (41) express the four transverse field components in terms of H_x. This completes Step 5.

Proceeding now to **Step 6**, we can obtain a wave equation in H_x by taking the y derivative of (38) and the z derivative of (39) and substituting both in (35). This yields

$$-\gamma H_x - \frac{1}{\gamma - YZ_{yz}} \left(\frac{\partial^2 H_x}{\partial y^2} + \frac{\partial^2 H_x}{\partial z^2} \right) = 0 \tag{42}$$

or

$$\frac{\partial^2 H_x}{\partial y^2} + \frac{\partial^2 H_x}{\partial z^2} + \gamma(\gamma - YZ_{yz})H_x = 0 \tag{43}$$

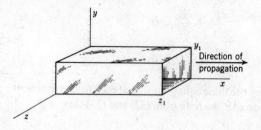

Figure 13-10 Coordinates for hollow rectangular waveguide.

Putting $k^2 = \gamma(\gamma - YZ_{yz})$ reduces (43) to

$$\frac{\partial^2 H_x}{\partial y^2} + \frac{\partial^2 H_x}{\partial z^2} + k^2 H_x = 0 \tag{44}$$

This is a partial differential equation of the second order and first degree. It is a scalar wave equation in H_x. It applies to a TE wave in a guide of any cross-sectional shape. This completes Step 6.

Step 7 is to find a solution of (44) that satisfies the boundary conditions for the waveguide under consideration, which is a hollow rectangular type, as shown in Fig. 13-10. The width of the guide is z_1, and the height is y_1. Assuming that the walls are perfectly conducting, the tangential component of **E** must vanish at the guide surface. Thus, at the sidewalls E_y must be zero, and at the top and bottom surfaces E_z must be zero. The problem now is to find a solution of (44) subject to these boundary conditions. The method of separation of variables can be used in obtaining the solution. Thus, H_x in (44) is a function of y and z. Hence we may seek a solution of the form

$$H_x = YZ \tag{45}$$

where $Y =$ a function of y only, that is, $Y = f(y)$
$Z =$ a function of z only†

Substituting (45) in (44) gives

$$Z\frac{d^2Y}{dy^2} + Y\frac{d^2Z}{dz^2} + k^2YZ = 0 \tag{46}$$

Dividing by YZ to separate variables gives

$$\frac{1}{Y}\frac{d^2Y}{dy^2} + \frac{1}{Z}\frac{d^2Z}{dz^2} = -k^2 \tag{47}$$

---

† Sometimes the notation $f(y)$ or $Y(y)$ is used to represent a function of y only and $f(z)$ or $Z(z)$ a function of z only. However, to simplify notation, the symbols Y and Z are used in Eqs. (45) to (55), inclusive, to indicate functions only of y or z, respectively. Y and Z in these equations should not be confused with admittance and impedance, for which these symbols are also used.

The first term is a function of y alone, the second term is a function of z alone, while k^2 is a constant. For the two terms (each involving a different independent variable) to equal a constant requires that each term be a constant. Thus we can write

$$\frac{1}{Y}\frac{d^2Y}{dy^2} = -A_1 \tag{48}$$

and

$$\frac{1}{Z}\frac{d^2Z}{dz^2} = -A_2 \tag{49}$$

where A_1 and A_2 are constants. It follows that

$$A_1 + A_2 = k^2 \tag{50}$$

Equations (48) and (49) each involve but one independent variable. A solution of (48) is

$$Y = c_1 \sin b_1 y \tag{51}$$

Substituting (51) in (48) yields

$$b_1 = \sqrt{A_1} \tag{52}$$

Hence (51) is a solution provided (52) is fulfilled. Another solution is

$$Y = c_2 \cos b_1 y \tag{53}$$

If (51) and (53) are each a solution for Y, their sum is also a solution, or

$$Y = c_1 \sin \sqrt{A_1}\, y + c_2 \cos \sqrt{A_1}\, y \tag{54}$$

In the same manner a solution may be written for Z as

$$Z = c_3 \sin \sqrt{A_2}\, z + c_4 \cos \sqrt{A_2}\, z \tag{55}$$

Substituting (54) and (55) into (45), we obtain the solution for H_x as

$$H_x = c_1 c_3 \sin \sqrt{A_1}\, y \sin \sqrt{A_2}\, z + c_2 c_3 \cos \sqrt{A_1}\, y \sin \sqrt{A_2}\, z$$
$$+ c_1 c_4 \sin \sqrt{A_1}\, y \cos \sqrt{A_2}\, z + c_2 c_4 \cos \sqrt{A_1}\, y \cos \sqrt{A_2}\, z \tag{56}$$

Substituting (56) in (40) and (41) and introducing the boundary conditions that $E_y = 0$ at $z = 0$ and $z = z_1$ and $E_z = 0$ at $y = 0$ and $y = y_1$, we find that only the last term of (56) can satisfy the boundary conditions and then only provided that

$$\sqrt{A_1} = \frac{n\pi}{y_1} \tag{57}$$

and

$$\sqrt{A_2} = \frac{m\pi}{z_1} \tag{58}$$

where m and n are integers $(0, 1, 2, 3, \ldots)$. They may be equal to the same integer or to different integers. The solution for H_x now assumes the form

$$H_x(y, z) = H_0 \cos \frac{n\pi y}{y_1} \cos \frac{m\pi z}{z_1} \tag{59}$$

where $H_0 = c_2 c_4 = $ a constant. If (59) is multiplied by a constant factor, it is still a solution. That is, the factor should not involve y or z although it may involve x and the time t. Accordingly, (59) may be multiplied by the exponential factor in (9) since this gives the variation assumed for the fields with respect to x and t. The complete solution for H_x then becomes

$$H_x(y, z, x, t) = H_0 \cos \frac{n\pi y}{y_1} \cos \frac{m\pi z}{z_1} e^{-\gamma x} \tag{60}$$

This completes Step 7. To perform **Step 8**, (60) is substituted into (38) through (41), giving the solutions for the transverse field components as

$$H_y = \frac{\gamma H_0}{k^2} \frac{n\pi}{y_1} \sin \frac{n\pi y}{y_1} \cos \frac{m\pi z}{z_1} e^{-\gamma x} \tag{61}$$

$$H_z = \frac{\gamma H_0}{k^2} \frac{m\pi}{z_1} \cos \frac{n\pi y}{y_1} \sin \frac{m\pi z}{z_1} e^{-\gamma x} \tag{62}$$

$$E_y = \frac{\gamma Z_{yz} H_0}{k^2} \frac{m\pi}{z_1} \cos \frac{n\pi y}{y_1} \sin \frac{m\pi z}{z_1} e^{-\gamma x} \tag{63}$$

$$E_z = -\frac{\gamma Z_{yz} H_0}{k^2} \frac{n\pi}{y_1} \sin \frac{n\pi y}{y_1} \cos \frac{m\pi z}{z_1} e^{-\gamma x} \tag{64}$$

Equations (60) to (64), to which may be added $E_x = 0$, are the solutions we have sought for the field components of a TE mode in a hollow rectangular guide of width z_1 and height y_1. This completes Step 8.†

Turning our attention now to an interpretation of the solutions for the field components, let us consider the significance of the integers m and n. It is apparent that for $m = 1$ and $n = 0$ we have only three field components H_x, H_z, and E_y and, further, that each of these components has no variation with respect to y but each has a half-cycle variation with respect to z. For example, E_y has a sinusoidal variation across the guide (in the z direction), being a maximum in the center and zero at the walls, and has no variation as a function of y.

If $m = 2$, there is a variation of two half-cycles (full-cycle variation) of each field component as a function of z. When $n = 1$, there is a half-cycle variation of each field component with respect to y. Hence we may conclude that the value of m or n indicates the number of half-cycle variations of each field component with respect to z and y, respectively. Each combination of m and n values represents a different field configuration or mode in the guide. Since we are dealing here with TE modes, it is convenient to designate them by adding the subscript mn so that, in general, any TE mode can be designated by the notation TE_{mn}, where m is the

† Note that to be explicit in (60) through (64) we could write $H_x(y, z, x, t)$ instead of H_x, $H_y(y, z, x, t)$ instead of H_y, $H_z(y, z, x, t)$ instead of H_z, etc., but for simplicity we do not and, in accord with phasor notation, the time factor ($e^{j\omega t}$) is omitted.

number of half-cycle variations in the z direction (usually taken as the larger transverse dimension of the guide) and n is the number of half-cycle variations in the y direction (usually taken as the smaller transverse dimension of the guide).

Case 1: TE$_{10}$ mode For this mode $m = 1$ and $n = 0$, and we have, as mentioned above, only three components E_y, H_x, and H_z that are not zero. The six field components for the TE$_{10}$ mode are

$$E_x = 0 \qquad \text{TE mode requirement}$$

$$E_y = \frac{\gamma Z_{yz} H_0}{k^2} \frac{\pi}{z_1} \sin \frac{\pi z}{z_1} e^{-\gamma x}$$

$$E_z = 0 \tag{65}$$

$$H_x = H_0 \cos \frac{\pi z}{z_1} e^{-\gamma x}$$

$$H_y = 0$$

$$H_z = \frac{\gamma H_0}{k^2} \frac{\pi}{z_1} \sin \frac{\pi z}{z_1} e^{-\gamma x}$$

The variation of these components as a function of z is portrayed in Fig. 13-11a. There is no variation with respect to y. This mode has the longest cutoff wavelength of any higher-order mode, and hence the lowest frequency of transmission in a hollow rectangular waveguide must be in the TE$_{10}$ mode. In Fig. 13-12a the field

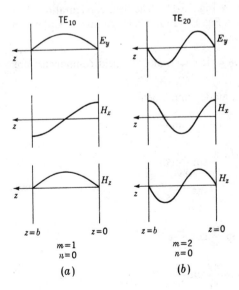

Figure 13-11 Variation of field components for TE$_{10}$ and TE$_{20}$ modes in a hollow rectangular waveguide. (Wave traveling out of page.)

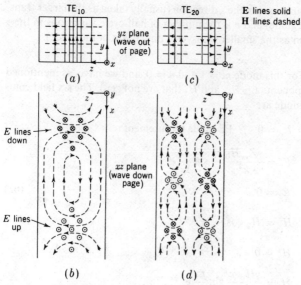

Figure 13-12 Field configurations for TE_{10} and TE_{20} modes in a hollow rectangular waveguide.

configuration of the TE_{10} mode is illustrated for a guide cross section and in Fig. 13-12b for a longitudinal section of the guide (top view).

Case 2: TE_{20} **mode** The variation of the field components as a function of z for the TE_{20} mode ($m = 2, n = 0$) is shown is Fig. 13-11b. The field configuration for a TE_{20} mode is shown in cross section in Fig. 13-12c and in longitudinal section (top view) in Fig. 13-12d.

Case 3: TE_{11} **mode** For this mode $m = 1$, $n = 1$, and the field components are given by

$$E_x = 0 \qquad \text{TE mode requirement}$$

$$E_y = \frac{yZ_{yz}H_0}{k^2} \frac{\pi}{z_1} \cos \frac{\pi y}{y_1} \sin \frac{\pi z}{z_1} e^{-\gamma x}$$

$$E_z = -\frac{\gamma Z_{yz}H_0}{k^2} \frac{\pi}{y_1} \sin \frac{\pi y}{y_1} \cos \frac{\pi z}{z_1} e^{-\gamma x}$$

$$H_x = H_0 \cos \frac{\pi y}{y_1} \cos \frac{\pi z}{z_1} e^{-\gamma x} \qquad (66)$$

$$H_y = \frac{\gamma H_0}{k^2} \frac{\pi}{y_1} \sin \frac{\pi y}{y_1} \cos \frac{\pi z}{z_1} e^{-\gamma x}$$

$$H_z = \frac{\gamma H_0}{k^2} \frac{\pi}{z_1} \cos \frac{\pi y}{y_1} \sin \frac{\pi z}{z_1} e^{-\gamma x}$$

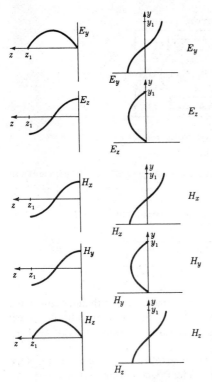

Figure 13-13 Variation of field components for TE_{11} mode in a square waveguide. (Wave traveling out of page.)

For this mode five field components have a value, only E_x being everywhere and always zero. The variation of the five field components with respect to z and y is shown in Fig. 13-13. It is assumed that the guide has a square cross section ($y_1 = z_1$). The field configuration for the TE_{11} mode in a square guide is illustrated in cross section (end view) in Fig. 13-14a and in longitudinal section (side view) in Fig. 13-14b.

The solution we have obtained tells us what modes are possible in the hollow rectangular waveguide. However, the particular mode or modes that are actually present in any case depend on the guide dimensions, the method of exciting the guide, and the irregularities or discontinuities in the guide. The resultant field in the guide is equal to the sum of the fields of all modes present.

Returning now to a consideration of the general significance of the solution, we have from (50), (57), and (58) that

$$\left(\frac{n\pi}{y_1}\right)^2 + \left(\frac{m\pi}{z_1}\right)^2 = k^2 \tag{67}$$

From (43), (37), and (19), k^2 is given by

$$k^2 = \gamma^2 - j\omega\mu(\sigma + j\omega\epsilon) \tag{68}$$

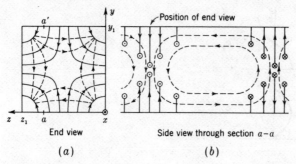

End view

Side view through section $a-a$

(a)

(b)

Figure 13-14 Field configurations for TE_{11} mode in a square waveguide. **E** lines are solid, and **H** lines are dashed.

Assuming a lossless dielectric medium in the guide, we can put $\sigma = 0$. Then equating (67) and (68) and solving for γ yields

$$\gamma = \sqrt{\left(\frac{n\pi}{y_1}\right)^2 + \left(\frac{m\pi}{z_1}\right)^2 - \omega^2\mu\epsilon} \qquad (69)$$

At sufficiently low frequencies the last term in (69) is smaller than the sum of the first two terms under the square-root sign. It follows that for this condition γ is real and therefore that the wave is attenuated. Under this condition it is said that the wave (or mode) is not propagated.

At sufficiently high frequencies the last term in (69) is larger than the sum of the first two terms under the square-root sign. Under this condition γ is imaginary, and therefore the wave is propagated without attenuation.

At some intermediate frequency the right-hand side of (69) is zero, and hence $\gamma = 0$. This frequency is called the *cutoff frequency* for the mode under consideration. At frequencies higher than cutoff this mode propagates without attenuation, while at frequencies lower than cutoff the mode is attenuated.

To summarize the three cases:

1. At low frequencies, ω small, γ real, guide opaque, wave does not propagate
2. At an intermediate frequency, ω intermediate, $\gamma = 0$, transition condition (cutoff)
3. At high frequencies, ω large, γ imaginary, guide transparent, wave propagates

Referring to (69), it is to be noted that $\sqrt{\omega^2\mu\epsilon}$ is equal to the phase constant β_0 for a wave traveling in an unbounded medium of the same dielectric material as fills the guide. Thus we can write

$$\gamma = \sqrt{k^2 - \beta_0^2} \qquad (m^{-1}) \qquad (70)$$

where $\beta_0 = \sqrt{\omega^2\mu\epsilon} = 2\pi/\lambda_0$ = phase constant in unbounded medium
λ_0 = wavelength in unbounded medium
$k = \sqrt{(n\pi/y_1)^2 + (m\pi/z_1)^2}$

Thus, at frequencies higher than cutoff $\beta_0 > k$, and

$$\gamma = \sqrt{k^2 - \beta_0^2} = j\beta \tag{71}$$

where $\beta = 2\pi/\lambda = \sqrt{\beta_0^2 - k^2}$ = phase constant in guide, rad m^{-1}
λ = wavelength in guide, m

At sufficiently high frequencies ($\beta_0 \gg k$) we note that the phase constant β in the guide approaches the phase constant β_0 in an unbounded medium. On the other hand, at frequencies less than cutoff $\beta_0 < k$, and

$$\gamma = \sqrt{k^2 - \beta_0^2} = \alpha \tag{72}$$

where α is the attenuation constant.

At sufficiently low frequencies ($\beta_0 \ll k$) we note that the attenuation constant α approaches a constant value k.

At the cutoff frequency, $\beta_0 = k$, and $\gamma = 0$. Thus, at cutoff

$$\omega^2 \mu\epsilon = \left(\frac{n\pi}{y_1}\right)^2 + \left(\frac{m\pi}{z_1}\right)^2 \tag{73}$$

It follows that the *cutoff frequency* is

$$f_c = \frac{1}{2\sqrt{\mu\epsilon}} \sqrt{\left(\frac{n}{y_1}\right)^2 + \left(\frac{m}{z_1}\right)^2} \qquad \text{(Hz)} \tag{74}$$

and the *cutoff wavelength* is

$$\boxed{\lambda_{oc} = \frac{2\pi}{\sqrt{(n\pi/y_1)^2 + (m\pi/z_1)^2}} = \frac{2}{\sqrt{(n/y_1)^2 + (m/z_1)^2}} \qquad \text{(m)}} \tag{75}$$

where λ_{oc} is the wavelength in an unbounded medium at the cutoff frequency (or, more concisely, the *cutoff wavelength*).† *Equations (74) and (75) give the cutoff frequency and cutoff wavelength for any* TE$_{mn}$ *mode in a hollow rectangular guide.* For instance, the cutoff wavelength of a TE$_{10}$ mode is

$$\lambda_{oc} = 2z_1 \tag{76}$$

This is identical with the value found in Sec. 13-3 since $z_1 = b$.

At frequencies above cutoff ($\beta_0 > k$)

$$\beta = \sqrt{\beta_0^2 - k^2} = \sqrt{\omega^2 \mu\epsilon - \left(\frac{n\pi}{y_1}\right)^2 - \left(\frac{m\pi}{z_1}\right)^2} \tag{77}$$

---

† Note that $k = 2\pi/\lambda_{oc}$. If this value of k is introduced, (71) can be used to relate λ, λ_0, and λ_{oc} when $\lambda_0 < \lambda_{oc}$.

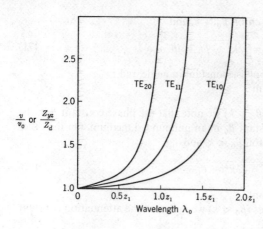

Figure 13-15 Relative phase velocity v/v_0 or relative transverse impedance Z_{yz}/Z_d as a function of the wavelength λ_0 for TE modes in a hollow square guide (height y_1 equal to width z_1).

It follows that the *phase velocity v_p in the guide* is equal to

$$v_p = \frac{\omega}{\beta} = \frac{v_0}{\sqrt{1 - (n\lambda_0/2y_1)^2 - (m\lambda_0/2z_1)^2}} \quad \text{(m s}^{-1}) \quad (78)$$

or

$$v_p = \frac{v_0}{\sqrt{1 - (\lambda_0/\lambda_{oc})^2}} \quad (79)$$

where $v_0 = 1/\sqrt{\mu\epsilon}$ = phase velocity in unbounded medium ($= 300$ Mm s^{-1} for air)

λ_0 = wavelength in unbounded medium

λ_{oc} = cutoff wavelength

The ratio v/v_0 as a function of the wavelength λ_0 is shown in Fig. 13-15 for several TE modes in a hollow waveguide of square cross section ($y_1 = z_1$).

In the above analysis there is no attenuation whatsoever at frequencies above cutoff. This results from the assumption of perfectly conducting guide walls and a lossless dielectric medium filling the guide. However, if the walls are not perfectly conducting or the medium is not lossless, or both, there is attenuation.†

If the guide is filled with air, the dielectric loss is usually negligible compared with losses in the guide walls, so that the attenuation at frequencies greater than cutoff is mainly determined by the conductivity of the guide walls. The fact that the guide walls are not perfectly conducting means that the tangential component E_t of the electric field is not zero at the walls but has a finite value. However, for walls made of a good conductor, such as copper, E_t will generally be so small that the above analysis (based on $E_t = 0$) is not affected to any appreciable extent. However, as a result of the finite wall conductivity α is not zero. Thus, in most practical problems where the wall conductivity is high (but not infinite) the field configura-

---

† That γ may have both a real and an imaginary part at frequencies greater than cutoff can be shown by solving (68) for γ under these conditions, with σ not equal to zero. See also Sec. 13-8.

tion in the guide, the wavelength λ, the phase constant β, the phase velocity v, etc., can all be calculated with high accuracy on the assumption that the walls have infinite conductivity, as done earlier in this section. The small (but not zero) attenuation may then be calculated separately, using (13-8-7) to find the power lost per unit area in the guide wall, it being assumed that the **H**-field distribution is the same as with perfectly conducting walls.

Finally, let us determine the value of the transverse-wave impedance Z_{yz} for TE modes in a rectangular hollow guide. Thus, from (37)

$$Z_{yz} = \frac{j\omega\mu}{\gamma} \tag{80}$$

At frequencies higher than cutoff $\gamma = j\beta$; so

$$Z_{yz} = \frac{\omega\mu}{\beta} = \frac{Z_d}{\sqrt{1 - (\lambda_0/\lambda_{oc})^2}} \quad (\Omega) \tag{81}$$

where Z_d = intrinsic impedance of dielectric medium filling guide
　　　　$= \sqrt{\mu/\epsilon} = 376.7 \ \Omega$ for air
　　λ_0 = wavelength in unbounded medium
　　λ_{oc} = cutoff wavelength

The ratio of Z_{yz} (transverse-wave impedance) to Z_d (intrinsic impedance) as a function of the wavelength λ_0 is shown in Fig. 13-15 for several TE modes in a hollow waveguide of square cross section ($y_1 = z_1$).

Thus far only TE-mode waves have been considered. To find the field relations for transverse magnetic (TM) mode waves we proceed precisely as in the eight-step list given earlier in this section except that where TE appears we substitute TM and where E_x appears we substitute H_x, and vice versa. In the TM wave $H_x = 0$, and the longitudinal field component is E_x. This analysis will not be carried through here (see Prob. 13-1-6). However, it may be mentioned that (75) for the cutoff wavelength applies to both TE and TM waves, as does (78) for the phase velocity, but this is not the case with (81) for the transverse impedance (see Prob. 13-3-7). The notation for any TM mode, in general, is TM_{mn}, where m and n are integers $(1, 2, 3, \ldots)$. It is to be noted that neither m nor n may be equal to zero for TM waves. Thus, the lowest-frequency TM wave that will be transmitted by a rectangular waveguide is the TM_{11} mode.

We have seen that each mode of transmission in a waveguide has a particular cutoff wavelength, velocity, and impedance. When the frequency is high enough to permit transmission in more than one mode, the resultant field is the sum of the fields of the individual mode fields in the guide.

For example, suppose that a rectangular waveguide, as shown in cross section in Fig. 13-16a, is excited in the TE_{10} mode. The variation of E_y across the guide is sinusoidal, as shown in Fig. 13-16b. Suppose now that z_1 exceeds 1λ, so that the TE_{20} mode can also be transmitted.† If only the TE_{10} mode is excited, no TE_{20} will

---

† But $y_1 < \lambda_0/2$, so that no TE_{01} mode (**E** in z direction) is transmitted.

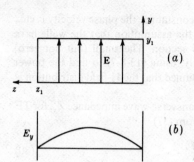

(a)

(b)

Figure 13-16 Rectangular waveguide with TE_{10} mode only.

appear provided that the guide is perfectly regular. However, in practice certain asymmetries and irregularities will be present, and these will tend to convert some of the TE_{10}-mode energy into TE_{20}-mode energy. Thus, if an asymmetrically located screw projects into the guide as in Fig. 13-17a, the total E_y field will tend to become asymmetrical, as suggested in Fig. 13-17b. This total field may be resolved into TE_{10} and TE_{20} components as shown in Fig. 13-17c and d. If both TE_{10} and TE_{20} modes can be transmitted, the field in the guide beyond the screw location will have energy in both modes. In effect the screw is a receiving antenna that extracts energy from the incident TE_{10}-mode wave and reradiates it so as to excite the TE_{20} mode. However, if the frequency is decreased so that only the TE_{10} wave can be transmitted, the asymmetric field (Fig. 13-17b) will exist only in the vicinity of the screw and farther down the guide the field will be entirely in the TE_{10} mode. To avoid the problems of multiple-mode transmission, a waveguide is usually operated so that only one mode is capable of transmission.† For instance, to ensure transmission only in the TE_{10} mode, z_1 must be less than 1λ and y_1 less than $\lambda/2$. But to allow transmission of the TE_{10} mode, z_1 must exceed $\lambda/2$. Hence z_1 must be between $\lambda/2$ and 1λ, and a value of 0.7λ is often used since this is well below 1λ and yet enough more than $\lambda/2$ so that the velocity and transverse impedance values are not too critical a function of frequency. We recall that at cutoff ($z_1 = \lambda/2$) the velocity and impedance approach infinite values. The height y_1 may be as small as desired without preventing transmission of the TE_{10} mode. Too small a value of y_1, however, increases attenuation (because of power lost in the guide walls) and also reduces the power-handling capabilities of the guide. It is often the practice to make $y_1 = z_1/2$. Many TE_{mn} and TM_{mn} modes of a rectangular guide for which $y_1 = z_1/2$ are shown in Fig. 13-18. The slant scale gives the frequency relative to the cutoff frequency for the dominant mode (TE_{10}). Thus, if the frequency is 3 times this value, we note from Fig. 13-18 that the following modes will be transmitted: TE_{10}, TE_{01}, TE_{20}, TE_{11}, TM_{11}, TE_{21}, and TM_{21}. An additional mode (TE_{30}) is at its cutoff frequency.

The relations derived in this section for a hollow rectangular waveguide (see Fig. 13-10) are summarized in Table 13-1.

---

† The lowest-frequency mode that a guide can transmit is called the *dominant mode*.

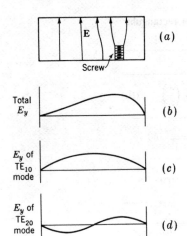

(a)

(b)

(c)

(d)

Figure 13-17 Rectangular waveguide with TE_{20} mode induced from TE_{10} mode by asymmetrically placed projection (screw).

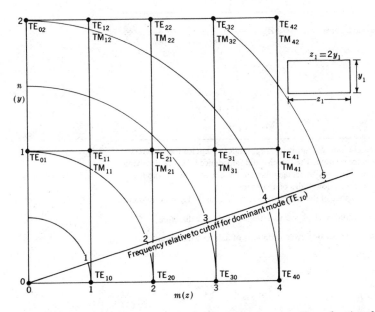

Figure 13-18 Possible TE and TM modes in a hollow rectangular waveguide as a function of frequency. At 3 times the cutoff frequency for the TE_{10} mode there are seven modes which will pass (see text at end of Sec. 13-4) and one mode (TE_{30}) at cutoff.

Table 13-1 Relations for TE_{mn} modes in hollow rectangular waveguides†

Name of relation	Relation
Cutoff frequency	$f_c = \dfrac{1}{2\sqrt{\mu\epsilon}} \sqrt{\left(\dfrac{n}{y_1}\right)^2 + \left(\dfrac{m}{z_1}\right)^2}$ (Hz)
Cutoff wavelength	$\lambda_{oc} = \dfrac{2}{\sqrt{(n/y_1)^2 + (m/z_1)^2}}$ (m)
Wavelength in guide	$\lambda_g = \dfrac{\lambda_0}{\sqrt{1 - (\lambda_0/\lambda_{oc})^2}}$ (m)
Phase velocity	$v_p = \dfrac{v_0}{\sqrt{1 - (n\lambda_0/2y_1)^2 - (m\lambda_0/2z_1)^2}}$ $= \dfrac{v_0}{\sqrt{1 - (\lambda_0/\lambda_{oc})^2}}$ $= \dfrac{v_0}{\sqrt{1 - (f_c/f)^2}}$ (m s^{-1}) where $v_0 = 1/\sqrt{\mu\epsilon}$
Transverse-wave impedance	$Z_{yz} = \dfrac{Z_d}{\sqrt{1 - (n\lambda_0/2y_1)^2 - (m\lambda_0/2z_1)^2}}$ $= \dfrac{Z_d}{\sqrt{1 - (\lambda_0/\lambda_{oc})^2}}$ $= \dfrac{Z_d}{\sqrt{1 - (f_c/f)^2}}$ (Ω) where $Z_d = \sqrt{\mu/\epsilon}$

† All the relations also apply to TM_{mn} modes except for the transverse-wave impedance relation. The velocity and impedance relations involving $(\lambda_0/\lambda_{oc})^2$ apply not only to rectangular guides but also to TE modes in hollow single-conductor guides of any shape.

The significance of the subscripts mn in TE_{mn} or TM_{mn} is as follows:

m = number of $\frac{1}{2}$-cycle variations of field in z direction

n = number of $\frac{1}{2}$-cycle variations of field in y direction

13-5 THE HOLLOW CYLINDRICAL WAVEGUIDE†

Consider the problem of describing wave propagation in a hollow (circular) cylindrical waveguide of radius r_0. This problem is most easily handled with a cylindrical coordinate system, as shown in Fig. 13-19. The procedure is similar to that used in the preceding section for the rectangular waveguide. We assume time-harmonic variation, perfectly conducting walls, and a lossless interior medium ($\sigma = 0$) containing no charge ($\rho = 0$).

Maxwell's two curl equations yield six scalar equations, and Maxwell's two divergence equations yield two scalar equations. In cylindrical coordinates these are as follows:

$$\frac{1}{r}\frac{\partial E_z}{\partial \phi} + \gamma E_\phi - ZH_r = 0 \tag{1}$$

$$-\gamma E_r - \frac{\partial E_z}{\partial r} - ZH_\phi = 0 \tag{2}$$

$$\frac{\partial E_\phi}{\partial r} + \frac{1}{r}E_\phi - \frac{1}{r}\frac{\partial E_r}{\partial \phi} - ZH_z = 0 \tag{3}$$

$$\frac{1}{r}\frac{\partial H_z}{\partial \phi} + \gamma H_\phi - YE_r = 0 \tag{4}$$

$$-\gamma H_r - \frac{\partial H_z}{\partial r} - YE_\phi = 0 \tag{5}$$

$$\frac{\partial H_\phi}{\partial r} + \frac{1}{r}H_\phi - \frac{1}{r}\frac{\partial H_r}{\partial \phi} - YE_z = 0 \tag{6}$$

$$\frac{\partial E_r}{\partial r} + \frac{E_r}{r} + \frac{1}{r}\frac{\partial E_\phi}{\partial \phi} - \gamma E_z = 0 \tag{7}$$

$$\frac{\partial H_r}{\partial r} + \frac{H_r}{r} + \frac{1}{r}\frac{\partial H_\phi}{\partial \phi} - \gamma H_z = 0 \tag{8}$$

where Z = series impedance = $-j\omega\mu$, $\Omega\,\text{m}^{-1}$
$\quad Y$ = shunt admittance = $j\omega\epsilon$, $\mho\,\text{m}^{-1}$
$\quad E_\phi = E_1 e^{-\gamma z}$
$\quad \gamma = \alpha + j\beta$ = propagation constant

These eight relations are the general equations for the steady-state field of a wave traveling in the z direction as expressed in cylindrical coordinates. At this point we can limit the problem to either a TE wave or a TM wave. Selecting the TE wave

---

† W. L. Barrow, Transmission of Electromagnetic Waves in Hollow Tubes of Metal, *Proc. IRE*, **24**:1298–1328 (October 1936); G. C. Southworth, Some Fundamental Experiments with Wave Guides, ibid., **25**:807–822 (July 1937).

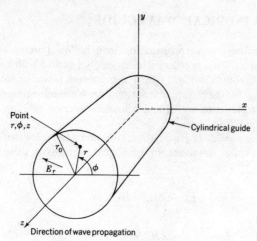

Figure 13-19 Coordinates for hollow cylindrical waveguide.

and proceeding in the same manner as for a rectangular waveguide following (13-4-27), we find the solution for the H_z component to be

$$H_z = c_1 c_3 \sin n\phi \, J_n(kr) + c_2 c_3 \cos n\phi \, J_n(kr)$$
$$+ c_1 c_4 \sin n\phi \, N_n(kr) + c_2 c_4 \cos n\phi \, N_n(kr) \quad (9)$$

where $k = \sqrt{\gamma^2 + ZY} = \sqrt{\gamma^2 + \omega^2 \mu \epsilon}$
      $n =$ integer

Equation (9) is a solution for H_z, but any term or combination of terms is also a solution. The appropriate solution must satisfy the boundary conditions, which are that $E_\phi = 0$ at $r = r_0$ and that the fields in the guide are finite.† The Neumann function becomes infinite at $r = 0$ and so is not a suitable solution. Thus, (9) can be reduced to

$$H_z = H_0(\cos n\phi + j \sin n\phi) J_n(kr) \quad (10)$$

where $$H_0 = c_2 c_3 = \frac{c_1 c_3}{j}$$

When the sine term is dropped, (10) simplifies to

$$H_z = H_0 \cos n\phi \, J_n(kr) \quad (11)$$

This choice means that H_z is a maximum where $\phi = 0$ and $\phi = \pi$ and is zero where $\phi = \pi/2$ and $3\pi/2$ provided $n \neq 0$. If we had dropped the cosine term instead of the sine term, the conditions for H_z would be rotated 90° in the guide. Assuming that

---

† The boundary condition $E_\phi = 0$ at $r = r_0$ can also be expressed

$$\hat{n} \times \mathbf{E} = 0 \quad \text{at } r = r_0$$

the field orientation is arbitrary, either solution would suffice (but both would be required for a circularly polarized wave in the guide). The six components are†

$$E_r = \frac{n\gamma H_0 Z_{r\phi}}{k^2 r} \sin n\phi \, J_n(kr) e^{-\gamma z} = Z_{r\phi} H_\phi \tag{12}$$

$$E_\phi = \frac{\gamma H_0 Z_{r\phi}}{k^2} \cos n\phi \, \frac{dJ_n(kr)}{dr} e^{-\gamma z} = -Z_{r\phi} H_r \tag{13}$$

$$E_z = 0 \qquad \text{TE-mode requirement} \tag{14}$$

$$H_r = \frac{-\gamma H_0}{k^2} \cos n\phi \, \frac{dJ_n(kr)}{dr} e^{-\gamma z} \tag{15}$$

$$H_\phi = \frac{n\gamma H_0}{k^2 r} \sin n\phi \, J_n(kr) e^{-\gamma z} \tag{16}$$

$$H_z = H_0 \cos n\phi \, J_n(kr) e^{-\gamma z} \tag{17}$$

Alternative expressions for E_ϕ and H_r can be written using the recurrence relations (see Appendix, Sec. A-10):

$$\frac{dJ_n(kr)}{dr} = k\left[\frac{n}{kr} J_n(kr) - J_{n+1}(kr)\right] \tag{18}$$

Equations (12) to (17) are general expressions for a TE-mode wave in a cylindrical guide. However, the boundary condition $E_\phi = 0$ at $r = r_0$ has not been imposed. Applying this condition requires that

$$\frac{dJ_n(kr)}{dr} = 0 \qquad \text{at } r = r_0 \tag{18a}$$

and hence that

$$k = \frac{k'_{nr}}{r_0} \tag{19}$$

where k'_{nr} is the rth root of the derivative of the nth-order Bessel function. The situation is illustrated in Fig. 13-20a for $n = 1$. The first three roots of the derivative of the first-order Bessel function occur at $kr_0 = k'_{nr} = 1.84, 5.33$, and 8.54. Thus, the roots are $k'_{11} = 1.84$, $k'_{12} = 5.33$, and $k'_{13} = 8.54$, corresponding to wave modes TE_{11}, TE_{12}, and TE_{13} in the guide. In general, a TE mode in a cylindrical guide is designated TE_{nr}, where the subscript n indicates the order of the Bessel function and r indicates the rank of the root. For TE_{01} and TE_{02} modes, $n = 0$ and the roots of (18) for this case are $k'_{nr} = 3.832$ and 7.016. These are the same as the zeros of $J_1(kr)$. See Fig. 13-20a.

† In (12) and (13)

$$Z_{r\phi} = \frac{E_r}{H_\phi} = -\frac{E_\phi}{H_r} = -\frac{Z}{\gamma} = \frac{j\omega\mu}{\gamma} \qquad (\Omega)$$

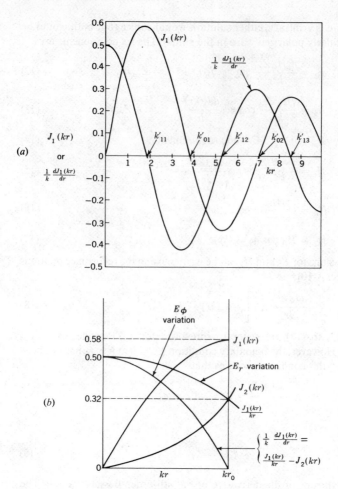

Figure 13-20 (a) First-order Bessel function and its derivative as a function of kr. (b) Bessel function relations involved in determining E_r and E_ϕ variation for TE_{11} mode.

Referring to (13), the variation of E_ϕ with r (at $\phi = 0$) for the TE_{11} mode is as shown in Fig. 13-20b. From (12) the variation of E_r with r (at $\phi = 90°$) is as given by the curve for $J_1(kr)/kr$. The amplitudes of the two field components, E_r and E_ϕ, can thus be represented by curves, as in Fig. 13-21a, or by arrows, as in Fig. 13-21b. Note that E_ϕ is a maximum at $\phi = 0°$ and $180°$ and zero at $\phi = 90°$ and $270°$ while E_r is a maximum at $\phi = 90°$ and $270°$ and zero at $\phi = 0°$ and $180°$. Note also that at the center of the guide ($r = 0$) E_r and E_ϕ both have the same direction and amplitude. When the E_r and E_ϕ components are added, the total electric field for the TE_{11} in the cylindrical guide is as suggested by the heavy arrows in Fig. 13-21b with

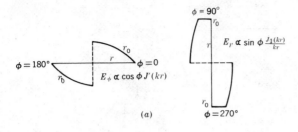

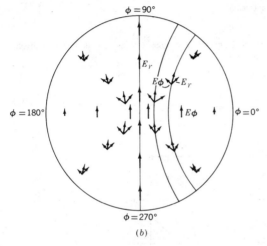

Figure 13-21 (a) Variation of electric field components E_r and E_ϕ with radial distance r for TE_{11} mode in cylindrical waveguide. (b) Electric field vectors for TE_{11} mode in cylindrical waveguide with component fields E_r and E_ϕ.

the direction of the total electric field as indicated by the curved lines (see also Fig. 13-22a). In a similar manner the transverse magnetic field configuration for the TE_{11} mode can be deduced from the variation of H_r and H_ϕ with r [from (15) and (16)], with the result shown in Fig. 13-22a.

The solution we have obtained for the TE_{11} mode is but one of an infinite number of possible wave modes in a cylindrical guide. The particular mode or modes that are actually present in any case depend on the guide dimensions, the method of exciting the guide, and discontinuities in the guide. The resultant field in the guide is equal to the sum of the fields of all modes present. The field configuration for the TE_{01} mode is illustrated in Fig. 13-22b.

Thus far only TE modes in a cylindrical guide have been considered. To find the field relations for TM-mode waves we let $H_z = 0$ in (1) to (8) and express the remaining field components in terms of the longitudinal field E_z. A wave equation in E_z is developed and solved to satisfy the boundary condition, which requires that

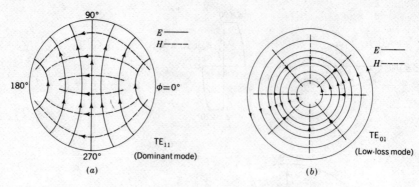

Figure 13-22 Electric field lines (solid) and magnetic field lines (dashed) in hollow cylindrical waveguide for (a) TE_{11} (dominant) mode and (b) TE_{01} (low-loss) mode.

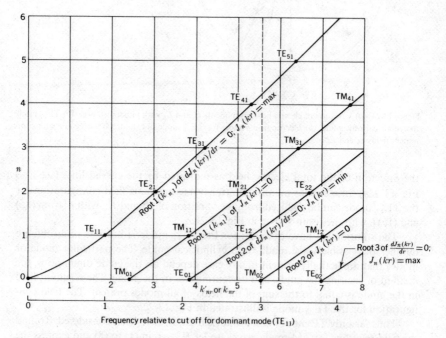

Figure 13-23 Possible TE and TM modes in a hollow cylindrical waveguide as a function of frequency. (*After C-T Tai.*) At 3 times the cutoff frequency for the TE_{11} mode there are nine modes which will pass (see text paragraph preceding Sec. 13-6) and one mode (TM_{02}) at cutoff.

Table 13-2 Cylindrical waveguide modes

Mode designation†	Eigenvalues		Cutoff wavelength, λ_{oc}
	k'_{nr}	k_{nr}	
TM_{01}		2.405	$2.61r_0$
TE_{01} (low loss)	3.832		$1.64r_0$
TM_{02}		5.520	$1.14r_0$
TE_{02}	7.016		$0.89r_0$
TE_{11} (dominant)	1.840		$3.41r_0$
TM_{11}		3.832	$1.64r_0$
TE_{12}	5.330		$1.18r_0$
TM_{12}		7.016	$0.89r_0$
TE_{21}	3.054		$2.06r_0$
TM_{21}		5.135	$1.22r_0$
TE_{22}	6.706		$0.94r_0$
TE_{31}	4.201		$1.49r_0$
TM_{31}		6.379	$0.98r_0$
TE_{41}	5.318		$1.18r_0$
TM_{41}		7.588	$0.83r_0$
TE_{51}	6.416		$0.98r_0$

† The subscripts nr as in TE_{nr} or k_{nr} have the following significance:

$n = n$th-order Bessel function

$r =$ order of root of nth-order Bessel function

$J_n(kr) = 0$ at $r = r_0$. The first three roots $k_{nr}(= kr_0)$ for the case $n = 1$ are $k_{11} = 3.832$, $k_{12} = 7.016$, and $k_{13} = 10.173$. The fields for TM modes are given in Prob. 13-2-10.

The roots (or eigenvalues) for the TM modes written k_{nr} (unprimed) correspond to zero values of the Bessel function $J_n(kr)$, whereas the roots for the TE modes written k'_{nr} (primed) correspond to zero values of the derivative (with respect to r) of the Bessel function. The relations for TE and TM modes are displayed in Fig. 13-23 and their numerical values listed in Table 13-2.

As seen in Fig. 13-23, the TE_{01} mode should logically be designated the TE_{02} mode since it refers to the *second* root of the derivative of the Bessel function. Likewise the TE_{02} mode should properly be designated TE_{03}. The first root is $k'_{nr} = 0$, corresponding to a trivial (nonexistent) mode. This inconsistency in the nomenclature for cylindrical guides has been pointed out by C-T Tai,† but the common and incorrect designations persist and Table 13-2 and Fig. 13-23 conform to this common usage.

† C-T Tai, On the Nomenclature of TE_{01} modes in a Cylindrical Waveguide, *Proc. IRE*, **49**:1442–1443 (September 1961).

Let us now examine the conditions necessary for propagation inside a cylindrical waveguide. Substituting (19) into $k = \sqrt{\gamma^2 + \omega^2\mu\epsilon}$ and solving for the propagation constant, we obtain

$$\gamma = \pm\sqrt{\left(\frac{k'_{nr}}{r_0}\right)^2 - \omega^2\mu\epsilon} = \alpha + j\beta \qquad (20)$$

As in the rectangular guide, there are three conditions:

1. At low frequencies, ω small, γ real, guide opaque (wave does not propagate)
2. At an intermediate frequency, ω intermediate, $\gamma = 0$, transition condition (cutoff)
3. At high frequencies, ω large, γ imaginary, guide transparent (wave propagates)

Putting $\gamma = 0$ in (20), we find for the *cutoff frequency* and *cutoff wavelength*

$$f_c = \frac{1}{2\pi\sqrt{\mu\epsilon}}\frac{k'_{nr}}{r_0} \qquad \text{(Hz)} \qquad (21)$$

$$\lambda_{oc} = \frac{2\pi r_0}{k'_{nr}} \qquad \text{(m)} \qquad (22)$$

For the TE_{11} mode $k'_{nr} = k'_{11} = 1.84$, so that $\lambda_{oc} = 2\pi r_0/1.84 = 3.41 r_0$. Thus, the cutoff wavelength for the TE_{11} mode corresponds to a wavelength 3.41 times the radius of the guide. The cutoff wavelengths for various modes in a cylindrical guide are listed in Table 13-2.

At frequencies above cutoff

$$\beta = \sqrt{\omega^2\mu\epsilon - \left(\frac{k'_{nr}}{r_0}\right)^2} \qquad \text{(rad m}^{-1}\text{)} \qquad (23)$$

From (23) and (22) we get for the *wavelength in the guide* (in z direction)

$$\lambda_g = \frac{\lambda_0}{\sqrt{1 - (\lambda_0/\lambda_{oc})^2}} \qquad \text{(m)} \qquad (24)$$

where λ_0 = wavelength in unbounded medium of same type that fills guide, m
λ_{oc} = cutoff wavelength, m

For the *phase velocity in the guide* ($v_p = f\lambda_g$) we obtain

$$v_p = \frac{\omega}{\beta} = \frac{v_0}{\sqrt{1 - (\lambda_0/\lambda_{oc})^2}} \qquad \text{(m s}^{-1}\text{)} \qquad (25)$$

where $v = 1/\sqrt{\mu\epsilon}$

Equations (24) and (25) are identical to those derived earlier for the rectangular waveguide (see Table 13-1). They also apply to waves in hollow guides of any cross section.

It will be noted that the roots k'_{nr} (also called *eigenvalues*) are not regularly spaced, in contrast to the case for the rectangular guide. Table 13-2 gives the roots and cutoff wavelengths for some TE and TM modes in a cylindrical guide. These modes are also illustrated in Fig. 13-23. The TE_{11} mode will propagate at a lower frequency than any other mode (including TM modes) and hence is called the *dominant mode* for a cylindrical guide. The TE_{01} mode is of interest because of its low attenuation characteristics in practical guides having finite wall conductivity. For this mode the attenuation decreases monotonically with increasing frequency (see Sec. 13-8).

Referring to Fig. 13-23, we note that if the frequency is 3 times that required to pass the dominant mode (TE_{11}), the guide will pass the following modes (to the left of the dashed line in Fig. 13-23): TE_{11}, TM_{01}, TE_{21}, TM_{11}, TE_{01}, TE_{31}, TM_{21}, TE_{12}, and TE_{41}. An additional mode (TM_{02}) is at its cutoff frequency.

13-6 HOLLOW WAVEGUIDES OF OTHER CROSS SECTION

In earlier sections we considered rectangular and cylindrical waveguides. These are only two of an infinite variety of forms in which single-conductor hollow wave-guides may be made. For example, the waveguide could have an elliptical† cross section, as in Fig. 13-24d, or a reentrant‡ cross section, as in Fig. 13-24f.

All these forms and many others may be regarded as derivable from the rectangular type (Fig. 13-24a). Thus the square cross section (Fig. 13-24b) is a special case of the rectangular guide. By bending out the walls the square guide can be transformed to the circular shape (Fig. 13-24c). By flattening the circular guide the elliptical form of Fig. 13-24d is obtained. On the other hand, by bending the top and bottom surfaces of the rectangular waveguide inward the form shown in Fig. 13-24e is obtained. A still further modification is the reentrant form with central ridge in Fig. 13-24f. The value of regarding these as related forms is that often certain properties of a guide of a particular shape can be interpolated approximately from the known properties of waveguides of closely related shape.

For example, the longest wavelength that the square guide (Fig. 13-24b) will transmit is equal to $2b$. This is for the TE_{10} mode. This information can be used to predict with fair accuracy the longest wavelength that a circular guide can transmit. Thus, if the cross-sectional area of the square guide is taken equal to the area of the circular guide,

$$b^2 = \pi \left(\frac{d}{2}\right)^2 \tag{1}$$

† L. J. Chu, Electromagnetic Waves in Hollow Elliptic Pipes of Metal, *J. Appl. Phys.*, **9** (September 1938).

‡ S. B. Cohn, Properties of Ridge Wave Guide, *Proc. IRE*, **35**: 783–789 (August 1947).

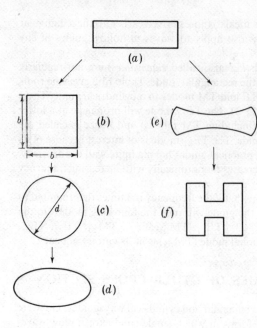

Figure 13-24 Forms of hollow single-conductor waveguides showing how square-shaped (b), cylindrically-shaped (c), and elliptically-shaped (d), waveguides may be derived from a rectangular type as at (a). Types (e) and (f) may also be derived from (a).

where d is the diameter of the circular guide. Now $\lambda_{oc} = 2b$ for the square waveguide or $b = \lambda_{oc}/2$, and so we have

$$\left(\frac{\lambda_{oc}}{2}\right)^2 = \pi\left(\frac{d}{2}\right)^2 \tag{2}$$

or $$\lambda_{oc} = \sqrt{\pi}\, d = 1.77d = 3.54r \tag{3}$$

as the cutoff wavelength for a circular waveguide of diameter d or radius r. This approximate value is only 4 percent greater than the exact value of $3.41r$ (see Table 13-2).

The procedure for carrying out a complete analysis of the properties of a waveguide of any shape is formally the same as in the preceding sections for the rectangular and circular waveguides. It is usually most convenient, however, to set up the equations in a coordinate system such that the waveguide surfaces can be specified by a fixed value of a coordinate. Thus, as we have seen, a rectangular guide is conveniently handled with rectangular coordinates, the guide surfaces being specified by $y = 0$, $y = y_1$, $z = 0$, and $z = z_1$. Likewise, a circular waveguide is readily analyzed using cylindrical coordinates, the guide surface being specified by $r = r_0$. Referring to Fig. 13-24d and e, these shapes can be analyzed using elliptical-hyperbolic coordinates. However, if the guide surface cannot be specified in a simple manner, the application of the boundary condition ($E_t = 0$) may be very difficult.

13-7 ATTENUATION AT FREQUENCIES LESS THAN CUTOFF

It has been shown that at frequencies less than cutoff, waves are not transmitted through hollow single-conductor guides but are attenuated. Let us now calculate the magnitude of this attenuation. From (13-4-72) the attenuation constant for a rectangular guide at frequencies less than cutoff is

$$\alpha = \sqrt{\left(\frac{n\pi}{y_1}\right)^2 + \left(\frac{m\pi}{z_1}\right)^2 - \left(\frac{2\pi}{\lambda_0}\right)^2} \tag{1}$$

Noting (13-4-75), we can reexpress this as

$$\alpha = \frac{2\pi}{\lambda_0} \sqrt{\left(\frac{\lambda_0}{\lambda_{oc}}\right)^2 - 1} = \beta_0 \sqrt{\left(\frac{\lambda_0}{\lambda_{oc}}\right)^2 - 1} \quad (\text{Np m}^{-1}) \tag{2}$$

where λ_0 = wavelength in unbounded medium, m

λ_{oc} = cutoff wavelength, m

The attenuation constant α as given in (2) applies not only to rectangular guides but to hollow single-conductor guides of any cross-sectional shape.

If the frequency is much less than cutoff ($\lambda_0 \gg \lambda_{oc}$), (2) reduces to the approximate relation

$$\alpha \approx \frac{2\pi}{\lambda_{oc}} \quad (\text{Np m}^{-1}) \tag{3}$$

where λ_{oc} is the cutoff wavelength in meters. Since, in dealing with voltage, 1 Np = 8.69 dB,†

$$\alpha \approx \frac{2\pi \times 8.69}{\lambda_{oc}} = \frac{54.6}{\lambda_{oc}} \quad (\text{dB m}^{-1}) \tag{4}$$

Example A certain waveguide has a cutoff wavelength λ_{oc} of 100 mm. Find the attenuation per meter along the guide for an applied wavelength λ_0 of 1 m.

SOLUTION Since $\lambda_0 \gg \lambda_{oc}$, (3) or (4) can be used, yielding

$$\alpha = 20\pi \ \text{Np m}^{-1}, \text{ or } 546 \ \text{dB m}^{-1}$$

This is a very high rate of attenuation, the applied field falling to a negligible value in a very short distance.

---

† Note that a 1-neper (1 Np) attenuation means a reduction to $1/e$ of the original value. Conversely, an increase of 1 Np means an increase to e ($=2.7183$) times the original value. Hence, for voltages 1 Np is equal to $20 \log e = 8.69$ dB.

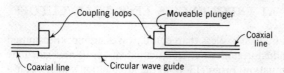

Figure 13-25 Attenuator for use at frequencies less than cutoff.

A simple attenuator operating at frequencies less than cutoff is illustrated in longitudinal section in Fig. 13-25. A metal tube, acting as a waveguide, has loops arranged at each end, as shown, to couple from coaxial transmission lines into and out of the waveguide. One of the loops is mounted on a movable plunger so that the distance between the loops is variable. If the applied wavelength λ_0 is much longer than the cutoff wavelength λ_{oc} of the guide and the loops are not in too close proximity, the attenuation is as given by (3) or (4). For instance, if $\lambda_{oc} = 100$ mm and λ_0 is much greater (1 m or more), the attenuation increases 5.46 dB per centimeter of outward movement of the plunger. This type of attenuator is very useful but has the disadvantage of a high insertion loss, i.e., a large initial attenuation when inserted in a coaxial line. Since $\beta = 0$, there is no change in phase with change in plunger position.

13-8 ATTENUATION AT FREQUENCIES GREATER THAN CUTOFF

If the waveguide has perfectly conducting walls and the medium filling the guide is lossless, there is no attenuation at frequencies greater than cutoff. Thus, $\alpha = 0$, and from (13-4-71) we have for a rectangular guide that

$$\gamma = j\beta = \sqrt{\left(\frac{n\pi}{y_1}\right)^2 + \left(\frac{m\pi}{z_1}\right)^2 - \left(\frac{2\pi}{\lambda_0}\right)^2} \tag{1}$$

or

$$\beta = \frac{2\pi}{\lambda_0} \sqrt{1 - \left(\frac{\lambda_0}{\lambda_{oc}}\right)^2} = \beta_0 \sqrt{1 - \left(\frac{\lambda_0}{\lambda_{oc}}\right)^2} \tag{2}$$

The phase constant β as given in (2) applies not only to rectangular guides but to hollow single-conductor guides of any cross-sectional shape.

The behavior of the phase constant β for this case and for the case discussed in Sec. 13-7 is compared on the composite graph in Fig. 13-26. Here the propagation constant γ is shown as ordinate vs. λ_0 in an unbounded medium as abscissa. The real part of $\gamma(=\alpha)$ is plotted as the solid curve above the x axis and the imaginary part $(=\beta)$ as the dashed curve below the axis. At very short wavelengths $(\lambda_0 \rightarrow 0)$, α is zero, and β approaches an infinite value that is equal to β_0 for an unbounded medium. As λ_0 increases, β decreases until at cutoff $(\lambda_0 = \lambda_{oc})$ β is zero. At still longer wavelengths, β remains zero, but α does not. At sufficiently long wavelengths $(\lambda_0 \gg \lambda_{oc})$, α approaches a value of $2\pi/\lambda_{oc}$ as indicated. This diagram applies to lossless hollow single-conductor guides with cross sections of any shape.

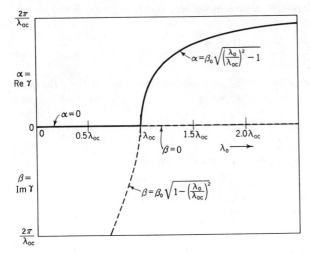

Figure 13-26 Composite graph showing attenuation constant α and phase constant β for a lossless hollow single-conductor waveguide as a function of the wavelength λ_0 in an unbounded medium. At wavelengths less than the cutoff wavelength (λ_{oc}) there is no attenuation ($\alpha = 0$) but a phase shift β. At wavelengths longer than cutoff there is no phase shift ($\beta = 0$) but an attenuation α.

Actual guides are not lossless, so that α is not zero for $\lambda_0 < \lambda_{oc}$, as indicated in Fig. 13-26. However, for air-filled guides of a good conducting material, such as copper, β is substantially as indicated for $\lambda_0 < \lambda_{oc}$, while α is small but not necessarily negligible. To calculate α, we note (see Fig. 13-27) that the average power in the guide varies with the distance x in the direction of transmission as given by

$$P = P_0 e^{-2\alpha x} \quad \text{(W)} \tag{3}$$

where P_0 = average power at reference point ($x = 0$), W

x = distance in direction of transmission through guide, m

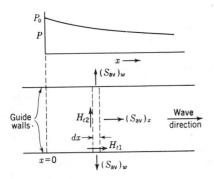

Figure 13-27 Power lost in walls of waveguide results in attenuation.

The factor 2 in the exponent is present because power is proportional to field squared. It follows that

$$\alpha = \frac{1}{2} \frac{-dP/dx}{P} \quad (\text{Np m}^{-1}) \tag{4}$$

In (4), $-dP/dx$ represents the decrease in power per unit distance along the guide at a particular location, while P is the power transmitted through the guide at that location.†

Thus, in words the attenuation constant in nepers per unit distance is expressed by

$$\alpha = \frac{\text{power lost per unit distance}}{\text{twice the power transmitted}}$$

If the medium filling the guide is lossless, the decrease in power per unit distance is equal to the power lost per unit distance in the walls of the guide. This is

$$-\frac{dP}{dx} = \frac{1}{dx} \iint (S_{av})_w \, ds = \int (S_{av})_w \, dl \tag{5}$$

where $(S_{av})_w$ is the average Poynting vector into the wall (average with respect to time). The surface integral in (5) is taken over a strip of length dx of the interior surface of the waveguide (Fig. 13-27). The line integral in (5) is taken around the inside of the guide (same path as that for strip). In general, the average Poynting vector is

$$\mathbf{S}_{av} = \tfrac{1}{2} \text{Re} \, \mathbf{E} \times \mathbf{H}^* \tag{6}$$

Since $\mathbf{E}$ and $\mathbf{H}$ are normal, the magnitude of the average Poynting vector into the conducting wall medium is (see Fig. 13-27)

$$(S_{av})_w = \tfrac{1}{2} \text{Re} \, H_{t1} H_{t1}^* Z_c = \tfrac{1}{2} |H_{t1}|^2 \, \text{Re} \, Z_c \tag{7}$$

where $|H_{t1}|$ = absolute value (or magnitude) of component of $\mathbf{H}$ tangent to conducting surface of guide walls

$\text{Re} \, Z_c$ = real part of intrinsic impedance of conducting wall medium
$= \sqrt{\mu\omega/2\sigma}$

Introducing (7) in (5) yields

$$-\frac{dP}{dx} = \frac{\text{Re} \, Z_c}{2} \int |H_{t1}|^2 \, dl \tag{8}$$

Now the power traveling through the guide (in x direction) is

$$P = \iint (S_{av})_x \, ds \quad (\text{W}) \tag{9}$$

---

† It is to be noted that the attenuation in this case is due to an actual power loss (joule heating of guide walls), whereas at frequencies less than cutoff no joule-heating effect is involved, the attenuation being due to the inability of the guide to transmit the higher-order mode (wave reflected).

where in this case the surface integral is taken over the guide cross section. It follows that

$$P = \tfrac{1}{2} \operatorname{Re} Z_{yz} \iint |H_{t2}|^2 \, ds \qquad (10)$$

where $|H_{t2}|$ = absolute value of component of **H** tangent to a cross-sectional plane through guide (Fig. 13-27)

$\operatorname{Re} Z_{yz}$ = real part of transverse impedance of guide

Therefore, the attenuation constant α is, in general, given by

$$\alpha = \frac{\operatorname{Re} Z_c \int |H_{t1}|^2 \, dl}{2 \operatorname{Re} Z_{yz} \iint |H_{t2}|^2 \, ds} \qquad (\text{Np m}^{-1}) \qquad (11)$$

where $\operatorname{Re} Z_c$ = real part of intrinsic impedance of guide walls (conductor)

$\operatorname{Re} Z_{yz}$ = real part of transverse impedance of guide

$|H_{t1}|$ = absolute value of component of **H** tangent to conducting surface of guide walls (integrated around interior surface of guide)

$|H_{t2}|$ = absolute value of component of **H** tangent to plane of cross section through guide (integrated over cross-sectional area) (Fig. 13-27)

Equation (11) applies to any mode in any guide. For each mode the attenuation constant must be calculated using (11), with values of $|H_{t1}|$ and $|H_{t2}|$ corresponding to the field distribution for that mode. If the guide walls are of good conducting material, we may assume, with but little error, that the **H**-field distribution used in (11) is the same as for perfectly conducting walls. The following example illustrates an application of (11) to a simple problem.

Example Find the attenuation constant for a 300-MHz TEM wave in an infinite-parallel-plane transmission line with a spacing between planes of 100 mm. The planes or walls are made of copper, and the medium between the planes is air.

SOLUTION For a TEM wave the transverse impedance equals the intrinsic impedance; so (11) becomes

$$\alpha = \frac{2 \operatorname{Re} Z_c \int_0^{y1} |H_{t1}|^2 \, dy}{2 \operatorname{Re} Z_d \int_0^{y1} \int_0^{z1} |H_{t2}|^2 \, dz \, dy} \qquad (12)$$

where y_1 = arbitrary distance along conducting wall (see Fig. 13-2)

z_1 = spacing between walls, m

$\operatorname{Re} Z_c$ = real part of intrinsic impedance of conducting walls, Ω

$\operatorname{Re} Z_d$ = real part of intrinsic impedance of dielectric medium between walls (Z_d is entirely real for lossless medium)

The integral with H_{t1} involves the power lost in one wall of the line. The total power lost in both walls is twice this; hence the factor 2 in the numerator. For a TEM wave **H** is everywhere parallel to the walls and normal to the direction of propagation, so that both H_{t1} and H_{t2} are perpendicular to the page instead of as suggested in Fig. 13-27. It follows that $|H_{t1}| = |H_{t2}| =$ a constant. Hence, (12) reduces to

$$\alpha = \frac{\text{Re } Z_c y_1}{\text{Re } Z_d y_1 z_1} = \frac{\text{Re } Z_c}{z_1 \text{ Re } Z_d} \tag{13}$$

For copper at 300 MHz, Re $Z_c = 4.55$ mΩ, while for air Re $Z_d = 376.7$ Ω. Therefore

$$\alpha = 1.2 \times 10^{-4} \text{ Np m}^{-1}$$

or

$$\alpha = 1.04 \times 10^{-3} \text{ dB m}^{-1}$$

Thus, the attenuation amounts to about 1 dB km^{-1}.

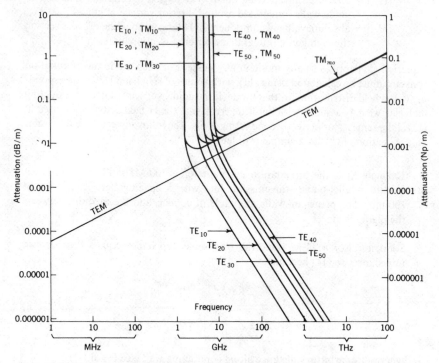

Figure 13-28 Attenuation α in decibels and nepers per meter as a function of frequency for various modes in an air-filled infinite-parallel-plane waveguide of copper with 100-mm spacing between planes. Note that the TEM mode and the TM_{mo} modes shown cannot occur in a single hollow rectangular guide but the TE_{mo} modes can. At frequencies below cutoff for the TE_{10} mode (1.5 GHz) only the TEM mode is transmitted but many TE and TM modes can be transmitted above cutoff and at sufficiently high frequencies the TE modes have the lowest attenuation.

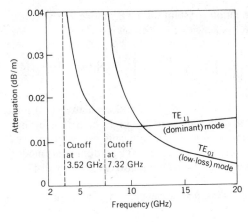

Figure 13-29 Attenuation α in decibels per meter as a function of frequency for TE_{01} (low-loss) and TE_{11} (dominant) modes in a circular copper waveguide of 50.8-mm inside diameter.

The attenuation in decibels per meter (and in nepers per meter) as a function of frequency for a number of modes in an air-filled infinite-parallel-plane guide of copper is shown in Fig. 13-28. The spacing between planes is 100 mm. We note that there is no cutoff for the TEM mode and attenuation decreases for this mode with decreasing frequency. However, at 100 GHz the TE_{10} mode has the lowest attenuation although the TE_{20} and higher-order modes can also be transmitted. The TE_{20} and higher-order modes all have higher attenuation than the TE_{10} mode.

Table 13-3 Rectangular and circular cylindrical waveguide parameters

Symbol	Name	Equation (rectangular or cylindrical)		
$\alpha\dagger$	Attenuation constant:			
	Below cutoff frequency	$\alpha = \dfrac{2\pi}{\lambda_0}\sqrt{\left(\dfrac{\lambda_0}{\lambda_{oc}}\right)^2 - 1}$ (Np m^{-1})		
	Above cutoff frequency	$\alpha = \dfrac{\operatorname{Re} Z_c \int |H_{t1}|^2\, dl}{2\operatorname{Re} Z_{yz}\iint |H_{t2}|^2\, ds}$ (Np m^{-1})		
$\beta\dagger$	Phase constant	$\beta = \sqrt{\left(\dfrac{2\pi}{\lambda_0}\right)^2 - k^2}$ (rad m^{-1})		
v_p	Phase velocity	$v_p = \dfrac{\omega}{\beta} = \dfrac{v_0}{\sqrt{1-(\lambda_0/\lambda_{oc})^2}}$ (m s^{-1})		
		Rectangular	Cylindrical	
$Z_{yz},\ Z_{r\phi}$	Transverse-wave impedance	$Z_{yz} = \dfrac{Z_d}{\sqrt{1-(\lambda_0/\lambda_{oc})^2}}$	$Z_{r\phi} = \dfrac{Z_d}{\sqrt{1-(\lambda_0/\lambda_{oc})^2}}$	(Ω)
		TE_{mn} mode only	TE_{nr} mode	
λ_{oc}	Cutoff wavelength	$\lambda_{oc} = \dfrac{2}{\sqrt{(n/y_1)^2 + (m/z_1)^2}}$	$\lambda_{oc} = \dfrac{2\pi r_0}{k'_{nr}}$ or $\dfrac{2\pi r_0}{k_{nr}}$	(m)
		TE_{mn} or TM_{mn} mode	TE_{nr} mode (k'_{nr}), TM_{nr} mode (k_{nr})	

$\dagger\ \gamma$ = propagation constant = $\alpha + j\beta$.

It may be shown that the attenuation constant for a TE_{nr} mode in a circular cylindrical waveguide of radius a at frequencies above cutoff is

$$\alpha = \frac{\text{Re } Z_c}{a \text{ Re } Z_d \sqrt{1 - (f_c/f)^2}} \left[\left(\frac{f_c}{f} \right)^2 + \frac{n^2}{(k'_{nr})^2 - n^2} \right] \quad (\text{Np m}^{-1}) \qquad (14)$$

For copper $\text{Re } Z_c = \sqrt{\pi \mu / \sigma} \sqrt{f} = 2.63 \times 10^{-7} \sqrt{f}$ (Ω), and evaluating (14) for TE_{01} and TE_{11} modes yields the attenuation curves of Fig. 13-29. Note that for the TE_{01} (low-loss) mode the attenuation decreases monotonically as the frequency increases. This trend also occurs for any TE_{0r} mode. The TE_{11} (dominant) mode has a minimum at $f/f_c = 3.1$. A minimum at this f/f_c ratio also occurs with all TE_{nr} modes for which $n \neq 0$.

Parameters of rectangular and cylindrical waveguides are summarized in Table 13-3.

13-9 WAVEGUIDE DEVICES

Several basic waveguide devices are discussed in this section, namely, *terminations*, *power dividers*, and *guide-to-line transitions*. Reference should be made to more specialized books for detailed treatments of these and other waveguide devices.†

A matched *termination* for a rectangular or circular waveguide is shown in cross section in Fig. 13-30a. A card of resistance material is placed transversely in the guide $\lambda/4$ from a metal plate capping the end of the guide. The situation here is similar to that discussed in Sec. 10-16 for the terminated wave. For zero reflection it is necessary only that the card have a resistance per square equal to the transverse-wave impedance of the guide. The termination of Fig. 13-30a will be matched at the design frequency for which the card-to-end-plate distance is $\lambda/4$ but not at adjacent frequencies. This termination is a narrowband device. To provide a broadband termination a wedge of resistance material can be used, as suggested in Fig. 13-30b. The length of the wedge should be of the order of a wavelength or more.

In applications where waveguides feed two or more antennas a *power divider* may be required to divide the power in a predetermined ratio. Figure 13-30c shows a power divider for a rectangular waveguide (with TE_{10} mode) delivering twice the power to the lower branch as compared to the upper. The division is achieved by inserting a thin septum which divides the guide cross-sectional area in the ratio 2:1. Note that the septum is perpendicular to the direction of the electric field vector **E** so that it does not disturb the field configuration in the guide. The height of the guide to the right of the septum in each branch is increased gradually over a distance of several wavelengths back to the standard height h. If both branches are

---

† N. Marcuvitz, "Waveguide Handbook," McGraw-Hill Book Company, New York, 1949; J. C. Slater, "Microwave Electrons," D. Van Nostrand Company, Inc., New York, 1950.

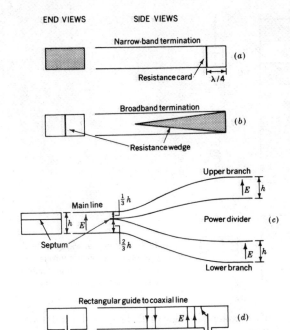

Figure 13-30 (a) and (b) Waveguide terminations, (c) power divider, and (d) and (e) waveguide-to-coaxial-line transitions. The arrows with E indicate the direction of the electric fields in the guides.

connected into nonreflecting loads or antennas, there will be no reflection from this power divider and the device can be used over a broad band of frequencies.

At some point in most waveguide systems it is necessary to convert to TEM-mode transmission on a coaxial line. Two *waveguide-to-coaxial-line transitions* are shown in Fig. 13-30d and e. The one in Fig. 13-30d provides a transition from a TE_{10} mode in a rectangular guide to a coaxial line, while the one in Fig. 13-30e provides a transition from a TM_{01} mode in a circular waveguide to a coaxial line.

13-10 WAVEGUIDE IRIS THEORY

A partial opening, or iris, in a waveguide presents a discontinuity which is analogous to placing a susceptance across a transmission line. If the iris is located the proper distance from a load or other discontinuity, and if the susceptance of the iris is correct, the iris can act as a matching device.

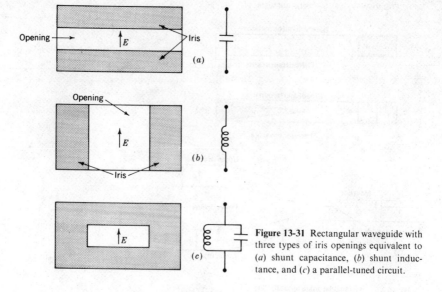

Figure 13-31 Rectangular waveguide with three types of iris openings equivalent to (a) shunt capacitance, (b) shunt inductance, and (c) a parallel-tuned circuit.

Consider the cross section of a rectangular guide (with TE$_{10}$ mode) and iris or slot openings as in Fig. 13-31. It can be shown† that the normalized susceptance presented by the iris is given by

$$B = j2Z^2 \frac{\iint \mathbf{H}_t \cdot \mathbf{H}_t' \, ds}{\iint \mathbf{E}_t \cdot \mathbf{E}_t' \, ds} \qquad (1)$$

where $\mathbf{E}_t, \mathbf{H}_t$ = traverse field components of dominant mode in guide (without iris)

$\mathbf{E}_t', \mathbf{H}_t'$ = transverse components of the actual field at iris

The quantity Z is the characteristic impedance of the guide.‡ The numerator is integrated over the metallic surface of the iris, and the denominator is integrated over the opening. The field components E_t and H_t can be calculated simply from the guide geometry, but the determination of the actual field components (E_t' and H_t') is more difficult since they are the resultant of many higher-order modes which exist in the proximity of the iris.

If the opening is small compared to the metal area of the iris, the denominator in (1) is small and the susceptance B is large. As the opening is reduced to zero, B approaches infinity. Conversely, a large opening gives a small B, and as the opening becomes complete (no iris at all), B goes to zero.

---

† Slater, ibid., p. 118.

‡ The term *characteristic impedance* usually implies a two-conductor transmission line but it can be extended to the dominant mode in a waveguide. For example, in a rectangular waveguide with TE$_{10}$ mode it is equal to the ratio of the maximum voltage across the guide to the wall current.

For the iris of Fig. 13-31a with full horizontal slot the susceptance is capacitive (positive), and the effect is like a capacitor across a transmission line. For the iris of Fig. 13-31b with full vertical slot the susceptance is inductive (negative), and the slot acts like an inductance across a transmission line. For the iris of Fig. 13-31c there are both capacitive and inductive susceptances, and if the dimensions are chosen properly, the total, or net, susceptance can be made zero. In this case the effect is like a parallel-tuned circuit across a transmission line, and the device provides a bandpass-filtering action with no reflection (complete transmission) at the resonant or design frequency.

13-11 INTRINSIC, CHARACTERISTIC, AND WAVE IMPEDANCES

We have dealt with three types of impedances: intrinsic, characteristic, and wave. *Intrinsic impedance* refers to the ratio of the phasor fields E and H for a plane (TEM) wave in an unbounded medium.

For a medium with complex μ and ϵ,

$$\text{Intrinsic impedance} = Z = \frac{E}{H} = \sqrt{\frac{\omega\mu'' + j\omega\mu'}{\sigma + j\omega(\epsilon' - j\epsilon'')}} \quad (\Omega) \quad (1)$$

For a conducting medium with real μ and ϵ,

$$\text{Intrinsic impedance} = Z = \frac{E}{H} = \sqrt{\frac{j\omega\mu}{\sigma + j\omega\epsilon}} \quad (\Omega) \quad (2)$$

For a nonconducting medium with real μ and ϵ,

$$\text{Intrinsic impedance} = Z_d = \frac{E}{H} = \sqrt{\frac{\mu}{\epsilon}} \quad (\Omega) \quad (3)$$

For free space,

$$\text{Intrinsic impedance} = Z_0 = \frac{E}{H} = \sqrt{\frac{\mu_0}{\epsilon_0}} = 376.7 \ \Omega \text{ per square} \quad (4)$$

For a good conductor ($\sigma \gg \omega\epsilon$)

$$\text{Intrinsic impedance} = Z_c = \frac{E}{H} = \sqrt{\frac{\mu\omega}{\sigma}} \underline{/45°} \quad (\Omega) \quad (4a)$$

The *characteristic impedance* refers to the ratio of the (phasor) voltage V to the (phasor) current I on an infinite two-conductor transmission line. The voltage V is equal to the integral of the electric field **E** along a path between the conductors, while the current I is equal to the integral of the magnetic field **H** around one of the conductors (Ampère's law). The characteristic impedance of a two-conductor line can also be expressed in terms of the series resistance R, series inductance L, the shunt conductance G, and shunt capacitance C. Thus, for a two-conductor line

$$\text{Characteristic impedance} = Z_0 = \frac{V}{I} = \frac{\int \mathbf{E} \cdot d\mathbf{l}}{\oint \mathbf{H} \cdot d\mathbf{l}} = \sqrt{\frac{R + j\omega L}{G + j\omega C}} \quad (\Omega) \quad (5)$$

For a transmission-line cell the characteristic impedance V/I equals the intrinsic impedance E/H.

The *wave impedance* refers to the ratio of an electric field component to a magnetic field component at the same point of the same wave. For a TEM wave the wave impedance is the same as the intrinsic impedance, but for higher-order modes, as in a hollow single-conductor waveguide, there can be as many wave impedances as there are combinations of electric and magnetic field components. In the preceding sections we have confined our attention to the *transverse-wave impedance*, e.g., the ratio of E_y to H_z for a TE mode in a rectangular guide (both E_y and H_z are transverse to the direction of wave propagation x). Thus,

$$\text{Transverse-wave impedance} = Z_{yz} = \frac{E_y}{H_z} \quad (\Omega) \quad (6)$$

In a cylindrical waveguide the transverse-wave impedance equals $Z_{r\phi}(= E_r/H_\phi)$. The transverse-wave impedance of a waveguide is a function of the intrinsic impedance of the medium filling the guide and the guide dimensions as given in Tables 13-1 and 13-3. As the transverse guide dimensions become very large compared to the wavelength, the transverse-wave impedance of the guide approaches the intrinsic impedance of the medium.

Whereas the characteristic impedance is basically a circuit quantity ($= V/I$), the intrinsic impedance and the wave impedance are field or wave quantities involving the ratios of electric to magnetic fields.

13-12 WAVES TRAVELING PARALLEL TO A PLANE BOUNDARY

In Sec. 13-8 we considered the attenuation due to the power lost in the walls of a waveguide. In this section some of the phenomena associated with this power loss or power flow are discussed in more detail.

Consider the plane boundary between two media shown in Fig. 13-32a, assuming that medium 1 is air and medium 2 is a perfect conductor. From the boundary condition that the tangential component of the electric field vanishes at the surface of a perfect conductor, the electric field of a TEM wave traveling parallel to the boundary must be exactly normal to the boundary, as portrayed in the figure. However, if medium 2 has a finite conductivity σ, there will be a tangential electric field E_x at the boundary, and, as a result, the electric field of a wave traveling along the boundary has a *forward tilt*, as suggested in Fig. 13-32b. From the continuity relation for tangential electric fields, the field on both sides of the boundary is E_x.

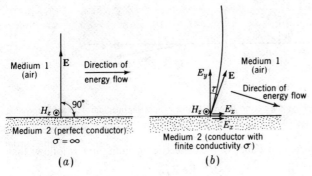

Figure 13-32 TEM wave traveling to right (*a*) along surface of perfectly conducting medium and (*b*) along surface of medium with finite conductivity.

The direction and magnitude of the power flow per unit area are given by the Poynting vector. The average value of the Poynting vector is

$$S_{av} = \tfrac{1}{2} \operatorname{Re} \mathbf{E} \times \mathbf{H}^* \qquad (\text{W m}^{-2}) \qquad (1)$$

At the surface of the conducting medium (Fig. 13-32*b*) the power into the conductor is in the negative *y* direction, and from (1) its average value per unit area is†

$$S_y = -\tfrac{1}{2} \operatorname{Re} E_x H_z^* \qquad (2)$$

The space relation of E_x, H_z (or H_z^*), and S_y is shown in Fig. 13-33*a*. But

$$\frac{E_x}{H_z} = Z_c \qquad (3)$$

where Z_c is the intrinsic impedance of the conducting medium, so that (2) can be written

$$S_y = -\tfrac{1}{2} H_z H_z^* \operatorname{Re} Z_c = -\tfrac{1}{2} H_{z0}^2 \operatorname{Re} Z_c \qquad (4)$$

where $H_z = H_{z0} e^{-j\xi - \gamma x}$

$H_z^* = H_{z0} e^{j\xi + \gamma x}$ = complex conjugate of H_z

ξ = phase lag of H_z with respect to E_x

The relation for the Poynting vector in (4) is the same as given in (13-8-7).

At the surface of the conducting medium (Fig. 13-32*b*) the power per unit area flowing parallel to the surface (*x* direction) is

$$S_x = \tfrac{1}{2} \operatorname{Re} E_y H_z^* \qquad (5)$$

The space relation of E_y, H_z (or H_z^*), and S_x is illustrated by Fig. 13-33*b*. But

$$\frac{E_y}{H_z} = Z_d \qquad (6)$$

† The component of the average Poynting vector in the *y* direction is $(S_{av})_y$, but to simplify notation we shall write S_y for $(S_{av})_y$.

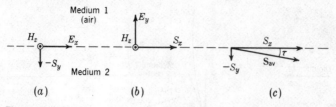

Figure 13-33 Fields and Poynting vector at surface of a conducting medium with wave traveling parallel to surface.

where Z_d is the intrinsic impedance of the dielectric medium (air). It follows that

$$S_x = \tfrac{1}{2}H_{z0}^2 \operatorname{Re} Z_d \qquad (7)$$

The total average Poynting vector is then

$$\mathbf{S}_{av} = \hat{\mathbf{x}}S_x + \hat{\mathbf{y}}S_y = \frac{H_{z0}^2}{2}(\hat{\mathbf{x}} \operatorname{Re} Z_d - \hat{\mathbf{y}} \operatorname{Re} Z_c) \qquad (8)$$

The relation of $\mathbf{S}_{av}$ to its x and y components is illustrated in Fig. 13-33c. It is to be noted that the average power flow (per unit area) is not parallel to the surface but inward at an angle τ. This angle is also the same as the angle of forward tilt of the average electric field (see Fig. 13-32b). If medium 2 were perfectly conducting, τ would be zero.

It is of interest to evaluate the tilt angle τ for a couple of practical situations. This is done in the following examples.

Example 1 Find the forward tilt angle τ for a vertically polarized 3-GHz wave traveling in air along a sheet of copper.

SOLUTION From (8) the tilt angle τ is given by

$$\tau = \tan^{-1}\frac{\operatorname{Re} Z_c}{\operatorname{Re} Z_d} \qquad (9)$$

At 3 GHz, we have for copper that $\operatorname{Re} Z_c = 14.4$ mΩ. The intrinsic impedance of air is independent of frequency ($\operatorname{Re} Z_d = 376.7\ \Omega$). Thus

$$\tau = \tan^{-1}\frac{1.44 \times 10^{-2}}{376.7} = 0.0022°$$

Although τ is not zero in the above example, it is very small, so that **E** is nearly normal to the copper surface and **S** nearly parallel to it. This small value of tilt is typical at most air-conductor boundaries but accounts for the power flow into the conducting medium. If the conductivity of medium 2 is very low, or if it is a dielectric medium, τ may amount to a few degrees. Thus the forward tilt of a vertically polarized radio wave propagating along poor ground is sufficient to produce an appreciable horizontal electric field component. In the Beverage, or wave, antenna this horizontal component is utilized to induce emfs along a horizontal wire oriented parallel to the direction of transmission of the wave.

In contrast to Example 1, in which medium 2 is copper, the following example considers the case of freshwater as medium 2.

Example 2 Find the forward tilt angle τ for a vertically polarized 3-GHz wave traveling in air along the surface of a smooth freshwater lake.

SOLUTION At 3 GHz the conduction current in freshwater is negligible compared with the displacement current, so that the lake may be regarded as a dielectric medium of relative permittivity $\epsilon_r = 80$. Thus

$$\tau = \tan^{-1} \frac{1}{\sqrt{80}} = 6.4°$$

In this case the forward tilt of 6.4° is sufficient to be readily detected by a direct measurement of the direction of the electric field.

The angle τ discussed above is an average value. In general, the instantaneous direction of the electric field varies as a function of time. In the case of a wave in air traveling along a copper sheet, E_y and E_x are in phase octature (45° phase difference), so that at one instant of time the total field $\mathbf{E}$ may be in the x direction and $\frac{1}{8}$ period later it will be in the y direction. (See Fig. 13-34a.) With time, the locus of the tip of $\mathbf{E}$ describes a cross-field ellipse as portrayed in Fig. 13-34b for a 3-GHz wave in air traveling along a copper sheet (in the x direction) as in Example 1. The ellipse is not to scale, the abscissa values being magnified 5,000 times. The positions

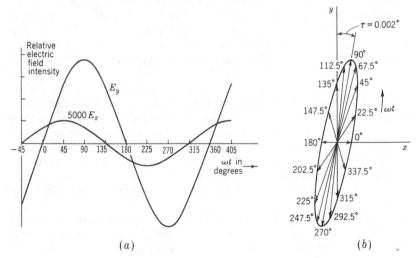

(a)   (b)

Figure 13-34 (a) Magnitude variation with time of E_y and E_x components of $\mathbf{E}$ in air at the surface of a copper region for a 3,000-MHz TEM wave traveling parallel to the surface. (b) Resultant values of $\mathbf{E}$ (space vector) at 22.5° intervals over one cycle, illustrating elliptical cross field at the surface of the copper region. The wave is traveling to the right. Note that although $\mathbf{E}$ has x and y components (cross field) it is linearly polarized (in y-z plane).

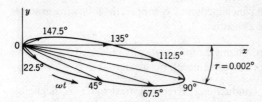

Figure 13-35 Poynting vector in air at a point on the surface of a copper region for a 3,000-MHz TEM wave traveling along the surface (to right). The Poynting vector is shown at 22.5° intervals over one-half cycle. The ordinate values are magnified 5,000 times compared with the abscissa values.

of **E** for various values of ωt are indicated. The variation of the instantaneous Poynting vector for this case is shown in Fig. 13-35. Here the ordinate values are magnified 5,000 times. It is to be noted that the tip of the Poynting vector travels around the ellipse twice per cycle.

Whereas copper has a complex intrinsic impedance, freshwater, at the frequency considered in Example 2, has a real intrinsic impedance. It follows that the E_x and E_y components of the total field **E** are in time phase so that the cross-field ellipse in this case collapses to a straight line (linear cross field) with a forward tilt of 6.4°.

13-13 OPEN WAVEGUIDES

In the previous section we have seen that a wave traveling along an air-conductor or air-dielectric boundary has a longitudinal (E_x) component of the electric field, resulting in a forward tilt of the total electric field. Hence the Poynting vector is not entirely parallel to the boundary but has a component directed from the air into the adjacent medium, as suggested in Fig. 13-33c. This tends to keep the energy in the wave from spreading out and to concentrate it near the surface, resulting in a *bound wave*, or *surface wave*. The phase velocity of such a bound wave is always less than the velocity in free space. Although the field of this guided wave extends to infinity, such a large proportion of the energy may be confined within a few wavelengths of the surface that the surface can be regarded as an open type of waveguide. It should be noted, however, that even though the forward-tilt effect is present along all finitely conducting surfaces, the bound wave may be of negligible importance without a launching device of relatively large dimensions (several wavelengths across) to initiate the wave. If the surface is perfectly smooth and perfectly conducting, the tangential component of the electric field vanishes, there is no forward tilt of the electric field, and there is no tendency whatever for the wave to be bound to the surface.

In 1899 Sommerfeld† showed that a wave could be guided along a round wire of finite conductivity. Zenneck‡ pointed out that for similar reasons a wave traveling along the earth's surface would tend to be guided by the surface.

---

† A. Sommerfeld, Forpflanzung elektrodynamischer Wellen an einem zylindrischen Leiter, *Ann. Phys. u. Chem.*, **67**:233 (December 1899).

‡ J. Zenneck, Uber die Fortpflanzung ebener elektromagnetischer Wellen langs einer ebenen Leiterflache und ihre Beziehung zur drahtlosen Telegraphie, *Ann. Phys.*, (4), **23**:846–866 (Sept. 20, 1907).

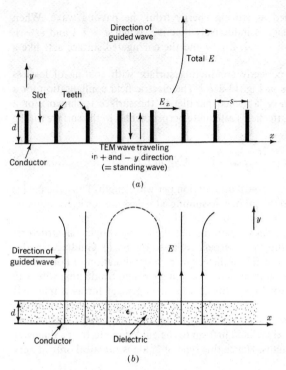

Figure 13-36 (a) Wave guided by corrugated surface. (b) TM_0 wave guided by conductor with dielectric coating of thickness d.

The guiding action of a flat conducting surface can be enhanced by adding corrugations or a dielectric coating or slab. If the dielectric slab is sufficiently thick, it can act alone as an effective *nonmetallic guide*. The characteristics of a number of these open waveguides are discussed in this chapter.

Consider a perfectly conducting flat surface of infinite extent with transverse conducting corrugations, as in Fig. 13-36a. The corrugations have many teeth per wavelength ($s \ll \lambda$). The slots between the teeth can support a TEM wave traveling vertically with electric field component E_x. Thus, each slot acts like a short-circuited section of a parallel-plane two-conductor transmission line of length d. Assuming lossless materials, the impedance Z presented to a wave traveling vertically downward onto the corrugated surface is a pure reactance, or

$$Z = jZ_d \tan \frac{2\pi\sqrt{\epsilon_r}d}{\lambda_0} \quad (\Omega) \qquad (1)$$

where Z_d = intrinsic impedance of medium filling the slots, Ω
$\quad \epsilon_r$ = relative permittivity of the medium, dimensionless
$\quad \lambda_0$ = free-space wavelength, m
$\quad d$ = depth of the slots, m

The slots may be regarded as storing energy from the passing wave. When $2\pi\sqrt{\epsilon_r}\,d/\lambda_0 < 90°$, the surface is inductively reactive. When $d = \lambda/4$ and $\epsilon_r = 1$, $Z = \infty$. When $d = \lambda/2$ and $\epsilon_r = 1$, $Z = 0$ and the corrugated surface acts like a conducting sheet.

Consider next a flat perfectly conducting surface with coating of lossless dielectric of thickness d, as in Fig. 13-36b.† The electric field configuration for a TM$_0$ (dominant) mode wave launched parallel to the surface is shown. For a sufficiently thick coating d, the fields attenuate perpendicular to the surface (in the y direction) as $e^{-\alpha y}$, where

$$\alpha = \frac{2\pi}{\lambda_0}\sqrt{\epsilon_r - 1} \qquad (\text{Np m}^{-1}) \qquad (2)$$

For $\epsilon_r = 2$ this gives over 50-dB attenuation per wavelength. Thus, the field is effectively confined close to the surface. Assuming a lossless dielectric, the attenuation in the x direction will be due entirely to radiation.

In the preceding paragraphs we have discussed guiding by open flat structures of infinite extent in the z direction (perpendicular to the page). Guiding can also occur along round wires of metal, of dielectric, or a combination of both. As an example, let us consider the guiding action of a single round conducting wire with dielectric coating. As shown by Goubau,‡ this arrangement forms a relatively efficient open type of waveguide. However, to initiate the guided wave along the wire with good efficiency requires a relatively large launching device, its function being to excite a mode, closely related in form to the guided mode, over a diameter of perhaps several wavelengths. Hence this type of guide is practical only at very high frequencies.

A dielectric-coated single-wire waveguide with typical dimensions is illustrated in Fig. 13-37. The dielectric coat consists of a layer of enamel of relative permittivity $\epsilon_r = 3$ having a thickness of only 0.0005λ. The wire diameter is 0.02λ. The configuration of the electric field lines in the launcher and along the wire guide is suggested in the figure. The mode on the wire is a TM type, but it is like a plane TEM wave to a considerable distance from the wire.

Wires wound in the form of long helices are also effective single-conductor open-type waveguides. Helix diameters as large as 0.4λ can be used successfully.

For efficient transmission of energy by a guiding system the attenuation should be small. With two-conductor transmission lines this requires that the series resistance R and the shunt conductance G be small. The conductor separation must also be small compared to the wavelength in order that radiation losses be negligible. Under these conditions the fields vary as $1/r^2$, where r is the distance perpendicular to the line and the power density varies as $1/r^4$. Thus, most of the power flow is close to the line (see Fig. 10-59b). Waves carried in a single hollow conducting

---

† S. A. Schelkunoff, Anatomy of "Surface Waves," *IRE Trans. Antennas Propag.*, **AP-7**:S133–139 (December 1959); R. F. Harrington, "Time-Harmonic Electromagnetic Fields," p. 168, McGraw-Hill Book Company, New York, 1961.

‡ G. Goubau, Surface Waves and Their Application to Transmission Lines, *J. Appl. Phys.*, **21**:1119–1128 (November 1950).

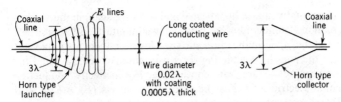

Figure 13-37 Single coated-wire open waveguide, or "G string." (*After Goubau.*)

waveguide will be unattenuated if the guide walls are perfectly conducting and the material filling the guide is lossless. Perfectly conducting walls also prevent any radiation from the guide. With open guides losses due to radiation tend to become significant, and modes which confine the power flow close to the guiding surface are desirable for efficient transmission. High attenuation of the fields perpendicular to guides is also important to reduce coupling or cross talk between adjacent transmission systems.

13-14 DIELECTRIC SHEET WAVEGUIDE

The guides we have been discussing are totally or partially metallic. Let us consider now guides which are entirely of dielectric material such as an infinite dielectric sheet of thickness d as suggested in Fig. 13-38. The sheet extends infinitely far in the x and z directions, and its permittivity ϵ_1 is greater than the permittivity ϵ_2 of the medium above and below.

In Chaps. 10 and 12 we discussed wave propagation into dielectric media with the wave entering as from above in Fig. 13-38. Now let us consider wave propagation in the sheet in the x direction (to right) due to a TEM wave launched into the sheet from the left. This wave can be largely confined inside the sheet by multiple reflections provided its angle of incidence θ_i with respect to the upper and lower surfaces is less than the critical angle θ_{ic}. Under this condition, as discussed in Sec.

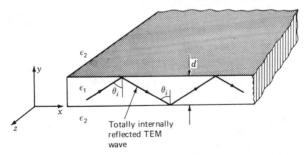

Figure 13-38 Section of dielectric sheet of infinite extent (in x and z directions) and of thickness d (in y direction). Zigzag line shows path of TEM wave propagating inside the sheet (in x direction) by total internal reflection.

12-2, the wave will be totally internally reflected and will propagate in the x direction via a zigzag path in a manner like that between two infinite conducting sheets, one on the upper and the other on the lower surface. However, there are important differences between the parallel conducting sheet waveguide of Sec. 13-3 and the dielectric sheet waveguide we are now considering. For perfectly conducting sheets the tangential electric field must be zero at the sheets ($E_x = E_z = 0$ at $y = 0$ and $y = d$) and zero outside. For the dielectric sheet, the field is not zero at the surfaces and it extends outside, theoretically to infinity. But as we noted in Example 2 of Sec. 12-2, this external field may attenuate very rapidly away from the sheet, indicating that it is tightly bound to the sheet.

At first glance one might suppose that any wave for which $\theta_i > \theta_{ic}$ will propagate in the dielectric sheet. However, because of interference, waves will actually propagate only at certain angles. Referring to Fig. 13-39, consider two rays, 1 and 2, belonging to the same TEM wave and incident at an angle $\theta_i > \theta_{ic}$. This requires that

$$\theta_i > \theta_{ic} = \sin^{-1} \sqrt{\frac{\epsilon_2}{\epsilon_1}} \tag{1}$$

where ϵ_1 = permittivity of sheet, F m^{-1}
ϵ_2 = permittivity of medium above and below sheet, F m^{-1}

with $\epsilon_1 > \epsilon_2$.

The ray paths are shown by solid lines and the constant-phase fronts by dashed lines. The necessary condition for wave propagation in the sheet is that the phase length a of ray 2 and the phase length b of ray 1, including phase shifts on reflection, be equal or equal plus or minus an integral number of 2π radians.† Thus, in symbols the requirement is

$$\frac{2\pi}{\lambda_0} \eta_1 (b - a) + \phi = 2\pi n \tag{2}$$

where η_1 = index of refraction of medium 1 ($= \sqrt{\epsilon_{1r}}$)
ϕ = phase shift on reflection from surface
λ_0 = free space wavelength, m
n = integer ($= 0, 1, 2, 3, \ldots$)

or

$$\frac{2\pi\eta_1 d}{\lambda_0} \left[\frac{1}{\cos \theta_i} - \sin \theta_i \left(\tan \theta_i - \frac{1}{\tan \theta_i} \right) \right] + \phi = 2\pi n \tag{3}$$

which can be reduced to

$$\frac{4\pi\eta_1 d \cos \theta}{\lambda_0} + \phi = 2\pi n \tag{4}$$

† Dietrich Marcuse, "Theory of Dielectric Optical Waveguides," Academic Press, Inc., New York, 1974.

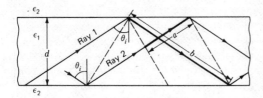

Figure 13-39 Geometry of wave reflection in dielectric sheet for permitted angles of reflection.

Restricting our attention to waves with **E** perpendicular to the plane of incidence (**E** in z direction), we have from (12-2-21) that the reflection coefficient for $\theta_i > \theta_{ic}$ is

$$\rho_\perp = \frac{\cos\theta_i - j\sqrt{\sin^2\theta_i - (\epsilon_2/\epsilon_1)}}{\cos\theta_i + j\sqrt{\sin^2\theta_i - (\epsilon_2/\epsilon_1)}} = 1\underline{/\phi} \tag{5}$$

where $\phi = -2\tan^{-1}(\sqrt{\sin^2\theta_i - (\epsilon_2/\epsilon_1)}/\cos\theta_i)$

Thus, the reflection coefficient $\rho_\perp$ has unit magnitude and a phase shift ϕ. Introducing (5) into (4),

$$\frac{4\pi\eta_1 d\cos\theta_i}{\lambda_0} - 2\pi n = 2\tan^{-1}\frac{\sqrt{\sin^2\theta_i - (\epsilon_2/\epsilon_1)}}{\cos\theta_i} \tag{6}$$

or

$$\tan\left(\frac{2\pi\eta_1 d\cos\theta_i}{\lambda_0} - \pi n\right) = \frac{\sqrt{\eta_1^2\sin^2\theta_i - \eta_2^2}}{\eta_1\cos\theta_i} \tag{7}$$

where η_1 = index of refraction of sheet ($= \sqrt{\epsilon_{1r}}$)
η_2 = index of refraction of medium above and below sheet ($= \sqrt{\epsilon_{2r}}$)
d = thickness of sheet, m
θ_i = angle of incidence, rad or deg
λ_0 = free space wavelength, m
n = integer ($= 0, 1, 2, 3, \ldots$)

To illustrate the significance of (7) consider the following example.

Example: Dielectric sheet waveguide A sheet of dielectric has a thickness $d = 10$ mm and index of refraction $\eta_1 = 1.5$. The medium above and below is air ($\eta_2 = 1$). The electric field is parallel to the sheet (in z direction in Fig. 13-38 with wave progressing in x direction). Find the angles of incidence θ_i if the wavelength $\lambda_0 = 10$ mm.

SOLUTION We note that the field **E**, although parallel to the sheet, is perpendicular to the plane of incidence. From (1) or (12-2-22) the critical angle

$$\theta_{ic} = \sin^{-1}\sqrt{\frac{\epsilon_2}{\epsilon_1}} = \sin^{-1}\frac{\eta_2}{\eta_1} = \sin^{-1}\frac{1}{1.5} = 41.8° \tag{8}$$

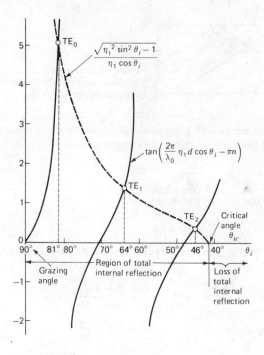

Figure 13-40 Solution of Eq. (7) for permitted angles of reflection in dielectric sheet of index of refraction $\eta_1 = 1.5$ and 1λ thick situated in air (Example 1). The three θ_i values (81, 64, and 46°) where the two curves intersect are the three solutions, or eigenvalues, for which Eq. (7) is satisfied. These angles correspond to three transverse electric modes: TE$_0$, TE$_1$, and TE$_2$.

Evaluating the left-hand side of (7) for angles of incidence θ_i greater than 41.8° yields the solid curves of Fig. 13-40. Evaluating the right-hand side of (7) in the same way yields the dashed curve. The three θ_i values, where the two curves intersect, are the three solutions, or eigenvalues, for which equation (7) is satisfied corresponding to $\theta_i = 46°$, $64°$, and $81°$. These angles correspond to three transverse electric modes in the sheet: TE$_0$, TE$_1$, and TE$_2$. All may be present simultaneously, but if the thickness d is decreased or the wavelength λ increased (or both), fewer solutions (eigenvalues) or modes will be possible. However, one solution (for the TE$_0$ mode) will always exist so that, at least in theory, waves can propagate to zero frequency.

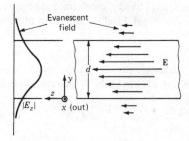

Figure 13-41 End view of sheet (from x direction) with graph (at left) and arrows (at right) suggesting variation of E_z with respect to y for TE$_0$-mode wave propagating out of page. Note that there is a spillover (evanescent) field above and below the sheet.

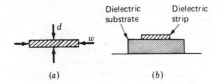

Figure 13-42 (*a*) End view of dielectric strip guide of thickness *d* and width *w*. (*b*) Dielectric strip on dielectric substrate of lower index of refraction as employed for light transmission in integrated circuits.

The variation of the field E_z across the sheet (in y direction) is shown in Fig. 13-41 for the TE_0 mode. The maximum field is at the center of the sheet with a spill-over (evanescent) field above and below which attenuates rapidly with distance above and below the sheet.

In addition to the transverse electric (TE) modes we have discussed, other modes are also possible, such as transverse magnetic (TM) modes with **H** transverse (in z direction).

If the sheet (of thickness d) is reduced to a width w in the z direction ($w \gg d$), it forms a dielectric strip waveguide or transmission line, as in Fig. 13-42*a*, with properties like those discussed above for the infinite sheet. Dielectric strips are often used as light guides in integrated circuits, the strip being mounted on a dielectric substrate of smaller index of refraction than that of the strip as suggested in Fig. 13-42*b*.

13-15 DIELECTRIC FIBER AND ROD WAVEGUIDES†

With Sec. 13-14 on the dielectric sheet waveguide as background, let us consider next the dielectric cylinder waveguide, which is usually referred to as a *fiber* or a *rod* depending on its diameter. Although the basic principles of the cylindrical dielectric guide are similar to those for the sheet, there are significant differences.

At or near optical wavelengths the dielectric cylinder guide can be physically small or threadlike in diameter. Such guides, called *optical fibers*, consist typically of a transparent core fiber of glass of index of refraction η_1 surrounded by a transparent glass sheath, or cladding, of slightly lower index η_2 with both enclosed in an opaque protective jacket. See Fig. 13-43. A typical core fiber of 25 μm is as fine as a human hair and can carry over a thousand two-way voice communication channels with an attenuation as small as 1 dB/km at light or infrared wavelengths ($\frac{1}{2}$ to 1 μm).

Figure 13-44 shows a cross section through the axis of an optical-fiber core of index of refraction η_1 with cladding of index η_2. A ray entering the core from an

† D. Hondros and P. Debye, Elektromagnetische Wellen an dielektrischen Drahten, *Ann. Phys.*, **32**:465–476 (1910); R. M. Whitmer, Fields in Non-metallic Guides, *Proc. IRE*, **36**:1105–1109 (September 1948).

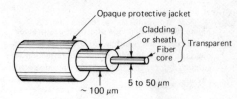

Figure 13-43 Optical-fiber guide with transparent core, transparent cladding, or sheath, and opaque protective jacket. Typical dimensions are indicated.

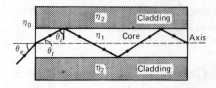

Figure 13-44 Wave entering fiber-optic core on axis will be trapped and propagate by total internal reflection if the entry angle θ_e is less than a critical value.

external medium of index η_0 at an angle θ_e will make an angle θ_t with respect to the axis inside the core. The relation between the angles, as given by Snell's law, is

$$\sin \theta_t = \frac{\eta_0}{\eta_1} \sin \theta_e \qquad (1)$$

The ray continuing in the core will be incident on the core-cladding boundary at an angle θ_i. If $\theta_i > \theta_{ic}$, where θ_{ic} is the critical angle, the ray will be totally internally reflected and continue to propagate inside the core.† From (1) and

$$\sin \theta_{ic} = \frac{\eta_2}{\eta_1} \qquad (2)$$

we have

$$\sin \theta_e = \frac{\eta_1}{\eta_0} \sin \theta_t = \frac{\eta_1}{\eta_0} \sin(90° - \theta_{ic}) = \frac{\eta_1}{\eta_0} \cos \theta_{ic} \qquad (3)$$

or

$$\sin \theta_e = \frac{\sqrt{\eta_1^2 - \eta_2^2}}{\eta_0} \qquad (4)$$

where θ_e = entrance angle on axis of core, rad or deg
  η_1 = index of refraction of core, dimensionless
  η_2 = index of refraction of cladding, dimensionless
  η_0 = index of refraction of external medium, dimensionless

For air as the external medium ($\eta_0 = 1$), (4) reduces to

$$\sin \theta_e = \sqrt{\eta_1^2 - \eta_2^2} \qquad (5)$$

For core index $\eta_1 = 1.5$, cladding index $\eta_2 = 1.485$ and external medium of air, $\theta_e = 12.2°$. Thus, any ray entering the end of the fiber core on axis with $\theta_e < 12.2°$

† N. S. Kapany, "Fiber Optics," Academic Press, Inc., New York, 1967.

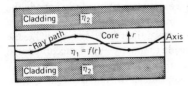

Figure 13-45 By gradually decreasing the index of refraction η_1 of the optical-fiber core as a function of radius r, the ray path becomes a smooth undulating curve which does not reach the core-cladding boundary so that the wave propagates as though in an unbounded medium.

will be trapped inside (totally internally reflected) and propagate by multiple reflections down the inside of the core. Although many modes may propagate in a fiber under suitable conditions, there is one mode (as in a dielectric sheet) for which no cutoff exists. Thus, if

$$\lambda_0 > \frac{2\pi a \sqrt{\eta_1^2 - \eta_2^2}}{k_{01}} = \frac{2\pi a \eta_1 \cos \theta_{ic}}{k_{01}} \quad (\text{m}) \tag{6}$$

where λ_0 = free space wavelength, m

a = core radius, m

η_1 = core index of refraction

η_2 = cladding index of refraction

k_{01} = 2.405 = first root of zeroth-order Bessel function (J_0) (see Table 13-2)

θ_{ic} = critical angle of incidence at core-cladding boundary

then only one mode propagates and the fiber is a single-mode guide.

If the index of refraction of the core fiber decreases continuously as a function of radius, it is possible to change the path from an angular zigzag as in Fig. 13-44 to a smooth undulating curve which does not reach the core boundary, as suggested in Fig. 13-45. Under these conditions the wave propagates as though in an un-bounded optical medium.

A typical optical-fiber communication link is illustrated in Fig. 13-46 with a laser or light-emitting diode (LED) as the transmitter and a phototransistor or other photosensitive device as the receiver.

In typical optical fibers, lowest attenuation occurs in the 700- to 1100-nm range. This is in the infrared region. The light wavelengths to which the human eye are sensitive are nominally from 400 nm (violet light) to 700 nm (red light). See the electromagnetic spectrum chart in Chap. 1.

Optical fibers usually have core diameters of 5 to 50 μm, so that they are many light or infrared wavelengths in diameter and losses due to radiation are small. As the diameter is decreased or wavelength increased, radiation losses increase. For moderate indices of refraction (~ 1.5) rods or fibers greater than 1λ in diameter are predominantly guides (most energy inside and radiation small), while rods or

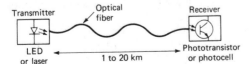

Figure 13-46 Optical-fiber communication link with laser or light-emitting-diode (LED) transmitter and photo-transistor or other photosensitive device as receiver.

fibers less than 1λ in diameter have most of the energy traveling along outside the dielectric with radiation becoming important. Thus, if the dielectric cylinder or rod is gradually tapered down from a diameter of more than 1λ to less than 1λ, the guiding action will shift from energy mostly inside to mostly outside accompanied by radiation in the direction of the rod so that it behaves as an end-fire antenna.† Because tapered dielectric rod antennas for centimeter wavelengths have often been made of polystyrene, they are called *polyrod* antennas.

A detailed analysis of wave propagation in dielectric cylinders is a complex topic and beyond the scope of this introductory treatment. (See footnote references to this section and Sec. 13-14 for more details.)

13-16 RETINAL OPTIC FIBERS

The retina of a human eye contains a bundle of more than 100 million optical fibers, each acting as both a light waveguide and also a photon detector (receptor).‡ There are two classes of these fibers: the *cones*, occupying the central area of the retina, and the more numerous *rods* in the outer surrounding regions. Almost all of the cones are individually connected by nerve transmission lines (axons) to the brain, where signal processing and image formation occurs, and it is with the cones that fine details (such as the book printing you are reading) can be distinguished. The rods provide much poorer resolution of image details but can give better vision at low light levels because of their higher sensitivity and the fact that many rods may be connected in parallel to a single axon-brain line. The rods also provide peripheral vision.

Figure 13-47*a* is a cross section of a human eye showing lens, retina, and optic nerve to the brain. Figure 13-47*b* is an enlarged view of a section of the retina, which is a transparent medium containing rods, cones, cells, and dendrites. It has an opaque backing called the *pigmented layer*. Figure 13-47*c* is a still more enlarged view of a single cone. The narrow ends of the rods and cones, called *outer segments*, are of the order of 1 μm in diameter and about 20 times as long. The outer segments have an index of refraction η_1 of about 1.39 with the cladding or surrounding (interstitial) medium index η_2 a few percent less. These index values are very close to those employed in typical commercial optical fibers ($\eta_1 = 1.46$, $\eta_2 = 1.44$), but the diameter of the outer segments is less (1.5 to 2λ), so that the outer segments apparently have more radiating action than a standard commercial optical fiber.

The nucleus of a cone or rod acts as a lens, concentrating light into the interior, where it travels through both segments by total internal reflection. Any light photons not absorbed in the outer segment pass out the far end and impinge on the opaque pigmented layer. In human beings the pigmented layer absorbs light, preventing any reflection, but in night-hunting animals, such as cats, the pigmented layer is replaced by a tapetum which is highly reflective so that light not

---

† In practice, bends, nicks, irregularites, and discontinuities also induce radiation.

‡ All electromagnetic waves are transmitted in quantum units, called *photons*, of energy equal to hf, where h = Planck's constant ($=6.63 \times 10^{-34}$ J s) and f = frequency (Hz).

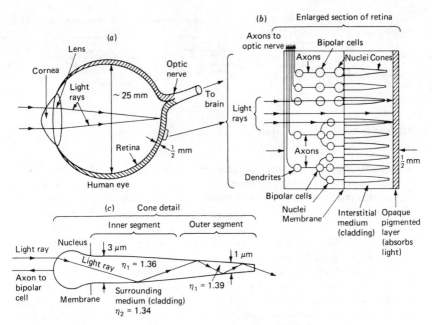

Figure 13-47 (a) Section of human eye, (b) enlarged section of retina, and (c) details of cone. The retina contains over 100 million optical fibers (rods and cones) which act as both polyrod antennas and photon detection devices. Normal vision wavelengths cover the range from 400 to 700 nm, which is approximately half the diameter of the outer segments of the rods and cones. The diagrams are simplified and somewhat schematic.

absorbed on its way to the tapetum is reflected back to the rods and cones. This gives cats a 6-dB dark vision advantage over humans.

It is to be noted that although the index of refraction of the rods and cones varies somewhat with position, it is always greater than that of the cladding or surrounding medium, which is a necessary condition for total internal reflection.

When photons are absorbed by molecules in the outer segment, a current is initiated which flows back to the bipolar cell, which fires the impulse that travels to the brain via the axons and dendrites. Thus, a rod or cone might be described as similar to an end-fire (polyrod) antenna (with unity front-to-back ratio) which is also equipped with sensitive detectors that convert the light photon frequencies (10^{15} Hz) to near dc impulses for transmission to and processing by the brain.

13-17 CAVITY RESONATORS

The purpose of transmission lines and waveguides is to transmit electromagnetic energy efficiently from one point to another. A *resonator*, on the other hand, is an energy storage device. As such it is equivalent to a resonant circuit element. At low

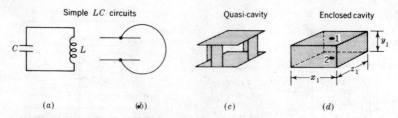

Simple LC circuits Quasi-cavity Enclosed cavity

Figure 13-48 Evolution of (enclosed) cavity resonator from simple LC circuit.

frequencies a parallel-connected capacitor and inductor, as in Fig. 13-48a, form a resonant circuit. To make this combination resonate at shorter wavelengths the inductance and capacitance can be reduced, as in Fig. 13-48b. Parallel straps reduce the inductance still further, as in Fig. 13-48c. The limiting case is the completely enclosed rectangular box, or *cavity resonator*, shown in Fig. 13-48d. In this cavity resonator the maximum voltage is developed between points 1 and 2 at the center of the top and bottom plates.

Resonators can also be constructed using sections of open- or short-circuited transmission lines, as in Fig. 13-49. The type at (a) uses a two-conductor transmission line, while that at (b) uses a coaxial line. The disadvantage of the two-conductor (open) type is that there can be a small but significant loss due to radiation. In the resonators of Fig. 13-49 the fields are in the TEM mode, whereas in the cavity resonator of Fig. 13-48d the fields must be in higher-order modes.

The basic principle of a cavity resonator was described in connection with the pure standing wave of Sec. 10-17. Here the energy oscillates back and forth from entirely electric to entirely magnetic twice per cycle. Let us now consider the case of a rectangular cavity resonator in more detail and determine the *resonant frequency* and Q. It is convenient to begin by recalling the situation for a TE_{m0}-mode wave in a hollow rectangular waveguide. Referring to Fig. 13-50, let a TE_{m0}-mode wave traveling in the $-x$ direction be incident on a conducting plate across the guide at $x = 0$, producing a pure standing wave in the guide. This standing wave is the resultant of two traveling waves of equal amplitude traveling in the negative x

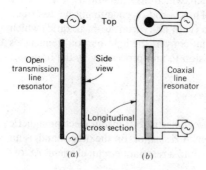

Figure 13-49 Resonator consisting of (a) two-conductor open transmission line and (b) enclosed coaxial line.

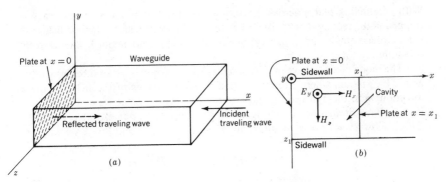

Figure 13-50 (*a*) Perspective view of rectangular waveguide closed with plate at $x = 0$ and (*b*) cross-sectional top view with additional plate at $x = x_1$ trapping wave inside the cavity.

direction (incident wave) and in the positive x direction (reflected wave). The fields of these traveling waves (with time shown explicitly) are given by

$$\overset{\bullet}{E}_y = \frac{j\beta Z_{yz}}{k_z} H_0 \sin k_z z \, e^{j(\omega t \pm \beta x)} \tag{1}$$

$$H_x = H_0 \cos k_z z \, e^{j(\omega t \pm \beta x)} \tag{2}$$

$$H_z = \frac{j\beta}{k_z} H_0 \sin k_z z \, e^{j(\omega t \pm \beta x)} \tag{3}$$

where $k_z = m\pi/z_1$. The plus sign in the exponential refers to a wave traveling in the $-x$ direction and the minus sign to one traveling in the $+x$ direction. Adding the fields of the two traveling waves to obtain the fields of the standing wave, we obtain

$$E_y = \frac{-j\beta Z_{yz}}{k_z} H_0 \sin k_z z (e^{j\beta x} - e^{-j\beta x}) e^{j\omega t} \tag{4}$$

$$= \frac{2\beta Z_{yz}}{k_z} H_0 \sin k_z z \sin \beta x e^{j\omega t} \tag{5}$$

Inserting another conducting plate across the guide at $x = x_1$ requires that $\beta = k_x = l\pi/x_1$. Noting that the transverse-wave impedance $Z_{yz} = \omega\mu/\beta = \omega\mu/k_x$, we get

$$E_y = \frac{2\omega\mu}{k_z} H_0 \sin k_x x \sin k_z z \, e^{j\omega t} \tag{6}$$

Proceeding in like manner for the magnetic field components, we get

$$H_x = -2H_0 \sin k_x x \cos k_z z \, e^{j[\omega t + (\pi/2)]} \tag{7}$$

$$H_z = \frac{2k_x}{k_z} H_0 \cos k_x x \sin k_z z \, e^{j[\omega t + (\pi/2)]} \tag{8}$$

With conducting plates across the waveguide at $x = 0$ and $x = x_1$ the wave is trapped in the rectangular cavity. We note that the electric and magnetic fields are in time-phase quadrature ($\pi/2$ in exponent for H_x and H_z but not for E_y), as is typical of a standing wave.

The mode of a TE wave in a rectangular cavity would in general be designated as a TE_{lmn} mode, where l refers to (half-cycle) variations of the fields in the x direction, m in the z direction, and n in the y direction. Since we assumed $n = 0$ in the above discussion, the designation appropriate to our example would be TE_{lm0}. Now $k^2 = k_z^2 = \gamma^2 + \omega^2\mu\epsilon$, but $\gamma^2 = -\beta^2$ ($\alpha = 0$) and $\beta = k_x$. Thus,

$$k_z^2 = -k_x^2 + \omega^2\mu\epsilon = -k_x^2 + (2\pi f)^2 \frac{1}{(f\lambda)^2}$$

so

$$\lambda = \frac{2}{\sqrt{(l/x_1)^2 + (m/z_1)^2}} \qquad (9)$$

where λ is the *resonant wavelength* and x_1 and z_1 are the resonator dimensions. (See Fig. 13-50b.) For example, the resonant wavelength for a TE_{110} mode in a square-box resonator ($x_1 = z_1$) is given by

$$\lambda = \frac{2}{\sqrt{2/x_1^2}} = 1.41x_1 \qquad (m) \qquad (10)$$

Thus, the resonant wavelength is equal to the diagonal of the square box, as suggested in Fig. 13-51a. The resonant frequency is given by $f = c/\lambda = (3 \times 10^8)/\lambda$ Hz. The electric and magnetic field configuration in the resonator is as indicated in Fig. 13-51b and c. There is no variation in the y direction ($n = 0$). The wave inside the cavity is equivalent to a TEM-mode wave reflected at 45° angles, as in Fig. 13-51a, assuming that the box is infinitely long in the y direction (perpendicular to the page in Fig. 13-51a). In principle the y dimension is noncritical for the mode being considered ($n = 0$), but in practice too large a value of y could permit modes with field variations in the y direction ($n \neq 0$).

At one instant of time ($t = 0$) the energy is all in electric form, as suggested in Fig. 13-52a, with accumulation of positive and negative charges on the top and bottom surfaces of the cavity, as in Fig. 13-52b. One-quarter cycle later ($t = T/4$) the energy is all in magnetic form (Fig. 13-52c) with electric currents flowing down the sidewalls, as in Fig. 13-52d.

To find the Q of the cavity resonator we note that by definition

$$Q = 2\pi \frac{\text{total energy stored}}{\text{decrease in energy in 1 cycle}}$$

The energy situation as a function of time is presented in Fig. 13-53. To get the total energy stored we can integrate the electric energy density w_e ($= \frac{1}{2}\epsilon_0 E_y^2$) over the interior volume of the cavity when E_y is a maximum, and to get the decrease in energy we can integrate the average power into the walls of the cavity and multiply

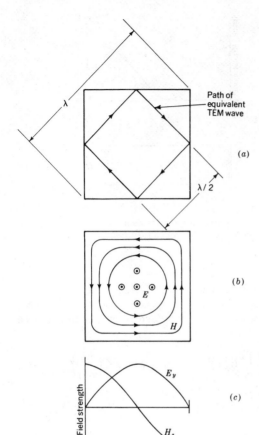

Path of equivalent TEM wave

λ

$\lambda/2$

(a)

(b)

E

H

Field strength

E_y

(c)

H_z

Figure 13-51 (a) The resonant wavelength of a square cavity is equal to the diagonal distance. The path of the equivalent TEM wave is also shown. (b) Electric and magnetic field configuration in square cavity resonator with TE_{110} mode. (c) Variation of E_y and H_z field components across centerline of cavity.

this by the period $T(= 1/f)$. It is assumed that the medium filling the cavity is air or vacuum. Thus, we have

$$Q = \frac{2\pi W}{T(-dW/dt)} = \frac{2\pi \iiint w_e \, dv}{T\frac{1}{2} \operatorname{Re} Z_c \iint |H_t|^2 \, ds} \tag{11}$$

where H_t is the magnetic field component tangent to the cavity wall. Noting (6) and performing the integration for the total energy stored gives

$$W = W_e = \tfrac{1}{2}\epsilon_0 \left(\frac{2\omega\mu_0 H_0}{k_z}\right)^2 \int_0^{x_1} \int_0^{y_1} \int_0^{z_1} \sin^2 k_x \sin^2 k_z z \, dx \, dy \, dz$$

$$= \tfrac{1}{8}\epsilon_0 x_1 y_1 z_1 \left(\frac{2\omega\mu_0 H_0}{k_z}\right)^2 = \frac{\mu_0 H_0^2 x_1 y_1 z_1}{2}\left[\left(\frac{lz_1}{mx_1}\right)^2 + 1\right] \qquad (J) \tag{12}$$

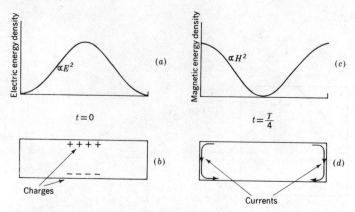

Figure 13-52 Electric and magnetic energy along centerline of cavity resonator. At time $t = 0$ all the energy is in the electric form, (a) and (b), while one-quarter period later ($t = T/4$) all the energy is in the magnetic form, (c) and (d).

The factor $-dW/dt$ in (11) is the power lost in the walls. It is convenient to calculate this as the power lost in three pairs of faces: two perpendicular to the x direction, two perpendicular to the y direction (top and bottom), and two perpendicular to the z direction. Thus,

$$P_x = 2 \times \tfrac{1}{2} \operatorname{Re} Z_c \int_0^{y_1} \int_0^{z_1} H_z^2 \, dy \, dz = 2H_0^2 y_1 z_1 \operatorname{Re} Z_c \left(\frac{l z_1}{m x_1}\right)^2 \qquad (13)$$

$$P_y = H_0^2 x_1 z_1 \operatorname{Re} Z_c \left[\left(\frac{l z_1}{m x_1}\right)^2 + 1\right] \qquad (14)$$

$$P_z = 2H_0^2 x_1 y_1 \operatorname{Re} Z_c \qquad (15)$$

where Z_c is the intrinsic impedance of the conducting material forming the walls of the cavity.

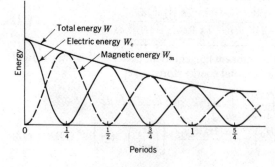

Figure 13-53 Decrease of stored energy with time in a resonator. The energy oscillates from all electric energy at one instant to all magnetic energy $\tfrac{1}{4}$ period later.

Substituting these results in (11), we have for the Q of a rectangular cavity with a TE_{lm0} mode

$$Q = \frac{\mu_0 \pi f}{Re\ Z_c} x_1 y_1 z_1 \frac{(lz_1/mx_1)^2 + 1}{2(lz_1/mx_1)^2 y_1 z_1 + [(lz_1/mx_1)^2 + 1]x_1 z_1 + 2x_1 y_1} \quad (16)$$

For a square cavity ($x_1 = z_1$) and a TE_{110} mode

$$Q = \frac{2\mu_0 \pi f}{Re\ Z_c} \frac{x_1 y_1 z_1}{2(y_1 z_1 + x_1 z_1 + x_1 y_1)}$$

$$= \frac{2\mu_0 \pi f}{Re\ Z_c} \frac{\text{volume of cavity}}{\text{interior surface area of cavity}} \quad (17)$$

or

$$Q = \frac{2}{\delta} \frac{\text{volume of cavity}}{\text{interior surface area of cavity}} \quad (18)$$

where $\delta = 2\ Re\ Z_c/\omega\mu_0 = 1/e$ depth of penetration. For copper $\delta = 6.6 \times 10^{-2}/\sqrt{f}$. If $x_1 = z_1 = 100\ mm$ and $y_1 = 50\ mm$, we have that the resonant wavelength $\lambda = 141\ mm$ and $Q = 17,500$ (dimensionless).

Methods for coupling of cavity resonators are illustrated in Fig. 13-54 for a square cavity with TE_{110} mode. Couplings to coaxial lines are shown in Fig. 13-54a. These involve an electric probe at the center of the cavity ($\mathbf{E} = max$) and a current loop at the wall ($\mathbf{H} = max$). Coupling to a hollow rectangular waveguide can be via a hole in the cavity wall, as in Fig. 13-54b. Coupling to an electron beam can be accomplished with holes in the top and bottom surfaces of the cavity, as in

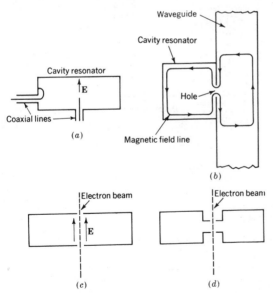

Figure 13-54 Cross sections through rectangular cavity resonators showing couplings (a) to coaxial lines, (b) to a hollow rectangular waveguide, and (c) and (d) to electron beams.

Fig. 13-54*c*. Here the electrons move parallel to the electric field where it is a maximum. To reduce the transit time for the electrons the cavity may be modified as in Fig. 13-54*d*.

13-18 MODES

Different modes on transmission lines and waveguides (TEM, TE_{10}, TE_{01}, TE_{11}, TM_{01}, TM_0, and TE_0) and in cavities (TE_{110} and TM_{01}) are shown for comparison in Fig. 13-55.

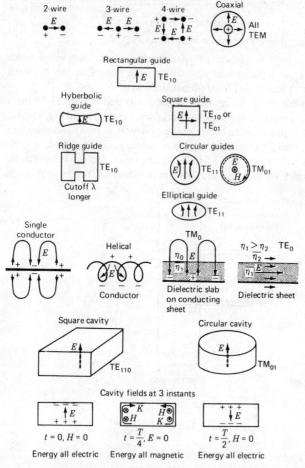

Figure 13-55 From top to bottom: Electromagnetic waves on transmission lines, in hollow waveguides, on open guides, and in cavities. On the open guides (third row up) wave propagation is left to right; however for the dielectric sheet the wave direction is out of the page (same as for the waveguides and transmission lines above).

PROBLEMS†

Group 13-1: Secs. 13-1 through 13-4. Parallel plane guide and hollow rectangular guide

★13-1-1. **Rectangular guide** A rectangular air-filled waveguide has a cross section of 50 mm by 100 mm. If the frequency is 1.6 times the cutoff frequency, find (a) cutoff wavelength λ_{oc} for the dominant mode and (b) phase velocity in the guide.

13-1-2. **Rectangular guide** A rectangular air-filled waveguide has a cross section of 10 mm by 100 mm. If the wavelength is more than 150 mm, find (a) mode or modes transmitted and (b) phase velocity in the guide at a wavelength of 150 mm. Express in terms of c (velocity of light).

★13-1-3. **Square waveguide** What modes are passed at wavelengths longer than 80 mm for a square waveguide 100 mm on a side?

13-1-4. **Rectangular waveguide** An air-filled hollow rectangular waveguide has cross-sectional dimensions $y_1 = 40$ mm and $z_1 = 50$ mm. (a) At frequencies below 8 GHz what modes will this guide transmit of the TE and TM type? (b) Find the ratio of the wave velocity in the guide to the velocity in free space for each of the modes if $f = 1.7 f_c$.

13-1-5. **Rectangular waveguide. Reflection angle** Find the reflection angle θ of the TE_{10} mode for a rectangular waveguide of 40 mm by 80 mm cross section with E parallel to the short side at a wavelength of (a) 120 mm, (b) 135 mm, and (c) 150 mm.

13-1-6. **Rectangular guide. TM wave** Show that the field components for a TM wave in a hollow rectangular single-conductor waveguide (see Fig. 13-10) are given by

$$E_x = E_0 \sin\frac{n\pi y}{y_1} \sin\frac{m\pi z}{z_1} e^{-\gamma x}$$

$$E_y = \frac{-\gamma E_0 n\pi}{k^2 y_1} \cos\frac{n\pi y}{y_1} \sin\frac{m\pi z}{z_1} e^{-\gamma x}$$

$$E_z = \frac{-\gamma E_0 m\pi}{k^2 z_1} \sin\frac{n\pi y}{y_1} \cos\frac{m\pi z}{z_1} e^{-\gamma x}$$

$$H_x = 0 \qquad \text{TM mode condition}$$

$$H_y = \frac{-\gamma Y_{yz} E_0}{k^2} \frac{m\pi}{z_1} \sin\frac{n\pi y}{y_1} \cos\frac{m\pi z}{z_1} e^{-\gamma x}$$

$$H_z = \frac{\gamma Y_{yz} E_0}{k^2} \frac{n\pi}{y_1} \cos\frac{n\pi y}{y_1} \sin\frac{m\pi z}{z_1} e^{-\gamma x}$$

•

Group 13-2: Secs. 13-5 through 13-8. Cylindrical guide and attenuation above and below cutoff

★13-2-1. **Cylindrical guide** An air-filled cylindrical waveguide has a diameter of 100 mm. (a) Find the cutoff wavelength for the low-loss TE_{01} mode. (b) At 99 percent of this wavelength, what other modes will be passed?

13-2-2. **Cylindrical guide** (a) An air-filled cylindrical waveguide has a diameter $d = 75$ mm. At frequencies below 5 GHz, what modes will this guide transmit of the TE and TM type? (b) Find the ratio of the wave velocity in the guide to the velocity in free space for each of the modes if $f = 1.2 f_c$.

13-2-3. **Cylindrical guide. Attenuation** An air-filled cylindrical waveguide has a diameter $d = 50$ mm. (a) Find the cutoff frequencies for the following modes: $TM_{01}, TM_{02}, TM_{11}, TM_{12}, TE_{01}, TE_{02}, TE_{11}$, and TE_{12}. (b) Find the ratio of the wave velocity in the guide to the velocity in free space for each of the modes if $f = 1.3 f_c$. (c) If the waveguide is of copper, find the attenuation in decibels per meter for each of the modes at $1.3 f_c$.

† Answers to starred problems are given in Appendix C.

13-2-4. Cylindrical guide. Power Show that the average power transmitted by a perfectly conducting cylindrical guide operating in the TE_{11} mode is given by

$$P_{av} = \frac{\omega\mu\beta|H_0|^2 r_0^4}{82}$$

13-2-5. Parallel plane guide. TE attenuation Show that the attenuation constant for a TE_{10} wave at frequencies above cutoff in an infinite-prallel-plane guide is

$$\alpha = \frac{2 \text{ Re } Z_c}{d \text{ Re } Z_d} \frac{(\lambda_0/2d)^2}{\sqrt{1 - (\lambda_0/2d)^2}} \quad (\text{Np m}^{-1})$$

where Re Z_c = real part of intrinsic impedance of wall medium (conductor), Ω
Re Z_d = real part of intrinsic impedance of medium-filling guide (dielectric), Ω
d = wall spacing, m
λ_0 = wavelength in unbounded medium, m

13-2-6. Parallel plane guide. TE attenuation Show that the attenuation constant for a TE_{m0} wave at frequencies above cutoff in an infinite-parallel-plane guide of spacing d is

$$\alpha = \frac{\text{Re } Z_c}{d \text{ Re } Z_d} \frac{2(\lambda_0/\lambda_{oc})^2}{\sqrt{1 - (\lambda_0/\lambda_{oc})^2}}$$

where λ_{oc} is the cutoff wavelength.

★**13-2-7. Parallel-plane line. TEM and TE attenuation** In an infinite-parallel-plane air-filled transmission line or guide of 15-mm spacing find the attenuation constant α for a TEM wave and a TE_{10} wave at 12 GHz. The planes are made of copper.

13-2-8. Parallel plane guide. TM attenuation In an infinite-parallel-plane guide show that at a wavelength λ_0, less than cutoff, the attenuation constant for a TM_{10} wave is

$$\alpha' = \frac{2\alpha}{\sqrt{1 - (\lambda_0/2b)^2}}$$

where α = attenuation constant for TEM wave
b = parallel-plane spacing

★**13-2-9. Hollow guide. Attenuation** What is the attenuation constant α in decibels per meter for a hollow single-conductor waveguide at an applied frequency 0.85 times the lowest cutoff frequency for the guide?

13-2-10. Cylindrical guide. TM fields Show that the field components for a TM wave in a hollow cylindrical waveguide (see Fig. 13-19) are given by

$$E_r = -\frac{\gamma E_0}{k^2} \cos n\phi \frac{dJ_n(kr)}{dr} e^{-\gamma z}$$

$$E_\phi = -\frac{n\gamma E_0}{k^2 r} \sin n\phi J_n(kr) e^{-\gamma z}$$

$$E_z = E_0 \cos n\phi J_n(kr) e^{-\gamma z}$$

$$H_z = 0 \quad \text{TM mode condition}$$

$$H_r = \frac{n\gamma E_0 Y_{r\phi}}{k^2 r} \sin n\phi J_n(kr) e^{-\gamma z}$$

$$H_\phi = -\frac{n\gamma E_0}{k^2} \cos n\phi \frac{dJ_n(kr)}{dr} e^{-\gamma z}$$

★13-2-11. Strip line. Power and attenuation A 500-MHz TEM traveling wave in a strip transmission line has an electric field $E = 500$ mV m^{-1} rms. The line consists of a copper strip 10 mm wide separated 1 mm from a flat copper ground plane by a 1-mm-thick ribbon of polystyrene ($\epsilon_r = 2.7$). (a) Find the average power in the direction of transmission. (b) Find the average power into the copper walls per meter of length. (c) What is the attenuation constant in decibels per meter? (d) What is the characteristic impedance? (e) What is the wave velocity?

13-2-12. Rectangular guide. Attenuation Show that the attenuation constant for a TE$_{m0}$ wave at frequencies above cutoff in a hollow rectangular waveguide of height y_1 and width z_1 is

$$\alpha = \frac{2 \operatorname{Re} Z_c[(\lambda_0/\lambda_{oc})^2 + z_1/2y_1]}{z_1 \operatorname{Re} Z_d \sqrt{1 - (\lambda_0/\lambda_{oc})^2}}$$

13-2-13. Rectangular guide. Attenuation An air-filled rectangular waveguide has cross-sectional dimensions of 50 mm by 100 mm. At a wavelength of 150 mm, with E parallel to the short side, find the attenuation in decibels per meter for the following modes: (a) TE$_{10}$, (b) TE$_{01}$, and (c) TE$_{11}$. Take the guide wall conductivity as infinite.

★13-2-14. Strip line. Attenuation A microstrip transmission line 25 mm wide is situated on a 5-mm-thick substrate ($\epsilon_r = 9$, $\mu_r = 1$, and $\sigma = 0$). The strip and ground plane have a conductivity $\sigma = 10^6$ ℧ m^{-1}. If the electric field $E = 2$ V m^{-1} (rms) and the frequency is 300 MHz, find (a) power transmitted and (b) decibels per meter attenuation.

13-2-15. Rectangular guide. Attenuation An air-filled hollow rectangular waveguide has cross-sectional dimensions $y_1 = 40$ mm and $z_1 = 80$ mm. (a) Find the cutoff frequencies for the following modes: TEM, TE$_{10}$, TE$_{20}$, TE$_{01}$, TE$_{02}$, TE$_{11}$, TE$_{21}$, and TE$_{12}$. (b) Find the ratio of the wave velocity in the guide to the velocity in free space for each of the modes if $f = 1.5f_c$. (c) If the waveguide is of copper, find the attenuation in decibels per meter for each of the higher-order modes at $1.5f_c$.

13-2-16. Rectangular guide. Power Show that the average power transmitted by a perfectly conducting rectangular guide operating in the TE$_{10}$ mode is given by

$$P_{av} = \frac{\omega\mu\beta|H_0|^2 y_1 z_1^3}{4\pi^2}$$

•

Group 13-3: Secs. 13-9 through 13-16. Waveguide devices, impedance, surface waves, dielectric fiber and rod guides

★13-3-1. Surface wave powers A 100-MHz wave is traveling parallel to a copper sheet ($|Z_c| = 3.7 \times 10^{-3}\,\Omega$) with E ($= 100$ V m^{-1} rms) perpendicular to the sheet. Find (a) Poynting vector (W m^{-2}) parallel to sheet, (b) Poynting vector into the sheet, and (c) Poynting vector for same wave incident normally on sheet instead of traveling along it.

13-3-2. Surface wave powers A 100-MHz wave is traveling parallel to a conducting sheet for which $|Z_c| = 0.02\,\Omega$. If E is perpendicular to the sheet and equal to 150 V m^{-1} (rms), find (a) watts per square meter traveling parallel to sheet and (b) watts per square meter into sheet.

★13-3-3. Surface wave power A plane 3-GHz wave in air is traveling parallel to the boundary of a conducting medium with H parallel to the boundary. The constants for the conducting medium are $\sigma = 10^7$ ℧ m^{-1} and $\mu_r = \epsilon_r = 1$. If the traveling-wave rms electric field $E = 75$ mV m^{-1}, find the average power per unit area lost in the conducting medium.

13-3-4. Surface wave current sheet A TEM wave is traveling in air parallel to the plane boundary of a conducting medium. Show that if $K = \rho_s v$, where K is the sheet-current density in amperes per meter, ρ_s is the surface charge density in coulombs per square meter, and v the velocity of the wave in meters per second, it follows that $K = H$, where H is the magnitude of the **H** field of the wave.

13-3-5. Rectangular guide TE_{10} impedance In a hollow rectangular guide with TE_{10} mode show that the ratio of the voltage V between the top and bottom of the guide (at the middle) to the longitudinal current I on the upper or lower inside surface is an impedance given by

$$Z = \frac{V}{I} = \frac{\pi y_1}{2z_1} Z_{yz}$$

where y_1 = height of guide
z_1 = width of guide
Z_{yz} = transverse impedance

13-3-6. Rectangular-guide TE impedance Show that the transverse impedance of a TE wave in a rectangular guide is equal to $Z_{yz} = Z_d(\lambda/\lambda_0)$, where Z_d is the intrinsic impedance of the medium, λ the wavelength in the guide, and λ_0 the wavelength in an unbounded medium of the same material that fills the guide.

13-3-7. Rectangular-guide TM impedance Show that the transverse impedance of a TM wave in rectangular guide is equal to

$$Z_{yz} = Z_d \sqrt{1 - \left(\frac{n\lambda_0}{2y_1}\right)^2 - \left(\frac{m\lambda_0}{2z_1}\right)^2} = Z_d \sqrt{1 - \left(\frac{\lambda_0}{\lambda_{oc}}\right)^2}$$

13-3-8. Rectangular-waveguide group velocity Show that the group velocity u in a hollow rectangular single-conductor waveguide (equal to the velocity of energy transport) is given by

$$u = v_0 \sqrt{1 - \left(\frac{n\lambda_0}{2y_1}\right)^2 - \left(\frac{m\lambda_0}{2z_1}\right)^2}$$

$$= v_0 \sqrt{1 - \left(\frac{\lambda_0}{\lambda_{oc}}\right)^2}$$

where λ_0 = wavelength in unbounded medium
λ_{oc} = cutoff wavelength

Both λ_0 and λ_{oc} should be distinguished from λ, the wavelength in the guide. Note that it follows that $uv = v_0^2$, where v is the phase velocity in the guide and v_0 is the phase velocity in an unbounded medium.

13-3-9. Coated surface wave power Show that for a dielectric-coated conductor, as in Fig. 13-36b, the ratio of the power transmitted in the dielectric P_d to the power transmitted in the air P_a is given by

$$\frac{P_d}{P_a} = \frac{\cos \beta d}{\sin^3 \beta d} (\sin 2\beta d - 2\beta d)$$

where d is the thickness of the dielectric coating.

★13-3-10. Coated surface wave cutoff. A perfectly conducting flat sheet of large extent has a dielectric coating ($\epsilon_r = 3$) of thickness $d = 5$ mm. Find the cutoff frequency for the TM_0 (dominant) mode and its attenuation per unit distance.

★13-3-11. Infrared fiber guide A fiber guide has a core of index 1.53 and cladding of index 1.51. For $\lambda = 1 \, \mu$m find the maximum angle θ_e at which rays will enter the fiber and be trapped.

★13-3-12. Infrared fiber core diameter (a) Find the core diameter required for a single mode of propagation in a fiber guide at the infrared wavelength of 1.1 μm if the core index is 1.54 and the cladding index is 1.535. (b) Find the maximum entrance angle θ_e.

★13-3-13. Infrared fiber guide attenuation A fiber guide of complex index $1.54\sqrt{1 - j10^{-10}}$ operates at a wavelength of 1 μm. $\mu_r = 1$ and $\sigma = 0$. Find the dB attenuation per kilometer.

Group 13-4: Secs. 13-17 through 13-18. Cavity resonators

13-4-1. Square cavity A square cavity resonator of x mm on a side has a depth of $x/4$. At a wavelength $\lambda_0 = 10$ mm find for the TE_{110} mode (a) x for resonance and (b) Q if the $1/e$ depth of penetration in the cavity walls is 1 μm.

★**13-4-2. Cylindrical cavity** A cylindrical cavity resonator of height equal to its radius operates at $\lambda_0 = 10$ mm in the TM_{02} cylindrical guide mode. Find (a) the cavity diameter d and (b) Q if $\sigma = 40$ M℧ m^{-1} for the cavity walls.

13-4-3. Gold-plated cylindrical cavity Repeat Prob. 13-4-2 for the case where the cavity walls are gold-plated.

★**13-4-4. Square cavity** Repeat Prob. 13-4-1 for the case where $\sigma = 40$ M℧ m^{-1} for the cavity walls.

13-4-5. Gold-plated square cavity An air-filled cavity resonator operates in the TE_{110} mode. The cavity is square, $x_1 = z_1 = 80$ mm, with height $y_1 = 35$ mm. The cavity is made of copper and is gold-plated inside. Find (a) resonant frequency, (b) resonant wavelength, and (c) Q.

★**13-4-6. Silver-plated cylindrical cavity** A cavity resonator is constructed of a short section of cylindrical tubing of diameter d closed at both ends by flat plates separated by $d/2$. The cavity is made of brass and is silver-plated inside. (a) What is the required value of d if the cavity is to operate in its dominant mode at K band (10 mm)? (b) What is Q?

13-4-7. Comparison of Qs Compare the Q values attainable with a parallel inductor-capacitor circuit, a short-circuited section of two-conductor line, a short-circuited section of coaxial line, a cavity resonator, a quartz crystal, an ammonia molecule, and a cesium atom (hyperfine transition of ground state of cesium 133).

Group 13-5: Practical applications

13-5-1. Waveguide bandwidth. Ridge or H guide A standard size of L-band rectangular waveguide has inside dimensions of 83 mm by 166 mm. (a) Over what frequency range can the guide be operated in only the TE_{10} mode? (b) Is it desirable to operate over this full range? (c) To extend the frequency range of the guide, ridges can be installed along the top and bottom surfaces (where E is maximum) in a manner similar to that shown in Fig. 13-24f for the reentrant or H-shaped guide. By suitable choice of dimensions of the ridges the ratio of cutoff frequencies between TE_{10} and TE_{20} modes can be increased to as much as 4 to 1 as compared to the 2 to 1 ratio for a standard rectangular guide. Give ridge dimensions and expected usable bandwidth.

★**13-5-2. Guide-horn transition with corrugations** To reduce the side lobes of a horn antenna (see Chap.14) its interior may be corrugated to prevent waves from diffracting around its edge (or surface currents from flowing around the edge and back over the outside). A circular waveguide-fed horn with corrugations is shown in cross section in Fig. P13-5-2. In the transition section the corrugation depth changes

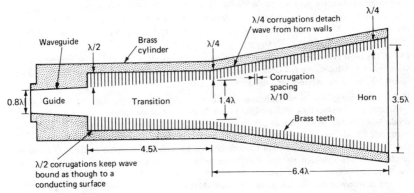

Figure P13-5-2 Cross section of circular waveguide-fed corrugated horn with corrugated transition.

from $\lambda/2$, where the corrugations act like a conducting surface, to $\lambda/4$, where the corrugations act like an open region. The corrugations are equivalent to sections of transmission line, a shorted $\lambda/2$ line presenting a short circuit while a shorted $\lambda/4$ line presents an open circuit. Calculate the surface impedance for corrugations (a) $\lambda/2$ deep, (b) $\frac{3}{8}\lambda$ deep, and (c) $\lambda/4$ deep. The slots are air-filled. (d) For the dimensions shown, how much longer is the cutoff wavelength of the guide for the TE_{11} mode? (e) How much longer is the cutoff wavelength for the TM_{01} mode? (f) What is the useful bandwidth of the device? (g) How critical is the $\lambda/10$ spacing between corrugations? [Re corrugated guide and horn see T. S. Chu, R. W. Wilson, R. W. England, D. A. Gray, and W. E. Legg, Crawford Hill 7-m Millimeter Wave Antenna, *Bell Sys. Tech. Jour.*, **57**:1257–1288 (May–June 1978).]

★**13-5-3. Tunnel communications** (a) A communications service using mobile units wants communication to be maintained even when its radio-equipped automobiles and trucks are in a vehicular tunnel. If the smallest tunnel diameter encountered is 5 m, what is the lowest frequency which can be employed? (b) What could be installed in the tunnel to permit communication at lower frequencies? No frequency converters are permitted. *Hint*: The tunnel is a waveguide. What can be installed so it can transmit TEM waves?

★**13-5-4. Waves through windows** A metal-clad building has window openings 1.5 m by 0.5 m. What is the longest wavelength which can readily penetrate into the interior of the building?

13-5-5. K-band waveguide The wavelength region around 10 mm is commonly referred to as K band, 30 mm as X band, and 100 mm as S band. Design a rectangular waveguide for transmitting TE_{10}-mode K-band waves with minimum attenuation and with minimum change in velocity over the wavelength range 9 to 11 mm. The guide should be incapable of passing any modes other than the TE_{10} (such as TE_{01}) in this wavelength range and any waves at all at wavelengths greater than 15 mm. Your result should be in the form of a drawing with dimensions.

13-5-6. Electronic oven leakage Radio-frequency dielectric heaters are widely used in many industrial processes, particularly for injection molding of plastics. The heating results from the dielectric hysteresis. Another application of this heating effect is for cooking using microwave ovens. The oven consists of a cavity resonator connected to a microwave power source. Design such a microwave oven operating at 2,450 MHz with 750-W radio-frequency power and an oven volume of 200 mm by 400 mm by 400 mm. (a) What maximum rate of temperature rise will be possible for 3 kg of dielectric material with complex permittivity $\epsilon_r = 4 + j1$ if its specific heat is 5 J g^{-1} °C^{-1}? (b) If the oven has a window covered by a metal screen, give the screen specifications, material, wire size, and hole size required to keep radiation loss through the screen to less than 10 mW cm^{-2}. Also is it important that the screen wires be bonded at all contact points? (c) Describe the standing-wave mode or modes in the oven and what means (such as fan or paddles) can be used to "stir" the standing-wave pattern. (d) How can the edges of the oven door be sealed to prevent radio-frequency leakage?

13-5-7. Lunar communication Discuss the possibilities of using dielectric-slab surface-wave modes for radio communication on the moon over long distances (1,000 km or more). Note that the moon has no ionosphere. See, for example, W. W. Salisbury and D. L. Fernald, Post-occultation Reception of Lunar Ship Endeavour Radio Transmission, *Nature*, **234**:95 (Nov. 12, 1971); also A. F. Wickersham, Jr., Generation, Detection and Propagation on the Earth of HF and VHF Radio Surface Waves, *Nature*, **230**:125–130 (April 5, 1971).

13-5-8. Gigawatt guide Design a waveguide for handling gigawatt power levels with minimum attenuation over a 20-km distance at 1 GHz. Consider using a cylindrical guide in a mode or modes which are of much higher order number than the dominant mode.

13-5-9. Lines and guides for all applications Select a suitable type of transmission line or waveguide for frequencies, applications, and line lengths as follows: (a) dc low-voltage measurement, length 5 m, (b) dc 1-MV buried power line, length 10 km, (c) 60-Hz low-voltage flexible power ($\sim$10-kW) line, 50 m, (d) 500-Hz to 5-kHz bandwidth audio circuit, ordinary telephone application over 50 km distance with 100 such circuits bundled as a compact cable with minimum cross talk, (e) 100-kHz two-conductor 600-Ω line from 100-kW transmitter to antenna, distance 1 km, (f) 2-MHz buried 50-Ω line from 50-kW transmitter to antenna, distance 500 m, (g) 5- to 25-MHz 50-Ω line from shortwave antenna to receiver, distance 500 m, (h) 54- to 890-MHz 300-Ω line from receiving antenna to TV set through low electromagnetic noise, distance 50 m, (*i*) same as (*h*) but through high noise, distance 200 m, (*j*) 1.42-GHz line or guide with 6-MHz bandwidth between radio-telescope antenna and receiver, distance 10 m, lowest possible loss, (*k*) 3-GHz (S band) line or guide with 10-MHz bandwidth from 100-kW radar transmitter to antenna, distance 10 m, (*l*) 30-GHz (K band) line or guide with 10-GHz bandwidth between radio-telescope antenna and receiver, distance 10 m, (*m*) 100-GHz line or guide with 10-GHz bandwidth for data transmission, between amplifiers, distance 5 km. Some of the types to choose from are as follows: two-conductor parallel or twisted pair, two conductors in shield, coaxial line (dielectric-filled), coaxial line (air-filled), strip line, and waveguide (rectangular or circular) in dominant or higher-order modes. Specify conductor or guide dimensions and material in all cases. Also explain reasons for selection, including attenuation, cost, etc.

13-5-10. Diffraction effect? Hold your hand between your eye and a lamp or bright area with your forefinger close to but not quite touching your thumb. Explain the phenomenon you observe when you look at the finger-thumb gap. *Hints*: Consider higher-order modes between two parallel boundaries. Consider diffraction. This effect may be noted whenever you view a lighted area through a narrow slit.

FOURTEEN

ANTENNAS AND RADIATION

14-1 INTRODUCTION

In discussing transmission lines and waveguides in Chap. 13 our attention was focused on guiding energy along the system. Little consideration was given to *radiation*, i.e., the loss of energy into free space. While transmission lines or waveguides are usually made so as to minimize radiation, antennas are designed to radiate (or receive) energy as effectively as possible.

A two-wire transmission line is shown in Fig. 14-1a, connected to a radiofrequency generator (or transmitter). Along the uniform part of the line, energy is guided as a plane TEM-mode wave with little loss. The spacing between wires is assumed to be a small fraction of a wavelength. At the right, the transmission line is opened out. As the separation approaches the order of a wavelength or more, the wave tends to be radiated so that the opened-out line acts like an antenna which launches a free-space wave. The currents on the transmission line flow out on the antenna and end there, but the fields associated with them keep on going. To be more explicit, the region of transition between a guided wave and a free-space wave may be defined as an antenna.

We have described the antenna as a transmitting device. As a receiving device the definition is turned around, and an antenna is the region of transition between a free-space wave and a guided wave. Thus, *an antenna is a transition device, or transducer, between a guided wave and a free-space wave or vice versa.* Stated in another way, an antenna is a device which interfaces a circuit and space.

Thus, referring to Fig. 14-1b, the antenna appears from the transmission line as a two-terminal circuit element with an *impedance Z* having a resistive component called the *radiation resistance R_r*, while from space, the antenna is characterized by its *radiation pattern* or patterns involving field quantities.

The radiation resistance R_r has nothing to do with any resistance in the antenna proper but is a resistance coupled from space via the antenna pattern to the antenna terminals.

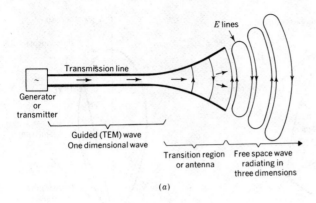

(a)

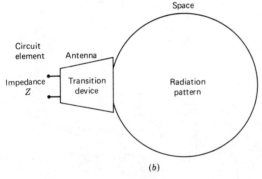

(b)

Figure 14-1 (a) The antenna as a region of transition between a guided wave and a free-space wave or vice versa. (b) Schematic representation of antenna as transition device between a circuit impedance and a radiation field. The antenna interfaces electrons on conductors and photons in space. The eye is another such device.

Associated with the radiation resistance is also an *antenna temperature T_A*. For lossless antennas this temperature has nothing to do with the physical temperature of the antenna proper but is related to the temperature of distant regions of space coupled to the antenna via its radiation pattern. Actually, the antenna temperature is not so much an inherent property of the antenna as it is a parameter which depends on the temperature of the regions the antenna is "looking at." In this sense, a receiving antenna may be regarded as a remote-sensing, temperature-measuring device.

Both the radiation resistance R_r and its temperature T_A are simple scalar quantities. The radiation patterns, on the other hand, are three-dimensional quantities involving the variation of field or power (proportional to the field squared) as a function of the spherical coordinates θ and ϕ. Figure 14-2a shows a field pattern with radial distance r proportional to the field intensity in the direction

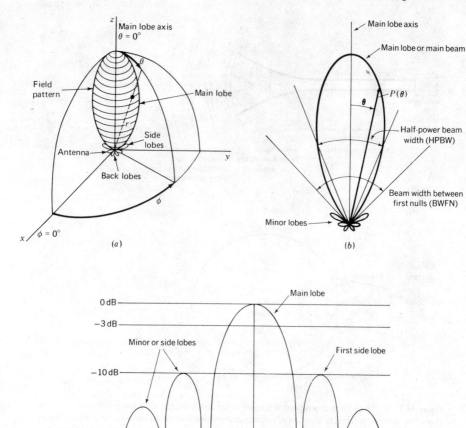

Figure 14-2 (*a*) Antenna field pattern with coordinate system. (*b*) Antenna power pattern in polar coordinates (linear scale). (*c*) Antenna pattern in rectangular coordinates and decibel (logarithmic) scale. Patterns (*b*) and (*c*) are the same.

θ, ϕ. The pattern has its *main lobe* (maximum radiation) in the *z* direction ($\theta = 0$) with *minor* (side and back) *lobes* in other directions.

To completely specify the radiation pattern with respect to field intensity and polarization requires three patterns:

$E_\theta(\theta, \phi)$, the θ component of E as a function of θ and ϕ
$E_\phi(\theta, \phi)$, the ϕ component of E as a function of θ and ϕ
$\delta(\theta, \phi)$, the phase difference of $E_\theta(\theta, \phi)$ and $E_\phi(\theta, \phi)$ as a function of θ and ϕ

For many purposes, a pattern of the Poynting vector $S(\theta, \phi)$ or a *normalized power pattern*, $P_n(\theta, \phi)$, is useful.† Thus,

$$P_n(\theta, \phi) = \frac{S(\theta, \phi)}{S(\theta, \phi)_{\max}} \qquad (1)$$

where $S(\theta, \phi) = [E_\theta^2(\theta, \phi) + E_\phi^2(\theta, \phi)]/Z_0$
   $S(\theta, \phi_{\max})$ = maximum value of $S(\theta, \phi)$

Any of these field or power patterns can be presented in three-dimensional spherical coordinates, as in Fig. 14-2a, or by plane cuts through the main-lobe axis. Two such cuts at right angles, called the *principal plane patterns* (as in the xz and yz planes in Fig. 14-2a) may suffice for a single field component, and if the pattern is symmetrical around the z axis, one cut is sufficient.

Figure 14-2b is a principal plane pattern in polar coordinates, and to show the minor lobes in more detail the same pattern is presented in Fig. 14-2c in rectangular coordinates on a logarithmic or decibel scale. If the pattern is symmetrical, the three-dimensional pattern would be a figure-of-revolution of Fig. 14-2b around the main-lobe axis similar to the pattern in Fig. 14-2a.

Although the radiation pattern characteristics of an antenna involve three-dimensional vector fields for a full representation, there are several simple scalar parameters which can provide all of the pattern information required for most engineering applications. These are the *beam solid angle* Ω_A, the *directivity D* (or *gain G*), and the *effective aperture* A_e.‡

The *beam solid angle* is given by the integral of the normalized power pattern over a sphere (4π sr), or

$$\Omega_A = \iint\limits_{4\pi} P_n(\theta, \phi)\, d\Omega \qquad \text{(sr)} \qquad (2)$$

where $d\Omega = \sin\theta\, d\theta\, d\phi$ (see Fig. 2-24).

This solid angle can often be described approximately in terms of angles subtended by the half-power points in the two principal planes, and a useful approximation involving these angles is that

$$\Omega_A \approx \theta_{\mathrm{HP}}\,\phi_{\mathrm{HP}} \qquad \text{(sr)} \qquad (3)$$

where θ_{HP} and ϕ_{HP} are the half-power beam widths in the two principal planes

The *directivity* of an antenna is given by the ratio of the maximum to the average Poynting vector. Thus,

$$D = \frac{S(\theta, \phi)_{\max}}{S(\theta, \phi)_{\mathrm{av}}} = \frac{1}{4\pi}\iint\limits_{4\pi} \frac{S(\theta, \phi)}{S(\theta, \phi)_{\max}}\, d\Omega = \frac{4\pi}{\Omega_A} \qquad (4)$$

---

† It is to be noted that the radiation resistance, the antenna temperature, and the radiation patterns are functions of the frequency. In general, the patterns are also functions of the distance at which they are measured.

‡ Distinguish carefully between Ω_A for beam solid angle and Ω for ohms.

The smaller the solid beam angle (or half-power beam width), the greater the directivity.

If an antenna has a main lobe with both HPBWs = 20°, its directivity is approximately

$$D = \frac{4\pi \text{ (sr)}}{\Omega_A \text{ (sr)}} = \frac{41{,}253 \text{ (deg}^2)}{\theta_{HP}^\circ \phi_{HP}^\circ} = \frac{41{,}253 \text{ (deg}^2)}{20° \times 20°}$$

$$= 103 = 20.1 \text{ dB (dB above isotropic)} \qquad (5)\dagger$$

which means that the antenna radiates a power in the direction of the main lobe which is 103 times as much as would be radiated by a nondirectional (isotropic) antenna with the same power input.

For an antenna which is isotropic or radiates the same in all directions $[P_n(\theta, \phi) = 1]$, $\Omega_A = 4\pi$ and $D = 1$. This is the smallest directivity an antenna can have. For more directive antennas, Ω_A is less than 4π and D is greater than unity.

The *gain G* equals the directivity for lossless antennas. If there is loss, the gain is less than the directivity. Thus, the gain $G = kD$, where k = efficiency factor $(0 \le k \le 1)$.

The *effective aperture* will be described with the antenna operating as a receiving device. Thus, if a plane wave with Poynting vector S is incident normally on the aperture of a horn antenna, as in Fig. 14-3, the received power P is dependent on the collecting, or effective, aperture of the antenna and is given by $P = SA_e$, from which the effective aperture is given by

$$A_e = \frac{P}{S} \quad (\text{m}^2) \qquad (5a)$$

where P = received power, W
 S = incident Poynting vector, W m^{-2}

For a large horn antenna with funnel-like collecting aperture, its effective aperture A_e is less than its physical aperture (width w times height h), but for antennas in general the effective aperture may be greater or less than the physical aperture, or cross section. This is discussed further in Sec. 14-22.

It will be shown later that the effective aperture A_e over which the antenna collects energy from a passing wave is also related to the beam solid angle Ω_A by the relation

$$A_e = \frac{\lambda^2}{\Omega_A} \quad (\text{m}^2) \qquad (6)$$

where λ = wavelength, m

Since both the antenna radiation resistance and the antenna radiation patterns are determined by the antenna's current distribution, the radiation resistance is related to the radiation pattern. To show this in an elementary way, consider the

$\dagger$ 1 sr $= \left(\frac{180}{\pi}\right)^2 (\text{deg}^2) = 3{,}282.81 \text{ deg}^2$, so 4π sr $= 41{,}253$ square degrees (deg^2).

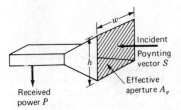

Figure 14-3 Incident Poynting vector (watts per square meter) collected by horn over its effective aperture A_e yields received power P.

power radiated by an antenna as given by the integral of the Poynting vector over a surface enclosing the antenna. Referring to Fig. 14-2a and from (1) and (2), we have

$$\text{Power radiated} = \iint_{4\pi} S(\theta, \phi) ds = S(\theta, \phi)_{max} r^2 \iint_{4\pi} P_n(\theta, \phi) d\Omega$$

$$= S(\theta, \phi)_{max} r^2 \Omega_A \qquad (7)$$

Now from the conservation of power, the power radiated (for a lossless antenna) should be equal to the power input to the antenna or equal to $I^2 R_r$, where I is the current at the antenna terminals and R_r is the radiation resistance. Equating power in to power out,

$$I^2 R_r = S(\theta, \phi)_{max} r^2 \Omega_A \qquad (W) \qquad (8)$$

Thus, the radiation resistance is seen to be a function of the pattern parameter Ω_A as given by

$$R_r = \frac{S(\theta, \phi)_{max} r^2}{I^2} \Omega_A \qquad (\Omega) \qquad (9)$$

As an illustration, consider an antenna with an isotropic (nondirectional) field pattern given by $E_\phi = 0$ and

$$E_\theta = \frac{Ik}{r} \qquad (V \ m^{-1} \ rms) \qquad (10)$$

where I = rms terminal current, A
r = distance, m
k = constant,

Thus, from (9) and (10)

$$R_r = \frac{E_\theta^2 r^2 \Omega_A}{Z_0 I^2} = \frac{k^2}{Z_0} \Omega_A \qquad (\Omega) \qquad (11)$$

where Z_0 = intrinsic impedance of space = 376.7 Ω

Since the antenna is isotropic $[P_n(\theta, \phi) = 1]$, $\Omega_A = 4\pi$, and if, for example, $k = 60 \ \Omega$, the radiation resistance

$$R_r = \frac{3{,}600 \ 4\pi}{376.7} = 120 \ \Omega \qquad (12)$$

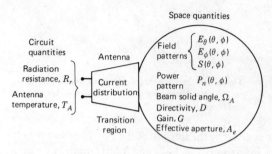

Figure 14-4 Schematic diagram of basic antenna parameters, illustrating the duality of an antenna, being a circuit device (with a resistance and temperature) on the one hand and a space device (with radiation patterns, beam angles, directivity, gain and aperture) on the other.

To summarize: the antenna, or transition region between a guided and free-space wave, has a current distribution which links, on the one hand, a circuit device with radiation resistance and temperature and, on the other hand, a space device having radiation patterns with associated parameters of beam solid angle, directivity, gain, and effective aperture. These interrelations are illustrated schematically in Fig. 14-4.

The quantities mentioned in this brief introduction will be treated in more detail later in the chapter with discussions of a variety of antenna types. These include dipole, loop, helix, reflector, lens, array, and interferometer antennas. The short dipole is a basic form and is considered first (in Sec. 14-3).

14-2 RETARDED POTENTIALS

In dealing with antennas or radiating systems the propagation time is a matter of great importance. Thus, if an alternating current is flowing in the short element in Fig. 14-5, the effect of the current is not felt instantaneously at the point P, but only after an interval equal to the time required for the disturbance to propagate over the distance r.

Accordingly, instead of writing the current I as

$$I = I_0 \cos \omega t \tag{1}$$

which implies instantaneous propagation of the effect of the current, we can introduce the time of propagation (or retardation time†), as done by Lorentz, and write

$$[I] = I_0 \cos \omega \left(t - \frac{r}{c} \right) \tag{2}$$

where $[I]$ is called the *retarded current*. The brackets [ ] may be added, as here, to indicate explicitly that the current is retarded.

Equation (2) is a statement of the fact that the disturbance at a time t and at a distance r from the element is caused by a current $[I]$ that occurred at an earlier time $t - (r/c)$. The time difference r/c is the interval required for the disturbance to travel the distance r, where c is the velocity of light (300 Mm s^{-1}).

---

† Called *retardation time* because the phase of the wave at P is retarded with respect to the phase of the current in the element by an angle $\omega r/c$.

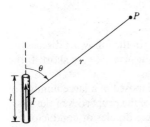

Figure 14-5 Short current-carrying element.

It is to be noted that we dealt with retarded quantities in Chap. 10 in connection with wave propagation, although the term *retarded* was not used. For example, in Chap. 10 a solution of the wave equation is given that involves $\cos(\omega t - \beta x)$, which is similar in form to the trigonometric function in (2) since†

$$\cos \omega\left(t - \frac{r}{c}\right) = \cos(\omega t - \beta r) \tag{3}$$

where $\beta\ (=\omega/c - 2\pi/\lambda)$ is the phase constant.

In complex form (2) is‡

$$[I] = I_0\, e^{j\omega[t - (r/c)]} = I_0\, e^{j(\omega t - \beta r)} \qquad \text{(A)} \tag{4}$$

In the more general situation we may write for the retarded-current density at a point

$$[\mathbf{J}] = \mathbf{J}_0\, e^{j\omega[t - (r/c)]} = \mathbf{J}_0\, e^{j(\omega t - \beta r)} \qquad (\text{A m}^{-2}) \tag{5}$$

Introducing this value of current density in (5-22-10) for the vector potential, we obtain a *retarded vector potential* that is applicable in time-varying situations where the distances involved are significant in terms of the wavelength. That is, the retarded vector potential is

$$[\mathbf{A}] = \frac{\mu_0}{4\pi}\int_v \frac{[\mathbf{J}]}{r}\, dv = \frac{\mu_0}{4\pi}\int_v \frac{\mathbf{J}_0\, e^{j\omega[t - (r/c)]}}{r}\, dv \qquad (\text{Wb m}^{-1}) \tag{6}$$

Likewise the *scalar potential* V can be put in the retarded form

$$[V] = \frac{1}{4\pi\epsilon_0}\int_v \frac{[\rho]}{r}\, dv \tag{7}$$

where $[V]$ = retarded scalar potential, V

$$[\rho] = \rho_0\, e^{j\omega[t - (r/c)]} = \text{retarded charge density, C m}^{-3}$$

---

† The expression $\cos(\omega t - \beta x)$ in Chap. 10 refers to a plane wave traveling in the x direction. The relation $\cos\omega[t - (r/c)]$ or $\cos(\omega t - \beta r)$ refers to a spherical wave traveling in the radial direction. An important point of difference between a plane and a spherical wave is that a plane wave suffers no attenuation (in a lossless medium) but a spherical wave does because it expands over a larger and larger region as it propagates.

‡ It is understood in this and the next section that the instantaneous value of current is given by the real (Re) part of the exponential expressions such as in (4), (5), (6), and (7); that is, the expressions are phasors but with $e^{j\omega t}$ explicitly stated.

14-3 THE SHORT DIPOLE ANTENNA

A short linear conductor is often called a short *dipole*. In the following discussion, a short dipole is always of finite length even though it may be very short. If the dipole is vanishingly short, it is an infinitesimal dipole.

Any linear antenna may be regarded as being composed of a large number of short dipoles connected in series. Thus, a knowledge of the properties of the short dipole is useful in determining the properties of longer dipoles or conductors of more complex shape such as are commonly used in practice.

Let us consider a short dipole like that shown in Fig. 14-6a. The length l is very short compared with the wavelength ($l \ll \lambda$). Plates at the ends of the dipole provide capacitative loading. The short length and the presence of these plates result in a uniform current I along the entire length l of the dipole. The dipole may be energized by a balanced transmission line, as shown. It is assumed that the transmission line does not radiate, and its presence will therefore be disregarded. Radiation from the end plates is also considered to be negligible. The diameter d of the dipole is small compared with its length ($d \ll l$). Thus, for purposes of analysis we may consider that the short dipole appears as in Fig. 14-6b. Here it consists simply of a thin conductor of length l with a uniform current I and point charges q at the ends. According to (4-13-6), the current and charge are related by

$$\frac{\partial q}{\partial t} = I \tag{1}$$

Let us now proceed to find the fields everywhere around a short dipole. Let the dipole of length l be placed coincident with the z axis and with its center at the origin, as in Fig. 14-7. At any point P the electric field has, in general, three components, E_θ, E_ϕ, and E_r, as shown. It is assumed that the medium surrounding the dipole is air or vacuum.

From (8-16-8) the electric field intensity **E** at any point P is expressed by

$$\mathbf{E} = -\nabla V - \frac{\partial \mathbf{A}}{\partial t} \quad (\text{Vm}^{-1}) \tag{2}$$

where V = electric scalar potential at point P, V
$\quad$ **A** = vector potential at point P, Wb m^{-1}

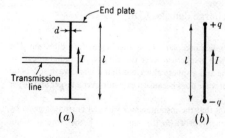

Figure 14-6 (*a*) Short-dipole antenna fed by two-conductor transmission line and (*b*) its equivalent.

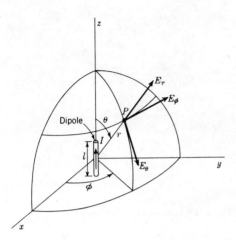

Figure 14-7 Relation of dipole antenna to coordinates.

From (5-22-2) the magnetic field **H** at any point P is

$$\mathbf{H} = \frac{1}{\mu_0}\, \mathbf{\nabla} \times \mathbf{A} \qquad (\mathrm{A\ m^{-1}}) \tag{3}$$

where μ_0 = permeability of air $(400\pi \ \mathrm{nH\ m^{-1}})$
$\mathbf{A}$ = vector potential at point P, Wb m^{-1}

If the scalar potential V and the vector potential **A** at the point P are known, the electric and magnetic fields **E** and **H** at P can be determined by means of (2) and (3). Since we are interested in the fields not only at points near the dipole but also at distances that are comparable to and larger than the wavelength, we must use the retarded potentials given in (14-2-6) and (14-2-7). Thus we have

$$\mathbf{E} = \mathbf{\nabla}[V] - \frac{\partial [\mathbf{A}]}{\partial t} = -\mathbf{\nabla}[V] - j\omega[\mathbf{A}] \qquad (\mathrm{V\ m^{-1}}) \tag{4}$$

and

$$\mathbf{H} = \frac{1}{\mu_0}\, \mathbf{\nabla} \times [\mathbf{A}] \qquad (\mathrm{A\ m^{-1}}) \tag{5}$$

where

$$[V] = \frac{1}{4\pi\epsilon_0} \int \frac{\rho_0\, e^{j\omega[t-(r/c)]}}{r}\, dv \qquad (\mathrm{V}) \tag{5a}$$

$$[\mathbf{A}] = \frac{\mu_0}{4\pi} \int_v \frac{\mathbf{J}_0\, e^{j\omega[t-(r/c)]}}{r}\, dv \qquad (\mathrm{Wb\ m^{-1}}) \tag{5b}$$

The electric and magnetic fields due to any configuration of currents and charges are given by (4) and (5), where the retarded scalar potential $[V]$ is a

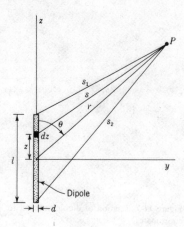

Figure 14-8 Geometry for short dipole.

quantity that depends only on the charges (stationary) and the retarded vector potential [**A**] is a quantity that depends only on the currents. Equation (5) indicates that the magnetic field **H** depends only on the currents, while (4) indicates that the electric field **E** depends on *both* the currents and the charges. However, it will be shown later that in determining the radiation field (at large distances from a current and charge distribution) only the currents need be considered. Since the retarded potentials will be used exclusively in the following development, the brackets will be omitted for the sake of simplicity, it being understood that the potentials are retarded.

We shall now proceed to find the electric and magnetic fields everywhere from a short dipole by first determining the retarded vector and scalar potentials and then substituting these values in (4) and (5) and performing the indicated operations.

Referring to Fig. 14-7 or 14-8, the current is entirely in the z direction. Hence, it follows that the retarded vector potential has only a z component. Its value is

$$A_z = |\mathbf{A}| = \frac{\mu_0 I_0}{4\pi} \int_{-l/2}^{l/2} \frac{e^{j(\omega t - \beta s)}}{s} \, dz \tag{6}$$

where I_0 = amplitude (peak value in time) of current (same at all points along dipole), A

μ_0 = permeability of free space = 400π nH m^{-1}

dz = element of length of conductor, m

ω = radian frequency ($= 2\pi f$, where f = frequency, Hz)

t = time, s

s = distance from dz to point P (see Fig. 14-8), m

β = phase constant, rad m^{-1} ($= 2\pi/\lambda$)

If the distance from the dipole is large compared with its length ($r \gg l$) and if the wavelength is large compared with the length ($\lambda \gg l$), we can put $s = r$ and neglect the magnitude and phase differences of the contributions from the different parts of the wire.† Thus (6) becomes

$$A_z = \frac{\mu_0 I_0 l e^{j(\omega t - \beta r)}}{4\pi r} \tag{7}$$

The electric charge is confined to the ends of the dipole; so turning our attention to the retarded scalar potential, we find its value to be

$$V = \frac{q_0}{4\pi\epsilon_0} \left(\frac{e^{j(\omega t - \beta s_1)}}{s_1} - \frac{e^{j(\omega t - \beta s_2)}}{s_2} \right) \quad \text{(V)} \tag{8}$$

where q_0 = amplitude (peak value in time) of charge at ends of dipole, C
   s_1 = distance from upper end of dipole to P, m
   s_2 = distance from lower end of dipole to P, m

From (1)

$$q = \int I \, dt = \frac{I}{j\omega} \tag{9}$$

where $q = q_0 e^{j(\omega t - \beta s)}$ = retarded charge, C
   $I = I_0 e^{j(\omega t - \beta s)}$ = retarded current, A

It follows that $q_0 = I_0/j\omega$ so that (8) can be reexpressed

$$V = \frac{I_0}{4\pi\epsilon_0 j\omega} \left(\frac{e^{j(\omega t - \beta s_1)}}{s_1} - \frac{e^{j(\omega t - \beta s_2)}}{s_2} \right) \tag{10}$$

When $r \gg l$, the lines of length s_1 and s_2 from the ends of the dipole to the point P may be considered parallel, as shown in Fig. 14-9, so that $s_1 = r - (l/2)\cos\theta$ and $s_2 = r + (l/2)\cos\theta$. Substituting these into (10) and clearing fractions yields

$$V = \frac{I_0 e^{j(\omega t - \beta r)}}{4\pi\epsilon_0 j\omega}$$

$$\times \frac{[r + (l/2)\cos\theta]\exp(j\tfrac{1}{2}\beta l \cos\theta) - [r - (l/2)\cos\theta]\exp(-j\tfrac{1}{2}\beta l \cos\theta)}{r^2}$$

$$\tag{11}$$

where the term $(l^2 \cos^2\theta)/4$ in the denominator has been neglected in comparison with r^2 since $r \gg l$. By de Moivre's theorem, (11) becomes

$$V = \frac{I_0 e^{j(\omega t - \beta r)}}{4\pi\epsilon_0 j\omega r^2} \left[\left(\cos\frac{\beta l \cos\theta}{2} + j \sin\frac{\beta l \cos\theta}{2} \right)\left(r + \frac{l}{2}\cos\theta \right) \right.$$

$$\left. - \left(\cos\frac{\beta l \cos\theta}{2} - j \sin\frac{\beta l \cos\theta}{2} \right)\left(r - \frac{l}{2}\cos\theta \right) \right] \tag{12}$$

† If r is large compared with l but λ is not large compared with l, we may put $s = r$ in the denominator in (6) and neglect the differences in magnitude. However, in such cases we should retain s in the exponential expression since the difference in phase of the contributions may be significant.

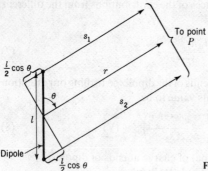

Figure 14-9 Relations for short dipole when $r \gg l$.

Since it is assumed that the wavelength is much greater than the length of the dipole ($\lambda \gg l$),

$$\cos \frac{\beta l \cos \theta}{2} = \cos \frac{\pi l \cos \theta}{\lambda} \approx 1 \qquad (13)$$

and

$$\sin \frac{\beta l \cos \theta}{2} \approx \frac{\beta l \cos \theta}{2} \qquad (14)$$

Introducing (13) and (14) into (12) reduces the expression for the scalar potential to

$$V = \frac{I_0 l e^{j(\omega t - \beta r)} \cos \theta}{4\pi\epsilon_0 c} \left(\frac{1}{r} + \frac{c}{j\omega} \frac{1}{r^2} \right) \qquad (\text{V}) \qquad (15)\dagger$$

where I_0 = amplitude (peak value in time) of current, A

l = length of dipole, m

ω = radian frequency ($= 2\pi f$, where f = frequency, Hz)

β = phase constant, rad m^{-1}($= 2\pi/\lambda$)

t = time, s

θ = angle between dipole and radius vector of length r to point P, dimensionless

ϵ_0 = permittivity of free space = 8.85 pF m^{-1}

c = velocity of light = 300 Mm s^{-1}

j = complex operator = $\sqrt{-1}$

r = distance from center of dipole to point P, m

Equation (15) gives the retarded scalar potential and (7) the retarded vector potential at a distance r and at an angle θ from a short dipole. The only restrictions are that $r \gg l$ and $\lambda \gg l$. Before substituting these values in (4) and (5), let us express **E** and **A** in polar coordinates. Thus, in polar coordinates (see Fig. 14-7),

$$\mathbf{E} = \hat{\mathbf{r}} E_r + \hat{\boldsymbol{\theta}} E_\theta + \hat{\boldsymbol{\phi}} E_\phi \qquad (16)$$

† Note that $1/\epsilon_0 c = \mu_0 c = \sqrt{\mu_0/\epsilon_0} = 376.7 \, \Omega$.

Figure 14-10 Resolution of vector potential into A_r and A_θ components.

and

$$\mathbf{A} = \hat{\mathbf{r}}A_r + \hat{\boldsymbol{\theta}}A_\theta + \hat{\boldsymbol{\phi}}A_\phi \tag{17}$$

In our case **A** has only a z component, so that $A_\phi = 0$, and from Fig. 14-10

$$A_r = A_z \cos\theta \tag{17a}$$

$$A_\theta = -A_z \sin\theta \tag{17b}$$

In polar coordinates we also have for the gradient of the scalar potential

$$\nabla V = \mathbf{r}\frac{\partial V}{\partial r} + \hat{\boldsymbol{\theta}}\frac{1}{r}\frac{\partial V}{\partial \theta} + \hat{\boldsymbol{\phi}}\frac{1}{r\sin\theta}\frac{\partial V}{\partial \phi} \tag{18}$$

It follows from (4) and the above relations that the components of **E** are

$$E_r = -j\omega A_r - \frac{\partial V}{\partial r} = -j\omega A_z \cos\theta - \frac{\partial V}{\partial r} \tag{19}$$

$$E_\theta = -j\omega A_\theta - \frac{1}{r}\frac{\partial V}{\partial \theta} = j\omega A_z \sin\theta - \frac{1}{r}\frac{\partial V}{\partial \theta} \tag{20}$$

$$E_\phi = -j\omega A_\phi - \frac{1}{r\sin\theta}\frac{\partial V}{\partial \phi} = -\frac{1}{r\sin\theta}\frac{\partial V}{\partial \phi} \tag{21}$$

Now introducing the value of A_z from (7) and V from (15) into these equations, we find that $E_\phi = 0$ (since V is independent of ϕ, so that $\partial V/\partial\phi = 0$) and also that

$$\boxed{E_r = \frac{I_0\,l e^{j(\omega t - \beta r)}\cos\theta}{2\pi\epsilon_0}\left(\frac{1}{cr^2} + \frac{1}{j\omega r^3}\right)} \tag{22}$$

and

$$\boxed{E_\theta = \frac{I_0\,l e^{j(\omega t - \beta r)}\sin\theta}{4\pi\epsilon_0}\left(\frac{j\omega}{c^2 r} + \frac{1}{cr^2} + \frac{1}{j\omega r^3}\right)} \tag{23}$$

Turning our attention now to the magnetic field, we can calculate it by (5). In polar coordinates the curl of **A** is

$$\nabla \times \mathbf{A} = \frac{\hat{\mathbf{r}}}{r\sin\theta}\left[\frac{\partial(\sin\theta\,A_\phi)}{\partial\theta} - \frac{\partial(A_\theta)}{\partial\phi}\right]$$
$$+ \frac{\hat{\boldsymbol{\theta}}}{r\sin\theta}\left[\frac{\partial A_r}{\partial\phi} - \frac{\partial(r\sin\theta)A_\phi}{\partial r}\right] + \frac{\hat{\boldsymbol{\phi}}}{r}\left[\frac{\partial(rA_\theta)}{\partial r} - \frac{\partial A_r}{\partial\theta}\right] \tag{24}$$

Since $A_\phi = 0$, the first and fourth terms of (24) are zero. From (7), (17a), and (17b) we note that A_r and A_θ are independent of ϕ, so that the second and third terms of (24) are also zero. Thus, since only the last two terms in (24) contribute, $\nabla \times \mathbf{A}$ has only a ϕ component. Introducing A_r and A_θ into (24), performing the indicated operations, and substituting this result into (5), we find that $H_r = H_\theta = 0$ and that

$$H_\phi = |\mathbf{H}| = \frac{I_0 \, l e^{j(\omega t - \beta r)} \sin \theta}{4\pi} \left(\frac{j\omega}{cr} + \frac{1}{r^2} \right) \qquad (25)$$

Thus the electric and magnetic fields from the dipole have only three components, E_r, E_θ, and H_ϕ. The components E_ϕ, H_r, and H_θ are everywhere zero.

When r is very large, the terms involving $1/r^2$ and $1/r^3$ in (22), (23), and (25) can be neglected in comparison with terms in $1/r$. Thus, in the *far field* E_r is negligible, and we have effectively only two field components, E_θ and H_ϕ, given by

$$E_\theta = \frac{j\omega I_0 \, l e^{j(\omega t - \beta r)} \sin \theta}{4\pi\epsilon_0 c^2 r} = j \frac{30 I_0 \beta l}{r} e^{j(\omega t - \beta r)} \sin \theta \qquad (26)$$

$$H_\phi = \frac{j\omega I_0 \, l e^{j(\omega t - \beta r)} \sin \theta}{4\pi c r} = j \frac{I_0 \beta l}{4\pi r} e^{j(\omega t - \beta r)} \sin \theta \qquad (27)$$

Taking the ratio of E_θ to H_ϕ as given by (26) and (27), we obtain for air or vacuum

$$\frac{E_\theta}{H_\phi} = \frac{1}{\epsilon_0 c} = \sqrt{\frac{\mu_0}{\epsilon_0}} = 376.7 \ \Omega \qquad (28)$$

This is the intrinsic impedance of free space.

It is to be noted that E_θ and H_ϕ are in time phase in the far field. Thus, $\mathbf{E}$ and $\mathbf{H}$ in the far field of the spherical wave from the dipole are related in the same manner as in a plane traveling wave. Both are also proportional to $\sin \theta$. That is, both are a maximum when $\theta = 90°$ and a minimum when $\theta = 0$ (in the direction of the dipole axis). This variation of E_θ (or H_ϕ) with angle can be portrayed by a *field pattern* as in Fig. 14-11, the length ρ of the radius vector being proportional to the value of the far field (E_θ or H_ϕ) in that direction from the dipole. The pattern in Fig. 14-11a is one-half of a three-dimensional pattern and illustrates that the fields are a function of θ but are independent of ϕ. The pattern in Fig. 14-11b is two-dimensional and represents a cross section through the three-dimensional pattern. The three-dimensional far-field pattern of the short dipole is doughnut-shaped, while the two-dimensional pattern has the shape of a figure eight.

From (22), (23), and (25) we note that for a small value of r the electric field has two components, E_r and E_θ, both of which are in time-phase quadrature with the magnetic field H_ϕ. Thus, in the *near field*, $\mathbf{E}$ and $\mathbf{H}$ are related as in a standing wave. At intermediate distances, E_θ and E_r can approach time-phase quadrature

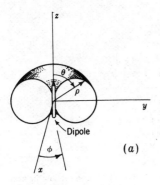

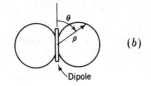

Figure 14-11 (*a*) Three-dimensional and (*b*) two-dimensional field pattern of far field (E_θ or H_ϕ) from a short dipole.

with each other so that the total electric field vector rotates in a plane parallel to the direction of propagation and containing the dipole, exhibiting the phenomenon of cross field.

In the far field the energy flow is real. That is, the energy flow is always radially outward. This energy is radiated. As a function of angle it is maximum at the equator ($\theta = 90°$). In the near field the energy flow is largely reactive. That is, energy flows out and back twice per cycle without being radiated. There is also angular energy flow (in the θ direction). This energy picture is suggested by Fig. 14-12, where the arrows represent the direction of energy flow at successive instants.†

Let us now consider the situation at very low frequencies. This will be referred to as the *quasi-stationary* case. Noting that $I_0 = j\omega q_0$, we can express the field components as

$$E_r = \frac{q_0\, l e^{j(\omega t - \beta r)} \cos \theta}{2\pi\epsilon_0} \left(\frac{j\omega}{cr^2} + \frac{1}{r^3} \right) \tag{29}$$

$$E_\theta = \frac{q_0\, l e^{j(\omega t - \beta r)} \sin \theta}{4\pi\epsilon_0} \left(-\frac{\omega^2}{c^2 r} + \frac{j\omega}{cr^2} + \frac{1}{r^3} \right) \tag{30}$$

$$H_\phi = \frac{I_0\, l e^{j(\omega t - \beta r)} \sin \theta}{4\pi} \left(\frac{j\omega}{cr} + \frac{1}{r^2} \right) \tag{31}$$

† The instantaneous direction and time rate of energy flow per unit area is given by the instantaneous Poynting vector ($=\mathbf{E} \times \mathbf{H}$).

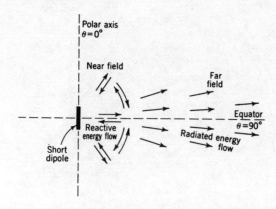

Figure 14-12 Energy flow in near and far regions of short dipole.

As the frequency approaches zero ($\omega \to 0$), the terms with ω in the numerator can be neglected. Also $e^{j(\omega t - \beta r)}$ approaches unity. Thus, for the quasi-stationary (or dc) case† the field components become

$$E_r = \frac{q_0 l \cos\theta}{2\pi\epsilon_0 r^3} \tag{32}$$

$$E_\theta = \frac{q_0 l \sin\theta}{4\pi\epsilon_0 r^3} \tag{33}$$

$$H_\phi = \frac{I_0 l \sin\theta}{4\pi r^2} \tag{34}$$

The electric field components, (32) and (33), are the same as (2-15-4) and (2-15-5) for a static electric dipole, while the magnetic field component H_ϕ in (34) is equivalent to (5-3-3) for a current element. Since these fields vary as $1/r^2$ or $1/r^3$, they are effectively confined to the vicinity of the dipole and radiation is negligible. At high frequencies in the far field, however, we note from (26) and (27) that the fields (E_θ and H_ϕ) vary as $1/r$. These fields are radiated and hence are often called the *radiation fields* of the dipole.

The expressions for the fields from a short dipole, developed above, are summarized in Table 14-1. In the table the restriction applies that $r \gg l$ and $\lambda \gg l$. The three field components not listed are everywhere zero; that is, $E_\phi = H_r = H_\theta = 0$.

If we had been interested only in the far field, the development following (7) could have been much simplified. The scalar potential V does not contribute to the far field, so that both **E** and **H** can be determined from **A** alone. Thus, from (4), E_θ and H_ϕ of the far field can be obtained very simply from $E_\theta = |\mathbf{E}| = -j\omega A_\theta$ and $H_\phi = |\mathbf{H}| = E_\theta/Z_0 = -(j\omega/Z_0)A_\theta$, where $Z_0 = \sqrt{\mu_0/\epsilon_0} = 376.7\ \Omega$. Or H_ϕ may be obtained directly from (5) and E_θ from this. Thus

$$H_\phi = |\mathbf{H}| = \frac{1}{\mu_0}|\nabla \times \mathbf{A}| \tag{35}$$

† For this case the wavelength is very large ($\lambda \to \infty$) so that $\lambda >>> l$. We also have $r \gg l$ and hence in *this* case $\lambda \gg r$.

Table 14-1 Fields of a short dipole

Component	General expression	Far field	Quasi-stationary
E_r	$\dfrac{I_0 l e^{j(\omega t - \beta r)} \cos\theta}{2\pi\epsilon_0}\left(\dfrac{1}{cr^2} + \dfrac{1}{j\omega r^3}\right)$	0	$\dfrac{q_0 l \cos\theta}{2\pi\epsilon r^3}$
E_θ	$\dfrac{I_0 l e^{j(\omega t - \beta r)} \sin\theta}{4\pi\epsilon_0}\left(\dfrac{j\omega}{c^2 r} + \dfrac{1}{cr^2} + \dfrac{1}{j\omega r^3}\right)$	$\dfrac{j60\pi I_0 e^{j(\omega t - \beta r)} \sin\theta}{r}\dfrac{l}{\lambda}$	$\dfrac{q_0 l \sin\theta}{4\pi\epsilon r^3}$
H_ϕ	$\dfrac{I_0 l e^{j(\omega t - \beta r)} \sin\theta}{4\pi}\left(\dfrac{j\omega}{cr} + \dfrac{1}{r^2}\right)$	$\dfrac{jI_0 e^{j(\omega t - \beta r)} \sin\theta}{2r}\dfrac{l}{\lambda}$	$\dfrac{I_0 l \sin\theta}{4\pi r^2}$

and, neglecting terms in $1/r^2$,

$$E_\theta = |\mathbf{E}| = ZH_\phi = \frac{Z}{\mu_0}|\nabla \times \mathbf{A}| \tag{36}$$

Referring to (26) for the far E_θ field of a short dipole, it is instructive to separate the expression into its six basic factors. Thus,

$$E_\theta = 60\pi \quad I_0 \quad \frac{l}{\lambda} \quad \frac{1}{r} \quad je^{j(\omega t - \beta r)} \quad \sin\theta \tag{37}$$

$$\text{Magni-}\quad \text{Current}\ \text{Length}\ \text{Distance}\quad \text{Phase}\quad \text{Pattern}$$
$$\text{tude}$$

where 60π is a constant (*magnitude*) factor, I_0 is the dipole *current*, l/λ is the dipole *length* in terms of wavelengths, $1/r$ is the *distance* factor, $je^{j(\omega t - \beta r)}$ is the *phase* factor, and $\sin\theta$ is the *pattern* factor giving the variation of the field with angle. In general, the expression for the field of any antenna will involve these six factors.

The field relations in Table 14-1 are those for a short dipole. Longer linear antennas or large antennas of other shape may be regarded as being made up of many such short dipoles. Hence the fields of these larger antennas can be obtained by integrating the field contributions from all the small dipoles making up the antenna (see Sec. 14-9).

14-4 RADIATION RESISTANCE OF A SHORT DIPOLE

By taking the surface integral of the average Poynting vector over any surface enclosing an antenna the total power radiated by the antenna is obtained. Thus

$$P = \int_s \mathbf{S}_{av} \cdot d\mathbf{s} \quad \text{(W)} \tag{1}$$

where P = power radiated, W
$\mathbf{S}_{av}$ = average Poynting vector, W m^{-2}

The simplest surface we might choose is a sphere with the antenna at the center. Since the far-field equations for an antenna are simpler than the near-field relations, it will be to our advantage to make the radius of the sphere large compared with the dimensions of the antenna. In this way the surface of the sphere lies in the far field, and only the far-field components need be considered.

Assuming no losses, the power radiated by the antenna is equal to the average power delivered to the antenna terminals. This equals $\frac{1}{2}I_0^2 R$, where I_0 is amplitude (peak value in time) of the current at the terminals and R is the *radiation resistance* appearing at the terminals. Thus $P = \frac{1}{2}I_0^2 R$, and the radiation resistance is

$$R = \frac{2P}{I_0^2} \quad (\Omega) \tag{2}$$

where P is the radiated power in watts.

Let us now carry through the calculation, as outlined above, in order to find the radiation resistance of a short dipole. The power radiated is

$$P = \int \mathbf{S}_{av} \cdot d\mathbf{s} = \frac{1}{2} \int_s \text{Re} \, (\mathbf{E} \times \mathbf{H}^*) \cdot d\mathbf{s} \tag{3}$$

In the far field only E_θ and H_ϕ are not zero, so that (3) reduces to

$$P = \frac{1}{2} \int_s \text{Re} \, E_\theta H_\phi^* \, \hat{\mathbf{r}} \cdot d\mathbf{s} \tag{4}$$

where $\hat{\mathbf{r}}$ is the unit vector in the radial direction. Thus the power flow in the far field is entirely radial (normal to surface of sphere of integration). But $\hat{\mathbf{r}} \cdot d\mathbf{s} = ds$; so

$$P = \frac{1}{2} \int_s \text{Re} \, E_\theta H_\phi^* \, ds \tag{5}$$

where E_θ and H_ϕ^* are complex, H_ϕ^* being the complex conjugate of H_ϕ. Now $E_\theta = H_\phi Z$; so (5) becomes

$$P = \frac{1}{2} \int_s \text{Re} \, H_\phi H_\phi^* Z \, ds = \frac{1}{2} \int_s |H_\phi|^2 \, \text{Re} \, Z \, ds \tag{6}$$

Since $\text{Re} \, Z = \sqrt{\mu_0/\epsilon_0}$ and $ds = r^2 \sin \theta \, d\theta \, d\phi$,†

$$P = \frac{1}{2} \sqrt{\frac{\mu_0}{\epsilon_0}} \int_0^{2\pi} \int_0^{\pi} |H_\phi|^2 \, r^2 \sin \theta \, d\theta \, d\phi \tag{7}$$

---

† Since $\sqrt{\mu_0/\epsilon_0} = E_\theta/H_\phi \approx 120\pi$, we may also write

$$P = \frac{1}{240\pi} \int_0^{2\pi} \int_0^{\pi} |E_\theta|^2 r^2 \sin \theta \, d\theta \, d\phi$$

where the angles θ and ϕ are as shown in Fig. 14-7 and $|H_\phi|$ is the absolute value (or amplitude) of the H field. From (14-3-27) this is

$$|H_\phi| = \frac{\omega I_{av} l \sin \theta}{4\pi c r} \tag{8}$$

Substituting this into (7), we have

$$P = \frac{1}{32} \sqrt{\frac{\mu_0}{\epsilon_0}} \left(\frac{\beta I_{av} l}{\pi} \right)^2 \int_0^{2\pi} \int_0^\pi \sin^3 \theta \, d\theta \, d\phi \tag{9}$$

Upon integration (9) becomes

$$P = \sqrt{\frac{\mu_0}{\epsilon_0}} \frac{(\beta I_{av} l)^2}{12\pi} \quad (\text{W}) \tag{10}$$

This is the power radiated by the short dipole where I_{av} is the average current on the dipole.

Substituting the power P from (10) into (2) yields for the *radiation resistance of the short dipole*

$$R = \sqrt{\frac{\mu_0}{\epsilon_0}} \frac{(\beta l)^2}{6\pi} \left(\frac{I_{av}}{I_0} \right)^2 \quad (\Omega) \tag{11}$$

Since $\sqrt{\mu_0/\epsilon_0} = 376.7 \approx 120\pi \ \Omega$, (11) reduces to

$$R = 20(\beta l)^2 = 80\pi^2 \left(\frac{l}{\lambda} \right)^2 \left(\frac{I_{av}}{I_0} \right)^2 \quad (\Omega) \tag{12}$$

Example: Radiation resistance of short vertical antenna Calculate the radiation resistance of a broadcast antenna which is a 30-m-high, base-fed vertical conductor or mast operating over a perfectly conducting ground at a frequency of 500 kHz.

SOLUTION The wavelength $\lambda = c/f = 3 \times 10^8/5 \times 10^5 = 600$ m, and so the height is $\frac{30}{600} = \frac{1}{20}\lambda$.

As discussed in Sec. 7-16, the mast has an image as suggested in Fig. 14-13. The radiation resistance of the mast antenna (measured between terminal 1 and ground) is one-half the radiation resistance of a dipole antenna of twice the height with no ground present (measured between terminals 1 and 2).

The antenna current must be zero at the top of the mast, tapering almost linearly from its value at the terminals. Thus, the average current is one-half the terminal current I_0 and from (12)

$$R_r = \frac{1}{2} \times 80\pi^2 \left(\frac{l}{\lambda} \right)^2 \frac{1}{4} \quad (\Omega)$$

where $l = 60$ m

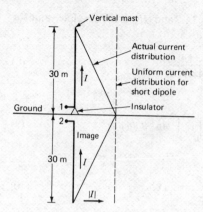

Figure 14-13 Electrically short vertical mast antenna above perfectly conducting ground with its image.

Thus, the 30-m mast has a radiation resistance of

$$R_r = \tfrac{1}{2} \times 80\pi^2 \left(\frac{60}{600}\right)^2 \tfrac{1}{4} \approx 1 \ \Omega$$

If there is any heat loss in the antenna due to finite conductivity or to losses in the associated dielectric structure, an equivalent *loss resistance* R_{loss} will appear and the terminal resistance will be given by

$$R = R_{\text{loss}} + R_r \qquad (\Omega) \qquad (13)$$

where R_{loss} = loss resistance, Ω
R_r = radiatior. resistance, Ω

Suppose, for instance, that $R_{\text{loss}} = 1 \ \Omega$ for the antenna of the above example. The terminal resistance is then $2\Omega \, (= 1 + 1)$. The *antenna efficiency k* is

$$k = \frac{\text{power radiated}}{\text{power input}} = \frac{R_r}{R_r + R_{\text{loss}}} = \frac{1}{1+1} = 50\% \qquad (13a)$$

A taller antenna with larger radiation resistance would be more efficient provided R_{loss} remained small.

The radiation resistance of antennas other than the short dipole can be calculated as above provided the far field is known as a function of angle. Thus, from (2) and (6) the *radiation resistance at the terminals of an antenna* is given by

$$R_r = \frac{120\pi}{I_0^2} \int_s |H|^2 \, ds = \frac{120\pi}{I_0^2} \int_s \frac{|E|^2}{Z_0^2} \, ds \qquad (\Omega) \qquad (14)$$

where $|H|$ = amplitude of far H field, A m^{-1}
$|E|$ = amplitude of far E field, V m^{-1}
I_0 = amplitude of terminal current, A
Z_0 = intrinsic impedance of space $= 376.7 \ \Omega$

If we integrate the complex Poynting vector $(=\frac{1}{2}\mathbf{E} \times \mathbf{H}^*)$ over a surface enclosing an antenna, we shall obtain, in general, both a real part equal to the power radiated and an imaginary part equal to the reactive power. Whereas the real part, or radiated power, is the same for *any* surface enclosing the antenna, the imaginary, or reactive, power obtained depends on the location and shape of the surface enclosing the antenna. For a large surface lying only in the far field the reactive power is zero, but for a surface lying in the near field it may be of considerable magnitude. In the case of a very thin linear antenna, it turns out that if the surface of integration is collapsed so as to coincide with the surface of the antenna, the complex power so obtained divided by the square of the terminal current yields the terminal impedance $R + jX$, where R is the radiation resistance.

14-5 EFFECTIVE APERTURE, DIRECTIVITY, AND GAIN

A useful way of describing the pattern of an antenna is in terms of the angular width of the main lobe at a particular level. Referring to Fig. 14-2b, the angle at the half-power level or *half-power beam width* (HPBW)† is usually the one given. The *beam width between first nulls* (BWFN) is also sometimes used.

The antenna power pattern as a function of angle can be expressed as the radial component of the average Poynting vector multiplied by the square of the distance r at which it is measured. Thus, regardless of the wave polarization, we have from (11-3-12) and (11-3-13) that

$$P(\theta, \phi) = S_r r^2 = \frac{1}{2} \frac{E^2(\theta, \phi)}{Z_0} r^2 = \frac{1}{2}H^2(\theta, \phi)Z_0 r^2 \qquad (\text{W sr}^{-1}) \qquad (1)$$

where S_r = radial component of Poynting vector at distance r, W m^{-2}
$E(\theta, \phi)$ = total transverse electric field as a function of angle, V m^{-1}
$H(\theta, \phi)$ = total transverse magnetic field as a function of angle, A m^{-1}
r = distance from antenna to point of measurement, m
Z_0 = intrinsic impedance of medium, Ω per square

It is assumed that the medium is lossless. In the far field the Poynting vector is entirely radial (fields entirely transverse), and E and H vary as $1/r$, so that $P(\theta, \phi)$ is independent of distance. The quantity $P(\theta, \phi)$ is often called the *radiation intensity* (W sr^{-1} or W rad^{-2}). Dividing $P(\theta, \phi)$ by its maximum value $[P(\theta, \phi)_{\text{max}}]$, we obtain the *normalized antenna power pattern*,

$$P_n(\theta, \phi) = \frac{P(\theta, \phi)}{P(\theta, \phi)_{\text{max}}} \qquad (\text{dimensionless}) \qquad (2)$$

A normalized antenna power pattern is shown in Fig. 14-14a with pattern maximum coinciding with the direction $\theta = 0$. A highly significant way of describing an antenna pattern is to integrate the normalized power pattern with respect to angle

---

† Also called the *3-dB beam width*.

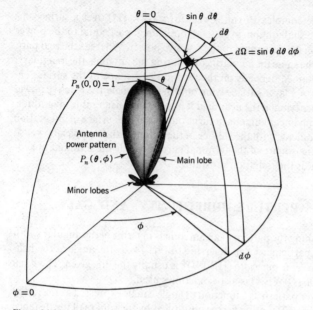

Figure 14-14a Antenna power pattern with maximum aligned with the $\theta = 0$ direction (zenith).

and obtain a *solid angle*. Thus, integrating $P_n(\theta, \phi)$ over 4π sr (rad²), we get the total pattern solid angle or *beam solid angle*

$$\Omega_A = \int_{\phi=0}^{\phi=2\pi} \int_{\theta=0}^{\theta=\pi} P_n(\theta, \phi) \sin\theta \, d\theta \, d\phi = \iint_{4\pi} P_n(\theta, \phi) \, d\Omega \quad \text{(sr)} \qquad (3)$$

Although the term *total pattern solid angle* would be more explicit, the term *beam solid angle* is customarily used for Ω_A.

The *beam solid angle* Ω_A is the angle through which all the power from a transmitting antenna would stream if the power (per unit solid angle) were constant over this angle and equal to the maximum value. It turns out that for typical symmetrical patterns this angle is *approximately* equal to the square of the half-power beam width (HPBW). See Fig. 14-14b, also (14-6.6-6).

Integrating over the main lobe gives the *main-lobe solid angle*†

$$\Omega_M = \iint_{\substack{\text{main} \\ \text{lobe}}} P_n(\theta, \phi) \, d\Omega \quad \text{(sr)} \qquad (4)$$

† In patterns for which no clearly defined minimum exists the extent of the main lobe may be somewhat indefinite, and an arbitrary level such as −20 dB can be used to delineate it.

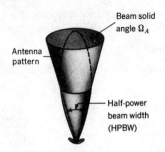

Figure 14-14b Antenna power pattern and its relation to the beam solid angle.

It follows that the *minor-lobe solid angle* Ω_m is given by the difference of the beam (total pattern) solid angle and the main-lobe solid angle. That is,

$$\Omega_m = \Omega_A - \Omega_M \quad \text{(sr)} \tag{5}$$

If the antenna has no minor lobes ($\Omega_m = 0$), then $\Omega_A = \Omega_M$. For an isotropic antenna (radiation same in all directions) $P_n(\theta, \phi) = 1$ for all θ and ϕ, and $\Omega_A = 4\pi$.†

Another important antenna parameter is the *directivity D*, which may be defined as the ratio of the maximum radiation intensity (antenna transmitting) to the average radiation intensity, or

$$D = \frac{\text{maximum radiation intensity}}{\text{average radiation intensity}} = \frac{P(\theta, \phi)_{max}}{P_{av}} \quad \text{(dimensionless)} \tag{6}$$

The average radiation intensity is given by the total power radiated W divided by 4π sr. The total power radiated is equal to the radiation intensity $P(\theta, \phi)$ integrated over 4π sr. Hence,

$$D = \frac{P(\theta, \phi)_{max}}{W/4\pi} = \frac{4\pi P(\theta, \phi)_{max}}{\iint\limits_{4\pi} P(\theta, \phi)d\Omega}$$

$$= \frac{4\pi}{\iint\limits_{4\pi} [P(\theta, \phi)/P(\theta, \phi)_{max}]d\Omega} = \frac{4\pi}{\iint\limits_{4\pi} P_n(\theta, \phi)d\Omega} \tag{7}$$

We note that the denominator of the last member of (7) is the beam solid angle as given by (3); so

$$D = \frac{4\pi}{\Omega_A} \quad \text{(dimensionless)} \tag{8}$$

Thus, the *directivity* of an antenna is equal to the solid angle of a sphere (4π sr) divided by the antenna beam solid angle Ω_A. In this relation, directivity is derived

---

† The *beam efficiency* ϵ_M may be defined as $\epsilon_M = \Omega_M/\Omega_A$.

from the pattern. The directivity is a unique dimensionless quantity. It indicates how well the antenna concentrates power into a limited solid angle; the smaller the angle, the larger the directivity.

Example 1 Calculate the directivity of an isotropic antenna (uniform radiation intensity in all directions).

SOLUTION For an isotropic antenna $P_n(\theta, \phi) = 1$ and $\Omega_A = 4\pi$; so

$$D = \frac{4\pi}{\Omega_A} = 1 \tag{9}$$

Thus, *the directivity of an isotropic antenna is unity. This is the smallest value the directivity can attain.*

Example 2 Calculate the directivity of a short dipole.

SOLUTION The short dipole has only an H_ϕ component of the magnetic field as given by (14-3-27). Thus, its normalized power pattern is

$$P_n(\theta, \phi) = \frac{H_\phi^2(\theta, \phi)}{H_\phi^2(\theta, \phi)_{\max}} = \sin^2 \theta \tag{10}$$

and noting (3) and (8),

$$D = \frac{4\pi}{\int_0^{2\pi}\int_0^\pi \sin^3 \theta \, d\theta \, d\phi} = \frac{3}{2} \tag{11}$$

Hence, the directivity of a short dipole is $\frac{3}{2}$. That is, the maximum radiation intensity is 1.5 times as much as if the power were radiated uniformly in all directions.

Directivity is based entirely on the shape of the far- (or radiated-) field pattern. The antenna efficiency is not involved. However, the power gain, or simply the *gain*, of an antenna does involve the efficiency. It is defined as follows:

$$\text{Gain} = G = \frac{\text{maximum radiation intensity}}{\substack{\text{maximum radiation intensity from a refer-} \\ \text{ence antenna with the same power input}}} \tag{12}$$

Any convenient type of antenna may be taken as the reference. If the reference antenna is a lossless isotropic type (radiation intensity uniform in all directions), the gain (designated G_0) is given by

$$G_0 = \frac{P'_m}{P_0} \tag{13}$$

where P'_m = maximum radiation intensity from antenna under consideration
P_0 = radiation intensity from a lossless (100 percent efficient) isotropic antenna with same power input

Since P'_m is related to the radiation intensity P_m of a 100 percent efficient antenna by the antenna efficiency factor k, same as k in (14-4-13a),

$$G_0 = \frac{kP_m}{P_0} = kD \tag{14}$$

Thus, the gain of an antenna over a lossless isotropic type equals the directivity if the antenna is 100 percent efficient ($k = 1$) but is less than the directivity if any losses are present in the antenna ($k < 1$).

The gain G_0 (over an isotropic source) when expressed in decibels can be written dBi (*decibels over isotropic*) to distinguish it from the gain over a reference $\lambda/2$ antenna, as discussed in Sec. 14-19. The gain equals the directivity D if the antenna is lossless.

The directivity D is never less than unity. Its value must lie between 1 and infinity ($1 \le D \le \infty$). On the other hand, the gain (G or G_0) may lie between 0 and infinity.

A transmitting antenna radiates energy. A receiving antenna, on the other hand, collects energy. In this connection it is often useful to consider that the receiving antenna possesses an aperture or area over which it extracts energy from a passing radio wave, the power collected being equal to the incident Poynting vector (W m^{-2}) times the area (m^2).

Thus, suppose that a receiving antenna is immersed in the field of a plane traveling wave, as suggested in Fig. 14-15a. The antenna is terminated in a load of impedance $Z_L = R_L + jX_L$. Let the aperture A of the antenna be defined as the ratio of the received power (power delivered to load) to the power density (or Poynting vector) of the incident wave. The received power is equal to $I^2 R_L$, where I is the terminal current. Therefore

$$A = \frac{I^2 R_L}{S} = \frac{\text{received power}}{\text{power density of incident wave}} \tag{15}$$

where A = aperture, m^2

$\quad\quad I$ = rms terminal current, A

$\quad\quad S$ = Poynting vector (or power density) of incident wave, W m^{-2}

$\quad\quad R_L$ = load resistance, Ω

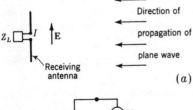

(a)

(b)

Figure 14-15 (a) Terminated receiving antenna immersed in field of plane traveling wave and (b) equivalent circuit.

Replacing the antenna by its equivalent, or Thévenin generator having an equivalent emf $\mathscr{V}$ and impedance Z_A ($= R_A + jX_A$), we may draw the equivalent circuit shown in Fig. 14-15b. The terminal current I is

$$I = \frac{\mathscr{V}}{Z_L + Z_A} \quad \text{(A)} \tag{16}$$

where $\mathscr{V}$ = rms emf induced by passing wave, V
$\quad Z_L$ = load impedance, Ω
$\quad Z_A$ = antenna impedance, Ω

When (16) is substituted into (15), it follows that

$$A = \frac{\mathscr{V}^2 R_L}{S[(R_A + R_L)^2 + (X_A + X_L)^2]} \tag{17}$$

where R_A = antenna resistance, Ω
$\quad X_A$ = antenna reactance, Ω
$\quad R_L$ = load resistance, Ω
$\quad X_L$ = load reactance, Ω

The emf $\mathscr{V}$ will be greatest when the antenna is oriented for maximum response. Under these conditions the maximum power will be transferred to the load when $X_L = -X_A$ and $R_L = R_r$. It is assumed that the antenna is lossless, so that R_A is entirely radiation resistance ($R_A = R_r$). Under these conditions we obtain the maximum aperture, known as the *effective aperture A_e*, as given by

$$\boxed{A_e = \frac{\mathscr{V}^2 R_L}{4SR_r^2} = \frac{\mathscr{V}^2}{4SR_r} \quad \text{(m}^2\text{)}} \tag{18}$$

The *effective aperture A_e has a unique, simply defined value for all antennas.*

Example 3 Find the effective aperture of a short dipole with uniform current.

SOLUTION Referring to (18), we need to know $\mathscr{V}$, S, and R_r. The emf induced in the short dipole is a maximum when the dipole is parallel to the incident electric field **E**. Hence

$$\mathscr{V} = El \quad \text{(V)} \tag{19}$$

The Poynting vector is

$$S = \frac{E^2}{Z_0} \quad \text{(W m}^{-2}\text{)} \tag{20}$$

where Z_0 is the intrinsic impedance of medium (air or vacuum)($= \sqrt{\mu_0/\epsilon_0}$). From (14-4-11) the radiation resistance is

$$R_r = \sqrt{\frac{\mu_0}{\epsilon_0}} \frac{(\beta l)^2}{6\pi} \quad \text{(}\Omega\text{)} \tag{21}$$

Substituting these values for $\mathscr{V}$, S, and R_r into (18), gives the effective aperture of a short dipole:

$$A_e = \frac{3}{8\pi} \lambda^2 = 0.119\lambda^2 \qquad (22)$$

Thus, regardless of how small the dipole is, it can collect power over an aperture of $0.119\lambda^2$ and deliver it to its terminal impedance or load. It is assumed here that the dipole is lossless. However, in practice, losses are present because of the finite conductivity of the dipole conductor, so that the actual aperture is less ($R_A = R_r + R_{\text{loss}}$).

Consider next that an antenna with effective aperture A_e has the beam solid angle Ω_A, as suggested in Fig. 14-16. If the field E_a is constant over the aperture, the power radiated is

$$P = \frac{E_a^2}{Z} A_e \qquad \text{(W)} \qquad (23)$$

where Z is the intrinsic impedance of the medium. Let the field at a radius r be E_r. Then the power radiated is given by

$$P = \frac{E_r^2}{Z} r^2 \Omega_A \qquad \text{(W)} \qquad (24)$$

Equating (23) and (24) and substituting $E_r = E_a A_e / r\lambda$ [see (14-7-5)] yields the important relation

$$\boxed{\lambda^2 = A_e \Omega_A \qquad (\text{m}^2)} \qquad (25)$$

where λ = wavelength, m
A_e = effective aperture, m^2
Ω_A = beam solid angle, sr

According to (25), the product of the effective aperture and the beam solid angle is equal to the wavelength squared. If A_e is known, we can determine Ω_A (or vice versa) at a given wavelength.

From (25) and (8) it follows that

$$\boxed{D = \frac{4\pi}{\lambda^2} A_e} \qquad (26)$$

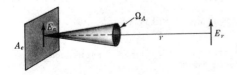

Figure 14-16 Radiation from aperture A_e with uniform field E_a.

Here the directivity is based on the aperture. Aperture is discussed further in Sec. 14-22. Three expressions have now been given for the directivity D. They are

$$D = \frac{P(\theta, \phi)_{max}}{P_{av}} \qquad \text{from pattern} \tag{27}$$

$$D = \frac{4\pi}{\Omega_A} \qquad \text{from pattern} \tag{28}$$

$$D = \frac{4\pi}{\lambda^2} A_e \qquad \text{from aperture†} \tag{29}$$

14-6 ARRAY THEORY

Much of antenna theory (and almost all of array theory) involves little more than the proper addition of the *field* contributions from all parts of an antenna. We must deal with the field (not power) since we must include both magnitude *and* phase.

14-6.1 Two Isotropic Point Sources

Consider two isotropic point sources separated by a distance d, as in Fig. 14-17a. A point source is an idealization representing here an isotropic radiator occupying zero volume. By reciprocity (see Sec. 14-17) the pattern of arrays of such sources (transmitting case) will be identical with the pattern when the array is used as a receiving antenna.

Let the two point sources be identical (in amplitude) and in the same phase. Assume also that both sources have the same polarization; i.e., let both be linearly polarized with **E** perpendicular to the page. When the reference point for phase is taken halfway between the sources, the far field in the direction θ is given by

$$E = E_2 e^{j\psi/2} + E_1 e^{-j\psi/2} \tag{1}$$

where E_1 = far electric field at distance r due to source 1
  E_2 = far electric field at distance r due to source 2
  $\psi = \beta d \cos \theta = (2\pi \, d/\lambda) \cos \theta$

The quantity ψ is the phase-angle difference between the fields of the two sources as measured along the radius-vector line at the angle θ (see Fig. 14-17a). When $E_1 = E_2$, we have

$$E = 2E_1 \frac{e^{j\psi/2} + e^{-j\psi/2}}{2} = 2E_1 \cos \frac{\psi}{2} \tag{2}$$

For a spacing of $\lambda/2$ the field pattern is as shown in Fig. 14-17b.

---

† Here A_e is independent of ohmic losses. Thus, the gain G of the antenna is the same as given by (29) for a lossless antenna but is less if losses are present.

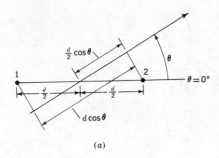

(a)

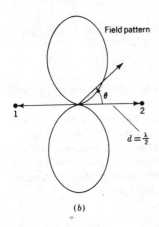

Field pattern

(b)

Figure 14-17 (a) Two isotropic point sources separated by a distance d and (b) field pattern when sources are of equal amplitude and in phase with $\lambda/2$ separation.

If the reference for phase for the two sources in Fig. 14-17 had been taken at source 1 (instead of midway between the sources), the resultant far field pattern would be

$$E = E_1 + E_2\, e^{j\psi} \tag{3}$$

and for $E_1 = E_2$

$$E = 2E_1 \cos\frac{\psi}{2}\, e^{j\psi/2} = 2E_1 \cos\frac{\psi}{2} \angle\ \psi/2 \tag{4}$$

The field (amplitude) pattern is the same as before, but the phase pattern is not. This is because the reference was taken at the *phase center* (midpoint of array) in developing (2) but at one end of the array in developing (4).

14-6.2 Pattern Multiplication

We assumed above that each point source was isotropic (completely nondirectional). If the individual point sources have directional patterns which are identical, the resultant pattern is given by (14-6.1-2) [or (14-6.1-4)], where E_1 is now also a

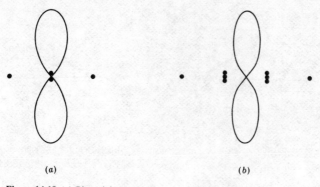

Figure 14-18 (*a*) Binomial array with source amplitudes $1:2:1$ and (*b*) binomial array with source amplitudes $1:3:3:1$. The spacing between sources is $\lambda/2$.

function of angle $[E_1 = E(\theta)]$. The pattern $E(\theta)$ may be called the *primary pattern* and $\cos \psi/2$ the *secondary pattern* or *array factor*. This is an example of the *Principle of Pattern Multiplication*, which may be stated more generally as follows: *The total field pattern of an array of nonisotropic but similar sources is the product of the individual source pattern and the pattern of an array of isotropic point sources each located at the phase center of the individual source with the relative amplitude and phase of the source, while the total phase pattern is the sum of the phase patterns of the individual sources and the array of isotropic point sources.*†

14-6.3 Binomial Array

From (14-6.1-2) the relative far-field pattern of two identical in-phase isotropic point sources spaced $\lambda/2$ apart is given by

$$E = \cos\left(\frac{\pi}{2} \cos \theta\right) \tag{1}$$

For convenience we have set $E_1 = \frac{1}{2}$ in (14-6.1-2) so that the pattern is normalized. This pattern, as shown in Fig. 14-17b, has no minor lobes. If a second identical array of two sources is placed $\lambda/2$ from the first, the arrangement shown in Fig. 14-18a is obtained. The two sources at the center should be superimposed but are shown separated for clarity. By the principle of pattern multiplication, the resultant pattern is given by

$$E = \cos^2\left(\frac{\pi}{2} \cos \theta\right) \tag{2}$$

† It is assumed that the pattern of the individual source is the same when it is in the array as when it is isolated.

Table 14-2 Pascal's triangle

			1			
		1		1		
	1		2		1	
1		3		3		1
1	4		6		4	1
1	5	10		10	5	1
1	6	15	20	15	6	1

as shown in Fig. 14-18a. If this three-source array with amplitudes $1:2:1$ is arrayed with an identical one at a spacing $\lambda/2$, the arrangement of Fig. 14-18b is obtained with the pattern

$$E = \cos^3\left(\frac{\pi}{2}\cos\theta\right) \tag{3}$$

as shown in Fig. 14-18b. This four-source array has amplitudes $1:3:3:1$, and it also has no minor lobes. Continuing this process, it is possible to obtain a pattern with arbitrarily high directivity and no minor lobes if the amplitudes of the sources in the array correspond to the coefficients of a binomial series.† These coefficients are conveniently displayed by Pascal's triangle (Table 14-2). Each internal integer is the sum of the adjacent ones above. The pattern of the array is then

$$E = \cos^{n-1}\left(\frac{\pi}{2}\cos\theta\right) \tag{4}$$

where n is the total number of sources.

Although the above array has no minor lobes, its directivity is less than that of an array of the same size with equal-amplitude sources. In practice most arrays are designed as a compromise between these extreme cases (binomial and uniform).

14-6.4 Array with n Sources of Equal Amplitude and Spacing

The binomial array is a nonuniform array. For a uniform array with n isotropic sources of equal amplitude and spacing, as in Fig. 14-19, the far field is

$$E = E_0(1 + e^{j\psi} + e^{j2\psi} + \cdots + e^{j(n-1)\psi}) \tag{1}‡$$

or

$$E = E_0 \sum_{n=1}^{N} e^{j(n-1)\psi} \tag{2}$$

where $\psi = \beta d\cos\theta + \delta$
$\quad d$ = spacing between sources
$\quad \delta$ = progressive phase difference between sources

---

† J. S. Stone, U.S. Pats. 1,643,323 and 1,715,433.

‡ This is a polynomial of degree $n-1$, as can be seen by putting $e^{j\psi}$ equal to a unit phasor r. Thus (1) becomes

$$\frac{E}{E_0} = 1 + r + r^2 + \cdots + r^{n-1}$$

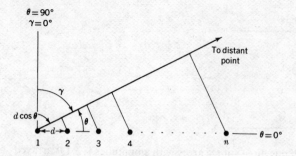

Figure 14-19 Array of n isotropic point sources of equal amplitude and spacing.

Multiplying (1) by $e^{j\psi}$ yields

$$Ee^{j\psi} = E_0(e^{j\psi} + e^{j2\psi} + e^{j3\psi} + \cdots + e^{jn\psi}) \tag{3}$$

Subtracting (3) from (1), we have

$$E = E_0 \frac{1 - e^{jn\psi}}{1 - e^{j\psi}} = E_0 \frac{\sin (n\psi/2)}{\sin (\psi/2)} \angle (n - 1)\frac{\psi}{2} \tag{4}$$

If the center of the array is chosen as the reference for phase, instead of source 1, the phase angle, $(n - 1)\psi/2$, is eliminated. If the sources are nonisotropic but similar, E_0 represents the primary or individual source pattern, while $\sin (n\psi/2)/\sin (\psi/2)$ is the array factor.

For isotropic sources and the center of the array as reference for phase, the pattern is

$$E = E_0 \frac{\sin (n\psi/2)}{\sin (\psi/2)} \tag{5}$$

As $\psi \to 0$, (5) reduces to

$$E = nE_0 \tag{6}$$

This is the maximum value of the field. It is n times the field from a single source. In the direction of the maximum the condition $\psi = 0$ or $\beta d \cos \theta = -\delta$ is satisfied. Dividing (5) by (6) yields the *normalized field pattern*

$$\boxed{E_n = \frac{E}{nE_0} = \frac{1}{n}\frac{\sin (n\psi/2)}{\sin (\psi/2)}} \tag{7}$$

Referring to (4), the null directions of the pattern occur for $e^{jn\psi} = 1$ provided $e^{j\psi} \neq 1$. This requires that $n\psi = \pm 2k\pi$, or

$$\pm \frac{2k\pi}{n} = \beta d \cos \theta_0 + \delta \tag{8}$$

or

$$\theta_0 = \cos^{-1}\left[\left(\pm \frac{2k\pi}{n} - \delta\right)\frac{\lambda}{2\pi d}\right] \tag{9}$$

where θ_0 = null angle

$k = 1, 2, 3, \ldots$ (but $k \neq mn$, where $m = 1, 2, 3, \ldots$)

14-6.5 Array with n Sources of Equal Amplitude and Spacing: Broadside Case

In the direction of the pattern maximum the condition $\beta d \cos \theta = -\delta$ must be satisfied. For a broadside array (maximum at $\theta = 90°$) the sources must be in phase ($\delta = 0$). When θ is replaced by its complementary angle γ (see Fig. 14-19), the null angles are given by

$$\gamma_0 = \sin^{-1}\left(\pm \frac{k\lambda}{nd}\right) \tag{1}$$

If the array is large, so that $nd \gg k\lambda$ (and γ_0 is small),

$$\gamma_0 = \frac{k}{nd/\lambda} \approx \frac{k}{L/\lambda} \tag{2}$$

where L = length of array, m

$L = (n - 1)d \approx nd$ if n is large

The first nulls (γ_{01}) occur when $k = 1$. Hence, the beam width between first nulls (BWFN) is

$$\text{BWFN} = 2\gamma_{01} \approx \frac{2}{L/\lambda} \text{ rad} = \frac{114.6°}{L/\lambda} \tag{3}$$

The more commonly used parameter is the half-power beam width (HPBW), which is about one-half (more nearly 0.44) the BWFN of a long uniform broadside array. Thus,

$$\boxed{\text{HPBW} \approx \frac{\text{BWFN}}{2} = \frac{1}{L/\lambda} \text{ rad} = \frac{57.3°}{L/\lambda}} \tag{4}$$

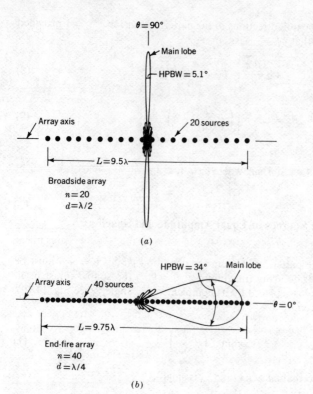

(a)

(b)

Figure 14-20 (a) Broadside array of 20 point sources of equal amplitude and $\lambda/2$ spacing with field pattern; (b) (ordinary) end-fire array of 40 point sources of equal amplitude and $\lambda/4$ spacing with field pattern. Note that both arrays are of approximately the same length (9.5 vs. 9.75λ).

The field pattern of a broadside array of 20 sources of equal amplitude spaced $\lambda/2$ apart is shown in Fig. 14-20a. The BWFN is 11.5°, and the HPBW is 5.1°. The three-dimensional pattern is a (disk-shaped) figure of revolution obtained by rotating the pattern of Fig. 14-20a around the array axis.

14-6.6 Array with n Sources of Equal Amplitude and Spacing: End-Fire Case

For an end-fire array (maximum at $\theta = 0°$) the condition $\beta d \cos \theta = -\delta$ requires that the progressive phase difference δ between sources be $-\beta d$. From (14-6.4-8)

$$\cos \theta_0 - 1 = \pm \frac{k}{nd/\lambda} \tag{1}$$

or

$$\frac{\theta_0}{2} = \sin^{-1}\left(\pm \sqrt{\frac{k}{2nd/\lambda}}\right) \tag{2}$$

For a long array $(nd \gg k\lambda)$

$$\theta_0 = \pm \sqrt{\frac{2k}{nd/\lambda}} \simeq \sqrt{\frac{2k}{L/\lambda}} \qquad (3)$$

where $L = (n - 1)d \approx nd$ if n is large. The first nulls (θ_{01}) occur when $k = 1$. Hence,

$$\boxed{\text{BWFN} = 2\theta_{01} \simeq \pm 2\sqrt{\frac{2}{L/\lambda}} \text{ rad} = \pm 114.6\sqrt{\frac{2}{L/\lambda}} \text{ deg}} \qquad (4)$$

The field pattern of an end-fire array of 40 sources spaced $\lambda/4$ apart is shown in Fig. 14-20b. The BWFN is 52°, and the HPBW† is 34°. The three-dimensional pattern is a (cigar-shaped) figure of revolution obtained by rotating the pattern of Fig. 14-20b around the array axis.

Although the two arrays in Fig. 14-20 are approximately the same length (broadside array 9.5λ, end-fire array 9.75λ), we note that the beam width is much smaller for the broadside array (5° as compared to 34°) but only in the plane shown in the figure. The pattern of the broadside array is disk-shaped with a small beam width in the plane of the figure, but the beam width is 360° in a plane perpendicular to the page. On the other hand, the pattern of the end-fire array is ellipsoidal, or cigar-shaped, with the same pattern in both principal planes (in plane of page and perpendicular to page through array axis), and, as will be shown below (compare Examples 1 and 2), the directivity of the end-fire array is greater.

From (14-5-28) the directivity of an antenna is given by

$$D = \frac{4\pi}{\Omega_A} \qquad (5)$$

An exact evaluation of the beam solid angle Ω_A requires an integration as in (14-5-3). However, a very simple *approximation*‡ can be made using the HPBWs in the two principal planes as follows:

$$\boxed{D \simeq \frac{4\pi}{\theta_{\text{HP}}\phi_{\text{HP}}} = \frac{41,253}{\theta_{\text{HP}}^\circ \phi_{\text{HP}}^\circ}} \qquad (6)$$

where θ_{HP} = HPBW in θ plane, rad
       ϕ_{HP} = HPBW in ϕ plane, rad
       θ_{HP}° = HPBW in θ plane, deg
       ϕ_{HP}° = HPBW in ϕ plane, deg

---

† This HPBW is calculated exactly from (14-6.4-7). For a long uniform ordinary end-fire array HPBW $\approx \frac{2}{3}$ BWFN, while for a long uniform broadside array HPBW $\approx \frac{1}{2}$ BWFN, as given in (14-6.5-4).

‡ Minor lobes are ignored, and $\theta_{\text{HP}}\phi_{\text{HP}}$ is only approximately equal to the beam solid angle. Hence (6) may be off by 1 dB or so.

In connection with (6) note that

$$4\pi \text{ sr} = 4\pi \text{ rad}^2 = 4\pi\left(\frac{180}{\pi}\right)^2 \text{ deg}^2 = 41{,}253 \text{ deg}^2 \qquad (7)$$

Example 1 For the 9.5λ broadside array discussed above, with $\theta_{HP}^\circ = 5.1^\circ$ and $\phi_{HP}^\circ = 360^\circ$, calculate the approximate directivity.

SOLUTION From (6)

$$D \approx \frac{41{,}253}{5.1^\circ \times 360^\circ} = 22.5 \text{ (or 13.5 dB)} \qquad (8)$$

Example 2 For the 9.75λ end-fire array discussed above, with $\theta_{HP}^\circ = \phi_{HP}^\circ = 34^\circ$, calculate the approximate directivity.

SOLUTION From (6)

$$D \approx \frac{41{,}253}{34^\circ \times 34^\circ} = 35.7 \text{ (or 15.5 dB)} \qquad (9)$$

This directivity is 2 dB more than for the broadside array of approximately the same length.

14-6.7 Graphical Representation of Phasor Addition of Fields

At the beginning of Sec. 14-6 it was stated that much of array theory is basically little more than the proper addition of the field contributions from all parts of an antenna. This can be emphasized by analyzing two simple antenna arrays using only graphical methods.

Example 1: Broadcast station array with heart-shaped pattern Find the radiation pattern in the horizontal plane for an array of two short vertical antennas spaced $\lambda/4$ apart on an east-west line and fed currents of equal magnitude and phase quadrature (current in eastern antenna lagging current in western antenna by 90°).

SOLUTION Figure 14-21a is a plan view with phasor arrows indicating a phase $\psi = 0^\circ$ for antenna 1 (western antenna) and phase $\psi = 270^\circ$ for antenna 2 (eastern antenna).† To find the field radiated to the east by the array, imagine that we start at antenna 1 and travel eastward, riding with the field of a radiated

---

† The magnitude of the arrows (phasors) is considered constant as a function of distance (even though fields do attenuate with distance) because we are interested in the difference of the field components at a distant point. Thus, when observing at a distance of thousands of wavelengths, there is a negligible difference in magnitude from sources separated $\lambda/4$, as above, and this effect can be ignored. However, the *phase difference* cannot be neglected.

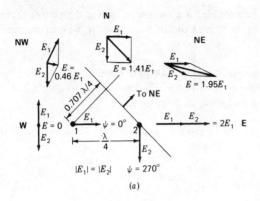

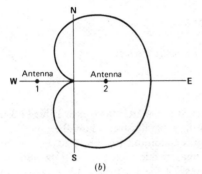

Figure 14-21 (*a*) Plan view of broadcast array of two vertical antennas with $\lambda/4$ spacing and quadrature phase ($\psi = 0°$ for antenna 1 and $\psi = 270°$ for antenna 2.) The phase angle ψ advances counterclockwise (ccw). Phasor addition is shown for fields in several directions from the array. (*b*) The field pattern is heart-shaped with a maximum to the east, a null to the west, and down 3 dB north and south.

wave like a surfer rides a breaker. The phase of the field of the wave we are riding does not change, but by the time we have traveled $\lambda/4$ and arrived at antenna 2, $\frac{1}{4}$ period has elapsed, so that the current in antenna 2 will have advanced 90° and so its phase $\psi = 0°$ (the same as that of the field we are riding). Thus, the fields of antenna 1 and 2 add or reinforce to the east (E) so $E(E) = 2E_1$, where $E_1 =$ field from either antenna alone.

Now imagine that we start at antenna 2 with a field of phase $\psi = 270°$ and travel west to antenna 1. By the time we arrive, the phase of antenna 1 will have advanced from $\psi = 0°$ to $\psi = 90°$ and, therefore, will be in phase opposition to the field we are riding from antenna 2, so that the total field to the west is zero [$E(W) = 0$].

Next, starting at antenna 1 and traveling northeast (NE), we travel 0.707 $\lambda/4$ to arrive at a perpendicular line through antenna 2, and by this time the phase of antenna 2 has advanced 0.707 × 90° = 63.6°, so that its phase is $\psi = 333.6°$, or $-26.4°$. Adding the fields vectorially $E(NE) = 1.95E_1$. In a similar way, $E(N) = 1.41E_1$ and $E(NW) = 0.46E_1$. By symmetry $E(S) = E(N)$, $E(SE) = E(NE)$, and $E(SW) = E(NW)$. Plotting these fields yields the cardioid (heart-shaped) pattern of Fig. 14-21*b*.

This pattern might be suitable for a broadcast station desiring to service an area to the east while protecting a distant station to the west operating on the same frequency.

Example 2: Six-source broadside array Calculate the pattern of a broadside array of six isotropic point sources with $d = \lambda/2$ and draw the phasor (addition) diagrams for the maxima and nulls of the pattern.

SOLUTION The pattern is given by (14-6.4-5) with $\delta = 0°$. From (14-6.4-8) the nulls (roots) occur when

$$\pm \frac{2k\pi}{6} = \frac{2\pi}{\lambda} \frac{\lambda}{2} \cos \theta = \pi \cos \theta \tag{1}$$

or

$$\cos \theta = \pm \frac{k}{3} \tag{2}$$

The minor-lobe maxima occur *approximately* when

$$\psi = \pm \frac{(2k + 1)\pi}{n} \tag{3}$$

where $k = 1, 2, 3, \ldots$. The values of k, θ, and ψ are related as in Table 14-3, and the column "Phasor addition" in the table illustrates how the unit phasors from each source add for the various conditions (see also Sec. 14-6.7). Note that in the maximum direction all phasors add in line while in the null directions the phasors close on themselves. The field pattern of the array is shown in Fig. 14-22.

In this example and also in other cases considered in this section the total field is n times the field from a single source ($E = nE_1$) in the direction of the maximum radiation. This is the ordinary situation. However, it should be noted that it is possible to have a total $E < nE_1$ in the direction of maximum radiation. This condition is sometimes used to obtain increased directivity.

14-6.8 Simple Two-Element Interferometer

The interferometer is a specialized type of antenna used widely for radio-astronomy observations. It provides a good example of pattern multiplication and array theory.

Consider two short-dipole antennas with spacing s connected in phase, as in Fig. 14-23. From (14-6.1-2) the field pattern is given by

$$E = 2E_1 \cos \frac{\psi}{2} \tag{1}$$

where $\psi = \beta s \cos \theta$

Table 14-3 Parameters for linear array of six isotropic in-phase point sources of equal amplitude and spacing $d = \lambda/2$. Field pattern is in Fig. 14-22.

Condition	k	θ	ψ	Phasor addition
Maximum	0	$\pm 90°$	0	$E = 6$ ($\theta = \pm 90°$) Unit phasors
Null	1	$\pm 70.5°$	$\pm\frac{\pi}{3}(\pm 60°)$	$E = 0$ ($\theta = 70.5°$) $\psi = 60°$ / ($\theta = -70.5°$) $\psi = -60°$
Maximum of minor lobe	—	$\pm 60°$	$\pm\frac{\pi}{2}(\pm 90°)$	$E = 1.41$ ($\theta = 60°$) $\psi = 90°$ / ($\theta = -60°$) $-90°$
Null	2	$\pm 48.2°$	$\pm\frac{2}{3}\pi(\pm 120°)$	$E = 0$ ($\theta = 48.2°$) $\psi = 120°$ / ($\theta = -48.2°$) $\psi = -120°$
Maximum of minor lobe	—	$\pm 33.7°$	$\pm\frac{5}{6}\pi(\pm 150°)$	$E = 1.04$ ($\theta = 33.7°$) $\psi = 150$ / ($\theta = -33.7°$) $-150°$
Null	3	$0°, 180°$	$\pm\pi(\pm 180°)$	$E = 0$ ($\theta = 0°$) $\psi = 180°$ / ($\theta = 180°$) $\psi = -180°$

By pattern multiplication E_1 is the field pattern of the short dipole [$=\sin\theta$, see (14-3-37)], and $\cos(\psi/2)$ is the array factor. Hence, the complete interferometer pattern is

$$E = 2 \sin\theta \cos\left(\frac{\pi s}{\lambda}\cos\theta\right) \qquad (2)$$

When θ is replaced by its complementary angle γ, the first null occurs when $(\pi s/\lambda)\sin\gamma_{01} = \pi/2$ or

$$\gamma_{01} = \sin^{-1}\frac{1}{2s/\lambda} \qquad (3)$$

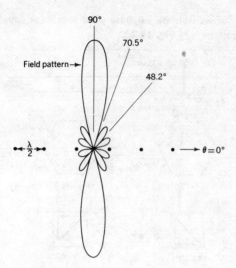

Figure 14-22 Field pattern of array of six isotropic in-phase point sources of equal amplitude with $\lambda/2$ spacing (see Table 14-3 for graphical addition of fields in different directions).

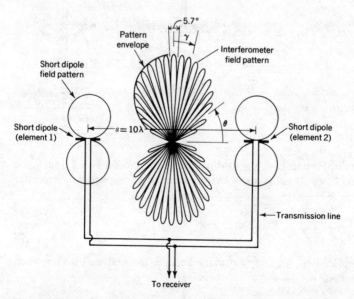

Figure 14-23 Simple two-element interferometer using short dipoles spaced 10λ showing patterns of the individual (short-dipole) elements and of the interferometer.

If we assume that the spacing s is many wavelengths, the BWFN is then

$$\text{BWFN} = 2\gamma_{01} = \frac{1}{s/\lambda} \text{ rad} = \frac{57.3}{s/\lambda} \text{ deg} \tag{4}$$

This BWFN is one-half the value for a uniform array of n sources as given by (14-6.5-3). The interferometer field pattern is illustrated in Fig. 14-23 for $s = 10\lambda$. The BWFN of this interferometer pattern is, from (4), equal to 5.7°. The envelope of the multilobed pattern is the same shape as the short-dipole pattern ($= \sin \theta$).

In radio-astronomy applications an interferometer is used for receiving radiation from celestial radio objects as they drift across the sky through the antenna pattern. The detection and location of the objects are facilitated by the small beam widths of the lobes of the pattern. We have assumed that the object detected is of small angular extent (or much less than 5.7° in this example). If not, the pattern is modified (nulls filled in). This effect enables one to deduce the angular size of objects.†

14-7 CONTINUOUS APERTURE DISTRIBUTION

In the preceding pages we have dealt with arrays of point sources, i.e., arrays of a finite number of sources separated by finite distances. Now let us consider a *continuous* array of point sources, i.e., an array of an infinite number of point sources separated by infinitesimal distances. By Huygens' principle (Sec. 12-4) a continuous array of point sources is equivalent to a continuous field (or current-sheet distribution). In this way we can extend array analysis to include the radiation patterns of field distributions across apertures of electromagnetic horns or parabolic dish antennas.

Consider the aperture distribution of length a in the y direction and width x_1 in the x direction (perpendicular to page), as suggested in Fig. 14-24. The magnitude of the field over the aperture is a function of y but not of x. The field is in the same phase over the entire aperture. The infinitesimal field at a distance from an elemental aperture area $dx\,dy$ is, from $E = -j\omega A$ and (14-3-7),‡

$$dE = -j\omega[dA_x] = \frac{-j\omega\mu E(y)}{4\pi rZ} e^{-j\beta r} dx\,dy \qquad (\text{V m}^{-1}) \tag{1}$$

where $[A_x] = x$ component of retarded vector potential [see (14-2-6)], Wb m^{-1}

$E(y) =$ electric field distribution as function of y across aperture, V m^{-1}

$Z =$ intrinsic impedance of medium, Ω

---

† This discussion is much simplified. For a more detailed discussion see, for example, J. D. Kraus, "Radio Astronomy," chap. 6, Cygnus-Quasar Books, Powell, Ohio 43065, 1982.

‡ Note that $E(y)/Z = E_x(y)/Z = H_y = J_x z_1$, where z_1 is the thickness of the equivalent current sheet in the aperture. Thus, in (14-3-7) let $I_0 = J_x z_1\,dy$ and $l = dx$.

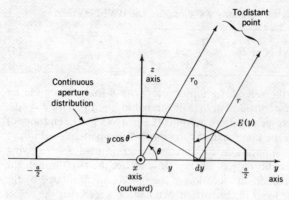

Figure 14-24 Aperture of width a and amplitude distribution $E(y)$.

The total field in the θ direction at a distance r_0 from the center of the aperture is obtained by integrating (1) over the aperture (extent a by x_1). If $r \gg a$, then $r = r_0 - y \cos \theta$ and

$$E(\theta) = \frac{-j\omega\mu x_1 e^{-j\beta r_0}}{4\pi r_0 Z} \int_{-a/2}^{+a/2} E(y) \exp(j\beta y \cos \theta) dy \qquad (2)$$

The magnitude of $E(\theta)$ is then†

$$E(\theta) = \frac{x_1}{2r_0 \lambda} \int_{-a/2}^{+a/2} E(y) \exp(j\beta y \cos \theta) dy \qquad (3)$$

For a uniform aperture distribution $[E(y) = E_a]$ (3) reduces to

$$E(\theta) = \frac{x_1 E_a}{2r_0 \lambda} \int_{-a/2}^{+a/2} \exp(j\beta y \cos \theta) dy \qquad (4)$$

On axis ($\theta = 90°$) we have

$$E(\theta = 90°) = \frac{E_a a x_1}{2r_0 \lambda} = \frac{E_a A_e}{2r_0 \lambda} \qquad (5)$$

where $A_e = ax_1$ is the effective aperture (square meters). If the aperture is unidirectional (maximum radiation at $\theta = 90°$ but none at $\theta = 270°$), the value of $E(\theta)$ should be doubled. This is the value which was used in deriving (14-5-25). Integrating (4) gives

$$E(\theta) = \frac{E_a A_e}{2r_0 \lambda} \frac{\sin\left[(\beta a/2) \cos \theta\right]}{(\beta a/2) \cos \theta} \qquad (6)$$

†Note that $\omega\mu/4\pi Z = 1/2\lambda$.

This gives the far-field pattern as a function of θ for a continuous, uniform, in phase distribution of field intensity E_a over a rectangular aperture of area A_e with length a (in the y direction).

Referring to the array of n isotropic sources of equal amplitude and spacing discussed earlier in Sec. 14-6.4, we recall that the field pattern was given in (14-6.4-5) as

$$E(\theta) = E_0 \frac{\sin n\psi/2}{\sin \psi/2} \tag{7}$$

where $\psi = \beta d \cos \theta + \delta$

E_0 = field from individual source

For the broadside case, $\delta = 0$, and (7) becomes

$$E(\theta) = E_0 \frac{\sin [(n\beta d/2) \cos \theta]}{\sin [(\beta d/2) \cos \theta]} \tag{8}$$

Let a' = length of array. Then if n and a' are large, $a' = (n - 1)d \approx nd$. Restricting our attention to angles near $90°$ (main beam direction), we can rewrite (8) as

$$E(\theta) = nE_0 \frac{\sin [(\beta a'/2) \cos \theta]}{(\beta a'/2) \cos \theta} \tag{9}$$

Comparing this expression with (6), we see that for a large aperture and at angles near the main beam, the field pattern of a continuous aperture is identical with the pattern of an n source broadside array of same length ($a \doteq a'$). Also the fields have the same absolute value provided $nE_0 = E_a A_e/2r_0 \lambda$.

14-8 FOURIER TRANSFORM RELATIONS BETWEEN THE FAR-FIELD PATTERN AND THE APERTURE DISTRIBUTION[†]

A one-dimensional aperture distribution $E(y)$ and its far-field pattern $E(\theta)$ are reciprocal *Fourier transforms*. For a finite aperture one of these transforms may be written

$$E(\theta) = \int_{-a/2}^{+a/2} E(y) \exp (j\beta y \cos \theta) \, dy \tag{1}$$

This is identical with (14-7-3) except for a constant factor.[‡] Examples of the far-field patterns $E(\theta)$ of several field distributions $E(y)$ are presented in Fig. 14-25.

[†]H. G. Booker and P. C. Clemmow, The Concept of an Angular Spectrum of Plane Waves and Its Relation to That of Polar Diagram and Aperture Distribution, *Proc. IEE London*, (3)**97**:11–17 (January 1950).

[‡]Equation (14-7-3) is an absolute relation, whereas (1) is relative.

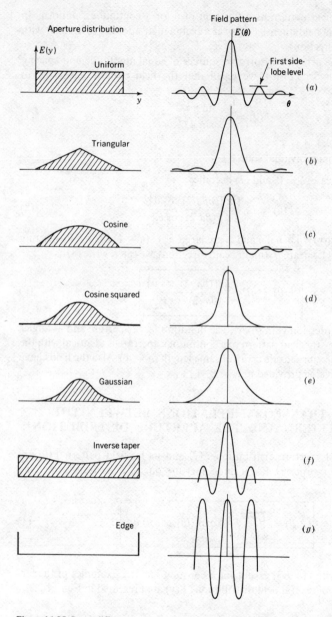

Figure 14-25 Seven different antenna aperture distributions with associated far-field patterns.

Taking the uniform aperture distribution and its pattern (Fig. 14-25*a*) as reference, we find that the tapered distributions (triangular and cosine) have larger beam widths but smaller minor lobes, while the most gradually tapered distributions (cosine squared and gaussian) have still larger beam widths but no minor lobes. On the other hand, an inverse taper, as in Fig. 14-25*f*, yields a smaller beam width but larger minor lobes than for the uniform distribution. Carrying the inverse taper to its extreme limit results in the edge distribution of Fig. 14-25*g*, which is equivalent to that of a two-element interferometer. The pattern in this case has a beam width one-half that of the uniform distribution but side lobes equal in amplitude to the main (central) lobe.

A useful property of (1) is that an aperture distribution can be taken as the sum of two or more component distributions with the resulting pattern equal to the sum of the transforms of these distributions.

14-9 LINEAR ANTENNAS

In Sec. 14-3 the dipole antenna was analyzed for the case where its length was small compared to the wavelength (say $\lambda/10$ or less). Let us now consider linear dipole antennas of any length L. It is assumed that the current distribution is sinusoidal. This is a satisfactory approximation provided the conductor diameter is small, say $\lambda/200$ or less.

Referring to Fig. 14-26, we shall calculate the far-field pattern for a symmetrical thin linear center-fed dipole antenna of length L. It will be done by regarding the antenna as being made up of series of elemental short dipoles of length dy and current I and integrating their contribution at a large distance. The form of the

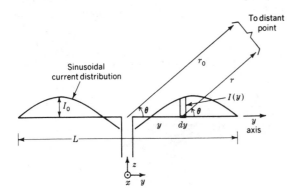

Figure 14-26 Geometry for linear center-fed dipole antenna of length L with sinusoidal current distribution.

current distribution on the antenna is found from experimental measurements to be given approximately by

$$I = I_0 \sin\left[\frac{2\pi}{\lambda}\left(\frac{L}{2} \pm y\right)\right] \quad \text{(A)} \tag{1}$$

where I_0 is the value of the current at the current maximum point, and where $(L/2) + y$ is used when $y < 0$ and $(L/2) - y$ is used when $y > 0$. Referring to Table 14-1 for the fields of a short dipole, we find the far (electric) field of an antenna of length L to be

$$E_\theta = k \sin\theta \int_{-L/2}^{L/2} I \exp(j\beta y \cos\theta)\, dy \tag{2}$$

where $k = \{j60\pi \exp[j(\omega t - \beta r_0)]\}/r_0\lambda$. Substituting (1) for I in (2) and integrating yields

$$E_\theta = \frac{j60[I_0]}{r_0}\left\{\frac{\cos[(\beta L\cos\theta)/2] - \cos(\beta L/2)}{\sin\theta}\right\} \tag{3}$$

where $[I_0] = I_0 \exp\{j\omega[t - (r_0/c)]\}$ is the retarded current. The shape of the far-field pattern is given by the factor in braces in (3). For a $\lambda/2$ dipole antenna this factor reduces to

$$E_\theta(\theta) = \frac{\cos[(\pi/2)\cos\theta]}{\sin\theta} \tag{4}$$

with the shape shown in Fig. 14-27a. This pattern is only slightly more directional than the $\sin\theta$ pattern of an infinitesimal or short dipole (shown dashed). The HPBW of the $\lambda/2$ antenna is 78°, compared with 90° for a short dipole.

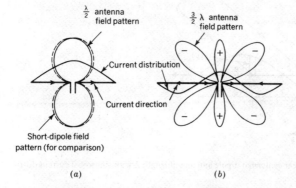

(a)   (b)

Figure 14-27 (a) Far-field pattern of $\lambda/2$ center-fed dipole antenna and (b) far-field pattern of 1.5λ center-fed dipole antenna.

For a 1.5λ dipole the pattern factor is

$$E_\theta(\theta) = \frac{\cos\left(\frac{3}{2}\pi\cos\theta\right)}{\sin\theta} \tag{5}$$

as shown in Fig. 14-27b. With the midpoint of the antenna as phase center, the phase of the far field (at a constant r_0) shifts 180° at each null, the relative phase of the lobes being indicated by the plus and minus signs. The three-dimensional pattern is a figure of revolution of the one shown in Fig. 14-27a and b with the dipole as axis (y axis in Fig. 14-26).

Example Find the directivity of a λ/2 linear dipole.

SOLUTION For the λ/2 dipole we have from (4) and (14-5-7) that

$$D = \frac{4\pi}{\iint P_n(\theta,\phi)\,d\Omega} = \frac{4\pi}{2\pi\displaystyle\int_0^\pi \frac{\cos^2[(\pi/2)\cos\theta]}{\sin^2\theta}\sin\theta\,d\theta} = 1.64\ (=2.15\text{ dBi}) \tag{6}$$

This is 1.64/1.5 = 1.09 (=0.4 dB) greater than the directivity of an infinitesimal or short dipole. Note: The integration of (6) involves sine (Si) and cosine (Ci) integrals.

To find the radiation resistance R of the λ/2 dipole the average Poynting vector is integrated over a large sphere, yielding the power radiated. This power is then equated to I^2R, where I is the rms terminal current. Performing this calculation for the λ/2 dipole yields

$$R = 30\ \text{Cin}\ 2\pi = 30 \times 2.44 = 73\ \Omega \tag{7}$$

where Cin is the modified cosine integral. This is the resistance presented by a λ/2 dipole to the transmission-line terminals (Fig. 14-27a). The antenna impedance Z also contains a positive reactance of 42.5 Ω ($Z = 73 + j42.5\ \Omega$). To make the antenna resonant ($X = 0$), the length can be shortened a few percent. However, this also results in a slight reduction of the radiation resistance (to approximately 70 Ω). The radiation resistance for the 1.5λ linear dipole is about 100 Ω.

14-10 FIELDS OF λ/2 DIPOLE ANTENNA

To illustrate the configuration of the fields radiated by an antenna, the electric field lines of a λ/2 antenna are shown in Fig. 14-28 at four successive instants of time. Although the fields were computer-generated (and plotted) for a prolate spheroid with λ/2 interfocal distance, they are substantially the same as for a thin λ/2 center-fed dipole† except in the immediate proximity of the antenna. At time $t = 0$ (Fig.

---

† For the thin λ/2 dipole the E_θ component would be as obtained from an equation like (14-9-2) but one which holds at small distances (see General expression in Table 14-1). Similarly calculating an E_r component makes it possible to determine the total field at any point.

Electric field lines

Equivalent
$\lambda/2$ dipole

Spheroid

$t = 0$ (a)

$t = \frac{1}{100} T$ (b)

$t = \frac{1}{4} T$ (c)

$t = \frac{1}{2} T$ (d)

14-28a) the antenna current is zero, and the charge at the ends of the dipole is a maximum. One-eighth of a period later ($t = T/8$) the current has started to flow, as suggested in Fig. 14-28b. At $t = T/4$ (Fig. 14-28c) the current has reached its peak value, and the charge at the ends is zero. At $t = \frac{3}{8}T$ (Fig. 14-28d) the current continues to flow in the same direction but with reduced magnitude. At $t = T/2$ the current is again zero, and the charge is a maximum at the ends. The field configuration at $t = T/2$ is identical with that shown in Fig. 14-28a except that the arrows and signs are reversed. At $t = \frac{5}{8}T$ the current has started to flow in the reverse direction (up), and the field is identical to that shown in Fig. 14-28b except that all arrows are reversed. Note that the lines in Fig. 14-28 are the actual field lines, not the field pattern (field intensity vs. angle). Compare with Fig. 14-27a with the dipole turned vertically.

Referring to all parts of Fig. 14-28, we note that a constant-phase point, such as P, moves radially outward approximately $\lambda/2$ in $\frac{1}{2}$ period (traveling approximately at the velocity of light, c). However, careful measurements show that P moves faster than c close to the antenna, like a higher-order-mode wave confined between the parallel plates of a waveguide, but farther out the velocity becomes exactly equal to c.† Figure 14-28 is incomplete in that the magnetic field lines are not shown. They consist of circular loops concentric with the antenna and perpendicular to the plane of the figure. From the standpoint of power flow the situation is similar to that shown by the arrows in Fig. 14-12 for the short dipole with maximum power radiated outward in the equatorial plane ($\theta = 90°$). This is real power flow. There is also reactive (nonradiating) power flow, oscillating back and forth close to the antenna, as suggested in Fig. 14-12.

14-11 TRAVELING-WAVE ANTENNAS‡

In Sec. 14-9 on the linear antenna a sinusoidal current distribution was assumed. This type of current distribution is closely approximated if the antenna is thin (conductor diameter $< \lambda/200$) and provided the antenna is *unterminated*.

---

†This may be inferred from the fact that the distances between similar phase points (P, P', P'') are more than $\lambda/2$ apart near the antenna but are exactly $\lambda/2$ apart farther out.

‡For a more detailed discussion of traveling-wave antennas, see, for example, C. H. Walter, "Traveling Wave Antennas," Dover Publications, Inc., New York, 1972.

Figure 14-28 Electric field configuration for a $\lambda/2$ antenna at four instants of time: (a) $t = 0$, (b) $t = T/8$, (c) $t = T/4$, and (d) $t = \frac{3}{8}T$, where $T =$ period. Note outward movement of constant-phase points P, P', and P'' as time advances. The strength of the electric field is proportional to the density of the lines. (*From "A Resonant Spheroid Radiator," produced at the Ohio State University for the National Committee for Electrical Engineering Films; Project Initiator, Prof. Edward M. Kennaugh; diagrams courtesy of Prof. John D. Cowan, Jr.*)

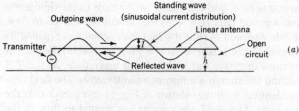

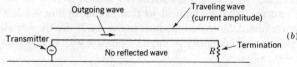

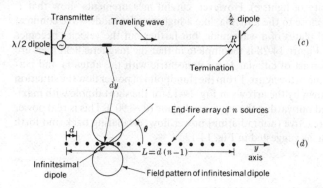

Figure 14-29 Long linear antennas (*a*) without and (*b*) and (*c*) with termination; (*d*) equivalent end-fire array of isotropic point sources.

Consider a thin linear antenna several wavelengths long situated a distance *h* above a perfectly conducting ground plane, as in Fig. 14-29*a*, with a transmitter connected at the left end. The antenna–ground-plane combination acts like an open-circuited transmission line. The transmitter sends a traveling wave down the line (to the right), which is reflected at the open end, causing an equal-amplitude wave to travel back (to the left). As a result, a standing wave is set up on the line with a sinusoidal current distribution. The current is zero at the open end and is a maximum $\lambda/4$ from the open end, as suggested in Fig. 14-29*a* (see also Sec. 10-6). This is the situation assumed in Sec. 14-9 on the linear dipole antenna.

Let us now connect a resistance *R* equal to the characteristic resistance of the antenna–ground-plane transmission line between the open end of the antenna and the ground, as in Fig. 14-29*b*. This converts the antenna to a *terminated* type. There is an outgoing traveling wave (to the right) but (ideally) no reflected wave.

The current amplitude is constant over the length of the antenna provided losses are negligible and there is a progressive uniform phase change along the antenna. An arrangement for establishing a traveling wave on a long linear antenna remote from the ground is suggested in Fig. 14-29c. The midpoint of a $\lambda/2$ antenna is used as a ground equivalent.

The pattern of a linear antenna with traveling wave can be readily determined if we regard the antenna as being made up of an array of a large number of series-connected infinitesimal dipoles operating as an end-fire array, as in Fig. 14-29d. Thus, from Secs, 14-6.6 and 14-6.4 the far electric field pattern of such an array is given by

$$E_0(\theta) = E_0 \frac{\sin (n\psi/2)}{n \sin (\psi/2)} \tag{1}$$

where $\psi = \beta d \cos \theta - \beta d = \beta d(\cos \theta - 1)$. The first factor (E_0) represents the pattern of the infinitesimal elements of the array, which have the same pattern as a short dipole ($E_0 = \sin \theta$). The second factor is the normalized pattern of an array of n isotropic sources. By the principle of pattern multiplication the resultant pattern $[E_\theta(\theta)]$ is the product of these two factors. For large L [$=d(n - 1) \approx nd$], resulting from a large n and small d, (1) can be expressed approximately as

$$E_\theta(\theta) = \sin \theta \frac{\sin [(\beta L/2)(1 - \cos \theta)]}{(\beta L/2)(1 - \cos \theta)} \tag{2}$$

For example, the pattern of a 9.75λ linear antenna with traveling wave is shown in Fig. 14-30a as the product of the pattern of an infinitesimal dipole and the pattern of an end-fire array of 40 isotropic sources with $\lambda/4$ spacing and uniform progressive phase shift $\delta = -\beta d = -\pi/2$. This is the same array as in Sec. 14-6.6 (see also Fig. 14-20b). The pattern of the array of 40 isotropic sources has a maximum at $\theta = 0°$ (end-fire direction). However, the infinitesimal dipoles making up the array have a null at $\theta = 0°$ and a maximum at $\theta = 90°$. The resulting (product) pattern is of conical shape with cross section as shown in Fig. 14-30a (pattern at right). The first minor lobe is only 4.7 dB down. The large minor lobes are characteristic of long linear antennas. By contrast, the first minor lobe of the isotropic-source array is 13.5 dB down (see Fig. 14-30a, center pattern).

The maximum radiation for the 9.75λ linear antenna is at $\theta \approx 15°$. By arranging two such linear antennas in the form of a V with about 30° included angle, as in Fig. 14-30b, a undirectional antenna is obtained which can be conveniently fed by a balanced two-conductor transmission line. The resulting pattern of this terminated V antenna is suggested in Fig. 14-30b (pattern at right).

The vertical patterns of these antennas when operated parallel to a ground plane, as in Fig. 14-29b, can be obtained by multiplying the patterns of the isolated antenna by that of a pair of isotropic sources, one at the center of the antenna and the other (as its image) an equal distance below the ground plane (see Sec. 14-19.1).

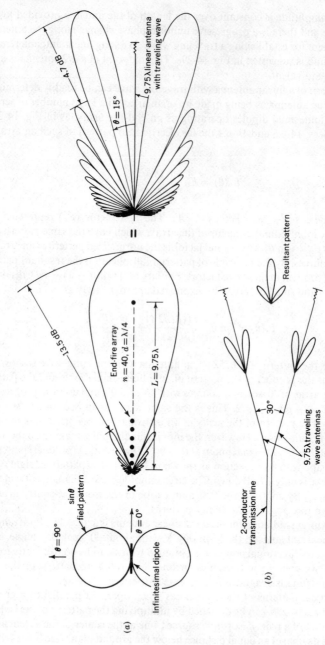

Figure 14-30 (a) Far-field pattern of 9.75λ linear (terminated) antenna as obtained by pattern multiplication of short-dipole pattern with pattern of end-fire array of isotropic point sources. (b) Combination of two antennas of (a) to form a V antenna.

14-12 THE SMALL LOOP ANTENNA

Consider a small circular loop antenna of diameter d, as in Fig. 14-31a. Its dimensions are assumed small compared with the wavelength ($d \ll \lambda$), so that the current is of constant amplitude and phase around the loop. This small circular loop may be approximated by a square loop of equal area $[s^2 = \pi(d/2)^2]$, as suggested in Fig. 14-31b. In this way the loop can be analyzed as four short dipoles whose properties we considered in Sec. 14-3. Let the square loop be oriented with respect to the coordinates as in Fig. 14-31b and c. The field in the yz plane is produced by the radiation from the two short dipoles 2 and 4. Short dipoles 1 and 3 do not contribute since their fields are *exactly* equal and in opposite phase at all angles in the yz plane. Since the short dipoles (2 and 4) are nondirectional in the yz plane, the field pattern in this plane is the same as for two isotropic point sources in opposite phase. Thus, the electric field pattern is given by

$$E(\theta) = E_2 e^{-j\psi/2} - E_4 e^{j\psi/2} \tag{1}$$

where $E_2 = E_4$ is the electric field from the individual short dipole. Assuming $s \ll \lambda$, it follows that

$$E(\theta) = -jE_2 \beta s \sin \theta \tag{2}$$

Introducing the value of the far field E_2 of a short dipole from Table 14-1, we get

$$E(\theta) = \frac{60\pi Il}{r\lambda} \beta s \sin \theta \tag{3}$$

where l is the length of the short dipole. But in our case $l = s$ and $s^2 = A$ ($=$ area of loop); so

$$E(\theta) = \frac{120\pi^2 I}{r} \frac{A}{\lambda^2} \sin \theta \qquad (\text{V m}^{-1}) \tag{4}$$

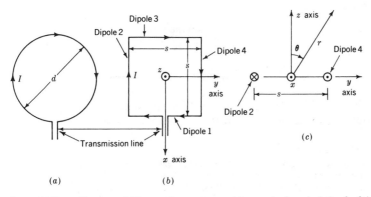

Figure 14-31 (a) Circular and (b) square loop antennas; (c) geometry for calculating far field.

This is the value of the electric field from a small loop antenna of area A (square or circular†). It is the same in any plane perpendicular to the plane of the loop. The approximation (square for circular loop) becomes exact as the loop dimensions become very small.

14-13 THE HELICAL-BEAM ANTENNA‡

A helix with a circumference of about 1λ acts as an end-fire beam antenna with maximum radiation in the axial direction. The radiation is circularly polarized on axis, and the directivity of the helix increases linearly with the length of the helix; i.e., doubling the length doubles the directivity.

The helical-beam antenna is conveniently fed with a coaxial transmission line and square or circular ground, as suggested in Fig. 14-32a. The important dimensions of the helix are its diameter D, circumference C, spacing between turns S, turn length l, and pitch angle α. These are related as in Fig. 14-32b. When the helix circumference ($C = \pi D$) is approximately 1λ, the instantaneous charge and current distribution on a single turn will be as suggested in the end view of Fig. 14-32c, with the resultant **E** field as indicated. As the wave travels outward on the helix, the electric field vector rotates, producing circular polarization on axis. The helical-beam antenna acts as a traveling-wave antenna without requiring a termination. The current distribution is essentially that of a single outward-traveling wave of uniform amplitude; the wave reflected back toward the feed point from the open end of the helix is very small.

The pattern of a helical-beam antenna can be readily calculated if it is regarded as an end-fire array of n isotropic sources, each source representing one turn. By the principle of pattern multiplication the actual helix pattern is then the product of this array pattern and the pattern of one turn. The single-turn pattern may be approximated as a cosine function, $E(\theta) = \cos\theta$. This is an oversimplification, but the single-turn pattern has little effect, especially for long helices, because the array pattern is so much sharper than the single-turn pattern. Thus, we may write for the field pattern

$$E(\theta) = \cos\theta \, \frac{\sin(n\psi/2)}{\sin(\psi/2)} \tag{1}$$

where

$$\psi = \beta S \cos\theta - \frac{\beta L}{p} \tag{2}$$

We note that (1) differs from the linear traveling-wave antenna pattern as given in (14-11-2) in that the single-turn helix pattern ($\cos\theta$) has a maximum on the axis (in

---

† Or any other shape.

‡ J. D. Kraus, Helical Beam Antenna, *Electronics*, **20**:109–111 (April 1947); also J. D. Kraus, Helical Beam Antennas for Wide Band Applications, *Proc. IRE* , **36**:1236–1242 (October 1948). For extensive treatment see J. D. Kraus, "Antennas," chap. 7, McGraw-Hill Book Company, New York, 1950.

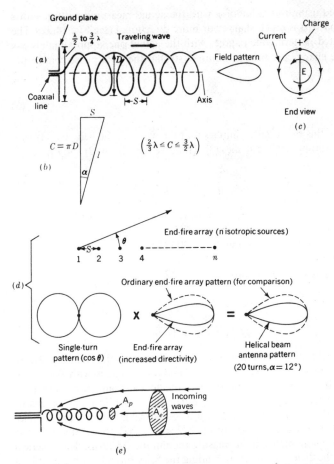

Figure 14-32 (a) and (b) Dimensions for helical-beam antenna. (c) End view of helix with instantaneous direction of rotating electric field (E). (d) Far-field pattern of helical-beam antenna obtained by pattern multiplication of single-turn pattern with end-fire array of isotropic sources with increased directivity. The pattern for an ordinary end-fire array of the same length is shown for comparison. (e) Helix acts like a lens, refracting, or bending, incoming waves into itself from an effective aperture A_e which is larger than its cross section or physical aperture A_p.

the same direction as the array maximum) while for the linear antenna the infinitesimal dipole pattern has a *null* on the axis (in the direction of the array maximum), resulting in the conical pattern of Fig. 14-30a and less directivity.

Since the radiation is circularly polarized on axis and symmetrical, (1) is the pattern for both components, that is, $E_\theta(\theta)$ and $E_\phi(\theta)$.

If all fields added in phase on axis, we would have the ordinary end-fire condition, which requires $\psi = 0, \pm 2\pi, \pm 4\pi$, etc. In our case $\psi = -2\pi$. Comparing a pattern calculated using this value of ψ with the actual measured pattern shows that the actual pattern is much sharper or more directional (see Fig. 14-32d). To obtain a calculated pattern which agrees with the actual one, approximately π/n additional phase change per turn is required, so that the total phase change per turn is[†]

$$\psi = -2\pi - \frac{\pi}{n} \tag{3}$$

The fulfillment of this requirement implies that the wave on the helix travels with a velocity which automatically satisfies (3). Thus, we have from (2) and (3)

$$-2\pi - \frac{\pi}{n} = \beta S - \frac{\beta l}{p} \tag{4}$$

where $p = v/c$ = relative phase velocity of wave on helix, dimensionless
 v = velocity of wave on helix, m s^{-1}
 c = velocity of light or radio waves = 300 Mm s^{-1}
 l = length of turn, m
 S = spacing between helix turns, m

Solving (4) for p, we get

$$p = \frac{1/\lambda}{(S/\lambda) + (2n + 1)/2n} \tag{5}$$

In a typical case where the helix dimensions are $C = \lambda$, $\alpha = 12°$, and $n = 20$, we obtain a phase velocity $p = 0.82$. This calculated value agrees with the actual measured value. It is a remarkable property of the helical-beam antenna that the relative phase velocity p of the wave traveling out on the helix adjusts automatically, so that increased directivity is obtained over a considerable range of pitch angles ($5° < \alpha < 20°$) and over a 2 to 1 bandwidth ($\frac{3}{4}\lambda < C < \frac{3}{2}\lambda$) provided $n > 3$. The directivity can be made higher by increasing the number of turns. The practical limit here is that the dielectric structure holding the helix should not be capable of supporting other modes to any appreciable extent. The patterns shown in Fig. 14-32d are those of a 20-turn helix with $\alpha = 12$ and an ordinary end-fire array of the same length with 20 sources each with a $\cos \theta$ pattern.

The approximate directivity of a helical-beam antenna is given by[‡]

$$\boxed{D = 15\left(\frac{C}{\lambda}\right)^2 \frac{nS}{\lambda}} \tag{6}$$

[†] Hansen and Woodyard have shown that the additional π/n phase change results in increased directivity; W. W. Hansen and J. R. Woodyard, A New Principle in Directional Antenna Design, *Proc. IRE*, **26**:333–345 (March 1938).

[‡] Kraus, "Antennas," p. 197.

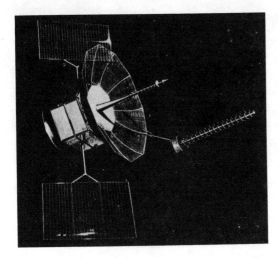

Figure 14-33 Fleetsatcom (communication) satellite in Clarke or geostationary orbit 36,000 km above the earth's equator. The parabolic antenna with helix feed is for transmitting (down-link to earth) and the single larger helix is for receiving (up-link from earth). Four such satellites are in orbit at 90° spacing around the Clarke orbit for complete global coverage. (Actually a minimum of three at 120° spacing could do it.) (*Courtesy TRW Corporation.*)

For a 20-turn helix with typical parameters $C = \lambda$ and $\alpha = 12°$ the directivity $D = 64$ (or 18 dB).

With a pitch angle $\alpha = 12°$, $S = 0.213\lambda$, so there are nearly 5 turns per wavelength, and the 20-turn helix is 4.3λ long ($= nS = 20 \times 0.213\lambda$). An ordinary endfire array of this length has less than one-fourth of this directivity. Because of this four-fold directivity, or gain, improvement of a helical-beam antenna over an ordinary end-fire antenna of the same length, the helical-beam antenna can be called a *super-gain* antenna.

The helical-beam antenna behaves like a lens, collecting waves from an effective aperture A_e which is larger than its cross section or physical aperture A_p, as in Fig. 14-32e.

The high gain, large bandwidth, simplicity, and circular polarization† of the helical-beam antenna have made it indispensable for space-communication applications. For example, the communications satellite shown in Fig. 14-33 has two helical-beam antennas, one for reception from the earth (up-link) and the other as a feed antenna for a parabola for transmitting to the earth (down-link). Four of these satellites in Clarke, or geostationary, orbit at a height of 36,000 km above the earth's equator provide global coverage for communication services. (See Probs. 10-6-1, 14-5-13, and 14-6-8.)

---

† With linearly polarized antennas, the plane of polarization of the wave from the transmitting antenna should be aligned with the receiving antenna [$MM_a = 0$ in (11-4-4)]. When the planes are perpendicular, the received power goes to zero ($MM_a = 180°$). Because of the rotation of a satellite or spacecraft or because of Faraday rotation in the ionosphere, the plane of polarization may rotate and cause signal fading and dropouts. This fading is overcome with circularly polarized antennas.

Table 14-4 Array beam widths and directivities†

Type of array	BWFN	HPBW	Directivity
Broadside	$\dfrac{115°}{L/\lambda}$	$\dfrac{51°}{L/\lambda}$	$2L/\lambda$
End fire (ordinary)	$\dfrac{162°}{\sqrt{L/\lambda}}$	$\dfrac{108°}{\sqrt{L/\lambda}}$	$3L/\lambda$
Helical-beam antenna (end fire with increased directivity)	$\dfrac{115°}{(C/\lambda)\sqrt{L/\lambda}}$	$\dfrac{52°}{(C/\lambda)\sqrt{L/\lambda}}$	$15\left(\dfrac{C}{\lambda}\right)^2\dfrac{L}{\lambda}$
Square-aperture broadside array (unidirectional)	$\dfrac{115°}{L/\lambda}$	$\dfrac{51°}{L/\lambda}$	$13(L/\lambda)^2$

† L = array length for end-fire arrays or side length for broadside array (relations are less accurate for $L < \lambda$)
C = circumference of helix
λ = wavelength

14-14 BEAM WIDTH AND DIRECTIVITY OF ARRAYS

In the preceding sections we have discussed the directional properties of a number of types of uniform arrays, both broadside and end-fire. Table 14-4 summarizes approximate relations for the BWFN, the HPBW, and directivity for a uniform linear broadside array and an ordinary end-fire array of isotropic sources and for a helical-beam antenna (end fire with increased directivity). In addition the béam widths and directivity are listed for a unidirectional square-aperture broadside array with uniform distribution.

14-15 SCANNING ARRAYS

In Secs. 14-6.5 and 14-6.6 we considered broadside and end-fire arrays. In the broadside array all sources are in phase ($\delta = 0$), while in the (ordinary) end-fire array the progressive phase difference between sources is $-\beta d$, where d is the spacing between sources. Actually the beam maximum can be tilted to any arbitrary direction $\theta_{\max}$ if the phase difference δ satisfies the relation

$$\delta = -\beta d \cos \theta_{\max} \qquad (1)$$

For a broadside array $\theta_{\max} = 90°$ and $\delta = 0$. For an end-fire array $\theta_{\max} = 0°$ and $\delta = -\beta d$. To direct the maximum radiation at $45°$ ($\theta_{\max} = 45°$) requires that $\delta = -0.707\beta d$. By controlling the phase difference between elements of an array the beam angle can be adjusted or scanned without turning the antenna. Such arrays are sometimes called *scanning arrays* or *antennas*.

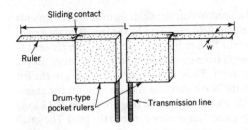

Figure 14-34 A frequency-independent antenna made of adjustable roll-up pocket rulers set so that L equals $\lambda/2$ at the operating frequency.

14-16 FREQUENCY-INDEPENDENT ANTENNAS

A frequency-independent antenna may be defined as one for which the impedance and pattern (and hence directivity) remain constant as a function of the frequency.

Consider the arrangement in Fig. 14-34, which consists of an adjustable $\lambda/2$ dipole made with two drum-type pocket rulers. If L is adjusted to approximately $\lambda/2$ at the frequency of operation, the impedance and pattern remain the same. Strictly speaking, the element thickness or width w should also be adjusted, but this effect is small and for many purposes may be negligible. This simple antenna illustrates the requirement that *the antenna should expand or contract in proportion to the wavelength in order to be frequency-independent*, or if the antenna structure is not mechanically adjustable as above, the size of the active or radiating region should be proportional to the wavelength.

Rumsey† proposed that an antenna which is defined only in terms of angles would be frequency-independent.‡ This requirement can be met with a variety of physical structures such as a pair of concentric equiangular spirals.

The log-periodic antenna§ is another example of a frequency-independent antenna, and one employing dipoles¶ is shown in Fig. 14-35. The dipole lengths increase along the antenna so that the included angle α is a constant, and the lengths and spacings of adjacent elements are scaled so that

$$\frac{l_{n+1}}{l_n} = \frac{s_{n+1}}{s_n} = k \tag{1}$$

where k is a constant. At a wavelength near the middle of the operating range, radiation occurs primarily from the central region of the antenna, as suggested in Fig. 14-35. The elements in this active region are about $\lambda/2$ long, and the phasing is (as indicated by the arrows) below the elements. By the time the field radiated to the left from element 8 arrives at element 7, the phase of 7 has advanced 90° and its field adds in phase to that from element 8. When these fields from elements 8 and 7

---

† V. H. Rumsey, "Frequency Independent Antennas," Academic Press, Inc., New York, 1966.

‡ Such antennas possess a wide instantaneous bandwidth, whereas the antenna of Fig. 14-34 has a narrow bandwidth but is tunable.

§ So called because the logarithm of (1) yields a constant logarithmic increment for each step or period in the structure.

¶ D. E. Isbell, *IRE Trans. Antennas Propag.*, **AP-8**:260–267 (1960).

arrive at element 6, its phase has advanced so that the fields from all three elements add in phase, producing a large resultant field radiating to the left, as suggested at the bottom of Fig. 14-35. Examining the field radiated to the right from element 6, we find that it combines with the fields from elements 7 and 8 to produce only a small resultant field radiating to the right. Thus, maximum radiation is to the left or in the backward direction from the wave entering the antenna on the transmission, or feed, line. Elements 9, 10, and 11 are in the neighborhood of 1λ long and carry only small currents (they present a large impedance to the line). The small currents in elements 9, 10, and 11 mean that the antenna is effectively truncated at the right of the active region. Any small fields from elements 9, 10, and 11 also tend to cancel in both forward and backward directions. However, some radiation may occur broadside since the currents are approximately in phase. The elements at the left (1, 2, 3, etc.) are less than $\lambda/2$ long and present a large capacitative reactance to the line. Hence, currents in these elements are small and radiation is small. Thus, under the condition assumed above that the wavelength is near the center of the operating range, only the middle portion of the antenna is active, while the left and right ends are relatively inactive. If the wavelength is increased, the active region moves to the right, and if the wavelength is decreased, the active region moves to

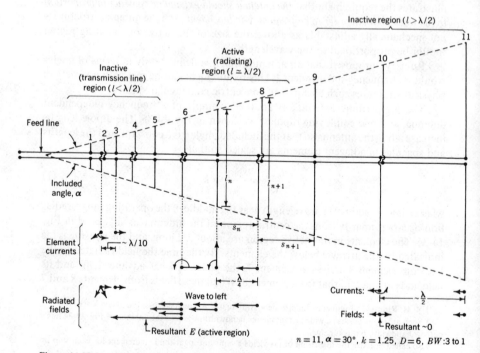

Figure 14-35 Log-periodic frequency-independent antenna of 8-dBi gain with 11 dipoles showing active (central) region and inactive regions (left and right ends).

the left. Thus, we see that at any given frequency only a fraction of the antenna is used (where the dipoles are about $\lambda/2$ long). At the short-wavelength limit of the bandwidth only 15 percent of the length may be used, while at the long wavelength limit a larger fraction is used but still less than 50 percent. For the log-periodic antenna of Fig. 14-35 the included angle $\alpha = 30°$, $k = 1.25$, the directivity is about 6 ($=8$ dB), and the bandwidth about 3 to 1.

14-17 RECIPROCITY

By reciprocity it follows that the *pattern, directivity, aperture, and terminal impedance of an antenna are the same transmitting or receiving.* In general, however, the current distribution on an antenna is not the same for transmission as for reception.

To show that reciprocity applies to antennas consider the two antennas 1 and 2 (Fig. 14-36a) in a medium that is linear, passive, and isotropic. Let a zero-impedance transmitter of frequency f be connected to the terminals of antenna 1 (Fig. 14-36a), producing a current I_1 and inducing a voltage V_{21} at the (open) terminals of antenna 2. Now let the transmitter be transferred to the terminals of antenna 2 (Fig. 14-36b), producing a current I_2 and inducing a voltage V_{12} at the open terminals of antenna 1. Since any four-terminal network can be reduced to an equivalent T section, the antenna arrangement of Fig. 14-36a and b can be replaced by the network of Fig. 14-36c. Since reciprocity is easily demonstrated for this network, it follows that

$$\frac{V_{21}}{I_1} = \frac{V_{12}}{I_2} \tag{1}$$

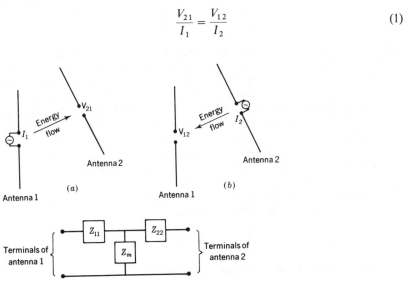

Figure 14-36 (*a*) and (*b*) Reciprocity between two antennas. (*c*) Equivalent circuit.

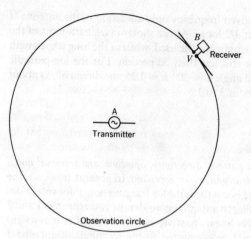

Figure 14-37 Pattern measurement on observation circle illustrating reciprocity for antenna patterns.

and that reciprocity applies to antennas. In words, the reciprocity theorem for antennas states that *if a current I_1 at the terminals of antenna 1 induces a voltage V_{21} at the (open) terminals of antenna 2 and a current I_2 at the terminals of antenna 2 induces a voltage V_{12} at the (open) terminals of antenna 1, then if $I_2 = I_1$, $V_{12} = V_{21}$.* In Fig. 14-36c Z_{11} is the self-impedance of antenna 1, Z_{22} is the self-impedance of antenna 2, and Z_m is the mutual impedance of antennas 1 and 2 as given by

$$Z_m = Z_{12} = \frac{-V_{12}}{I_2} = \frac{-V_{21}}{I_1} = Z_{21} \tag{2}$$

Pattern If all media involved are linear, passive, and isotropic, reciprocity holds and it follows that *the transmitting and receiving patterns of an antenna are identical.* Referring to Fig. 14-37, let the pattern of antenna A (transmitting) be measured by observing its field with a receiving antenna B at points around an observation circle having A as its center.† The voltage at the terminals of antenna B is a measure of the field at the observation circle. If the measurement procedure is reversed by interchanging transmitter and receiver so that antenna B transmits and antenna A receives, it follows from reciprocity that the pattern of antenna A, observed by moving antenna B, will be identical to that obtained when antenna A is transmitting.

Directivity and aperture Directivity was discussed in Sec. 14-5 for a transmitting antenna. In general, the directivity is equal to

$$D = \frac{4\pi}{\iint P_n(\theta, \phi) \, d\Omega} \tag{3}$$

where $P_n(\theta, \phi)$ is the normalized power pattern.

† The same result is obtained by keeping antenna B fixed and rotating antenna A about its center point.

Thus, D depends only on the shape of the power pattern. As shown in the preceding subsection, an antenna has the same pattern for both transmission and reception. Hence, the directivity determined with an antenna's transmitting pattern is identical with the directivity determined with the antenna's receiving pattern. Accordingly, the term *directivity* can be applied to both transmitting and receiving antennas, the directivity of an antenna being the same for both situations.

According to (14-5-29), the effective aperture A_e of an antenna is equal to the directivity of the antenna times a constant. Hence, the term *effective aperture* may be applied to both transmitting and receiving antennas, it being understood that the effective aperture of a transmitting antenna is the same as its effective aperture when receiving.

Impedance When an antenna is transmitting, it may be excited at only one point. However, when used for reception, the antenna is excited over its entire extent by the received wave.† As a consequence the current distribution on the antenna is, in general, not the same for transmission and reception. However, the antenna always behaves as the same circuit regardless of the mode of excitation, so that the impedance of an antenna is the same for transmission and reception. This means that if the terminal impedance of a transmitting antenna is Z_T, the load impedance required for maximum power transfer when the antenna is receiving is equal to its complex conjugate Z_T^*.

That Z_T must be the same for transmission and reception can also be seen by considering a circuit or network of many meshes. Although the currents in the network are dependent on the location or locations of the emfs, the circuit impedances are independent of the distribution of the emfs.

14-18 SELF-IMPEDANCE AND MUTUAL IMPEDANCE AND ARRAYS OF DIPOLES

The impedance presented by an antenna to a transmission line can be represented by a two-terminal network. This is illustrated in Fig. 14-38, in which the antenna is replaced by its equivalent impedance connected to the transmission-line terminals. This impedance is called the *terminal* or *driving-point impedance* Z_T. If the antenna is isolated, i.e., remote from other objects, and is lossless, its terminal impedance is the same as the *self-impedance* of the antenna ($Z_T = Z_{11}$). This impedance has a real part equal to the *radiation resistance* R_{11} and an imaginary part called the *self-reactance* X_{11}. If there are nearby objects, such as other active antennas, the terminal impedance will be modified to include contributions involving the mutual impedance between the antennas and currents flowing in them.

In Sec. 14-9 the radiation resistance of a thin $\lambda/2$ dipole was calculated and found to be equal to 73 Ω. The method used involved a calculation of the (real)

---

† Furthermore, the manner in which the receiving antenna is excited depends on the direction of the incident wave.

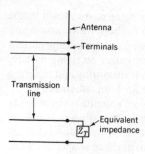

Figure 14-38 Transmission line with antenna and with equivalent lumped (terminal) impedance Z_T.

power radiated by integrating the Poynting vector over a large sphere (in far field). No reactance term resulted from this calculation. To include the antenna reactance requires integration (in effect) of the Poynting vector over a closed surface collapsed down so that it just fits over the antenna conductors. In this way the reactive (non-radiating) power circulating back and forth near the antenna enters into the integration, and, in general, this will result in a reactance term. Such calculations are tedious and will be omitted here.[†] As stated following (14-9-7), the self-impedance of a thin linear $\lambda/2$ center-fed dipole is given by

$$Z_{11} = R_{11} + jX_{11} = 73 + j42.5 \ \Omega \qquad (1)$$

The terminal impedance of a thin linear center-fed dipole antenna as a function of the frequency is shown by the spiral solid line in the impedance diagram of Fig. 14-39.[‡] The length-to-diameter ratio (l/d ratio) for this antenna (solid curve) is 2,000. With change in frequency the impedance varies over a wide range of values. At frequencies where $1/\lambda \approx 0.5$ the resistance is small, and the reactance changes rapidly from large negative to large positive values with increase in frequency (increase in l/λ). At resonance $l/\lambda = 0.48$ and $Z_{11} = 70 + j0 \ \Omega$. At higher frequencies where $l/\lambda \approx 1$ the resistance is several thousand ohms, and the reactance may also be of comparable magnitude. Another (second) resonance occurs when $l/\lambda = 0.93$, and for this condition $Z_{11} = 3,300 + j0 \ \Omega$. At still higher frequencies the impedance values again become small, and another (third) resonance occurs when $l/\lambda = 1.48$. For this resonance $Z_{11} = 100 + j0/\Omega$. With further increase in frequency the antenna impedance goes through further resonances. At the odd-numbered resonances the resistance is small, and at the even-numbered it is high. If the antenna thickness is increased (l/d ratio smaller), the impedance variation with frequency is smaller. This is illustrated by the dashed spiral curve in Fig. 14-39, which is for a center-fed dipole with l/d ratio of 60. The resistance values of this dipole at its first and third resonances are not much different from those for the thinner dipole but the resistance values at the second and fourth

† For a detailed discussion see, for example, Kraus, "Antennas," chaps. 9 and 10.

‡ Erik Hallén, Admittance Diagrams for Antennas and the Relation between Antenna Theories, *Harv. Univ., Cruft Lab. Tech. Rep.* 46, June 1, 1948.

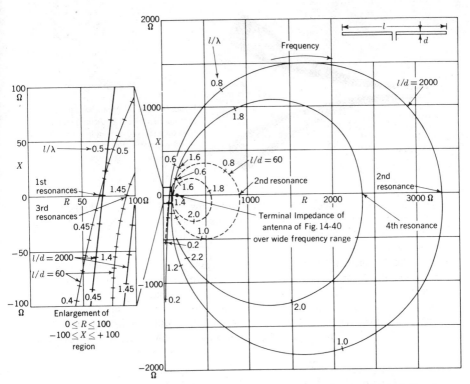

Figure 14-39 Self-impedance diagram (R, X) as a function of frequency (expressed in l/λ ratios indicated along curves) for linear antennas with length-diameter (l/d) ratios of 60 (dashed) and 2,000 (solid). Resonance ($X = 0$) occurs where the impedance curve crosses the horizontal axis. The diagram at the left is an enlargement of the region of first and third resonances near the origin of the main diagram. (*After Hallén.*)

resonances are greatly reduced. As the antenna thickness is increased further, the impedance variation becomes still less.

Basic broadband antenna A simple explanation for this tendency follows from a consideration of the *V*-type antenna shown in Fig. 14-40. If a two-conductor transmission line is gradually opened out and the ratio of conductor spacing D to diameter d kept constant, the characteristic impedance of the line will also be constant. Hence, there will be no reflected wave on the line provided none occurs at the end, and this will be essentially the case if D is sufficiently large ($D > \lambda$) at the end of the structure. In the opened-out region of the thick conductors the transmission line acts as an antenna, and most of the power is radiated in a beam to the right. Accordingly, the antenna terminal impedance will be equal to the characteristic resistance of the transmission line, and this value will remain substantially

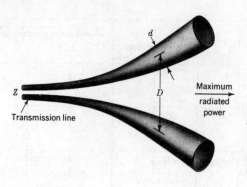

Figure 14-40 Basic broad-band directional antenna in which a two-conductor transmission line is opened out into an antenna. By keeping ratio D/d constant the arrangement presents a constant impedance Z over a wide frequency range.

constant at all frequencies for which the above conditions hold. Thus, for an antenna like that in Fig. 14-40, the impedance spiral reduces to a single point representing the characteristic impedance (all resistance) of the opened-out transmission line.

The arrangement of Fig. 14-40 may be regarded as a *basic broadband antenna*. If a two-conductor transmission line with uniform diameter conductors is opened out, as in Fig. 14-41a, it will behave like the basic broadband antenna at higher frequencies but not at lower frequencies where D is less than 1λ since the outgoing wave will have been only partially radiated and there will be reflection and diffraction at the ends of the conductors. To improve the low-frequency performance,

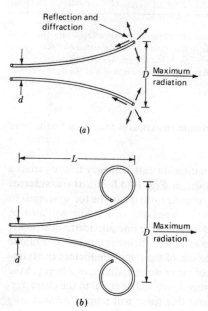

Figure 14-41 Broad-band directional antennas with opened-out transmission lines. (*a*) Simple form. (*b*) Design with curled ends to improve low-frequency performance. L should be at least $\lambda/2$ at the lowest frequency.

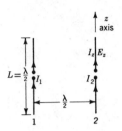

Figure 14-42 Two-parallel linear $\lambda/2$ center-fed dipole antennas with $\lambda/2$ spacing.

the ends of the conductors can be curled back as in Fig. 14-41b, extending operation to frequencies for which D is as small as $\lambda/4$. The high-frequency performance will deteriorate when the spacing d (at the input end) becomes too large (as $\lambda/4$).

Thus, the useful bandwidth of the antenna is determined by the ratio D/d, which, in practice, may be quite large (50 to 1 or more). Whereas the directivity of a log-periodic antenna (Sec. 14-16) is relatively constant over its bandwidth of 4 to 1, the directivity of this antenna increases with frequency.

Mutual impedance Let us consider the mutual impedance of the two antennas of Fig. 14-42. The mutual impedance is given by

$$Z_m = Z_{12} = Z_{21} = \frac{-V_{21}}{I_1} = \frac{-1}{I_1 I_2} \int_0^L I_z E_z \, dz \tag{2}$$

where I_z, E_z = current and field at distance z along antenna 2 induced by current in antenna 1

I_1, I_2 = terminal currents of antennas 1 and 2, respectively

The integration is over the length L of antenna 2. The evaluation of (2) will be omitted here. However, for the case of two parallel thin linear $\lambda/2$ center-fed antennas spaced $\lambda/2$, as in Fig. 14-42,

$$Z_m = Z_{12} = Z_{21} = -13 - j29 \, \Omega \tag{3}$$

The mutual impedance of two parallel linear $\lambda/2$ antennas as a function of spacing is shown by the spiral curves of Fig. 14-43.† The solid spiral curve is for antenna elements with an l/d ratio of 11,000. The dashed spiral curve is for an l/d ratio of 73. As the spacing s between the antennas increases, the mutual impedance rapidly spirals in around the origin ($R_m = X_m = 0$). It is to be noted that whereas the self-resistance of an antenna is always positive, the mutual resistance may be either positive or negative.

The mutual impedance Z_m of two antennas is usually a complicated function of the size, separation, and orientation of the antennas. However, in the case of two short-dipole antennas the relation is relatively simple and illustrates clearly the

† C-T Tai, Coupled Antennas, *Proc. IRE.* **36**:487–500 (April 1948); P. S. Carter, Circuit Relations in Radiating Systems and Applications to Antenna Problems, ibid., **20**:1004–1041 (June 1932).

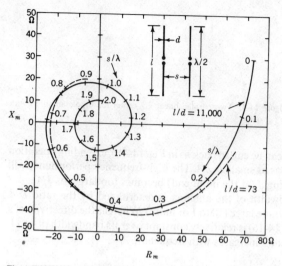

Figure 14-43 Mutual-impedance diagram (R_m, X_m) as a function of element spacing s/λ for two parallel $\lambda/2$ center-fed dipoles with length-diameter (l/d) ratios of 73 (dashed) and 11,000 (solid). (*After Tai and Carter.*)

effect of these parameters. Referring to Fig. 14-44, the mutual impedance of the two dipoles is given by

$$Z_m = \frac{60\pi l^2}{s\lambda} (\sin \theta \sin \theta')(\sin \beta s + j \cos \beta s) \tag{4}$$

Note that the first factor is a *magnitude factor*, the second an *orientation factor*, and the third a *periodic, or complex, factor* with real and imaginary parts. Note also that when the separation distance $s = n\lambda/4$ and n is odd, Z_m is real; but if n is even, Z_m is imaginary. Both dipoles are in the same plane and it is assumed that $l \ll \lambda$ and $l \ll s$ and also that the current is of uniform magnitude and constant phase on each dipole.

Two-element array With self-impedance and mutual impedance much of the field theory of antenna arrays can be reduced to a circuit theory of networks as illustrated by the following example.

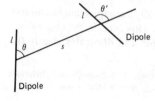

Figure 14-44 Two short dipoles of length l, separation s, and orientation angles θ and θ' for mutual-impedance equation.

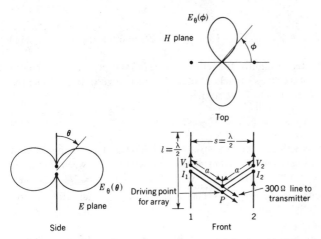

Figure 14-45 Broadside array of two thin $\lambda/2$ center-fed dipoles with $\lambda/2$ spacing showing field patterns in both principal planes.

Consider an array of two identical parallel thin linear center-fed dipole antenna elements which are to be driven with equal in-phase currents and fed by a two-conductor 300-Ω transmission line. This can be accomplished by the arrangement of Fig. 14-45. The voltage at the terminals of element 1 is

$$V_1 = I_1 Z_{11} + I_2 Z_{12} \tag{5}$$

where I_1 = terminal current in element 1
$\quad I_2$ = terminal current in element 2
$\quad Z_{11}$ = self-impedance of element 1
$\quad Z_{12}$ = mutual impedance between elements

Likewise, we have for the terminal voltage of element 2

$$V_2 = I_2 Z_{22} + I_1 Z_{21} \tag{6}$$

where Z_{22} is the self-impedance of element 2. We require $I_1 = I_2$, and we also know that $Z_{11} = Z_{22}$ and $Z_{12} = Z_{21}$. Therefore,

$$V_1 = I_1(Z_{11} + Z_{12}) = V_2 \tag{7}$$

and the terminal impedance Z_1 of element 1 and Z_2 of element 2 are equal and given by

$$Z_1 = \frac{V_1}{I_1} = Z_{11} + Z_{12} = Z_2 \tag{8}$$

For the case where $l = \lambda/2$ and $s = \lambda/2$

$$Z_{11} = 73 + j43 \ \Omega \tag{9}$$

$$Z_{12} = -13 - j29 \ \Omega \tag{10}$$

Therefore,

$$Z_1 = Z_{11} + Z_{12} = 73 - 13 + j(43 - 29) = 60 + j14\,\Omega \tag{11}$$

If each element is shortened slightly, the self-reactance ($43\,\Omega$) may be reduced more than the mutual reactance so as to yield a total reactance of zero. In so doing the self-resistance will be reduced to $70\,\Omega$, so that

$$Z_1 = 57 + j0\,\Omega \tag{12}$$

In order that the array will present a resistive load of $300\,\Omega$ at the driving point of the array (point P in Fig. 14-45) the transmission-line sections between P and the element terminals can be constructed so that their length $a = \lambda/4$ and their characteristic impedance $Z_0 = \sqrt{57 \times 600} = 185\,\Omega$. Thus, each $\lambda/4$ section transforms the element terminal resistance of 57 to $600\,\Omega$ and the two lines in parallel yield a driving-point impedance $Z = 300 + j0\,\Omega$.

The above array consisting of two parallel thin $\lambda/2$ antenna elements fed with equal in-phase currents has maximum radiation broadside. The principal-plane patterns are as shown in Fig. 14-45 and are given by:

E plane:
$$E_\theta(\theta) = 2kI_1 \frac{\cos\left[(\pi/2)\cos\theta\right]}{\sin\theta} \tag{13}$$

H plane:
$$E_\theta(\phi) = 2kI_1 \cos\frac{\pi\cos\phi}{2} \tag{14}$$

The shape of the E-plane pattern is the same as for a $\lambda/2$ linear antenna, as given by (14-9-4), while the shape of the H-plane pattern is the same as for two equal isotropic in-phase point sources spaced $\lambda/2$ apart, as given by (14-6.1-2) with $d = \lambda/2$.

The directivity of the above array can be calculated from (14-5-7) by integrating the power pattern over 4π sr. An alternative method of calculating the directivity will now be described which employs the self-resistance and mutual resistance of the antenna elements.

Assuming no heat losses, the power input to the array (Fig. 14-45) is given by

$$P = 2I_1^2(R_{11} + R_{12}) \tag{15}$$

Hence, the individual element currents are

$$I_1 = I_2 = \sqrt{\frac{P}{2(R_{11} + R_{12})}} \tag{16}$$

If the same power is supplied to a single isolated $\lambda/2$ antenna, its current is

$$I_0 = \sqrt{\frac{P}{R_{11}}} \tag{17}$$

In the H plane the pattern of the single $\lambda/2$ antenna is nondirectional and is given by†

$$E_{HW}(\phi) = kI_0 = k\sqrt{\frac{P}{R_{11}}} \tag{18}$$

The maximum radiation from the array occurs in the direction $\phi = 90°$ (Fig. 14-45). Hence, the gain of the array over the single $\lambda/2$ antenna as reference is given by the ratio of (14) to (18) with $\phi = 90°$, or

$$\boxed{\text{Array gain (field)} = \frac{2kI_1}{kI_0} = \sqrt{\frac{2R_{11}}{R_{11} + R_{12}}}} \tag{19}$$

Now $R_{11} = 70\,\Omega$ and $R_{12} = -13\,\Omega$. Therefore,

$$\text{Array gain} = \sqrt{\frac{2 \times 70}{70 - 13}} = 1.57 \quad (= 3.9 \text{ dB})$$

This gain is with respect to a $\lambda/2$ antenna (with same power input) as reference. The directivity of a linear $\lambda/2$ (dipole) antenna has been calculated earlier as 1.64 ($=$ gain of $\lambda/2$ antenna over isotropic source). Therefore, the directivity of the array of Fig. 14-45 is

$$D = 1.57^2 \times 1.64 = 4\,(= 6 \text{ dBi})‡$$

This is the same directivity that would be obtained by integrating the power pattern of the array over 4π sr, and yet we have carried out no such integration, at least not *explicitly*. However, such an integration or its equivalent is *implicit* in the self-resistance and mutual resistance.

14-19 REFLECTOR AND LENS ANTENNAS

In this section a variety of reflector antennas is considered. The lens-type antenna is also discussed. It is convenient to treat both reflector and lens types together since the design of both involves the principle of equality of path length.

Figure 14-46 displays six types of reflector antennas. In Fig. 14-46a the reflector is an infinite flat sheet, but in (b) the sheet is of finite extent. The infinite flat sheet can be analyzed by simple image theory, but the finite one requires the addition of diffraction theory. In Fig. 14-46c the reflector has degenerated to a thin linear element, and this arrangement is analyzed by simple array and circuit theory. Corner reflector antennas are shown in Fig. 14-46d and e. In one case the corner reflector is active (contains driven element), and in the other case the corner reflector is passive. A parabolic reflector antenna is shown in Fig. 14-46f.

---

† The subscript HW of E in (18) stands for *half-wavelength antenna*.
‡ Decibels over isotropic.

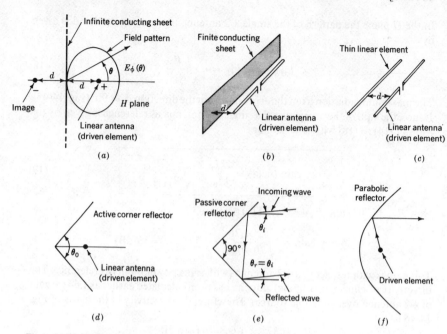

Figure 14-46 Linear antenna (*a*) with infinite reflecting sheet and (*b*) with sheet of finite extent. In (*c*) the reflecting sheet has degenerated to a thin linear element. (*d*) Corner reflector with (active) driven element and (*e*) passive corner reflector (no driven element). (*f*) Driven element and parabolic reflector.

14-19.1 Infinite-Flat-Sheet Reflector

In Fig. 14-46*a* a linear antenna is situated a distance *d* from an infinite flat perfectly conducting metal sheet, or ground plane. This combination can be analyzed simply by replacing the ground plane by an image of the antenna at a distance 2*d* from the antenna (see Fig. 7-17). The pattern in the θ plane (*H* plane)† is that of two equal isotropic point sources spaced a distance 2*d* apart and in opposite phase, or

$$E_\phi(\theta) = \sin(\beta d \cos\theta) \tag{1}$$

If $d = \lambda/4$, the pattern is as suggested in Fig. 14-46*a* with maximum at $\theta = 0°$.

14-19.2 Finite-Flat-Sheet Reflector

When the reflecting sheet or ground plane is reduced in size, the analysis is less simple. Consider, as suggested in Fig. 14-46*b*, that the flat conducting sheet is reduced in width to $2\frac{1}{4}\lambda$ and that the driven element is a $\lambda/2$ dipole spaced $\frac{3}{8}\lambda$ out from the sheet. The arrangement is shown in cross section in Fig. 14-47*a*.

---

† In the *E* plane the pattern is that of an isolated antenna multiplied by (1).

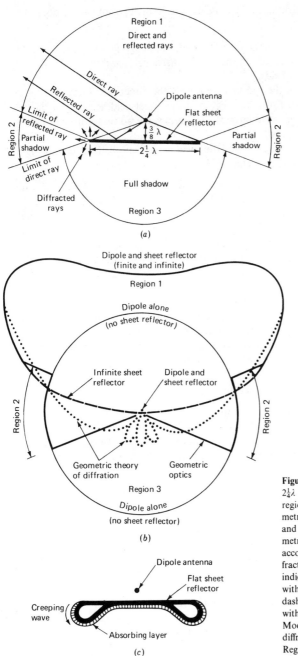

Figure 14-47 (*a*) Dipole antenna with $2\frac{1}{4}\lambda$ flat-sheet reflector with three regions of radiation according to geometric optics. (*b*) Field pattern of dipole and sheet reflector according to geometric optics (heavy solid line), and according to geometric theory of diffraction (dotted line). The solid circle indicates the field from the dipole alone with no reflecting sheet present and the dashed line gives the pattern for dipole with an infinite sheet reflector. (*c*) Modification of edges of sheet to reduce diffracted back-side radiation (in Region 3).

According to geometric optics, the fields can be divided into three zones or regions:

In *Region 1* (above or in front of the sheet) the radiated field is given by the resultant of the field coming directly from the dipole (direct field) and the field reflected off the sheet (reflected field).

In *Region 2*, to the sides of the sheet, there is only the direct field from the dipole (no reflection), so that this is the region of the direct field. Region 2 is in the shadow of the reflected wave but not the direct wave, so that it may be called a region of partial shadow.

In *Region 3* (below or behind the sheet) the sheet acts as an obstacle, producing a full shadow (no direct or reflected fields).

The field pattern by geometric optics is shown by the heavy solid line in Fig. 14-47b, with abrupt steps at the limits of direct and reflected rays. The pattern for the dipole alone (no sheet present and for the same current) is shown by the solid circle. According to the geometric theory of diffraction (GTD), the transitions will be less abrupt and there will be diffraction around the edges as though originating from a weak line source along each edge, resulting in the pattern shown by the dotted line. For comparison, the dashed line shows the pattern for an infinite sheet or ground plane.

To reduce the diffracted backside radiation coming from the sharp edges of the sheet, rolled edges may be added as in Fig. 14-47c, with absorbing material (complex μ and ϵ) added to attenuate waves creeping around the edges. With this modification the sheet width is increased slightly, but a pattern is obtained which largely eliminates backside radiation (in Region 3).

14-19.3 Thin Reflectors and Directors

When the reflecting sheet degenerates into a thin linear conductor, as in Fig. 14-46c, the arrangement can be analyzed by circuit and array theory as follows.† Let the active (driven) element be designated element 1 and the passive (degenerate flat sheet) element be designated element 2, as in Fig. 14-48a. The circuit relations for this arrangement are

$$\boxed{\begin{aligned} V_1 &= I_1 Z_{11} + I_2 Z_{12} \\ 0 &= I_2 Z_{22} + I_1 Z_{21} \end{aligned}} \tag{1}$$

where V_1 = terminal voltage of element 1

I_1 = terminal current of element 1

I_2 = terminal current of element 2

Z_{11} = self-impedance of element 1

Z_{22} = self-impedance of element 2 (if it were opened at center and measured in isolated situation)

$Z_{12} = Z_{21}$ = mutual impedance of elements 1 and 2

† G. H. Brown, Directional Antennas, *Proc. IRE*, **25**: 78–145 (January 1937).

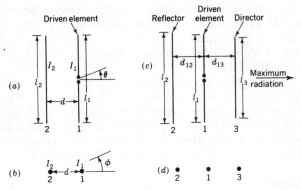

Figure 14-48 (*a*) and (*b*) Driven element with single parasitic (passive) element and (*c*) and (*d*) driven element with parasitic reflector and director.

From (1)

$$I_2 = -I_1 \frac{Z_{12}}{Z_{22}} = -I_1 \frac{|Z_{12}| \angle \tan^{-1}(X_{12}/R_{12})}{|Z_{22}| \angle \tan^{-1}(X_{22}/R_{22})} \tag{2}$$

The field pattern $E_\theta(\phi)$ (*H* plane) is then

$$E_\theta(\phi) = k(I_1 + I_2 \underline{/\beta d \cos \phi}) \tag{3}$$

where I_2 is as given in (2) and k is a constant. In general, I_1 and I_2 will be of different magnitude, so that the pattern cannot be reduced to the simple forms used in Sec. 14-6.1. Also note that, in general, $Z_{22} \neq Z_{11}$. Suppose that the driven element 1 is approximately $\lambda/2$ long ($l_1 \approx \lambda/2$) and that l_2 is *about* the same length but not necessarily exactly the same. Referring to Fig. 14-39, we note that the reactance of a thin linear element varies rapidly as a function of frequency when $l/\lambda \approx 0.5$, going from positive to negative reactance values as the length is reduced. Thus, if the length of element 2 is more than $\lambda/2$ ($l_2/\lambda > 0.5$), it has a positive self-reactance; while if it is somewhat shorter (say $l_2/\lambda < 0.48$), it has a negative reactance. Calculating the pattern $E_\theta(\phi)$ as above shows that in the former case (element 2 reactance positive) element 2 acts as a *reflector* and maximum radiation from the array is to the right ($\theta = \phi = 0$, Fig. 14-48*a* and *b*). On the other hand, if l_2 is shorter, so that its self-reactance is negative, the maximum radiation is to the left ($\theta = \phi = 180°$) and element 2 acts as a *director*. It is assumed that the interelement spacing is small, say 0.1 to 0.2λ.

An extension of the arrangement of Fig. 14-48*a* and *b* is to add a third (passive) element as in Fig. 14-48*c* and *d*. Let element 2 be made inductively reactive (X_{22} positive) and element 3 capacitively reactive (X_{33} negative). In this way element 2 acts as a reflector and element 3 as a director. For thin linear elements typical dimensions are as follows: driven-element length $l_1 = 0.48\lambda$ (to make $X_{11} = 0$), reflector-element length $l_2 = 0.50\lambda$, director-element length $l_3 = 0.46\lambda$, and interelement spacings $d_{12} \approx d_{13} \approx 0.1\lambda.$† The gain of such an array is about 5 (= 7 dB) as compared with a reference $\lambda/2$ dipole and the directivity $D = 5 \times 1.64 = 8.2$ (= 9 dBi = gain over isotropic source). With a single reflector (or director) element as in Fig. 14-48a and b the maximum possible gain is about 5 dB with respect to a $\lambda/2$ dipole ($D = 7$ dBi).

In arrays like those discussed above, where one element is driven and the others are passive, it is customary to refer to the passive elements as *parasitic* elements. Thus, the above reflector and director elements are parasitic elements. Arrays may be constructed with larger numbers of parasitic elements. For example, Yagi‡ has built arrays with many parasitic director elements, and arrays with a number of parasitic elements are commonly referred to as *Yagi* or *Yagi-Uda antennas.*§

14-19.4 Corner Reflectors ¶

A corner reflector consists of two flat reflecting sheets intersecting at an angle, as in Fig. 14-46d. When the corner angle is 90°, the sheets meet at right angles, forming a square-corner reflector. With a driven element placed as shown, the arrangement is an effective (active) directional antenna for a wide range of corner angles ($0 < \theta_0 < 180°$). Without a driven element a square corner is an effective (passive) wave reflector over a wide range of angles of incidence ($0 < \theta_i < \pm 45°$). The reflected wave is directed back along the path direction of the incoming wave, as illustrated in Fig. 14-46e. Thus, the passive corner reflector acts as a *retro-reflector.*

The active corner reflector (with driven element) can be readily analyzed using the method of images for corner angles equal to $180°/n$, where $n = 1, 2, 3, \ldots$. It will be convenient to illustrate the method for a *square corner*, as in Fig. 14-49. For this case there are three images, as indicated. The driven element and all images carry equal currents. Driven element 1 and image element 4 are in phase, while image elements 2 and 3 are in opposite phase (with respect to 1 and 4). The field pattern $E_\phi(\theta)$ (H plane) is given by

$$E_\phi(\theta) = 2kI_1 \left| \cos(\beta s \cos \theta) - \cos(\beta s \sin \theta) \right| \tag{1}$$

where s is the corner-to-driven-element spacing. The maximum radiation is in the direction $\theta = 0°$ (along corner bisector) for all values of $s < \lambda$.

The terminal voltage for the driven element is

$$V_1 = I_1 Z_{11} + I_1 Z_{14} - 2I_1 Z_{12} \tag{2}$$

where Z_{11} = self-impedance of element 1 (driven element)

Z_{12} = mutual impedance between elements 1 and 2 or between 1 and 3

Z_{14} = mutual impedance between elements 1 and 4

---

† The exact lengths are a weak function of the element thickness. Thus, compare solid and dashed curves in the *RX* diagrams of Figs. 14-39 and 14-43.

‡ H. Yagi, Beam Transmission of Ultra-short Waves, *Proc. IRE*, **16**:7v5-740 (June 1928).

§ S. Uda and Y. Mushiake, "Yagi-Uda Antenna," Research Institute of Electrical Communication, Tohoku University, Sendai, Japan, 1954.

¶ J. D. Kraus, The Corner Reflector Antenna, *Proc. IRE*, **28**:513-519 (November 1940); J. D. Kraus, U.S. Patent 2,270,314, Jan. 20, 1942.

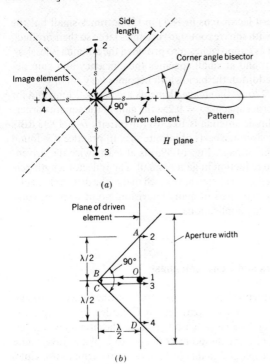

Figure 14-49 (*a*) Square-corner reflector with driven element analyzed as driven element with three image elements. (*b*) Path lengths in corner reflector.

Similar expressions can be written for the voltage at the terminals of each of the image elements. Then for a power P to the driven element (power to each image is also P)† we have

$$I_1 = \sqrt{\frac{P}{R_{11} + R_{14} - 2R_{12}}} \tag{3}$$

The pattern for the isolated driven element (no reflector present) is

$$E_\phi(\theta) = k\sqrt{\frac{P}{R_{11}}} \tag{4}$$

Substituting (3) in (1) and taking the ratio of (1) to (4) yields the gain over the reference (driven) element (in direction $\theta = 0$) as

$$\text{Gain over reference} = 2\sqrt{\frac{R_{11}}{R_{11} + R_{14} - 2R_{12}}}\,|(\cos \beta s - 1)| \tag{5}$$

† But the actual total power supplied to the corner-reflector antenna is that supplied to the driven element ($= P$).

At small spacings s, the pattern factor ($\cos \beta s - 1$) in (5) becomes small but the coupling or resistance factor (with square root sign) becomes large so their product, or gain, approaches a constant as s approaches zero provided the antenna is lossless. In practice losses are not zero and at small spacings the efficiency and gain decreases (see Sec. 14-4) and in addition the bandwidth becomes very small.[†]

If the driven element is $\lambda/2$ long, the image elements are also $\lambda/2$ long and for a typical driven-element-to-corner spacing $s = 0.25\lambda$ the gain of the square corner-reflector antenna over a $\lambda/2$ dipole antenna is 10.2 dB (directivity $D = 12.35$ dBi). The above analysis assumes reflecting sheets of infinite size. In practice it is found that if the side length (in plane of page, Fig. 14-49a) is at least twice the driven-element-to-corner spacing, the reduction in gain is small. The reflector length perpendicular to the page should equal or somewhat exceed that of the driven element.[‡] The corner reflector is very widely used in many television, point-to-point communication, and radio-astronomy applications.

14-19.5 Parabolic Reflectors and Lens Antennas

Parabolic reflectors and lens antennas may be considered together since the design of both involves the principle of equality of path length. Consider, for example, the parabolic reflector shown in Fig. 14-50a. We wish to convert the circular wavefront from the driven element or source at the point O (focus) to a plane wavefront.[§] From geometric optics the *Principle of Equality of Path Length* (Fermat's principle) requires that the electrical length along all ray paths between a pair of wavefronts be equal.[¶] This means here that the electrical distance $OAO = OBC$ or that $2L = R + R \cos \theta$, from which[‖]

$$R = \frac{2L}{1 + \cos \theta} \qquad (2)$$

This is the equation of the required surface. It is a *parabola* with focus at O.

[†] For a more detailed discussion of spacing, efficiency, and bandwidth see Kraus, "Antennas," McGraw-Hill Book Company, New York, 1950, sec. 11-5.

[‡] For a detailed discussion of corner-reflector parameters see Kraus, "Antennas," pp. 328–336.

[§] A wavefront is an equiphase surface; i.e., the phase is the same at all points on the surface.

[¶] For a more detailed discussion see, for example, S. Silver, "Microwave Antenna Theory and Design," McGraw-Hill Book Company, New York, 1949.

[‖] More explicitly we should write

$$2L + \tfrac{1}{2} = R + R \cos \theta + \tfrac{1}{2} \qquad (1)$$

where the distances are expressed in wavelengths. Waves reflected at the parabolic surface undergo a phase reversal (180° phase change) which is equivalent to a path length of $\lambda/2$. Hence the terms with $1/2$ in (1).

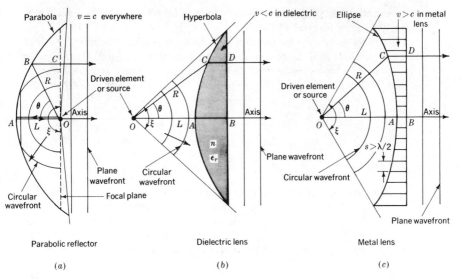

Figure 14-50 (*a*) Parabolic-reflector, (*b*) dielectric-lens, and (*c*) metal-lens antenna with path lengths used in analysis.

Consider next the dielectric lens of Fig. 14-50*b* with driven element or source at O. The principle of equality of path length requires that $OAB = OCD$ or that $R = \eta(R \cos \theta - L) + L$, from which

$$R = \frac{(\eta - 1)L}{\eta \cos \theta - 1} \tag{3}$$

where η is the index of refraction ($\eta = \sqrt{\epsilon_r}$ for nonmagnetic media). Equation (3) gives the required shape of the curved lens surface. It is a *hyperbola*.

We note that with the parabolic reflector, $\eta = 1$ everywhere and the wave velocity $v = c$. For the dielectric lens, however, $\eta > 1$ in the lens, and hence $v < c$. Therefore, the electrical distance, or phase change, is greater per unit distance in the lens than in the air outside, and this is accounted for in (3). A general expression for the electrical length is

$$\text{Electrical length} = \beta_0 \int \eta \, dl = \frac{2\pi}{\lambda_0} \int \eta \, dl \tag{4}$$

where λ_0 is the free-space wavelength and the integration is over the physical path length. In (2) and (3) we have applied (4) in incremental form with $\eta = 1$ in air and

$\eta = \sqrt{\epsilon_r}$ in the dielectric lens. The integral form of (4) is required only if η is a continuously variable function of distance.

Finally, let us analyze the metal-plate lens of Fig. 14-50c. The lens consists of a stack of parallel metal sheets spaced somewhat more than $\lambda/2$ and hence capable of transmitting a higher-mode wave (**E** perpendicular to page), as in a waveguide (see Sec. 13-3).[†] For equality of electrical path length we must have $OAB = OCD$ of $R + (L - R \cos \theta)\eta = L$, from which

$$R = \frac{(1 - \eta)L}{1 - \eta \cos \theta} \tag{5}$$

This relation is identical with (3) for the dielectric lens. However, in the metal-plate lens $v > c$ and $\eta < 1$, and so the form of (5) is used in order to keep both numerator and denominator positive. For this case ($\eta < 1$), (5) is the equation of an *ellipse*. Metal-plate lens antennas with **E** parallel to the page (and metal plates parallel to the page) are also possible. They are analyzed in the same way as the lens of Fig. 14-50c.

This discussion of the antennas of Fig. 14-50 applies both to cylindrical geometry (all surfaces same as shown with line source perpendicular to page) and to spherical geometry (surfaces given by figure of revolution around axes) with point source. As a generalization, all the antennas of Fig. 14-50 may be regarded as *wave transformers* which convert a cylindrical or spherical wave to a plane wave or vice versa.

14-19.6 Some Comments on Corner Reflectors vs. Parabolic Reflectors

The driven element or source for the parabola (and also lens) antennas should be directional, and for greatest operating efficiency most of its pattern angle should be included within the angle ξ subtended by the reflector (or lens) (Fig. 14-50). This puts an additional requirement on the primary or feed antenna used at the focus of a parabolic reflector. This requirement is not present in the corner reflector, where the driven element is a simple linear antenna with omnidirectional pattern (in plane of page, Fig. 14-49).

It should be noted that the principal advantages of the corner reflector are for small apertures. When the aperture width (Fig. 14-49b) is less than 2λ, a corner reflector may be simpler and more practical than a parabolic reflector, but for larger apertures the parabola is better. For small apertures the actual shape of the reflector becomes of secondary importance since the ray-path differences involved are small.

[†] W. E. Kock, Metal Lens Antennas, *Proc. IRE*, **34**:828–836 (November 1946).

It is instructive to apply the principle of equality of path length to the corner reflector. Referring to the square-corner reflector of Fig. 14-49b with driven-element-to-corner spacing of $\lambda/2$, the path lengths of the four principal rays to a point at an infinitesimal distance to the right of the plane of the driven element are

$$\text{Length (ray 1)} = 0\lambda$$

$$\text{Length (ray 2)} = OA + \tfrac{1}{2} = \tfrac{1}{2} + \tfrac{1}{2} = 1\lambda$$

$$\text{Length (ray 3)} = OB + CO + \tfrac{1}{2} - \tfrac{1}{2} = \tfrac{1}{2} + \tfrac{1}{2} + \tfrac{1}{2} - \tfrac{1}{2} = 1\lambda$$

$$\text{Length (ray 4)} = OD + \tfrac{1}{2} = \tfrac{1}{2} + \tfrac{1}{2} = 1\lambda$$

Thus, all four rays are in the same phase.

14-20 SLOT AND COMPLEMENTARY ANTENNAS

If a $\lambda/2$ slot is cut in a large flat metal sheet and a transmission line connected to the points FF, as in Fig. 14-51a, the arrangement will radiate effectively because of currents flowing on the sheet. The analysis of such a *slot antenna* is greatly facilitated by considering the slot's *complementary antenna*. Thus, the antenna which is complementary to the slot of Fig. 14-51a is the dipole of Fig. 14-51b. The metal and air regions of the slot are interchanged for the dipole. According to Booker's theory,[†] the pattern of the slot of Fig. 14-51a is identical in shape to that of the dipole of Fig. 14-51b except that the electric field will be vertically polarized for the slot and horizontally polarized for the dipole. Further, the terminal impedance Z_s of the slot is related to the terminal impedance Z_d of the dipole by the intrinsic

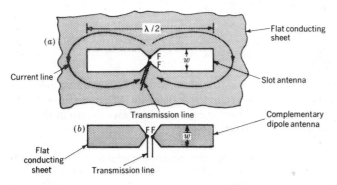

Figure 14-51 (a) Slot antenna and (b) complementary dipole antenna. Both are $\lambda/2$ long and have a width w. They are fed at the points FF.

[†] H. G. Booker, Slot Aerials and Their Relation to Complementary Wire Aerials, *J. IEEE (London)*, **93**, part IIIA, no. 4 (1946).

impedance Z_0 ($= 376.7 \, \Omega$) of free space as follows:

$$\boxed{Z_s = \frac{Z_0^2}{4Z_d} = \frac{35{,}481}{Z_d} \quad (\Omega)} \tag{1}$$

Thus, knowing the properties of the dipole enables us to predict the properties of the complementary slot. For example, let the width of the dipole and slot of Fig. 14-51 be reduced to a very small fraction of a wavelength so that the dipole qualifies as a thin $\lambda/2$ linear dipole with $Z_d = 73 + j42.5$ (see Sec. 14-18). According to (1), the terminal impedance of the complementary slot antenna will be

$$Z_s = \frac{35{,}481}{73 + j42.5} = 363 - j211 \, \Omega$$

14-21 HORN ANTENNAS

A horn antenna may be regarded as a flared-out (or opened-out) waveguide. A pyramidal horn fed by a rectangular waveguide is shown in Fig. 14-52a. The function of the horn is to produce a uniform phase front with a larger aperture than that of the waveguide and hence greater directivity. The principle of equality of path length is applicable to the horn design but with a different emphasis. Instead of specifying that the wave over the plane of the horn mouth be *exactly* in phase, this requirement is relaxed to one where the phase may deviate but by less than a specified amount.† From the geometry of Fig. 14-52b we have $\cos \theta = l/(l + \delta)$, $\sin \theta = h/[2(l + \delta)]$, $\tan \theta = h/2l$, from which

$$l = \frac{h^2}{8\delta} \qquad \delta \text{ small} \tag{1}$$

$$\theta = \tan^{-1} \frac{h}{2l} = \cos^{-1} \frac{l}{l + \delta} \tag{2}$$

In the E plane of the horn, it is customary to allow no more than 72° phase difference ($\delta < \lambda/5$) (deviation $\pm 36°$). However, in the H plane the difference can be larger, say 135° ($\delta < 3\lambda/8$), without undesirable effects since E goes to zero at the horn edges (boundary condition $E_t = 0$ satisfied).

Example Determine the length l, width w, and half-angles θ and ϕ (in E and H planes, respectively) of a pyramidal electromagnetic horn as in Fig. 14-52 for which the mouth height $h = 10\lambda$. The horn is fed by a rectangular waveguide with TE_{10} mode as in the figure.

---

† This could be stated as the *principle of inequality of path length*, where the inequality does not exceed a specified amount.

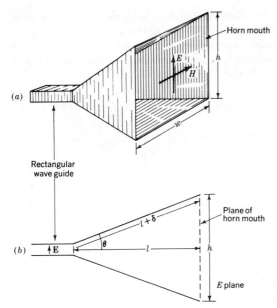

(a)

Rectangular
wave guide

(b)

Figure 14-52 (*a*) Horn antenna showing field orientation in mouth and (*b*) dimensions used in analysis.

SOLUTION Taking $\delta = \lambda/5$ in the E plane, we have from (1) that $l = h^2/8\delta = 100\lambda/(\frac{8}{5}) = 62.5\lambda$ and half the flare angle (in the E plane) is

$$\theta = \tan^{-1} \frac{h}{2l} = \tan^{-1} \tfrac{10}{125} = 4.6°$$

Taking $\delta = 3\lambda/8$ in the H plane gives

$$\phi = \cos^{-1} \frac{62.5}{62.5 + 0.375} = 6.3°$$

and the width w of the horn in the H plane is

$$w = 2l \tan \phi = 2 \times 62.5 \times \tan 6.3° = 13.7\lambda$$

14-22 APERTURE CONCEPT

According to (14-5-29), the *effective aperture* A_e of an antenna is related to its directivity D by

$$D = \frac{4\pi}{\lambda^2} A_e \qquad (1)$$

The effective aperture is a unique quantity for any antenna. For a short-dipole antenna it may be larger than its physical cross section or aperture. However, for antennas like horns or parabolic reflectors which have large aperture dimensions in terms of wavelength, the effective aperture is always less than the physical aperture. Ideally if the field is exactly uniform and in phase over the aperture and there are no thermal losses, the *effective aperture A_e* is equal to the *physical aperture A_p*. Under these conditions the maximum directivity D_m is achieved, or

$$D_m = \frac{4\pi}{\lambda^2} A_p \tag{2}$$

The ratio of (1) to (2) gives the *aperture efficiency*

$$\epsilon_{ap} = \frac{A_e}{A_p} \tag{3}$$

A uniform field distribution has the disadvantage that minor lobes are large (see Fig. 14-25). Although minor lobes are less for a tapered field distribution (zero minor lobes for cosine-squared or gaussian distributions), the effective aperture and directivity are less. One of the problems of antenna design is to achieve a suitable compromise, so that the aperture efficiency is as large as possible for a given side-lobe level. An empirical approach to the problem is to postulate a number of aperture field distributions and to select the best efficiency and side-lobe combination. The pattern with side lobes can be calculated from (14-8-3), while the aperture efficiency can be evaluated from

$$\epsilon_{ap} = \frac{E_{av}^2}{(E^2)_{av}} \tag{4}$$

where E = field at any point in aperture
E_{av} = average of E over aperture
$(E^2)_{av}$ = average of E^2 over aperture

It is assumed that E is exactly in phase over the aperture and that there are no thermal losses.†

Example Calculate the aperture efficiency of a rectangular aperture antenna (x_0 by y_0) with a uniform field in the y direction and a linearly tapered field in the x direction going from E_0 at the centerline ($x = x_0/2$) to 0 at both edges (triangular taper as in Fig. 14-25b).

SOLUTION We have $E(x) = E_0 [x/(x_0/2)]$, and $E(x)_{av} = E_0/2$. Also

$$[E^2(x)]_{av} = \frac{1}{x_0/2} \int_0^{x_0/2} E^2(x) \, dx = \frac{E_0^2}{(x_0/2)^3} \int_0^{x_0/2} x^2 \, dx = \frac{E_0^2}{3}$$

---

† This discussion is highly simplified. For a more complete treatment see Kraus, "Radio Astronomy," pp. 212–223.

Hence the aperture efficiency is

$$\epsilon_{ap} = \frac{[E(x)_{av}]^2}{[E^2(x)]_{av}} = \frac{\frac{1}{4}}{\frac{1}{3}} = \frac{3}{4}$$

A cosine rather than a linear type of taper is common in practice for large apertures, the field going from a maximum at the center to about 0.3 times the center value at the edges (down about 10 dB).

The aperture concept is useful in other ways. For example, it can be used to help decide between a very long end-fire antenna or a broadside array of several shorter end-fire antennas and, for the latter, to determine what spacing should be used between the antennas.

14-23 FRIIS FORMULA AND RADAR EQUATION

Consider the communication circuit of Fig. 14-53a, consisting of a transmitter T with antenna of effective aperture A_{et} and receiver R with antenna of effective aperture A_{er}. The distance between transmitting and receiving antennas is r. If the

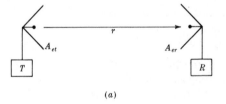

(a)

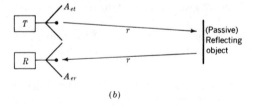

(b)

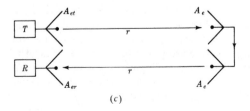

(c)

Figure 14-53 (a) Single transmission path with parameters used in Friis transmission formula. (b) and (c) Double-path geometry used in obtaining radar equation.

transmitter power P_t were radiated by an isotropic source, the power received per unit area at the receiving antenna would be

$$S = \frac{P_t}{4\pi r^2} \tag{1}$$

and the power available to the receiver would be

$$P_r' = SA_{er} \tag{2}$$

The transmitting antenna has an effective aperture A_{et} and hence a directivity $D = 4\pi A_{et}/\lambda^2$, so that with the antenna the power available at the receiver is

$$P_r = \frac{P_r' 4\pi A_{et}}{\lambda^2} \tag{3}$$

Substituting (1) in (2) and (2) in (3) gives

$$P_r = \frac{P_t A_{er}}{4\pi r^2} \frac{4\pi A_{et}}{\lambda^2}$$

or

$$\boxed{\frac{P_r}{P_t} = \frac{A_{er} A_{et}}{r^2 \lambda^2}} \quad \text{(dimensionless)} \tag{4}$$

where P_r = received power, W
P_t = transmitted power, W
A_{er} = effective aperture of receiving antenna, m^2
A_{et} = effective aperture of transmitting antenna, m^2
r = distance between receiving and transmitter antennas, m
λ = wavelength, m

This is the *Friis transmission formula*, which gives the ratio of the received to transmitted power for a direct path.

The following worked example illustrates an application of the Friis transmission formula and also the effects of ground reflection and polarization.

Example A TV broadcasting station radiates a power of 1 kW from an antenna on a 200-m tower rising above a flat perfectly conducting ground. The antenna is omnidirectional in the horizontal plane with a 10° half-power beam width in the vertical plane. If the wavelength is 1 m, what is the optimum height for a receiving antenna at a distance of 4 km if the transmitting and receiving antennas are (a) vertically polarized, (b) horizontally polarized, and (c) circularly polarized? (d) What is the received power for the three cases? The vertical and horizontally polarized antennas are log-periodic types with 6-dBi gain and the circularly polarized antenna is a 6-turn helical-beam antenna with 12.5° pitch angle and spacing between turns of 0.22λ.

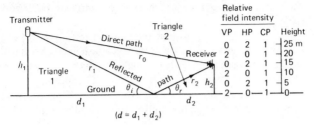

Figure 14-54 Broadcast transmitter with transmission to receiver via direct path and via ground reflection for three cases: vertical (VP), horizontal (HP), and circular polarization (CP). Relative received field intensities vs. height are shown at right for geometry of worked example. Drawing is not to scale.

SOLUTION To satisfy the boundary condition that the tangential field component be zero at the ground, the reflected horizontally polarized wave must be reversed in phase from the incident horizontally polarized wave. Therefore, at zero height h_2 at the receiving location (see Fig. 14-54) the received power will be zero since the direct and reflected fields cancel. For vertical polarization, however, the incident and reflected fields are in phase, giving twice the field intensity and 4 times the Poynting vector (power proportional to field squared).

In general, as indicated in Fig. 14-54, horizontal polarization will produce zero field and vertical polarization will produce doubled field at all heights for which

$$r_1 + r_2 - r_0 = n\lambda \qquad (5)$$

where $n = 0, 1, 2, 3, \ldots$

Further, vertical polarization will produce zero field and horizontal polarization will produce doubled field for all heights for which

$$r_1 + r_2 - r_0 = n\frac{\lambda}{2} \qquad (6)$$

where $n = 1, 3, 5, \ldots$

These heights are intermediate to those for condition (5), as indicated in Fig. 14-54.

If the transmitting antenna is circularly polarized (say, right-handed), then a circularly polarized receiving antenna will be responsive only to waves arriving via one path (the direct path if the receiving antenna is right-handed or the reflected path if the receiving antenna is left-handed). For circular polarization the field intensity is unity and independent of height. Although the field is half the maximum possible field for either vertical or horizontal polarization, there are no zero response heights for circular polarization (see Fig. 14-54).

Since the angle of incidence (θ_i) equals the angle of reflection (θ_r) of the ground reflected wave, triangles 1 and 2 are similar and the heights (h_2) at which the fields are zero (or doubled) are readily calculated from the geometry using (5) or (6) (see Prob. 14-5-15) with the result that the difference of the direct and reflected path lengths $(r_1 + r_2 - r_0)$ is given by $2h_1h_2/d$. Thus, the heights for maximum vertically polarized field are given by integral multiples of

$$h_2 = \frac{d\lambda}{2h_1} = \frac{4 \times 10^3 \times 1}{2 \times 200} = 10 \text{ m} \qquad (6a)$$

with heights for maximum horizontally polarized field at 5, 15, 25, ... m, as shown in Fig. 14-54.

From the Friis transmission formula, the received power via one path is found to be

$$P_r = \frac{P_t A_{et} A_{er}}{r^2 \lambda^2} = \frac{10^3 \times 0.32 \times 0.91}{16 \times 10^6 \times 1} = 18 \text{ } \mu\text{W}$$

where P_t = transmitted power = 1 kW

A_{et} = effective aperture of transmitting antenna

$$= \frac{\lambda^2}{4\pi} \frac{41,253}{360 \times 10} = 0.91 \text{ m}^2$$

A_{er} = effective aperture of receiving antenna

$$= \frac{\lambda^2}{4\pi} D = \frac{1^2 \times 4}{4\pi} = 0.32 \text{ m}^2$$

$r = 4 \times 10^3$ m

$\lambda = 1$ m

The power received via both direct and reflected paths for linear polarization is 4 times this, or 72 μW as its maximum value at certain critical heights and between 72 μW and zero at other heights.

For the circularly polarized case, the directivity of the receiving antenna

$$D = 15 \frac{nS}{\lambda} = 15 \times 6 \times 0.22 = 20$$

and

$$A_{er} = \frac{\lambda^2 D}{4\pi} = 1.6 \text{ m}^2$$

so that the received power for circular polarization is

$$P_r(CP) = \frac{1.6}{0.32} \, 18 = 90 \ \mu W$$

at all heights.

In calculating these powers we have not taken into account the fact that the actual path length is slightly more than 4 km. However, for cases where the receiving antenna height is less than that of the transmitting antenna, the difference is very small and the actual received power is within 1 percent or within 0.05 dB of the above values.

Consider now the situation in Fig. 14-53b, where a passive reflecting object, such as a flat sheet or a single passive corner reflector, constitutes part of the transmission path. For convenience let the reflecting object be regarded as a pair of antennas connected as in Fig. 14-53c. Applying the Friis formula in two steps, we obtain

$$\frac{P_r}{P_t} = \frac{A_{er} A_{er} A_e^2}{r^4 \lambda^4} \qquad \text{(dimensionless)} \qquad (7)$$

where A_e = effective aperture of reflecting object, m^2

r = distance from transmitting (or receiving) antenna to reflecting object, m

The remainder of the symbols in (7) are the same as for (4). This is a radar equation for the ratio of the received to transmitted power. If the receiving and transmitting antennas are identical so $A_{et} = A_{er} = A$, we have

$$\frac{P_r}{P_t} = \frac{A^2 A_e^2}{r^4 \lambda^4} \qquad (8)$$

The aperture A_e is the *effective scattering aperture* of the reflecting object.† If the reflecting object is a large, flat conducting sheet oriented for specular reflection, $A_e = A_p$, where A_p is the physical aperture.

For most objects the scattering aperture is much less than the cross-sectional area because the objects tend to scatter or reradiate isotropically. Making this change (i.e., assuming isotropic scattering), we get the *radar equation*

$$\boxed{\frac{P_r}{P_t} = \frac{A^2 \sigma}{4\pi \lambda^2 r^4}} \qquad (9)$$

---

† It is assumed that the angle subtended by the reflecting object is small compared to the antenna beam width.

where σ is the *radar cross section* (m^2).[†] The radar cross section (σ) is equal to the ratio of the scattered power (assumed isotropic) to the incident Poynting vector (or power density). Thus

$$\sigma = \frac{\text{scattered power}}{\text{incident power density}} = \frac{4\pi r^2 S_r}{S_{\text{inc}}} \qquad (10)$$

where S_r = scattered power density at distance r, W m^{-2}
S_{inc} = power density incident on object, W m^{-2}

For a large perfectly reflecting metal sphere of radius a, the radar cross section is equal to the physical cross section (πa^2). For imperfectly reflecting spheres, the radar cross section is smaller. For example, at meter wavelengths, the radar cross section of the moon is about 0.1 of the physical cross section.

14-24 RADIO TELESCOPES, ANTENNA TEMPERATURE, SYSTEM TEMPERATURE, REMOTE SENSING, AND RESOLUTION[‡]

The noise power per unit bandwidth available at the terminals of a resistor of resistance R and temperature T (Fig. 14-55a) is given by[§]

$$w = kT \qquad (1)$$

where w = power per unit bandwidth, W Hz^{-1}
k = Boltzmann's constant = 1.38×10^{-23} J K^{-1}
T = absolute temperature of resistor, K

The temperature T in (1) may be called the *noise temperature*.

If the resistor is replaced by a lossless resonant antenna of radiation resistance R (Fig. 14-55b), the impedance presented at the terminals is unchanged. However, the noise power will not be the same unless the antenna is receiving from a region at the temperature T.

[†] We have

P_r = (scattered power) × (fraction of sphere subtended by receiving antenna)
= (power density incident on object) × σ × (fraction of sphere subtended by receiving antenna)
= $(P_t 4\pi A/4\pi r^2 \lambda^2) \times \sigma \times (A/4\pi r^2)$
= $P_t A^2 \sigma / 4\pi \lambda^2 r^4$

We can also derive (9) from (8) as follows: Referring to (8) the aperture of the object is set equal to the radar cross section as regards incident power [$A_e(\text{inc}) = \sigma$], but it is set equal to the aperture of an isotropic antenna as regards scattered power [$A_e(\text{sc}) = \lambda^2/4\pi$]. Hence, we can put $A_e^2 = A_e(\text{inc})A_e(\text{sc}) = \sigma\lambda^2/4\pi$ in (8), which gives (9).

[‡] For a more complete treatment see Kraus, "Radio Astronomy," pp. 97–104.
[§] H. Nyquist, Thermal Agitation of Electric Charge in Conductors, *Phys. Rev.*, **32**: 110–113 (1928).

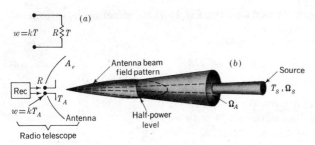

Figure 14-55 (*a*) Noise power of resistance R at room temperature T and (*b*) noise power of antenna radiation resistance R at temperature T_A due to distant regions in antenna beam.

In a radio-telescope antenna operating at centimeter wavelengths, the beam may be directed at regions of the sky which are at effective temperatures close to absolute zero (0 K, or $-273°C$). The noise temperature T_A of the antenna is equal to this sky temperature and *not* to the physical temperature of the antenna structure. Thus, for a radio-telescope antenna the noise power per unit bandwidth is given by†

$$ w = kT_A \qquad (\text{W Hz}^{-1}) \qquad (2) $$

where T_A is the *antenna* (*noise*) *temperature*, or temperature of the antenna radiation resistance, determined by the sky temperature at which the antenna beam is directed. Thus, a radio-telescope antenna (and receiver) may be regarded as a radiometer (or temperature-measuring device) for determining the temperature of distant regions of space coupled to the system through the radiation resistance of the antenna. It has been assumed that the antenna has no thermal losses and that all the antenna pattern is directed at the sky (negligible side and back lobes).

In receiving with a radio telescope it is convenient to express the received power as a (power) *flux density*. Thus, dividing the received power per unit bandwidth by the effective aperture A_e of the antenna (Fig. 14-55b) yields the flux density S, or

$$ S = \frac{w}{A_e} = \frac{kT_A}{A_e} \qquad (\text{W m}^{-2}\,\text{Hz}^{-1}) \qquad (3) $$

Note that the units for the flux density ($\text{W m}^{-2}\,\text{Hz}^{-1}$) are the same as for the Poynting vector per unit bandwidth, so that we may regard the flux density as a measure of the Poynting vector (per unit bandwidth) received from distant celestial sources of radio emission.‡ Most of these sources radiate waves which are completely unpolarized; i.e., they have polarization incoherence. Since any antenna (regardless of its polarization characteristics) can receive only half the

---

† The total power received is $kT_A B$ (W), where B = receiver bandwidth (Hz).

‡ The unit of flux density in radio-astronomy measurements is the jansky (Jy) = 10^{-26} W m^{-2} Hz^{-1}.

incident power of an unpolarized wave (see Fig. 11-7), the actual flux density should be twice that given in (3), or

$$S = \frac{2kT_A}{A_e} \tag{4}$$

where S = source flux density, $W\ m^{-2}\ Hz^{-1}$
k = Boltzmann's constant $= 1.38 \times 10^{-23}\ J\ K^{-1}$
T_A = antenna (noise) temperature, K
A_e = effective aperture of telescope antenna, m^2

The value S in (4) results from power received over the entire antenna pattern. If the celestial source extent Ω_s is small compared to the antenna-beam solid angle Ω_A and the antenna beam is aligned with the source, the observed flux density is as given in (3). Hence, it is said that this relation applies to *point sources*.

If the angular size of the source is small (compared to Ω_A) and its magnitude is known, it is possible to determine the source temperature T_s very simply from the relation

$$T_A = \frac{\Omega_S}{\Omega_A} T_S \quad \text{or} \quad T_S = \frac{\Omega_A}{\Omega_S} T_A \tag{5}$$

where Ω_A = antenna beam solid angle, sr (see Fig. 14-55b)
Ω_S = source solid angle, sr (see Fig. 14-55b)
T_A = antenna (noise) temperature, K
T_S = source temperature, K

It is important to note that the antenna temperature T_A has nothing to do with the physical temperature of the antenna provided it is lossless.

Example: Temperature of Mars The antenna temperature for the planet Mars measured with the U.S. Naval Research Laboratory† 15-m radio telescope at 31.5 mm wavelength was 0.24 K. Mars subtended an angle of 0.005° at the time of measurement, and the antenna HPBW = 0.116°. Find the temperature of Mars.

SOLUTION From (5) the temperature of Mars is given by

$$T_S = \frac{\Omega_A}{\Omega_S} T_A \approx \frac{0.116^2}{\pi(0.005^2/4)} 0.24 = 164\ K$$

This temperature is less than the infrared temperature of the sunlit side (250 K), implying that the radio emission may originate further below the Martian surface than the infrared radiation.

---

† C. H. Mayer, T. P. McCullough, and R. M. Sloanaker, Observations of Mars and Jupiter at a Wavelength of 3.15 cm, *Astrophys. J.*, **127**:11–16 (January 1958).

System temperature and remote sensing Any antenna may be regarded as a remote sensing device because its antenna temperature is an indication of the temperature of the objects in its beam. Thus, a centimeter-wavelength radio telescope antenna directed up at the empty sky may have as low a temperature as a few degrees kelvin. The minimum possible antenna temperature is 3 K, which is the residual temperature of the primordial fireball which created the universe. If any radio sources are in the beam, the temperature will be higher. At meter wavelengths radiation from our galaxy may reach temperatures of thousands of degrees (K). The temperatures we are discussing are thermal (noise) temperatures like those of a perfect emitting-absorbing object called a *blackbody*. A hot carbon block or black metal sphere filling the beam of a receiving antenna will ideally produce an antenna temperature equal to the block or sphere's thermal (thermometer-measured) temperature. However, the oscillating currents of a transmitting antenna can simulate a temperature of millions of degrees even though the transmitting antenna is at normal outdoor temperature. It is said that the transmitting antenna (and its currents) have an *equivalent blackbody (or noise) temperature* of millions of degrees.

For example, suppose that an AM broadcast signal from a transmitting antenna produces a field intensity of 10 μV m^{-1} at a receiving antenna with an effective aperture of 10 m^2 connected to a receiver of bandwidth $B = 10$ kHz. The received power is

$$W = \frac{E^2}{Z_0} A_e = \frac{10^{-10}}{377} 10 = 2.65 \times 10^{-12} \text{ W}$$

From (2) the equivalent blackbody (noise) temperature of the receiving antenna is

$$T_A = \frac{W}{kB} = \frac{2.65 \times 10^{-12}}{1.38 \times 10^{-23} \times 10^4} = 1.9 \times 10^7 \text{ K}$$

Thus, a radio signal of only 10 μV m^{-1} field intensity can produce an equivalent receiving antenna temperature of 19 million kelvin degrees. Such temperatures are so large compared to what is called the *system temperature* of a receiving system that the system temperature may be of little concern. However, radio telescopes and other remote sensing systems operate at such high sensitivity (low signal levels) that the system temperature is extremely important.

Consider the receiving system in Fig. 14-56 with an antenna of temperature T_A and receiver of equivalent (noise) temperature T_R, connected by a transmission line at a physical (thermometer) temperature T_{LP}. Looking toward the receiver from the antenna terminals, the receiver temperature T_R would be observed if the transmission line were lossless, but for a line with efficiency ϵ the the equivalent temperature observed is T_R/ϵ, where $\epsilon = 1$ for a lossless line and $\epsilon = 0$ for zero efficiency.† The transmission line at temperature T_{LP} is both an emitter and absorber and contributes an observed temperature of $T_{LP}[(1/\epsilon) - 1]$.

---

† $\epsilon = e^{-\alpha l}$, where α = attenuation constant (Np m^{-1}) and l = length of transmission line (m)

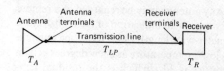

Figure 14-56 Antenna, transmission line, and receiver for system temperature determination.

The total or system temperature observed at the antenna terminals is then the sum of the above temperatures plus the antenna temperature T_A, or

$$T_{sys} = T_A + T_{LP}\left(\frac{1}{\epsilon} - 1\right) + \frac{T_R}{\epsilon} \quad \text{(K)} \quad (6)$$

Thus, for a receiving system with an antenna temperature $T_A = 50$ K, a receiver with temperature $T_R = 100$ K, a transmission line at the ambient temperature $T_{LP} = 300$ K, and a line efficiency $\epsilon = 0.9$, the system temperature†

$$T_{sys} = 50 + 300\left(\frac{1}{0.9} - 1\right) + \frac{1}{0.9}100 = 194 \text{ K}$$

The system temperature is a determining factor in the sensitivity of a radio telescope or other remote sensing system. The sensitivity, or *minimum detectable temperature* ΔT_{min}, is equal to the rms noise temperature ΔT_{rms} of the system, as given by

$$\Delta T_{min} = \frac{k' T_{sys}}{\sqrt{Bt}} = \Delta T_{rms} \quad \text{(K)} \quad (7)$$

where k' = system constant (order of unity), dimensionless
T_{sys} = system temperature (sum of antenna, line and receiver temperatures), K
ΔT_{rms} = rms noise temperature, K
B = predetection bandwidth of receiver, Hz
t = postdetection time constant of receiver, s

The *criterion of detectability* is that the incremental increase in antenna temperature ΔT_A due to a radio source be equal to or exceed ΔT_{min}, that is,

$$\Delta T_A \geq \Delta T_{min} \quad (8)$$

and the *signal-to-noise (S/N) ratio* is then

$$\frac{S}{N} = \frac{\Delta T_A}{\Delta T_{min}} \quad (9)$$

Example 1: Minimum detectable flux density The Ohio State University 110 by 21 m radio telescope has an effective aperture of 1,200 m² and a system temperature of 50 K at 1,415 MHz. The radio-frequency bandwidth is

---

†See noise temperature–noise figure conversion chart in Appendix sec. A-14.

100 MHz, the output time constant is 10 s, and the system constant is 2.2. Find the minimum detectable flux density.

SOLUTION From (7) the minimum detectable temperature is given by

$$\Delta T_{min} = \frac{k' T_{sys}}{\sqrt{Bt}} = \frac{2.2 \times 50}{\sqrt{100 \times 10^6 \times 10}} = 0.0035 \text{ K}$$

From (4) the minimum detectable flux density is

$$\Delta S_{min} = \frac{2k\Delta T_{min}}{A_e} = \frac{2 \times 1.38 \times 10^{-23} \times 0.0035}{1,200}$$

$$= 8.1 \times 10^{-29} \text{ W m}^{-2} \text{ Hz}^{-1}$$

or $\Delta S_{min} = 8.1 \text{ mJy}$

By repeating observations and averaging, the minimum can be further reduced ($\propto \sqrt{1/n}$, where n = number of observations). In an all-sky survey at 1,415 MHz with the Ohio State University radio telescope, shown in Fig. 14-57, about 20,000 radio sources were detected at flux densities above 0.18 Jy.

Example 2: Effect of Ground pickup on antenna temperature A radio-telescope antenna of 500 m² effective aperature is directed at the zenith. Calculate the antenna temperature, assuming that the sky temperature is uniform and equal to 10 K. Take the ground temperature equal to 300 K and assume that half the minor-lobe solid beam angle is in the back direction. The wavelength is 200 mm and the beam efficiency is 0.7.

SOLUTION Referring to Fig. 14-2a, consider that the xy plane represents the ground and that the main lobe is pointing straight up at the zenith (z direction) with sky temperature = 10 K. The side lobes are also directed at the sky but the back lobes are directed into the ground at a temperature of 300 K. This ground temperature of 300 K ($=27°C = 80°F$) is a typical summer-day outdoor temperature.

The beam efficiency of an antenna is the ratio of solid angle Ω_M in the main beam (or lobe) to the total beam solid angle Ω_A. Thus,

$$\text{Beam efficiency} = \frac{\Omega_M}{\Omega_A} = 0.7 \qquad (10)$$

so $\Omega_M = 0.7\Omega_A$ and the minor-lobe solid angle $\Omega_m = 0.3 \ \Omega_A$. A more general expression for the antenna temperature T_A than (5) is

$$T_A = \frac{1}{\Omega_A} \iint T(\theta, \phi) P_n(\theta, \phi) \, d\Omega \qquad (11)$$

where T = temperature of sky (or ground)

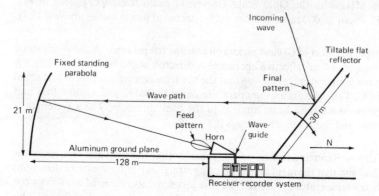

Figure 14-57 The Ohio State University radio telescope, shown in the air view and cross section, consists of a 110 m × 21 m section of a paraboloid of revolution (at left) with a 104 m × 30 m tiltable flat reflector (at right). The two reflectors are joined by a 15,000-m² aluminum ground plane which acts as a guiding surface. Incoming radio waves are reflected by the tiltable flat reflector into the paraboloid, which in turn focuses the waves into a cluster of horn antennas. Rectangular waveguides operating in the TE_{10} mode then convey the received energy to low-noise-temperature receivers for 1,415 and 2,650 MHz (21- and 11-cm λ).

At 2,650 MHz the antenna gain is over 60 dB with beamwidths of 0.07° in right ascension (east-west coordinate) and 1/3° in declination (north-south coordinate). The telescope is a survey instrument. For a given flat reflector setting a 1/6° zone in declination is scanned through 360° in right ascension each day with the earth's rotation. By moving the flat reflector to a slightly different setting, adjacent declination zones are scanned on successive days, and in this way the entire sky is surveyed.

With this telescope 20,000 celestial radio objects were located, mapped and catalogued. Among them are the most distant known objects in the universe (specifically OH471 and OQ172) at distances of 15 billion light-years or more.

Integrating in three steps, we have

$$\text{Sky contribution} = 10 \times 0.7\Omega_A$$
$$\text{Side-lobe contribution} = 10 \times \tfrac{1}{2} \times 0.3\Omega_A$$
$$\text{Back-lobe contribution} = 300 \times \tfrac{1}{2} \times 0.3\Omega_A$$

and
$$T_A = \frac{1}{\Omega_A}(10 \times 0.7 + 10 \times \tfrac{1}{2} \times 0.3 + 300 \times \tfrac{1}{2} \times 0.3)\Omega_A$$

$$= 7 + 1.5 + 45 = 53.5 \text{ K}$$

Note that 45 of the 53.5 K, or 84 percent, of the antenna temperature results from the back-lobe pickup from the ground. With no back lobes the antenna temperature could ideally be only 10 K, so that in this example the back lobes are very detrimental to the telescope sensitivity. It is for this reason that radio-telescope antennas are usually designed to reduce back- and side-lobe response to a minimum. Antennas for receiving Clarke-orbit communication satellites have identical requirements.

The effective aperture and wavelength information in the example is superfluous to the problem but the values are consistent with those of a typical radio-telescope antenna.

The concept of remote sensing is particularly applicable to radio telescopes whether they are ground-based and pointed at the sky for observing celestial objects or on an aircraft or satellite and pointed at the surface of the earth.

Consider the radio astronomy situation of Fig. 14-58a, in which the antenna beam is completely subtended by a celestial source of temperature T_s with an intervening absorbing-emitting cloud of temperature T_c. With no cloud present, T_A equals T_s, but with the cloud it may be shown that the observed antenna temperature

$$T_A = T_c(1 - e^{-\tau_c}) + T_s e^{-\tau_c} \quad \text{(K)} \tag{12}$$

where τ_c = absorption coefficient of cloud ($=0$ for no absorption and $=\infty$ for infinite absorption). Thus, knowing T_s and τ_c, the cloud's equivalent blackbody temperature can be determined.

Now, referring to Fig. 14-58b, let us reverse the situation and use the radio telescope to observe the earth (at temperature T_e) from a satellite in space with the antenna beam completely subtended by a large forest at a temperature T_f. The satellite antenna temperature is then

$$T_A = T_f(1 - e^{-\tau_f}) + T_e e^{-\tau_f} \quad \text{(K)} \tag{13}$$

where τ_f = absorption coefficient of the forest. Knowing T_e and τ_f, the temperature of the forest can be determined, or knowing T_e and T_f, the absorption coefficient can be deduced. It is by such a technique that the whole earth can be surveyed and much information obtained about temperatures of land and water areas and from the absorption coefficients about the nature of the surface cover.

If the antenna-line-receiver combination is viewed from the receiver terminals, as in Fig. 14-58c, and compared with Figs, 14-58a and b, we note that the analogy between space links and transmission lines as discussed in Chap. 10 extends here to the concept of temperature. Thus, the emitting-absorbing line and antenna of (c) are like the astronomical path of (a) with emitting-absorbing interstellar cloud and

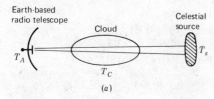

(a)

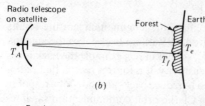

(b)

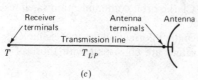

(c)

Figure 14-58 (a) Earth-based radio telescope remote-sensing celestial source through intervening interstellar cloud. (b) Radio telescope on satellite remote-sensing the earth through forest. (c) Receiver detecting antenna output through transmission line. The cloud, forest, and transmission line have analogous emitting-absorbing properties.

celestial source, or the remote earth-sensing path of (b) with the emitting-absorbing forest and the earth. The analogy may be emphasized by comparing (12) and (13) with (6) (for the antenna-receiver), rewritten for the temperature as seen from the receiver terminals so that the equations for the three situations have the identical form as follows:

Antenna looking at celestial source
$$T_A = T_c(1 - e^{-\tau_c}) + T_s e^{-\tau_c} \quad \text{[Fig. 14-58(a)]} \quad (14)$$

Antenna looking at earth
$$T_A = T_f(1 - e^{-\tau_f}) + T_e e^{-\tau_f} \quad \text{[Fig. 14-58(b)]} \quad (15)$$

Receiver looking at antenna
$$T = T_{LP}(1 - e^{-\alpha l}) + T_A e^{-\alpha l} \quad \text{[Fig. 14-58(c)]} \quad (16)$$

where T_A = antenna temperature, K
T_c = cloud temperature, K
τ_c = cloud absorption coefficient, dimensionless
T_s = celestial source temperature, K
T_f = forest temperature, K
τ_f = forest absorption coefficient, dimensionless
T_e = temperature of earth, K
T_{LP} = transmission line physical temperature, K
α = transmission line attenuation constant, Np m^{-1}
l = length of transmission line, m

Note that *system* temperature should be referred to the antenna terminals as in (6) and not to the receiver terminals as in (16). Thus, if the line is completely lossy ($e^{-\alpha l} = \epsilon = 0$), (6) gives an infinite system temperature which is correct, meaning that the system has no sensitivity whatever. But with this condition, a system temperature viewed from the receiver terminals, as in (16), would equal the temperature T_{LP} of the line plus the receiver temperature, a completely misleading result since it indicates that the system still has sensitivity.

Resolution The resolution of a radio-telescope antenna is inversely proportional to its beam width, the narrower the beam width the greater the ability of the telescope to distinguish between two closely-spaced celestial objects.

Example 1: Resolution of eye and radio telescope The pupil diameter of the human eye is about 3 mm. What is the diameter required for a circular-dish radio telescope antenna to provide the same resolution at 2-cm wavelength as the eye does at light wavelengths?

SOLUTION The Rayleigh resolution, or the ability to distinguish between two closely-spaced objects, is given by BWFN/2. Taking the light $\lambda = \frac{1}{2}\,\mu m$, the resolution of the eye from (14-6.5-4) is approximately

$$\frac{\text{BWFN}}{2} = \frac{57.3°}{L/\lambda} = \frac{57.3°}{3\,\text{mm}/\frac{1}{2}\,\mu m} = 0.01°$$

meaning that the eye can just distinguish between two objects separated by 1 m (like a pair of automobile headlights) at a distance of 5 km.

For the same resolution at 2-cm wavelength

$$L = \frac{57.3°}{\text{BWFN}/2}\,\lambda = \frac{57.3°}{0.01°}\,0.02 = 115\,\text{m}$$

Thus, to match the resolving power of the eye requires a radio telescope 115 m in diameter operating at a wavelength of 2 cm. At longer wavelengths the diameter would need to be proportionately greater.

Note: (14-6.5-4) applies to a rectangular aperture. The pupil and the dish are circular apertures and a better angle for the resolution would be 60 or 65 instead of 57.3°. This difference, however, does not affect this example since it involves only a ratio.

Example 2: Resolution and directivity How many objects can a radio telescope with directivity D resolve?

SOLUTION For most antennas the Rayleigh resolution ($=$ BWFN/2) is approximately equal to the HPBW. For a symmetrical single-lobe pattern the solid beam angle $\Omega_A \simeq$ HPBW². Hence, the solid beam angle Ω_A is approximately equal to the solid angle for which an object can be just resolved, and since the whole sky contains 4π sr, the number of objects N_r a radio telescope can resolve in the entire sky is given by†

$$N_r = \frac{4\pi}{\Omega_A} = D \qquad \text{(dimensionless)} \qquad (17)$$

---

† N_r is, of course, an idealization which assumes that the sources are smaller than the beam width of the telescope, are uniformly distributed, etc.

Thus, the directivity of an antenna may be regarded as having the significance that it is a measure of the number of objects the antenna can resolve in the whole sky.

14-25 TABLE OF ANTENNA AND ANTENNA SYSTEMS RELATIONS

Directivity, $D = \dfrac{4\pi A_e}{\lambda^2} = \dfrac{4\pi}{\Omega_A} = \dfrac{P(\theta, \phi)_{max}}{P_{av}}$ (dimensionless)

Directivity (approx), $D = \dfrac{4\pi \, (\text{sr})}{\theta_{HP}\phi_{HP}(\text{sr})} = \dfrac{41{,}253 \, (\text{deg}^2)}{\theta^\circ_{HP} \phi^\circ_{HP}}$

Directivity of short dipole, $D = 1.5$ (1.76 dBi)

Directivity of $\lambda/2$ dipole, $D = 1.64$ (2.15 dBi)

Directivity of helical beam antenna, $D = 15\left(\dfrac{C}{\lambda}\right)^2 \dfrac{nS}{\lambda}$ (dimensionless)

Gain, $G = kD$ (dimensionless)

Effective aperture, $A_e = \lambda^2/\Omega_A$ (m^2)

Aperture efficiency, $\epsilon_{ap} = \dfrac{A_e}{A_p} = \dfrac{E^2_{av}}{(E^2)_{av}}$ (dimensionless)

Beam solid angle, $\Omega_A = \displaystyle\iint\limits_{4\pi} P_n(\theta, \phi)\, d\Omega$ (sr)

Beam solid angle (approx.), $\Omega_A = \theta_{HP}\phi_{HP}(\text{sr}) = \theta^\circ_{HP}\phi^\circ_{HP}(\text{deg}^2)$

Beam width (long broadside array), HPBW $\approx \dfrac{\lambda}{L/\lambda}$ (rad) $= \dfrac{57.3^\circ}{L/\lambda}$

Radiation resistance, $R_r = \dfrac{E^2 r^2 \Omega_A}{Z_0 I^2}$ (Ω)

Radiation resistance of short dipole, $R_r = 80\pi^2 \left(\dfrac{l}{\lambda}\right)^2 \left(\dfrac{I_{av}}{I_0}\right)^2$ (Ω)

Self-impedance of $\lambda/2$ dipole, $R_r + jX = 73 + j42.5\ \Omega$

Slot-dipole impedance relation, $Z_s = \dfrac{Z_0^2}{4Z_d}$ (Ω)

Array factor (n sources of equal amplitude and spacing),

$$E_n = \frac{1}{n}\frac{\sin{(n\psi/2)}}{\sin{(\psi/2)}}\quad \text{(dimensionless)}$$

where $\psi = \beta d \cos\theta + \delta$ (rad or deg)

Continued

Friis formula, $\dfrac{P_r}{P_t} = \dfrac{A_{er} A_{et}}{r^2 \lambda^2}$ (dimensionless)

Radar equation, $\dfrac{P_r}{P_t} = \dfrac{A^2 \sigma}{4\pi \lambda^2 r^4}$ (dimensionless)

Nyquist power, $w = kT$ (W Hz^{-1})

Flux density, $S = \dfrac{2kT_A}{A_e}$ (W m^{-2} Hz^{-1})

Minimum detectable flux density, $\Delta S_{min} = \dfrac{2k\,\Delta T_{min}}{A_e}$ (W m^{-2} Hz^{-1})

System temperature, $T_{sys} = T_A + T_{LP}\left(\dfrac{1}{\ell} - 1\right) + \dfrac{T_R}{\ell}$ (K)

Minimum detectable temperature, $\Delta T_{min} = \dfrac{k' T_{sys}}{\sqrt{Bt}} = \Delta T_{rms}$ (K)

Antenna temperature (through emitting-absorbing cloud),

$$T_A = T_c(1 - e^{-\tau_c}) + T_s e^{-\tau_c}   \quad (K)$$

Signal-to-noise ratio, $\dfrac{S}{N} = \dfrac{\Delta T_A}{\Delta T_{min}}$ (dimensionless)

Resolution angle $\simeq \dfrac{\text{BWFN}}{2}$

Number of resolvable objects, $N_r = \dfrac{4\pi}{\Omega_A} = D$

PROBLEMS†

Group 14-1: Secs. 14-1 through 14-5. Introduction to antenna parameters, retarded potentials, short dipole, radiation resistance, aperture, directivity, and gain

★**14-1-1. Isotropic antenna. Resistance** An omnidirectional (isotropic) antenna has a field pattern given by $E = 10\ I/r$ (V m^{-1}), where I = terminal current (A) and r = distance (m). Find the radiation resistance.

14-1-2. Short dipole For a thin center-fed dipole $\lambda/15$ long find (a) directivity D, (b) gain G, (c) effective aperture A_e, (d) beam solid angle Ω_A, and (e) radiation resistance R_r. The antenna tapers linearly from its values at the terminals to zero at its ends. The loss resistance is 1 Ω.

★**14-1-3. Conical pattern** An antenna has a conical field pattern with uniform field for zenith angles (θ) from 0 to 60° and zero field from 60 to 180°. Find exactly (a) beam solid angle and (b) directivity. The pattern is independent of the azimuth angle (ϕ).

★**14-1-4. Conical pattern** An antenna has a conical field pattern with uniform field for zenith angles (θ) from 0 to 45° and zero field from 45 to 180°. Find exactly (a) beam solid angle (b) directivity and (c) effective aperture. (d) Find radiation resistance if the field $E = 5$ V m^{-1} at a distance of 50 m for a terminal current $I = 2$ A (rms). The pattern is independent of the azimuth angle (ϕ).

† Answers to starred problems are given in Appendix C.

★**14-1-5. Directional pattern in θ and ϕ** An antenna has a uniform field pattern for zenith angles (θ) between 45 and 90° and for azimuth (ϕ) angles between 0 and 120°. If $E = 3$ V m^{-1} at a distance of 500 m from the antenna and the terminal current is 5 A, find the radiation resistance of the antenna. $E = 0$ except within the angles given above.

14-1-6. Directional pattern in θ and ϕ An antenna has a uniform field $E = 2$ V m^{-1} (rms) at a distance of 100 m for zenith angles between 30 and 60° and azimuth angles ϕ between 0 and 90° with $E = 0$ elsewhere. The antenna terminal current is 3 A (rms). Find (a) directivity, (b) effective aperture, and (c) radiation resistance.

★**14-1-7. Directional pattern with back lobe** The field pattern of an antenna varies with zenith angle (θ) as follows: E_n ($= E_{normalized}$) $= 1$ between $\theta = 0°$ and $\theta = 30°$ (main lobe), $E_n = 0$ between $\theta = 30°$ and $\theta = 90°$, and $E_n = \frac{1}{3}$ between $\theta = 90°$ and $\theta = 180°$ (back lobe). The pattern is independent of azimuth angle (ϕ). (a) Find the exact directivity. (b) If the field equals 8 V m^{-1} (rms) for $\theta = 0°$ at a distance of 200 m for a terminal current $I = 4$ A (rms), find the radiation resistance.

14-1-8. Short dipole The radiated field of a short-dipole antenna with uniform current is given by $|E| = 30\beta l(I/r) \sin\theta$, where $l = $ length, $I = $ current, $r = $ distance, and $\theta = $ pattern angle. Find the radiation resistance.

Group 14-2: Secs. 14-6 through 14-8. Array theory, continuous apertures, and Fourier transform relations
14-2-1. Broadside arrays. $\lambda/2$ spacing The following BASIC programs provide antenna field pattern graphs in polar and rectangular coordinates for an array of 4 sources as illustrated in Fig. P14-2-1. Using modifications of these programs, produce graphs of the patterns for a larger number of in-phase sources spaced $\lambda/2$ apart, such as (a) 6 sources, (b) 8 sources, and (c) 12 sources.

```
BROADSIDE ARRAY N = 4      d = λ/2
POLAR PLOT
10 HOME
20 HGR
30 HCOLOR=3
40 FOR A = .02 TO 6.26 STEP .01
50 R=15 * SIN(6.28 * COS(A))/SIN(1.57 * COS(A))
60 HPLOT 138 + R * COS(A),79 + R * SIN(A)
70 NEXT A
RECTANGULAR PLOT (POLAR PLOT STEPS 10 60 70 OMITTED)
60 HPLOT A * 30,R + 75
70 NEXT A
NOTE THAT SINGULAR POINTS FOR WHICH DENOMINATOR IN 50 IS ZERO ARE AVOIDED
BY SUITABLE SELECTION OF LIMITS IN 40
```

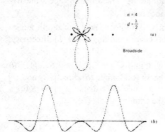

Figure P14-2-1 Field pattern of broadside array of four sources in polar and rectangular coordinates.

14-2-2. End-fire arrays. $\lambda/2$ spacing The following BASIC programs provide antenna field pattern graphs in polar and rectangular coordinates for an array of 4 sources as illustrated in Fig. P14-2-2. Using modifications of these programs, produce graphs of the patterns for a larger number of sources spaced $\lambda/2$ apart with end-fire phasing, such as (a) 5 sources, (b) 6 sources, (c) 8 sources, and (d) 12 sources.

END-FIRE ARRAY N = 4   d = $\lambda/2$

POLAR PLOT

```
10 HOME
20 HGR
30 HCOLOR=3
40 FOR A = .02 TO 3.12 STEP .01
50 R=15 * SIN(6.28 * (COS(A) - 1))/SIN(1.57 * (COS(A) - 1))
60 HPLOT 138 + R * COS(A),79 + R * SIN(A)
61 HPLOT 138 + R * (-COS(A)),79 + R * SIN(A)
70 NEXT A
```

RECTANGULAR PLOT (POLAR PLOT STEPS 10 60 61 70 OMITTED)

```
60 HPLOT A * 30,R + 75
61 HPLOT (A+3.16) * 30,-R + 75
70 NEXT A
```

NOTE THAT SINGULAR POINTS FOR WHICH DENOMINATOR IN 50 IS ZERO ARE AVOIDED BY SUITABLE SELECTION OF LIMITS IN 40

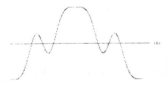

Figure P14-2-2 Field pattern of end-fire array of four sources in polar and rectangular coordinates.

14-2-3. End-fire arrays. $\lambda/4$ spacing Repeat Prob. 14-2-2 for the case where the spacing is $\lambda/4$ instead of $\lambda/2$. The patterns in this case will be unidirectional instead of bidirectional as with $\lambda/2$ spacing.

14-2-4. Two-element interferometer Using a computer as in the above problems, produce graphs of the field patterns of two-isotropic in-phase sources with spacings of (a) 8λ, (b) 16λ, and (c) 32λ.

14-2-5. Two sources in phase quadrature Two isotropic point sources of equal amplitude but in phase quadrature are spaced $\lambda/4$ apart. Plot a graph of the field pattern which will be a cardioid (heart shaped). A computer solution is suggested.

★14-2-6. Two sources in phase Two isotropic point sources of equal amplitude and same phase are spaced 2λ apart. (a) Plot a graph of the field pattern. (b) Tabulate the angles for maxima and nulls.

14-2-7. Two sources in opposite phase Two isotropic sources of equal amplitude and opposite phase have 1.5λ spacing. Find the angles for all maxima and nulls.

★14-2-8. Seven short dipoles. 4-dB angle A linear broadside (in-phase) array of 7 short dipoles has a separation of 0.35λ between dipoles. Find the angle from the maximum field for which the field is 4 dB down (to nearest $0.1°$).

14-2-9. Three unequal sources Three isotropic in-line sources have $\lambda/4$ spacing. The middle source has 3 times the current of the end sources. If the phase of the middle source is $0°$, the phase of one end source is $+90°$, and if the other end source is $-90°$, make a graph of the normalized field pattern.

14-2-10. Square array Four identical short dipoles (perpendicular to page) are arranged at the corners of a square $\lambda/2$ on a side. The upper left and lower right dipoles are in the same phase while the two dipoles at the other corners are in the opposite phase. If the direction to the right (x direction) corresponds to $\phi = 0°$, find the angles ϕ for all maxima and minima of the field pattern in the plane of the page.

14-2-11. Long broadside array Show that the HPBW of a long uniform broadside array is given (without approximation) by $50.8°/L_\lambda$, where $L_\lambda = L/\lambda$ = length of array in wavelengths.

★**14-2-12. Rectangular aperture. Cosine taper** An antenna with rectangular aperture $x_1 y_1$ has a uniform field in the y direction and a cosine field distribution in the x direction (zero at edges, maximum at center). If $x_1 = 16\lambda$ and $y_1 = 8\lambda$, calculate (a) aperture efficiency and (b) directivity.

14-2-13. Rectangular aperture. Cosine tapers Repeat Prob. 14-2-12 for the case where the aperture field has a cosine distribution in both x and y directions.

★**14-2-14. 20λ line source. Cosine-square taper** (a) Calculate and plot the far-field pattern of a continuous in-phase line source 20λ long with cosine-squared field distribution. (b) What is the HPBW?

14-2-15. Triangle array Three isotropic point sources of equal amplitude are arranged at the corners of an equilateral triangle, as in Fig. P14-2-15. If all sources are in phase, determine and plot the far-field pattern.

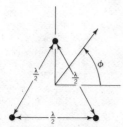

Figure P14-2-15 Three sources in triangle.

14-2-16. Square array Four isotropic point sources of equal amplitude are arranged at the corners of a square, as in Fig. P14-2-16. If the phases are as indicated by the arrows, determine and plot the far-field pattern. A graphical solution may be used for this problem and Prob. 14-2-15 (see Fig. 14-21 and Table 14-3).

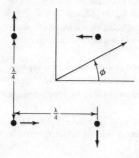

Figure P14-2-16 Four sources in square.

14-2-17. Interferometer. Pattern multiplication An interferometer antenna consists of two square broadside in-phase apertures with uniform field distribution. (*a*) If the apertures are 10 λ square and are separated 60λ on centers, calculate and plot the far-field pattern to the first null of the aperture pattern. (*b*) How many lobes are contained between first nulls of the aperture pattern? (*c*) What is the effective aperture? (*d*) What is the HPBW of the central interferometer lobe? (*e*) How does this HPBW compare with the HPBW for the central lobe of two isotropic in-phase point sources separated 60λ?

Group 14-3: Secs. 14-9 through 14-16. Linear antennas, $\lambda/2$ dipole, traveling-wave antennas, loops, helices, scanning arrays, and frequency-independent antennas

14-3-1. Collinear array of three $\lambda/2$ dipoles An antenna array consists of three in-phase collinear thin $\lambda/2$ dipole antennas each with sinusoidal current distribution and spaced $\lambda/2$ apart. The current in the center dipole is twice the current in the end dipoles (binomial array). (*a*) Calculate and plot the far-field pattern. (*b*) What is the HPBW? (*c*) How does this HPBW value compare with the HPBW for a binomial array of three isotropic point sources spaced $\lambda/2$ apart?

14-3-2. Terminated V. Traveling wave (*a*) Calculate and plot the far-field pattern of a terminated-V antenna with 5λ legs and 45° included angle. (*b*) What is the HPBW?

14-3-3. Square loop A square loop antenna is 1λ on a side. If the current is uniform and in phase around the loop, calculate and plot the far-field pattern in a plane normal to the plane of the loop and parallel to one side.

14-3-4. Circular loop A circular loop antenna with uniform in-phase current has a diameter D. Find (*a*) far-field pattern (calculate and plot), (*b*) radiation resistance, and (*c*) directivity for the following three cases: (1) $D = \lambda/3$, (2) $D = 0.75\lambda$, and (3) $D = 2\lambda$.

14-3-5. Small-loop resistance (*a*) Using a Poynting vector integration, show that the radiation resistance of a small loop is equal to $320\pi^4(A/\lambda)^2\,\Omega$, where A = area of loop. (*b*) Show that the effective aperture of an isotropic antenna equals $\lambda^2/4\pi$.

14-3-6. 10-turn helix A conductor is wound into a right-handed helix of 10 turns with diameter of 100 mm and turn spacing of 20 mm. Calculate and plot the far-field pattern and describe the polarization state if the helix is fed with radio-frequency power at a frequency of (*a*) 3 MHz and (*b*) 1 GHz.

14-3-7. 30-turn helix A helical-beam antenna has 30 turns, $\lambda/3$ diameter, and $\lambda/5$ turn spacing. The helix is right-handed. (*a*) Calculate and plot the far-field pattern. (*b*) What is the HPBW? (*c*) What is the directivity? (*d*) What is the polarization state?

14-3-8. Helices. Left and right Two identical helical-beam antennas, one left-handed and the other right-handed, are arranged as in Fig. P14-3-8. What is the polarization of the radiation to the right if the two helices are fed (*a*) in phase and (*b*) in opposite phase?

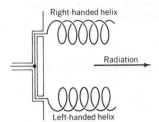

Figure P14-3-8 Left- and right-handed helices.

$\star$**14-3-9. Crossed dipoles for CP and other states** Two $\lambda/2$ dipoles are crossed at 90°. If the two dipoles are fed with equal currents, what is the polarization of the radiation perpendicular to the plane of the dipoles if the currents are (*a*) in phase, (*b*) phase quadrature (90° difference in phase), and (*c*) phase octature (45° difference in phase)?

14-3-10. Eight-source scanning array A linear broadside array has eight sources of equal amplitude and $\lambda/2$ spacing. Find the progressive phase shift required to swing the beam (a) 5°, (b) 10°, and (c) 15° from the broadside direction. (d) Find BWFN when all sources are in phase.

14-3-11. 24-dipole scanning array A linear array consists of an in-line configuration of 24 $\lambda/2$ dipoles spaced $\lambda/2$. The dipoles are fed with equal currents but with an arbitrary progressive phase shift δ between dipoles. What value of δ is required to put the main-lobe maximum (a) perpendicular to the line of the array (broadside condition), (b) 25° from broadside, (c) 50° from broadside, and (d) 75° from broadside? (e) Calculate and plot the four field patterns in polar coordinates. (f) Discuss the feasibility of this arrangement for a scanning array by changing feed-line lengths to change δ or by keeping the array physically unchanged but changing the frequency. What practical limits occur in both cases?

14-3-12. Three-helix scanning array Three identical right-handed helical-beam antennas spaced 1.5λ apart are arranged in a broadside array as in Fig. P14-3-12. (a) If the outer two helices rotate on their axes in opposite directions while the center helix is fixed, determine the angle ϕ of the main lobe with respect to the broadside direction and describe how ϕ varies as the helices rotate. (b) What is the maximum scan angle ϕ? (c) What is the main-lobe HPBW as a function of ϕ?

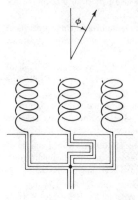

Figure P14-3-12 Three-helix array for beam scanning.

★**14-3-13. Four-tower broadcast array** A broadcast array has four identical vertical towers arranged in an east-west line with a spacing d and progressive phase shift δ. Find (a) d and (b) δ so that there is a maximum field at $\phi = 45°$ (northeast) and a null at $\phi = 90°$ (north). There can be other nulls and maxima, but no maximum can exceed the one at 45°. The distance d must be less than $\lambda/2$.

14-3-14. Log-periodic antenna Design a log-periodic antenna of the type of Fig. 14-35 to operate at frequencies between 50 and 250 MHz. Make a drawing with dimensions in meters.

Group 14-4: Secs. 14-17 through 14-22. Reciprocity, self-impedance, mutual impedance, lenses, reflectors, directors, diffraction, slots, horns, and aperture concept

14-4-1. Two short dipoles. Impedance Two short-dipole antennas $\lambda/10$ long with 1-Ω loss resistance are oriented parallel to each other (side by side) and situated 1λ apart. Find (a) the mutual impedance and (b) the terminal impedance.

★**14-4-2. Matched feed for $\lambda/2$ dipoles** Two parallel center-fed $\lambda/2$ dipoles spaced $\lambda/2$ apart are to be fed equal currents in phase opposition from a 400-Ω transmission line. If each antenna has a terminal impedance of $25 + j0\ \Omega$, find (a) length and (b) impedance of the matching sections needed between the transmission line and the dipoles in order that the transmission line is matched. (c) If the impedance of one of the dipoles alone (isolated) is $35 + j0\ \Omega$, find the gain of the array in decibels above isotropic.

14-4-3. Horizontal dipole above ground A thin $\lambda/2$ dipole is parallel to a flat, perfectly conducting ground and a distance s above it, as in Fig. P14-4-3. (a) Calculate and plot the gain of the dipole above ground with respect to a single $\lambda/2$ dipole in free space for s values from zero to $\lambda/2$. Confine consideration to the zenith direction (perpendicular to ground). Assume zero losses. (b) Repeat (a) for dipole loss resistance $R_L = 1\ \Omega$.

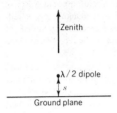

Ground plane   **Figure P14-4-3** Dipole above ground.

14-4-4. Reflecting lens A dielectric lens of index η is backed by a perfectly conducting sheet as in Fig. P14-4-4. Show that this lens-reflector will bring incoming rays to a focus at F if the distance R from the focus to a point on the lens is given by

$$R = \frac{(\eta - 1)2L}{(2\eta - 1)\cos\theta - 1}$$

where $L =$ focal distance (see Fig. P14-4-4)

Note that since a ray traverses this lens twice, the lens thickness is about half that of an ordinary one-way transmission lens. See J. D. Kraus, Some Unique Reflector-Type Antennas, *IEEE Ants. and Prop. Society Newsletter*, **24**(2): 10–12, April 1982.

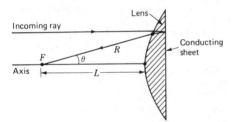

Figure P14-4-4 Reflecting lens.

14-4-5. Parabolic reflector field (a) Show that the variation of field across the aperture of a paraboloidal reflector with isotropic source is proportional to $1/[1 + (r/2L)^2]$, where $r =$ radial distance from axis of paraboloid and $L =$ focal distance. (b) If the parabola extends to the focal plane and the feed is isotropic over the hemisphere subtended by the parabola, calculate the aperture efficiency.

14-4-6. Corner reflector. $\lambda/4$ to driven element A square-corner reflector has a spacing of $\lambda/4$ between the driven $\lambda/2$ element and the corner. Show that the directivity $D = 12.5\ \text{dB}$.

14-4-7. Corner reflector. $\lambda/2$ to driven element A square-corner reflector has a driven $\lambda/2$ element $\lambda/2$ from the corner. (a) Calculate and plot the far-field pattern in both principal planes. (b) What are the HPBWs in the two principal planes? (c) What is the terminal impedance of the driven element? (d) Calculate the directivity two ways: (1) from impedances of driven and image dipoles and (2) from HPBWs, and compare. Assume perfectly conducting sheet reflectors of infinite extent.

14-4-8. Mutual impedance for large separation Show that the maximum mutual impedance Z_m of two antennas separated by a large distance is $Z_m = (\sqrt{D_1 D_2}\sqrt{R_1 R_2})/2\pi r_\lambda$, where D_1 = directivity of antenna 1, D_2 = directivity of antenna 2, R_1 = radiation resistance of antenna 1, R_2 = radiation resistance of antenna 2, and $r_\lambda = r/\lambda$ = separation of antennas (wavelengths) (see Fig. 14-36). It is assumed that the receiving antenna is terminated for maximum power transfer.

14-4-9. Terminal impedance (a) Show by means of an equivalent network that at the terminals of a receiving antenna (see Fig. 14-36) the equivalent, or Thévenin, generator has an impedance $Z_{22} - (Z_m^2/Z_{11})$ and an emf $\mathcal{V}_1 Z_m/Z_{11}$, where Z_{11} = self-impedance of transmitting antenna, Z_{22} = self-impedance of receiving antenna, Z_m = mutual impedance, and $\mathcal{V}_1$ = emf applied to terminals of transmitting antenna. (b) What load impedance connected to the terminals of the receiving antenna results in the maximum power transfer?

14-4-10. Corner retroreflector Consider the three mutually perpendicular rectangular coordinate planes for which $x = 0$, $y = 0$, and $z = 0$ and their intersection at the origin. If three flat conducting sheets are situated in these planes, a three-dimensional square-corner reflector is formed. The arrangement may be regarded as a cluster of eight back-to-back corner reflectors each covering one octant ($4\pi/8$ sr). To distinguish this reflector from a corner reflector with driven element it may be called a passive corner reflector. (a) Show that a plane wave incident on this passive reflector from any direction is reflected back in the same direction (producing retroreflection). (b) Show that if the dimensions of this *retroreflector* are large compared to the wavelength, the effective scattering aperture equals the physical aperture. These reflectors are often used to provide a large radar return. Thus, a small boat carrying such a reflector can increase its radar cross section or echo area, making its presence more prominent on radar screens.

14-4-11. $\lambda/2$ slots Two $\lambda/2$-slot antennas, as in Fig. 14-51a, are arranged end to end in a large conducting sheet with a spacing of 1λ between centers. If the slots are fed with equal in-phase voltages, calculate and plot the far-field pattern in the two principal planes. Note that the H plane coincides with the line of the slots.

★14-4-12. Boxed slot The complementary dipole of a slot antenna has a terminal impedance $Z = 90 + j10 \, \Omega$. If the slot antenna is boxed in so that it radiates only in one half-space, what is the terminal impedance of the slot antenna? The box adds no shunt susceptance at the terminals.

★14-4-13. Pyramidal horn (a) Determine the length l, width w, and half-angles in E and H planes for a pyramidal electromagnetic horn for which the mouth height $h = 8\lambda$. The horn is fed with a rectangular waveguide with TE_{10} mode, as in Fig. 14-52. Take $\delta = \lambda/10$ in the E plane and $\delta = \lambda/4$ in the H plane. (b) What are the HPBWs in both E and H planes? (c) What is the directivity? (d) What is the aperture efficiency?

14-4-14. Self-resistance and mutual resistance Explain why the mutual *resistance* of two antennas can be both positive and negative but the self-resistance of a single antenna can only be positive.

14-4-15. Corrugated horn gain Referring to Prob. 13-5-2, calculate (a) HPBW and (b) gain G of the corrugated horn, noting that the corrugations reduce the field to a low value at the teeth. (c) Compare these values with those if the horn section is removed at the transition so that the opening is 1.4λ in diameter.

Group 14-5: Secs. 14-23 through 14-24. Friis formula, radar equation, radio telescope, temperature, remote sensing, and resolution

★14-5-1. Spacecraft link over 100 Mm Two spacecraft are separated by 100 Mm. Each has an antenna with $D = 1,000$ operating at 2.5 GHz. If craft A's receiver requires 1 pW for a 20-dB signal-to-noise ratio, what transmitter power is required on craft B to achieve this signal-to-noise ratio?

14-5-2. Spacecraft link over 3 Mm Two spacecraft are separated by 3 Mm. Each has an antenna with $D = 200$ operating at 2 GHz. If craft A's receiver requires 1 pW for a 20-dB signal-to-noise ratio, what transmitter power is required on craft B to achieve this signal-to-noise ratio?

★14-5-3. Radar detection A radar receiver has a sensitivity of 10^{-12} W. If the radar antenna effective aperture is 1 m^2 and the wavelength is 10 cm, find the transmitter power required to detect an object with 5-m^2 radar cross section at a distance of 1 km.

★14-5-4. Radio telescope parameters The Ohio State University radio telescope operates at 2,650 MHz with the following parameters: system temperature 150 K, predetection bandwidth 100 MHz, postdetection time constant 5 s, system constant 2.2, and effective aperture of antenna 800 m^2. Find (a) minimum detectable temperature and (b) minimum detectable flux density. (c) If four records are averaged, what change results in (a) and (b)?

14-5-5. Remote-sensing antenna Design a 30-GHz antenna for an earth-resources remote-sensing satellite to measure earth-surface temperatures with 1-km^2 resolution from a 300-km orbital height.

14-5-6. Jupiter signals Flux densities of 10^{-20} W m^{-2} Hz^{-1} are commonly received from Jupiter at 20 MHz. What is the power per unit bandwidth radiated at the source? Take the earth-Jupiter distance as 40 light-minutes and assume that the source radiates isotropically.

14-5-7. Red shifts. Powers Some radio sources have been identified with optical objects and the Doppler or red shift z ($= \Delta\lambda.\ \lambda = v\ c$) measured from an optical spectrum. Assume that the objects with larger red shift are more distant, according to the Hubble relation $R = v\ H_0$, where $R =$ distance in megaparsecs (1 megaparsec = 1 Mpc = 3.26×10^6 light-years), $v =$ velocity of recession of object in meters per second, and $H_0 =$ Hubble's constant = 75 km s^{-1} Mpc^{-1}. Determine the distance R to the following radio sources: (a) Cygnus A (prototype radio galaxy), $z = 0.06$; (b) 3C273 (quasistellar radio source, or quasar), $z = 0.16$; and (c) OQ172 (distant quasar), $z = 3.53$. The above sources have flux densities as follows at 3 GHz: Cygnus A, 600 Jy; 3C273, 30 Jy; OQ172, 2 Jy (1 Jy = 10^{-26} W m^{-2} Hz^{-1}). (d) Determine the radio power per unit bandwidth radiated by each source. Assume that the source radiates isotropically.

★14-5-8. Pulsar spin down The peak flux density of a pulsar is 25 Jy at 150 MHz. The pulsar rate is 30 s^{-1} with pulse duration one-fourth the pulse period. (a) What is the energy density per pulse (J m^{-2} Hz^{-1})? (b) If the pulsar is at a distance of 100 light-years, what is the energy radiated per pulse in a 20-MHz bandwidth? Assume that the source radiates isotropically and that 25 Jy is the average flux density over the bandwidth. (c) Assuming that the above radio-frequency radiation is the only energy lost by the pulsar, how long can it radiate at the above rate if it has one solar mass ($= 2 \times 10^{30}$ kg)? Assume that the pulsar diameter is 25 km and that its kinetic energy of rotation is transformed to radio-frequency power with 5 percent efficiency.

14-5-9. Earth station antenna temperature An earth station dish of 100-m^2 effective aperture is directed at the zenith. Calculate the antenna temperature assuming that the sky temperature is uniform and equal to 6 K. Take the ground temperature equal to 300 K and assume that one-third the minor-lobe beam area is in the back direction. The wavelength is 75 mm and the beam efficiency is 0.8.

14-5-10. Thompson scatter The alternating electric field of a passing electromagnetic wave causes an electron (initially at rest) to oscillate. This oscillation of the electron makes it equivalent to a dipole radiator. Show that the ratio of the power scattered per steradian to the incident Poynting vector is given by $(\mu_0 e^2 \sin\theta\ 4\pi m)^2$, where e and m are the charge and mass of the electron and θ is the angle of the scattered radiation with respect to the direction of the electric field $\mathbf{E}$ of the incident wave. This ratio times 4π is the radar cross section of the electron. Such reradiation is called *Thompson scatter*.

14-5-11. Thompson scatter radar A ground-based vertical-looking radar can be used to determine electron densities in the earth's ionosphere by means of *Thompson scatter* (see Prob. 14-5-10). The scattered-power radar return is proportional to the electron density. If a short pulse is transmitted by the radar, the back-scattered power as a function of time is a measure of the electron density as a function of height. Design a Thompson scatter radar operating at 430 MHz capable of measuring ionospheric electron densities with 1-km resolution in height and horizontal position to heights of 1 Mm. The radar should also be capable of detecting a minimum of 100 electrons at a height of 1 Mm. The design should specify radar peak power, pulse length, antenna size, and receiver sensitivity. See W. E. Gordon. Radar Backscatter from the Earth's Ionosphere, *IEEE Trans. Antennas Propag.*, **AP-12**:873–876 (December 1964).

★14-5-12. Critical frequency. MUF Layers may be said to exist in the earth's ionosphere where the ionization gradient is sufficient to refract radio waves back to the earth. [Although the wave actually may be bent gradually along a curved path in an ionized region of considerable thickness, a useful simplification for some situations is to assume that the wave is reflected as though from a horizontal perfectly conducting surface situated at a (virtual) height h.] The highest frequency at which this layer reflects a vertically incident wave back to the earth is called the *critical frequency* f_0. Higher frequencies at vertical incidence pass through. For waves at oblique incidence ($\phi > 0$ in Fig. P14-5-12) the *maximum usable frequency* (MUF) for point-to-point communication on the earth is given by MUF = $f_0/\cos \phi$, where ϕ = angle of incidence. The critical frequency $f_0 = 9\sqrt{N}$, where N = electron density (number m^{-3}). N is a function of solar irradiation and other factors. Both f_0 and h vary with time of day, season, latitude, and phase of the 11-year sunspot cycle. Find the MUF for (a) a distance $d = 1.3$ Mm by F$_2$-layer ($h = 325$ km) reflection with F$_2$-layer electron density $N = 6 \times 10^{11}$ m^{-3}, (b) a distance $d = 1.5$ Mm by F$_2$-layer ($h = 275$ km) reflection with $N = 10^{12}$ m^{-3}, and (c) a distance $d = 1$ Mm by sporadic E-layer ($h = 100$ km) reflection with $N = 8 \times 10^{11}$ m^{-3}. Neglect earth curvature.

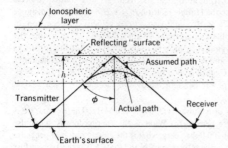

Figure P14-5-12 Communication path via reflection from ionospheric layer.

14-5-13. mUF for Clarke orbit satellites Stationary communication (relay) satellites are placed in the Clarke orbit at heights of about 36 Mm. This is far above the ionosphere, so that the transmission path passes completely through the ionosphere twice, as in Fig. P14-5-13. Since frequencies of 2 GHz and above are usually used, the ionosphere has little effect. The high frequency also permits wide bandwidths. If the ionosphere consists of a layer 200 km thick between heights of 200 and 400 km with a uniform electron density $N = 10^{12}$ m^{-3}, find the lowest frequency (or *minimum usable frequency*, mUF) which can be used with a communication satellite (a) for vertical incidence and (b) for paths 30° from the

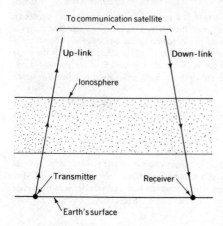

Figure P14-5-13 Communication path via geostationary Clarke-orbit relay satellite.

zenith. (c) For an earth station on the equator, what is the mUF for a satellite 15° above the eastern or western horizon?

14-5-14. S/N **equation** Derive the signal-to-noise ratio equation of Prob. 10-6-1.

14-5-15. Multiple-path transmission In the worked example involving the height of a receiving antenna above ground (see Fig. 14-54) confirm the first equality of Eq. (14-23-6a).

14-5-16. Polarization measurements Consider the following four antennas: horizontal $\lambda/2$ dipole, vertical $\lambda/2$ dipole, left-handed 8-turn helical-beam antenna, and right-handed 8-turn helical-beam antenna with four power responses P_x, P_y, P_L, and P_R, respectively. All antennas are oriented for maximum response to waves from a distant source. Each antenna can be switched in turn to a power-measuring unit to which it is properly matched. (a) Determine the Stokes parameters and also the parameters d, AR, and τ for eight different waves if the measured power responses are as tabulated below. Take the directivity of each $\lambda/2$ dipole as 1.6 and of each helical-beam antenna as 32. (b) What types of transmitting systems are required to produce waves of these types?

				Case				
	1	2	3	4	5	6	7	8
P_x	1	1	0.5	0.5	0.6	0.3	0.5	0.4
P_y	1	0.5	0	0.5	0.4	0.1	0.5	0.5
P_L	20	10	10	20	9	2	15	1
P_R	20	10	10	0	6	8	15	18

Group 14-6: Practical applications

14-6-1. Directional broadcast array (a) An AM broadcasting station is to be located south of the area it is to serve. Design an antenna for this station which gives a broad coverage to the north (from NW through N to NE) with reduced field intensity in other directions. However, to obtain Federal Communications Commission approval the pattern must have a null SE (135° from N) in order to protect another station on the same frequency in that direction. The antenna is to consist of an in-line array of $\lambda/4$ vertical elements oriented along a north-south line with equal spacing between elements. The minimum number of elements should be used. *Hint*: In plan view the problem reduces to a linear array of isotropic point sources. (b) Repeat part (a) with the additional requirement of another null to the west (90° from N). *Hint*: Apply pattern multiplication.

⋆**14-6-2. Dish gain** (a) An antenna manufacturer claims that its parabolic dish antenna of 4-m diameter operating at 20-cm wavelength has a gain of 38 dBi. Do you believe this? (b) If not, what is the maximum gain it could have? (c) If the antenna has the gain in (b) how many beam areas will it have in the whole sky?

⋆**14-6-3. Pocket radio** (a) What is the effective aperture of a pocket transistor-radio receiver at 1 MHz? Assume isotropic pattern. (b) At 1 km from a 10-kW, 1-MHz broadcast station what power would be available in this aperture? (c) What power is actually required by the receiver (approximately 1 μV in 300 Ω)? (d) What is the antenna efficiency? (e) Would the pocket unit be effective for transmitting at 1 MHz? (f) Would the above considerations be altered appreciably if one used the actual patterns instead of assuming isotropic patterns?

14-6-4. Police radar Design a CW Doppler radar operating at 30 GHz for measuring the velocity of automobiles at distances up to 1.5 km. The transmitter power is limited to 500 mW, and the receiver requires 10^{-10} W. Specify antenna size and the precision with which frequency shift must be measured in order to achieve velocity measurements accurate to 1 km h^{-1}. Take the radar cross section of an automobile as 30 percent of the physical cross section.

14-6-5. Moon link A radio link from the moon to the earth has a moon-based 5λ-long right-handed helical-beam antenna (as in Fig. 14-33) and a 2-W transmitter operating at 1.5 GHz. What should the polarization state and effective aperture be for the earth-based antenna in order to deliver 10^{-14} W to the receiver? Take the earth-moon distance as 1.27 light-seconds.

14-6-6. Venus radar (a) Design an earth-based radar system capable of delivering 10^{-15} W of peak echo power from Venus to a receiver. The radar is to operate at 2 GHz, and the same antenna is to be used for both transmitting and receiving. Specify the effective aperture of the antenna and the peak transmitter power. Take the earth-Venus distance as 3 light-minutes, the diameter of Venus as 12.6 Mm, and the radar cross section of Venus as 10 percent of the physical cross section. (b) If the system of (a) is used to observe the moon, what will the received power be? Take the moon diameter as 3.5 Mm and the moon radar cross section as 10 percent of the physical cross section.

14-6-7. Mars link (a) Design a two-way radio link to operate over earth-Mars distances for data and picture transmission with a Mars probe at 2.5 GHz with 5-MHz bandwidth. A power of 10^{-19} W Hz^{-1} is to be delivered to the earth receiver and 10^{-17} W Hz^{-1} to the Mars receiver. The Mars antenna must be no larger than 3 m in diameter. Specify effective aperture of Mars and earth antennas and transmitter power (total over entire bandwidth) at each end. Take earth-Mars distance as 6 light-minutes. (b) Repeat (a) for an earth-Jupiter link. Take the earth-Jupiter distance as 40 light-minutes.

★**14-6-8. Earth station radio telescope** Referring to the 2.6-m earth station dish shown in the frontispiece, find (a) minimum detectable temperature and (b) minimum detectable flux density at 4 GHz if the receiver temperature is 80 K, sky temperature 20 K, antenna aperture efficiency 55 percent, predetection bandwidth 1 GHz, system constant unity (total power), and postdetection time constant 4 s.

14-6-9. Earth station boiler Referring to the 2.6-m earth station dish shown in the frontispiece, determine how long it takes to bring a 2l tank of water at the focus to boiling temperature when the dish is pointed at the sun. Assume dish efficiency for this application as 20 percent. Take solar flux as 1 kW m^{-2} and initial water temperature as 25°C. Neglect re-radiation, convection, or conduction heat loss from the tank. If the dish is used as a radio receiving antenna and inadvertently pointed at the sun or if the dish is used as a radio telescope antenna and intentionally pointed at the sun (to measure the solar radio emission), what can be done to prevent heat damage to an rf preamplifier at the focus?

★**14-6-10. Solar interference to earth station** (a) Twice a year the sun passes through the apparent declination of the geostationary Clarke-orbit satellites, causing solar-noise interference to earth stations. If the equivalent temperature of the sun at 4 GHz is 50,000 K, find the sun's signal-to-noise ratio (in decibels) for an earth station with 3-m parabolic dish antenna at 4 GHz. Take the sun's diameter as 0.5° and the earth station system temperature as 100 K. (b) Compare this result with that for the carrier-to-noise ratio calculated in Prob. 10-6-1 for a typical Clarke-orbit TV transponder. (c) How long does the interference last? Note that the relation $\Omega_A = \lambda^2/A_e$ gives the solid beam angle in steradians and not in square degrees.

4-6-11. Quad-helix array Design a square array of four helical beam antennas with total directivity $D = 100$. Determine (a) the length L of each helix and (b) the spacing d between helix axes (as measured along the sides of square). The feed point impedance of a helical beam antenna may be anywhere between 50 and 150 Ω resistive depending on the geometry of the last turn (at feed end). (See J. D. Kraus, 50-ohm Input for Helical Beam Antennas, *IEEE Trans. Ants. Prop.*, **AP-25**, Nov. 1977, p. 913; also J. Kraus, "Our Cosmic Universe," Cygnus-Quasar Books, Powell, Ohio 43065, 1980, p. 252.) (c) Design a transmission line network to feed all helices in phase from a 50-Ω coaxial line. Note that since side-by-side helical beam antennas have very small mutual impedance, this may be neglected in the design.

UNITS, CONSTANTS, AND OTHER USEFUL RELATIONS

A-1 UNITS

Multiples and submultiples of the basic units are designated by the prefixes listed in Table 1. Note that in the SI system, multiples and submultiples are in steps of 10^3 or 10^{-3}. Quantities larger than 10^{18} or smaller than 10^{-18} are designated by the exponential form. For a discussion of dimensions and units see Secs. 1-2 to 1-4.

In Table 2 dimensions or quantities commonly used in electromagnetics are listed alphabetically under the headings Fundamental, Mechanical, Electrical, and Magnetic. In the *first* column the name of the dimension or quantity is given and in the *second* column the common symbol for designating it. In the *third* column (Description) the dimension is described in terms of the fundamental dimensions (mass, length, time, and electric current) or other secondary dimensions. The

Table 1

Prefix	Abbreviation	Magnitude	Derivation
exa (eks-a)	E	10^{18}	Greek *ex*, "beyond"
peta (pet-a)	P	10^{15}	Greek *petasos*, "outreach"
tera (těr′à)	T	10^{12}	Greek *teras*, "monster"
giga (jǐ′gà)	G	10^9	Latin *gigas*, "giant"
mega (měg′à)	M	10^6	Greek *megas*, "great"
kilo (kǐl′ò)	k	10^3	Greek *chilioi*, "a thousand"
milli (mǐl′ ǐ)	m	10^{-3}	Latin *mille*, "a thousand"
micro (mǐ′ krò)	μ	10^{-6}	Greek *mikros*, "small"
nano (năn′ ò)	n	10^{-9}	Greek *nanos*, "dwarf"
pico (pē′ kò)	p	10^{-12}	Spanish *pico*, "small quantity"
femto (fěm′tò)	f	10^{-15}	Danish *femten*, "fifteen"
atto (ăt′ tò)	a	10^{-18}	Danish *atten*, "eighteen"

fourth column (SI unit) lists the SI unit and abbreviation, and the *fifth* column gives equivalent units. The *last* column indicates the fundamental dimensions by means of the symbols M (mass), L (length), T (time), and I (electric current).

Table 2

Fundamental units					
Name of dimension or quantity	Symbol	Description	SI unit and abbreviation	Equivalent units	Dimension
Current	I	$\dfrac{\text{charge}}{\text{time}}$	ampere (A)	6.25 electron charges per second $= \dfrac{C}{s}$	I
Length	L, l		meter (m)	1,000 mm $= 100$ cm	L
Mass	M, m		kilogram (kg)	1,000 g	M
Time	T, t		second (s)	$\frac{1}{60}$ min $= \dfrac{1}{3,600}$ hr $= \dfrac{1}{86,400}$ day	T
Mechanical units					
Acceleration	a	$\dfrac{\text{velocity}}{\text{time}} = \dfrac{\text{length}}{\text{time}^2}$	$\dfrac{\text{meter}}{\text{second}^2}$ (m s^{-2})	$\dfrac{N}{kg}$	$\dfrac{L}{T^2}$
Area	A, a, s	length^2	meter^2 (m^2)		L^2
Energy or work	W	force $\times$ length $=$ power $\times$ time	joule (J)	N m $=$ W s $=$ V C $= 10^7$ ergs $= 10^8$ dynes mm	$\dfrac{ML^2}{T^2}$
Energy density	w	$\dfrac{\text{energy}}{\text{volume}}$	$\dfrac{\text{joule}}{\text{meter}^3}$ (J m^{-3})	$\dfrac{\frac{1}{100}\,\text{erg}}{\text{mm}^3}$	$\dfrac{M}{LT^2}$
Force	$\mathbf{F}$	mass $\times$ acceleration	newton (N)	$\dfrac{kg\ m}{s^2} = \dfrac{J}{m}$ $= 10^5$ dynes	$\dfrac{ML}{T^2}$
Frequency	f	$\dfrac{\text{cycles}}{\text{time}}$	hertz (Hz)	$\dfrac{\text{cycles}}{s}$	$\dfrac{1}{T}$
Impedance	Z	$\dfrac{\text{force}}{\text{mass} \times \text{velocity}}$	$\dfrac{\text{newton-second}}{\text{kilogram-meter}}$	$\dfrac{N\ s}{kg\ m}$	$\dfrac{1}{T}$

Table 2—(*Continued*)

			Mechanical units (*continued*)		
Name of dimension or quantity	Symbol	Description	SI unit and abbreviation	Equivalent units	Dimension
Length	L, l		meter (m)	1,000 mm $= 100$ cm	L
Mass	M, m		kilogram (kg)	1,000 g	M
Moment (torque)		force × length	newton-meter (N m)	$\dfrac{\text{kg m}^2}{\text{s}} = \text{J}$	$\dfrac{ML^2}{T^2}$
Momentum	$m\mathbf{v}$	mass × velocity $= $ force × time $= \dfrac{\text{energy}}{\text{velocity}}$	newton-second (N s)	$\dfrac{\text{kg m}}{\text{s}} = \dfrac{\text{J s}}{\text{m}}$	$\dfrac{ML}{T}$
Period	T	$\dfrac{1}{\text{frequency}}$	second (s)		T
Power	P	$\dfrac{\text{force} \times \text{length}}{\text{time}}$ $= \dfrac{\text{energy}}{\text{time}}$	watt (W)	$\dfrac{\text{J}}{\text{s}} = \dfrac{\text{N m}}{\text{s}}$ $= \dfrac{\text{kg m}^2}{\text{s}^3}$	$\dfrac{ML^2}{T^3}$
Time	T, t		second (s)	$\dfrac{1}{60} \min = \dfrac{1}{3,600} \text{hr}$ $= \dfrac{1}{86,400} \text{day}$	T
Velocity (velocity of light in vacuum $= 300$ Mm s^{-1})	$\mathbf{v}$	$\dfrac{\text{length}}{\text{time}}$	$\dfrac{\text{meter}}{\text{second}}$ (m s^{-1})		$\dfrac{L}{T}$
Volume	v	length^3	meter^3 (m^3)		L^3
			Electrical units		
Admittance	Y	$\dfrac{1}{\text{impedance}}$	mho (℧)	$\dfrac{\text{A}}{\text{V}} = \dfrac{\text{C}^2}{\text{J s}} = \text{S}\dagger$	$\dfrac{I^2 T^3}{ML^2}$
Capacitance	C	$\dfrac{\text{charge}}{\text{potential}}$	farad (F)	$\dfrac{\text{C}}{\text{V}} = \dfrac{\text{C}^2}{\text{J}} = \dfrac{\text{As}}{\text{V}}$ $= 9 \times 10^{11}$ cm (cgs esu)	$\dfrac{I^2 T^4}{ML^2}$

(*continued*)

Table 2—(*Continued*)

			Electrical units		

Name of dimension or quantity	Symbol	Description	SI unit and abbreviation	Equivalent units	Dimension
Charge	Q, q	current × time	coulomb (C)	6.25×10^{18} electron charges $=$ A s $= 3 \times 10^9$ cgs esu $= 0.1$ cgs emu	IT
Charge (volume) density	ρ	$\dfrac{\text{charge}}{\text{volume}} = \nabla \cdot \mathbf{D}$	coulomb meter^3 (C m^{-3})	$\dfrac{\text{A s}}{\text{m}^3}$	$\dfrac{IT}{L^3}$
Conductance	G	$\dfrac{1}{\text{resistance}}$	mho (℧)	$\dfrac{\text{A}}{\text{V}} = \dfrac{\text{C}^2}{\text{J s}} = \text{S†}$	$\dfrac{IT^3}{ML^2}$
Conductivity	σ	$\dfrac{1}{\text{resistivity}}$	mho meter (℧ m^{-1})	$\dfrac{1}{\Omega\,\text{m}}$	$\dfrac{I^2 T^4}{ML^3}$
Current	I, i	$\dfrac{\text{charge}}{\text{time}}$	ampere (A)	$\dfrac{\text{C}}{\text{s}} = 3 \times 10^9$ cgs esu $= 0.1$ cgs emu	I
Current density	J	$\dfrac{\text{current}}{\text{area}}$	ampere meter^2 (A m^{-2})	$\dfrac{\text{C}}{\text{s m}^2}$	$\dfrac{I}{L^2}$
Dipole moment	$\mathbf{p}(= q\mathbf{l})$	charge × length	coulomb-meter (C m)	A s m	LIT
Emf	$\mathscr{V}$	$\int \mathbf{E}_e \cdot d\mathbf{l}$	volt (V)	$\dfrac{\text{Wb}}{\text{s}} = \dfrac{\text{J}}{\text{C}}$	$\dfrac{ML^2}{IT^3}$
Energy density (electric)	w_e	$\dfrac{\text{energy}}{\text{volume}}$	joule meter^3 (J m^{-3})	$\dfrac{\frac{1}{100}\text{ erg}}{\text{mm}^3}$	$\dfrac{M}{LT^2}$
Field intensity	$\mathbf{E}$	$\dfrac{\text{potential}}{\text{length}} = \dfrac{\text{force}}{\text{charge}}$	volt meter (V m^{-1})	$\dfrac{\text{N}}{\text{C}} = \dfrac{\text{J}}{\text{C m}}$ $= \frac{1}{3} \times 10^{-4}$ cgs esu $= 10^6$ cgs emu	$\dfrac{ML}{IT^3}$
Flux	ψ	charge $= \iint \mathbf{D} \cdot d\mathbf{s}$	coulomb (C)	A s	IT

† See page 732.

Table 2—(*Continued*)

			Electrical units (*continued*)		
Name of dimension or quantity	Symbol	Description	SI unit and abbreviation	Equivalent units	Dimension
Flux density (displacement) (*D* vector)	**D**	$\dfrac{\text{charge}}{\text{area}}$	coulomb $\overline{\text{meter}^2}$ (C m^{-2})	$\dfrac{\text{A s}}{\text{m}^2} = \dfrac{\text{A}}{\text{m}^2 \text{ s}^{-1}}$	$\dfrac{IT}{L^2}$
Impedance	Z	$\dfrac{\text{potential}}{\text{current}}$	ohm (Ω)	$\dfrac{\text{V}}{\text{A}}$	$\dfrac{ML^2}{I^2T^3}$
Linear charge density	ρ_L	$\dfrac{\text{charge}}{\text{length}}$	coulomb $\overline{\text{meter}}$ (C m^{-1})	$\dfrac{\text{A s}^{-1}}{\text{m}}$	$\dfrac{IT}{L}$
Permittivity (dielectric constant) (for vacuum, $\epsilon_0 = 8.85$ pF $\approx 10^{-9}/36\pi$ F m^{-1})	ϵ	$\dfrac{\text{capacitance}}{\text{length}}$	farad $\overline{\text{meter}}$ (F m^{-1})	$\dfrac{\text{C}}{\text{V m}}$	$\dfrac{I^2T^4}{ML^3}$
Polarization	**P**	$\dfrac{\text{dipole moment}}{\text{volume}}$	coulomb $\overline{\text{meter}^2}$ (C m^{-2})	$\dfrac{\text{A s}}{\text{m}^2}$	$\dfrac{IT}{L^2}$
Potential	V	$\dfrac{\text{work}}{\text{charge}}$	volt (V)	$\dfrac{\text{J}}{\text{C}} = \dfrac{\text{N m}}{\text{C}} = \dfrac{\text{W s}}{\text{C}}$ $= \dfrac{\text{W}}{\text{A}} = \dfrac{\text{Wb}}{\text{s}}$ $= \dfrac{1}{300}$ cgs esu $= 10^8$ cgs emu	$\dfrac{ML^2}{IT^3}$
Poynting vector	**S**	$\dfrac{\text{power}}{\text{area}}$	watt $\overline{\text{meter}^2}$ (W m^{-2})	$\dfrac{\text{J}}{\text{s m}^2}$	$\dfrac{M}{T^3}$
Radiation intensity	P	$\dfrac{\text{power}}{\text{unit solid angle}}$	watt $\overline{\text{steradian}}$ (W sr^{-1})		$\dfrac{ML^2}{T^3}$
Reactance	X	$\dfrac{\text{potential}}{\text{current}}$	ohm (Ω)	$\dfrac{\text{V}}{\text{A}}$	$\dfrac{ML^2}{I^2T^3}$
Relative permittivity	ϵ_r	ratio $\dfrac{\epsilon}{\epsilon_0}$			dimensionless

(*continued*)

Table 2—(*Continued*)

			SI unit and abbrevia-tion	Equivalent units	Dimen-sion
Electrical units (*continued*)					
Name of dimension or quantity	Symbol	Description			
Resistance	R	$\dfrac{\text{potential}}{\text{current}}$	ohm (Ω)	$\dfrac{\text{V}}{\text{A}} = \dfrac{\text{J s}}{\text{C}^2}$ $= \frac{1}{9} \times 10^{-11}$ cgs esu $= 10^{-9}$ cgs emu	$\dfrac{ML^2}{I^2 T^3}$
Resistivity	S	resistance × length $= \dfrac{1}{\text{conductivity}}$	ohm-meter, (Ω m)	$\dfrac{\text{V m}}{\text{A}}$	$\dfrac{ML^3}{I^2 T^4}$
Sheet-current density	**K**	$\dfrac{\text{current}}{\text{length}}$	ampere meter (A m^{-1})	$\dfrac{\text{A}}{\text{m}^2} \times \text{m}$	$\dfrac{I}{L}$
Susceptance	B	$\dfrac{1}{\text{reactance}}$	mho (℧)	$\dfrac{\text{A}}{\text{V}} = \text{S}$	$\dfrac{I^2 T^3}{ML^2}$
Wavelength	λ	length	meter (m)		L
Magnetic units					
Dipole moment (magnetic)	m $(= Q_m l)$	pole strength × length $=$ current × area $= \dfrac{\text{torque}}{\text{magnetic flux density}}$	ampere-meter^2 (Am^2)	$\dfrac{\text{C m}^2}{\text{s}}$	IL^2
Energy density (magnetic)	w_m	$\dfrac{\text{energy}}{\text{volume}}$	joule meter^3 (J m^{-3})	$\dfrac{\frac{1}{100}\,\text{erg}}{\text{mm}^3}$	$\dfrac{M}{LT^2}$
Flux (magnetic)	ψ_m	$\iint \mathbf{B} \cdot d\mathbf{s}$	weber (Wb)	$\text{V s} = \dfrac{\text{n m}}{\text{A}}$ $= 10^8$ Mx‡ (cgs emu)	$\dfrac{ML^2}{IT^2}$
Flux density (*B* vector)	**B**	$\dfrac{\text{force}}{\text{pole}} = \dfrac{\text{force}}{\text{current moment}}$ $= \dfrac{\text{magnetic flux}}{\text{area}}$	tesla (T) $= \dfrac{\text{weber}}{\text{meter}^2}$ (Wb m^{-2})	$\dfrac{\text{V s}}{\text{m}^2} = \dfrac{\text{N}}{\text{A m}}$ $= 10^4$ G‡ (cgs emu)	$\dfrac{M}{IT^2}$
Flux linkage	Λ	flux × turns	weber-turn (Wb turn)		$\dfrac{ML^2}{IT^2}$

Table 2—(*Continued*)

			Magnetic units (*continued*)		
Name of dimension or quantity	Symbol	Description	SI unit and abbreviation	Equivalent units	Dimension
H field (H vector)	**H**	$\dfrac{\text{mmf}}{\text{length}}$	ampere meter (A m^{-1})	$\dfrac{\text{N}}{\text{Wb}} = \dfrac{\text{W}}{\text{V m}}$ $= 4\pi \times 10^{-3}$ Oe‡ (cgs emu) $= 400\pi$ gammas	$\dfrac{I}{L}$
Inductance	L	$\dfrac{\text{magnetic flux linkage}}{\text{current}}$	henry (H)	$\dfrac{\text{Wb}}{\text{A}} = \dfrac{\text{J}}{\text{A}^2}$ $= \Omega \text{ s}$ $= \frac{1}{9} \times 10^{-11}$ cgs esu $= 10^9$ cm (cgs emu)	$\dfrac{ML^2}{I^2T^2}$
Magnetization (magnetic polarization)	**M**	$\dfrac{\text{magnetic moment}}{\text{volume}}$	ampere meter (A m^{-1})	$\dfrac{\text{A m}^2}{\text{m}^3} = \dfrac{\text{A m}}{\text{m}^2}$	$\dfrac{I}{L}$
Mmf	F	$\int \mathbf{H} \cdot d\mathbf{l}$	ampere-turn (A turn)	$\dfrac{\text{C}}{\text{s}}$	
Permeability (for vacuum, $\mu_0 = 400\pi$ nH m^{-1})	μ	$\dfrac{\text{inductance}}{\text{length}}$	henry meter (H m^{-1})	$\dfrac{\text{Wb}}{\text{A m}} = \dfrac{\text{V s}}{\text{A m}}$	$\dfrac{ML}{I^2T^2}$
Permeance	$\mathscr{P}$	$\dfrac{\text{magnetic flux}}{\text{mmf}}$ $= \dfrac{1}{\text{reluctance}}$	henry (H)	$\dfrac{\text{Wb}}{\text{A}}$	$\dfrac{ML^2}{I^2T^2}$
Pole density	ρ_m	$\dfrac{\text{pole strength}}{\text{volume}}$ $= \dfrac{\text{current}}{\text{area}}$ $= \nabla \cdot \mathbf{H} = -\nabla \cdot \mathbf{M}$	ampere meter^2 (A m^{-2})		$\dfrac{I}{L^2}$
Pole strength	Q_m, q_m	current $\times$ length $= \iiint \rho_m \, dv$	ampere-meter (A m)	$\dfrac{\text{C m}}{\text{s}}$	IL
Potential (magnetic) (for **H**)	U	$\int \mathbf{H} \cdot d\mathbf{l}$	ampere (A)	$\dfrac{\text{J}}{\text{Wb}} = \dfrac{\text{W}}{\text{V}} = \dfrac{\text{C}}{\text{s}}$ $= \dfrac{4\pi}{10}$ Gb§ (cgs emu)	I

‡, § See page 732.

(*continued*)

Table 2—(*Continued*)

			SI unit and abbrevia-tion	Equivalent units	Dimen-sion
Name of dimension or quantity	**Symbol**	**Description**			
		Magnetic units (*continued*)			
Relative permeability	μ_r	ratio $\dfrac{\mu}{\mu_0}$			Dimen-sion-less
Reluctance	$\mathscr{R}$	$\dfrac{\text{mmf}}{\text{magnetic flux}}$ $= \dfrac{1}{\text{permeance}}$	$\dfrac{1}{\text{henry}}$ (H^{-1})	$\dfrac{\text{A}}{\text{Wb}}$	$\dfrac{I^2 T^2}{ML^2}$
Vector potential	**A**	current × permeability	$\dfrac{\text{Weber}}{\text{meter}}$ (Wb m^{-1})	$\dfrac{\text{H A}}{\text{m}} = \dfrac{\text{N}}{\text{A}}$	$\dfrac{ML}{IT^2}$

† S is the SI abbreviation for siemens, used often for mho.
‡ Mx, G, and Oe are SI abbreviations for maxwell, gauss, and oersted.
§ Gb is the SI abbreviation for gilbert.

A-2 CONSTANTS AND CONVERSIONS

Quantity	Symbol or abbre-viation	Nominal value	More accurate value†
Astronomical unit	AU	1.5×10^8 km	1.496
Boltzmann's constant	k	1.38×10^{-23} J K^{-1}	1.38062×10^{-23}
Earth mass		6.0×10^{24} kg	5.98×10^{24}
Earth radius (average)		6.37 Mm	
Electron charge	e	-1.60×10^{-19} C	-1.602
Electron rest mass	m	9.11×10^{-31} kg	9.10956×10^{-31}
Electron charge-to-mass ratio	e/m	1.76×10^{11} C kg^{-1}	1.758803×10^{11}
Flux density (power)	Jy‡	10^{-26} W m^{-2} Hz^{-1}	10^{-26} (by definition)
Foot	ft	0.30 m (1 m = 3.281 ft)	0.3048
Foot squared	ft^2	9.3×10^{-2} m^2 (1 m^2 = 10.76 ft^2)	9.290×10^{-2}
Light-second		300 Mm ($\approx$ 1 Gft)	
Light, velocity of	c	300 Mm s^{-1} ($\approx$ 1 Gft s^{-1})	299.7925
Light-year	LY	9.46×10^{12} km	9.4605×10^{12}
Logarithm (natural) base	e	2.72	2.71828

Table A-2—(*Continued*)

Quantity	Symbol or abbre- viation	Nominal value	More accurate value†
Logarithm, reciprocal of base	$1/e$	0.368	0.36788
Logarithm conversion		$\ln x = 2.3 \log x$ $\log x = 0.43 \ln x$	$\ln x = 2.3026 \log x$ $\log x = 0.4343 \ln x$
Mile (statute)	mi	1.61 km (1 km = 0.6214 mi)	1.60935
Moon distance (average)		380 Mm	
Moon mass		6.7×10^{22} kg	
Moon radius (average)		1.738 Mm	
Permeability of vacuum	μ_0	1,260 nH m^{-1}	400π (exact value)
Permittivity of vacuum	ϵ_0	8.85 pF m^{-1}	$8.854185 = 1/\mu_0 c^2$
Pi	π	3.14	3.1415927
Pi squared	π^2	9.87	9.8696044
Planck's constant	h	6.63×10^{-34} J s	6.62620×10^{-34}
Proton rest mass		1.67×10^{-27} kg	1.67261×10^{-27}
Radian	rad	57.3°	57.2958°
Space, impedance of	Z	376.7 ($\approx 120\pi$) Ω	$376.7304 = \mu_0 c$
Sphere, solid angle		12.6 sr	$4\pi = 12.5664$
		41,253 deg^2	41,252.96
Square degree	deg^2	3.05×10^{-4} sr	3.04617×10^{-4}
Steradian ($\approx$ square radian)	sr	3,283 deg^2	$(180/\pi)^2 = 3,282.806$
Sun, distance	AU	1.5×10^8 km	1.496×10^8
Sun mass	$M_\odot$	2.0×10^{30} kg	1.99×10^{30}
Sun radius (average)	$R_\odot$	700 Mm	695.3
	$\sqrt{2}$	1.414	1.41421
	$\sqrt{3}$	1.73	1.73205
	$\sqrt{10}$	3.16	3.16228

† Same units as nominal value. Regarding permittivity ϵ_0 and space impedance Z note that the values for these quantities are determined by the exact (definition) value of μ_0 and the measured value of c (velocity of light).

‡ Jy = jansky

A-3 TRIGONOMETRIC RELATIONS

$$\sin (x \pm y) = \sin x \cos y \pm \cos x \sin y$$

$$\cos (x \pm y) = \cos x \cos y \mp \sin x \sin y$$

$$\sin (x + y) + \sin (x - y) = 2 \sin x \cos y$$

$$\cos(x + y) + \cos(x - y) = 2\cos x \cos y$$

$$\sin(x + y) - \sin(x - y) = 2\cos x \sin y$$

$$\cos(x + y) - \cos(x - y) = -2\sin x \sin y$$

$$\sin 2x = 2\sin x \cos x$$

$$\cos 2x = \cos^2 x - \sin^2 x = 2\cos^2 x - 1 = 1 - 2\sin^2 x$$

$$\cos x = 2\cos^2 \tfrac{1}{2}x - 1 = 1 - 2\sin^2 \tfrac{1}{2}x$$

$$\sin x = 2\sin \tfrac{1}{2}x \cos \tfrac{1}{2}x$$

$$\sin^2 x + \cos^2 x = 1$$

$$\tan(x + y) = \frac{\tan x + \tan y}{1 - \tan x \tan y}$$

$$\tan(x - y) = \frac{\tan x - \tan y}{1 + \tan x \tan y}$$

$$\tan 2x = \frac{2\tan x}{1 - \tan^2 x}$$

$$\sin x = x - \frac{x^3}{3!} + \frac{x^5}{5!} - \frac{x^7}{7!} + \cdots$$

$$\cos x = 1 - \frac{x^2}{2!} + \frac{x^4}{4!} - \frac{x^6}{6!} + \cdots$$

$$\tan x = x + \frac{x^3}{3} + \frac{2x^5}{15} + \frac{17x^7}{315} + \frac{62x^9}{2{,}835} + \cdots$$

A-4 HYPERBOLIC RELATIONS

$$\sinh x = \frac{e^x - e^{-x}}{2} = x + \frac{x^3}{3!} + \frac{x^5}{5!} + \frac{x^7}{7!} + \cdots$$

$$\cosh x = \frac{e^x + e^{-x}}{2} = 1 + \frac{x^2}{2!} + \frac{x^4}{4!} + \frac{x^6}{6!} + \cdots$$

$$\tanh x = \frac{\sinh x}{\cosh x}$$

$$\coth x = \frac{\cosh x}{\sinh x} = \frac{1}{\tanh x}$$

$$\sinh(x \pm jy) = \sinh x \cos y \pm j\cosh x \sin y$$

$$\cosh(x \pm jy) = \cosh x \cos y \pm j\sinh x \sin y$$

$$\left.\begin{aligned}\cosh(jx) &= \tfrac{1}{2}(e^{+jx} + e^{-jx}) = \cos x \\ \sinh(jx) &= \tfrac{1}{2}(e^{+jx} - e^{-jx}) = j\sin x\end{aligned}\right\} \quad \text{de Moivre's theorem}$$

$$e^{\pm jx} = \cos x \pm j\sin x$$

$$e^{\pm jx} = 1 \pm jx - \frac{x^2}{2!} \mp j\frac{x^3}{3!} + \frac{x^4}{4!} \pm j\frac{x^5}{5} - \cdots$$

$$e^x = \cosh x + \sinh x$$

$$e^{-x} = \cosh x - \sinh x$$

$$e^x = 1 + x + \frac{x^2}{2!} + \frac{x^3}{3!} + \frac{x^4}{4!} + \cdots$$

$$\cosh x = \cos jx$$

$$j\sinh x = \sin jx$$

$$\tanh(x \pm jy) = \frac{\sinh 2x}{\cosh 2x + \cos 2y} \pm j\frac{\sin 2y}{\cosh 2x + \cos 2y}$$

$$\coth(x \pm jy) = \frac{\sinh 2x}{\cosh 2x - \cos 2y} \pm j\frac{\sin 2y}{\cosh 2x - \cos 2y}$$

A-5 LOGARITHMIC RELATIONS

$$\log_{10} x = \log x \qquad \text{common logarithm}$$

$$\log_e x = \ln x \qquad \text{natural logarithm}$$

$$\log_{10} x = 0.4343 \log_e x = 0.4343 \ln x$$

$$\ln x = \log_e x = 2.3026 \log_{10} x$$

$$e = 2.71828$$

$$dB = 10 \log(\text{power ratio}) = 20 \log(\text{voltage ratio})$$

$$1 \text{ Np (attenuation)} = \frac{1}{e} = 0.368 \text{ (voltage)} = -8.68 \text{ dB}$$

A-6 APPROXIMATION FORMULAS FOR SMALL QUANTITIES

(δ is a small quantity compared with unity.)

$$(1 \pm \delta)^2 = 1 \pm 2\delta$$

$$(1 \pm \delta)^n = 1 \pm n\delta$$

$$\sqrt{1 + \delta} = 1 + \tfrac{1}{2}\delta$$

$$\frac{1}{\sqrt{1 + \delta}} = 1 - \tfrac{1}{2}\delta$$

$$e^{\delta} = 1 + \delta$$

$$\ln(1 + \delta) = \delta$$

$$J_n(\delta) = \frac{\delta^n}{n!\,2^n} \qquad \text{for } |\delta| \ll 1$$

where J_n is Bessel function of order n. Thus

$$J_1(\delta) = \frac{\delta}{2}$$

A-7 SERIES

Binomial:

$$(x + y)^n = x^n + nx^{n-1}y + \frac{n(n-1)}{2!}x^{n-2}y^2 + \frac{n(n-1)(n-2)}{3!}x^{(n-3)}y^3 + \cdots$$

Taylor's:

$$f(x + y) = f(x) + \frac{df(x)}{dx}\frac{y}{1} + \frac{d^2f(x)}{dx^2}\frac{y^2}{2!} + \frac{d^3f(x)}{dx^3}\frac{y^3}{3!} + \cdots$$

A-8 SOLUTION OF QUADRATIC EQUATION

If $ax^2 + bx + c = 0$, then

$$x = \frac{-b \pm \sqrt{b^2 - 4ac}}{2a}$$

A-9 VECTOR IDENTITIES

(F, f, G, and g are scalar functions; $\mathbf{F}, \mathbf{G}$, and $\mathbf{H}$ are vector functions.)

$$\mathbf{F} \cdot \mathbf{G} = FG \cos \theta \qquad \text{scalar (or dot) product}$$

where $\theta = \cos^{-1}(\mathbf{F} \cdot \mathbf{G})/FG$

$$\mathbf{F} \times \mathbf{G} = \hat{\mathbf{n}}FG \sin \theta \qquad \text{vector (or cross) product}$$

where $\theta = \sin^{-1}(\mathbf{F} \times \mathbf{G})/\hat{n}FG$

$\hat{n}$ = unit vector normal to plane containing $\mathbf{F}$ and $\mathbf{G}$

$$\nabla \cdot (\nabla \times \mathbf{F}) = 0 \tag{1}$$

$$\nabla \cdot \nabla f = \nabla^2 f \tag{2}$$

$$\nabla \times \nabla f = 0 \tag{3}$$

$$\nabla(f + g) = \nabla f + \nabla g \tag{4}$$

$$\nabla \cdot (\mathbf{F} + \mathbf{G}) = \nabla \cdot \mathbf{F} + \nabla \cdot \mathbf{G} \tag{5}$$

$$\nabla \times (\mathbf{F} + \mathbf{G}) = \nabla \times \mathbf{F} + \nabla \times \mathbf{G} \tag{6}$$

$$\nabla(fg) = g\nabla f + f\nabla g \tag{7}$$

$$\nabla \cdot (f\mathbf{G}) = \mathbf{G} \cdot (\nabla f) + f(\nabla \cdot \mathbf{G}) \tag{8}$$

$$\nabla \times (f\mathbf{G}) = (\nabla f) \times \mathbf{G} + f(\nabla \times \mathbf{G}) \tag{9}$$

$$\nabla \times (\nabla \times \mathbf{F}) = \nabla(\nabla \cdot \mathbf{F}) - \nabla^2 \mathbf{F} \tag{10}$$

$$\diamondsuit \mathbf{F} = \nabla^2 \mathbf{F} = \hat{x}\nabla^2 F_x + \hat{y}\nabla^2 F_y + \hat{z}\nabla^2 F_z \tag{11}$$

$$\nabla \cdot (\mathbf{F} \times \mathbf{G}) = \mathbf{G} \cdot (\nabla \times \mathbf{F}) - \mathbf{F} \cdot (\nabla \times \mathbf{G}) \tag{12}$$

$$\mathbf{F} \cdot (\mathbf{G} \times \mathbf{H}) = \mathbf{G} \cdot (\mathbf{H} \times \mathbf{F}) = \mathbf{H} \cdot (\mathbf{F} \times \mathbf{G}) \tag{13}$$

$$\nabla \times (\mathbf{F} \times \mathbf{G}) = \mathbf{F}(\nabla \cdot \mathbf{G}) - \mathbf{G}(\nabla \cdot \mathbf{F}) + (\mathbf{G} \cdot \nabla)\mathbf{F} - (\mathbf{F} \cdot \nabla)\mathbf{G} \tag{14}$$

$$\nabla(\mathbf{F} \cdot \mathbf{G}) = (\mathbf{F} \cdot \nabla)\mathbf{G} + (\mathbf{G} \cdot \nabla)\mathbf{F} + \mathbf{F} \times (\nabla \times \mathbf{G}) + \mathbf{G} \times (\nabla \times \mathbf{F}) \tag{15}$$

In the following three relations the surface (left side of equation) encloses the volume (right side of equation)

$$\oint_s f\, d\mathbf{s} = \int_v \nabla f\, dv \tag{16}$$

$$\oint_s \mathbf{F} \cdot d\mathbf{s} = \int_v \nabla \cdot \mathbf{F}\, dv \qquad \text{divergence theorem} \tag{17}$$

$$\oint_s \hat{n} \times \mathbf{F}\, ds = \int_v \nabla \times \mathbf{F}\, dv \tag{18}$$

In the following two formulas the line integral (left-side of equation) is along a closed path which bounds an unclosed surface (right side of equation)

$$\oint f\, d\mathbf{l} = \int_s \hat{n} \times \nabla f\, ds \tag{19}$$

$$\oint \mathbf{F} \cdot d\mathbf{l} = \int_s (\nabla \times \mathbf{F}) \cdot d\mathbf{s} \qquad \text{Stokes' theorem} \tag{20}$$

A-10 RECURRENCE RELATIONS FOR BESSEL FUNCTIONS

Equations expressing Bessel functions or their derivatives in terms of Bessel functions of the same or different order are called *recurrence formulas* or *relations*. A few of these formulas are listed below for reference.

$$\frac{dJ_v(u)}{du} = \frac{v}{u} J_v(u) - J_{v+1}(u) \tag{1}$$

$$\frac{dJ_v(u)}{du} = -\frac{v}{u} J_v(u) + J_{v-1}(u) \tag{2}$$

$$\frac{dJ_v(u)}{du} = \tfrac{1}{2}[J_{v-1}(u) - J_{v+1}(u)] \tag{3}$$

$$J_v(u) = \frac{u}{2v} [J_{v+1}(u) + J_{v-1}(u)] \tag{4}$$

$$J_{n+1}(u) = \frac{2n}{u} J_n(u) - J_{n-1}(u) \tag{5}$$

From (1) we have for $v = n = 0$,

$$\frac{dJ_0(u)}{du} = -J_1(u) \tag{6}$$

That is, the slope of the $J_0(u)$ curve is equal to $-J_1(u)$.

A-11 COORDINATE DIAGRAMS

Diagrams for rectangular, cylindrical, and spherical coordinates are shown in Fig. A-1. The constructions for elemental volumes and lengths in the three systems are shown in Fig. A-2. See inside back cover for gradient, divergence, curl, and laplacian in rectangular, cylindrical, spherical, and general curvilinear coordinates.

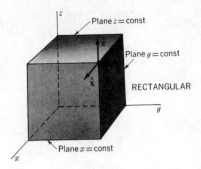

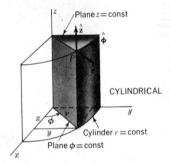

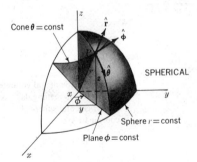

Figure A-1 Rectangular, cylindrical, and spherical coordinate systems.

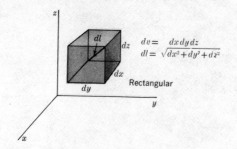

$$dv = dx\,dy\,dz$$
$$dl = \sqrt{dx^2 + dy^2 + dz^2}$$

Rectangular

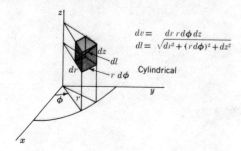

$$dv = dr\,r\,d\phi\,dz$$
$$dl = \sqrt{dr^2 + (r\,d\phi)^2 + dz^2}$$

Cylindrical

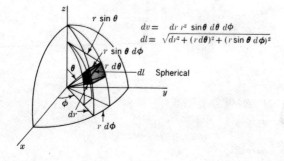

$$dv = dr\,r^2\,\sin\theta\,d\theta\,d\phi$$
$$dl = \sqrt{dr^2 + (r\,d\theta)^2 + (r\sin\theta\,d\phi)^2}$$

Spherical

Figure A-2 Elemental volumes and lengths in rectangular, cylindrical, and spherical coordinate systems.

A-12 TABLE OF DIELECTRIC MATERIALS

The following table presents data on the relative permittivity, dielectric strength, power factor, and resistivity of a number of common dielectric materials.† The power factor is given in percent at the frequencies 60 Hz, 1 MHz, and 100 MHz.

Material	Relative permittivity (relative dielectric constant)	Dielectric strength, megavolts/ meter	Power factor, percent			Resistivity, ohm-meters
			60 Hz	1 MHz	100 MHz	
Air (atmospheric pressure)	1.0006	3	—	—	—	—
Amber	3	91	0.1	0.56	—	10^{14}
Ammonia (liquid)	22	—	—	—	—	—
Bakelite	5	25	2	1	—	10^{14}
Cellulose acetate	7	12	7	5	5	—
Cellulose nitrate	5	—	7	5	—	—
Epoxy resin	3.7	16	—	2	3	10^{5}
Glass (plate)	6	30	—	0.5	—	10^{13}
Glycerine	50	—	—	3	5	—
Halowax	4	—	0.2	0.2	10	10^{11}
Mica	6	200	0.5	0.03	0.03	10^{15}
Oil (mineral)	2.2	15	0.01	0.01	0.04	10^{14}
Paper (impregnated)	3	50	1	4	7	—
Paraffin	2.1	10	0.02	0.02	0.02	10^{15}
Plywood	2.1	—	1.2	2.3	—	—
Polyethylene	2.2	47	0.02	0.02	0.02	10^{15}
Polystyrene	2.7	24	0.005	0.007	0.01	10^{16}
Polyvinyl	3.2	34	0.3	2	1	10^{15}
Quartz	5	40	0.09	0.02	0.02	10^{17}
Rubber	3	34	0.08	1.2	0.8	10^{13}
Rutile‡ (titanium dioxide, TiO_2)	89–173	—	—	0.03	—	—
Soil (dry sandy)	3.4	—	—	1.7	—	—
Sulfur	4	—	—	—	—	10^{15}
Teflon	2.1	59	0.05	0.02	0.02	10^{15}
Water (distilled)	81	—	—	4	0.5	10^{4}

‡ See footnote for Table 3-1.

† Much of this data was taken from "Tables of Dielectric Materials," vols. I–IV, by the Laboratory for Insulation Research of the Massachusetts Institute of Technology, Cambridge, Massachusetts, 1953.

A-13 SMITH CHART

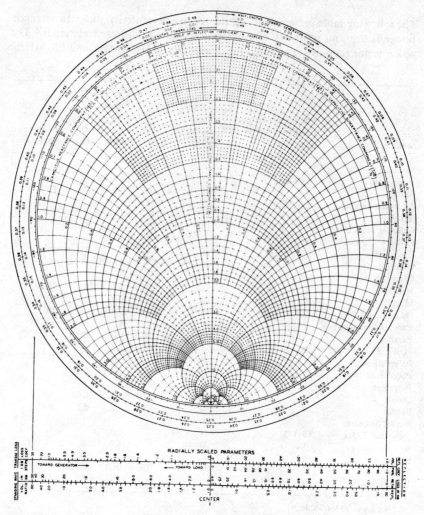

Figure A-3 Smith chart. (©. *Reproduced by permission of Phillip H. Smith. Charts are available from Analog Instruments Co., P.O. Box 808, New Providence, New Jersey, 07974. Regular charts are 72 percent larger than the one above.*)

A-14 NOISE-TEMPERATURE–NOISE-FIGURE CHART

$$T = (F - 1)T_0$$

where T = noise temperature, K
T_0 = 290 K
F = noise figure, dimensionless

$$F_{dB} = 10 \log F$$

where F_{dB} = noise figure, dB

$$F = \frac{T}{T_0} + 1 = 10^{F_{dB}/10} = \text{antilog} \frac{F_{dB}}{10}$$

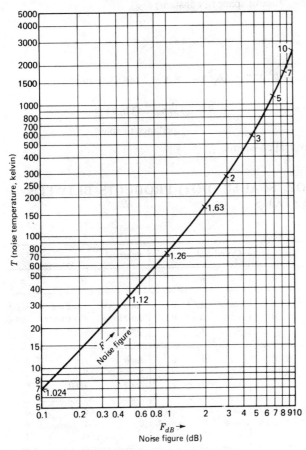

Figure A-4 Noise-temperature–noise-figure chart.

A-15 FUNDAMENTAL FORCE LAWS†

$$F = k_1 \frac{Q_1 Q_2}{r^2} \qquad \text{Two charges}$$

$$F = k_2 \frac{l I_1 I_2}{d} \qquad \text{Two parallel current-carrying wires}$$

Ratio:

$$k_1/k_2 = I^2 r^2 / Q^2 = (\text{distance/time})^2 = v^2$$

where v = wave velocity (a constant)

Product:

$$k_1 k_2 = F^2 r^2 / Q^2 I^2 = (\text{resistance})^2 = R^2$$

where R = intrinsic resistance of space (a constant)

Putting

$$k_1 = 1/4\pi\epsilon_0 \qquad \text{and} \qquad k_2 = \mu_0/4\pi$$

Ratio:

$$k_1/k_2 = 1/\mu_0\epsilon_0 = v^2 \text{ or } v = 1/\sqrt{\mu_0\epsilon_0} = c = \text{velocity of light}$$

Product:

$$k_1 k_2 = \mu_0/\epsilon_0 = R^2 \qquad \text{or} \qquad R = \sqrt{\mu_0/\epsilon_0} = 376.73 \ \Omega$$
$$= \text{impedance of space}$$

where c and R are universal constants.

A-16 UNIT VECTOR, SCALAR (DOT) PRODUCTS BETWEEN COORDINATE SYSTEMS

Table 1

		Rectangular			Cylindrical			Spherical		
$\cdot$		$\hat{\mathbf{x}}$	$\hat{\mathbf{y}}$	$\hat{\mathbf{z}}$	$\hat{\mathbf{r}}$	$\hat{\boldsymbol{\phi}}$	$\hat{\mathbf{z}}$	$\hat{\mathbf{r}}$	$\hat{\boldsymbol{\theta}}$	$\hat{\boldsymbol{\phi}}$
Rectangular	$\hat{\mathbf{x}}$	1	0	0	$\cos\phi$	$-\sin\phi$	0	$\sin\theta\cos\phi$	$\cos\theta\cos\phi$	$-\sin\phi$
	$\hat{\mathbf{y}}$	0	1	0	$\sin\phi$	$\cos\phi$	0	$\sin\theta\sin\phi$	$\cos\theta\sin\phi$	$\cos\phi$
	$\hat{\mathbf{z}}$	0	0	1	0	0	1	$\cos\theta$	$-\sin\theta$	0
Cylindrical	$\hat{\mathbf{r}}$	$\cos\phi$	$\sin\phi$	0	1	0	0	$\sin\theta$	$\cos\theta$	0
	$\hat{\boldsymbol{\phi}}$	$-\sin\phi$	$\cos\phi$	0	0	1	0	0	0	1
	$\hat{\mathbf{z}}$	0	0	1	0	0	1	$\cos\theta$	$-\sin\theta$	0
Spherical	$\hat{\mathbf{r}}$	$\sin\theta\cos\phi$	$\sin\theta\sin\phi$	$\cos\theta$	$\sin\theta$	0	$\cos\theta$	1	0	0
	$\hat{\boldsymbol{\theta}}$	$\cos\theta\cos\phi$	$\cos\theta\sin\phi$	$-\sin\theta$	$\cos\theta$	0	$-\sin\theta$	0	1	0
	$\hat{\boldsymbol{\phi}}$	$-\sin\phi$	$\cos\phi$	0	0	1	0	0	0	1

Note that the unit vectors $\hat{\mathbf{r}}$ in the cylindrical and spherical systems are *not* the same.

† From lecture by Erik Hallén at Ohio State University, Aug. 25, 1954.

Table 2

Rectangular-cylindrical product in rectangular coordinates			
$\cdot$	$\hat{\mathbf{x}}$	$\hat{\mathbf{y}}$	$\hat{\mathbf{z}}$
$\hat{\mathbf{r}}$	$\dfrac{x}{\sqrt{x^2+y^2}}$	$\dfrac{y}{\sqrt{x^2+y^2}}$	0
$\hat{\boldsymbol{\phi}}$	$\dfrac{-y}{\sqrt{x^2+y^2}}$	$\dfrac{x}{\sqrt{x^2+y^2}}$	0
$\hat{\mathbf{z}}$	0	0	1

Example: $\hat{\boldsymbol{\phi}} \cdot \hat{\mathbf{y}} = \cos\phi = \dfrac{x}{\sqrt{x^2+y^2}}$

Rectangular-spherical product in rectangular coordinates			
$\cdot$	$\hat{\mathbf{x}}$	$\hat{\mathbf{y}}$	$\hat{\mathbf{z}}$
$\hat{\mathbf{r}}$	$\dfrac{x}{\sqrt{x^2+y^2+z^2}}$	$\dfrac{y}{\sqrt{x^2+y^2+z^2}}$	$\dfrac{z}{\sqrt{x^2+y^2+z^2}}$
$\hat{\boldsymbol{\theta}}$	$\dfrac{xz}{\sqrt{x^2+y^2}\sqrt{x^2+y^2+z^2}}$	$\dfrac{yz}{\sqrt{x^2+y^2}\sqrt{x^2+y^2+z^2}}$	$-\dfrac{\sqrt{x^2+y^2}}{\sqrt{x^2+y^2+z^2}}$
$\hat{\boldsymbol{\phi}}$	$-\dfrac{y}{\sqrt{x^2+y^2}}$	$\dfrac{x}{\sqrt{x^2+y^2}}$	0

Example: $\hat{\mathbf{x}} \cdot \hat{\mathbf{r}} = \sin\theta \cos\phi = \dfrac{x}{\sqrt{x^2+y^2+z^2}}$

A-17 COMPONENT TRANSFORMATIONS

Rectangular to Cylindrical

$$A_r = A_x \frac{x}{\sqrt{x_2 + y^2}} + A_y \frac{y}{\sqrt{x^2 + y^2}}$$

$$A_\phi = -A_x \frac{y}{\sqrt{x^2 + y^2}} + A_y \frac{x}{\sqrt{x^2 + y^2}}$$

$$A_z = Z_z$$

Rectangular to Spherical

$$A_r = A_x \frac{x}{\sqrt{x^2 + y^2 + z^2}} + A_y \frac{y}{\sqrt{x^2 + y^2 + z^2}} + A_z \frac{z}{\sqrt{x^2 + y^2 + z^2}}$$

$$A_\theta = A_x \frac{xz}{\sqrt{x^2 + y^2}\sqrt{x^2 + y^2 + z^2}} + A_y \frac{yz}{\sqrt{x^2 + y^2}\sqrt{x^2 + y^2 + z^2}}$$

$$- A_z \frac{\sqrt{x^2 + y^2}}{\sqrt{x^2 + y^2 + z^2}}$$

$$A_\phi = -A_x \frac{y}{\sqrt{x^2 + y^2}} + A_y \frac{x}{\sqrt{x^2 + y^2}}$$

Cylindrical to Rectangular

$$A_x = A_r \cos \phi - \acute{A}_\phi \sin \phi$$

$$A_y = A_r \sin \phi + A_\phi \cos \phi$$

$$A_z = A_z$$

Spherical to Rectangular

$$A_x = A_r \sin \theta \cos \phi + A_\theta \cos \theta \cos \phi - A_\phi \sin \phi$$

$$A_y = A_r \sin \theta \sin \phi + A_\theta \cos \theta \sin \phi + A_\phi \cos \phi$$

$$A_z = A_r \cos \theta - A_\theta \sin \theta$$

BIBLIOGRAPHY

The following list includes references for supplemental reading plus a few visual aids for classroom instruction.

Balanis, C. A.: "Antenna Theory Analysis and Design," Harper & Row Publishers, Incorporated, New York, 1982.

Davidson, C. W.: "Transmission Lines for Communications," John Wiley & Sons, Inc., New York, 1978.

Edminister, J. A.: "Theory and Problems of Electromagnetics," McGraw-Hill Book Company, New York, Schaum's Outline Series, 1979.

Elliot, R. S.: "Electromagnetics," McGraw-Hill Book Company, New York, 1966. Provides excellent historical insight.

Enoch, J. M., and F. L. Tobey (eds.): "Vertebrate Photo-receptor Optics," Springer Verlag New York Inc., New York, 1981.

Geddes, L. A., and L. E. Baker: "Principles of Applied Biomedical Instrumentation," 2d ed., John Wiley & Sons, Inc., New York, 1975.

Halliday, D., and R. Resnick: "Physics," John Wiley & Sons, Inc., New York, 1962.

Harrington, R. F.: "Time-harmonic Electromagnetic Fields," McGraw-Hill Book Company, New York, 1961.

Hayt, W. H., Jr.: "Engineering Electromagnetics," 4th ed., McGraw-Hill Book Company, New York, 1981.

Johnson, R. C. (ed.): "Jasik's Antenna Engineering Handbook," 2d ed., McGraw-Hill Book Company, New York, 1984.

Karplus, W. J., and W. J. Soroka: "Analog Methods: Computation and Simulation," 2d ed., McGraw-Hill Book Company, New York, 1959.

Kraus, J. D.: "Antennas," McGraw-Hill Book Company, New York, 1950.

Kraus, J. D.: "Electromagnetics," 1st ed., McGraw-Hill Book Company, New York, 1953. Includes sections on field mapping and on boundary value problems not in later editions.

Kraus, J. D., and K. R. Carver: "Electromagnetics," 2d ed., McGraw-Hill Book Company, New York, 1973. Contains chapters on particles and plasmas and on moving systems and space-time not included in third edition.

Kraus, J. D.: "Radio Astronomy," McGraw-Hill Book Company, New York, 1966. Cygnus-Quasar Books, Powell, Ohio 43065, 1982.

Kraus, J. D.: "Antennas and wave propagation phenomena," Lecture-demonstration. Includes demonstration of Heinrich Hertz's original radio transmitter and receiver. 70-minute color video tape, 1982. Station WOSU-TV, The Ohio State University, Columbus, Ohio 43210.

Midwinter, J. E.: Optical Fiber Systems: Low Loss Links, *IEEE Spectrum*, March 1980, p. 45.

Moore, A. D.: "Fundamentals of Electrical Design," McGraw-Hill Book Company, New York, 1927. A classic.

Plonsey, R., and D. G. Fleming: "Bioelectric Phenomena," McGraw-Hill Book Company, New York, 1969.

Ramo, S., J. R. Whinnery, and R. Vanduzer: "Fields and Waves in Communication Electronics," 3d ed., John Wiley & Sons, Inc., New York, 1965.

"Reference Data for Radio Engineers," H. W. Sams, Indianapolis, Indiana, 6th ed., 1975. Engineering formulas and data for transmission lines, waveguides, antennas, propagation, properties of materials, etc.

Schelkunoff, S. A.: "Electromagnetic Waves," D. Van Nostrand Company, Inc., Princeton, N.J., 1943.

Schelkunoff, S. A.: "Applied Mathematics for Engineers and Scientists," D. Van Nostrand Company, Inc., Princeton, N.J., 1948.

Shive, J. N.: "Similarities in Wave Behavior," 26-minute black and white, 16 mm film, 1960. Excellent demonstration of wave phenomena. A classic. Good supplement to Chap. 10. Available through Bell Telephone Co. local business offices.

Skilling, H. H.: "Fundamentals of Electric Waves," John Wiley & Sons, Inc., New York, 1948. A classic.

Skitek, G. G., and S. V. Marshall: "Electromagnetic Concepts and Applications," Prentice-Hall, Inc., Englewood Cliffs, N.J., 1982.

Skolnik, M. I.: "Introduction to Radar Systems," 2d ed., McGraw-Hill Book Company, New York, 1980.

Stratton, J. A.: "Electromagnetic Theory," McGraw-Hill Book Company, New York, 1941.

Stutzman, W. L., and G. A. Thiele: "Antenna Theory and Design," John Wiley & Sons, Inc., New York, 1981.

Williams, H. J.: "The Formation of Ferromagnetic Domains," 45-minute, 16 mm sound and color film, 1959. Good supplement to Chap. 6. Available through Bell Telephone Co. local offices.

ANSWERS TO STARRED PROBLEMS

Chapter 1

1-2-1. Velocity, L/T, meters/second; work, ML^2/T^2, joules; ratio, dimensionless

1-2-2. (*a*) Charge, IT, coulombs; (*b*) potential, ML^2/T^3I, volts; (*c*) electric field intensity, ML/T^3I, volts/meter; (*d*) potential, ML^2/T^3I, volts; (*e*) constant, ML^3/T^4I^2, meters/farad; (*f*) force, ML/T^2, newtons; (*g*) current density, I/L^2, amperes/meter^2; (*h*) force, ML/T^2, newtons

1-2-3. (*a*) $\mathscr{E}$; (*b*) $\mathscr{E}$; (*c*) $\mathscr{E}T/L$ or ML/T; (*d*) $\mathscr{E}/T$; (*e*) $\mathscr{E}/LIT$ or ML/IT^3; (*f*) $\mathscr{E}/IL^2$ or M/IT^2; (*g*) $\mathscr{E}/L^3$

1-2-4. (*a*) Yes; (*b*) no

Chapter 2

2-1-1. $\mathbf{F} = 3.18\underline{/45°}\ \mu N$

2-1-3. 857 mm

2-1-4. $\mathbf{E} = \hat{\mathbf{y}}11.6\,\text{GN C}^{-1}$

2-1-6. $\mathbf{E} = \hat{\mathbf{z}}11.52\,\text{V m}^{-1}$

2-1-8. $E = 117.6\,\text{V m}^{-1}$

2-1-13. 173 TC

2-2-1. $V = 99.6\,\text{mV}$

2-2-3. (*a*) $V = 0$; (*b*) $\mathbf{E} = 2.25\underline{/135°}\ \text{V m}^{-1}$

2-2-7. $V = 8.66\,\text{V}$

2-2-9. (*a*) $V = 690\,\text{mV}$; (*b*) $\mathbf{E} = -\hat{\mathbf{x}}\,917\text{mV m}^{-1}$

2-2-11. $\hat{\mathbf{n}} = (\hat{\mathbf{x}}2 + \hat{\mathbf{y}}3 + \hat{\mathbf{z}}6)/7$

2-2-15. (*a*) $V = 7.45\,\text{V}$; (*b*) $\mathbf{E} = \hat{\mathbf{z}}5.27\,\text{V m}^{-1}$

2-2-17. (*a*) $V = 4.5\,\text{V}$; (*b*) $\mathbf{E} = \hat{\mathbf{r}}2.25\,\text{V m}^{-1}$

2-2-23. $\mathbf{E} = \hat{\mathbf{x}}2xy^2z^3 + \hat{\mathbf{y}}2x^2yz^3 + \hat{\mathbf{z}}3x^2y^2z^2\,\text{V m}^{-1}$

2-2-26. $V = 2\,\text{V}$; $\mathbf{E} = 2\sqrt{2}\,\underline{/45°}\ \text{V m}^{-1}$

2-3-2. $T = QlE\sin\theta$

2-3-6. $0°$

2-3-8. $(\hat{\mathbf{x}}3 - \hat{\mathbf{y}}6 + \hat{\mathbf{z}})/6.78$

2-3-11. (*a*) $r = 900\,\mu\text{m}$; (*b*) $V = 159\,\text{V}$

2-3-20. $\psi = 770\,\text{pC}$

Chapter 3

3-1-1. (a) $P = 1.67$ C m^{-2}; (b) $Ql = 0.17$ C m
3-1-4. (a) $E_n = D/\sqrt{2}\epsilon_0$; (b) $E_t = D/3\sqrt{2}\epsilon_0$

3-2-2. $E = 727$ V m^{-1}
3-2-8. No.
3-2-10. (a) $V_1/V_0 = 0.625$; (b) $C_1/C_0 = 1.6$
3-2-14. (a) $L = 155$ mm; (b) $L = 150$ mm; (c) $V_a = 51.780$ V, $V_b = 52,670$ V; (d) $V = 45,970$ V; (e) $V = 60,000$ V
3-2-18. 7.2 m^3
3-2-19. (a) 300 kV; (b) 1.5 MV; (c) 600 kV and 3 MV
3-2-24. (a) $W = Q^2/8\pi\epsilon_0 R$; (b) $2R$

3-3-2. (b) $V = 23.8$ V; (d) $C/l = 48.3$ pF m^{-1}
3-3-4. $V = 37.5$ V
3-3-6. (a) $C/l = 55.2$ pF m^{-1}; (b) $E = 3.33$ kV m^{-1}; (c) $E = 794$ V m^{-1}

3-4-1. $\nabla \cdot \mathbf{D} = 6 + 2z$ C m^{-3}
3-4-4. $D = \rho r/3$ C m^{-2} inside r_1; $\nabla \cdot \mathbf{D} = \rho$ inside r_1; $D = \rho r_1^3/3r^2$ outside r_1; $\nabla \cdot \mathbf{D} = 0$ outside r_1

3-5-1. (a) $C = 15.8$ fF; (b) $V = 24$ kV
3-5-6. $q = -mg/NE$
3-5-13. (a) $W = 1.1$ TJ; (b) $V = 11.3$ GV; (c) $E = 2.26$ MV m^{-1}; (d) within 25 percent

Chapter 4

4-1-5. (a) $\rho = 13.7$ GC m^{-3}; (b) $\mu = 4.16 \times 10^{-3}$ m^2 V^{-1} s^{-1}
4-1-8. 11 kΩ km^{-1}
4-1-11. $K = 318$ A m^{-1}

4-2-5. (a) When $R_0 = 4.5$ Ω, $I = 1$ A; when $R_0 = 0$ Ω, $I = 3.25$ A; (b) when $R_0 = 6.5$ Ω, $I = 0.5$ A; when $R_0 = 0$ Ω, $I = 1$ A

4-3-5. (a) $R = 1.97$ $\mu\Omega$; (b) 1.13; (c) current flowing from side 1 to side 2 has a longer average path length in the L-shaped bar.
4-3-9. $R/l = 17.6$ MΩ km^{-1}
4-3-10. (a) $R = 107$ $\mu\Omega$; (b) $R = 63$ $\mu\Omega$; (c) 171 mm
4-3-11. (a) ~ 8 mm
4-3-16. (a) $E = 100$ mV m^{-1}; (b) $E = 910$ mV m^{-1} at 6.3° from normal
4-3-20. (a) $R = 17.9$ Ω; (b) $R = 1.40$ Ω
4-3-21. (a) $R = 1.33$ Ω; (b) $R = 0.188$ Ω
4-3-24. (a) $0.8/\sigma l$ Ω, $\sigma = \infty$; (b) $1.25/\sigma l$ Ω, $\sigma = 0$
4-3-25. (a) $R = 13.1$ $\mu\Omega$; (b) $R = 12.9$ $\mu\Omega$; (c) $R = 14.3$ $\mu\Omega$; (d) current flowing from side 1 to side 2 has a longer average path length for case (a) than (b) but shorter than (c).

4-4-2. (a) $R = 39$ Ω; (b) 0.36 V

Chapter 5

5-1-1. $I = 3.18$ A
5-1-2. (a) $B = 848$ nT
5-1-5. (a) $\mathbf{T}/l = \mathbf{z}Ik$ N; (b) ccw; (c) $\mathbf{T}/l = 0$
5-1-8. $\mathbf{T} = \hat{\phi}41.7$ fN m
5-1-10. 71.6°

5-1-20. (a) $\psi_m = 9.73$ Wb; (b) $\psi_m = 2\pi k r_0$ Wb
5-1-23. $B = 2\pi\mu_0 I/4\pi r_0(e + 1)$

5-2-2. (a) $L = 5.26$ mH; (b) $w_m = 10.05$ J m^{-3}; (c) $w_m Al$
5-2-4. (a) $J = \hat{z}2$ A m^{-2}; (b) $I = 40.5$ A; (c) $\psi_m = 40.5$ Wb
5-2-7. (a) $B = 62.52$ μT; (b) $B = 31.38$ μT; (c) $B = 62.15$ μT
5-2-11. (a) $E(P_2) = 379$ V m^{-1}; (b) $E(P_3) = 30.6$.V m^{-1}; (c) $H(P_2) = 2.37$ A m^{-1}; (d) $H(P_3) = 220$ mA m^{-1}; (e) $C/d = 33$ pF m^{-1}; (f) $L/d = 330$ nH m^{-1}; (g) $D(P_2) = 3.35$ nC m^{-2}; (h) $D(P_3) = 271$ pC m^{-2}; (i) $B(P_2) = 2.98$ μT; (j) $B(P_3) = 277$ nT; (k) $w_e(P_2) = 636$ nJ m^{-3}; (l) $w_e(P_3) = 4.14$ nJ m^{-3}; (m) $w_m(P_2) = 3.53$ μJ m^{-3}; (n) $w_n(P_3) = 30.5$ nJ m^{-3}; (o) $Q/d = 330$ pC m^{-1}; (p) $\psi_m/d = 89$ nWb m^{-1}; (q) $I = 271$ mA
5-2-14. (a) $H = 0$; (b) $H = 318$ A m^{-1}; (c) $H = 167$ A m^{-1}

5-3-1. $J = \hat{z}3kr$ A m^{-2}
5-3-3. $B = 7.07$ μT
5-3-6. $\nabla \times F = -\hat{x}2y + \hat{y}2x$; circle with center at origin

5-4-1. (a) $v = 1$ Mm s^{-1}; (b) $R = 500$ mm
5-4-4. (a) $W = 14$ MeV; (b) $W = 14$ MeV; (c) $W = 7$ MeV; (d) $n_\alpha = 388$ r, $n_p = 777$ r, $n_d = 388$ r
5-4-6. $R =$ (a) 20.8 m; (b) 1.04 km; (c) 10.4 Mm
5-4-8. $W = (\text{Re}B)^2/2m$
5-4-14. $R = 2$ mm
5-4-16. (a) $R = 500$ m; (b) $R = \infty$; (c) $E = 10$ MV m^{-1}

5-5-1. (a) $V_d = 1.4$ kV; (b) $B = 2.82$ mT
5-5-3. $V = 25$ kV
5-5-9. $I = 257$ A

Chapter 6

6-1-3. (a) $L/l = 10.05$ μH m^{-1}; (b) 0.5%
6-1-4. (a) $\mu_0[(N^2\pi a^2/l^2) + (1/2\pi) \ln b/a]$; (b) $\mu_0[(\mu_r N^2\pi a^2/l^2) + (1/2\pi) \ln b/a]$
6-1-8. $F_{\max} = (Q_m l)^2(\mu - \mu_0/\mu + \mu_0)\mu_0(3/32\pi d^4)$;
$\quad T = (Q_m l)^2(\mu - \mu_0/\mu + \mu_0)\mu_0(\sin \theta/16\pi d^3)\sqrt{\cos^2 \theta + \tfrac{1}{4} \sin^2 \theta}$

6-3-1. $(BH)_{\max} = -5$ kJ m^{-3}
6-3-2. $w_m = H_1^2(\mu_i/2 + 2aH_1/3)$ J m^{-3}

6-4-4. 5.1 $\pm$ 0.5 m south, 6.7 $\pm$ 1.7 m deep
6-4-6. 0.889

6-5-2. $NI = 2.4$ kA-turns
6-5-4. $B = 327$ mT
6-5-6. $F = 78.3$ kN
6-5-9. $\psi_m = 1$ mWb
6-5-11. (c) 200 MA m

6-6-5. ~ 60 kA-turns for each coil
6-6-6. (a) $m = 9.49 \times 10^{22}$ A m^2; (b) $B_{eq} = 95.2$ nT, $B_{polar} = 190.4$ nT
6-6-7. (a) $B = 20$ μT; (b) 9%; (c) $B = 10$ μT; (d) 12%

Chapter 7

7-1-2. (a) $V = 6.67$ V
7-1-8. $V = 1.94$ V
7-1-9. $V = 570$ mV

7-2-1. $V = V_1 \sin(\pi\alpha/2\theta)$ (α is the angle from $V = 0$ plate)

7-2-3. (a) $K_\phi = -3B_0 \sin \theta/2\mu_0$ A m^{-1}

7-2-5. $V_0 = \dfrac{-E_0 \cos \phi}{r} \dfrac{[\epsilon(b^2 + a^2)(r^2 - b^2) + \epsilon_0(b^2 - a^2)(r^2 + b^2)]}{\epsilon(b^2 + a^2) + \epsilon_0(b^2 - a^2)}$

$V_i = \dfrac{-2E_0 b^2 \epsilon_0 \cos \phi(r^2 - a^2)}{r[\epsilon(b^2 + a^2) + \epsilon_0(b^2 - a^2)]}$

7-2-9. (a) $V = [(a/r)^3 - 1]E_0 r \cos \theta$; (b) $\rho_s = 3\epsilon_0 E_0 \cos \theta$ (on sphere), $\rho_s = \epsilon_0 E_0[1 - (a/r)^3]$ (on sheet)

7-2-11. $V = \rho_0(x_1^3 - x^3)/6\epsilon x_1$

7-2-12. (a) $V = (\rho_L/\epsilon) \ln [(x + \sqrt{x^2 + x_1^2})/x_1]$; (b) $C = 0.094\epsilon A/x_1$

7-2-13. $C = 150$ fF

Chapter 8

8-1-2. (a) $\mathscr{V} = \pi NBR^2$ (V); (b) $T = \frac{1}{2}IBR^2$ (N m); (c) cw

8-1-6. (a) $V_{av} = 8NRIB$ (V); (b) $T_{av} = 4IRlB/\pi$ (N m); (c) ccw

8-1-10. $\mathscr{V} = 78.5$ mV

8-1-12. $\mathscr{V} = 2vB\sqrt{R^2 - v^2t^2}$ (V)

8-1-13. (a) cw; (b) ccw

8-1-21. (a) $\mathscr{V} = 1$ V; (b) $\mathscr{V} = 1$ V

8-1-22. (a) $V = -312$ V; (b) $V = -50.3$ V; (c) $V = -362.3$ V

8-1-26. (a) $\mathscr{V} = \pi f B_0 R^2 \cos 4\omega t$; (b) $\mathscr{V} = 1,777$ V

8-2-1. $M = (\mu_0 l/2\pi) \ln r_2/r_1$ (H)

8-2-3. (a) $M/L = (\mu_0/2\pi)\{\ln [1 + \sqrt{1 + (s/L)^2}] - \sqrt{1 + (s/L)^2} + s/L\}$ H m^{-1}; (b) $\mathscr{V}/L = \mu Iv/2\pi s$ V m^{-1}

8-2-4. (a) $M = (\mu Nl/2\pi) \ln c/b$ (H); (b) $V = (\omega\mu NlI_0/2\pi)(\cos \omega t) \ln c/b$ (V)

8-3-1. $P = 1.34$ kW

8-3-2. 86.6 W

8-3-4. (a) $I = 15.3$ A; (b) $P = 1.06$ kW

8-3-6. $P = 5.22$ W

8-4-1. (a) 400; (b) parallel to **B**

8-4-3. $P = 2.85$ kW

8-4-5. (a) $I = 50$ kA; (b) $j5$ GΩ

8-4-7. (a) $K = 2.39$ MA m^{-1}; (b) $\mathscr{V} = 5.65$ V

8-4-9. (a) $J_d = 0.7$ A m^{-2}; (b) $V = 1.57$ V; (c) $J_d = 5.76$ A m^{-2}; $V = 13.7$ V

8-4-11. (a) $I = 265$ mA; (b) $P = 701$ mW; (c) $P_{max} = 844$ mW; (e) yes; (f) yes

Chapter 9

9-1-5. Because of no isolated magnetic charges

9-1-8. $J_d = \frac{1}{2}\omega^2\epsilon_0 B_0 r \sin \omega t$ A m^{-2}

Chapter 10

10-1-5. 2.6

10-2-1. (a) $\rho = 0.62 \underline{/83°}$; (b) VSWR = 4.26; (c) $14 - j20$ Ω; (d) 0.115λ; (e) 213 Ω

10-2-5. $d = 0.342\lambda$, $Z_1 = 229$ Ω

10-2-9. $d_1 = 0.304\lambda$, $d_2 = 0.106\lambda$

10-2-11. $d_1 = 0.021\lambda, d_2 = 0.106\lambda$
10-2-13. $d_1 = 0.195\lambda, d_2 = 0.150\lambda$
10-2-16. $A = 0.33\lambda, B = 0.044\lambda$
10-2-18. $d_1 = 0.035\lambda, d_2 = 0.106\lambda$
10-2-21. (a) $P = 833\ \mu\text{W}$; (b) $V_{max} = V_{min} = 500\ \text{mV}$

10-3-7. $u = 13.9\ \text{Mm s}^{-1}$

10-4-3. $t = 77.7\ \text{mm}$
10-4-4. (a) $t = 54.8\ \text{mm}$; (b) $\epsilon_r = 1.87$
10-4-7. $398\ \mu\text{m}$
10-4-9. $1.64\ \text{dB}$
10-4-11. (a) $377\ \Omega$; (b) $274\ \text{mm}$

10-5-1. $-9.9\ \text{dB}$
10-5-3. (a) $Z_0 = 337.2\ \Omega$; (b) $f = 60\ \text{MHz}$
10-5-8. (a) $S = 1.4\ \text{kW m}^{-2}$; (b) $P = 3.95 \times 10^{26}\ \text{W}$; (c) $E = 726.5\ \text{V m}^{-1}$; (d) $t = 8.33\ \text{min}$
10-5-10. (a) $S = 10\ \text{pW m}^{-2}$; (b) $E = 61.4\ \mu\text{V m}^{-1}$; (c) $P = 2.83\ \text{TW}$

10-6-1. (a) $S/N = 12.2\ \text{dB}$; (b) $d = 2.3\ \text{m}$; (c) $d = 1\ \text{m}$
10-6-5. (a) $S = 5.51\ \text{aW m}^{-2}$; (b) $E = 45\ \text{nV m}^{-1}$; (c) $t = 1.27\ \text{s}$; (d) $4.14 \times 10^6\ \text{photons m}^{-2}\text{s}^{-1}$

Chapter 11

11-1-1. (a) $AR = -2.33$; (b) $\tau = -45°$; (c) RH
11-1-3. (a) AR = 5; (b) RH; (c) $S = 69\ \text{mW m}^{-2}$
11-1-5. (a) AR = 6; (b) $\tau = 0°$; (c) RH
11-1-9. $P = 14.9\ \text{kW}$

11-2-2. $P = 4.635\ \text{W}$
11-2-4.

		Antenna		
		1	2	3
	a	0.50	0.50	0.50
Wave	b	0.50	0.25	0.35
	c	0.50	0.20	0

11-2-8. $d =$ (a) 0%; (b) 100%; (c) 50%; (d) 70.7%; (e) 100%

Chapter 12

12-1-2. AR = 2.62
12-1-3. (a) AR = 1.34; (b) AR = ∞; (c) 24.1°
12-1-5. AR = (a) 1.43; (b) ∞
12-1-6. AR = 3.74

Chapter 13

13-1-1. (a) $\lambda_{oc} = 200\ \text{mm}$; (b) $v_{ph} = 384\ \text{Mm s}^{-1}$
13-1-3. $TE_{10}, TE_{01}, TE_{02}, TE_{20}, TE_{11}, TM_{11}, TE_{12}, TM_{12}, TE_{21}, TM_{21}$
13-2-1. (a) $\lambda_{oc} = 82\ \text{mm}$; (b) $TM_{01}, TE_{11}, TM_{11}, TE_{21}$
13-2-7. TEM: $\alpha = 5.09\ \text{mNp m}^{-1}$; TE_{10}: $\alpha = 12.8\ \text{mNp m}^{-1}$
13-2-9. $\alpha = 8.42/r_0\ \text{dB m}^{-1}$, where r_0 = radius
13-2-11. (a) $P = 10.9\ \text{nW}$; (b) $P/l = 550\ \text{pW m}^{-1}$; (c) $\alpha = 0.22\ \text{dB m}^{-1}$
13-2-14. (a) $P = 3.98\ \mu\text{W}$; (b) $0.46\ \text{dB m}^{-1}$

13-3-1. $S =$ (a) 26.5 W m^{-2}; (b) 182 μW m^{-2}; (c) 728 μW m^{-2}
13-3-3. $S = 1.35$ nW m^{-2}
13-3-10. $f_c = 0$, $\alpha = 8.89/\lambda_0$ Np m^{-1}
13-3-11. $\theta_e = 14.3°$
13-3-12. (a) $d = 6.8$ μm; (b) $\theta_e = 7.13°$
13-3-13. 4.2 dB km^{-1}

13-4-2. (a) $d = 17.5$ mm; (b) $Q = 9544$
13-4-4. (a) $d = 7.07$ mm; (b) $Q = 2,565$
13-4-6. (a) $d = 7.7$ mm; (b) $Q = 5,178$

13-5-2. (a) 0; (b) $-j377$ Ω; (c) ∞; (d) 1.36; (e) 1.04
13-5-3. (a) $\lambda_{oc} = 35.2$ MHz; (b) run a wire along the inside of the tunnel so its acts as a coaxial line and extend wire beyond end of tunnel to act as an antenna
13-5-4. $\lambda = 4$ m

Chapter 14

14-1-1. $R_r = 3.33$ Ω
14-1-3. (a) $\Omega_A = \pi$ sr; (b) $D = 4$
14-1-4. (a) $\Omega_A = 1.84$ sr; (b) $D = 6.83$; (c) $A_e = 0.543$ λ^2; (d) $R_r = 76.3$ Ω
14-1-5. $R_r = 354$ Ω
14-1-7. (a) $D = 8.16$; (b) $R_r = 654$ Ω

14-2-6. Max: 0°, 180°, $\pm 60°$, $\pm 90°$, $\pm 120°$
Nulls: $\pm 41.4°$, $\pm 75.5°$, $\pm 104.5°$, $\pm 138.6°$
14-2-8. 78.3°
14-2-12. (a) $\epsilon_{ap} = 81.1\%$; (b) $D = 31.2$ dB
14-2-14. (b) HPBW $= 4.2°$

14-3-9. (a) LP, $\tau = 45°$; (b) CP; (c) EP, AR $= 2.41$, $\tau = 45°$
14-3-13. (a) $d = 0.354\lambda$; (b) $\delta = -\pi/2$

14-4-2. (a) $\lambda/4$; (b) 141 Ω; (c) 6.6 dBi
14-4-12. $Z = 778 - j86$ Ω
14-4-13. (a) $l = 80\lambda$, $\theta_E = 2.86°$, $\phi = 4.52°$, $w = 12.7\lambda$; (b) HPBW(E) = 7°, HPBW(H) = 5.3°

14-5-1. $P = 110$ W
14-5-3. $P = 25$ mW
14-5-4. (a) $\Delta T_{min} = 0.015$ K; (b) $\Delta S_{min} = 52$ mJy; (c) $\Delta T_{min} = 0.008$ K, $\Delta S_{min} = 26$ mJy
14-5-8. (a) $w = 2.08 \times 10^{-27}$ J m^{-2} Hz^{-1}; (b) $W = 4.68 \times 10^{17}$ J; (c) $t = 5.2 \times 10^{13}$ years
14-5-12. (a) MUF $= 15.6$ MHz; (b) MUF $= 26.1$ MHz

14-6-2. (a) No; (b) 36 dB; (c) 3,948
14-6-3. (a) $A_e = 7,162$ m^2; (b) $P_r = 5.7$ W; (c) $P_{reqd} = 3.3$ fW; (e) no; (f) no
14-6-8. (a) $T_{min} = 1.58$ mK; (b) $S_{min} = 1.5$ Jy
14-6-10. (a) $S/N = 15.7$ dB; (b) 3.5 dB more than for TV carrier; (c) ~ 6 min
14-6-11. (a) $L = 1.7\lambda$; (b) $d = 1.4\lambda$

INDEX

Abramowitz, M., 526n.
Absolute potential, 22
Absolute zero, 118
Absorbing sheet, 490 (Prob. 10-4-14)
Absorption coefficient of forest, 709
Acceptors, 121
Aircraft indicator for bank and pitch, 55 (Prob. 2-4-1)
Amber, 1, 11
Ampère, André Marie, 2, 3, 162
Ampere (unit), 113, 155
  definition of, 6
Ampere-turns, 174
Ampère's law, 170, 366
Ampère's theory, 216
Amplifier, GaAsFET, 433
Analog computer solution, 294−297
Analog field plotter, 296−297
Angle:
  of incidence, 513
  of reflection, 513
  of refraction, 511
Angle factor, 36
Anistropic materials, 57
Answers to problems, 749−754
Antarctica, 493 (Prob. 10-6-4)
Antenna:
  aperture, effective, 638
    (See also Aperture)
  array (see Arrays)
  broad-band, basic, 677−679
  directivity, 635
  efficiency, 632
  fields, computer-calculated, 659−661
  gain, 636
  impedance, self- and mutual, 675−683, 712
  reactance, 638
  resistance, 638
  temperature, 493 (Prob. 10-6-1), 613, 703, 707
  thickness, 676
Antenna and antenna system relations, table of, 712
Antennas, 612−713
  complementary, 693−694
  current distribution on, 658

Antennas (Cont.):
  dipole, 657
  Fourier transform relation for, 655−657
  frequency-independent, 671−673
    active region, 671
    equiangular spirals, 671
    log-periodic, 671−673
    pocket ruler, 671, 672
  half-wave dipole (see Half-wave dipole)
  helical-beam, 666−670
    (See also Helical-beam antennas)
  horn, 694−695
    pyramidal, 694−695
  interferometer (see also Interferometer)
  isotropic, 636
  lens (see Lens antenna)
  linear, 657−659
  loop, 665−666
  mutual impedance of, 679−680, 720 (Prob. 14-4-8)
  parabolic dish, 653
  phase center of, 659
  pocket ruler, 671, 672
  polarization state, 503, 507
  prolate spheroid, 659−660
  radio telescope, 703
    (See also Radio telescope)
  receiving, 612, 637
  reflector, 683−693
    corner, 688−689
    finite-flat-sheet, 684−686
    infinite-flat-sheet, 684
    parabolic, 689−691
    thin, 686−688
  resonance of, 659
  retroreflector, 720 (Prob. 14-4-10)
  satellite, 492, 669
    (See also Satellite topics)
  short dipole (see Short dipole)
  slot, 693−694
  super-gain, 669
  transmitting, 612, 637
  traveling-wave, 661−664
  V, 663, 664, 677−678
  Yagi-Uda, 688

Antiferromagnetic materials, 214
Aperture:
  effective, 616, 633, 638, 695, 712
  efficiency, 696, 712
  physical, 616, 696
  scattering, 701
Aperture concept, 695−697
Aperture distribution, 657
  continuous, 653−655
Approximation formulas, 735−736
Array:
  factor, 712
  gain, 683
  long broadside, 716 (Prob. 14-2-11)
  theory, 640
Arrays:
  beam width of, 670
  binomial, 642−643
  collinear, 717 (Prob. 14-3-1)
  continuous, 653
  directivity of, 670
  with *n* sources, 643−648
    broadside case, 645−646
    end-fire case, 646, 648
  null directions, 645
  phase center, 641
  scanning, 670
  of six-point sources, 650
  of two isotropic point sources, 640−641
  uniform, 643
Artificial dielectric, 104 (Prob. 3-1-6)
Astronomical unit, 732
Asymmetric coaxial line, 106 (Prob. 3-3-4), 203
  (Prob. 5-2-11)
Atom as electromechanical system, 346
Attenuation:
  at frequencies greater than cutoff, 573−578
  at frequencies less than cutoff, 571−572
Attenuation constant, 393, 575
Attenuation for frequencies less than cutoff,
  572
Auxiliary equation, 392
Auxiliary line, 487 (Prob. 10-2-24)
Axial bright spot, 529
Axial ratio, 495
Axon, 475,−476, 597

B on solenoid axis, 203 (Prob. 5-2-8)
Backlobe, antenna, 614
Bailin, Louis, xvi
Baker, L. E., 747
Balance of forces, 209 (Prob. 5-5-9)
Balanced transmission line, 620
Balanis, C. A., 747
Balloon:
  charged, 55 (Prob. 5-3-21)
  contracting, 357 (Prob. 8-1-18)
Bandwidth, 420−424
  of helical beam antenna, 668, 669
  predetection, 706

Bar:
  with hole, 143 (Prob. 4-3-10)
  with notches, 278 (Prob. 6-4-2)
  with and without notches, 278 (Prob. 6-4-5)
Bar and strip, 143 (Prob. 4-3-11)
Bar magnet, 213
  magnetization, 277 (Prob. 6-1-9)
  translational force on, 277 (Prob. 6-1-6)
Barrier surface, 175
Barrow, W. L., 543*n*., 561*n*.
BASIC computer programs, 53 (Prob. 2-2-32), 714
  (Prob. 14-2-1), 715 (Prob. 14-2-2)
Beam:
  efficiency, 635*n*.
  solid angle, 615, 634, 712
  widths, 633, 670, 712
    half power, 633
    3-dB, 633*n*.
Bennett, P. I., 146 (Prob. 4-4-6)
Bessel function, 303, 564, 738
Betatron, 205 (Prob. 5-4-2)
BH product, 251
Binomial array, 642−643
Binomial series, 424, 643
Biomedical topics:
  axon, 475−476, 597
    active transmission line, 475
    dendrites, 475, 597
    myelin sheath, 475
    neurons, 475
    sciatic nerve, 475
    soma, 475
    synapses, 475
  bone, field in, 110 (Prob. 3-5-16)
  chest resistance, 146 (Prob. 4-4-5)
  defibrillator, 146 (Prob. 4-4-6)
  eye, 596−597
    bipolar cell, 597
    cones, 596
    outer segments, 596, 597
    pigmented layer, 596, 597
    retina, 596, 597
    rods, 596
    tapetum, 596
  fibrillation, 146 (Prob. 4-4-6)
  heart: defillibrator, 146 (Prob. 4-4-6)
    dipole field, 93−94, 110 (Prob. 3-5-15)
Biot-Savart law, 152−153
Bipolar cell, 597
Birks, J. B., 58*n*.
Blackbody radiator, 491 (Prob. 10-5-12)
Blackbody temperature, 705
Boltzmann's constant, 491 (Prob. 10-5-12), 704,
  732
Booker, H. G., 655*n*., 693*n*.
Born, M., 500*n*., 509*n*.
Bound wave, 586
Boundary:
  between conductor and dielectric, 67
  between transmission lines, 456
  between two dielectrics, 66

Boundary conditions;
  coaxial line, 304
  parallel-plate capacitor, 308
  rectangular waveguide, 548
  two conducting planes, 299
Boundary relations, 63−68, 234−237, 350−351
  electric and magnetic fields, table of, 351
  electric fields, table of, 68
  magnetic fields, table of, 238
Boundary-value problems, 282
Bounded fields, 282
Boxed slot, 720 (Prob. 14-4-12)
Braking field, 207 (Prob. 5-4-16)
Breakdown:
  of capacitor, 72
  for sphere, 105 (Prob. 3-2-19)
Breakwater, 541
Brewster angle, 518−519
Brillouin, Leon, 436n.
Broadcast array, 648−650, 718 (Prob. 14-3-13)
Broadside arrays, 714 (Prob. 14-2-1)
Brown, G. H., 686n.
Buried pipeline, 278 (Prob. 6-4-4)
Burnside, Walter, xvi
Bushing:
  capacitor, 105 (Prob. 3-2-14)
  high voltage, 105 (Prob. 3-2-14)

Can, potential in:
  cylindrical, 315 (Prob. 7-1-3)
  tall and short, 317 (Prob. 7-1-12)
Candela, definition of, 6
Capacitance, 69
  cell, 76
  of earth, 104 (Prob. 3-2-3)
  earth-ionosphere, 104 (Prob. 3-2-4)
  energy, 106 (Prob. 3-2-21)
  per unit length, 85, 388
Capacitive load, 410
Capacitor, 69, 359 (Prob. 8-2-7)
  with air gap, 104 (Prob. 3-2-1)
  breakdown, 72
  displacement current, 377 (Prob. 9-1-9)
  energy in, 74−77, 104 (Prob. 3-2-11), 105
    (Prob. 3-2-18)
  field-cell, 89
  parallel-plate, 69−70
  plates at angle, 318 (Prob. 7-2-1)
  for radio transmitter, 108 (Prob. 3-5-1)
  with space charge, 308−310
  three-section, 104 (Prob. 3-2-2)
  unequal, 105 (Prob. 3-2-20)
  wedge-shaped, 104 (Prob. 3-2-6)
Carrier-to-noise ratio *(C /N)*, 493 (Prob. 10-6-1)
Carver, Keith R., xvi, 747
Catérpillar, 438
Cathode, 308
Cathode-ray tube (CRT):
  electric deflection, 206 (Prob. 5-4-12)
  magnetic deflection, 207 (Prob. 5-4-12)

Cavities, 276 (Prob. 6-1-5)
  cylindrical, 609 (Prob. 13-4-2)
  in dielectric, 104 (Prob. 3-1-3)
  silver plated, 609 (Prob. 13-4-6)
Cavity resonators, 597−604
  coupling to, 603
Celestial source, 709
Cell (*see* Field cell)
Chalmers, J. A., 55 (Prob. 2-4-1)
Chan, L. C., 364 (Prob. 8-4-12)
Characteristic impedance, 581
  of coaxial and two-wire lines, 402
  by dc measurement, 399
  of field-cell transmission line, 396
  of transmission lines, table of, 395
Characteristic resistance, 394
Charge, shell of, 42−44
Charge and mass of particles, table of, 197
Charge density, 28
  linear, 29
  surface, 29
  volume, 29
Charge density and mobility for copper, 140 (Prob.
  4-1-5)
Charge-free boundary, 65
Charged particles:
  in electric fields, 192−195
  in magnetic fields, 195−196
Charges:
  in gases, 120
  in liquids, 120
  in plasmas, 120
Chawla, B. R., 316n.
Chest resistance, 146 (Prob. 4-4-5)
Child, D. C., 310n.
Child-Langmuir space-charge law, 310
Chu, L. J., 543n., 569n.
Chu, T. S., 610 (Prob. 13-5-2)
Circuit and field quantities, table of, 376
Circuit and field relations, table of, 483
Circuit and field theory, applications of, 366−368
Circuit application of Poynting vector, 472−475
Circuit quantities, 396
Circuits, 534
Circular loop, 717 (Prob. 14-3-4)
Circular polarization, 495−498
Circular-square loop, 200 (Prob. 5-1-14)
Circularly polarized wave, 509 (Prob. 11-1-1)
  at Brewster angle, 531 (Prob. 12-1-3)
  at 45°, 531 (Prob. 12-1-2)
  at 50°, 531 (Prob. 12-1-7)
  for more power, 509 (Prob. 11-1-7)
Circularly polarized wave and prism, 531 (Prob.
  12-1-6)
Circularly polarized wave reflection and
  transmission, 509 (Prob. 11-1-4)
Cladding, 593
Clarke, Arthur C., 480
Clarke orbit, 669
Clarke-orbit satellite, 480−481, 492 (Prob.
  10-6-1), 709, 722 (Prob. 14-5-13)

Clemmow, P. C., 655n.
Closure current for relay, 279 (Prob. 6-5-8)
Coaxial line, 304–307, 401
  asymmetric, 106 (Prob. 3-3-4), 203 (Prob.
    5-2-11)
  with ferrous-titanate, 276 (Prob. 6-1-3)
  flexible, 108 (Prob. 3-5-3)
  with helical inner conductor, 276 (Prob. 6-1-4),
    485
  inductance of, 167, 204 (Prob. 5-2-15)
  RG-59/U, 493 (Prob. 10-6-2)
  skin effect, 204 (Prob. 5-2-16)
  with square outer conductor, 108 (Prob. 3-3-2),
    400
Coaxial transmission line, 81, 134, 320 (Prob.
    7-3-1), 367, 391
Coercive force, 245
Coherency matrix, 509
  power, 510 (Prob. 11-2-9)
Cohn, S. B., 569n.
Collinear array, 717 (Prob. 14-3-1)
Collins, Stuart, xvi
Columbium, 118
Communication satellite (see Satellite topics)
Comparison:
  of circuits, lines, and guides, 534–535
  of electric and magnetic circuits, table of, 258
  of electric and magnetic field relations for time-
    changing situations, table of, 353
  of field equations, table of, 274
Compass deflection, 276 (Prob. 6-1-2)
Compass needle, 147
Complementary antennas, 693–694
Complementary solution to Poisson's equation, 307
Complex conjugate, 465
Complex power, 633
Complex quantity, 344
Computer methods:
  analog, 294–297
  attenuation on lossy line, 490 (Prob. 10-4-12)
  broadside arrays, 714 (Prob. 14-2-1)
  end-fire arrays, 715 (Prob. 14-2-2, 14-2-3)
  field map of 8 point charges, 50–51 (Prob.
    2-2-28)
  field map of 3 point charges, 49–50 (Prob.
    2-2-27)
  potential plots, 53 (Prob. 2-2-32)
  two-element interferometer, 715 (Prob. 14-2-4)
  two sources in phase quadrature, 715 (Prob.
    14-2-5)
Computer programs:
  BASIC, 53 (Prob. 2-2-32), 714 (Prob. 14-2-1),
    715 (Prob. 14-2-2)
  FORTRAN IV, 293, 453
Condenser, 69n.
Conductance, 118
  per unit length, 388
Conducting media, 388, 447–454
  Laplace's equation for, 138–139
Conducting sheet between two conducting planes,
    297–303

Conducting shell, 44–47
Conduction current, 340, 445
Conductivity, 118
  super-, 118
  table of, 123
Conductor, 11, 44, 112, 120, 445–447
  cell, 135–137
  charge density for, 140 (Prob. 4-1-4)
  cylindrical, over ground plane, 312–314
  infinite linear, 153–154
  at interface, images, 320 (Prob. 7-2-14)
  junction, 129
  parallel to sheet, 145 (Prob. 4-3-22)
  with shield, 277 (Prob. 6-2-6)
  stack, 144 (Prob. 4-3-13)
Conductor-conductor boundary, 134–135, 144
    (Prob. 4-3-15)
Conductor-insulator boundary, 131–133
Cones, 596
  concentric, 318 (Prob. 7-2-2)
Conformal transformation, 320 (Prob. 7-3-2)
Conical pattern, 713 (Prob. 14-1-3)
Conservation of charge, 131n.
Conservative field, 25
Constant-conductance line, 418
Constant-phase point, 439
Constants:
  common media, table of, 447
  table of, 732–733
  for transmission lines and waves, table of, 484
Constitutive relations, 274, 373
Contact pressure, 280 (Prob. 6-6-4)
Continuity, 129
Continuity relation, 131, 373
Continuous aperture distribution, 653–655
Continuous array, 653
Convolution, 532 (Prob. 12-2-1)
Cooper, John, xvi
Coordinate diagrams, 738–740
Copper, table of penetration depths, wavelength,
    velocity, and refractive index, 451
Copper-germanium comparison, 141
    (Prob. 4-2-3)
Core fiber, 593
Corner reflector, 688–689, 692–693, 719 (Prob.
    14-4-6)
Corner retroreflector, 720 (Prob. 14-4-10)
Cornu's spiral, 526
Corona discharge, 73
Corrugated horn, 720 (Prob. 14-4-15)
Corrugated surface, 587
Cosine integrals, 659
Cosmic ray, 206 (Prob. 5-4-10)
Coulomb, Charles A. de, 2, 3, 12
Coulomb force, 14
Coulomb's law, 12
Coulomb's Law Committee, 59n.
Coupling factor, wave-to-antenna, 507
Coupling to cavity resonators, 603
Cowan, J. D., Jr., xvi, 661
Criterion of detectability, 706

Critical angle, 515
  of incidence, 595
Critical frequency, 722 (Prob. 14-5-12)
Cross-field ellipse, 585
Cross product, 159
Cross section, radar, 702
Crossed dipoles, 717 (Prob. 14-3-9)
Crystals, 58n.
Cube of charge, 108 (Prob. 3-4-7)
Cubical box, 315 (Prob. 7-1-2)
Curie temperature, 214
Curl, 179−186
Curl meter, 183
Current, 112−114
  continuity relation, 131
  density of, 114
  distribution on antenna, 658
  induced, 322
  loop, 213
  mapping, 135−138
  model, 219, 225
    magnetization, 275
  retarded, 618
  sheet, 235
  solenoidal, 129
Current-carrying conductor over ground plane, 314
Current-carying loop, 155−156
Current-carrying wire, 148−149
Curvilinear square, 88, 471
Cutoff frequency:
  cylindrical guide, 568
  rectangular guide, 555
Cutoff wavelength, 538
  cylindrical guide, 568
  rectangular guide, 555
Cyclotron, 205 (Prob. 5-4-4)
Cyclotron frequency, 197n.
Cyclotron magnet, 280 (Prob. 6-6-5)
Cylinder of charge, 80
Cylindrical guide, 605 (Prob. 13-2-1)
  attenuation, 605 (Prob. 13-2-3)
  modes, table of, 567
  power, 606 (Prob. 13-2-4)
  TM fields, 606 (Prob. 13-2-10)
  (See also Waveguide, cylindrical)

D'Alembert's equation, 382
Davidson, C. W., 747
DC machine, 354 (Prob. 8-1-6)
DC measurement, 399
DC transmission line, 427
Debye, P., 347n., 593n.
Degaussing, 281 (Prob. 6-6-7)
Degaussing earth, 281 (Prob. 6-6-8)
Degaussing ship, 281 (Prob. 6-6-7)
Degree of polarization, 506
Del operator, 32, 185
Delay line, 485 (Prob. 10-1-6)
Delay time, 424

Demagnetization, 252−254
  by reversals, 252
Demagnetization curve, 250
  of permanent magnet, 278 (Prob. 6-3-1)
Dendrites, 475, 597
Deperming, 252
Depth of penetration, 449
Detection:
  of automobiles, 364 (Prob. 8-4-12)
  of dielectric anomalies, 364 (Prob. 8-4-12)
  of pipes, 364 (Prob. 8-4-12)
Deuteron, proton, and alpha particle, 205 (Prob. 5-4-3)
Diamagnetic materials, 214
Dicke, R. H., 431
Dielectric, 11, 57−58, 120, 445−447
  artificial, 104 (Prob. 3-1-6)
  class A, 57n.
Dielectric-conductor boundary, 64
Dielectric constant, 57n.
Dielectric guide, 587
Dielectric hysteresis, 346, 359 (Probs. 8-3-1, 8-3-2, 8-3-3), 610 (Prob. 13-5-6)
Dielectric-lens antenna, 691
Dielectric materials, table of, 741
Dielectric sandwich, 104 (Prob. 3-1-5)
Dielectric sheet waveguide, 589−593
Dielectric slab, 103 (Prob. 3-1-1)
Dielectric strength, 72−74
  table of, 74
Differential permeability, 338
Diffraction, 511
  behind conducting strip, 528
  over mountain, 527
  of waves, 524
Diffraction effects, 611 (Prob. 13-5-10)
Dimensional analysis, 8, 9
Dimensional balance, 9, 19 (Prob. 1-2-4)
Dimensions, 5
Diode, 122, 308
Dipole:
  center-fed, 657
  electric (see Electric dipole)
  magnetic (see Magnetic dipoles)
  short: fields, table of, 629
    quasi-stationary case, 627
    radiation fields, 628
    (See also Short dipole)
Dipole antenna, half-wave, 659−661
Dipole fields, 274
Dipole moment, 275
Dipole patterns, 627
Directional broadcast array, 723 (Prob. 14-6-1)
Directional coupler, 419−420
Directional pattern, 714 (Prob. 14-1-5)
Directivity, 615, 633, 635, 636
  approximate formula for, 647
  of isotropic antenna, 636
  relation to gain, 637
  of short dipole, 636
  table of, 670

Directors, 686–688
Discontinuity, 129
Dish antenna, parabolic, 653
Dish gain, 723 (Prob. 14-6-2)
Disk charge, 54 (Prob. 2-3-12)
Dispersive medium, 437, 541
Displacement current, 340, 445
Displacement-current density, 359 (Prob. 8-2-6)
Distance factor, 36
Distributed quantities, 388
Divergence, 185
   in capacitor, 99–102
   compared with curl, 186–188
   of **D**, 94–97
   of **J**, 130
Divergence operator, 97
Divergence theorem, 99
Dixon, Robert S., 292n.
Domains, 214, 238
Dominant mode, 558n.
   for cylindrical guide, 596
Donors, 121
Doping, 121
Downlink, TV, 492 (Prob. 10-6-1), 533 (Prob. 12-3-1)
Drift velocity, 113
Driving-point impedance, 675
Drude-Debye relations, 360 (Prob. 8-3-7)
Duct line, 478
Dyadic Green's function, 543n.

Earth inductor, 360 (8-4-1)
Earth-moon force, 48 (Prob. 2-1-13)
Earth station:
   antenna, 493 (Prob. 10-6-1), iv
   antenna temperature, 721 (Prob. 14-5-9)
   boiler, 724 (Prob. 14-6-9)
   radio telescope, 724 (Prob. 14-6-8)
Echo signals (see Ghost images)
Echo strength, 424
Eddy currents, 340
Edison, Thomas A., 4
Edminister, J. A., 747
Effective aperture, 616, 633, 638, 695, 712
Effective radiated power, 492 (Prob. 10-6-1)
Effective scattering aperture, 701
Efficiency:
   antenna, 632
   aperture, 696
   beam, 635n.
Eigenvalues, 567, 569, 592
Eight charges at cube corners, 47 (Prob. 2-1-8)
Eight point charges, maps of, 50–51 (Prob. 2-2-28)
Einstein, Albert, 3, 4
Electret, 59n.
Electric and magnetic field relations, comparison of, 174
Electric circuit, 116
Electric current, 112–114

Electric dipole, 26, 27, 35–37
Electric-dipole moment, 35, 36
Electric eel, 146 (Prob. 4-4-7)
Electric field cell, 263
Electric field intensity, 14–16
   for insulated wire, 318 (Prob. 7-2-5)
Electric field strength (see Electric field intensity)
Electric flux, 38–42
   over closed surface, 40–42
   density, 38–39, 62, 94–97
      boundary relations for, 65
      definition of, 38
      divergence of, in capacitor, 99–102
      relation of, to electric field intensity, 38–39, 62
   scalar potential, 20–22
   susceptibility, 62n.
   tubes of, 38
      figure, 39
Electromagnet for lifting, 280 (Prob. 6-6-3)
Electromagnetic spectrum, 5
Electromagnetics, history of, 1–5
Electromechanical system, 346
Electromotive force (emf), 125
Electron drift velocity, 140 (Prob. 4-1-2)
Electron excursion distance, 140 (Prob. 4-1-3)
Electron gun, 207 (Prob. 5-5-1)
Electron propulsion, 209 (Prob. 5-5-10)
Electron volt, 194
Electronic oven leakage, 610 (Prob. 13-5-6)
Electrostatic press, 105 (Prob. 3-2-17)
Electrostatic spray painting, 109 (Prob. 3-5-7)
Elektron, 1
Elliot, R. S., 747
Ellipse, 692
   cross-field, 585
   polarization, 497, 500–503
Elliptical polarization, 495–498
Elliptically polarized wave, 509 (Prob. 11-1-2)
   oblique incidence, 519–524
   power, 509 (Prob. 11-1-9)
Emde, F., 526n.
Emf (electromotive force), 125
Emf-producing field, 124
Empire State Building, 14
End-fire array, 646–648, 715 (Prob. 14-2-2)
Energy, 10 (Prob. 1-2-3)
   in bar magnet, 277 (Prob. 6-2-2)
   in capacitor, 74–77, 104 (Prob. 3-2-11), 105 (Prob. 3-2-18)
   between current sheets, 277 (Prob. 6-2-3)
   density, 76, 223, 467
   in inductor, 231–232
   in magnet, 248–249
   product, 251
   reactive, 469
   relations, 464–470
   in solenoid, 277 (Prob. 6-2-1)
   stored, 469
   velocity, 490 (Prob. 13-5-2)
Enoch, J. M., 747

Equality of path length, principle of, 690
Equation numbering, 8
Equiangular spiral antenna, 671
Equipotential contour, 25
Equipotential line, 22
Equivalence of point and line charges, 48 (Prob. 2-2-8)
Equivalent conductivity, 348
Equivalent current sheet, 163
Euler's identity, 344
Evanescent field, 592
Evanescent waves, 517
Expanding circular loop, 357 (Prob. 8-1-23)
Expanding square loop, 357 (Prob. 8-1-24)
Exponential line, 487 (Prob. 10-2-22)
Extragalactic radio source, 491 (Prob. 10-5-11)
Eye, human, 597
   (*See also* Biomedical topics, eye)

Far field, 626, 627
Faraday, Michael, 2, 3, 148, 154, 214, 322, 323
Faraday disk generator, 360 (Prob. 8-4-2)
   in superconducting solenoid, 362 (Prob. 8-4-7)
Faraday rotation, 669n.
Faraday's ice pail experiment, 44n.
Faraday's law, 322−323
Fell, Charles, xvi
Fermat's principle, 690
Fernald, D. L., 610 (Prob. 13-5-7)
Ferrimagnetic materials, 214
Ferrite-titanate medium, 462−464
Ferrites, 214
Ferromagnetic materials, 210, 214
   ac behavior of, 338−339
Ferromagnetism, 238−339
Fiber core, 593
Fiber core diameter, 608 (Prob. 13-3-12)
Fiber guide attenuation, 608 (Prob. 13-3-13)
Fiber-optic core, 594
Fiber waveguides, 593−596
Fibrillation, 146 (Prob. 4-4-6)
Field, 14
   applied, 44
   of bar magnet, 277 (Prob. 6-1-7)
   in bone, 110 (Prob. 3-5-16)
   of cube, 49 (Prob. 2-2-18)
   of disk, 49 (Prob. 2-2-15)
   evanescent, 592
   lamellar, 188
   of sphere, 49 (Prob. 2-2-17)
   of square, 49 (Prob. 2-2-16)
   (*See also entries beginning with the word:* Field)
Field and circuit theory, relation between, 365
Field angles at interface, 277 (Prob. 6-2-4)
Field cell, 88−93, 176−178
   capacitance, 76
   capacitor, 89
   magnetic, 259
   of the same kind, 90, 260
   transmission line, 176, 388, 391

Field direction from loop, 198 (Prob. 5-1-10)
Field distributions, 77−78
Field-effect transistor (FET), 433
Field lines, 25, 87n.
Field-map quantities, table of, 266
Field mapping, 77
   for currents in conductors, 135−138
   for electric fields, 86−93
   for magnetic fields, 258−263
Field maps, 86−93
   comparison of, 263−267
   properties of, 91−92
Field pattern, 614, 626
Field relations, general, 351−352
Fields of half-wave dipole, 659−661
Fields of short dipole, table of, 629
Finite-flat-sheet reflector, 684−686
Finite line of charge, field of, 78−80
Five waves, power, 510 (Prob. 11-2-8)
Fleetsatcom satellite, 669
Fleming, D. G., 748
Flip coil, 357 (Prob. 8-1-17)
Floating bar, 355 (Prob. 8-1-9)
Flood, Walter, xvi
Flux:
   electric (*see* Electric flux)
   magnetic (*see* Magnetic flux)
Flux-cutting law (*see* Motional-induction law)
Flux density, 38−39, 65
   due to conductor, 200 (Prob. 5-1-21)
   electric (*see* Electric flux, density)
   from loop, 198 (Prob. 5-1-7)
   magnetic, 149, 156−157
   of radio source, 703, 713
Flux linkage, magnetic, 165
Flux tubes, electric, 38−39, 87
Fluxmeter, 240
FM modulation, 493 (Prob. 10-6-1)
Footprint, 492 (Prob. 10-6-1)
Force:
   per charge, 15
   coercive, 245
   between conductors, 154, 155, 276 (Prob. 6-1-1)
   between current elements, 200 (Prob. 5-1-24)
   driving, 347
   electric, 13
   gravitational, 14
   on loop, 357 (Prob. 8-1-19)
   Lorentz, 160
   magnetic, 148−151
   magnetic gap, 270−272
   per unit length, 154
Force relations, 373
Forest, absorption coefficient of, 709
FORTRAN IV program, 293, 453
Forward tilt of wave, 582
Four charges in a plane, 47 (Prob. 2-1-6)
Four-turn coil, 325
Fourier transform relations between pattern and aperture, 655−657

Fourth power effect between magnets, 277 (Prob. 6-1-11)
Franklin, Benjamin, 2, 3, 73
Free charge, 95
Frequency-independent antennas, 671–673
Frequency of revolution, 197
Fresnel cosine integral, 526
Fresnel integrals, 526
Fresnel reflection coefficient, 514, 519
Fresnel sine integral, 526
Friis formula, 697–698, 712
Fringing field, 320 (Prob. 7-3-2
    magnetic, 148
Fuller, Buckminster, 494 (Prob. 10-6-7)
Fundamental dimensions, 5
Fundamental force laws, 744
Fundamental units, 6, 7

GaAsFET amplifier, 433
Gabor, D., $529n$.
Gain, 616, 633, 712
    antenna, 636
    array, 683
    relation to directivity, 637
Gap force, magnetic, 270–272
Gapless circuit, 267–268
Gapless iron ring, 279 (Prob. 6-5-1)
Garbacz, Robert, xvi
Gauss, Karl F., 2, 3, 40
Gauss (unit), 732
Gauss' law, 40–41, 94
    for magnetic fields, 158
Gauss' theorem (see Divergence theorem)
Geddes, L. A., 747
Generator-motor, 354 (Prob. 8-1-6)
Geometric mean, 421
Geometrical optics, 529–531
Geometrical theory of diffraction (GTD), 529
Geostationary satellites, 492 (Prob. 10-6-1)
Germanium semiconductor, 121
Ghost images, 533 (Prob. 12-3-1)
Gigawatt guide, 610 (Prob. 13-5-8)
Gilbert, William, 2, 3
Gilbert (unit), 732
Gold-plated cylindrical cavity, 609 (Prob. 13-4-3)
Gold-plated square cavity, 609 (Prob. 13-4-5)
Gordon, W. E., 721 (Prob. 14-5-11)
Goubau, G., 588
Gradient, 185
    of electric potential, 32
    in rectangular coordinates, 32–35
Grand Unified Theory, 4, 276
Graphical field mapping (see Mapping, graphical)
Graphical vector addition, 18
Graphite cone, 141 (Prob. 4-1-10)
Graphite shell with polar caps, 143 (Prob. 4-3-8)
Graphite slab, 142 (Prob. 4-3-6)
Graphite washer, 141 (Prob. 4-1-9)
Gravitational field of earth shell, 54 (Prob. 2-3-13)
Gravity, 1

Gray, D. A., 610 (Prob. 13-5-2)
Greenland, 493 (Prob. 10-6-4)
Green's function, dyadic, $543n$.
Ground reflection, effect of, 698
Ground resistance, 146 (Prob. 4-4-1)
Group velocity, 436, 437, 540
GTD (geometrical theory of diffraction), 529
Guide, 534
    fiber, 593–596
    helical, 588
    open wire, 588
    rod, 593–596
    (See also Waveguide)
Guide, ridge, bandwidth, 609 (Prob. 13-5-1)
Guide horn with corrugations, 609 (Prob. 13-5-2)
Guide-to-line transitions, 578
Gyro frequency, $197n$.

Half-cylindrical tube, 319 (Prob. 7-2-10)
Half-power beam width (HPBW), 633
Half-wave dipole:
    directivity of, 659
    fields of, 659–661
    radiation resistance of, 659
Half-wave slots, 720 (Prob. 14-4-11)
Hall, Edwin H., 208 (Prob. 5-5-8)
Hall effect, 208 (Prob. 5-5-8)
Hallén, E., $676n$.
Halliday, D., 747
Hansen, W. W., $668n$.
Hansen-Woodyard increased directivity, $668n$.
Hard magnetization, 242
Harmonics, 302
Harrington, R. F., $588n$., 747
Hayt, W. H., Jr., 747
Heart defibrillator, 146 (Prob. 4-4-6)
Heart dipole, 93–94, 110 (Prob. 3-5-15)
Heaviside, Oliver, $465n$.
Heaviside's condition, 395
Helical-beam antennas, 498, 666–669
    bandwidth of, 668, 669
    directivity of, 668
    pattern of, 666, 667
    phase velocity on, 668
    use of, in space communication, 669
Helical coil (see Solenoid)
Helical path of proton, 205 (Prob. 5-4-7)
Helices, left and right, 717 (Prob. 14-3-8)
Helix:
    ten-turn, 717 (Prob. 14-3-6)
    thirty-turn, 717 (Prob. 14-3-7)
Helmholtz, Hermann von, 208 (Prob. 5-5-7)
Helmholtz pair of coils, 208 (Prob. 5-5-7)
Hemisphere on conducting sheet, 319 (Prob. 7-2-9)
Henry, Joseph, 2, 3
Henry (unit), 165
Herman, Paul, 453
Hertz, Heinrich, 2, 3, 373, 747
High-voltage bushing, 105 (Prob. 3-2-14)

High-voltage dc transmission line, 106 (Prob. 3-3-1)
High-voltage line, field under, 108 (Prob. 3-5-5)
Higher-order mode, 379, 534
Hill, M. L., 55 (Prob. 2-4-1)
History of electromagnetics, 1−5
Hodge, Daniel, xvi
Hodgkin, A. L., 476
Holes in semiconductors, 120
Hollow conductor, internal inductance, 204 (Prob. 5-2-17)
Hollow cylindrical waveguide, 561−569
Hollow guide, attenuation, 606 (Prob. 13-2-9)
Hollow rectangular waveguide, 543−560
Hologram, 529
Holography, 529
Homogeneity, 56, 57
Hondros, D., 593n.
Horn antenna, 609 (Prob. 13-5-2), 694−695
Human eye, 597
    (See also Biomedical topics, eye)
Huxley, A. F., 476
Huygens, C., 524n.
Huygens' principle, 524, 525, 653
Hydraulic analog for electric circuit, 127
Hyperbola, 691
Hyperbolic relations, 734−735
Hysteresis, 245−248
Hysteresis loop, 245
    major, 245
    minor, 339

Ice, permittivity of, 360 (Prob. 8-3-7)
Ice caps, 493 (Prob. 10-6-4)
Idaho potato, 362 (Prob. 8-4-8)
Idealness, 14
Image:
    charges, 311
    of conductor, 314
    current, 314
    theory of, 85, 311−312
Impedance:
    antenna, 675−683
        mutual, 679−680
        self-, 675−679
    characteristic, 582
    intrinsic, 582
    of media, 395−402
    of transmission lines, 395−402
    transverse, 557
    transverse-wave, 582
    wave, 582
Incident wave, 440
Increased directivity condition, Hansen-Woodyard, 668n.
Incremental permeability, 339
Index of refraction, 435, 488 (Prob. 10-3-1)
Indicating meter, torque, 199 (Prob. 5-1-13)
Induced charges, 44
Induced current, 323

Induced emf, 326
Induced field, 44
Induced magnetization, 227, 239
Inductance, 164−165
    average, 338
    of coaxial line, 167
    of coil and wire, 359 (Prob. 8-2-4)
    definition of, 165
    self- and mutual, 335−338
    of solenoid, 166
    in terms of energy, 202 (Prob. 5-2-1)
    of toroid, 166
    of transmission line, 377 (Prob. 9-1-11)
    of two-conductor line, 168
    per unit length, 388
Induction:
    examples of, 328−332
    general case, 327−328
    in loop, 356 (Prob. 8-1-13)
    of loop: motion and B change, 329
        motion only, 328−329
    for moving conductor, 353 (Prob. 8-1-2)
    of moving strip, 330
    relations, 328
    of rotating loop, 331−332
    of stationary loop, 353 (Prob. 8-1-1)
Inductive load, 410
Inductor, 164−165
    coaxial line, 167
    definition of, 164
    energy in, 231−232
    solenoidal, 165−166
    toroidal, 166−167
    two-conductor line, 167−168
Infinite-flat-sheet reflector, 684
Infinite line of charge, field of, 80
Infinite-parallel-plane guide, 535−543
Infinite square trough, 289−292
Infinitesimal dipole (see Short dipole)
Infinitesimal permeability, 338
Infrared fiber guide, 608 (Prob. 13-3-11)
Ingling, Carl, xvi
Initial-magnetization curve, 242
Initial permeability, 242
Insulators, 11
Interfaces, waves at, 454−464
Interferometer, 717 (Prob. 14-2-17)
    as radio telescope antenna, 650, 653
    two-element, 650−653
Interferometer pattern, 653
Internal reflection, 592
International System of Units (SI), 6−7
Intrinsic impedance, 581
    of vacuum, 396
Inverse-square relation, 54 (Prob. 2-3-15)
Iris, waveguide, 579−581
Iron crystal, 239
Iron filings, 211
Iron ring, 279 (Prob. 6-5-4)
    flux, 279 (Prob. 6-5-9)
    with gap, 224, 279 (Prob. 6-5-2)

Iron ring (*Cont.*):
  gap force, 279 (Prob. 6-5-6)
  with spoke, 279 (Prob. 6-5-10)
Isbell, D. E., 671*n.*
Isotropic antenna, 636, 713 (Prob. 14-1-1)
Isotropic source (*see* Isotropic antenna)
Isotropy, 56−57
Iterative method (*see* Point-by-point method)

Jahnke, E., 526*n.*
Jansky (unit), 733
Japanese high-speed railroad, 279 (Prob. 6-6-1)
Jensen, H. J., 494 (Prob. 10-6-8)
Johnson, R. C., 747
Jones, V. C., 146 (Prob. 4-4-6)
Joule, James P., 3
Joule heating, 340, 574
Joule's law, 115
Junction, power loss, 490 (Prob. 10-4-13)
Jupiter signals, 721 (Prob. 14-5-6)

K-band waveguide, 610 (Prob. 13-5-5)
Kapany, N. S., 594*n.*
Karplus, W. J., 747
Keeper, soft iron, 252
Kelvin, Lord, 5
Kelvin, definition of, 6
Kennaugh, E. M., xvi, 661
Kilogram, definition of, 6
Kinetic energy, 308
Kirchhoff's current law, 129−130, 294
Kirchhoff's voltage law, 125, 258
Ko, Hsien C., xvi, 507*n.*
Kock, W. E., 692*n.*
Kouyoumjian, Robert, xvi
Kraus, J. D., 495*n.*, 524*n.*, 532 (Prob. 12-2-1),
  653*n.*, 666*n.*, 668*n.*, 676*n.*, 688*n.*, 689*n.*,
  696*n.* 702*n.*, 747

L-shaped bar, 278 (Prob. 6-4-3)
L-shaped conductor, 316 (Prob. 7-1-9)
L-shaped copper bar, 142 (Prob. 4-3-3)
L-shaped vs. straight bar (Prob. 6-4-6)
Lamellar field, 25, 188
Laminations, 340
Langmuir, I., 310*n.*
Laplace's equation, 103, 282
  for conducting media, 138−139
  in cylindrical coordinates, 303
  mean value, 315 (Probs. 7-1-5, 7-1-6)
  mean value significance, 315 (Prob. 7-1-1)
  for parallel-plate capacitor, 284−286
  in rectangular coordinates, 283−284
  solution with digital computer, 292−294
  in spherical coordinates, 303
Laplacian, 185
Laplacian operator, 103
Latitude on Poincaré sphere, 500

Laws:
  Ampère's, 170, 366
  Biot-Savart, 152−153
  Child-Langmuir, 310
  Coulomb's, 12
  Faraday's, 322−323
  Gauss', 40−41, 94
    for magnetic fields, 158
  Joule's, 115
  Kirchhoff's current, 129−130, 294
  Kirchhoff's voltage, 125, 258
  Lenz's, 323
  Newton's: second law, 195
    of Universal Gravitation, 47 (Prob. 2-1-11)
  Ohm's, 114, 258, 365
    at a point, 118, 373
  Plank's radiation, 491 (Prob. 10-5-12)
  Snell's: of reflection, 513
    of refraction, 514
Leakage field through slot, 319 (Prob. 7-2-7)
  at various distances, 319 (Prob. 7-2-8)
Leakage resistance to ground, 141 (Prob. 4-1-8)
Legendre function, 303
Legg, W. E., 610 (Prob. 13-5-2)
Leith, E. N., 529*n.*
Lens antenna, 689
  dielectric, 691−692
  metal plate, 692
Lenz, Heinrich F., 323
Lenz's law, 323
Levis, Curt, xvi
Levitation of charged ball, 47 (Prob. 2-1-3)
Light, velocity of, 387
Lightning conductor impedance, 361 (Prob. 8-4-5)
Lightning stroke current, 146 (Prob. 4-4-4)
Lightning rod, 73
Linear polarization, 495−497
Linearly polarized (LP) wave at angle, 531 (Prob.
  12-1-1)
Lines:
  of charge, field of, 82−84
  open-circuited, 405
  with series resistors, 487 (Prob. 10-2-21)
  short-circuited, 405
  with small R and G, (Prob. 10-4-2)
  transmission, 534
  with two loads, 486−487 (Prob. 10-2-20)
Line impedance, 404
Line integral, 23
  around closed path, 24
Line reflections, 488 (Prob. 10-3-3)
Line source, pattern of, 716 (Prob. 14-2-14)
Linear antennas, 657−659
Linear charge density, 29
Linear conductor, force on, 198 (Prob. 5-1-4)
Linear medium, 56, 57
Linear-motor vehicle, 363 (Prob. 8-4-10)
Linear polarization, 495−497
Linearly independent solution, 307
Lines and guides for all applications, 610 (Prob.
  13-5-9)

Linkage, magnetic flux, 165
Liquid hydrogen, 118
Load impedance, 403
Log-periodic antenna, 671, 718 (Prob. 14-3-14)
Logarithmic relations, 735
Longitude on Poincaré sphere, 500
Loop, 212−213
  antennas, 665−666
  atomic current, 210
  current-carrying, 155−156
  fields, 274
  hysteresis, 245
  saturation, 245
  torque on, 161−162, 199 (Prob. 5-1-11)
Lorentz, H. A., 160
Lorentz force, 160, 195, 326
Loss resistance, 632
Loss tangent, 349
Lossless line, 485 (Prob. 10-1-3)
Lossless medium, 387
Lossy lines, 447−454, 490 (Prob. 10-4-12)
Lossy medium, 462, 489 (Prob. 10-4-6)
Lossy sheet, 489 (Prob. 10-4-9)
Lossy slab, 489 (Prob. 10-4-8)
Lossy transmission line, 452
Low-frequency cutoff, 535
Low-pass filter, 451
Lunar communication, 610 (Prob. 13-5-7)
Lunar occultation, 532 (Prob. 12-2-1)

McCullough, T. P., 704n.
Magnesia, 1
Magnet:
  bar, 211−213, 217
  domains in, 239
  effect on current-carrying wire, 148−151
  energy in, 248−249
  with gap, 279 (Prob. 6-5-7)
  materials for, 251−252
  permanent *see* Permanent magnet)
Magnetic bottle, 206 (Prob. 5-4-9)
Magnetic circuit, 254−258
  with air gap, 269−270
Magnetic dipoles, 210−213, 216−218
Magnetic field:
  in air gap, 269
    figure, 261
  boundary relations for, 235
  cell, 259−260, 263
  definition of, 169−170
  of earth, 281 (Prob. 6-6-6)
  energy density in, 233
  H vector, 169
  H, tangential component of, 235
  mapping, 258−263
  relation of, to magnetic flux density, 169−170, 221
  of solid cylindrical conductor, 171
Magnetic flux, 156−157
  over closed surface, 157−158

Magnetic flux (*Cont.*):
  tubes of, 158
Magnetic flux density, 149, 156−157
  boundary relations for, 234
  of coaxial line, 167
  of current carrying loop, 155−156
  due to current distribution, 156
  definition of, 156
  of infinite linear conductor, 153−154
  of iron ring, 225−227
  relation of, to magnetic field, 169−170, 221
  relation of, to magnetization, 221
  of solenoid, 162−164, 203 (Prob. 5-2-8), 229−231
  of toroid, 166, 223−225
  of two-conductor line, 168
Magnetic flux linkage, 165
Magnetic gap force, 270−272
Magnetic levitation, 279 (Prob. 6-6-1), 355 (Prob. 8-1-9)
Magnetic materials, 214−215
  antiferromagnetic, 214
  diamagnetic, 214
  ferrimagnetic, 214
  ferromagnetic, 214
  paramagnetic, 214
  superparamagnetic, 214
Magnetic moment, 162, 213
Magnetic monopole, 276
  force, 279 (Prob. 6-5-11)
Magnetic potential function (*see* Magnetostatic potential)
Magnetic saturation, 239, 242
Magnetic sheet current density, 219−220
Magnetic susceptibility, 222n.
Magnetic vector potential, 187−192
Magnetic vectors, 220−231
Magnetization, 217−218
  current model, 275
  easy, 238, 242
  induced, 227, 239
  permanent, 227, 239
  pole model, 275
  table of, 275
Magnetization curve, 240−245
  magnetized rod, 218
Magnetizing force, 240
Magnetomotance (*see* Magnetomotive force)
Magnetomotive force (mmf), 173
Magnetostatic potential, 172−176
Magnitude factor, 680
Main lobe, 614
Main-lobe solid angle, 634
Major hysteresis loop, 245
Major lobe, 614
Mapping, graphical:
  current, 135−138
  electric field, 86−93
  magnetic field, 258−263
Marconi, Guglielmo, 3
Marcuse, Dietrich, 590n.

Marcuvitz, N., 578n.
Mars link, 724 (Prob. 14-6-7)
Marshall, S. V., 748
Matched line, 408
Matching:
  with double-stub, 486 (Prob. 10-2-13)
  with one wave and six antennas, 510 (Prob. 11-2-3)
  with λ/4 transformer, 423
  by three waves and three antennas, 510 (Prob. 11-2-4)
Matching plate, 489 (Prob. 10-4-4)
Matthias, B. T., 118n.
Maximum energy product, 251
Maximum permeability, 242
Maximum relative permeability, 215
Maximum usable frequency (MUF), 722 (Prob. 14-5-12)
Maxwell, James Clerk, 2, 3, 371n.
Maxwell (unit), 732
Maxwell's equations, 172, 365, 371
  from Ampère's law, 343, 371
  in differential form, table of, 376
  from Faraday's law, 325, 334, 372
  in free space, 374
  from Gauss' law, 97−98, 372, 373
  for harmonically varying fields, 374
  in integral form, table of, 375
  involving curl, 185
  involving divergence, 97−98, 158
Maxwell's theory, objection to, 377 (Prob. 9-1-6)
Mayer, C. H., 704n.
Measurable quantity, 59
Mechanical moment (see Torque)
Mechanical system, equivalent, 346, 347
Megaparsec, 721 (Prob. 14-5-7)
Mercury pool, 354 (Prob. 8-1-5)
Mercury pump, 209 (Prob. 5-5-11)
Mesh, 371
Metal detector, 364 (Prob. 8-4-12)
Metal-plate lens antenna, 691, 692
Meter, definition of, 6
Mho, 118
Microstrip line, 391, 397, 398, 485 (Probs. 10-1-2, 10-1-5)
Microwave circuit, 532 (Prob. 12-3-1)
Microwave oven, 610 (Prob. 13-5-6)
Midwinter, J. E., 748
Millikan's oil-drop experiment, 109 (Prob. 3-5-6)
Mine sweeper, 208 (Prob. 5-5-5)
Minimum detectable flux density, 713
Minimum detectable temperature, 706, 713
Minimum usable frequency (mUF), 722 (Prob. 14-5-13)
Minor hysteresis loop, 339
Minor lobe, 614
Minor-lobe solid angle, 635
Mirror point, 205 (Prob. 5-4-8)
Mismatch, double-stub, 486 (Prob. 10-2-14)
Mmf (magnetomotive force), 173
Mobility, 113, 120

Modes:
  on guides, 604
  higher-order, 534
  on lines, 604
Modified cosine integral, 659
Moffatt, D. L., 364 (Prob. 8-4-12)
Moment:
  current, 148, 149
  electric-dipole, 35, 36
  magnetic, 162, 213
  mechanical (see Torque)
Monochromatic transmitter, 503
Montgomery, C. G., 431
Moon link, 724 (Prob. 14-6-5)
Moore, A. D., 55 (Prob. 2-4-1), 297, 748
Moore, R. K., 425n.
Motional induction, 327, 328
Motional-induction law, 327
Motor, linear (see Linear-motor vehicle)
Motor equations, 149
Motor force, 148
Moving arm, 357 (Prob. 8-1-22)
Moving conductor in magnetic field, 326−327
Moving loop, 356 (Prob. 8-1-11)
Moving wire, 356 (Prob. 8-1-12)
MUF (see Maximum usable frequency)
mUF (see Minimum usable frequency)
Multilobed pattern, 653
Multiple-path transmission, 723 (Prob. 14-5-15)
Mushiake, Y., 688n.
Mutual impedance, 338n.
  of antenna, 679−680, 720 (Prob. 14-4-8)
Mutual inductance, 336
  of concentric loops, 359 (Prob. 8-2-2)
  of parallel wires, 359 (Prob. 8-2-3)
  of wire and loop, 358 (Prob. 8-2-1)
Myelin sheath, 475

N-type semiconductors, 121
Nabla, 32
Near field, 626−627
Negative magnetic pole (see South pole)
Nepers, 484, 571n.
Neumann function, 303, 562
Neumann's inductance formula, 377 (Prob. 9-1-10)
Neurons, 475
Newton, Isaac, 3
Newton (unit), 13n.
Newton's Law of Universal Gravitation, 47 (Prob. 2-1-11)
Newton's second law, 195
Niobium, 118
Noise-figure chart, 743
Noise power, 703
Noise temperature, 702
  chart of, 743
Nondispersive media, 437
Nonisotropic (see Anisotropic materials)
Nonlinear elements, 114
Nonmagnetic materials, 214

Nonmetallic guide, 587
Normal component of flux density **B**, 234
Normal magnetization curve, 247
Normalized antenna power pattern, 633
Normalized field pattern, 644
Normalized power pattern, 615
Normalized Stokes parameters for seven wave-polarization states, table of, 506
Normally dispersive medium, 437
North pole, magnetic, 150, 210−211
North-seeking pole, 210
Notation, how to read, 7
Number of resolvable objects, 713
Numbering, equation, 8
Nyquist, H., 702n.
Nyquist power, 713

Oblique incidence, 511−524
  elliptical polarization, 519−524
  parallel case, 518
  perpendicular case, 512
Octopole, 37
Oersted, Hans Christian, 2, 3, 147
Oersted (unit), 732
Ohio State University radio telescope, 706−708
Ohm, Georg Simon, 2, 3, 114
Ohmmeter, 399
Ohm's law, 114, 258, 365
  at a point, 118, 373
Ohms per square, 460
Oil-drop experiment, Millikan's, 109 (Prob. 3-5-6)
One-dimensional device, 319 (Prob. 7-2-11)
1/e depths, 489 (Prob. 10-4-5)
Onnes, H. K., 118
Open-circuited line, 408
Open waveguides, 586−589
Optical-fiber communication link, 595
Optical fibers, 593
Ordinary permeability, 338
Orientation factor, 680
Orthogonality, 25
Outer segments, 596
Outwardness, 16−17
Oven, miocrowave, 610 (Prob. 13-5-6)
Over-mountain TV, 533 (Prob. 12-3-2)
Overland TV, 532 (Prob. 12-3-1)
Overlapping bars, 146 (Prob. 4-3-25)

P-type semiconductors, 121
Pacht, Erich, xvi
Paddle wheel, 183
Parabola, 690
Parabolic line charge, 53 (Prob. 2-2-31)
Parabolic-reflector antennas, 691
Parabolic reflector field, 719 (Prob. 14-4-5)
Paraffin slab, 100
Parallel circuit, 377 (Prob. 9-1-7)
Parallel conductors, force on, 198 (Prob. 5-1-3)

Parallel-connected iron blocks, 257
Parallel plane guide, TE attenuation, 606 (Prob. 13-2-5)
Parallel-plate capacitor, 69−70, 284−286
  divergence of **D** and **P** in, 99−102
  Laplace's equation for, 284−286
  problems involving, 104−105
  with space charge, 308−310
Parallel polarization, 518−519
Parallel wires with sliders, 358 (Prob. 8-1-28)
Paramagnetic materials, 214
Parameters for linear array of six isotropic in-phase point sources, table of, 651
Parasitic element, 689
Partial differential equation, 383
Partial polarization, 503−509
Particle:
  acceleration of, 205 (Prob. 5-4-1)
  radius of, 196, 207 (Prob. 5-4-14)
Particular solution to Poisson's equation, 307
Pattern:
  antenna radiation, 612
  far field, 627
  field, 614, 626
  interferometer, 653
  multilobed, 653
  near-field, 626
  polar, 614
  power, 615
  principal-plane, 615
  primary, 642
  radiation, 612, 614
  rectangular, 614
  short-dipole, 653
Pattern multiplication, 641−642
Pendulum, 355 (Prob. 8-1-10)
Penetration depths, wavelength, velocity, and refractive index for copper, table of, 451
Periodic factor, 680
Permanent magnet, 250−251
  with gap, 272−273
Permanent magnetic materials, table of, 252
Permanent magnetization, 227, 239
Permeability, 151, 176, 177
  differential, 338
  incremental, 339
  infinitesimal, 338
  initial, 242
  maximum, 242
  ordinary, 339
  relative, 215
    table of, 216
  of vacuum, 152
Permeance, 255−258
  of bar, 278 (Prob. 6-4-1)
Permitted angles of reflection, 592
Permittivity:
  of air, 12
  of artificial dielectric, 104 (Prob. 3-1-6)
  of crystals, 58n.
  of ice, 360 (Prob. 8-3-7)

Permittivity (*Cont.*):
  relative, 57
    table of, 58
  of titanates, 58*n*.
  of vacuum, 12, 57
Perpendicular polarization, 512
Perpendicularly polarized wave, 512
Peters, Leon, Jr., xvi, 364 (Prob. 8-4-12)
Phase center, 641
Phase constant, 393
Phase factor, 394
Phase shift, 441
  via hole, 489 (Prob. 10-4-3)
Phase velocity, 384, 437, 450, 540
  in guide, 568
    rectangular, 556
  relative, 434
  of standing wave, 489 (Prob. 10-3-9)
  on transmission line, 395
Phasor addition of fields, 648–651
Phasors, 344–346
Phosphorus, 121
Photons, 491 (Prob. 10-5-13), 596*n*.
Physical aperture, 616, 696
Physical optics, 524–529
Physical optics approximation, 529
Pie-shaped loop, 358 (Prob. 8-1-26)
Piezoelectric, 110 (Prob. 3-5-16)
Pigmented layer, 596
Pipeline, buried, 278 (Prob. 6-4-4)
Pipeline resistance, 143 (Prob. 4-3-9)
Planck's constant, 491 (Prob. 10-5-12), 733
Planck's radiation law, 491 (Prob. 10-5-12)
Plane of incidence, 511
Plane wave, 381
  diffraction of, 525
Plessner, K. W., 58*n*.
Plonsey, R., 748
Plumb bob over mercury pool, 354 (Prob. 8-1-5)
Pocket radio, 723 (Prob. 14-6-3)
Pocket rulers, 671
Poincaré, H., 500*n*.
Poincaré sphere, 500–503
Point-by-point method, 287–289
Point charges, computer plots, 53 (Prob. 2-2-32)
Point relation, 371
Point source (*see* Isotropic antenna)
Point-to-wave correspondence, 525
Poisson's equation, 103, 189, 275, 282, 307
Polarization:
  circular, 495–498
  degree of, 506
  of dielectric, 59–63
  elliptical, 495, 498
  incoherent, 504
  left circular, 497
  linear, 495–497
  parallel, 518–519
  table of, 275
  unpolarized, 504
  volume charge density, 101

Polarization (*Cont.*):
  wave, 380
  (*See also* entries beginning with the word:
    Polarization)
Polarization ellipse, 497, 500–503
Polarization fading, 669*n*.
Polarization matching, 503
Polarization measurements, 723 (Prob. 14-5-16)
Polarization state, 500, 501
  antenna, 503, 507
  wave, 506
Polarized atom, 60
Polarizing angle, 519
Pole:
  north, 150, 210–211
  south, 150
Pole faces, 227
Pole model, 217, 227, 229
  magnetization, 275
Pole pieces, displacement current, 377 (Prob.
    9-1-8)
Pole strength, 150
Poles, magnetic, 150
Police radar, 723 (Prob. 14-6-4)
Polyrod antennas, 596
Polystyrene-air interface, 514
Positive set, 16
Positiveness, 16, 17
Potato, baked, 362 (Prob. 8-4-8)
Potential:
  absolute, 22
  electric scalar: definition of, 20*n*., 189*n*.
    of linear charge distribution, 30
    of quadrupole, 54 (Prob. 2-3-17)
    of surface charge distribution, 31
    of volume charge distribution, 31, 102
  magnetostatic, 172–176
  retarded, 618–619
  vector (*see* Vector potential)
Potential difference, 21
  rise in, 25
Pounder, E. R., 360 (Prob. 8-3-7)
Powdered-iron cores, 340
Power:
  complex, 633
  through curvilinear square, 471
  dissipated by dielectric, 349
  for partially polarized wave, 509
  reactive, 633 (Prob. 11-2-1)
  per square, 490 (Prob. 10-5-2)
Power divider, 486 (Prob. 10-2-20), 487 (Prob.
    10-2-23), 578
  waveguide, 578
Power factor, 349, 360 (Prob. 8-3-6)
Power flow:
  reactive, 661
  real, 661
  on transmission line, 470–472
Power-flow lines, 473
Power-flow map, 471
Power-line spillover, 361 (Probs. 8-4-3, 8-4-4)

Power pattern, 614
  normalized, 615
  in polar coordinates, 614
  in rectangular coordinates, 614
  relation to Poynting vector, 615
Power per unit area, average, 466
Power relations, 115, 464–470
Poynting, J. H., 465n.
Poynting vector, 465, 574
  average, 466
  circuit application of, 472–475
  for circular polarization, 509 (Prob. 11-1-8)
  flow lines of, 474
  instantaneous, 465
  for polarized wave, 498–500
  for traveling wave, 490 (Prob. 10-5-5)
Prefixes for units, 7
Press, electrostatic, 105 (Prob. 3-2-17)
Primary pattern, 642
Primary winding, 335
Principal plane pattern, 615
Principle of equality of path length, 690
Principle of inequality of path length, 694n.
Principle of pattern multiplication, 642
Principle of superposition:
  of electric fields, 17
  of electric potential, 29
Prolate spheroid antenna, 659–660
Propagation constant, 393, 404
Proton radius of curvature, 206 (Prob. 5-4-10)
Pulsar spin down, 721 (Prob. 14-5-8)
Pulse, square, 430
Pulses, 424–428
  on transmission line, 431
Purcell, E. M., 431
Pyramidal horn, 720 (Prob. 14-4-13)

Q, 598, 600
  comparison of, 609 (Prob. 13-4-7)
Quad-helix array, 724 (Prob. 14-6-11)
Quadratic equation, 736
Quadrupole, 37, 54 (Prob. 2-3-17)
Quantum theory, 491 (Prob. 10-5-13)
Quarter ring, 48 (Prob. 2-1-14)
Quarter-wave plate, 458
Quarter-wave transformer, 420–424, 486 (Prob. 10-2-17)
  double section, 423
  single section, 420–421
  three section, 424
  wave reflections on, 428–431
Quartz crystal, 58n.
Quasi-conductor, 446
Quasi-stationary field, 627

Radar cross section, 702
Radar detection, 721 (Prob. 14-5-3)
Radar equation, 701, 713
Radiation, 490 (Prob. 10-5-3)
  antenna and, 612–618

Radiation intensity, 633
Radiation pattern, 612, 614
Radiation resistance, 478, 612, 630, 632, 675, 712
Radio, pocket, 723 (Prob. 14-6-3)
Radio link, 480–482
Radio loop, 362–363 (Prob. 8-4-9)
Radio telescope, 709
  of Ohio State university, 706–708
  parameters of, 721 (Prob. 14-5-4)
  resolution of, 711
  sensitivity of, 706
Radiometer, 703
Radius:
  of curvature, 205 (Prob. 5-4-6)
  ratio, 205 (Prob. 5-4-5)
Ramo, S., 748
Ratio $\sigma/\omega\epsilon$, 446
Rationalized SI units, 7
Ray paths, 529
Rayleigh, Lord, 278 (Prob. 6-3-2), 543n.
Rayleigh relation, 278 (Prob. 6-3-2)
Rayleigh resolution, 711
Reactive energy, 469
Reactive power, 633
Receiver temperature, 493 (Prob. 10-6-1)
Receiving antenna, 612, 637
Reciprocal henry, 255
Reciprocity, 673–675
  directivity, 674–675
  impedance, 675
  pattern, 674
Reciprocity theorem, 338, 674
Rectangular and circular cylindrical waveguide parameters, table of, 577
Rectangular aperture, 716 (Prob. 14-2-12)
Rectangular guide, 605 (Prob. 13-1-1)
  attenuation, 607 (Prob. 13-2-12)
  impedance, 608 (Prob. 13-3-5)
  power, 607 (Prob. 13-2-16)
  TM wave, 605 (Prob. 13-1-6)
Rectangular trough, 317 (Prob. 7-1-13)
  with curved electrode, 317 (Prob. 7-1-14)
Rectangular-wavelength group velocity, 608 (Prob. 13-3-8)
Recurrence relations, 738
Red shifts, 721 (Prob. 14-5-7)
Reflected wave, 440, 461, 511, 512
Reflecting lens, 719 (Prob. 14-4-4)
Reflection, 511
  internal, 592
Reflection and transmission coefficients, table of, 407
Reflection angle, 605 (Prob. 13-1-5)
Reflection coefficient, 407, 408, 444, 455, 591
  for current, 403
  Fresnel, 514, 519
  for voltage, 403
Reflector (see Antennas, reflector)
Refracted wave, 511n.
Refraction, angle of, 511n., 512

Relative permeability, 215
 table of, 216
Relative phase velocity, 434
Relativistic effect, 194
Reluctance, 254–258, 337
Remanence, 245
Remote sensing, 702, 705, 709, 721 (Prob.
 14-5-5)
Residual flux density, 245
Resistance, 114–115
 antenna, 638
 loss, 632
 radiation, 478, 612, 630, 632, 675, 712
 of simple geometries, 135–138
 in terms of power, 140 (Prob. 4-1-1)
Resistance-measurement method, 400
Resistance paper, 296
Resistivity, 116–117
Resistor, 116
Resistor and power flow, 491 (Prob. 10-5-14)
Resnick, R., 747
Resolution, 702
 of eye, 711
 of radio telescope, 711
Resolution and directivity, 711
Resolution angle, 713
Resonant frequency, 598
 of atomic dipole, 347, 348
Resonant wavelength, 600
Resonator, 430, 597
Response, system, 508
Resultant wave, 538
Retardation time, 618n.
Retarded current, 618, 658
Retarded potentials, 618–619
Retarded scalar potential, 619
Retarded vector potential, 619
Retentivity, 245
Retina, 597
Retinal optic fibers, 596–597
Retroreflector, corner, 720 (Prob. 14-4-10)
RF loop, 362 (Prob. 8-4-9)
RF shielding, 493 (Prob. 10-6-3)
Richmond, Jack, xvi
Ridge guide, 609 (Prob. 13-5-1)
Ridge guide bandwidth, 609 (Prob 13-5-1)
Right circular polarization, 497
Right-hand rule, 147, 148
 figure, 148
Right-handedness, 16, 17
Ring:
 of charge, 49 (Prob. 2-2-13)
 composite, 279 (Prob. 6-5-3)
Rod waveguides, 593–596
Rods, 596
Root, rank of, 563
Rotatable bar magnet, 355 (Prob. 8-1-8)
Rowland, Henry, 361 (Prob. 8-4-6)
Rowland disk, 361 (Prob. 8-4-6)
Rowland ring, 241
Rumsey, V. H., 671n.

Rural ground, electrical behavior of, 446–447
Rutile, 58

S parameters, 488 (Prob. 10-3-2)
Salisbury, W. W., 610 (Prob. 13-5-7)
Sams, H. W., 748
Satellite topics:
 attenuation of 3-dimensional radio link, 481
 carrier-to-noise (C/N) ratio, 493 (Prob. 10-6-1)
 Clarke orbit, 669
 Clarke orbit satellite, 480–481, 492 (Prob.
  10-6-1), 709, 722 (Prob. 14-5-13)
 communication satellite, 669
 downlink, satellite TV, 492 (Prob. 10-6-1), 533
  (Prob. 12-3-1), 722 (Prob. 14-5-13)
 earth station, radio telescope, iv, 493 (Prob.
  10-6-1), 724 (Prob. 14-6-9), 724 (Prob.
  14-6-10)
 effective radiated power (ERP), 492 (Prob.
  10-6-1)
 Fleetsatcom satellite, 669
 FM-modulated signals, 493 (Prob. 10-6-1)
 footprint, 492 (Prob. 10-6-1)
 geostationary satellites, 492 (Prob. 10-6-1)
 helical antenna, use on satellites, 669
 Mars link, 724 (Prob. 14-6-7)
 moon link, 724 (Prob. 14-6-5)
 mUF for Clarke orbit satellites, 722 (Prob.
  14-5-13)
 radio-link transmission line, 480–482
 signal-to-noise ratio (S/N), 481, 492 (Prob.
  10-6-1), 706, 713
 solar interference to earth station, 724 (Prob.
  14-6-10)
 transponder, satellite TV, 492 (Prob. 10-6-1)
 TV downlink, 492 (Prob. 10-6-1)
Saturation, magnetic, 239, 242
Saturation loop, 245
Scalar potential, 189n.
 electric, 20–22, 352
  (See also Potential, electric scalar)
 magnetic, 173
  (See also Potential, magnetostatic)
 retarded, 619
Scalar products, 744–745
Scalar wave equation, 382
Scanning array, 670, 718 (Prob. 14-3-10)
Scarborough, J. B., 287n.
Scattering aperture, effective, 701
Scattering parameter, 432
Schelkunoff, S. A, 588n., 748
Schulman, J. H., 58n.
Schultz, Clarence, xvi
Sciatic nerve, 475
Seawater, electrical behavior of, 446
Second, definition of, 6
Secondary dimensions, 5
Secondary units, 6–7
Secondary winding, 335
Self-impedance of antenna, 675–679, 712

Self-inductance, 336
Seliga, Thomas, xvi
Semiconductor junction, 319 (Prob. 7-2-12)
  capacitance, 320 (Prob. 7-2-13)
Semiconductor mobilities, table of, 121
Semiconductors, 120−122
  germanium, 141 (Prob. 4-2-2)
  germanium and silicon, 121
Sensitivity of radio telescope, 706
Separation of variables, 283, 284, 299, 303, 305,
  548
Septum, waveguide, 578
Series, 736
Series circuit, 126, 141 (Prob. 4-2-5), 368−370
Series-connected iron blocks, 257
Series matching section, 485 (Prob. 10-2-5)
Series resistors in line, 487 (Prob. 10-2-21)
Seven short dipoles, 715 (Prob. 14-2-8)
Shearing line, 273
Sheath, 594
Shell of charge, 42−44, 54 (Probs. 2-3-10, 2-3-12)
  with hole, 53 (Prob. 2-2-30)
Shielding, 476−480
  from lightning, 110 (Prob. 3-5-14)
Shive, J. N., 748
Shoe-shaped block, 144 (Prob. 4-3-20)
Short-circuited line, 405
Short dipole, 620−629, 713 (Prob. 14-1-2)
  effective aperture of, 638
  energy flow of, 627
  far fields of, 626, 627
  fields of, 625−626
  near fields of, 626−627
  radiation resistance of, 629−633
  scalar potential for, 623
  vector potential for, 622
Short vertical antenna, 631−632
Shunt admittance, 389, 546
Shunt resistance of transmission line, 142 (Prob.
  4-3-2)
SI (International System of Units), 6−7
Side lobe, 614
Siemens (unit), 118, 732n.
Signal-to-noise ratio (S/N), 481, 492 (Prob.
  10-6-1), 706, 713
Silicon, 121
  cube of, 141 (Prob. 4-2-4)
Silver, S., 690n.
Silver paint, 296
Sinclair, G., 503n.
Sine integrals, 659
Single quarter-wave transformer, 423
Single-stub matching, 485 (Prob. 10-2-8)
Single-stub tuner, 415
Single-stub VSWR, 485 (Prob. 10-2-7)
Single-wire transmission line, 85−86
Singular point, 27
Six-source broadside array, 650
Skilling, H. H., 183n., 748
Skin effect, 204 (Prob. 5-2-15), 450
Skitek, G. G., 748

Skolnik, M. I., 748
Sky survey, 707
Sky temperature, 703, 707
Slater, J. C., 424n., 578n.
Sloanaker, R. M., 704n.
Slot antennas, 693−694
Slotted lione, 410
Smith, P. H., 412n.
Smith chart, 412−419, 488 (Prob. 10-2-27), 742
Smoke precipitator, 208 (Prob. 5-5-3)
Snell's law;
  of reflection, 513
  of refraction, 514
Solar interference, 724 (Prob. 14-6-10)
Solar power, 491 (Prob. 10-5-8)
Solar waves, 491 (Prob. 10-5-10)
Solenoid, 162−164, 202 (Prob. 5-2-2), 212−213
  air-filled, 218−220
  force on, 203 (Prob. 5-2-9)
  inductance of, 166
  magnetic flux density, 162−164, 203 (Prob.
    5-2-8), 229−231
Solenoidal currents, 129
Solenoidal field, 188
Solenoidal flux tubes, 158
Solid angle, 634
  beam, 615, 634, 712
  main-lobe, 634
  minor-lobe, 635
Soma, 475
Sommerfeld, A., 586
Soroka, W. J., 747
Source temperature, 704
South pole, magnetic, 150
Southworth, G. C., 561n.
Space and time quadrature, 468
Space charge, 307
Space-charge limitation, 308
Space cloth, 459−462
Space paper, 459
Spacecraft link, 720 (Prob. 14-5-1)
Spaceship near moon, 493 (Prob. 10-6-5)
Spangenberg, K. R., 308n.
Sparking, 72
Sphere:
  gaps, 110 (Prob. 3-5-10)
  Poincaré, 500−503
  with uniform **B** inside, 318 (Prob. 7-2-3)
Spherical current density, 142 (Prob. 4-3-1)
Spherical sheet current, 318 (Prob. 7-2-4)
Spherical trigonometry, 500
Spherical wave, 626
Spheroid radiator, 661
Split washer, 202 (Prob. 5-2-6)
Spoke:
  moving, 353−354 (Prob. 8-1-2)
  single, 353 (Prob. 8-1-2)
Spoke and rim, 357 (Prob. 8-1-25)
Spring constant, 346
Square array, 716 (Probs. 14-2-10, 14-2-16)
Square cavity, 609 (Prob. 13-4-1)

Square cavity resonator, 601
Square-corner reflector, 688, 690
Square enclosure, 316 (Prob. 7-1-8)
Square loop, 717 (Prob. 14-3-3)
of charge, 49 (Prob. 2-2-14)
Square resistor with cylindrical core, 145 (Prob. 4-3-24)
Square trough, infinite, 289−292
Square waveguide, 605 (Prob. 13-1-3)
Stabilized magnet, 273
Standing wave, 439−445, 469, 489 (Prob. 10-3-8)
pure, 441
Standing-wave envelope, 442, 443
Standing-wave pattern, 610 (Prob. 13-5-6)
Standing-Wave Ratio (SWR), 406, 443
(*See also* Voltage Standing-Wave Ratio)
State, polarization (*see* Polarization state)
Static electric field, 24
Staticness, 14
Stationary envelope, 442
Stationary loops, 356 (Prob. 8-1-13)
Stegun, I. A., 526n.
Stokes, G., 504
Stokes parameters, 504−509, 509 (Prob. 11-2-2), 510 (Prob. 11-2-7)
normalized, 506
for polarized wave, 510 (Prob. 11-2-6)
Stokes' theorem, 334
Stone, J. S., 643n.
Stored energy, 469
Stratton, J. A., 436, 748
Strip guide, 593
Strip line, 106 (Prob. 3-3-5), 391, 397
attenuation, 607 (Prob. 13-2-14)
nonplanar, 106 (Prob. 3-3-6)
Strip transmission line, 315 (Prob. 7-1-4)
Strong forces, 1
Stub:
matching, 417
short-circuited, 415
Stutzman, W. L., 748
Submarines, communication with, 451
Submerged transmitter, 532 (Prob. 12-2-2)
Sulfur slab, 71, 103 (Prob. 3-1-2)
Sunshine Scientific Co., 297n.
Super-gain antenna, 669
Superconducting cable, nanosecond response, 494 (Prob. 10-6-8)
Superconducting line, 493 (Prob. 10-6-6)
Superconducting world grid, 494 (Prob. 10-6-7)
Superconductivity, 118
Superparamagnetic materials, 214
Superposition of fields, principle of, 17
Superposition of potential, principle of, 29
Suppressed carrier, 438
Surface charge, 143 (Prob. 4-3-12)
Surface charge density, 29
Surface wave, 515, 586
Surface wave current sheet, 607 (Prob. 13-3-4)
Surface wave cutoff, 608 (Prob. 13-3-10)

Surface wave power, 607 (Prob. 13-3-1), 608 (Prob. 13-3-9)
Susceptibility, 275
electric, 62n.
magnetic, 222n.
Symbols, how to read, 7
Synapse, 475
Synchrotron, 206 (Prob. 5-4-11)
System temperature, 702, 705−706, 713

Tables:
antenna and antenna system relations, 712
array beam widths and directivities, 670
boundary relations: for electric and magnetic fields, 351
for electric fields, 68
for magnetic fields, 238
characteristic impedance of transmission lines, 395
charge and mass of particles, 197
circuit and field quantities, 376
circuit and field relations, 483
conductivities, 123
constants, 732−733
for some common media, 447
for transmission lines and waves, 484
cylindrical waveguide modes, 567
dielectric materials, 741
dielectric strengths, 74
dipole fields, 629
electric and magnetic circuits, comparison of, 258
electric and magnetic field relations, comparison of, 174, 274, 353
field-map quantities, 266
hollow rectangular waveguide modes, 560
Maxwell's equations, 374−376
in differential form, 376
in integral form, 375
penetration depths, wavelength, velocity, and refractive index for copper, 451
permanent magnetic materials, 252
peremeabilities, 216
peremittivities, 58
rectangular and circular cylindrical waveguide parameters, 577
reflection and transmission coefficients, 407
semiconductor mobilities, 121
short dipole fields, 629
six-source array parameters, 651
Stokes parameters for seven wave-polarization states, 506
terminated transmission line impedance, 406
units, 726−732
Tai, C-T, xvi, 543n., 566, 567n., 679n.
Taper for wide bandwidth, 424
Tapetum, 596
TE (transverse electric) mode, 536
TE_{10} mode, 551
TE_{11} mode, 552−553

TE_{20} mode, 551, 552
TE_{mn} modes in hollow rectangular waveguides, table of, 560
Teledeltos paper, 296, 399
TEM (Transverse ElectroMagnetic) field, 367
TEM modes, 379
TEM space waves, 379
TEM wave, 380
Temperature:
  antenna, 493 (Prob. 10-6-1), 613, 703
    effect of ground on, 707
  of Mars, 704
  minimum detectable, 706, 713
  noise, 702, 743
  sky, 703, 707
  source, 704
  system, 702, 705–706, 713
Ten waves, 510 (Prob. 11-2-5)
  Stokes parameters, 510 (Prob. 11-2-10)
Terminal impedance, 720 (Prob. 14-4-9)
Terminated line, 485 (Prob. 10-2-1)
Terminated transmission line, 402–407
Terminated V, 717 (Prob. 14-3-2)
Terminations, waveguide, 578
Tesla, Nikola, 4
Test charge, 20
Thales of Miletus, 1, 3, 11
Theoretical quantity, 59
Theory of images, 85, 311–312
Thévenin generator, 638
Thiele, G. A., 748
Thompson scatter, 721 (Probs. 14-5-10, 14-5-11)
Three point charges, map of, 49 (Prob. 2-2-27)
Three unequal sources, 716 (Prob. 14-2-9)
Thundercloud, 55 (Prob. 2-4-1)
  current, 146 (Prob. 4-4-3)
  energy in, 110 (Prob. 3-5-13)
  field under, 108 (Prob. 3-5-4)
Tilt angle:
  of polarization ellipse, 497
  of wave, 521, 584
Time-changing fields, 322–353
Time-phase quadrature, 626
TM (Transverse Magnetic) mode, 368
Tobey, F. L., 747
Toroid, 221
  inductance of, 166, 359 (Prob. 8-2-5)
  iron-cored, 228
Toroidal coil, 335
  with gap, 224
Torque, 151, 212, 274
  on dipole, 53 (Prob. 2-3-2)
  on loop, 161–162
  on solenoid, 164
Total internal reflection, principle of, 515
Trace, 508
Track-guided vehicle design, 363 (Prob. 8-4-10)
Transducer, 612
Transformations, component, 746
Transformer:
  and bandwidth, 420–424

Transformer (*Cont.*):
  double quarter-wave, 423
  single quarter-wave, 423
  spherical-to-plane wave, 692
Transformer induction, 328
Transformer induction equation, 325
Transformer matching, 486 (Prob. 10-2-12)
Transients, 424–428
Transistors, 122
Transition device, 578, 612
Transmission coefficient, 455, 521
  for current, 407
  for voltage, 406–407
Transmission line, 378, 379
  active, 475
  balanced, 620
  cell, 176, 263
  characteristic impedance of, 392–395
  coaxial, 81, 134, 391
  dielectric rod, 379
  distortionless, 395
  equivalent circuit, 389
  field under, 208 (Prob. 5-5-6)
  field-cell, 391
  high-voltage dc, 106 (Prob. 3-3-1)
  impedance, 395–402
    table of, 406
  inductance, 377 (Prob. 9-1-11)
  lossless, 402
  matching, 412–419, 423, 486 (Prob. 10-2-13)
  passive, 475
  phase velocity on, 395
  power flow on, 470–472
  radio-link, 480–482
  single-wire, 85–86
  of square cross section, 315 (Prob. 7-1-7)
  strip, 315 (Prob. 7-1-4)
    (*See also* Strip line)
  terminated, 402–407
  two-wire (*see* Two-wire line)
  uniform, 392
Transmitted wave, 440, 511
Transmitting antenna, 612, 637
Transponder, 492 (Prob. 10-6-1)
Transverse electric (TE) mode, 536
Transverse ElectroMagnetic (TEM) field, 367
Transverse ElectroMagnetic modes, 379
Transverse ElectroMagnetic space waves, 379
Transverse ElectroMagnetic wave, 380
Transverse field components, 550
Transverse Magnetic (TM) mode, 368
Transverse-wave impedance, 547, 557, 582
Traveling wave, 439–445, 467
  on antenna, 662
  energies, 490 (Prob. 10-5-7)
  Poynting vector, 490 (Prob. 10-5-5)
Traveling-wave antenna, 661–664
  terminated, 662
  unterminated, 661
Triangle array, 716 (Prob. 14-2-15)
Trigonometric relations, 733–734

Trigonometry, spherical, 500
Troughs, potential in, 317 (Prob. 7-1-11)
Tubes:
    of current, 128−129
    of electric flux, 38−39, 87
Tubular conductor, torque on, 198 (Prob. 5-1-5)
Tuner:
    double-stub, 417, 486 (Probs. 10-2-13, 10-2-19)
    single-stub, 415
Tunnel communications, 610 (Prob. 13-5-3)
TV downlinks, 492 (Prob. 10-6-1), 533 (Prob. 12-3-1)
TV screen, 207 (Prob. 5-5-2)
Twin-line, 108 (Prob. 3-5-2)
Two-conductor line (see Two-wire line)
Two-element array, 680−683
Two-element interferometer, 650−653, 715 (Prob. 14-2-4)
Two hemispheres, 318 (Prob. 7-2-6)
Two isotropic point sources, 640−641
Two loops, torque, 198 (Prob. 5-1-8), 202 (Prob. 5-1-25)
Two-part capacitor, 104 (Prob. 3-2-12)
Two-port junction, 432
Two sources:
    in opposite phase, 715 (Prob. 14-2-7)
    in phase, 715 (Prob. 14-2-6)
    in phase quadrature, 715 (Prob. 14-2-5)
Two-wire line, 390, 391
    capacitance of, 85
    characteristic impedance of, 401−402
    inductance of, 168
    potential of, 82−83

U-shaped electromagnet, 280 (Prob. 6-6-2)
Uda, S., 688n.
Uniform array, 643
Uniform transmission line, 392
Unique solution, 103, 286
Uniqueness, 286−287
Units, 6−7, 725
    table of, 726−732
Unpolarized wave, 504
    at angle, 531 (Prob. 12-1-9)
Upatnieks, J., 529n.
Urban ground, electrical behavior of, 446

V antenna, 663, 664, 677−678
Vacuum tube, 308
Van de Graaf generator, 110 (Prob. 3-5-9)
Vanduzer, R., 748
Vector identities, 736−737
Vector notation, full, 19
Vector potential:
    magnetic, 187−192
    retarded, 619
Vector products, 744−745
    (See also Cross product)

Vehicles:
    linear motor for, 363 (Prob. 8-4-10)
    radio-equipped, 493 (Prob. 10-6-4)
    space (see Spacecraft link; Spaceship near moon)
Vehicular tunnel, 610 (Prob. 13-5-3)
Velocities, phase and group, 489 (Prob. 10-3-6)
Velocity, 383
    energy, 490 (Prob. 10-5-4), 540
    group, 436, 437, 540
    of light, 387
    phase (see Phase velocity)
Venus radar, 724 (Prob. 14-6-6)
Vertically polarized wave, 380
Vibrational modes, 348
Volta, Alessandro, 2, 3
Voltage reflection coefficient, 403
Voltage Standing-Wave Ratio (VSWR), 406, 443
Voltmeter:
    digital (DVM), 295
    high-resistance, 295
Volume charge density, 29
Von Hippel, A. R., 58n.

Wainer, E., 58n.
Walter, C. H., 661n.
Water generator, 363 (Prob. 8-4-11)
Water plume, 541
Water trough, 181
Watt, James, 3
Wave, 378
    polarized: circularly, 495
        completely, 505
        elliptically, 495
        linearly, 495
        partially, 504
    reflected, 461
    resultant, 538
    spherical, 626
    surface, 515, 586
    TEM, 380
    transmitted, 461
    unpolarized, 505
    (See also entries beginning with the word: Wave)
Wave absorption, 459−464
Wave equation, 380−390, 544, 548
    for conductors, 447
    for dielectrics, 382
    general development of, 482
    for rectangular waveguide, 548
    scalar, 382
    solution, 383
    for transmission line, 390
Wave impedance, 582
Wave path, 536, 537
Wave polarization, 380
Wave reflection method, 428−431
Wave reflections on quarter-wave transformer, 428−431

Wave-to-antenna angle, 508
Wave-to-antenna coupling factor, 507
Wave transformers, 692
Wavefronts, 529, 537
Waveguide, 534
  coated conductor, 588
  corrugated, 587
  cylindrical, 561−569
    modes, chart of, 566
  dielectric sheet, 589−593
  discontinuities in, 553
  elliptical, 569−570
  G-string, 589
  nonmetallic, 587
  open surface, 586
  phase velocity in, 568
  rectangular, 543−560
    modes, chart of, 559
  reentrant, 569−570
  ridge, 569−570
  screw in, 558, 559
  square, 569−570
  wavelength in, 568
  (*See also* Guide)
Waveguide devices, 578−579
Waveguide iris, 579−581
Waveguide power dividers, 578
Waveguide terminations, 578
Waveguide transitions, 578
Wavelength, 383, 387
Wavelength change, 531 (Prob. 12-1-8)

Wavelength in guide, 568
Waves parallel to plane boundary, 582−586
Waves through windows, 610 (Prob. 13-5-4)
Weak forces, 1
Weber, Wilhelm E., 3
Weber (unit), 157
Weed, Herman, xvi
Wells, H. G., 5
West, R., 58*n*.
Whinnery, J. R., 748
Wickersham, A. F., Jr., 610 (Prob. 13-5-7)
Williams, H. J., 748
Wilson, R. W., 610 (Prob. 13-5-2)
Wire:
  of hyperbolic shape, 200 (Prob. 5-1-23)
  of parabolic shape, 200 (Prob. 5-1-22)
Wire loop, 323
Wolf, E., 500*n*., 509*n*.
Woodyard, J. R., 668*n*.
WOSU-TV, 747

Xerographic machine, 55 (Prob. 2-4-2)

Yagi, H., 688*n*.
Yagi-Uda antennas, 688

Zenneck, J., 586
Zonal harmonics, solid, 303

Spherical Coordinates

$$\mathbf{\nabla}f = \hat{\mathbf{r}}\,\frac{\partial f}{\partial r} + \hat{\boldsymbol{\theta}}\,\frac{1}{r}\frac{\partial f}{\partial \theta} + \hat{\boldsymbol{\phi}}\,\frac{1}{r\sin\theta}\frac{\partial f}{\partial \phi}$$

$$\mathbf{\nabla}\cdot\mathbf{A} = \frac{1}{r^2}\frac{\partial}{\partial r}\,r^2 A_r + \frac{1}{r\sin\theta}\frac{\partial}{\partial \theta}\,(A_\theta \sin\theta) + \frac{1}{r\sin\theta}\frac{\partial A_\phi}{\partial \phi}$$

$$\mathbf{\nabla}\times\mathbf{A} = \hat{\mathbf{r}}\,\frac{1}{r\sin\theta}\left[\frac{\partial}{\partial \theta}\,(A_\phi \sin\theta) - \frac{\partial A_\theta}{\partial \phi}\right] + \hat{\boldsymbol{\theta}}\,\frac{1}{r}\left(\frac{1}{\sin\theta}\frac{\partial A_r}{\partial \phi} - \frac{\partial}{\partial r}\,rA_\phi\right)$$

$$+ \hat{\boldsymbol{\phi}}\,\frac{1}{r}\left(\frac{\partial}{\partial r}\,rA_\theta - \frac{\partial A_r}{\partial \theta}\right)$$

$$\mathbf{\nabla}^2 f = \frac{1}{r^2}\frac{\partial}{\partial r}\left(r\,\frac{\partial f}{\partial r}\right) + \frac{1}{r^2\sin\theta}\frac{\partial}{\partial \theta}\left(\sin\theta\,\frac{\partial f}{\partial \theta}\right) + \frac{1}{r^2\sin^2\theta}\frac{\partial^2 f}{\partial \phi^2}$$

General Curvilinear Coordinates

$$\mathbf{\nabla}f = \hat{\mathbf{x}}_1\,\frac{1}{h_1}\frac{\partial f}{\partial x_1} + \hat{\mathbf{x}}_2\,\frac{1}{h_2}\frac{\partial f}{\partial x_2} + \hat{\mathbf{x}}_3\,\frac{1}{h_3}\frac{\partial f}{\partial x_3}$$

$$\mathbf{\nabla}\cdot\mathbf{A} = \frac{1}{h_1 h_2 h_3}\left(\frac{\partial}{\partial x_1}\,h_2 h_3 A_1 + \frac{\partial}{\partial x_2}\,h_3 h_1 A_2 + \frac{\partial}{\partial x_3}\,h_1 h_2 A_3\right)$$

$$\mathbf{\nabla}\times\mathbf{A} = \hat{\mathbf{x}}_1\,\frac{1}{h_2 h_3}\left[\frac{\partial}{\partial x_2}\,h_3 A_3 - \frac{\partial}{\partial x_3}\,h_2 A_2\right] + \hat{\mathbf{x}}_2\,\frac{1}{h_3 h_1}\left(\frac{\partial}{\partial x_3}\,h_1 A_1 - \frac{\partial}{\partial x_1}\,h_3 A_3\right)$$

$$+ \hat{\mathbf{x}}_3\,\frac{\partial}{h_1 h_2}\left(\frac{\partial}{\partial x_1}\,h_2 A_2 - \frac{\partial}{\partial x_2}\,h_1 A_1\right)$$

$$\mathbf{\nabla}^2 f = \frac{1}{h_1 h_2 h_3}\left[\frac{\partial}{\partial x_1}\left(\frac{h_2 h_3}{h_1}\frac{\partial f}{\partial x_1}\right) + \frac{\partial}{\partial x_2}\left(\frac{h_3 h_1}{h_2}\frac{\partial f}{\partial x_2}\right) + \frac{\partial}{\partial x_3}\left(\frac{h_1 h_2}{h_3}\frac{\partial f}{\partial x_3}\right)\right]$$

	h_1	h_2	h_3	x_1	x_2	x_3
Rectangular	1	1	1	x	y	z
Cylindrical	1	r	1	r	ϕ	z
Spherical	1	r	$r\sin\theta$	r	θ	ϕ

GRADIANT, DIVERGENCE, CURL AND LAPLACIAN IN RECTANGULAR, CYLINDRICAL, SPHERICAL, AND GENERAL CURVILINEAR COORDINATES

(See also Appendixes A-16 and A-17.)

Rectangular Coordinates

$$\nabla f = \hat{\mathbf{x}}\,\frac{\partial f}{\partial x} + \hat{\mathbf{y}}\,\frac{\partial f}{\partial y} + \hat{\mathbf{z}}\,\frac{\partial f}{\partial z}$$

$$\nabla \cdot \mathbf{A} = \frac{\partial A_x}{\partial x} + \frac{\partial A_y}{\partial y} + \frac{\partial A_z}{\partial z}$$

$$\nabla \times \mathbf{A} = \hat{\mathbf{x}}\left(\frac{\partial A_z}{\partial y} - \frac{\partial A_y}{\partial z}\right) + \hat{\mathbf{y}}\left(\frac{\partial A_x}{\partial z} - \frac{\partial A_z}{\partial x}\right) + \hat{\mathbf{z}}\left(\frac{\partial A_y}{\partial x} - \frac{\partial A_x}{\partial y}\right) = \begin{vmatrix} \hat{\mathbf{x}} & \hat{\mathbf{y}} & \hat{\mathbf{z}} \\ \dfrac{\partial}{\partial x} & \dfrac{\partial}{\partial y} & \dfrac{\partial}{\partial z} \\ A_x & A_y & A_z \end{vmatrix}$$

$$\nabla^2 f = \frac{\partial^2 f}{\partial x^2} + \frac{\partial^2 f}{\partial y^2} + \frac{\partial^2 f}{\partial z^2}$$

Cylindrical Coordinates

$$\nabla f = \hat{\mathbf{r}}\,\frac{\partial f}{\partial r} + \hat{\boldsymbol{\phi}}\,\frac{1}{r}\frac{\partial f}{\partial \phi} + \hat{\mathbf{z}}\,\frac{\partial f}{\partial z}$$

$$\nabla \cdot \mathbf{A} = \frac{1}{r}\frac{\partial}{\partial r}\,rA_r + \frac{1}{r}\frac{\partial A_\phi}{\partial \phi} + \frac{\partial A_z}{\partial z}$$

$$\nabla \times \mathbf{A} = \hat{\mathbf{r}}\left(\frac{1}{r}\frac{\partial A_z}{\partial \phi} - \frac{\partial A_\phi}{\partial z}\right) + \hat{\boldsymbol{\phi}}\left(\frac{\partial A_r}{\partial z} - \frac{\partial A_z}{\partial r}\right) + \hat{\mathbf{z}}\,\frac{1}{r}\left(\frac{\partial}{\partial r}\,rA_\phi - \frac{\partial A_r}{\partial \phi}\right)$$

$$= \begin{vmatrix} \hat{\mathbf{r}}\,\dfrac{1}{r} & \hat{\boldsymbol{\phi}} & \hat{\mathbf{z}}\,\dfrac{1}{r} \\ \dfrac{\partial}{\partial r} & \dfrac{\partial}{\partial \phi} & \dfrac{\partial}{\partial z} \\ A_r & rA_\phi & A_z \end{vmatrix}$$

$$\nabla^2 f = \frac{1}{r}\frac{\partial}{\partial r}\left(r\frac{\partial f}{\partial r}\right) + \frac{1}{r^2}\frac{\partial^2 f}{\partial \phi^2} + \frac{\partial^2 f}{\partial z^2} = \frac{\partial^2 f}{\partial r^2} + \frac{1}{r}\frac{\partial f}{\partial r} + \frac{1}{r^2}\frac{\partial^2 f}{\partial \phi^2} + \frac{\partial^2 f}{\partial z^2}$$